Student Solutions Manual

Kevin Bodden   Randy Gallaher

Experiencing
SECOND EDITION Introductory and Intermediate Algebra

JOANNE THOMASSON   BOB PESUT

Upper Saddle River, NJ 07458

Senior Acquisitions Editor: Paul Murphy
Editorial Assistant: Eileen Nee
Assistant Managing Editor: John Matthews
Production Editor: Wendy A. Perez
Supplement Cover Manager: Paul Gourhan
Supplement Cover Designer: Joanne Alexandris
Manufacturing Buyer: Ilene Kahn

© 2003 by Pearson Education, Inc.
Pearson Education, Inc.
Upper Saddle River, NJ 07458

All rights reserved. No part of this book may be reproduced in any form or by any means, without permission in writing from the publisher.

The author and publisher of this book have used their best efforts in preparing this book. These efforts include the development, research, and testing of the theories and programs to determine their effectiveness. The author and publisher make no warranty of any kind, expressed or implied, with regard to these programs or the documentation contained in this book. The author and publisher shall not be liable in any event for incidental or consequential damages in connection with, or arising out of, the furnishing, performance, or use of these programs.

Printed in the United States of America

10 9 8 7 6 5 4 3

ISBN 0-13-009444-7

Pearson Education Ltd., *London*
Pearson Education Australia Pty. Ltd., *Sydney*
Pearson Education Singapore, Pte. Ltd.
Pearson Education North Asia Ltd., *Hong Kong*
Pearson Education Canada, Inc., *Toronto*
Pearson Educación de Mexico, S.A. de C.V.
Pearson Education—Japan, *Tokyo*
Pearson Education Malaysia, Pte. Ltd.
Pearson Education, *Upper Saddle River, New Jersey*

Table of Contents

Discovery Box Answers	1
Chapter 1: Real Numbers	25
Chapter 2: Variables, Expressions, Equations and Formulas	52
Chapter 3: Relations, Functions and Graphs	71
Chapter 4: Linear Equations in One Variable	112
Chapter 5: Linear Equations and Functions	158
Chapter 6: Systems of Linear Equations	211
Chapter 7: Linear Inequalities	279
Chapter 8: Polynomial Functions	336
Chapter 9: Exponents and Polynomials	377
Chapter 10: Factoring	409
Chapter 11: Quadratic and Other Polynomial Equations and Inequalities	447
Chapter 12: Rational Expressions, Functions and Equations	512
Chapter 13: Radical Expressions, Functions and Equations	554
Chapter 14: Exponential and Logarithmic Functions and Equations	615
Calculator Appendix	649

Discovery Answers

Chapter 1

Discovery 1
1. a. 8
   b. 8
2. a. 12
   b. 12
3. a. −8
   b. 8
4. a. −12
   b. 12

To find the sum of two rational numbers with like signs, add the absolute values of the two addends. The sign of the sum is the like sign of the two addends.

Discovery 2
1. a. 4
   b. 4
2. a. −6
   b. 6
3. a. −4
   b. 4
4. a. 6
   b. 6

To find the sum of two rational numbers with unlike signs, subtract the absolute value of the smaller addend from the absolute value of the larger addend. The sign of the sum is the sign of the addend with the larger absolute value.

Discovery 3
1. a. 4
   b. 4
2. a. −6
   b. −6
3. a. −4
   b. −4
4. a. 6
   b. 6
5. a. 8
   b. 8
6. a. −12
   b. −12

To find the difference of two rational numbers, add the minuend to the opposite of the subtrahend.

Discovery 4
1. a. 6
   b. 6
2. a. 6
   b. 6
3. a. −6
   b. 6
4. a. −6
   b. 6

To find the product of two rational numbers with like signs, multiply the absolute values of the two factors. The product will be positive.

To find the product of two rational numbers with unlike signs, multiply the absolute values of the two factors. The product will be negative.

SSM: Experiencing Introductory and Intermediate Algebra

Discovery 5
1. a. 4
   b. 4
2. a. 4
   b. 4
3. a. −4
   b. 4
4. a. −4
   b. 4

To find the quotient of two rational numbers with like signs, divide the absolute value of the dividend by the absolute value of the divisor. The quotient will be positive.

To find the quotient of two rational numbers with unlike signs, divide the absolute value of the dividend by the absolute value of the divisor. The quotient will be negative.

Discovery 6
1. undefined
2. undefined
3. 0
4. 0
5. indeterminate

The quotient of any nonzero rational number divided by 0 will be undefined.

The quotient 0 divided by any nonzero rational number will be 0.

The quotient of 0 divided by 0 will be indeterminate.

Discovery 7
1. −72
2. 72
3. −72
4. 72

The product of three or more rational numbers will be positive when the number of negative factors is even. The product of three or more rational numbers will be negative when the number of negative factors is odd.

Discovery 8
1. 7
2. −4
3. 1
4. 1

First, evaluate all multiplications and divisions in order from left to right. Then, evaluate all additions and subtractions in order from left to right.

Discovery 9
1. 10
2. −10
3. 0
4. −1
5. 1
6. indeterminate

The value of an exponential expression with an exponent of 1 is equal to the base number.

The value of an exponential expression with a nonzero base and an exponent of 0 is 1. The value of an exponential expression with a base of 0 and an exponent of 0 is indeterminate.

Discovery 10
1. a. 0.1
   b. 0.1
2. a. 0.01
   b. 0.01
3. a. 0.001
   b. 0.001
4. a. 10
   b. 10
5. a. 100
   b. 100
6. a. 1000
   b. 1000

To evaluate an exponential expression with a nonzero base and a negative integer exponent, rewrite the expression with a new base that is the reciprocal of the original base and a new exponent that is the opposite of the original exponent. Then evaluate the new expression.

Discovery 11
1. a. 20
   b. 20
2. a. 4
   b. 4

The product of a real number factor times the sum of two or more real number addends is equal to the sum of the products of the real number factor times each addend.

Chapter 3

Discovery 1

Ordered pairs will vary.
In quadrant I, both coordinates are positive; in quadrant II, the x-coordinate is negative and the y-coordinate is positive; in quadrant III, both coordinates are negative; and in quadrant IV, the x-coordinate is positive and the y-coordinate is negative.

Discovery 2

Ordered pairs will vary.
On the x-axis, the y-coordinate is always 0. On the y-axis, the x-coordinate is always 0.

Discovery 3

1. Not possible.

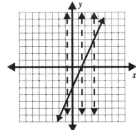

2. Not possible.

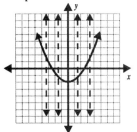

3. Not possible.

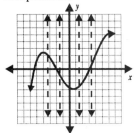

If a vertical line can be drawn through more than one point on the graph, the graph is not of a function.

4. Not a function. A vertical line can pass through more than one point of the graph.

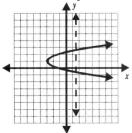

5. Not a function. A vertical line can pass through more than one point of the graph.

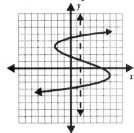

SSM: Experiencing Introductory and Intermediate Algebra

Discovery 4

1. $f(-2)=12$, $f(-1)=5$, $f(0)=0$, $f(1)=-3$, $f(2)=-4$, $f(3)=-3$, $f(4)=0$, $f(5)=5$, $f(6)=12$

2. The function values first **decrease** and then **increase** as the x-values increase.

3. Tracing the graph from left to right, we find that the graph first **falls** and then **rises**.

   If the graph falls as you move from left to right along the x-axis, then the graph is decreasing. If the graph rises, it is increasing.

4. The function has a relative **minimum**.

5. The graph has one **low** point.

   A continuous graph has a relative minimum at every low point and a relative maximum at every high point.

Chapter 4

Discovery 1

1.
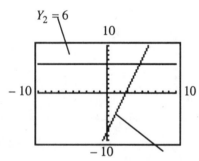

The graphs of both of the functions Y1 and Y2 are straight lines.

2.
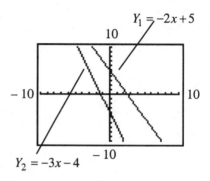

The graphs of both of the functions Y1 and Y2 are straight lines.

3. a.
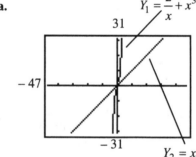

The graph of function Y1 is not a line. The graph of Y2 is a line.

Discovery 2

x	$3x-9$	$=-2x+6$		
0	-9	6	$-9<6$	$-9-6=-15$
1	-6	4	$-6<4$	$-6-4=-10$
2	-3	2	$-3<2$	$-3-2=-5$
3	0	0	$0=0$	$0-0=0$
4	3	-2	$3>-2$	$0-(-2)=2$

The solution of an equation is the value that when substituted for the variable causes the value of the expression on the left side of the equals sign to equal the value of the expression on the right side of the equals sign. Also, the solution of an equation is the value that when substituted for the variable causes the difference of the values of the expressions to be 0.

4

Discovery 3

x	(5x+4)−2(3x+1)	= 2(x−7)		
3	−1	−8	−1 > −8	−1 − (−8) = 7
4	−2	−6	−2 > −6	−2 − (−6) = 4
5	−3	−4	−3 > −4	−3 − (−4) = 1
6	−4	−2	−4 < −2	−4 − (−2) = −2
7	−5	0	−5 < 0	−5 − 0 = −5

The solution of an equation is between two integer values where the order of the values for the expression on the left side of the equals sign and the expression on the right side of the equals sign changes from less than to greater than or from greater than to less than. Also, the solution of an equation is between two integer values where the difference of the values of the expressions changes from positive to negative or from negative to positive.

Discovery 4

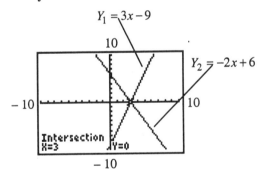

The value of the x-coordinate of the intersection point of the two lines is the solution of the equation.

The value of the y-coordinate of the intersection point of the two lines is the value of each expression when the equation is evaluated at the solution.

Discovery 5

1.

x	2x+5	= 2x+10		
0	5	10	5 < 10	5−10 = −5
1	7	12	7 < 12	7−12 = −5
2	9	14	9 < 14	9−14 = −5
3	11	16	11 < 16	11− 6 = −5
4	13	18	13 < 18	13−18 = −5

If the values of the expressions on the left side of the equals sign and on the rights side of the equals sign are never equal, and if their differences remain the same constant, then the equation has no solution.

2.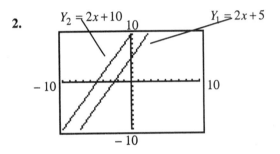

If the graphs of the functions defined by the expression on the left side of the equals sign and the expression on the right side of the equals sign never intersect (i.e. if the graphs are parallel), then the equation has no solution.

SSM: Experiencing Introductory and Intermediate Algebra

Discovery 6

1.

x	$2x+5$	$=(x+3)+(x+2)$		
0	5	5	5=5	5−5=0
1	7	7	7=7	7−7=0
2	9	9	9=9	9−9=0
3	11	11	11=11	11−11=0
4	13	13	13=13	13−13=0

If the values of the expressions on the left side of the equals sign and on the rights side of the equals sign are always equal, and thus if their differences are always 0, then the solution set of the equation is the set of all real numbers.

2.

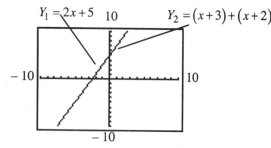

If the graphs of the functions defined by the expression on the left side of the equals sign and the expression on the right side of the equals sign are exactly the same (i.e. if the graphs coincide), then the solution set of the equation is the set of all real numbers.

Discovery 7

1.
$$7=7$$
$$\begin{array}{c|c} 7+(-2) & 7+(-2) \\ 5 & 5 \end{array}$$

2.
$$6+1=4+3$$
$$\begin{array}{c|c} 6+1+2 & 4+3+2 \\ 9 & 9 \end{array}$$

3.
$$6+1=4+3$$
$$\begin{array}{c|c} 6+1+(-2) & 4+3+(-2) \\ 5 & 5 \end{array}$$

If the same value is added to both sides of an equation, the result is an equivalent equation.

Discovery 8

1.
$$7=7$$
$$\begin{array}{c|c} 7(-2) & 7(-2) \\ -14 & -14 \end{array}$$

2.
$$6+1=4+3$$
$$\begin{array}{c|c} (6+1)2 & (4+3)2 \\ 14 & 14 \end{array}$$

3.
$$6+1=4+3$$
$$\begin{array}{c|c} (6+1)(-2) & (4+3)(-2) \\ -14 & -14 \end{array}$$

If both sides of an equation are multiplied by the same value, the result is an equivalent equation.

Discovery 9

$2x + 5 = 2x + 10$
$2x + 5 - 2x = 2x + 10 - 2x$ Subtract $2x$ from both sides.
$5 = 10$ Simplify.

When isolating the variable on one side of the equation if the variable is eliminated from the equation and the resulting statement is false (a contradiction), then the equation has no solution.

Discovery 10

$$2x + 5 = (x + 3) + (x + 2)$$
$$2x + 5 = 2x + 5 \quad \text{Simplify the right side.}$$
$$2x + 5 - 2x = 2x + 5 - 2x \quad \text{Subtract } 2x \text{ from both sides.}$$
$$5 = 5 \quad \text{Simplify.}$$

When isolating the variable on one side of the equation, if the variable is eliminated from the equation and the resulting statement is true (an identity) then the solution set of the equation is the set of all real numbers.

Discovery 11

1.

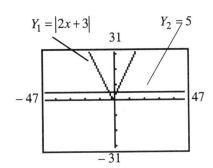

2.

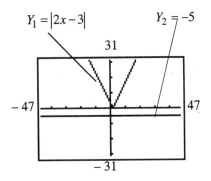

The graphs of the functions defined by absolute value expressions (i.e. Y1) are V-shaped graphs.

The graphs of the functions defined by constants (i.e. Y2) are horizontal lines.

3.

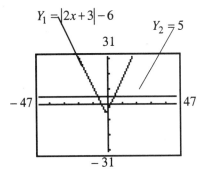

4.

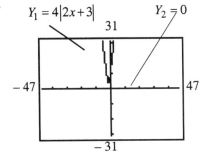

The graphs of the functions defined by absolute value expressions (i.e. Y1) are V-shaped graphs.

The graphs of the functions defined by constants (i.e. Y2) are horizontal lines.

Discovery 12

1. a.

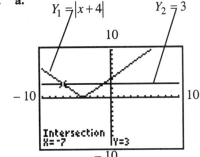

SSM: Experiencing Introductory and Intermediate Algebra

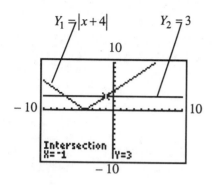

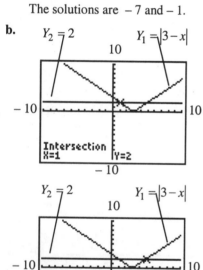

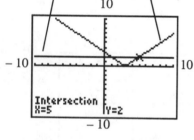

The solutions are −7 and −1.

b.

The solutions are 1 and 5.

2. a.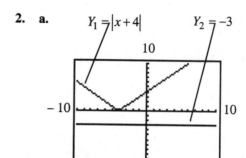

There is no intersection point, so this equation has no solution.

b.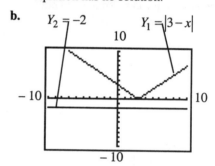

There is no intersection point, so this equation has no solution.

3. a.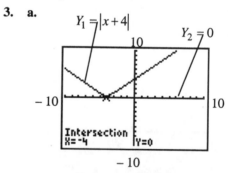

The solution is −4.

b. $Y_2 = 0$ $Y_1 = |3 - x|$

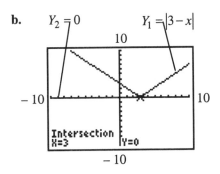

The solution is 3.

If, within a linear absolute-value equation, the absolute-value expression equals a positive number, then the equation will have two solutions.

If, within a linear absolute-value equation, the absolute-value expression equals a negative number, the equation will have no solution.

If, within a linear absolute-value equation, the absolute-value expression equals 0, then the equation will have one solution.

Chapter 5

Discovery 1

1.
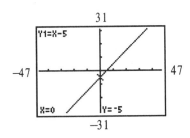

2.
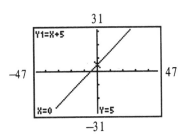

3.
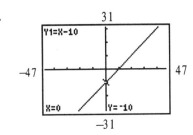

4.
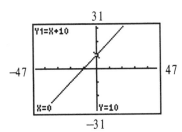

The y-coordinate of the y-intercept of the graph is the same as the constant term when the equation is solved for y.

Discovery 2

1.
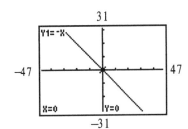

2.
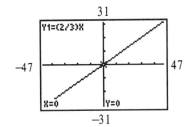

SSM: Experiencing Introductory and Intermediate Algebra

The graph of a linear equation in two variables written in standard form has one point that is both the x-intercept and y-intercept if the constant term is zero. The intercept is the origin, $(0,0)$.

The slope of a line connecting two points is the difference of the y-coordinates, divided by the difference of the x-coordinates (in the same order).

Discovery 3

1. slope = $\frac{1}{2}$
2. slope = 2
3. slope = $-\frac{1}{3}$
4. slope = -3
5. slope = 0
6. slope is undefined
7. positive; rise
8. negative; fall
9. 0; horizontal
10. undefined; vertical
11. more

Discovery 4

1.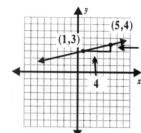

2. 1
3. 4
4. $4 - 3 = 1$
5. $5 - 1 = 4$
6. $\frac{1}{4}$

Discovery 5

1. a.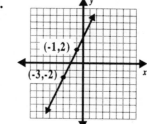

   b.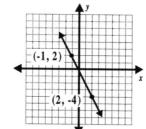

   c.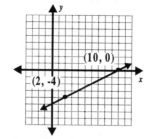

2. a. 2
   b. -2
   c. $\frac{1}{2}$

3. a. 2
   b. -2
   c. $\frac{1}{2}$

The slope of the line is the coefficient of the *x*-term in the equation when it is solved for *y*.

Discovery 6

1. a.
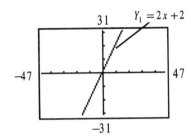

   b.
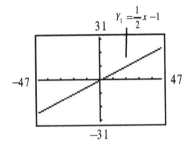

   c.
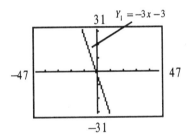

2. a. $m = 2; b = 2$
      $m = 2; b = 2$

   b. $m = \dfrac{1}{2}; b = -1$
      $m = \dfrac{1}{2}; b = -1$

   c. $m = -3; b = -3$
      $m = -3; b = -3$

3. coinciding

4. equal

5. equal

The graphs of two linear equations in two variables are coinciding if their corresponding equations, written in slope-intercept form, have the same slope, m, and the same y-coordinate of the y-intercept, b.

Discovery 7

1. a.
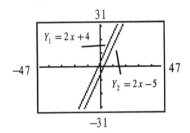

   b.
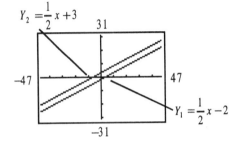

   c.
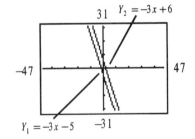

2. a. $m = 2; b = 4$
      $m = 2; b = -5$

   b. $m = \dfrac{1}{2}; b = -2$
      $m = \dfrac{1}{2}; b = 3$

   c. $m = -3; b = -5$
      $m = -3; b = 6$

3. parallel

4. equal

5. not equal

The graphs of two linear equations in two variables are parallel if their corresponding equations, written in slope-intercept form, have the same slope, m, and different y-coordinates for the y-intercept, b.

Discovery 8

1. a.

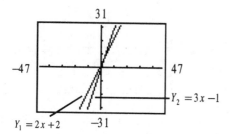

b.

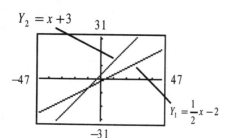

c.

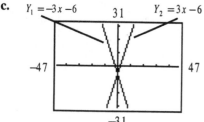

2. a. $m = 2; b = 2$
$m = 3; b = -1$

b. $m = \dfrac{1}{2}; b = -2$
$m = 1; b = 3$

c. $m = -3; b = -6$
$m = 3; b = -6$

3. intersecting

4. not equal

The graphs of two linear equations in two variables are intersecting if their corresponding equations, written in slope-intercept form, have different slopes.

Discovery 9

1. a.

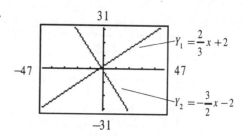

b.

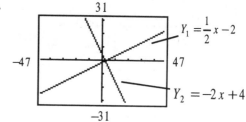

c.

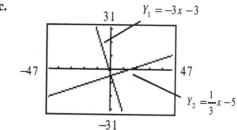

2. a. $m = \dfrac{2}{3}; b = 2$

$m = -\dfrac{3}{2}; b = -2$

b. $m = \dfrac{1}{2}; b = -2$

$m = -2; b = 4$

c. $m = -3; b = -3$

$m = \dfrac{1}{3}; b = -5$

3. intersecting and perpendicular

4. not equal

5. opposite

The graphs of two linear equations in two variables are perpendicular if their corresponding equations, written in slope-intercept form, have opposite reciprocal slopes.

Chapter 6

Discovery 1

1.

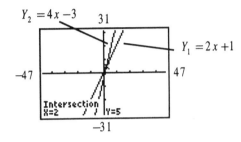

   a. $(2,5)$

   b. $(2,5)$

If the graphs of two linear equations in two variables intersect, then the coordinates of the point of intersection form the ordered pair solution to the system.

Discovery 2

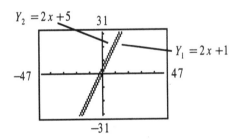

If the graphs of two linear equations in two variables do not intersect (i.e. they are parallel), then the system is inconsistent or has no solution.

Discovery 3

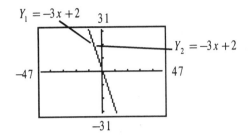

If the graphs of two linear equations in two variables are coinciding, then the system contains dependent equations and has an infinite number of solutions.

Discovery 4

Solve the first equation for y.
$y = 2x + 1$

Substitute $(2x+1)$ in for y in the second equation.
$$-4x + 2(2x+1) = 3$$
$$-4x + 4x + 2 = 3$$
$$2 = 3 \text{ (contradiction)}$$

If the substitution method yields a contradiction, then the system is inconsistent and has no solution.

Discovery 5

Substitute $(-3x+2)$ in for y in the first equation.
$$-3x - (-3x+2) = -2$$
$$-3x + 3x - 2 = -2$$
$$-2 = -2 \text{ (identity)}$$

If the substitution method yields an identity, then the system contains dependent equations and has an infinite number of solutions.

Discovery 6

Multiply both sides of the first equation by -2.
$$-2(-2x+y) = -2(1)$$

The new system is:
$$4x - 2y = -2$$
$$-4x + 2y = 3$$

SSM: Experiencing Introductory and Intermediate Algebra

Add the two equations together.
$0 = 1$ (contradiction)

If the elimination method yields a contradiction, then the system is inconsistent and has no solution.

Discovery 7

Write both equations in standard form.
$-3x - y = -2$
$3x + y = 2$
Add the two equations.
$0 = 0$ (identity)

If the elimination method yields an identity, then the system contains dependent equations and has an infinite number of solutions.

Chapter 7

Discovery 1

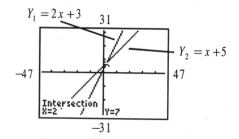

1. $(2, 7)$
2. $x = 2$
3. below; left; less than
4. $x \leq 2$

Discovery 2

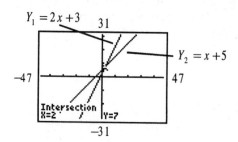

1. $(2, 7)$
2. $x = 2$
3. above; right; greater than
4. $x \geq 2$

Discovery 3

1. $ 10 < 12$

$10 + 2$	$12 + 2$
12	14

   This is true.

2. $ 10 < 12$

$10 + (-2)$	$12 + (-2)$
8	10

   This is true.

3. $ 10 < 12$

$10 - 2$	$12 - 2$
8	10

   This is true.

4. $ 10 < 12$

$10 - (-2)$	$12 - (-2)$
12	14

   This is true.

If you add or subtract the same quantity from both sides of a true inequality, the resulting inequality is also true.

Discovery 4

1. $$10 < 12$$

$10 \cdot 2$	$12 \cdot 2$
20	24

This is true.

2. $$10 < 12$$

$10 \cdot (-2)$	$12 \cdot (-2)$
-20	-24

This is not true.

3. $$10 < 12$$

$10 \div 2$	$12 \div 2$
5	6

This is true.

4. $$10 < 12$$

$10 \div (-2)$	$12 \div (-2)$
-5	-6

This is not true.

If you multiply or divide both sides of a true inequality by a positive number, the resulting inequality is also true. If you multiply or divide both sides of a true inequality by a negative number, the resulting inequality is not true. The direction of the inequality needs to be switched to make a true statement.

Discovery 5

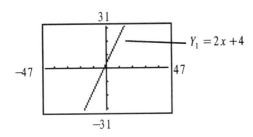

1. a. An infinite number of ordered pairs can be found. Some possible answers are:
   $(0,4)$ and $(1,6)$

   b. Yes, because of the equality.

2. a. An infinite number of ordered pairs can be found. Some possible answers are:
   $(-8,4)$ and $(3,19)$

   b. No.

3. a. An infinite number of ordered pairs can be found. Some possible answers are:
   $(-7,-22)$ and $(14,-8)$

   b. Yes.

The solution set for a "less than or equal to" linear inequality can be graphed by graphing the line for the related equation and shading all points below the line.

Discovery 6

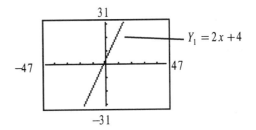

1. a. An infinite number of ordered pairs can be found. Some possible answers are:
   $(0,4)$ and $(1,6)$

   b. No, because there is no equality.

2. a. An infinite number of ordered pairs can be found. Some possible answers are:
   $(-8,4)$ and $(3,19)$

   b. Yes.

3. a. An infinite number of ordered pairs can be found. Some possible answers are:
   $(-7,-22)$ and $(14,-8)$

   b. No.

The solution set for a "greater than" linear inequality can be graphed by graphing the line for the related equation with a dashed line and shading all points above the line.

Chapter 8

Discovery 1

1.
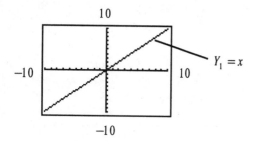

2.
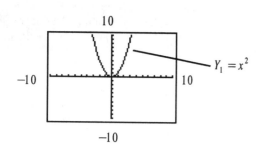

3.
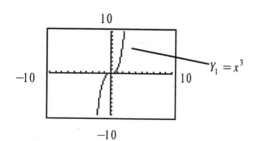

4.
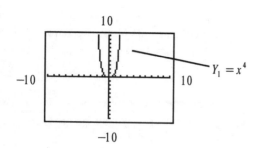

All the polynomials are functions. They all pass the vertical line test.

Discovery 2

1. to 6.
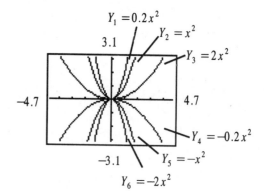

7. positive; upward

8. negative; downward

9. narrower

10. wider

Discovery 3

1.

x	y
−3	9
−2	4
−1	1
0	0
1	1
2	4
3	9

2.
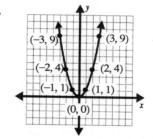

3. 1

4. 4

5. 9

Discovery 4

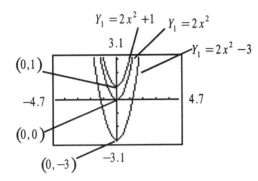

1. The y-coordinate of the y-intercept is the constant term.

2.

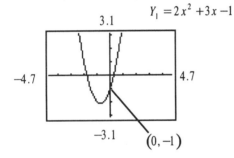

Chapter 9

Discovery 1

1. a. 1024
   b. 1024
2. a. 64
   b. 64
3. a. $\dfrac{27}{64}$
   b. $\dfrac{27}{64}$

To multiply exponential expressions with the same base, add the exponents.

4. x^7

Discovery 2

1. a. 64
   b. 64
2. a. 4
   b. 4
3. a. $\dfrac{9}{16}$
   b. $\dfrac{9}{16}$
4. a. Indeterminate
   b. Indeterminate

To divide exponential expressions with the same base, subtract the exponent in the denominator from the exponent in the numerator.

5. x^4

Discovery 3

1. a. 4096
   b. 4096
2. a. 256
   b. 256
3. a. $\dfrac{729}{4096}$
   b. $\dfrac{729}{4096}$

To raise an exponential expression to a power, multiply the powers.

4. x^{12}

SSM: Experiencing Introductory and Intermediate Algebra

Discovery 4

1. a. 512
   b. 512
2. a. 36
   b. 36
3. a. $\dfrac{27}{1000}$
   b. $\dfrac{27}{1000}$

Raising a product to a power is equivalent to the product of the factors raised to the power.

4. $x^4 y^4$

Discovery 5

1. $x^2 - 25$
2. $9x^2 - 1$
3. $x^2 - y^2$

The product of the sum and difference of the same two terms is the difference of the square of the first term and the square of the second term.

Discovery 6

1. $x^2 + 10x + 25$
2. $9x^2 - 6x + 1$
3. $x^2 + 2xy + y^2$
4. $x^2 - 2xy + y^2$

The square of a binomial that is the sum of two terms is the square of the first term, plus twice the product of the two terms, plus the square of the second term.

The square of a binomial that is the difference of two terms is the square of the first term, minus twice the product of the two terms, plus the square of the second term.

Chapter 10

Discovery 1

1. $x^3 + 125$
2. $x^3 - 125$

To factor a sum of two cubed terms, write a binomial factor and a trinomial factor. The binomial factor is the sum of the two terms being cubed. The trinomial factor is the square of the first term, minus the product of the two terms, plus the square of the second term.

To factor a difference of two cubed terms, write a binomial factor and a trinomial factor. The binomial factor is the difference of the two terms being cubed. The trinomial factor is the square of the first term, plus the product of the two terms, plus the square of the second term.

Discovery 2

1. $(x+3)(x+2)$; positive
2. $(x-3)(x-2)$; negative
3. $(x+3)(x-2)$; positive
4. $(x-3)(x+2)$; negative

If *c* is positive, then the factors both have the same sign as *b*. If *c* is negative, then the factors have different signs (the factor with the largest absolute value has the same sign as *b*).

Chapter 11

Discovery 1

1. a. 4.58258
   b. 4.58258
2. a. 2.44949
   b. 2.44949

To multiply square roots, multiply the radicands and take the square root of the product.

Discovery Boxes

Discovery 2
1. a. 1.73205
   b. 1.73205
2. a. 1.41421
   b. 1.41421

To divide square roots, divide the radicands and take the square root of the quotient.

Discovery 3
1. 49
2. 41
3. 0
4. −7

If the radicand, $b^2 - 4ac$, is a non-zero perfect square, there will be two rational solutions. If the radicand is 0, there will be one rational solution. If the radicand is positive (but not a perfect square) there will be two irrational solutions. If the radicand is negative, there will be no real solutions. (We are assuming the coefficients are rational)

Discovery 4
1.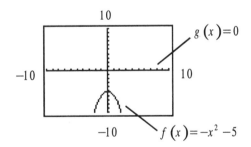
   a. No solution.
   b. No solution.
   c. All real numbers.
   d. All real numbers.

2.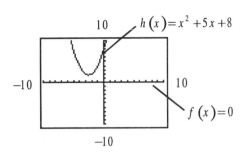
   a. All real numbers.
   b. All real numbers.
   c. No solution.
   d. No solution.

3.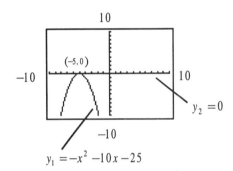
   a. No solution.
   b. $x = -5$
   c. $x \neq -5$
   d. All real numbers.

4.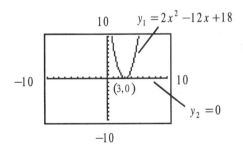
   a. $x \neq 3$
   b. All real numbers.
   c. No solution.
   d. $x = 3$

Chapter 12

Discovery 1

1. a. $f(x) = \dfrac{6}{x}$

x	$f(x)$
−3	−2
−2	−3
−1	−6
0	undefined
1	6
2	3
3	2

    Restricted values: $x = 0$

    b. $g(x) = \dfrac{x+2}{x-1}$

x	$g(x)$
−3	$\dfrac{1}{4}$
−2	0
−1	$-\dfrac{1}{2}$
0	−2
1	undefined
2	4
3	$\dfrac{5}{2}$

    Restricted values: $x = 1$

    c. $h(x) = \dfrac{x^2 - x - 2}{x - 2}$

x	$h(x)$
−3	−2
−2	−1
−1	0
0	1
1	2
2	undefined
3	4

    Restricted values: $x = 2$

2. a.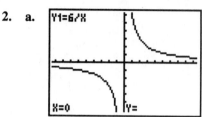

   b.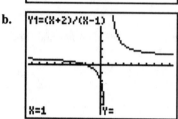

   c.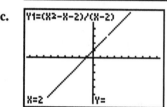

   When the cursor reaches the critical value, no value is displayed for y since it is undefined. Graphically there may be a "hole" in the graph or the function goes off to $\pm\infty$.

   The restricted values on the domain of a rational function are the numbers that result in undefined values for the rational expression.

   The restricted values of the function do not have corresponding function values.

   The restricted values can be found algebraically by setting the denominator equal to zero and solving the resulting equation for the independent variable.

Discovery 2

1. In viewing the table, the original expression is undefined when $x = 0$ and the simplified expression is defined for all x values.

2. The original expression has a restricted value of $x = 0$. The simplified expression has no restricted values.

3. When simplifying the original expression, we divided out an x in the numerator and denominator. A factor of x in the original expression is deleted in the simplified expression and prevents us from obtaining the indeterminate form $\frac{0}{0}$.

Chapter 13

Discovery 1

1. a. $f(x) = \sqrt{x}$

x	$f(x)$
−3	non-real number
−2	non-real number
−1	non-real number
0	0
1	1

    b. $g(x) = \sqrt[4]{x+1}$

x	$g(x)$
−3	non-real number
−2	non-real number
−1	0
0	1
2	1.1892

    c. $h(x) = \sqrt[3]{x}$

x	$h(x)$
−3	−1.4422
−2	−1.2599
−1	−1
0	0
1	1

    d. $f(x) = \sqrt[5]{x+1}$

x	$f(x)$
−3	−1.2457
−2	−1.1487
−1	−1
0	0
1	1

2. a. Restricted values: $x < 0$
    b. Restricted values: $x < -1$
    c. No restricted values.
    d. No restricted values.

3. The even roots have restricted values.

4.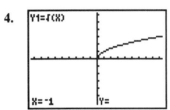
    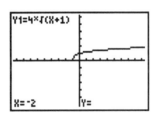

    When you reach the restricted values, there are no corresponding function values.

5. When viewing a table of values for a radical function, the restricted values are those numbers which result in non-real numbers when evaluating the radical expression.

6. When viewing the graph of a radical function, the restricted values are those numbers which have no corresponding function value.

7. We can determine the restricted values of a radical function by setting the radicand < 0 and solving the resulting inequality for the independent variable.

SSM: Experiencing Introductory and Intermediate Algebra

Discovery 2

1. a. 1.7321   b. 1.7321
2. a. 1.4422   b. 1.4422
3. a. 1.3161   b. 1.3161

To write a radical expression as an exponential expression, raise the radicand to a fractional power with 1 in the numerator and the radical index in the denominator.

Discovery 3

1. a. 8   c. 8
    b. 8
2. a. 16   c. 16
    b. 16
3. a. 16   c. 16
    b. 16
4. a. 16   c. 16
    b. 16

The base of the exponential expression becomes the radicand of the radical expression. The numerator in the exponent is the power to which the radicand, or the radical expression, is raised. The denominator in the exponent is the index of the radical expression.

Discovery 4

1. -1
2. $4 - x$

The factors contained square roots, but the product did not contain a square root.

Discovery 5

1. a. $-i$
    b. 1
    c. $-i$
    d. -1

2. $-i$
3. 1

Divide the power on *i* by 4. If the remainder is 1, the result is i; if the remainder is 2, the result is -1; if the remainder is 3, the result is $-i$; if the remainder is 0, the result is 1.

Chapter 14

Discovery 1

1.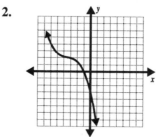

    Not possible.

2.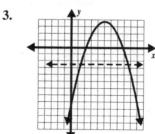

    Not possible.

    If it is possible to draw a horizontal line that crosses the graph of a function more than once, the function is not one-to-one.

3.

Discovery 2

1. a.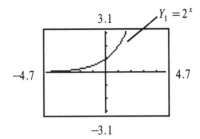

   b.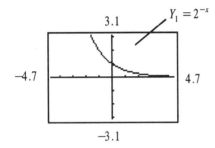

   c.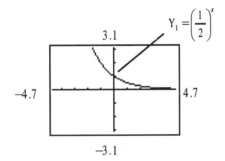

   d.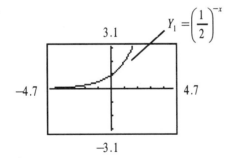

2. a and d; b and c

   An exponential function $f(x) = a^{-x}$, where $a > 1$, is equivalent to the function $f(x) = \left(\dfrac{1}{a}\right)^x$.

Discovery 3

1. a.–f.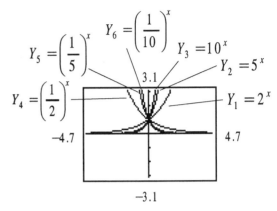

2. All real numbers.
3. All positive real numbers.
4. Yes.
5. $(0, 1)$
6. There are no x-intercepts.
7. Increasing; steeper
8. Decreasing; steeper

Discovery 4

1. a. 0
   b. 0
2. a. 1
   b. 1
3. a. 2
   b. 2
4. a. Undefined
   b. Undefined
5. a. Undefined
   b. Undefined
6. 0
7. 1

SSM: Experiencing Introductory and Intermediate Algebra

8. 2
9. Undefined
10. Undefined

Discovery 5

1. a. 1.0792
   b. 1.0792
   c. 2.3026
   d. 2.3026
2. a. −0.1249
   b. −0.1249
   c. −0.9163
   d. −0.9163
3. a. 1.9085
   b. 1.9085
   c. 3.4657
   d. 3.4657
4. The logarithm of a product is equal to the sum of the logarithms of the factors.
5. The logarithm of a quotient is equal to the difference of the logarithm of the numerator and the logarithm of the denominator.
6. The logarithm of a base raised to a power is the product of the power and the logarithm of the base.

Discovery 6

1. a. True
   b. True
   c. True

If a both sides of a true equation are written as exponential expressions using the same base and the original expressions as exponents, then the resulting equation is also true.

2. a. True
   b. True
   c. True

If we take the logarithm of both sides of a true equation, the resulting equation is also true (assuming both sides of the original equation are positive).

Chapter 1

1.1 Exercises

1. a. integer, rational
    b. natural, whole, integer, rational
    c. natural, whole, integer, rational
    d. rational
    e. rational

3. a. whole, integer, rational
    b. rational
    c. rational
    d. rational
    e. rational

5. $\frac{17}{3}$ is between 5 and 6.

7.

9. $-9 < -5$

11. $0 > -6$

13. $5 > 3$

15. $-\frac{1}{2} < -\frac{2}{5}$

17. $\frac{3}{7} < \frac{7}{10}$

19. $2\frac{3}{5} > -2\frac{3}{5}$

21. $1\frac{4}{5} = 1.8$

23. $-3.7 > -5.8$

25. $1.7 < 3.2$

27. $|-7| = 7$

29. $|-3.5| > -3.5$

31. $\left|\frac{1}{2}\right| = -\left(-\frac{1}{2}\right)$

33. $-|2| = -2$

35. $0.295 < 0.7$

37. $-5 < -1 \le 0$

39. $0.4 = \frac{2}{5}$

41. $|15.34| = 15.34$

43. $|-15.34| = 15.34$

45. $\left|-\left(-3\frac{1}{3}\right)\right| = \left|3\frac{1}{3}\right| = 3\frac{1}{3}$

47. $-|23| = -23$

49. $-|-23| = -23$

51. $-(15) = -15$

53. $-(-25) = 25$

55. $-\left(-\left(-\frac{3}{2}\right)\right) = -\left(\frac{3}{2}\right) = -\frac{3}{2}$

57. a. $-(26.97) = -26.97$; The 2001 return for the SSgA Growth & Income Fund was -26.97%.

    b. $|-26.97| = 26.97$; In 2001, the SSgA Growth & Income Fund had a loss of 26.97%.

59. 1st play: $45; 2nd play: $-\$115$; 3rd play: $200

61. -14 seconds

63. a. 5 points

SSM: Experiencing Introductory and Intermediate Algebra

b. -5 points

c. -2 points

65. $|-70| = 70$ degrees below zero Fahrenheit

1.1 Calculator Exercises

1.
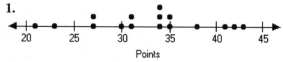
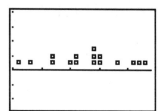

2.
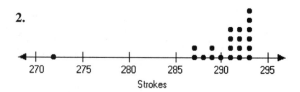
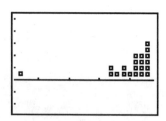

1.2 Exercises

1. $-7+9=2$
3. $-9+(-2)=-11$
5. $5+(-7)=-2$
7. $32+(-579)=-547$
9. $2.7+3.96=6.66$

11. $1.2+(-2.5)=-1.3$
13. $-2.73+4.1=1.37$
15. $-1.1+(-2.27)=-3.37$
17. $-\dfrac{3}{5}+\left(-\dfrac{1}{2}\right)=-\dfrac{6}{10}+\left(-\dfrac{5}{10}\right)=-\dfrac{11}{10}$
19. $-\dfrac{7}{9}+\dfrac{1}{6}=-\dfrac{14}{18}+\dfrac{3}{18}=-\dfrac{11}{18}$
21. $\dfrac{2}{3}+\left(-\dfrac{2}{9}\right)=\dfrac{6}{9}+\left(-\dfrac{2}{9}\right)=\dfrac{4}{9}$
23. $-2\dfrac{3}{4}+3\dfrac{2}{3}=-2\dfrac{9}{12}+3\dfrac{8}{12}=-\dfrac{33}{12}+\dfrac{44}{12}=\dfrac{11}{12}$
25. $-7-9=-7+(-9)=-16$
27. $-9-(-2)=-9+2=-7$
29. $5-(-13)=5+13=18$
31. $32-579=32+(-579)=-547$
33. $1.2-(-2.5)=1.2+2.5=3.7$
35. $-1.1-(-2.27)=-1.1+2.27=1.17$
37. $2.7-3.96=2.7+(-3.96)=-1.26$
39. $-\dfrac{3}{5}-\left(-\dfrac{1}{2}\right)=-\dfrac{3}{5}+\dfrac{1}{2}=-\dfrac{6}{10}+\dfrac{5}{10}=-\dfrac{1}{10}$
41. $-\dfrac{7}{9}-\dfrac{1}{6}=-\dfrac{7}{9}+\left(-\dfrac{1}{6}\right)=-\dfrac{14}{18}+\left(-\dfrac{3}{18}\right)=-\dfrac{17}{18}$
43. $\dfrac{2}{3}-\left(-\dfrac{7}{9}\right)=\dfrac{2}{3}+\dfrac{7}{9}=\dfrac{6}{9}+\dfrac{7}{9}=\dfrac{13}{9}$
45. $\dfrac{3}{7}-3=\dfrac{3}{7}+(-3)=\dfrac{3}{7}+\left(-\dfrac{21}{7}\right)=-\dfrac{18}{7}$
47. $-5-\left(-1\dfrac{4}{5}\right)=-5+1\dfrac{4}{5}=-5+\dfrac{9}{5}=-\dfrac{25}{5}+\dfrac{9}{5}$
$=-\dfrac{16}{5}=-3\dfrac{1}{5}$

49. $3\frac{2}{3} - 5 = 3\frac{2}{3} + (-5) = \frac{11}{3} + (-5) = \frac{11}{3} + \left(-\frac{15}{3}\right)$
$= -\frac{4}{3} = -1\frac{1}{3}$

51. $17 + (-23) + 0 + 13 = -6 + 0 + 13$
$= -6 + 13$
$= 7$

53. $-\frac{1}{2} + \frac{1}{3} + \left(-\frac{1}{4}\right) = -\frac{6}{12} + \frac{4}{12} + \left(-\frac{3}{12}\right)$
$= -\frac{2}{12} + \left(-\frac{3}{12}\right)$
$= -\frac{5}{12}$

55. $17 - (-23) - (-16) - 13 = 17 + 23 + 16 + (-13)$
$= 40 + 16 + (-13)$
$= 56 + (-13)$
$= 43$

57. $1.2 - (-2.31) - (-5.7) = 1.2 + 2.31 + 5.7$
$= 3.51 + 5.7$
$= 9.21$

59. $-\frac{1}{2} - \frac{1}{3} - \left(-\frac{1}{4}\right) = -\frac{1}{2} + \left(-\frac{1}{3}\right) + \frac{1}{4}$
$= -\frac{6}{12} + \left(-\frac{4}{12}\right) + \frac{3}{12}$
$= -\frac{10}{12} + \frac{3}{12}$
$= -\frac{7}{12}$

61. $1124 - (-924) - 2305 - (-1156) - (-109)$
$= 1124 + 924 + (-2305) + 1156 + 109$
$= 2048 + (-2305) + 1156 + 109$
$= -257 + 1156 + 109 = 889 + 109 = 1008$

63. $23 + 56 - 34 + (-12) - 68 - (-31)$
$= 23 + 56 + (-34) + (-12) + (-68) + 31$
$= 79 + (-34) + (-12) + (-68) + 31$
$= 45 + (-12) + (-68) + 31$
$= 33 + (-68) + 31 = -35 + 31 = -4$

65. $1\frac{1}{5} - 2\frac{3}{10} + \frac{4}{5} - \left(-\frac{7}{10}\right) + \left(-\frac{3}{5}\right)$
$= 1\frac{1}{5} + \left(-2\frac{3}{10}\right) + \frac{4}{5} + \frac{7}{10} + \left(-\frac{3}{5}\right)$
$= 1\frac{2}{10} + \left(-2\frac{3}{10}\right) + \frac{8}{10} + \frac{7}{10} + \left(-\frac{6}{10}\right)$
$= \frac{12}{10} + \left(-\frac{23}{10}\right) + \frac{8}{10} + \frac{7}{10} + \left(-\frac{6}{10}\right)$
$= -\frac{11}{10} + \frac{8}{10} + \frac{7}{10} + \left(-\frac{6}{10}\right)$
$= -\frac{3}{10} + \frac{7}{10} + \left(-\frac{6}{10}\right) = \frac{4}{10} + \left(-\frac{6}{10}\right)$
$= -\frac{2}{10} = -\frac{1}{5}$

67. $3.75 - 1.2 + (-1.09) - (-0.76) + 13.13$
$= 3.75 + (-1.2) + (-1.09) + 0.76 + 13.13$
$= 2.55 + (-1.09) + 0.76 + 13.13$
$= 1.46 + 0.76 + 13.13 = 2.22 + 13.13 = 15.35$

69. $29,786,000 + 2,881,000 = 32,667,000$; The 1998 projected population of California was 32,667,000.

71. $98 - (-90) = 98 + 90 = 188$; The range in temperature is $188°F$.

73. $130 - (-180) = 130 + 180 = 310$; The change in mean surface temperature is $310°C$.

75. $20,320 - (-282) = 20,320 + 282 = 20,602$; The range in elevation is 20,602 feet.

77. $\frac{1}{2} + \frac{1}{4} + 1 + 1 + \frac{1}{2} = \frac{2}{4} + \frac{1}{4} + \frac{4}{4} + \frac{4}{4} + \frac{2}{4} = \frac{13}{4} = 3\frac{1}{4}$;
The recipe calls for $3\frac{1}{4}$ cups of dry ingredients.

SSM: Experiencing Introductory and Intermediate Algebra

79. $-255 + 375 + (-575) + 1525 = 1070$; Karin is $1070 above her quota because 1070 is a positive number.

81. $1500 + 150 + 150 + 150 + (-75) + (-500) + (-200) + 12 = 1187$; The balance in Lindsay's account is $1187.

83. $19.95 + 19.95 + 19.95 + (-25.00) + (-59.27) + (-19.95) = -44.37$; Rosie's balance for the morning was $-\$44.37$. That is, Rosie lost $44.37 that morning.

85. $A = 8.4 - 1.1 - 0.7 - 0.8 = 5.8$ inches

1.2 Calculator Exercises

1. $\dfrac{55}{7} = 7\dfrac{6}{7}$

```
55/7
        7.857142857
Ans-7
        .8571428571
Ans▶Frac
                6/7
```

2. $-\dfrac{295}{113} = -2\dfrac{69}{113}$

```
295/113
        2.610619469
Ans-2
        .610619469
Ans▶Frac
              69/113
```

3. $\dfrac{1227}{487} = 2\dfrac{253}{487}$

```
1227/487
        2.519507187
Ans-2
        .5195071869
Ans▶Frac
              253/487
```

4. $-\dfrac{108}{19} = -5\dfrac{13}{19}$

```
108/19
        5.684210526
Ans-5
        .6842105263
Ans▶Frac
              13/19
```

1.3 Exercises

1. $-3 \cdot (-8) = 24$

3. $5 \cdot (-3) = -15$

5. $-2 \cdot 10 = -20$

7. $45 \cdot (-3) = -135$

9. $-32 \cdot (-4) = 128$

11. $0 \cdot (-15) = 0$

13. $(-1.7)(-0.2) = 0.34$

15. $(24.3)(0.3) = 7.29$

17. $(-0.25)(-50) = -12.5$

19. $\left(\dfrac{2}{5}\right)\left(\dfrac{25}{48}\right) = \dfrac{50}{240} = \dfrac{5}{24}$

21. $\left(1\dfrac{2}{3}\right)\left(-\dfrac{3}{4}\right) = \left(\dfrac{5}{3}\right)\left(-\dfrac{3}{4}\right) = -\dfrac{15}{12} = -\dfrac{5}{4} = -1\dfrac{1}{4}$

23. $\left(-\dfrac{4}{7}\right)\left(\dfrac{3}{16}\right) = -\dfrac{12}{112} = -\dfrac{3}{28}$

25. $15 \div (-3) = -5$

27. $-32 \div (-4) = 8$

29. $27 \div 3 = 9$

31. $0 \div (-15) = 0$

33. $26 \div (-0.13) = -200$

35. $-1.7 \div (-0.2) = 8.5$

37. $-2.7 \div (-2.7) = 1$

39. $\dfrac{5}{-25} = -0.2$

41. $\dfrac{0.88}{-1.1} = -0.8$

43. $\dfrac{2}{5} \div \dfrac{25}{48} = \dfrac{2}{5} \cdot \dfrac{48}{25} = \dfrac{96}{125}$

45. $1\dfrac{2}{3} \div \left(-\dfrac{3}{4}\right) = \dfrac{5}{3} \div \left(-\dfrac{3}{4}\right) = \dfrac{5}{3} \cdot \left(-\dfrac{4}{3}\right) = -\dfrac{20}{9} = -2\dfrac{2}{9}$

47. $-\dfrac{1}{3} \div \left(-\dfrac{3}{7}\right) = -\dfrac{1}{3} \cdot \left(-\dfrac{7}{3}\right) = \dfrac{7}{9}$

49. $-\dfrac{4}{7} \div \dfrac{3}{16} = -\dfrac{4}{7} \cdot \dfrac{16}{3} = -\dfrac{64}{21}$

51. $\dfrac{3}{5} \div 0 =$ undefined

53. $-\dfrac{2}{3} \div \left(-\dfrac{2}{3}\right) = -\dfrac{2}{3} \cdot \left(-\dfrac{3}{2}\right) = \dfrac{6}{6} = 1$

55. $\dfrac{8}{9} \div 4 = \dfrac{8}{9} \cdot \dfrac{1}{4} = \dfrac{8}{36} = \dfrac{2}{9}$

57. $14 \div \left(-\dfrac{1}{3}\right) = 14 \cdot \left(-\dfrac{3}{1}\right) = -\dfrac{42}{1} = -42$

59. $(-2)(-3)(-4)(-10)(20) = 6(-4)(-10)(20)$
$= -24(-10)(20)$
$= 240(20)$
$= 4800$

61. $\left(-\dfrac{1}{5}\right)\left(-\dfrac{2}{3}\right)\left(-\dfrac{4}{5}\right)\left(\dfrac{1}{2}\right) = \dfrac{2}{15}\left(-\dfrac{4}{5}\right)\left(\dfrac{1}{2}\right)$
$= -\dfrac{8}{75}\left(\dfrac{1}{2}\right)$
$= -\dfrac{8}{150}$
$= -\dfrac{4}{75}$

63. $(5.2)(-0.1)(-2.2) = -0.52(-2.2) = 1.144$

65. $(14)(0)(-35)(0)(-312) = 0$

67. $(-15)(4) \div (-3)(12) \div 3(-10)$
$= -60 \div (-3)(12) \div 3(-10)$
$= 20(12) \div 3(-10)$
$= 240 \div 3(-10)$
$= 80(-10)$
$= -800$

69. $(-3.3)(2.7) \div (-11)(0.6) = -8.91 \div (-11)(0.6)$
$= 0.81(0.6)$
$= 0.486$

71. $\left(\dfrac{2}{3}\right)\left(-\dfrac{5}{8}\right) \div \left(-\dfrac{5}{16}\right) = -\dfrac{10}{24} \div \left(-\dfrac{5}{16}\right)$
$= -\dfrac{10}{24} \cdot \left(-\dfrac{16}{5}\right)$
$= \dfrac{160}{120}$
$= \dfrac{4}{3}$

73. $15 + 9 \div 3 - 7 \cdot 6 \div 2 = 15 + 3 - 7 \cdot 6 \div 2$
$= 15 + 3 - 42 \div 2$
$= 15 + 3 - 21$
$= 18 - 21$
$= -3$

75. $2.7 + 5.6 - 16 \div 4 - 3 \cdot 2 = 2.7 + 5.6 - 4 - 3 \cdot 2$
$= 2.7 + 5.6 - 4 - 6$
$= 8.3 - 4 - 6$
$= 4.3 - 6$
$= -1.7$

77. $4.3(3) - 5(1.6) + 42.9 \div 3$
$= 12.9 - 5(1.6) + 42.9 \div 3$
$= 12.9 - 8 + 42.9 \div 3$
$= 12.9 - 8 + 14.3$
$= 4.9 + 14.3$
$= 19.2$

SSM: Experiencing Introductory and Intermediate Algebra

79. $0.22(-645) = -141.9$

    Sarah has $141.90 deducted every two weeks.

    $6(-141.9) = 851.4$

    Sarah has $851.40 deducted every twelve weeks.

81. 5 cents = 0.05; $0.05(40)\left(1\frac{1}{2}\right)(3) = 9$

    Sammy earns $9 for the work.

83. $50\left(8\frac{1}{2}\right)(4) = 1700$

    George drove approximately 1700 miles.

85. 59 cents = 0.59; $14(24)(0.59) = 198.24$

    The total retail value of the shipment is $198.24.

87. $659 \div 19.4 \approx 33.97$

    Al will use approximately 33.97 gallons of gas.

89. $47,500 \div 40 = 1187.5$

    1187 bottles can be filled completely.

91. $16,000,000 \text{ sec} \times \dfrac{1 \text{ min}}{60 \text{ sec}} \times \dfrac{1 \text{ hr}}{60 \text{ min}} \times \dfrac{1 \text{ day}}{24 \text{ hr}}$

    ≈ 185.19 day

    The event will kick off in about 185 days.

93. $335 \text{ CD} \times \dfrac{1 \text{ drawer}}{20 \text{ CD}} \times \dfrac{1 \text{ cabinet}}{3 \text{ drawer}} \approx 5.58$ cabinets

    Billie needs 6 cabinets in order to store her entire collection.

95. $\dfrac{\$4,040,374,000}{12 \text{ months}} \times \dfrac{3 \text{ month}}{1 \text{ quarter}} = \dfrac{\$1,010,093,500}{1 \text{ quarter}}$

    The expenditures were $1,010,093,500 per quarter.

97. $0.65(15,000) = 9750$

    A conservative investor should invest $9750 in U.S. stocks.

99. $20 \div 3 = 6\dfrac{2}{3}$

    The rate is $6\dfrac{2}{3}$ minutes per drawing.

$\left(6\dfrac{2}{3}\right)(17) = 113\dfrac{1}{3}$

It will take $113\dfrac{1}{3}$ minutes to plot the 17 drawings.

1.3 Calculator Exercises

1. $50 \div 5(2) = 20$

2. $50 \div 5 \cdot 2 = 20$

   ```
   50/5(2)
                  20
   50/5*2
                  20
   ```

3. $81 \div 9(-3) = -27$

4. $81 \div 9 \cdot (-3) = -27$

   ```
   81/9(-3)
                 -27
   81/9*(-3)
                 -27
   ```

5. $100 \div 25(4) = 16$

6. $100 \div 25 \cdot 4 = 16$

   ```
   100/25(4)
                  16
   100/25*4
                  16
   ```

7. $2 \div 2(5) = 5$

8. $2 \div 2 \cdot 5 = 5$

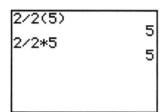

1.4 Exercises

1. $3^4 = 3 \cdot 3 \cdot 3 \cdot 3 = 81$

3. $(-3)^4 = (-3)(-3)(-3)(-3) = 81$

5. $-3^4 = -(3 \cdot 3 \cdot 3 \cdot 3) = -81$

7. $(-4)^3 = (-4)(-4)(-4) = -64$

9. $-4^3 = -(4 \cdot 4 \cdot 4) = -64$

11. $(-2.5)^2 = (-2.5)(-2.5) = 6.25$

13. $-\left(-\dfrac{3}{7}\right)^2 = -\left[\left(-\dfrac{3}{7}\right)\left(-\dfrac{3}{7}\right)\right] = -\dfrac{9}{49}$

15. $\left(1\dfrac{1}{3}\right)^3 = \left(\dfrac{4}{3}\right)^3 = \left(\dfrac{4}{3}\right)\left(\dfrac{4}{3}\right)\left(\dfrac{4}{3}\right) = \dfrac{64}{27} = 2\dfrac{10}{27}$

17. $\left(3\dfrac{2}{11}\right)^4 = \left(\dfrac{35}{11}\right)^4$
$= \left(\dfrac{35}{11}\right)\left(\dfrac{35}{11}\right)\left(\dfrac{35}{11}\right)\left(\dfrac{35}{11}\right)$
$= \dfrac{1,500,625}{14,641}$
$= 102\dfrac{7,243}{14,641}$
≈ 102.495

19. $0^8 = 0 \cdot 0 \cdot 0 \cdot 0 \cdot 0 \cdot 0 \cdot 0 \cdot 0 = 0$

21. $1^{32} = 1 \cdot 1 \cdot 1 \cdot 1 \cdot 1 \cdot 1 \cdot 1 \cdot 1 \cdot 1 \cdot 1 \cdot 1 \cdot 1 \cdot 1 \cdot 1 \cdot 1 \cdot 1$
$\cdot 1 \cdot 1 \cdot 1 \cdot 1 \cdot 1 \cdot 1 \cdot 1 \cdot 1 \cdot 1 \cdot 1 \cdot 1 \cdot 1 \cdot 1 \cdot 1 \cdot 1$
$= 1$

23. $(-1)^{29} = (-1)(-1)(-1)(-1)(-1)(-1)(-1)(-1)$
$(-1)(-1)(-1)(-1)(-1)(-1)(-1)(-1)$
$(-1)(-1)(-1)(-1)(-1)(-1)(-1)(-1)$
$(-1)(-1)(-1)(-1)(-1)$
$= -1$

25. $1256^1 = 1256$

27. $1256^0 = 1$

29. $-4721^0 = -\left(4721^0\right) = -1$

31. $(-325)^0 = 1$

33. $\sqrt{36} = 6$ because $6^2 = 36$

35. $\sqrt{256} = 16$ because $16^2 = 256$

37. $-\sqrt{25} = -5$ because $5^2 = 25$

39. $\sqrt{0.64} = 0.8$ because $0.8^2 = 0.64$

41. $-\sqrt{\dfrac{16}{9}} = -\dfrac{4}{3}$ because $\left(\dfrac{4}{3}\right)^2 = \dfrac{16}{3}$

43. $\sqrt{1} = 1$ because $1^2 = 1$

45. $-\sqrt{0} = 0 = 0$ because $0^2 = 0$

47. $\sqrt{-16}$ is not a real number

49. $\sqrt{10}$ is between $\sqrt{9} = 3$ and $\sqrt{16} = 4$;
$\sqrt{10} \approx 3.162$

51. $-\sqrt{3}$ is between $-\sqrt{4} = -2$ and $-\sqrt{1} = -1$;
$-\sqrt{3} \approx -1.732$

53. $\sqrt[3]{64} = 4$ because $4^3 = 64$

55. $\sqrt[3]{1728} = 12$ because $12^3 = 1728$

57. $\sqrt[3]{1234} \approx 10.726$

59. $\sqrt[3]{-125} = -5$ because $(-5)^3 = -125$

61. $\sqrt[3]{\dfrac{1}{8}} = \dfrac{1}{2}$ because $\left(\dfrac{1}{2}\right)^3 = \dfrac{1}{8}$

SSM: Experiencing Introductory and Intermediate Algebra

63.

65.

Description	Number line
All real numbers between −1.5 and 3, including −1.5	
All real numbers between −1.5 and 3, including 3	
All real numbers between −1.5 and 3, inclusive	
All real numbers between −1.5 and 3	
All real numbers greater than −1.5	
All real numbers less than or equal to 3	

67. $755^2 = 570,025$; The Great Pyramid covers $570,025 \text{ ft}^2$ of ground.

69. diagonal $= \sqrt{8^2 + 5^2} = \sqrt{64 + 25} = \sqrt{89} \approx 9.434$; Jennie will need about 9.434 ft of landscaping logs.

71. edge $= \sqrt[3]{3} \approx 1.442$; Each edge of the cube should be approximately 1.442 ft.

73. side $= \sqrt{12} \approx 3.464$; Each side of the table is approximately 3.464 ft.

75. side $= \sqrt{377,000} \approx 614.003$; Each side is approximately 614.003 ft.

77. radius $= \sqrt{\dfrac{19,400}{\pi}} \approx 78.583$; The radius is approximately 78.583 ft.

1.4 Calculator Exercises

Part 1

Number, n	Square, n^2	Cube, n^3
1	1	1
2	4	8
3	9	27
4	16	64
5	25	125
6	36	216
7	49	343
8	64	512
9	81	729
10	100	1000

Chapter 1: Real Numbers

11	121	1331
12	144	1728
13	169	2197
14	196	2744
15	225	3375
16	256	4096
17	289	4913
18	324	5832
19	361	6859
20	400	8000

1. $\sqrt{196} = 14$
2. $\sqrt{529} = 23$
3. $\sqrt[3]{4913} = 17$
4. $\sqrt[3]{9261} = 21$

Part 2

1. $\sqrt[5]{57,392} \approx 8.94891271$

2. $-\sqrt[4]{37,652} \approx -13.92986837$

3. $\sqrt[4]{\pi} \approx 1.331335364$

4. $\sqrt[6]{\frac{64}{729}} \approx 0.6666666667$

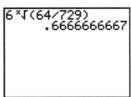

5. $\sqrt[7]{2.5} \approx 1.139852281$

6. $\sqrt[5]{-391.35393} = -3.3$

1.5 Exercises

1. $2^{-1} = \left(\frac{1}{2}\right)^1 = \frac{1}{2}$

3. $\left(\frac{1}{2}\right)^{-2} = 2^2 = 4$

5. $(-2)^{-1} = \left(-\frac{1}{2}\right)^1 = -\frac{1}{2}$

7. $\left(-\frac{1}{2}\right)^{-2} = (-2)^2 = 4$

SSM: Experiencing Introductory and Intermediate Algebra

Standard Notation	Scientific Notation	Calculator Notation
9. 23,450,000,000	2.345×10^{10}	2.345E10
11. -0.000006591	-6.591×10^{-6}	$-6.591\text{E}-6$
13. 3.6943	3.6493×10^{0}	3.6493E0
15. $-711,030$	-7.1103×10^{5}	$-7.1103\text{E}5$
17. 0.01966	1.966×10^{-2}	$1.966\text{E}-2$
19. -9.95	-9.95×10^{0}	$-9.95\text{E}0$
21. 27,000,000	2.7×10^{7}	2.7E7
23. -0.00030303	-3.0303×10^{-4}	$-3.0303\text{E}-4$
25. 1.26	1.26×10^{0}	1.26E0

27. $\$1.231 \times 10^{12} = \$1,231,000,000,000$

29. $1.231 \times 10^{12} - 3.517 \times 10^{11} = 8.793 \times 10^{11}$; The 1997 GNP of the United Kingdom exceeded the 1997 GNP of India by $\$8.793 \times 10^{11}$ or $\$879,300,000,000$.

31. Crimes: $13,175,000 = 1.3175 \times 10^{7}$

    Population: $268,008,000 = 2.68008 \times 10^{8}$

    Rate: $\dfrac{1.3175 \times 10^{7}}{2.68008 \times 10^{8}} \approx 4.9159 \times 10^{-2}$

    About 0.049159 crimes per person were committed in the U.S. in 1997.

33. Hispanic population: $30,250,000 = 3.025 \times 10^{7}$

    Total population: $270,561,000 = 2.70561 \times 10^{8}$

    $\dfrac{3.025 \times 10^{7}}{2.70561 \times 10^{8}} \approx 0.1118$

    In 1998, approximately 11.18% of the U.S. population was Hispanic.

35. $2.65 \times 10^{9} = 2,650,000,000$; Hallmark expected to sell 2,650,000,000 Christmas cards in 1996.

37. $6,000,000 \text{ ton} \times 2,000 \dfrac{\text{lb}}{\text{ton}} = 12,000,000,000 \text{ lb}$

    $= 1.2 \times 10^{10} \text{ lb}$

    Each pyramid at Giza weighs over 1.2×10^{10} or 12,000,000,000 pounds.

39. $4470 \times 5280 = 23601600 = 2.36016 \times 10^{7}$; The Great Wall of China is 2.36016×10^{7} feet long.

41. $4 \cdot (4.07 \times 10^{-10}) = 1.628 \times 10^{-9}$; The perimeter of one face of the silver crystal is 1.628×10^{-9} meters.

43. $\sqrt[3]{2.78 \times 10^{-29}} \approx 3.03 \times 10^{-10}$; The length of the edge of the cubic structure of the vanadium crystal is 3.03×10^{-10} meters.

1.5 Calculator Exercises

1. $3,000,000 \div 0.0005 = 6 \times 10^{9}$

   ```
   3000000/0.0005
                6E9
   ```

Chapter 1: Real Numbers

2. $(1.25 \times 10^5)(4.8 \times 10^{-9}) = 6 \times 10^{-4}$

```
1.25E5*4.8E-9
            6E-4
```

3. $(2.35 \times 10^{12}) - (3.87 \times 10^{11}) = 1.963 \times 10^{12}$

```
2.35E12-3.87E11
        1.963E12
```

4. $2.78 \times 10^{12} + 1.35 \times 10^{13} = 1.628 \times 10^{13}$

```
2.78E12+1.35E13
        1.628E13
```

1.6 Exercises

1. $\left(\dfrac{3}{8}\right)\left(\dfrac{5}{7} - \dfrac{1}{9}\right) = \left(\dfrac{3}{8}\right)\left(\dfrac{5}{7}\right) - \left(\dfrac{3}{8}\right)\left(\dfrac{1}{9}\right)$

3. $2.7(-1.5 + 3.2) = 2.7(-1.5) + 2.7(3.2)$

5. $\dfrac{217 - 175}{7} = \dfrac{217}{7} - \dfrac{175}{7}$

7. $-(15 + 19.3) = -15 + (-19.3) = -15 - 19.3$

9. $-\left(-\dfrac{6}{7} - \dfrac{5}{9}\right) = -\left[-\dfrac{6}{7} + \left(-\dfrac{5}{9}\right)\right]$

$\qquad = -\left(-\dfrac{6}{7}\right) + \left[-\left(-\dfrac{5}{9}\right)\right]$

$\qquad = \dfrac{6}{7} + \dfrac{5}{9}$

11. $-\left(1\dfrac{1}{7} - 2\dfrac{1}{5}\right) = -\left[1\dfrac{1}{7} + \left(-2\dfrac{1}{5}\right)\right]$

$\qquad = -\left(1\dfrac{1}{7}\right) + \left[-\left(-2\dfrac{1}{5}\right)\right]$

$\qquad = -1\dfrac{1}{7} + 2\dfrac{1}{5}$

13. $-(19.37 + 15.043) = -19.37 + (-15.043)$

$\qquad\qquad\qquad\quad = -19.37 - 15.043$

15. $15(17) + 15(23) = 15(17 + 23)$

17. $-3(14) - 3(21) = -3(14 + 21)$

19. $6(1.2) - 6(1.3) = 6(1.2 - 1.3)$

21. $3\left(\dfrac{2}{3}\right) - 9\left(\dfrac{2}{3}\right) = (3 - 9)\left(\dfrac{2}{3}\right)$

23. $-(6^2 - 12) + 5(-3 - 8) = -(36 - 12) + 5(-3 - 8)$

$\qquad\qquad\qquad\qquad\quad = -(24) + 5(-11)$

$\qquad\qquad\qquad\qquad\quad = -24 + (-55)$

$\qquad\qquad\qquad\qquad\quad = -79$

25. $[4(7 - 5) + 2] - 9 = [4(2) + 2] - 9$

$\qquad\qquad\qquad\quad = [8 + 2] - 9$

$\qquad\qquad\qquad\quad = 10 - 9$

$\qquad\qquad\qquad\quad = 1$

27. $-\sqrt{36 + 64} + 5 = -\sqrt{100} + 5 = -10 + 5 = -5$

29. $\dfrac{18 - 4^2 + 7}{2 + 1} = \dfrac{18 - 16 + 7}{2 + 1} = \dfrac{2 + 7}{3} = \dfrac{9}{3} = 3$

31. $-5(39 - 4^2) - 2(23 - 11)$

$= -5(39 - 16) - 2(23 - 11)$

$= -5(23) - 2(12)$

$= -115 - 24$

$= -139$

33. $4(15 - 8) + 31(14 - 11) = 4(7) + 31(3)$

$\qquad\qquad\qquad\qquad\quad = 28 + 93$

$\qquad\qquad\qquad\qquad\quad = 121$

SSM: Experiencing Introductory and Intermediate Algebra

35. $-|5.2 - 31.3 + 3.95| = -|-22.15|$
$ = -22.15$

37. $6^2 + 12 \div (-2) - 12 \cdot (-4)$
$= 36 + 12 \div (-2) - 12 \cdot (-4)$
$= 36 + (-6) - (-48)$
$= 36 + (-6) + 48$
$= 78$

39. $\left(\dfrac{2}{3}\right)^2 \div \left(\dfrac{1}{3} + \dfrac{1}{2}\right) \cdot \left(\dfrac{8}{9}\right) = \left(\dfrac{2}{3}\right)^2 \div \left(\dfrac{5}{6}\right) \cdot \left(\dfrac{8}{9}\right)$
$= \dfrac{4}{9} \div \left(\dfrac{5}{6}\right) \cdot \left(\dfrac{8}{9}\right)$
$= \dfrac{4}{9} \cdot \left(\dfrac{6}{5}\right) \cdot \left(\dfrac{8}{9}\right)$
$= \dfrac{64}{135}$

41. $\left(\dfrac{1}{5}\right) \cdot \left(\dfrac{15}{22}\right) \div \left(\dfrac{1}{11} + \dfrac{1}{33}\right) - \left(\dfrac{1}{3}\right)^2$
$= \left(\dfrac{1}{5}\right) \cdot \left(\dfrac{15}{22}\right) \div \left(\dfrac{4}{33}\right) - \left(\dfrac{1}{3}\right)^2$
$= \left(\dfrac{1}{5}\right) \cdot \left(\dfrac{15}{22}\right) \div \left(\dfrac{4}{33}\right) - \dfrac{1}{9}$
$= \left(\dfrac{3}{22}\right) \div \left(\dfrac{4}{33}\right) - \dfrac{1}{9}$
$= \left(\dfrac{3}{22}\right) \cdot \left(\dfrac{33}{4}\right) - \dfrac{1}{9}$
$= \dfrac{9}{8} - \dfrac{1}{9}$
$= \dfrac{73}{72}$

43. $100 - \left(24 + 7^2 - 5\right) \cdot 3 + 102 \div 2$
$= 100 - (24 + 49 - 5) \cdot 3 + 102 \div 2$
$= 100 - (68) \cdot 3 + 102 \div 2$
$= 100 - 204 + 51$
$= -53$

45. $2\{3[5 - 2(3 + 4)] + 9 \div 3\} - 9(-8)$
$= 2\{3[5 - 2(7)] + 9 \div 3\} - 9(-8)$
$= 2\{3[5 - 14] + 9 \div 3\} - 9(-8)$
$= 2\{3[-9] + 9 \div 3\} - 9(-8)$
$= 2\{-27 + 3\} - 9(-8)$
$= 2\{-24\} - 9(-8)$
$= -48 + 72$
$= 24$

47. $\dfrac{29 + 3 - 2^3}{2^2 + 2^3} = \dfrac{29 + 3 - 8}{4 + 8} = \dfrac{24}{12} = 2$

49. $15 - \sqrt{2^2 + 3 \cdot 7} = 15 - \sqrt{4 + 3 \cdot 7}$
$= 15 - \sqrt{4 + 21}$
$= 15 - \sqrt{25}$
$= 15 - 5$
$= 10$

51. $\dfrac{4 + 3 \cdot 9 - 5^2 - 2 \cdot 3}{6^2 - 5} = \dfrac{4 + 3 \cdot 9 - 25 - 2 \cdot 3}{36 - 5}$
$= \dfrac{4 + 27 - 25 - 6}{36 - 5}$
$= \dfrac{0}{31}$
$= 0$

53. $\dfrac{8^2 - 3 \cdot 12}{5 - 4 \cdot 6 + 2 \cdot 3^2 + 1^2} = \dfrac{64 - 3 \cdot 12}{5 - 4 \cdot 6 + 2 \cdot 9 + 1}$
$= \dfrac{64 - 36}{5 - 24 + 18 + 1}$
$= \dfrac{28}{0}$
$=$ undefined

55. $2.2[45.5 + 2.3(63 - 60)] = 2.2[45.5 + 2.3(3)]$
$= 2.2[45.5 + 6.9]$
$= 2.2[52.4]$
$= 115.28$
The recommended body weight of a woman who is 5 feet, 3 inches tall is 115.28 pounds.

Chapter 1: Real Numbers

57. $2.2[50+2.3(75-60)] = 2.2[50+2.3(15)]$
$= 2.2[50+34.5]$
$= 2.2[84.5]$
$= 185.9$

The recommended body weight of a man who is 6 feet, 3 inches tall is 185.9 pounds.

59. Current $= \dfrac{120}{400\sqrt{3}} \approx \dfrac{120}{400(1.732050808)}$
$\approx \dfrac{120}{692.820323} \approx 0.1732050808$

The current will be approximately 0.173 amps.

61. Speed $= 2\sqrt{5 \cdot 50} = 2\sqrt{250} \approx 2(15.8223883)$
≈ 31.6227766

The estimated speed is approximately 32 mph.

63. Speed $= 3.25\sqrt{300} \approx 3.25(17.32050808)$
≈ 56.29165125

The estimated speed is approximately 56 mph.

65. Flow rate $= (2.97)(1.5^2)\sqrt{40}$
$\approx (2.97)(2.25)(6.32455532)$
≈ 42.26384093

The flow rate is approximately 42 gallons per minute.

67. $\dfrac{-39.3+(-10.1)+(-22.5)+4.2+(-15.2)+(-30)}{6}$
$= \dfrac{-112.9}{6} \approx -18.81666667$

The average of the index returns was approximately -18.8%.

1.6 Calculator Exercises

1. $7-\{5[2+9(4-6)-12]+8(17-12 \div 6)\}$
$= 7-\{5[2+9(-2)-12]+8(17-12 \div 6)\}$
$= 7-\{5[2+(-18)-12]+8(17-2)\}$
$= 7-\{5[-28]+8(15)\} = 7-\{-140+120\}$
$= 7-\{-20\} = 7+20 = 27$

```
7-(5(2+9(4-6)-12
)+8(17-12/6))
              27
```

2. $\dfrac{4-[2(8+3)-5]}{5+2\{3[4-(8-2)]+11\}} = \dfrac{4-[2(11)-5]}{5+2\{3[4-6]+11\}}$
$= \dfrac{4-[2(11)-5]}{5+2\{3[-2]+11\}} = \dfrac{4-[22-5]}{5+2\{-6+11\}} = \dfrac{4-17}{5+2\{5\}}$
$= \dfrac{4-17}{5+10} = \dfrac{-13}{15} = -\dfrac{13}{15} \approx -0.8666666667$

```
(4-(2(8+3)-5))/(
5+2(3(4-(8-2))+1
1))
         -.8666666667
Ans▶Frac
               -13/15
```

Chapter 1 Section-By-Section Review

1. -15: integer, rational
  1000: natural, whole, integer, rational
  0: whole, integer, rational
  -2000: integer, rational
  13: natural, whole, integer, rational
  $-\dfrac{15}{17}$: rational
  12.97: rational
  $3\dfrac{5}{8}$: rational
  $\dfrac{12}{4} = 3$: natural, whole, integer, rational

37

SSM: Experiencing Introductory and Intermediate Algebra

2. a.

   (number line with points at -3, $-2\frac{1}{2}$, -1.1, 0, $\frac{3}{4}$, $1\frac{1}{2}$, 2.5, 4)

   b.

   (number line with open circle at -2 and closed circle at 3, segment shaded between)

3. $-7 > -9$

4. $\dfrac{3}{11} < \dfrac{1}{3}$

5. $\dfrac{27}{75} = 0.36$

6. $2035 > 491$

7. $12.304 < 12.344$

8. $-28 < 13$

9. $331 > -331$

10. $-1.34 < -1.04$

11. $-\dfrac{3}{4} < -\dfrac{5}{8}$

12. $|-10| = 10$

13. $-|10| < 10$

14. $-(-2) = |-2|$

15. $-100 \le -49 < -23$

16. $\dfrac{13}{17} > \dfrac{252}{529}$

17. $-\dfrac{1729}{50} = -34.58$

18. $-\left|-\dfrac{17}{33}\right| = -\left(\dfrac{17}{33}\right) = -\dfrac{17}{33}$

19. $|-(-67)| = |67| = 67$

20. $-(-32.698) = 32.698$

21. $-(-(-257)) = -(257) = -257$

22. 1st round: 0; 2nd round: -5; 3rd round: -2; 4th round: -3.

23. $-\$125{,}000{,}000$

24. $-27 + 13 = -14$

25. $-1123 + (-3406) = -4529$

26. $-5\dfrac{2}{9} + 4\dfrac{5}{9} = -\dfrac{47}{9} + \dfrac{41}{9} = -\dfrac{6}{9} = -\dfrac{2}{3}$

27. $\dfrac{102}{43} + \left(-\dfrac{29}{75}\right) = \dfrac{7650}{3225} + \left(-\dfrac{1247}{3225}\right) = \dfrac{6403}{3225}$

28. $235{,}407 + (-571{,}004) = -335{,}597$

29. $12.097 + 1.92 = 14.017$

30. $\dfrac{32}{77} + \dfrac{50}{99} = \dfrac{288}{693} + \dfrac{350}{693} = \dfrac{638}{693} = \dfrac{58}{63}$

31. $-3.57 + (-41.098) = -44.668$

32. $0 + (-123) = -123$

33. $15.28 + (-15.28) = 0$

34. $-35 - (-61) = -35 + 61 = 26$

35. $-\dfrac{3}{8} - \dfrac{4}{7} = -\dfrac{3}{8} + \left(-\dfrac{4}{7}\right) = -\dfrac{21}{56} + \left(-\dfrac{32}{56}\right) = -\dfrac{53}{56}$

36. $15.6 - 18 = 15.6 + (-18) = -2.4$

37. $0 - 3.97 = 0 + (-3.97) = -3.97$

38. $-2.3 - (-2.3) = -2.3 + 2.3 = 0$

39. $3\dfrac{5}{9} - 5\dfrac{2}{9} = 3\dfrac{5}{9} + \left(-5\dfrac{2}{9}\right)$
    $= \dfrac{32}{9} + \left(-\dfrac{47}{9}\right)$
    $= -\dfrac{15}{9}$
    $= -\dfrac{5}{3}$
    $= -1\dfrac{2}{3}$

40. $4{,}079{,}321 - 7{,}056{,}902$
$= 4{,}079{,}321 + (-7{,}056{,}902)$
$= -2{,}977{,}581$

41. $-473 - 2091 = -473 + (-2091) = -2564$

42. $-\dfrac{34}{75} - \left(-\dfrac{203}{275}\right) = -\dfrac{34}{75} + \dfrac{203}{275}$
$= -\dfrac{374}{825} + \dfrac{609}{825}$
$= \dfrac{235}{825}$
$= \dfrac{47}{165}$

43. $101.02 - (-23.9) = 101.02 + 23.9 = 124.92$

44. $553 - (-392) = 553 + 392 = 945$

45. $3.5 - 4.7 - (-8.2) - (-10.1) - 4.9 - (-16.7)$
$= 3.5 + (-4.7) + 8.2 + 10.1 + (-4.9) + 16.7$
$= -1.2 + 8.2 + 10.1 + (-4.9) + 16.7$
$= 7 + 10.1 + (-4.9) + 16.7$
$= 17.1 + (-4.9) + 16.7$
$= 12.2 + 16.7$
$= 28.9$

46. $1 - 2 - (-3) - (-4) - 5 - 6 - (-7) - 8 - (-9)$
$= 1 + (-2) + 3 + 4 + (-5) + (-6) + 7 + (-8) + 9$
$= -1 + 3 + 4 + (-5) + (-6) + 7 + (-8) + 9$
$= 2 + 4 + (-5) + (-6) + 7 + (-8) + 9$
$= 6 + (-5) + (-6) + 7 + (-8) + 9$
$= 1 + (-6) + 7 + (-8) + 9$
$= (-5) + 7 + (-8) + 9$
$= 2 + (-8) + 9$
$= (-6) + 9$
$= 3$

47. $\dfrac{3}{2} - \dfrac{1}{4} - \left(-\dfrac{8}{3}\right) - \dfrac{5}{6} - \left(-\dfrac{7}{2}\right) - 3$
$= \dfrac{3}{2} + \left(-\dfrac{1}{4}\right) + \dfrac{8}{3} + \left(-\dfrac{5}{6}\right) + \dfrac{7}{2} + (-3)$
$= \dfrac{18}{12} + \left(-\dfrac{3}{12}\right) + \dfrac{32}{12} + \left(-\dfrac{10}{12}\right) + \dfrac{42}{12} + \left(-\dfrac{36}{12}\right)$
$= \dfrac{15}{12} + \dfrac{32}{12} + \left(-\dfrac{10}{12}\right) + \dfrac{42}{12} + \left(-\dfrac{36}{12}\right)$
$= \dfrac{47}{12} + \left(-\dfrac{10}{12}\right) + \dfrac{42}{12} + \left(-\dfrac{36}{12}\right)$
$= \dfrac{37}{12} + \dfrac{42}{12} + \left(-\dfrac{36}{12}\right)$
$= \dfrac{79}{12} + \left(-\dfrac{36}{12}\right)$
$= \dfrac{43}{12}$

48. $31 - 16 + (-23) - 45 - (-37) + 52 + (-83)$
$= 31 + (-16) + (-23) + (-45) + 37 + 52 + (-83)$
$= 15 + (-23) + (-45) + 37 + 52 + (-83)$
$= -8 + (-45) + 37 + 52 + (-83)$
$= -53 + 37 + 52 + (-83)$
$= -16 + 52 + (-83)$
$= 36 + (-83)$
$= -47$

49. $3.7 - 6.83 + 5.5 + (-9.02) - 0.8 - (-15.2)$
$= 3.7 + (-6.83) + 5.5 + (-9.02) + (-0.8) + 15.2$
$= -3.13 + 5.5 + (-9.02) + (-0.8) + 15.2$
$= 2.37 + (-9.02) + (-0.8) + 15.2$
$= -6.65 + (-0.8) + 15.2$
$= -7.45 + 15.2$
$= 7.75$

SSM: Experiencing Introductory and Intermediate Algebra

50. $\dfrac{4}{5} - \dfrac{7}{3} + \dfrac{8}{9} - \left(-\dfrac{4}{15}\right) + \left(-\dfrac{2}{3}\right) + 3$

$= \dfrac{4}{5} + \left(-\dfrac{7}{3}\right) + \dfrac{8}{9} + \dfrac{4}{15} + \left(-\dfrac{2}{3}\right) + 3$

$= \dfrac{36}{45} + \left(-\dfrac{105}{45}\right) + \dfrac{40}{45} + \dfrac{12}{45} + \left(-\dfrac{30}{45}\right) + \dfrac{135}{45}$

$= -\dfrac{69}{45} + \dfrac{40}{45} + \dfrac{12}{45} + \left(-\dfrac{30}{45}\right) + \dfrac{135}{45}$

$= -\dfrac{29}{45} + \dfrac{12}{45} + \left(-\dfrac{30}{45}\right) + \dfrac{135}{45}$

$= -\dfrac{17}{45} + \left(-\dfrac{30}{45}\right) + \dfrac{135}{45}$

$= -\dfrac{47}{45} + \dfrac{135}{45}$

$= \dfrac{88}{45}$

51. $735.66 - 276.12 - 187.05 - 68.57 + 75 + 185 + 50 - 4.65 - 12$

$= 735.66 + (-276.12) + (-187.05) + (-68.57) + 75 + 185 + 50 + (-4.65) + (-12)$

$= 459.54 + (-187.05) + (-68.57) + 75 + 185 + 50 + (-4.65) + (-12)$

$= 272.49 + (-68.57) + 75 + 185 + 50 + (-4.65) + (-12)$

$= 203.92 + 75 + 185 + 50 + (-4.65) + (-12)$

$= 278.92 + 185 + 50 + (-4.65) + (-12)$

$= 463.92 + 50 + (-4.65) + (-12)$

$= 513.92 + (-4.65) + (-12)$

$= 509.27 + (-12)$

$= 497.27$

Cleta had a closing balance of $497.27.

52. $14{,}494 - (-282) = 14{,}494 + 282 = 14{,}776$
The range of elevation is 14,776 feet.

53. $(-13)(-6) = 78$

54. $(21)(-5) = -105$

55. $\left(-7\dfrac{2}{5}\right)\left(-\dfrac{5}{7}\right) = \left(-\dfrac{37}{5}\right)\left(-\dfrac{5}{7}\right) = \dfrac{185}{35} = \dfrac{37}{7} = 5\dfrac{2}{7}$

56. $(23.05)(-0.04) = -0.922$

57. $0 \cdot (-11) = 0$

58. $(-1) \cdot 25 = -25$

59. $\left(\dfrac{23}{171}\right)\left(\dfrac{33}{230}\right) = \dfrac{759}{39{,}330} = \dfrac{11}{570}$

60. $(-2.04)(-4.12) = 8.4048$

61. $2905 \cdot 1197 = 3{,}477{,}285$

62. $\dfrac{-13}{29} \cdot \dfrac{58}{169} = \dfrac{-754}{4901} = -\dfrac{2}{13}$

63. $765 \cdot (-835) = -638{,}775$

64. $(43{,}996)(0) = 0$

65. $(-32)(20)(-1)(5)(-2)(10)$
$= -640(-1)(5)(-2)(10)$
$= 640(5)(-2)(10)$
$= 3200(-2)(10)$
$= -6400(10)$
$= -64{,}000$

66. $(-1.1)(0.2)(-4)(-10)(-0.8)$
    $= -0.22(-4)(-10)(-0.8)$
    $= 0.88(-10)(-0.8)$
    $= -8.8(-0.8)$
    $= 7.04$

67. $\left(-\dfrac{1}{3}\right)\left(\dfrac{6}{7}\right)\left(-\dfrac{14}{15}\right)\left(-\dfrac{25}{8}\right) = -\dfrac{6}{21}\left(-\dfrac{14}{15}\right)\left(-\dfrac{25}{8}\right)$
    $= \dfrac{84}{315}\left(-\dfrac{25}{8}\right)$
    $= -\dfrac{2100}{2520}$
    $= -\dfrac{5}{6}$

68. $(-54)(21)(0)(32)(0)(-25) = 0$

69. $220 \div (-4) = -55$

70. $-78 \div (-3) = 26$

71. $\dfrac{5}{9} \div \left(\dfrac{-2}{3}\right) = \dfrac{5}{9} \cdot \left(\dfrac{3}{-2}\right) = \dfrac{15}{-18} = -\dfrac{5}{6}$

72. $-10.557 \div 2.3 = -4.59$

73. $0 \div 25 = 0$

74. $-2 \div 0 =$ undefined

75. $-13.7 \div (-1) = 13.7$

76. $-1{,}363{,}443 \div 2539 = -537$

77. $\dfrac{143{,}883}{657} = 219$

78. $\dfrac{-87}{121} \div \left(\dfrac{-58}{99}\right) = \dfrac{-87}{121} \cdot \left(\dfrac{99}{-58}\right) = \dfrac{-8613}{-7018} = \dfrac{27}{22}$

79. $\dfrac{65}{323} \div \dfrac{91}{247} = \dfrac{65}{323} \cdot \dfrac{247}{91} = \dfrac{16{,}055}{29{,}393} = \dfrac{65}{119}$

80. $7.32864 \div 1.056 = 6.94$

81. $(-25) \div (-5)(12)(-3) \div (-9)(-5)$
    $= 5(12)(-3) \div (-9)(-5)$
    $= 60(-3) \div (-9)(-5)$
    $= -180 \div (-9)(-5)$
    $= 20(-5)$
    $= -100$

82. $(4.2)(-3.2) \div (1.6)(0.2) \div (-0.4)(-2.2)$
    $= -13.44 \div (1.6)(0.2) \div (-0.4)(-2.2)$
    $= -8.4(0.2) \div (-0.4)(-2.2)$
    $= -1.68 \div (-0.4)(-2.2)$
    $= 4.2(-2.2)$
    $= -9.24$

83. $\left(\dfrac{5}{12}\right)\left(-\dfrac{6}{25}\right) \div \left(-\dfrac{3}{10}\right)\left(\dfrac{15}{17}\right) \div \left(-\dfrac{5}{13}\right)$
    $= -\dfrac{30}{300} \div \left(-\dfrac{3}{10}\right)\left(\dfrac{15}{17}\right) \div \left(-\dfrac{5}{13}\right)$
    $= -\dfrac{30}{300} \cdot \left(-\dfrac{10}{3}\right)\left(\dfrac{15}{17}\right) \div \left(-\dfrac{5}{13}\right)$
    $= \dfrac{300}{900}\left(\dfrac{15}{17}\right) \div \left(-\dfrac{5}{13}\right)$
    $= \dfrac{4500}{15300} \div \left(-\dfrac{5}{13}\right)$
    $= \dfrac{4500}{15{,}300} \cdot \left(-\dfrac{13}{5}\right)$
    $= -\dfrac{58{,}500}{76{,}500}$
    $= -\dfrac{13}{17}$

84. $17 + 31 - 20 \div 2 - 15 \cdot (-3)$
    $= 17 + 31 - 10 - 15 \cdot (-3)$
    $= 17 + 31 - 10 + 45$
    $= 48 - 10 + 45$
    $= 38 + 45$
    $= 83$

SSM: Experiencing Introductory and Intermediate Algebra

85. $-2(-4.8) - 5(1.7) + 9.2 = 9.6 - 5(1.7) + 9.2$
$= 9.6 - 8.5 + 9.2$
$= 1.1 + 9.2$
$= 10.3$

86. $0.83(125) = 103.75$; Of the $125 donated, $103.75 will go directly to program services.

87. $\frac{1}{7}(192,631,000) \approx 27,518,714.29$;

Approximately 27,518,714 of the 192,631,000 Americans 20 years and older could not find the United States on a world map.

88. $\frac{345}{18.7} \approx 18.44919786$; The average mileage for the trip was approximately 18.45 mpg.

89. $2^8 = 2 \cdot 2 \cdot 2 \cdot 2 \cdot 2 \cdot 2 \cdot 2 \cdot 2 = 256$

90. $(-3)^5 = (-3)(-3)(-3)(-3)(-3) = -243$

91. $(-3)^4 = (-3)(-3)(-3)(-3) = 81$

92. $1.2^2 = (1.2)(1.2) = 1.44$

93. $\left(1\frac{1}{3}\right)^3 = \left(\frac{4}{3}\right)^3 = \left(\frac{4}{3}\right)\left(\frac{4}{3}\right)\left(\frac{4}{3}\right) = \frac{64}{27} = 2\frac{10}{27}$

94. $0^{10} = 0 \cdot 0 \cdot 0 \cdot 0 \cdot 0 \cdot 0 \cdot 0 \cdot 0 \cdot 0 \cdot 0 = 0$

95. $1^{15} = 1 \cdot 1 \cdot 1 \cdot 1 \cdot 1 \cdot 1 \cdot 1 \cdot 1 \cdot 1 \cdot 1 \cdot 1 \cdot 1 \cdot 1 \cdot 1 \cdot 1 = 1$

96. $(-1)^{18} = (-1)(-1)(-1)(-1)(-1)(-1)(-1)(-1)(-1)(-1)(-1)(-1)(-1)(-1)(-1)(-1)(-1)(-1) = 1$

97. $(-1)^{21} = (-1)(-1)(-1)(-1)(-1)(-1)(-1)(-1)(-1)(-1)(-1)(-1)(-1)(-1)(-1)(-1)(-1)(-1)(-1)(-1)(-1) = -1$

98. $\left(-\frac{3}{4}\right)^6 = \left(-\frac{3}{4}\right)\left(-\frac{3}{4}\right)\left(-\frac{3}{4}\right)\left(-\frac{3}{4}\right)\left(-\frac{3}{4}\right)\left(-\frac{3}{4}\right)$
$= \frac{729}{4096}$

99. $\left(2\frac{1}{3}\right)^5 = \left(\frac{7}{3}\right)^5 = \left(\frac{7}{3}\right)\left(\frac{7}{3}\right)\left(\frac{7}{3}\right)\left(\frac{7}{3}\right)\left(\frac{7}{3}\right)$
$= \frac{16,807}{243} = 69\frac{40}{243}$

100. $(-15)^0 = 1$

101. $-15^0 = -(15^0) = -1$

102. $-15^1 = -15$

103. $3.079^1 = 3.079$

104. $\left(\frac{13}{23}\right)^0 = 1$

105. $1^0 = 1$

106. $0^0 =$ indeterminate

107. $1^1 = 1$

108. $0^1 = 0$

109. $\sqrt{81} = 9$ because $9^2 = 81$

110. $\sqrt{0.64} = 0.8$ because $0.8^2 = 0.64$

111. $\sqrt{\frac{9}{25}} = \frac{3}{5}$ because $\left(\frac{3}{5}\right)^2 = \frac{9}{25}$

112. $-\sqrt{49} = -7$ because $7^2 = 49$

113. $\sqrt{-16} =$ is not a real number

114. $\sqrt{1.2769} = 1.13$ because $1.13^2 = 1.2769$

115. $-\sqrt{470.89} = -21.7$ because $21.7^2 = 470.89$

116. $-\sqrt{\frac{576}{1369}} = -\frac{24}{37}$ because $\left(\frac{24}{37}\right)^2 = \frac{576}{1369}$

117. $\sqrt{15} \approx 3.872983346$

118. $\sqrt[3]{27} = 3$ because $3^3 = 27$

119. $\sqrt[3]{0.125} = 0.5$ because $0.5^3 = 0.125$

120. $\sqrt[3]{\dfrac{64}{729}} = \dfrac{4}{9}$ because $\left(\dfrac{4}{9}\right)^3 = \dfrac{64}{729}$

121. $\sqrt[3]{-27000} = -30$ because $(-30)^3 = -27000$

122. $-\sqrt[3]{10} \approx -2.15443469$

123. [Number line showing $-\sqrt{68}$, $-\sqrt{16}$, $-\sqrt{7.29}$, $\sqrt{9}$, π, $\sqrt{18}$, $\sqrt[3]{140}$, 9 plotted from -10 to 10]

124.

Description	Number line
All real numbers between -0.5 and 1, including -0.5	[closed dot at -0.5, open dot at 1]
All real numbers between -0.5 and 1, including 1	[open dot at -0.5, closed dot at 1]
All real numbers between -0.5 and 1, inclusive	[closed dots at -0.5 and 1]
All real numbers between -0.5 and 1	[open dots at -0.5 and 1]
All real numbers greater than or equal to 1	[closed dot at 1, ray right]
All real numbers less than or equal to -0.5	[closed dot at -0.5, ray left]

125. $\sqrt{729} = 27$; The diamond is 27 meters on each side.

126. $\sqrt[3]{91\dfrac{1}{8}} = \sqrt[3]{\dfrac{729}{8}} = \dfrac{9}{2} = 4\dfrac{1}{2}$; The box is $4\dfrac{1}{2}$ inches on each edge.

127. $4^{-1} = \left(\dfrac{1}{4}\right)^1 = \dfrac{1}{4}$

128. $\left(\dfrac{1}{4}\right)^{-1} = 4^1 = 4$

129. $\left(\dfrac{1}{4}\right)^{-2} = 4^2 = 16$

130. $(-4)^{-2} = \left(-\dfrac{1}{4}\right)^2 = \dfrac{1}{16}$

SSM: Experiencing Introductory and Intermediate Algebra

	Standard Notation	Scientific Notation	Calculator Notation
131.	0.000000189	1.89×10^{-7}	1.89E−7
132.	−27,085,000,000	-2.7085×10^{10}	−2.7085E10
133.	589,000,000,000	5.89×10^{11}	5.89E11
134.	−0.00007093	-7.093×10^{-5}	−7.093E−5

135. $682,100,000,000 = 6.821 \times 10^{11}$; The 1998 total U.S. export of goods was $\$6.821 \times 10^{11}$.

136. $9.119 \times 10^{11} = 911,900,000,000$; The 1998 total U.S. import of goods was $\$9.119 \times 10^{11}$.
$9.119 \times 10^{11} - 6.821 \times 10^{11} = 2.298 \times 10^{11}$;
In 1998, imports exceeded exports by $\$2.298 \times 10^{11} = \$229,800,000,000$.

137. $8 \times 10^{-6} = 0.000008$; The diameter of a red blood cell is 0.000008 meters.
$(8 \times 10^{-6}) \div 2 = 4 \times 10^{-6} = 0.000004$; The radius of a red blood cell is 4×10^{-6} meters or 0.000004 meters.

138. $-2.6(-1.9 + 3.2) = (-2.6)(-1.9) + (-2.6)(3.2)$

139. $\dfrac{5}{6}\left(-\dfrac{3}{5} - \dfrac{4}{15}\right) = \dfrac{5}{6}\left(-\dfrac{3}{5}\right) + \dfrac{5}{6}\left(-\dfrac{4}{15}\right)$

140. $\dfrac{1687 - 1372}{7} = \dfrac{1687}{7} - \dfrac{1372}{7}$

141. $-(2.7 + 3.09) = -2.7 + (-3.09) = -2.7 - 3.09$

142. $-[32 + (-51)] = -(32) + [-(-51)] = -32 + 51$

143. $21(18) + 21(42) = 21(18 + 42)$

144. $5(97) - 5(17) = 5(97 - 17)$

145. $5.6(18) + 5.6(-8) = 5.6[18 + (-8)]$

146. $121\left(\dfrac{3}{4}\right) - 161\left(\dfrac{3}{4}\right) = (121 - 161)\left(\dfrac{3}{4}\right)$

147. $41(-73 + 65) - (52 - 46) = 41(-8) - 6$
$= -328 - 6$
$= -334$

148. $-(27 - 4^2) - 16(-5 - 3)$
$= -(27 - 16) - 16(-5 - 3)$
$= -(11) - 16(-8)$
$= -11 - (-128)$
$= -11 + 128$
$= 117$

149. $\dfrac{191 + 104 - 11^2}{5^2 + 3^3 - 46} = \dfrac{191 + 104 - 121}{25 + 27 - 46} = \dfrac{174}{6} = 29$

150. $[15 + 21(14 - 18)] - 7^2 = [15 + 21(-4)] - 7^2$
$= [15 + (-84)] - 7^2$
$= -69 - 7^2$
$= -69 - 49$
$= -118$

151. $22 - \sqrt{9^2 - 45} + 5 = 22 - \sqrt{81 - 45} + 5$
$= 22 - \sqrt{36} + 5$
$= 22 - 6 + 5$
$= 21$

152. $\dfrac{8 \cdot 9 - 2 \cdot 6^2}{125 - 7^2} = \dfrac{8 \cdot 9 - 2 \cdot 36}{125 - 49}$
$= \dfrac{72 - 72}{125 - 49}$
$= \dfrac{0}{76}$
$= 0$

153. $\dfrac{78-3\cdot 17}{5^2-4\cdot 6-1^3} = \dfrac{78-3\cdot 17}{25-4\cdot 6-1}$

$= \dfrac{78-51}{25-24-1}$

$= \dfrac{27}{0}$

$=$ undefined

154. $2.2\left[0.5(184)-4.1\left(184^2 \div 70^2\right)\right]$

$= 2.2\left[0.5(184)-4.1(33856 \div 4900)\right]$

$\approx 2.2\left[0.5(184)-4.1(6.909387755)\right]$

$\approx 2.2[92-28.3284898]$

$\approx 2.2[63.6715102]$

≈ 140.0773224

The lean body weight is approximately 140 pounds.

155. $\dfrac{442+512+588+639}{4} = \dfrac{2181}{4} = 545.25$; The average earnings for the four periods shown is $545.25.

156. diagonal $= \sqrt{7^2+5^2} = \sqrt{49+25} = \sqrt{74}$;

lace $= 2\cdot 7+2\cdot 5+2\cdot \sqrt{74}$

$\approx 2\cdot 7+2\cdot 5+2\cdot 8.602325267$

$\approx 14+10+17.20465053$

≈ 41.20465053

Mary will need approximately 41.2 feet of lace.

Chapter 1 Chapter Review

	Standard Notation	Scientific Notation	Calculator Notation
1.	355,400,000,000,000,000	3.554×10^{17}	3.544E17
2.	-0.000092	-9.2×10^{-5}	-9.2E-5
3.	0.00000000000794	7.94×10^{-12}	7.94E-12
4.	$-68,760$	-6.876×10^4	-6.876E4

5. $42(73)-42(23)=42(73-23)$

6. $-31(18)-31(42)=-31(18+42)$

7. $75 > 59$

8. $3.54 < 3.65$

9. $\dfrac{31}{51} > \dfrac{10}{17}$

10. $-142 < -105$

11. $-\dfrac{21}{59} > -\dfrac{29}{59}$

12. $\dfrac{3}{8} = 0.375$

13. $-3.7 < 0.53$

14. $0 > -197$

15. $31 > 0$

16. $-\left|\dfrac{9}{25}\right| = -\dfrac{9}{25}$

17. $|-(-102)| = |102| = 102$

SSM: Experiencing Introductory and Intermediate Algebra

18. $-(0.085) = -0.085$

19. $-549 + (-908) = -1457$

20. $3.07 + (-2.9) = 0.17$

21. $-\dfrac{1}{8} + \left(-1\dfrac{1}{4}\right) = -\dfrac{1}{8} + \left(-\dfrac{5}{4}\right)$
    $= -\dfrac{1}{8} + \left(-\dfrac{10}{8}\right)$
    $= -\dfrac{11}{8}$
    $= -1\dfrac{3}{8}$

22. $-67{,}853 + 80{,}000 = 12{,}147$

23. $0.005 + 0.05 = 0.055$

24. $\dfrac{17}{65} + \dfrac{8}{91} = \dfrac{119}{455} + \dfrac{40}{455} = \dfrac{159}{455}$

25. $-576 - (-394) = -576 + 394 = -182$

26. $0.52 - 3 = 0.52 + (-3) = -2.48$

27. $0 - \dfrac{3}{11} = 0 + \left(-\dfrac{3}{11}\right) = -\dfrac{3}{11}$

28. $-4.07 - (-5.1) = -4.07 + 5.1 = 1.03$

29. $\dfrac{13}{25} - \dfrac{2}{5} = \dfrac{13}{25} + \left(-\dfrac{2}{5}\right) = \dfrac{13}{25} + \left(-\dfrac{10}{25}\right) = \dfrac{3}{25}$

30. $4500 - (-45) = 4500 + 45 = 4545$

31. $\left(-\dfrac{12}{55}\right)\left(-\dfrac{5}{6}\right) = \dfrac{60}{330} = \dfrac{2}{11}$

32. $\left(-4\dfrac{3}{7}\right)\left(\dfrac{14}{31}\right) = \left(-\dfrac{31}{7}\right)\left(\dfrac{14}{31}\right) = -\dfrac{434}{217} = -2$

33. $0 \cdot (-8.05) = 0$

34. $(-1)(-88) = 88$

35. $\left(\dfrac{17}{72}\right)\left(\dfrac{9}{34}\right) = \dfrac{153}{2448} = \dfrac{1}{16}$

36. $(-31.6)(-1.001) = 31.6316$

37. $4000 \cdot 1200 = 4{,}800{,}000$

38. $\dfrac{4}{13} \div \left(-\dfrac{12}{13}\right) = \dfrac{4}{13} \cdot \left(-\dfrac{13}{12}\right) = -\dfrac{52}{156} = -\dfrac{1}{3}$

39. $0 \div 0 =$ indeterminate

40. $-21 \div 0 =$ undefined

41. $56.8 \div (-1) = -56.8$

42. $-9.02 \div (-1.1) = 8.2$

43. $\dfrac{-8}{55} \div \left(\dfrac{-4}{17}\right) = \dfrac{-8}{55} \cdot \left(\dfrac{17}{-4}\right) = \dfrac{-136}{-220} = \dfrac{34}{55}$

44. $5.5 \div 2.2 = 2.5$

45. $\dfrac{184{,}008}{902} = 204$

46. $\dfrac{2}{5} \div \dfrac{56}{75} = \dfrac{2}{5} \cdot \dfrac{75}{56} = \dfrac{150}{280} = \dfrac{15}{28}$

47. $25 + 63 - 48 \div 3 - 22 \cdot (-8)$
    $= 25 + 63 - 16 - (-176)$
    $= 25 + 63 + (-16) + 176$
    $= 248$

48. $2 - \{4[3 + 2(7 - 5)] - 18 \div 3\}$
    $= 2 - \{4[3 + 2(2)] - 18 \div 3\}$
    $= 2 - \{4[3 + 4] - 18 \div 3\}$
    $= 2 - \{4[7] - 18 \div 3\}$
    $= 2 - \{28 - 6\}$
    $= 2 - 22$
    $= -20$

49. $-(34 - 7^2) - 21(-8 - 4)$
    $= -(34 - 49) - 21(-8 - 4)$
    $= -(-15) - 21(-12)$
    $= 15 + 252$
    $= 267$

50. $96 - \sqrt{11^2 - 21} - 36 = 96 - \sqrt{121 - 21} - 36$
    $= 96 - \sqrt{100} - 36$
    $= 96 - 10 - 36$
    $= 50$

51. $\dfrac{412+204-4^2}{5^2+35} = \dfrac{412+204-16}{25+35} = \dfrac{600}{60} = 10$

52. $\dfrac{10 \cdot 25 - 2 \cdot 5^3}{275 - 3^5} = \dfrac{10 \cdot 25 - 2 \cdot 125}{275 - 243}$
$= \dfrac{250 - 250}{275 - 243}$
$= \dfrac{0}{32}$
$= 0$

53. $\dfrac{35 - 6 \cdot 22}{6^3 - 4 \cdot 50 - 4^2} = \dfrac{35 - 6 \cdot 22}{216 - 4 \cdot 50 - 16}$
$= \dfrac{35 - 132}{216 - 200 - 16}$
$= \dfrac{-97}{0}$
$=$ undefined

54. $13^2 = 13 \cdot 13 = 169$

55. $(3.4)^7 = (3.4)(3.4)(3.4)(3.4)(3.4)(3.4)(3.4)$
≈ 5252.335014

56. $(-5)^3 = (-5)(-5)(-5) = -125$

57. $(-5)^4 = (-5)(-5)(-5)(-5) = 625$

58. $\left(-1\dfrac{2}{5}\right)^3 = \left(-\dfrac{7}{5}\right)^3$
$= \left(-\dfrac{7}{5}\right)\left(-\dfrac{7}{5}\right)\left(-\dfrac{7}{5}\right)$
$= -\dfrac{343}{125}$
$= -2\dfrac{93}{125}$

59. $(-1)^{24} = (-1)(-1)(-1)(-1)(-1)(-1)(-1)(-1)$
$(-1)(-1)(-1)(-1)(-1)(-1)(-1)(-1)$
$(-1)(-1)(-1)(-1)(-1)(-1)(-1)(-1)$
$= 1$

60. $(-1)^{35} = (-1)(-1)(-1)(-1)(-1)(-1)(-1)(-1)$
$(-1)(-1)(-1)(-1)(-1)(-1)(-1)(-1)$
$(-1)(-1)(-1)(-1)(-1)(-1)(-1)(-1)$
$(-1)(-1)(-1)(-1)(-1)(-1)(-1)(-1)$
$(-1)(-1)(-1)$
$= -1$

61. $3^4 = 3 \cdot 3 \cdot 3 \cdot 3 = 81$

62. $-\left(\dfrac{1}{3}\right)^5 = -\left(\dfrac{1}{3}\right)\left(\dfrac{1}{3}\right)\left(\dfrac{1}{3}\right)\left(\dfrac{1}{3}\right)\left(\dfrac{1}{3}\right) = -\dfrac{1}{243}$

63. $(-2.1)^4 = (-2.1)(-2.1)(-2.1)(-2.1) = 19.4481$

64. $22^0 = 1$

65. $(-22)^0 = 1$

66. $-22^0 = -1$

67. $22^1 = 22$

68. $-22^1 = -22$

69. $(-22)^1 = -22$

70. $0^0 =$ indeterminate

71. $0^1 = 0$

72. $(2,333,145)^0 = 1$

73. $(65.02)^1 = 65.02$

74. $(-0.84)^0 = 1$

75. $\left(\dfrac{79}{95}\right)^0 = 1$

76. $(-0.0004)^0 = 1$

77. $10^{-2} = \left(\dfrac{1}{10}\right)^2 = \dfrac{1}{100}$

78. $(-10)^{-2} = \left(-\dfrac{1}{10}\right)^2 = \dfrac{1}{100}$

SSM: Experiencing Introductory and Intermediate Algebra

79. $\left(\dfrac{1}{10}\right)^{-1} = 10^1 = 10$

80. $\left(-\dfrac{1}{10}\right)^{-1} = (-10)^1 = -10$

81. $\sqrt{2500} = 50$ because $50^2 = 2500$

82. $\sqrt{-400}$ is not a real number

83. $-\sqrt{\dfrac{441}{1444}} = -\dfrac{21}{38}$ because $\left(\dfrac{21}{38}\right)^2 = \dfrac{441}{1444}$

84. $\sqrt{1.44} = 1.2$ because $1.2^2 = 1.44$

85. $-\sqrt{\dfrac{5}{400}} \approx -0.1118033989$

86. $-\sqrt{2235.6} = -47.28213193$

87. $\sqrt[3]{-64} = -4$ because $(-4)^3 = -64$

88. $\sqrt{\dfrac{81}{16}} = \dfrac{9}{4}$ because $\left(\dfrac{9}{4}\right)^2 = \dfrac{81}{16}$

89. $-\sqrt[3]{274.625} = -6.5$ because $(-6.5)^3 = -274.625$

90. $-\sqrt[3]{25} = -2.924017738$

91. $\sqrt{2.56} = 1.6$ because $1.6^2 = 2.56$

92. $\sqrt[3]{-2\dfrac{5}{7}} \approx -1.394928199$

93. $\sqrt[3]{\dfrac{27}{64}} = \dfrac{3}{4}$ because $\left(\dfrac{3}{4}\right)^3 = \dfrac{27}{64}$

94. $-\sqrt{6.5536} = -2.56$ because $2.56^2 = 6.5536$

95. $7.3 - 38.6 + 5.5 + (-2.09) - 8 - (-2.51)$
$= 7.3 + (-38.6) + 5.5 + (-2.09) + (-8) + 2.51$
$= -31.3 + 5.5 + (-2.09) + (-8) + 2.51$
$= -25.8 + (-2.09) + (-8) + 2.51$
$= -27.89 + (-8) + 2.51$
$= -35.89 + 2.51$
$= -33.38$

96. $\dfrac{3}{7} - \dfrac{17}{21} + \dfrac{11}{14} - \left(-\dfrac{5}{6}\right) + \left(-\dfrac{35}{42}\right) + 7$
$= \dfrac{3}{7} + \left(-\dfrac{17}{21}\right) + \dfrac{11}{14} + \dfrac{5}{6} + \left(-\dfrac{35}{42}\right) + 7$
$= \dfrac{18}{42} + \left(-\dfrac{34}{42}\right) + \dfrac{33}{42} + \dfrac{35}{42} + \left(-\dfrac{35}{42}\right) + \dfrac{294}{42}$
$= -\dfrac{16}{42} + \dfrac{33}{42} + \dfrac{35}{42} + \left(-\dfrac{35}{42}\right) + \dfrac{294}{42}$
$= \dfrac{17}{42} + \dfrac{35}{42} + \left(-\dfrac{35}{42}\right) + \dfrac{294}{42}$
$= \dfrac{52}{42} + \left(-\dfrac{35}{42}\right) + \dfrac{294}{42}$
$= \dfrac{17}{42} + \dfrac{294}{42}$
$= \dfrac{311}{42}$

97. $(-55)(12)(-2)(9)(-3)(1)$
$= -660(-2)(9)(-3)(1)$
$= 1320(9)(-3)(1)$
$= 11{,}880(-3)(1)$
$= -35{,}640(1)$
$= -35{,}640$

98. $\left(-\dfrac{14}{22}\right)\left(\dfrac{8}{21}\right)\left(-\dfrac{11}{4}\right)\left(-\dfrac{3}{5}\right)$
$= -\dfrac{112}{462}\left(-\dfrac{11}{4}\right)\left(-\dfrac{3}{5}\right)$
$= \dfrac{1232}{1848}\left(-\dfrac{3}{5}\right)$
$= -\dfrac{3696}{9240}$
$= -\dfrac{2}{5}$

99. $(-11.2)(3.1)(0)(-9.4)(-1)(-7.5) = 0$

Chapter 1: Real Numbers

100. $(13.1)(-4.2) \div (2.62)(0.5) \div (-0.7)(-1.1)$
$= -55.02 \div (2.62)(0.5) \div (-0.7)(-1.1)$
$= -21(0.5) \div (-0.7)(-1.1)$
$= -10.5 \div (-0.7)(-1.1)$
$= 15(-1.1)$
$= -16.5$

101. $\left(\dfrac{9}{14}\right)\left(-\dfrac{7}{18}\right) \div \left(-\dfrac{3}{4}\right)\left(\dfrac{6}{7}\right) \div \left(-\dfrac{1}{21}\right)$
$= -\dfrac{63}{252} \div \left(-\dfrac{3}{4}\right)\left(\dfrac{6}{7}\right) \div \left(-\dfrac{1}{21}\right)$
$= -\dfrac{63}{252} \cdot \left(-\dfrac{4}{3}\right)\left(\dfrac{6}{7}\right) \div \left(-\dfrac{1}{21}\right)$
$= \dfrac{252}{756} \cdot \left(\dfrac{6}{7}\right) \div \left(-\dfrac{1}{21}\right)$
$= \dfrac{1512}{5292} \div \left(-\dfrac{1}{21}\right)$
$= \dfrac{1512}{5292} \div \left(-\dfrac{21}{1}\right)$
$= -\dfrac{31,752}{5292}$
$= -6$

102. $74 \le 75 < 76$

103. $\dfrac{1}{7} > -\dfrac{2}{5}$

104. $\dfrac{13}{25} = 0.52$

105.
a.
b.

106. $763 \div \left(6\dfrac{1}{2}\right) = 763 \div \left(\dfrac{13}{2}\right)$
$= 763 \div \left(\dfrac{2}{13}\right)$
$= \dfrac{1526}{13}$
$= 117\dfrac{5}{13}$
Approximately 117 credits run per minute.

107. $10 \cdot \left(\dfrac{3}{4}\right) = \dfrac{30}{4} = \dfrac{15}{2} = 7\dfrac{1}{2}$; It takes $7\dfrac{1}{2}$ gallons of water to fill ten 10-gallon hats.

108. $1012 - 430 = 582$; Captain Picard commanded 582 more people than Captain Kirk.
$430 - 127 = 303$; Captain Janeway commanded 303 fewer people than Captain Kirk.

109. $24 \cdot 31 \cdot 8 = 5952$; 5952 footballs will be needed.

110. $\dfrac{-213.63 + (-115.95) + (-31.32) + (-7.84)}{4}$
$= \dfrac{-368.74}{4}$
$= -92.185$
The average of the four indexes is -92.185.

111. $\sqrt[3]{1.303 \times 10^{-28}} \approx 5.06969079 \times 10^{-10}$; Each edge of the barium crystal cube will be approximately 5.070×10^{-10} meters long.

Chapter 1 Test

1. $\dfrac{4}{9} < \dfrac{5}{6}$

2. $-18 > -23$

3. $\dfrac{17}{25} = 0.68$

4. $\dfrac{1}{5} \le \dfrac{1}{4} < \dfrac{5}{16}$

5. $-2.3 > -3.2$

6. $\dfrac{3}{50} = 0.06$

7. $|-13.37| = 13.37$

8. $-|-20| = -20$

9. $59 + (-95) = -36$

10. $-\dfrac{17}{95} + \dfrac{4}{19} = -\dfrac{17}{95} + \dfrac{20}{95} = \dfrac{3}{95}$

11. $4.378 - 7.98 = 4.378 + (-7.98) = -3.602$

12. $\left(-\dfrac{3}{4}\right)\left(-\dfrac{8}{15}\right) = \dfrac{24}{60} = \dfrac{2}{5}$

13. $-6985 - (-2576) = -6985 + 2576 = -4409$

14. $45.78 \cdot (-1) = -45.78$

15. $(-37,562)(456) = -17,128,272$

16. $-413.9 + (-597.65) = -1011.55$

17. $-819 \div (-9) = 91$

18. $-\dfrac{44}{57} \div \dfrac{11}{19} = -\dfrac{44}{57} \cdot \dfrac{19}{11} = -\dfrac{836}{627} = -\dfrac{4}{3}$

19. $15.9 \div 0 = $ undefined

20. $0 \div (-53) = 0$

21. $2\dfrac{3}{7} \div 1\dfrac{3}{14} = \dfrac{17}{7} \div \dfrac{17}{14} = \dfrac{17}{7} \cdot \dfrac{14}{17} = \dfrac{238}{119} = 2$

22. $-12.05 - 2.4 = -12.05 + (-2.4) = -14.45$

23. $(-23)(-4)(0)(-17)(0)(-45) = 0$

24. $\left(-\dfrac{5}{6}\right)\left(-\dfrac{3}{7}\right)\left(\dfrac{1}{2}\right)\left(-\dfrac{2}{15}\right) = \dfrac{15}{42}\left(\dfrac{1}{2}\right)\left(-\dfrac{2}{15}\right)$
$= \dfrac{15}{84}\left(-\dfrac{2}{15}\right)$
$= -\dfrac{30}{1260}$
$= -\dfrac{1}{42}$

25. $4.3 + (-0.1) - (-2) + (-1.1)$
$= 4.3 + (-0.1) + 2 + (-1.1)$
$= 4.2 + 2 + (-1.1)$
$= 6.2 + (-1.1)$
$= 5.1$

26. $(-42)(22) \div 77(-4) \div (-16)$
$= -924 \div 77(-4) \div (-16)$
$= -12(-4) \div (-16)$
$= 48 \div (-16)$
$= -3$

27.
a. [number line showing -3.5, $-1\tfrac{1}{2}$, $\tfrac{1}{2}$, $2\tfrac{3}{4}$, 3, $\sqrt[3]{100}$, $\sqrt{25}$]

b. [number line with closed circle near -4 and open circle near 4]

28. $5,239,000,000,000,000 = 5.239 \times 10^{15}$

29. $-0.00000203 = -2.03 \times 10^{-6}$

30. $(1.5)^2 = (1.5)(1.5) = 2.25$

31. $\left(\dfrac{4}{3}\right)^4 = \left(\dfrac{4}{3}\right)\left(\dfrac{4}{3}\right)\left(\dfrac{4}{3}\right)\left(\dfrac{4}{3}\right) = \dfrac{256}{81}$

32. $1^9 = 1 \cdot 1 \cdot 1 \cdot 1 \cdot 1 \cdot 1 \cdot 1 \cdot 1 \cdot 1 = 1$

33. $0^{12} = 0 \cdot 0 \cdot 0 \cdot 0 \cdot 0 \cdot 0 \cdot 0 \cdot 0 \cdot 0 \cdot 0 \cdot 0 \cdot 0 = 0$

34. $(-4)^3 = (-4)(-4)(-4) = -64$

35. $(-4.008)^1 = -4.008$

36. $0^0 = $ indeterminate

37. $(-10)^{-1} = \left(-\dfrac{1}{10}\right)^1 = -\dfrac{1}{10}$

38. $(-3)^0 = 1$

39. $\sqrt{\dfrac{36}{121}} = \dfrac{6}{11}$ because $\left(\dfrac{6}{11}\right)^2 = \dfrac{36}{121}$

40. $-\sqrt{3.6} \approx -1.897366596$

41. $-[51.3-(-20.9)] = -[51.3+20.9]$
$= -[72.2]$
$= -72.2$

42. $\dfrac{-609+928}{29} = \dfrac{319}{29} = 11$

43. $\sqrt[3]{17^2 - 6\cdot 12 - 1} + 126 \div 9$
$= \sqrt[3]{289 - 6\cdot 12 - 1} + 126 \div 9$
$= \sqrt[3]{289 - 72 - 1} + 126 \div 9$
$= \sqrt[3]{216} + 126 \div 9$
$= 6 + 126 \div 9$
$= 6 + 14$
$= 20$

44. $\dfrac{2(5^2+3^2)-8^2-2^2}{3.65} = \dfrac{2(25+9)-64-4}{3.65}$
$= \dfrac{2(34)-64-4}{3.65}$
$= \dfrac{68-64-4}{3.65}$
$= \dfrac{0}{3.65}$
$= 0$

45. $(9.5)^2 - 68 = 90.25 - 68 = 22.25$; The garden will have 22.25 square feet of excess space.

46. $0.02(1.89) = 0.0378$; There are 0.0378 liters of fat in the milk.
$0.0378 \div 8 = 0.004725$; There are 0.004725 liters of fat in each serving.

47. $8.511\times 10^{12} - 5.934\times 10^{11} = 7.9176\times 10^{12}$; The GDP of the United States exceeded that of Russia by $\$7.9176\times 10^{12}$ or $\$7,917,600,000,000$.

48. If both numbers have the same sign (i.e. both positive or both negative), the product will be positive. If the signs of the two numbers are different (i.e. one positive and one negative), the product will be negative.

49. If the number of negatives is odd, the product will be negative. If the number of negatives is even, then the product will be positive.

Chapter 2

2.1 Exercises

For problems 1-20, let x, y, z = some number

1. Let p = total price of a gallon of milk
   $$\frac{3}{4}p$$

3. Let A = total amount
   $$\frac{A}{15}$$

5. $2.5x - \dfrac{19.59}{x}$

7. $12x - 25$

9. Let c = cost of chair in dollars
   $80 + 3c$

11. Let L = length of rectangle
    $$\frac{1}{3}(L-5)$$

13. $4 + 5x$

15. $(x+y) + 2x \cdot y$
    $x + y + 2xy$

17. $\dfrac{2x}{x+5}$

19. Let x = amount invested in first fund
    The amount he has left to invest is the total amount minus what he invested in the first fund. Thus, the expression for the amount he can invest in the second fund is $10,000 - x$

21. For $x = -3$
    a. $-x = -(-3) = 3$
    b. $-(-x) = -(-(-3)) = -(3) = -3$
    c. $x^2 = (-3)^2 = 9$
    d. $-x^2 = -(-3)^2 = -(9) = -9$

23. $3x + 5$ for $x = -5$
    $3(-5) + 5 = -15 + 5 = -10$

25. $3x + 5$ for $x = \dfrac{2}{3}$
    $3\left(\dfrac{2}{3}\right) + 5 = 2 + 5 = 7$

27. $3x + 5$ for $x = -2.7$
    $3(-2.7) + 5 = -8.1 + 5 = -3.1$

29. $18 - 3z$ for $z = -12.07$
    $18 - 3(-12.07) = 18 + 36.21 = 54.21$

31. $18 - 3z$ for $z = 23$
    $18 - 3(23) = 18 - 69 = -51$

33. $18 - 3z$ for $z = -\dfrac{5}{6}$
    $18 - 3\left(-\dfrac{5}{6}\right) = 18 + 3\left(\dfrac{5}{6}\right) = 18 + \dfrac{5}{2}$
    $= \dfrac{36}{2} + \dfrac{5}{2} = \dfrac{41}{2}$

35. $\dfrac{1}{2}bh$ for $b = 56$ and $h = 14$
    $\dfrac{1}{2}(56)(14) = (28)(14) = 392$

37. $\dfrac{1}{2}bh$ for $b = \dfrac{8}{5}$ and $h = \dfrac{16}{3}$
    $\dfrac{1}{2}\left(\dfrac{8}{5}\right)\left(\dfrac{16}{3}\right) = \left(\dfrac{4}{5}\right)\left(\dfrac{16}{3}\right) = \dfrac{64}{15}$

39. $\dfrac{1}{2}bh$ for $b = 6.8$ and $h = 4.2$
    $\dfrac{1}{2}(6.8)(4.2) = (3.4)(4.2) = 14.28$

41. $x^2 + 2x + 9$ for $x = -7$
    $(-7)^2 + 2(-7) + 9 = 49 - 14 + 9 = 44$

43. $x^2 + 2x + 9$ for $x = 2.5$
$(2.5)^2 + 2(2.5) + 9 = 6.25 + 5 + 9 = 20.25$

45. $\sqrt{4x^2 - 20x + 25} + 8$ for $x = 3$
$\sqrt{4(3)^2 - 20(3) + 25} + 8 = \sqrt{4(9) - 60 + 25} + 8$
$= \sqrt{36 - 35} + 8 = \sqrt{1} + 8 = 1 + 8 = 9$

47. $\sqrt{4x^2 - 20x + 25} + 8$ for $x = -4$
$\sqrt{4(-4)^2 - 20(-4) + 25} + 8$
$= \sqrt{4(16) + 80 + 25} + 8 = \sqrt{64 + 105} + 8$
$= \sqrt{169} + 8 = 13 + 8 = 21$

49. $|4.5 - 3.1b|$ for $b = 2$
$|4.5 - 3.1(2)| = |4.5 - 6.2| = |-1.7| = 1.7$

51. $|4.5 - 3.1b|$ for $b = -2$
$|4.5 - 3.1(-2)| = |4.5 + 6.2| = |10.7| = 10.7$

53. $\dfrac{-b + \sqrt{b^2 - 4ac}}{2a}$ for $a = 1, b = -2, c = -15$
$\dfrac{-(-2) + \sqrt{(-2)^2 - 4(1)(-15)}}{2(1)} = \dfrac{2 + \sqrt{4 + 60}}{2}$
$= \dfrac{2 + \sqrt{64}}{2} = \dfrac{2 + 8}{2} = \dfrac{10}{2} = 5$

55. $\dfrac{-b + \sqrt{b^2 - 4ac}}{2a}$ for $a = 2, b = 5, c = -12$
$\dfrac{-5 + \sqrt{(5)^2 - 4(2)(-12)}}{2(2)} = \dfrac{-5 + \sqrt{25 + 96}}{4}$
$= \dfrac{-5 + \sqrt{121}}{4} = \dfrac{-5 + 11}{4} = \dfrac{6}{4} = \dfrac{3}{2}$

57. Let a = number of adult tickets and let
c = number of child tickets
total $ raised = $6a + 2c$
When $a = 235$ and $c = 380$
total $ raised = $6(235) + 2(380)$
$= 1410 + 760 = 2170$
The total $ raised would be $2170.

59. Let x = number of pc systems and
y = pieces of ancillary equipment
monthly profit = revenue − costs
$= (200x + 50y) - 750 = 200x + 50y - 750$
When $x = 19$ and $y = 12$
monthly profit = $200(19) + 50(12) - 750$
$= 3800 + 600 - 750 = 3650$
Pablo would make a profit of $3650.

61. Let d = diameter of circle
circumference = $\pi \cdot d$
When $d = 12$ feet
circumference = $\pi(12) = 12\pi \approx 37.7$
The circumference of the pool would be roughly 37.7 feet.

63. Let d = number of days and m = miles driven
rental cost = $64.99d + 0.29m$
When $d = 8$ and $m = 1250$
rental cost = $64.99(8) + 0.29(1250) = 882.42$
The rental cost would be $882.42.

65. Let x = length of a side
area = x^2, perimeter = $4x$
When $x = 21$
area = $(21)^2 = 441$, perimeter = $4 \cdot 21 = 84$
The area would be 441 square inches and the perimeter would be 84 inches.

67. Let c = retail cost of coat
total cost = $c + 0.085c$
When $c = 149.00$
total cost of coat = $149.00 + 0.085(149.00) = 161.665$
The total cost of the coat would be $161.67.

SSM: Experiencing Introductory and Intermediate Algebra

2.1 Calculator Exercises

1. $a^2 + b^2$ for $a = 7, b = 9$
   $(7)^2 + (9)^2 = 49 + 81 = 130$

   ```
   7→A:9→B:(A+B)²
                256
   ```

2. $a^2 + b^2$ for $a = 0.8, b = 0.6$
   $(0.8)^2 + (0.6)^2 = 6.4 + 3.6 = 1$

   ```
   .8→A:.6→B:(A+B)²
                1.96
   ```

3. $a^2 + b^2$ for $a = \dfrac{9}{5}, b = \dfrac{12}{5}$

   $\left(\dfrac{9}{5}\right)^2 + \left(\dfrac{12}{5}\right)^2 = \dfrac{81}{25} + \dfrac{144}{25} = \dfrac{225}{25} = 9$

   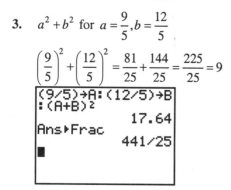

4. $a^2 + b^2$ for $a = 6, b = 5$
   $(6)^2 + (5)^2 = 36 + 25 = 61$

   ```
   6→A:5→B:(A+B)²
                121
   ```

5. $a^2 + b^2$ for $a = 0.4, b = 0.3$
   $(0.4)^2 + (0.3)^2 = 0.16 + 0.09 = 0.25$

   ```
   .4→A:.3→B:(A+B)²
                 .49
   ```

6. $a^2 + b^2$ for $a = \dfrac{15}{7}, b = \dfrac{8}{7}$

   $\left(\dfrac{15}{7}\right)^2 + \left(\dfrac{8}{7}\right)^2 = \dfrac{225}{49} + \dfrac{64}{49} = \dfrac{289}{49}$

   ```
   (15/7)→A:(8/7)→B
   :(A+B)²
            10.79591837
   Ans▶Frac
                 529/49
   ```

2.2 Exercises

1. $2x^2 - 6x + x + 12$
   a. Number of terms = 4
   b. Constant term(s): 12
   c. Variable term(s): $2x^2, -6x, x$
   d. Coefficients: $2, -6, 1, 12$
   e. Like terms: $-6x, x$

3. $3.4a - 11.2b - 0.3a$
   a. Number of terms: 3
   b. Constant term(s): none
   c. Variable term(s): $3.4a$, $-11.2b$, $-0.3a$
   d. Coefficients: $3.4, -11.2, -0.3$
   e. Like terms: $3.4a, -0.3a$

5. $3m(n-5) + 6(n-5)$
   a. Number of terms: 2

Chapter 2: Variables, Expressions, Equations, and Formulas

b. Constant term(s): none

c. Variable term(s): $3m(n-5), 6(n-5)$

d. Coefficients: 3, 6

e. Like terms: none

7. $x^2 + 3xy - y^2 + 7$

   a. Number of terms: 4

   b. Constant term(s): 7

   c. Variable term(s): $x^2, 3xy, -y^2$

   d. Coefficients: $1, 3, -1, 7$

   e. Like terms: none

9. $5x + 9 - 13x + 17 - 12 + 9x$
   $= 5x - 13x + 9x + 9 + 17 - 12$
   $= x + 14$

11. $2x^3 + 7x^2 - 2x + 8 - x^3 - 7x^2 + 3x - 2$
    $= 2x^3 - x^3 + 7x^2 - 7x^2 - 2x + 3x + 8 - 2$
    $= x^3 + x + 6$

13. $3.05a + 6.29b - 1.18a + 0.49b$
    $= 3.05a - 1.18a + 6.29b + 0.49b$
    $= 1.87a + 6.78b$

15. $\dfrac{1}{6}x + \dfrac{2}{9} - \dfrac{2}{3}x + \dfrac{5}{18}$
    $= \dfrac{1}{6}x - \dfrac{2}{3}x + \dfrac{2}{9} + \dfrac{5}{18}$
    $= \dfrac{1}{6}x - \dfrac{4}{6}x + \dfrac{4}{18} + \dfrac{5}{18}$
    $= -\dfrac{3}{6}x + \dfrac{9}{18} = -\dfrac{1}{2}x + \dfrac{1}{2}$

17. $6x^3 + 3x^2y - 5xy^2 + 3y^3 - 5x^2y + xy^2 + x^3 + 6y^3$
    $= 6x^3 + x^3 + 3x^2y - 5x^2y - 5xy^2 + xy^2 + 3y^3 + 6y^3$
    $= 7x^3 - 2x^2y - 4xy^2 + 9y^3$

19. $\dfrac{5x}{7} + \dfrac{y}{6} + \dfrac{5x}{6} - \left(\dfrac{-2y}{7}\right)$
    $= \dfrac{5x}{7} + \dfrac{5x}{6} + \dfrac{y}{6} + \dfrac{2y}{7} = \dfrac{30x}{42} + \dfrac{35x}{42} + \dfrac{7y}{42} + \dfrac{12y}{42}$
    $= \dfrac{65}{42}x + \dfrac{19}{42}y$

21. $5.3x + 1.4 + (3.4 - 1.7x)$
    $= 5.3x - 1.7x + 1.4 + 3.4$
    $= 3.6x + 4.8$

23. $(45x - 112) + (21x + 33)$
    $= 45x - 112 + 21x + 33$
    $= 45x + 21x - 112 + 33 = 66x - 79$

25. $(235 - 12y) - (307 + 31y)$
    $= 235 - 12y - 307 - 31y$
    $= -12y - 31y + 235 - 307 = -43y - 72$

27. $(x + y + 4z) - (2x - 5y + z)$
    $= x + y + 4z - 2x + 5y - z$
    $= x - 2x + y + 5y + 4z - z$
    $= -x + 6y + 3z$

29. $-(1.8y - 3.5z) + (4.1y - 2.7z)$
    $= -1.8y + 3.5z + 4.1y - 2.7z$
    $= -1.8y + 4.1y + 3.5z - 2.7z = 2.3y + 0.8z$

31. $(a+b) - (a+b) + (a+b) - (a+b)$
    $= a + b - a - b + a + b - a - b$
    $= a - a + a - a + b - b + b - b = 0$

SSM: Experiencing Introductory and Intermediate Algebra

33. $(-x+y)+(x-y)$
    $=-x+y+x-y=-x+x+y-y=0$

35. $(a+5b)-(-a+5b)$
    $=a+5b+a-5b=a+a+5b-5b=2a$

37. $-15(2x+3)$
    $=-15(2x)-15(3)=-30x-45$

39. $-4(-5z-14)$
    $=-4(-5z)-4(-14)=20z+56$

41. $2.2(3.5x-7.3)$
    $=2.2(3.5x)+2.2(-7.3)=7.7x-16.06$

43. $72\left(-\dfrac{7}{6}m+\dfrac{49}{72}\right)$
    $=72\left(-\dfrac{7}{6}m\right)+72\left(\dfrac{49}{72}\right)=-84m+49$

45. $-48\left(-\dfrac{5}{12}b+\dfrac{3}{16}\right)$
    $=-48\left(-\dfrac{5}{12}b\right)-48\left(\dfrac{3}{16}\right)=20b-9$

47. $\dfrac{36x+60}{12}$
    $=\dfrac{36x}{12}+\dfrac{60}{12}=3x+5$

49. $\dfrac{20.4b-3.4c}{-6.8}$
    $=\dfrac{20.4b}{-6.8}-\dfrac{3.4c}{-6.8}=-3b+\dfrac{1}{2}c$

51. $\dfrac{96a+24b-115c}{8}$
    $=\dfrac{96a}{8}+\dfrac{24b}{8}-\dfrac{115c}{8}=12a+3b-\dfrac{115}{8}c$

53. $11(-3a+2b-4c)-8(5a-7b+2c)$
    $=11(-3a)+11(2b)+11(-4c)$
    $\qquad-8(5a)-8(-7b)-8(2c)$
    $=-33a+22b-44c-40a+56b-16c$
    $=-33a-40a+22b+56b-44c-16c$
    $=-73a+78b-60c$

55. $-4.6(2x-5y)+9.9(5x-3y)$
    $=-4.6(2x)-4.6(-5y)+9.9(5x)+9.9(-3y)$
    $=-9.2x+23y+49.5x-29.7y$
    $=-9.2x+49.5x+23y-29.7y=40.3x-6.7y$

57. $\dfrac{3}{8}\left(-\dfrac{4}{9}p-\dfrac{2}{9}q\right)+\dfrac{2}{3}\left(\dfrac{7}{8}p-\dfrac{6}{7}q\right)$
    $=\dfrac{3}{8}\left(-\dfrac{4}{9}p\right)+\dfrac{3}{8}\left(-\dfrac{2}{9}q\right)+\dfrac{2}{3}\left(\dfrac{7}{8}p\right)+\dfrac{2}{3}\left(-\dfrac{6}{7}q\right)$
    $=-\dfrac{1}{6}p-\dfrac{1}{12}q+\dfrac{7}{12}p-\dfrac{4}{7}q$
    $=-\dfrac{1}{6}p+\dfrac{7}{12}p-\dfrac{1}{12}q-\dfrac{4}{7}q$
    $=-\dfrac{2}{12}p+\dfrac{7}{12}p-\dfrac{7}{84}q-\dfrac{48}{84}q=\dfrac{5}{12}p-\dfrac{55}{84}q$

59. $[15-2(3x+6y-10)+4x]+[6x+2(8y-12)]$
    $=[15-6x-12y+20+4x]+[6x+16y-24]$
    $=35-2x-12y+6x+16y-24=4x+4y+11$

61. $2[-5a+3(2b-4c)+15]-[7(2a+6b-c)+12]$
    $=2[-5a+6b-12c+15]-[14a+42b-7c+12]$
    $=-10a+12b-24c+30-14a-42b+7c-12$
    $=-10a-14a+12b-42b-24c+7c+30-12$
    $=-24a-30b-17c+18$

63. $6\{2[x+2(3y-4z)]-[x-y+3(y+2z)]\}$
    $=6\{2[x+6y-8z]-[x-y+3y+6z]\}$
    $=6\{2x+12y-16z-x+y-3y-6z\}$
    $=6\{x+10y-22z\}=6x+60y-132z$

Chapter 2: Variables, Expressions, Equations, and Formulas

65. $\dfrac{8(5a+7c)-6(2a+4c)}{4}$

$= \dfrac{8(5a+7c)}{4} - \dfrac{6(2a+4c)}{4}$

$= 2(5a+7c) - \dfrac{3}{2}(2a+4c)$

$= 10a + 14c - 3a - 6c$

$= 10a - 3a + 14c - 6c = 7a + 8c$

67. $\dfrac{2.6m + 3(1.2m - 2.6n) - (4.8m + 7.4n) - 1.6n}{3m + 2(-2m+1) + m}$

$= \dfrac{2.6m + 3.6m - 7.8n - 4.8m - 7.4n - 1.6n}{3m - 4m + 2 + m}$

$= \dfrac{2.6m + 3.6m - 4.8m - 7.8n - 7.4n - 1.6n}{3m - 4m + m + 2}$

$= \dfrac{1.4m - 16.8n}{2} = \dfrac{1.4m}{2} - \dfrac{16.8n}{2}$

$= 0.7m - 8.4n$

69. Let x = number of days and
    y = number of miles.
    Expense
    $= 21.99x + 0.29y + 2(120x + 50)$
    $= 21.99x + 0.29y + 240x + 100$
    $= 261.99x + 0.29y + 100$
    For $x = 4$ days and $y = 625$ miles,
    $= 261.99(4) + 0.29(625) + 100$
    $= 1047.96 + 181.25 + 100 = 1329.21$
    The total expense would be $1329.21.

71. Let h = number of hours.
    Cost $= 2(38h + 22) + 3(16h)$
    $= 76h + 44 + 48h = 124h + 44$
    For $h = 7$ hours,
    $= 124(7) + 44 = 868 + 44 = 912$
    The total cost would be $912.

73. Let c = number of square yards of carpet.
    Profit $= (20 + 1.55c) - (0.65c + 6.50)$
    $= 20 + 1.55c - 0.65c - 6.50$
    $= 13.50 + 0.90c$
    For $c = 250$ square yards

$= 13.50 + 0.90(250) = 13.50 + 225 = 238.5$
The profit would be $238.50.

75. Let w = number of ounces.
    Profit
    $= (5.00 + 1.50w) - (2.25)$
    $= 5.00 + 1.50w - 2.25 = 1.50w + 2.75$
    For $w = 4$ ounces,
    $= 1.50(4) + 2.75 = 6.00 + 2.75 = 8.75$
    The profit would be $8.75.

77. Let x = number of compact discs.
    Cost $= (9.99 + 0.08(9.99))x$
    $= (9.99 + 0.7992)x$
    $= 10.7892x$
    Change $= 50 - 10.7892x$
    For $x = 3$ compact discs,
    Cost $= 10.7892(3) = 32.3676$
    Change $= 50 - 10.7892(3) = 50 - 32.3676$
    $= 17.6324$
    Her cost would be $32.37 and her change would be $17.63.

2.2 Calculator Exercises

1. $12.078x + 2.093 - 17.42x - 13.9035$
   $= -5.342x - 11.8105$

   ```
   12.078-17.42
              -5.342
   2.093-13.9035
              -11.8105
   ```

2. $(2579x - 4302) - (1087x - 306)$
   $= 2579x - 4302 - 1087x + 306$
   $= 1492x - 3996$

   ```
   2579-1087
             1492
   -4302+306
             -3996
   ```

SSM: Experiencing Introductory and Intermediate Algebra

3. $\dfrac{10}{13}x - \dfrac{5}{52}y - \dfrac{17}{20}x - \dfrac{7}{13}y$

$= -\dfrac{21}{260}x - \dfrac{33}{52}y$

```
(10/13)-(17/20)▶
Frac
            -21/260
(-5/52)-(7/13)▶F
rac
             -33/52
```

4. $\left(2\dfrac{11}{25}\right)x + 5\dfrac{17}{30} - \left(3\dfrac{13}{15}\right)x - 3\dfrac{23}{75}$

$= -\dfrac{107}{75}x + \dfrac{113}{50}$

```
(2+11/25)-(3+13/
15)▶Frac
           -107/75
(5+17/30)-(3+23/
75)▶Frac
            113/50
```

5. $(1.0009x + 0.0004) - (0.0909x - 1.0031)$

$= 1.0009x + 0.0004 - 0.0909x + 1.0031$

$= 0.91x + 1.0035$

```
1.0009-.0909
               .91
.0004+1.0031
             1.0035
```

6. $-(935.3376x + 701.315) - (83.027x - 581.9534)$

$= -935.3376x - 701.315 - 83.027x + 581.9534$

$= -1018.3646x - 119.3616$

```
-935.3376-83.027
         -1018.3646
-701.315+581.953
4
           -119.3616
```

7. $3.995x + 12.083 - 2.995x - 9.083$

$= x + 3$

```
3.995-2.995
                 1
12.083-9.083
                 3
```

2.3 Exercises

1. Equation; equality.

3. Expression; no equality.

5. Equation; equality.

7. Equation; equality.

9. Expression; no equality.

11. $2(3) + 4 = 10$
    $6 + 4 = 10$
    $10 = 10$
    Since $10 = 10$, $x = 3$ is a solution.

13. $5(3) - 7 = 9$
    $15 - 7 = 9$
    $8 = 9$
    Since $8 \neq 9$, $y = 3$ is not a solution.

15. $6(3) + 5 = 3(3) + 17$
    $18 + 5 = 9 + 17$
    $23 = 26$
    Since $23 \neq 26$, $a = 3$ is not a solution.

17. $9(-2) - 23 = 6(-2) - 29$
    $-18 - 23 = -12 - 29$
    $-41 = -41$
    Since $-41 = -41$, $z = -2$ is a solution.

19. $.3((1) - 5) + 9 = 4(6(1) - 5) - 7$
    $3(-4) + 9 = 4(1) - 7$
    $-3 = -3$
    Since $-3 = -3$, $x = 1$ is a solution.

Chapter 2: Variables, Expressions, Equations, and Formulas

21. $2[3((8)-4)-6]+(8) = 3(8)$
    $2[3(4)-6]+8 = 24$
    $2[6]+8 = 24$
    $20 = 24$
    Since $20 \neq 24$, $x = 8$ is not a solution.

23. $(4)^2 + 5 = 33 - 3(4)$
    $16 + 5 = 33 - 12$
    $21 = 21$
    Since $21 = 21$, $x = 4$ is a solution.

For problems 25 to 35, let $x, y =$ a number

25. $x + 6 = 15$

27. Let $c =$ number of children and
    $a =$ number of adults.
    $2c - 5 = a + 2$

29. $x^2 - 21 = 100$

31. $2(x + 5^2) = x + 100$

33. $2(x+2) = 4 + 2x$

35. $17 + \dfrac{x}{2} = 4 + 3x$

37. Let $p =$ perimeter, $d =$ diameter, and
    $r =$ radius.
    $p = d + \pi r$

39. Let $I =$ interest, $A =$ compounded amount, and
    $P =$ principal.
    $I = A - P$

41. Let $n =$ number of nickels and
    $d =$ number of dimes.
    $n + d = 2n$

43. Let $x =$ amount invested in first account (\$) and
    $y =$ amount invested in other account (\$).
    $0.05x + 0.07y = 176$

45. Let $x =$ one angle and $y =$ another angle.
    $x = 3y + 10$

47. Let $H =$ magnetic intensity, $L =$ length,
    $N =$ number of turns, and $I =$ current.
    $H = \dfrac{N \cdot I}{L}$

2.3 Calculator Exercises

1. $10 + 42 \div 7x = 4x$ for $x = 3$

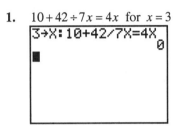

2. $10 + 42 \div (7x) = 4x$ for $x = 3$

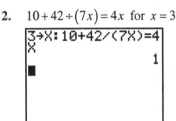

3. $10 + 42 \div 7x = 4x$
   $10 + \dfrac{42}{7}x = 4x$

4. $10 + 42 \div (7x) = 4x$
   $10 + \dfrac{42}{7x} = 2x$

5. Answers should be based on order of operations.

2.4 Exercises

1. $A = \dfrac{1}{2}bh = \dfrac{1}{2}(39)(24) = 468$ in^2
   $P = a + b + c = 39 + 40 + 25 = 104$ in

3. $A = L \cdot W = (6)(4) = 24$ cm^2
   $P = 2L + 2W = 2(6) + 2(4) = 12 + 8 = 20$ cm

5. $A = s^2 = \left(2\frac{1}{2}\right)^2 = 6.25$ ft^2
   $P = 4s = 4\left(2\frac{1}{2}\right) = 10$ ft

7. $A = b \cdot h = (68)(45) = 3060$ m^2
   $P = 2a + 2b = 2(53) + 2(68)$
   $= 106 + 136 = 242$ m

9. $A = \pi r^2 = \pi \left(5\frac{1}{4}\right)^2 \approx 86.6$ in^2
   $P = 2\pi r = 2\pi \left(5\frac{1}{4}\right) \approx 33.0$ in

11. $A = L \cdot W = (5)(4) = 20$
    cost $= 6.99(20) = 139.8$
    John will need to cover 20 yd^2 of space. It will cost him $139.80.

13. $A = s^2 = (52)^2 = 2704$
    $P = 4s = 4(52) = 208$
    The tablecloth covers 2704 square inches. Gretchen will need 208 inches of fringe material for the perimeter.

15. $A = \pi r^2 = \pi (16.5)^2 = 272.25\pi \approx 855.3$ mi^2
    $C = 2\pi r = 2\pi (16.5) = 33\pi \approx 103.7$ mi
    The bears roam an area of roughly 855.3 square miles. The circumference of this area is about 103.7 miles.

17. $A = L \cdot W = (100)(60) = 6000$ ft^2
    $P = 2L + 2W = 2(100) + 2(60)$
    $= 200 + 120 = 320$ ft
    A standard lot covered 6000 square feet and had a perimeter of 320 feet.

19. $V = L \cdot W \cdot H = (5)(3)(1) = 15$ ft^3
    $S = 2L \cdot W + 2W \cdot H + 2L \cdot H$
    $= 2(5)(3) + 2(3)(1) + 2(5)(1)$
    $= 30 + 6 + 10 = 46$ ft^2

21. $V = s^3 = (7.5)^3 = 421.875$ in^3
    $S = 6s^2 = 6(7.5)^2 = 337.5$ in^2

23. $V = \pi r^2 h = \pi (1.5)^2 (5) \approx 35.3$ in^3
    $S = 2\pi r^2 + 2\pi rh = 2\pi (1.5)^2 + 2\pi (1.5)(5)$
    ≈ 61.3 in^2
    The volume of the can is about 35.3 cubic inches and the surface area is about 61.3 square inches.

25. $V = \frac{4}{3}\pi r^3 = \frac{4}{3}\pi (10)^3 \approx 4189$ cm^3
    $S = 4\pi r^2 = 4\pi (10)^2 \approx 1257$ cm^2

27. $V = \frac{1}{3}\pi r^2 h = \frac{1}{3}\pi (.25)^2 (.75) \approx 0.049$ ft^3
    $S = \pi r \sqrt{r^2 + h^2} + \pi r^2$
    $= \pi (.25)\sqrt{(.25)^2 + (.75)^2} + \pi (.25)^2$
    ≈ 0.817 ft^2

29. $V = L \cdot W \cdot H = (4)(2)(2) = 16$ ft^3
    $S = 2LW + 2WH + 2LH$
    $= 2(4)(2) + 2(2)(2) + 2(4)(2)$
    $= 16 + 8 + 16 = 40$ ft^2
    $40 \div 20 = 2$
    The box will hold 16 cubic feet of toys. Jim painted 40 square feet of surface area which required 2 pints of paint.

31. $V = \frac{4}{3}\pi r^3 = \frac{4}{3}\pi (10)^3 \approx 4188.8$ in^3
    $S = 4\pi r^2 = 4\pi (10)^2 \approx 1256.6$ in^2
    The gas tank will hold about 4188.8 cubic inches of gas. The tank has a surface area of about 1256.6 square inches.

Chapter 2: Variables, Expressions, Equations, and Formulas

33. $V = s^3 = (18)^3 = 5832 \text{ in}^3$
    $S = 6s^2 = 6(18)^2 = 1944 \text{ in}^2$
    The doll case contains 5832 cubic inches of space. The surface area is 1944 square inches.

35. $V = \frac{1}{3}\pi r^2 h = \frac{1}{3}\pi(3.5)^2(7) \approx 89.80 \text{ in}^3$
    The paperweight has a volume of about 89.80 cubic inches.

37. $180° - (30° + 65°) = 180° - 95° = 85°$
    The third angle measures $85°$.

39. $r = \frac{d}{2} = \frac{95.25}{2} = 47.625$ inches
    The radius is 47.625 inches.
    $V = \frac{4}{3}\pi r^3 = \frac{4}{3}\pi(47.625)^3 \approx 452473.9 \text{ in}^3$
    $S = 4\pi r^2 = 4\pi(47.625)^2 \approx 28502.3 \text{ in}^2$
    The ball has a volume of about 452,473.9 cubic inches and a surface area of 28,502.3 square inches.
    $6900 \cdot (2)^2 = 27600 \text{ in}^2$
    The mirrors will cover 27,600 square inches of the surface area.

41. $90° - 65° = 25°$

43. $180° - 65° = 115°$

45. $180° - (33° + 68°) = 180° - 101° = 79°$

47. $90° - 50° = 40°$
    The pitch of the roof would measure $40°$.

49. $180° - 45° = 135°$
    The other angle formed by the guy wire would measure $135°$.

51. $F = \frac{9}{5}C + 32 = \frac{9}{5}(25) + 32 = 45 + 32 = 77$
    The Fahrenheit temperature is $77°F$.

53. $C = \frac{5}{9}(F - 32) = \frac{5}{9}(92 - 32) = \frac{100}{3} \approx 33.3$
    The high temperature, in Celsius, would be roughly $33.3°C$.

55. $F = \frac{9}{5}C + 32 = \frac{9}{5}(100) + 32 = 212$
    The boiling point of water at sea level is $212°F$.

57. $F = \frac{9}{5}C + 32 = \frac{9}{5}(95) + 32 = 203$
    The boiling point of water in Denver is $203°F$.

59. $C = \frac{5}{9}(F - 32) = \frac{5}{9}(-37.97 - 32)$
    $= -\frac{6997}{180} \approx -38.87$
    The melting point of mercury is about $-38.87°C$.

61. $C = \frac{5}{9}(F - 32) = \frac{5}{9}(84 - 32) = 28\frac{8}{9} \approx 28.9$
    The average high temperature is $28.9°C$.

63. $I = P \cdot r \cdot t = (2500)(0.065)(1) = 162.50$
    JoAnne paid $162.50 in simple interest.

65. $I = P \cdot r \cdot t = (500)(0.07)\left(\frac{1}{12}\right) \approx 2.92$
    $2.92 in simple interest would be earned.

67. $I = P \cdot r \cdot t = (1200)(0.08)\left(\frac{1}{2}\right) = 48$
    The simple interest would be $48.

69. $d = r \cdot t = (55)\left(9\frac{1}{2}\right) = 522.5$
    The distance covered was 522.5 miles.

71. $d = r \cdot t = (74.602)(3) = 223.806$
    He drove 223.806 miles in the first 3 hours.

2.4 Calculator Exercises

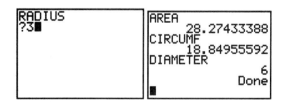

SSM: Experiencing Introductory and Intermediate Algebra

Chapter 2 Section-By-Section Review

1. Let $x =$ a number
    a. $4 + 55x$
    b. $(4 + x)55$

2. Let $x =$ a number
    a. $\frac{3}{4}(x + 35)$
    b. $\frac{3}{4}x + 35$

3. Let $x =$ a number
    a. $2x - 20$
    b. $20 - 2x$

4. Let $n =$ months
    Total cost $= 2500 + 275n$

5. Let $h =$ number of hours worked
    hourly rate $= \frac{650}{h}$

6. Let $k =$ number of customers who buy
    weekly pay $= 200 + 5.5k$

7. Let $x =$ liters of solution
    liters of antifreeze $= 0.40x$
    liters of water $= 0.60x$

8. Let $x =$ miles driven on first day
    miles on second day $= 758 - x$

9. Let $x =$ original price
    discount $= 0.25x$
    sale price $= x - 0.25x = 0.75x$

10. $-x = -\left(-\frac{2}{3}\right) = \frac{2}{3}$

11. $-(-x) = -\left(-\left(-\frac{2}{3}\right)\right) = -\frac{2}{3}$

12. $x^2 = \left(-\frac{2}{3}\right)^2 = \frac{4}{9}$

13. $-x^2 = -\left(-\frac{2}{3}\right)^2 = -\frac{4}{9}$

14. $|-x| = \left|-\left(-\frac{2}{3}\right)\right| = \left|\frac{2}{3}\right| = \frac{2}{3}$

15. $5x - 25 = 5(8) - 25 = 40 - 25 = 15$

16. $5x - 25 = 5(-4.6) - 25 = -23 - 25 = -48$

17. $\sqrt{a^2 + b^2} = \sqrt{(9)^2 + (12)^2} = \sqrt{81 + 144}$
    $= \sqrt{225} = 15$

18. $\sqrt{a^2 + b^2} = \sqrt{(4.5)^2 + (6)^2} = \sqrt{20.25 + 36}$
    $= \sqrt{56.25} = 7.5$

19. $\frac{12y - 84}{6y + 36} = \frac{12(9) - 84}{6(9) + 36} = \frac{108 - 84}{54 + 36} = \frac{24}{90} = \frac{4}{15}$

20. $\frac{12y - 84}{6y + 36} = \frac{12(-6) - 84}{6(-6) + 36} = \frac{-156}{0}$
    Undefined (division by 0).

21. $|2x^2 - 45x - 75| - 2$
    $= |2(0.1)^2 - 45(0.1) - 75| - 2$
    $= |.02 - 4.5 - 75| - 2$
    $= |-79.48| - 2 = 79.48 - 2 = 77.48$

22. $|2x^2 - 45x - 75| - 2 = |2(10)^2 - 45(10) - 75| - 2$
    $= |200 - 450 - 75| - 2 = |-325| - 2 = 325 - 2 = 323$

23. Let $L =$ length, $W =$ width, and $H =$ height
    a. volume $= L \cdot W \cdot H$
    b. volume $= (192)(126)(103) = 2,491,776$
    The building has a volume of 2,491,776 cubic feet (assuming a rectangular box).

	Algebraic Expression	Num. of terms	Constant terms	Variable terms	Coefficents	Like terms
24.	$3x-2y+4x+9y$	4	None	$3x,-2y,4x,9y$	$3,-2,4,9$	$3x$ and $4x$, $-2y$ and $9y$
25.	$2a^2-a+3a^2-5a^3$	4	None	$2a^2,-a,3a^2,-5a^3$	$2,-1,3,-5$	$2a^2,3a^2$
26.	$2.4x+5.1+6.2x$	3	5.1	$2.4x,6.2x$	$2.4,5.1,6.2$	$2.4x,6.2x$
27.	$4a(a+b)-b(a+b)$	2	None	$4a(a+b),-b(a+b)$	$4,-1$	None

28. $2.4z+1.7z-3.9z = (2.4+1.7-3.9)z = 0.2z$

29. $17x+51+26x-86-19x-7$
$= 17x+26x-19x+51-86-7$
$= 24x-42$

30. $\frac{3}{4}x+\frac{5}{8}y-\frac{2}{3}x+\frac{1}{4}y = \frac{3}{4}x-\frac{2}{3}x+\frac{5}{8}y+\frac{1}{4}y$
$= \frac{9}{12}x-\frac{8}{12}x+\frac{5}{8}y+\frac{2}{8}y$
$= \frac{1}{12}x+\frac{7}{8}y$

31. $15x^2-14xy+12y^2-23+42xy-7y^2+21-6x^2$
$= 15x^2-6x^2-14xy+42xy+12y^2-7y^2-23+21$
$= 9x^2+28xy+5y^2-2$

32. $(3a+4b)-(-2a-6b)+(a-b)-(-a+b)$
$= 3a+4b+2a+6b+a-b+a-b$
$= 3a+2a+a+a+4b+6b-b-b$
$= 7a+8b$

33. $-2(3.8a-4.7b) = -2(3.8a)-2(-4.7b)$
$= -7.6a+9.4b$

34. $-105\left(\frac{6}{7}x-\frac{4}{15}y\right) = -105\left(\frac{6}{7}x\right)-105\left(-\frac{4}{15}y\right)$
$= -90x+28y$

35. $2x(3x-17y) = 2x(3x)+2x(-17y)$
$= 6x^2-34xy$

36. $\frac{27a-36b+15c}{9} = \frac{27a}{9}-\frac{36b}{9}+\frac{15c}{9}$
$= 3a-4b+\frac{5}{3}c$

37. $12(7x+9y)+15(3x-7y)$
$= 12(7x)+12(9y)+15(3x)+15(-7y)$
$= 84x+108y+45x-105y$
$= 84x+45x+108y-105y$
$= 129x+3y$

38. $3[-2(x+3y)-5]-[3(2x+y)+16]$
$= 3[-2x-6y-5]-[6x+3y+16]$
$= -6x-18y-15-6x-3y-16$
$= -6x-6x-18y-3y-15-16$
$= -12x-21y-31$

39. $\dfrac{25(2a-6b)+5(4b-3c)+75}{4a-2(2a-3)-1}$
$= \dfrac{50a-150b+20b-15c+75}{4a-4a+6-1}$
$= \dfrac{50a-130b-15c+75}{5}$
$= \dfrac{50a}{5} - \dfrac{130b}{5} - \dfrac{15c}{5} + \dfrac{75}{5}$
$= 10a-26b-3c+15$

40. Let x = Tom's push-ups
Charles' push-ups: $2x-5$
Jim's push-ups: $x+7$
Total push-ups for all 3 boys
$= x+(2x-5)+(x+7) = x+2x-5+x+7$
$= 4x+2$
When $x = 45$,
$4x+2 = 4(45)+2 = 180+2 = 182$
The total number for all three boys will be 182 push-ups.

41. Let d = number of days and
m = number of miles
Total cost
$= 45d + 0.20m + 2(125d + 20d)$
$= 45d + 0.20m + 2(145d)$
$= 335d + 0.20m$
When $d = 4$ and $m = 345$,
$335d + 0.20m = 335(4) + 0.20(345) = 1409$.
The total cost for the trip will be $1409.

42. Let x = cans of peanuts and
y = cans of bridge mix
profit $= 2.75(x+y) - (1.25x + 1.50y)$
$= 2.75x + 2.75y - 1.25x - 1.50y$
$= 1.50x + 1.25y$

When $x = 220$ and $y = 480$,
profit $= 1.50(220) + 1.25(480) = 930$.
The school will make a profit of $930.

43. Let x = number of pins
change $= 20 - (5x)(1.06) = 20 - 5.3x$
When $x = 3$,
change $= 20 - 5.3(3) = 4.1$
Katie would receive $4.10 in change.

44. Let x = number of frames
net profit $= 12.00x - (52.65 + 3.25x)$
$= 8.75x - 52.65$
When $x = 32$,
net profit $= 8.75(32) - 52.65 = 227.35$.
Margaret's net profit would be $227.35.

45. Expression; not equality

46. Equation; equality

47. $15y - 35 = 12$ for $y = 3$
$15(3) - 35 = 12$
$45 - 35 = 12$
$10 = 12$
Since $10 \ne 12$, $y = 3$ is not a solution.

48. $8a + 2(3a-7) = 11a - 32$ for $a = -6$
$8(-6) + 2(3(-6)-7) = 11(-6) - 32$
$-48 + 2(-25) = -66 - 32$
$-98 = -98$
Since $-98 = -98$, $a = -6$ is a solution.

49. $x^3 - 25x = 2x^2 - 3x - 35$ for $x = 5$
$(5)^3 - 25(5) = 2(5)^2 - 3(5) - 35$
$125 - 125 = 50 - 15 - 35$
$0 = 0$
Since $0 = 0$, $x = 5$ is a solution.

50. $2(x - 3.4) = x - 5$ for $x = 1.8$
$2(1.8 - 3.4) = 1.8 - 5$
$2(-1.6) = -3.2$
$-3.2 = -3.2$
Since $-3.2 = -3.2$, $x = 1.8$ is a solution.

Chapter 2: Variables, Expressions, Equations, and Formulas

51. $x - \dfrac{2}{3} = -\dfrac{1}{9}$ for $x = \dfrac{4}{9}$

$\dfrac{4}{9} - \dfrac{2}{3} = -\dfrac{1}{9}$

$\dfrac{4}{9} - \dfrac{6}{9} = -\dfrac{1}{9}$

$-\dfrac{2}{9} = -\dfrac{1}{9}$

Since $-\dfrac{2}{9} \neq -\dfrac{1}{9}$, $x = \dfrac{4}{9}$ is not a solution.

52. Let $x =$ a number

$4 + 5x = 65 + \dfrac{x}{4}$

53. Let $x =$ pamphlets sold

$1500 = 1.35x$

54. Let $x =$ amt in first account and
$y =$ amt in second account
$1500 = 0.06x + 0.08y$

55. Let $x =$ amt in first account and
$y =$ amt in second account
$30{,}000 = x + y$

56. Let $f =$ frequency, $T =$ tension,
$m =$ mass per unit length, and $L =$ wire length

$f = \dfrac{\sqrt{T/m}}{2L}$

57. $A = \dfrac{1}{2}bh = \dfrac{1}{2}(26)(16) = 208$ m^2

$P = a + b + c = 22 + 26 + 20 = 68$ m

58. $A = L \cdot W = (44)(20) = 880$ in^2

$P = 2L + 2W = 2(44) + 2(20) = 128$ in

59. $A = s^2 = (15)^2 = 225$ cm^2

$P = 4s = 4(15) = 60$ cm

60. $A = b \cdot h = (10.0)(6.5) = 65.0$ m^2

$P = 2a + 2b = 2(7.0) + 2(10.0) = 34.0$ m

61. $A = \pi r^2 = \pi(12.2)^2 \approx 467.59$ ft^2

$C = 2\pi r = 2\pi(12.2) \approx 76.65$ ft

62. $V = L \cdot W \cdot H = (35)(9)(21) = 6615$ in^2

$S = 2LW + 2WH + 2LH$
$= 2(35)(9) + 2(9)(21) + 2(35)(21)$
$= 2478$ in^2

63. $V = s^3 = (14.6)^3 = 3112.136$ cm^3

$S = 6s^2 = 6(14.6)^2 = 1278.96$ cm^2

64. $V = \pi r^2 h = \pi(18)^2(54) \approx 54{,}965.3$ in^3

$S = 2\pi r^2 + 2\pi rh = 2\pi(18)^2 + 2\pi(18)(54)$
≈ 8143.01 in^2

65. $V = \dfrac{4}{3}\pi r^3 = \dfrac{4}{3}\pi(32.6)^3 \approx 145{,}124.72$ cm^3

$S = 4\pi r^2 = 4\pi(32.6)^2 \approx 13{,}355.04$ cm^2

66. $V = \dfrac{1}{3}\pi r^2 h = \dfrac{1}{3}\pi(4)^2(7) \approx 117.29$ cm^3

$S = \pi r\sqrt{r^2 + h^2} + \pi r^2$
$= \pi(4)\sqrt{4^2 + 7^2} + \pi(4)^2 \approx 151.58$ cm^2

67. $180° - 58° = 122°$

68. $90° - 58° = 32°$

69. $180° - (67° + 88°) = 180° - 155° = 25°$

70. $A = L \cdot W = (85)(60) = 5100$ ft^2

$P = 2L + 2W = 2(85) + 2(60) = 290$ ft

$290 - (3 + 35) = 290 - 38 = 252$ ft

Dan should order 5100 square feet of sod and 252 feet of fencing.

71. $A_{\text{garden}} = A_{\text{plot}} - A_{\text{tile}}$

$= \pi R^2 - \pi r^2 = \pi(R^2 - r^2)$

$= \pi(8^2 - 1^2) = 63\pi \approx 197.92$ ft^2

Amelia will have about 197.92 square feet for her garden.

SSM: Experiencing Introductory and Intermediate Algebra

72. 10-inch pizza:
$A = \pi r^2 = \pi(5)^2 \approx 78.54 \text{ in}^2$
14-inch pizza:
$A = \pi r^2 = \pi(7)^2 \approx 153.94 \text{ in}^2$
Since two 10-inch pizzas would have a total area of about 157 square inches, that would be the better deal.

73. $V = L \cdot W \cdot H = (8)(5)(2) = 80 \text{ ft}^3$
The truck will hold 80 cubic feet of mulch.

74. $S = 4\pi r^2 = 4\pi(3)^2 \approx 113.1 \text{ ft}^2$
The storage tank would have a surface area of about 113.1 square feet.

75. $I = P \cdot r \cdot t = (850)(0.125)(1) = 106.25$
$A = P + I = 850 + 106.25 = 956.25$
The interest on the loan will be $106.25. The total amount paid back will be $956.25.

76. $F = \frac{9}{5}C + 32 = \frac{9}{5}(50) + 32 = 122$
The temperature is $122°F$.

77. $C = \frac{5}{9}(F - 32) = \frac{5}{9}(80 - 32) = 26\frac{2}{3}$
The temperature is $26\frac{2}{3}°C$

78. $d = r \cdot t = (62)\left(5\frac{3}{4}\right) = 356.5$
The distance traveled was 356.5 miles.

Chapter 2 Chapter Review

	Algebraic Expression	Num. of terms	Constant terms	Variable terms	Coefficents	Like terms
1.	$12x + y - z + 23$	4	23	$12x, y, -z$	$12, 1, -1, 23$	None
2.	$3(a-2) + 5(b-4) + 75$	3	75	$3(a-2), 5(b-4)$	$3, 5, 75$	None
3.	$12 - 7x + 14x - 18 + x$	5	$12, -18$	$-7x, 14x, x$	$12, -7, 14, -18, 1$	12 and -18, $-7x$ and $14x$ and x
4.	$b^2 + 2b - 3b^2 + 6b + b^3$	5	none	$b^2, 2b, -3b^2, 6b, b^3$	$1, 2, -3, 6, 1$	b^2 and $-3b^2$, $2b$ and $6b$

5. $x^2 = (-18)^2 = 324$

6. $-x^2 = -(-18)^2 = -324$

7. $(-x)^2 = (-(-18))^2 = (18)^2 = 324$

8. $-(-x)^2 = -(-(-18))^2 = -(18)^2 = -324$

9. $\sqrt{12y + 20} - 16 = \sqrt{12(8) + 20} - 16$
$= \sqrt{116} - 16 \approx -5.22967$

10. $\sqrt{12y + 20} - 16 = \sqrt{12\left(-\frac{1}{3}\right) + 20} - 16$
$= \sqrt{16} - 16 = 4 - 16 = -12$

11. $\sqrt{12y + 20} - 16 = \sqrt{12(-5) + 20} - 16$
$= \sqrt{-40} - 16$
Not a real number.

Chapter 2: Variables, Expressions, Equations, and Formulas

12. $\dfrac{7x+84}{2x-3} = \dfrac{7(-12)+84}{2(-12)-3}$
    $= \dfrac{-84+84}{-36-3} = \dfrac{0}{-39} = 0$

13. $\dfrac{7x+84}{2x-3} = \dfrac{7(3)+84}{2(3)-3} = \dfrac{21+84}{6-3}$
    $= \dfrac{105}{3} = 35$

14. $\dfrac{7x+84}{2x-3} = \dfrac{7(1.5)+84}{2(1.5)-3} = \dfrac{10.5+84}{3-3} = \dfrac{94.5}{0}$
    Undefined (division by 0).

15. $-\dfrac{b}{2a} = -\dfrac{(-4)}{2(1)} = -(-2) = 2$

16. $-\dfrac{b}{2a} = -\dfrac{(5)}{2(2)} = -\dfrac{5}{4}$

17. $-\dfrac{b}{2a} = -\dfrac{(-3)}{2\left(\dfrac{1}{2}\right)} = -(-3) = 3$

18. $3x - 7 = x + 1$
    $3(4) - 7 = (4) + 1$
    $12 - 7 = 4 + 1$
    $5 = 5$
    Since $5 = 5$, $x = 4$ is a solution.

19. $5x + 17 = 10x + 5$
    $5(3) + 17 = 10(3) + 5$
    $15 + 17 = 30 + 5$
    $32 = 35$
    Since $32 \neq 35$, $x = 3$ is not a solution.

20. $3x + 17 = 2(x - 5)$
    $3(-27) + 17 = 2((-27) - 5)$
    $-81 + 17 = 2(-32)$
    $-64 = -64$
    Since $-64 = -64$, $x = -27$ is a solution.

21. $2.1x - 1.9 = 0.6x - 4.6$
    $2.1(-1.8) - 1.9 = 0.6(-1.8) - 4.6$
    $-3.78 - 1.9 = -1.08 - 4.6$
    $-5.68 = -5.68$
    Since $-5.68 = -5.68$, $x = -1.8$ is a solution.

22. $\dfrac{3}{4}\left(x - \dfrac{8}{9}\right) = \dfrac{1}{2}\left(x + \dfrac{2}{3}\right)$
    $\dfrac{3}{4}\left(5\dfrac{1}{3} - \dfrac{8}{9}\right) = \dfrac{1}{2}\left(5\dfrac{1}{3} + \dfrac{2}{3}\right)$
    $\dfrac{3}{4}\left(\dfrac{40}{9}\right) = \dfrac{1}{2}(6)$
    $\dfrac{10}{3} = 3$
    Since $\dfrac{10}{3} \neq 3$, $x = 5\dfrac{1}{3}$ is not a solution.

23. $3x^2 - 6x - 10 = x^2 + x - 5$
    $3(-5)^2 - 6(-5) - 10 = (-5)^2 + (-5) - 5$
    $75 + 30 - 10 = 25 - 5 - 5$
    $95 = 15$
    Since $95 \neq 15$, $x = -5$ is not a solution.

24. $12h + 9h - 4h = (12 + 9 - 4)h = 17h$

25. $6m + 22 - m - 12 + 3m$
    $= 6m - m + 3m + 22 - 12$
    $= 8m + 10$

26. $3x - 35 + 4y - 5x - 6y + 7x + 27 + 17y + 22x$
    $= 3x - 5x + 7x + 22x + 4y - 6y + 17y - 35 + 27$
    $= 27x + 15y - 8$

27. $3x^4 + 5x - 7x^2 + 12x^4 - 17x - 34x + x^3 - 1$
    $= 3x^4 + 12x^4 + x^3 - 7x^2 + 5x - 17x - 34x - 1$
    $= 15x^4 + x^3 - 7x^2 - 46x - 1$

28. $(6.2a + 5.3b) + (4.7a - 1.9b)$
    $= 6.2a + 5.3b + 4.7a - 1.9b$
    $= 6.2a + 4.7a + 5.3b - 1.9b$
    $= 10.9a + 3.4b$

29. $-(27y - 15) = -27y + 15$

30. $5g+8-(g+4) = 5g+8-g-4$
    $= 5g-g+8-4$
    $= 4g+4$

31. $(-2x+4y-7z)-(-x+6y+8z)$
    $= -2x+4y-7z+x-6y-8z$
    $= -2x+x+4y-6y-7z-8z$
    $= -x-2y-15z$

32. $50\left(\dfrac{11}{25}a+\dfrac{33}{50}b\right) = 22a+33b$

33. $\dfrac{104x-156y+30z}{13} = \dfrac{104x}{13}-\dfrac{156y}{13}+\dfrac{30z}{13}$
    $= 8x-12y+\dfrac{30}{13}z$

34. $\dfrac{-18x+24y-36z}{-6} = \dfrac{-18x}{-6}+\dfrac{24y}{-6}-\dfrac{36z}{-6}$
    $= 3x-4y+6z$

35. $4(3.9x-11.1y)+7(2.9x-0.7y)$
    $= 15.6x-44.4y+20.3x-4.9y$
    $= 15.6x+20.3x-44.4y-4.9y$
    $= 35.9x-49.3y$

36. $12[-3(2a-5b)+9]+8[-9(a+13)-6(b-12)]$
    $= 12[-6a+15b+9]+8[-9a-117-6b+72]$
    $= -72a+180b+108-72a-936-48b+576$
    $= -72a-72a+180b-48b+108-936+576$
    $= -144a+132b-252$

37. $\dfrac{14.4(2x+5)-21.6(5x-2)+7.2(x-1)}{3x-4(x+8)+x+34.4}$
    $= \dfrac{28.8x+72-108x+43.2+7.2x-7.2}{3x-4x-32+x+34.4}$
    $= \dfrac{-72x+108}{2.4} = \dfrac{-72x}{2.4}+\dfrac{108}{2.4}$
    $= -30x+45$

38. Let x = number of DVDs.
    change $= 100-19.95x$
    $= 100-19.95(3) = 40.15$
    Javan would have $40.15 in change.

39. Let x = pounds of apples and
    y = pounds of bananas.
    cost $= 0.89x+0.49y$
    $= 0.89(4)+0.49(2) = 4.54$
    Heather's total cost was $4.54.

40. Let x = a number.
    $x^2+2x = x+306$

41. Let x = hours spent.
    earnings $= 225+45x$
    $= 225+45(120) = 5625$
    Lakeetha will earn $5625.

42. Let n = number of weeks.
    savings $= 500+145n-15n$
    $= 500+130n$
    $= 500+130(15) = 2450$
    balance $= 2450-525 = 1825$
    After 15 weeks, Carmen would have $2450 in the account. If she withdrew $625 for that period, her balance would be $1825.

43. Let x = # of appliances sold by Beatrice.
    Marie: $2x-5$
    Ann: $\dfrac{1}{2}x$
    Magdalene: x
    $(100+25x)+[100+25(2x-5)]$
    $\quad +\left(100+25\left(\dfrac{x}{2}\right)\right)+(100+25x)$
    $= 100+25x+100+50x-125$
    $\quad +100+\dfrac{25x}{2}+100+25x$
    $= 275+\dfrac{225}{2}x$
    $= 275+\dfrac{225}{2}(20) = 2525$
    The total money earned was $2525.

44. $V = \pi r^2 h = \pi(10)^2(4.5)$
    $= 450\pi \approx 1413.7 \text{ ft}^3$
    There will be about 1413.7 cubic feet of water in the pool.

Chapter 2: Variables, Expressions, Equations, and Formulas

45. $C = \frac{5}{9}(F-32) = \frac{5}{9}(96-32) = 35\frac{5}{9} \approx 35.56$

The temperature is about $35.56°F$.

46. $A = P(1+r)^t = 18500(1+.055)^{10}$
$\approx 31,600.67$
You will have about $31,600.67.

47. $d = r \cdot t = 13(1.25) = 16.25$ miles
Randy traveled 16.25 miles.

48. $V = \pi r^2 h = \pi(1.625)^2(1.5) \approx 12.44 \text{ in}^3$
$S = 2\pi r^2 + 2\pi r h = 2\pi r(r+h)$
$= 2\pi(1.625)(1.625+1.5) \approx 31.91 \text{ in}^2$

The can has a volume of 12.44 cubic inches and a surface area of 31.91 square inches.

49. Let x = number of books.
profit $= 2.00x - (175.00 + 0.50x)$
$= 2.00x - 175.00 - 0.50x$
$= 1.50x - 175.00$
$= 1.50(250) - 175 = 200$

The store will make a profit of $200.

50. The velocity of the reaction is given by:
$v = \frac{ax}{x+k}$

Since speed is the absolute value of velocity, the speed would be:

reaction speed $= |v| = \left|\frac{ax}{x+k}\right|$

51. $V = \pi r^2 h = \pi(0.02)^2(4)$
$= 0.0016\pi \approx 0.005 \text{ in}^3$
The volume of the wire (i.e. the paper clip) is about 0.005 cubic inches.

52. Equation; equality.

53. Expression; no equality.

54. Expression; no equality.

55. Equation; equality.

56. $90° - 85° = 5°$
The complimentary angle is $5°$.

57. $180° - 43° = 137°$
The supplementary angle is $137°$.

58. $180° - (31° + 58°) = 180° - 89° = 91°$
The third angle is $91°$.

Chapter 2 Test

1. Let x = amt in first account.
amt in 2nd account = $3500 - x$

2. Let x = selling price.
tax $= 0.08x$
total = price + tax
$= x + 0.08x$
$= 1.08x$
$= 1.08(15) = 16.2$
The total cost would be $16.20.

3. Let L = length and W = width.
$L = 3W - 5$ (inches)

4. Let x = raw score, m = mean, z = std. score and s = std. deviation.
$z = \frac{x-m}{s}$

5. $\sqrt{4x^2 - 20x + 25} + 8$
$= \sqrt{4(2)^2 - 20(2) + 25} + 8$
$= \sqrt{1} + 8 = 9$

6. $\frac{-18x+54}{x-8} = \frac{-18(-1)+54}{(-1)-8}$
$= \frac{18+54}{-1-8} = \frac{72}{-9} = -8$

7. $x^2 = (-6)^2 = 36$

8. $-x^2 = -(-6)^2 = -(36) = -36$

9. $(-x)^2 = (-(-6))^2 = (6)^2 = 36$

10. 8 terms.

11. $y^3, -5y^2, 15y, 7y^2, 4y, 6y^3$

SSM: Experiencing Introductory and Intermediate Algebra

12. $-3, -12$

13. $1, -5, 15, -3, 7, -12, 4, 6$

14. y^3 and $6y^3$, $-5y^2$ and $7y^2$, $15y$ and $4y$, -3 and -12.

15. $\dfrac{2x}{3} + \dfrac{5y}{6} - \dfrac{8}{9} + \dfrac{x}{6} + \dfrac{7y}{9} + \dfrac{1}{3}$
    $= \dfrac{12x}{18} + \dfrac{15y}{18} - \dfrac{16}{18} + \dfrac{3x}{18} + \dfrac{14y}{18} + \dfrac{6}{18}$
    $= \dfrac{15x}{18} + \dfrac{29y}{18} - \dfrac{10}{18}$
    $= \dfrac{5}{6}x + \dfrac{29}{18}y - \dfrac{5}{9}$

16. $-(5p+2q)-(-9p+q)+(p+q)-(-p-q)$
    $= -5p - 2q + 9p - q + p + q + p + q$
    $= -5p + 9p + p + p - 2q - q + q + q$
    $= 6p - q$

17. $\dfrac{25x - 45}{-5} = \dfrac{25x}{-5} - \dfrac{45}{-5} = -5x + 9$

18. $5[2(x+3) - 4(2x+1)]$
    $= 5[2x + 6 - 8x - 4] = 5[-6x + 2]$
    $= -30x + 10$

19. $-7x - 4 = 6x + 9$
    $-7(-2) - 4 = 6(-2) + 9$
    $14 - 4 = -12 + 9$
    $10 = -3$
    Since $10 \neq -3$, $x = -2$ is not a solution.

20. $8x^2 + 40x + 45 = 2x^2 - 2x - 27$
    $8(-4)^2 + 40(-4) + 45 = 2(-4)^2 - 2(-4) - 27$
    $128 - 160 + 45 = 32 + 8 - 27$
    $13 = 13$
    Since $13 = 13$, $x = -4$ is a solution.

21. $V = L \cdot W \cdot H = (4)(2)(1.5) = 12 \text{ ft}^3$
    $S = 2LW + 2WH + 2LH$
    $= 2(4)(2) + 2(2)(1.5) + 2(4)(1.5)$
    $= 34 \text{ ft}^2$
    The toolbox has a volume of 12 cubic feet and a surface area of 34 square feet.

22. $A = P + I = P + P \cdot r \cdot t$
    $= 2000 + 2000(0.055)(40)$
    $= 6400$
    The account will contain $6400 in 40 years.

23. $180° - 78° = 102°$
    The supplementary angle is $102°$.

24. $A = L \cdot W = (6)(2.5) = 15 \text{ m}^2$
    $P = 2L + 2W = 2(6) + 2(2.5) = 17 \text{ m}$
    The rectangle would have an area of 15 square meters and a perimeter of 17 meters.

25. $F = \dfrac{9}{5}C + 32 = \dfrac{9}{5}(25) + 32 = 77$
    The temperature is $77°F$.

26. Wording may vary.
    One possible answer: The area is the amount of space enclosed by the boundaries, whereas the perimeter is the distance around the boundaries (i.e. the total length of the boundaries).

Chapter 3

3.1 Exercises

1. $y = 5x + 4$

x	$y = 5x + 4$	y
-2	$y = 5(-2) + 4$	-6
-1	$y = 5(-1) + 4$	-1
0	$y = 5(0) + 4$	4
1	$y = 5(1) + 4$	9
2	$y = 5(2) + 4$	14
3	$y = 5(3) + 4$	19

3. $y = \frac{3}{5}x - 2$

x	$y = \frac{3}{5}x - 2$	y
-15	$y = \frac{3}{5}(-15) - 2$	-11
-10	$y = \frac{3}{5}(-10) - 2$	-8
-5	$y = \frac{3}{5}(-5) - 2$	-5
0	$y = \frac{3}{5}(0) - 2$ $y = -2$	-2
5	$y = \frac{3}{5}(5) - 2$	1
10	$y = \frac{3}{5}(10) - 2$	4
15	$y = \frac{3}{5}(15) - 2$	7

5. $y = 2.3x + 1.6$

x	$y = 2.3x + 1.6$	y
-2	$y = 2.3(-2) + 1.6$	-3
-1	$y = 2.3(-1) + 1.6$	-0.7
0	$y = 2.3(0) + 1.6$	1.6
1	$y = 2.3(1) + 1.6$	3.9
2	$y = 2.3(2) + 1.6$	6.2

7. $y = \frac{1}{3}(x + 7)$

x	$y = \frac{1}{3}(x + 7)$	y
-1	$y = \frac{1}{3}(-1 + 7)$	2
-4	$y = \frac{1}{3}(-4 + 7)$	1
-7	$y = \frac{1}{3}(-7 + 7)$	0
-10	$y = \frac{1}{3}(-10 + 7)$	-1
-13	$y = \frac{1}{3}(-13 + 7)$	-2

9. $y = 6x - 8$

x	$y = 6x - 8$	y
-2	$y = 6(-2) - 8$	-20
-1	$y = 6(-1) - 8$	-14
0	$y = 6(0) - 8$	-8
1	$y = 6(1) - 8$	-2
2	$y = 6(2) - 8$	4

11. $y = \frac{2}{7}x - 2$

x	$y = \frac{2}{7}x - 2$	y
-14	$y = \frac{2}{7}(-14) - 2$	-6
-7	$y = \frac{2}{7}(-7) - 2$	-4
0	$y = \frac{2}{7}(0) - 2$	-2
7	$y = \frac{2}{7}(7) - 2$	0
14	$y = \frac{2}{7}(14) - 2$	2

SSM: Experiencing Introductory and Intermediate Algebra

13. $y = -4.6x + 2.1$

x	$y = -4.6x + 2.1$	y
-2	$y = -4.6(-2) + 2.1$	11.3
-1	$y = -4.6(-1) + 2.1$	6.7
0	$y = -4.6(0) + 2.1$	2.1
1	$y = -4.6(1) + 2.1$	-2.5
2	$y = -4.6(2) + 2.1$	-7.1

15. $y = \frac{1}{4}(3x - 2)$

x	$y = \frac{1}{4}(3x - 2)$	y
-6	$y = \frac{1}{4}(3(-6) - 2)$	-5
-2	$y = \frac{1}{4}(3(-2) - 2)$	-2
0	$y = \frac{1}{4}(3(0) - 2)$	$-\frac{1}{2}$
2	$y = \frac{1}{4}(3(2) - 2)$	1
6	$y = \frac{1}{4}(3(6) - 2)$	4

17. $y = 12x - 13$

x	$y = 12x - 13$	x
-4	$y = 12(-4) - 13$	-61
-2	$y = 12(-2) - 13$	-37
0	$y = 12(0) - 13$	-13
2	$y = 12(2) - 13$	11
4	$y = 12(4) - 13$	35

19. $z = \frac{1}{3}y + 5$

y	$z = \frac{1}{3}y + 5$	z
-6	$z = \frac{1}{3}(-6) + 5$	3
-3	$z = \frac{1}{3}(-3) + 5$	4
0	$z = \frac{1}{3}(0) + 5$	5
3	$z = \frac{1}{3}(3) + 5$	6
6	$z = \frac{1}{3}(6) + 5$	7

21. $a = 14.2b + 5.7$

b	$a = 14.2b + 5.7$	a
-2	$a = 14.2(-2) + 5.7$	-22.7
-1	$a = 14.2(-1) + 5.7$	-8.5
0	$a = 14.2(0) + 5.7$	5.7
1	$a = 14.2(1) + 5.7$	19.9
2	$a = 14.2(2) + 5.7$	34.1

23. $y = 2x^2 + 3x + 1$

x	$y = 2x^2 + 3x + 1$	y
-3	$y = 2(-3)^2 + 3(-3) + 1$	10
-2	$y = 2(-2)^2 + 3(-2) + 1$	3
-1	$y = 2(-1)^2 + 3(-1) + 1$	0
0	$y = 2(0)^2 + 3(0) + 1$	1
1	$y = 2(1)^2 + 3(1) + 1$	6
2	$y = 2(2)^2 + 3(2) + 1$	15
3	$y = 2(3)^2 + 3(3) + 1$	28

Chapter 3: Relations, Functions, and Graphs

25. $y = (2x-3)(3x+4)$

x	$y = (2x-3)(3x+4)$	y
–2	$y = (2(-2)-3)(3(-2)+4)$	14
–1	$y = (2(-1)-3)(3(-1)+4)$	–5
0	$y = (2(0)-3)(3(0)+4)$	–12
1	$y = (2(1)-3)(3(1)+4)$	–7
2	$y = (2(2)-3)(3(2)+4)$	10

27. $y = \dfrac{3x+7}{x-1}$

x	$y = \dfrac{3x+7}{x-1}$	y
–3	$y = \dfrac{3(-3)+7}{(-3)-1}$	$\dfrac{1}{2}$
–1	$y = \dfrac{3(-1)+7}{(-1)-1}$	–2
1	$y = \dfrac{3(1)+7}{(1)-1}$	undefined
3	$y = \dfrac{3(3)+7}{(3)-1}$	8

29. Let x = hours driven and y = miles traveled

  a. $y = 55x$

  b.

x	$y = 55x$	y
4	$y = 55(4)$	220
8	$y = 55(8)$	440
12	$y = 55(12)$	660
16	$y = 55(16)$	880
20	$y = 55(20)$	1100

If Chameeka drives 4, 8, 12, 16, or 20 hours, she will travel 220, 440, 660, 880, or 1100 miles respectively.

31. Let h = height (in feet) and V = volume

$V = l \cdot w \cdot h = 4 \cdot 2 \cdot h = 8h$ cubic feet.

h	$V = 8h$	V
1	$V = 8(1)$	8
3	$V = 8(3)$	24
5	$V = 8(5)$	40
7	$V = 8(7)$	56
9	$V = 8(9)$	72

33. Let t = number of years, P = principal, and I = interest (in dollars)

$I = P \cdot r \cdot t = 5000(0.045)t = 225t$

t	$I = 225t$	I
1	$I = 225(1)$	225
2	$I = 225(2)$	450
3	$I = 225(3)$	675
4	$I = 225(4)$	900
5	$I = 225(5)$	1125
6	$I = 225(6)$	1350
7	$I = 225(7)$	1575
8	$I = 225(8)$	1800
9	$I = 225(9)$	2025
10	$I = 225(10)$	2250
11	$I = 225(11)$	2475
12	$I = 225(12)$	2700

SSM: Experiencing Introductory and Intermediate Algebra

35. $I = \dfrac{V}{R} = \dfrac{9}{R}$

R	$I = \dfrac{9}{R}$	I
1	$I = \dfrac{9}{1}$	9
2	$I = \dfrac{9}{2}$	4.5
3	$I = \dfrac{9}{3}$	3
4	$I = \dfrac{9}{4}$	2.25
5	$I = \dfrac{9}{5}$	1.8
6	$I = \dfrac{9}{6}$	1.5
7	$I = \dfrac{9}{7}$	1.2857
8	$I = \dfrac{9}{8}$	1.125
9	$I = \dfrac{9}{9}$	1

37. $y = 1.2x + 4$

x	$y = 1.2x + 4$	ordered pair
−2	$y = 1.2(-2) + 4 = 1.6$	$(-2, 1.6)$
−1	$y = 1.2(-1) + 4 = 2.8$	$(-1, 2.8)$
0	$y = 1.2(0) + 4 = 4$	$(0, 4)$
1	$y = 1.2(1) + 4 = 5.2$	$(1, 5.2)$
2	$y = 1.2(2) + 4 = 6.4$	$(2, 6.4)$

39. $q = 1 - p$

p	$q = 1 - p$	ordered pair
$\dfrac{1}{6}$	$q = 1 - \left(\dfrac{1}{6}\right) = \dfrac{5}{6}$	$\left(\dfrac{1}{6}, \dfrac{5}{6}\right)$
$\dfrac{1}{5}$	$q = 1 - \dfrac{1}{5} = \dfrac{4}{5}$	$\left(\dfrac{1}{5}, \dfrac{4}{5}\right)$
$\dfrac{1}{4}$	$q = 1 - \dfrac{1}{4} = \dfrac{3}{4}$	$\left(\dfrac{1}{4}, \dfrac{3}{4}\right)$
$\dfrac{1}{3}$	$q = 1 - \dfrac{1}{3} = \dfrac{2}{3}$	$\left(\dfrac{1}{3}, \dfrac{2}{3}\right)$
$\dfrac{1}{2}$	$q = 1 - \dfrac{1}{2} = \dfrac{1}{2}$	$\left(\dfrac{1}{2}, \dfrac{1}{2}\right)$

41. Let s = length of side in inches
    $P = 4s$ inches

s	$P = 4s$	P
2	$P = 4(2) = 8$	$(2, 8)$
4	$P = 4(4) = 16$	$(4, 16)$
6	$P = 4(6) = 24$	$(6, 24)$
8	$P = 4(8) = 32$	$(8, 32)$
10	$P = 4(10) = 40$	$(10, 40)$

43. $(1990, 110), (1991, 66), (1992, 22),$
    $(1993, 83), (1994, 175), (1995, 198),$
    $(1996, 225), (1997, 244),$ and $(1998, 238)$

45. $R = \{(3, 15.8), (5, 17.8), (7, 19.8), (9, 21.8)\}$
    Domain: $\{3, 5, 7, 9\}$
    Range: $\{15.8, 17.8, 19.8, 21.8\}$

47. $S = \{(4, -3), (4, -1), (4, 1), (4, 3), ...\}$
    Domain: $\{4\}$
    Range: $\{-3, -1, 1, 3, ...\}$

49. $T = \{..., (2, -2), (2, -1), (2, 0), (2, 1), (2, 2), ...\}$
    Domain: $\{2\}$
    Range: $\{..., -2, -2, 0, 1, 2, ...\}$

51. $y = 4x - 5$

x	$y = 4x - 5$
2	$y = 4(2) - 5 = 3$
4	$y = 4(4) - 5 = 11$
6	$y = 4(6) - 5 = 19$

Domain: $\{2, 4, 6\}$

Range: $\{3, 11, 19\}$

53. $y = 6 - x$

x	$y = 6 - x$
0	$y = 6 - 0 = 6$
1	$y = 6 - 1 = 5$
2	$y = 6 - 2 = 4$
3	$y = 6 - 3 = 3$
4	$y = 6 - 4 = 2$
5	$y = 6 - 5 = 1$
6	$y = 6 - 6 = 0$

Domain: $\{0, 1, 2, 3, 4, 5, 6\}$

Range: $\{0, 1, 2, 3, 4, 5, 6\}$

55. $y = x^2 + 1$

x	$y = x^2 + 1$
0	$y = (0)^2 + 1 = 1$
0.5	$y = (.5)^2 + 1 = 1.25$
1	$y = (1)^2 + 1 = 2$
1.5	$y = (1.5)^2 + 1 = 3.25$
2	$y = (2)^2 + 1 = 5$
2.5	$y = (2.5)^2 + 1 = 7.25$
3	$y = (3)^2 + 1 = 10$
3.5	$y = (3.5)^2 + 1 = 13.25$
4	$y = (4)^2 + 1 = 17$

Domain: $\{0, 0.5, 1, 1.5, 2, 2.5, 3, 3.5, 4\}$

Range: $\{1, 1.25, 2, 3.25, 5, 7.25, 10, 13.25, 17\}$

57. $y = \sqrt{x - 2}$

x	$y = \sqrt{x - 2}$
2	$y = \sqrt{2 - 2} = 0$
3	$y = \sqrt{3 - 2} = 1$
6	$y = \sqrt{6 - 2} = 2$
11	$y = \sqrt{11 - 2} = 3$
18	$y = \sqrt{18 - 2} = 4$
27	$y = \sqrt{27 - 2} = 5$

Domain: $\{2, 3, 6, 11, 18, 27\}$

Range: $\{0, 1, 2, 3, 4, 5\}$

59. $y = \dfrac{6}{x - 2}$

x	$y = \dfrac{6}{x - 2}$
-3	$y = \dfrac{6}{-3 - 2} = -\dfrac{6}{5}$
-1	$y = \dfrac{6}{-1 - 2} = -2$
1	$y = \dfrac{6}{1 - 2} = -6$
3	$y = \dfrac{6}{3 - 2} = 6$
5	$y = \dfrac{6}{5 - 2} = 2$
7	$y = \dfrac{6}{7 - 2} = \dfrac{6}{5}$

Domain: $\{-3, -1, 1, 3, 5, 7\}$

Range: $\left\{-6, -2, -\dfrac{6}{5}, \dfrac{6}{5}, 2, 6\right\}$

SSM: Experiencing Introductory and Intermediate Algebra

61. $s = -16t^2 + 50$

t	$s = -16t^2 + 50$	s
0	$s = -16(0)^2 + 50$	50
0.5	$s = -16(0.5)^2 + 50$	46
1	$s = -16(1)^2 + 50$	34
1.5	$s = -16(1.5)^2 + 50$	14

When the diver has fallen for 0, 0.5, 1, and 1.5 seconds, his position above the water is 50, 46, 34, and 14 feet respectively.

63. $d = r \cdot t$

t	$d = 65t$	d	ordered pair (t, d)
1	$d = 65(1)$	65	$(1, 65)$
2	$d = 65(2)$	130	$(2, 130)$
3	$d = 65(3)$	195	$(3, 195)$
4	$d = 654$	260	$(4, 260)$

The distance traveled (in miles) after 1, 2, 3, and 4 hours is 65, 130, 195, and 260 miles respectively.

65. Let w = number of weeks and c = total cost.

a. $c = 225 + 175w$

b.

w	$c = 225 + 175w$	c
1	$c = 225 + 175(1)$	400
2	$c = 225 + 175(2)$	575
3	$c = 225 + 175(3)$	750
4	$c = 225 + 175(4)$	925

The total cost for renting the apartment for 1, 2, 3, or 4 weeks is $400, $575, $750, or $925 respectively.

3.2 Exercises

1. $A(-7, -5); B(-5, -7)$

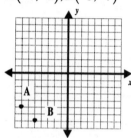

3. $E(4, 9); F(9, 4)$

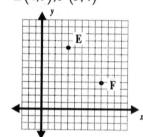

5. $I(-5, 5); J(5, -5)$

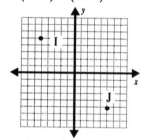

7. $M(2, -1); N(-1, 2)$

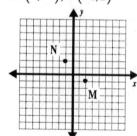

9. $Q(3,-1); R(-3,1)$

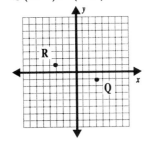

11. $U(-8,-2); V(8,2)$

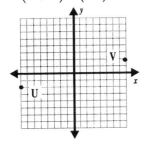

13. $A(1.2, 2.4)$

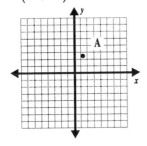

15. $C(-4.5, -2.6)$

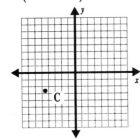

17. $E(-2.4, 2.1)$

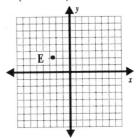

19. $G(1.8, -2.7)$

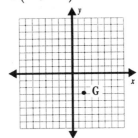

21. $(-21, 35)$

Coordinates are $(-,+)$ so Quadrant II

23. $(4, 96)$

Coordinates are $(+,+)$ so Quadrant I

25. $(-3, -19)$

Coordinates are $(-,-)$ so Quadrant III

27. $(0, -31)$

x-coordinate is 0 so on the y-axis

29. $(90, -100)$

Coordinates are $(+,-)$ so Quadrant IV

31. $(24, 0)$

y-coordinate is 0 so on the x-axis

33. $(-19, 0)$

y-coordinate is 0 so on the x-axis

35. $(0.05, 1.003)$

Coordinates are $(+,+)$ so Quadrant I

37. $(0, 3.7)$

   x-coordinate is 0 so on the y-axis

39. $\left(\dfrac{13}{27}, -\dfrac{11}{19}\right)$

   Coordinates are $(+,-)$ so Quadrant IV

41. $\left(-\dfrac{53}{100}, -\dfrac{39}{100}\right)$

   Coordinates are $(-,-)$ so Quadrant III

43. $\left(0, \dfrac{28}{51}\right)$

   x-coordinate is 0 so on the y-axis

45. $A(8,2), B(-9,7), C(0,0), D(-2,-3),$
   $E(0,4), F(5,-6), G(-5,0)$

47. $A = \{(-3,-3),(-2,-2),(-1,-1),(0,0),$
   $(1,1),(2,2),(3,3)\}$

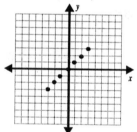

49. $C = \{(-4,4),(-2,2),(0,0),(2,-2),(4,-4)\}$

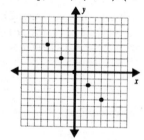

51. $E = \{(5,-3),(5,-1),(5,1),(5,3)\}$

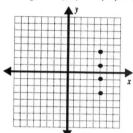

53.

x	y
-2	-1
-1	0
0	1
1	2
2	3

55.

x	y
-2	6
-1	-5
0	-4
1	-3
2	-2
3	-1

57. $y = 12x - 15$ for $x = \{-1, 0, 1\}$

x	y
-1	-27
0	-15
1	-3

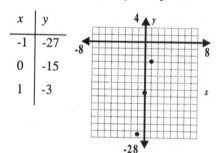

59. $y = \frac{1}{2}x + 3$ for $x = \{-4, -2, 0, 2, 4\}$

x	y
-4	1
-2	2
0	3
2	4
4	5

61. $y = -10x + 9$
Possible points:

x	y
0.5	4
1	-1
1.5	-6

63. $y = -\frac{2}{3}x - 2$
Possible points:

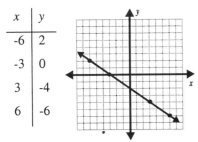

x	y
-6	2
-3	0
3	-4
6	-6

65. $y = 2x^2 - 5$
Possible points:

x	y
-2	3
-1	-3
0	-5
1	-3
2	3

67. $y = |2x|$
Possible points:

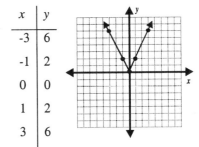

x	y
-3	6
-1	2
0	0
1	2
3	6

69. *Domain:* All real numbers.
Range: All real numbers ≤ -1.

71. *Domain:* All real numbers.
Range: All real numbers ≥ 3.

73. *Domain:* $0 \leq x \leq 40$
Range: roughly 37% to 62%
The domain represents the time period between 1960 and 2000 when the data was collected. The range represents the percent of total votes cast for the Democratic candidate in the given year.

75. a. $s = -16t^2 + 100$

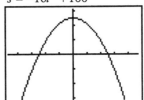

b. Domain: $0 \leq x \leq 2.5$
Range: $0 \leq y \leq 100$

3.2 Calculator Exercises

Part A

1. $y = 2x - 3$

   ZDecimal

   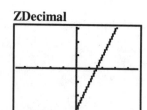

   ZStandard

   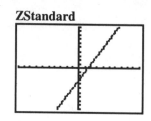

   ZInteger

   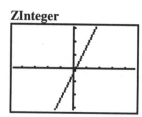

   Without making any adjustments, the standard window appears to be best. It shows the main features of the graph more clearly (the two intercepts in particular). The integer setting would possibly be better if we zoomed in.

2. $y = 2x^2 - 3x + 1$

   Zdecimal

   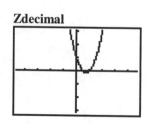

   ZStandard

   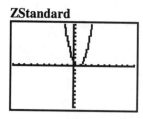

   Zinteger

   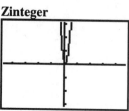

   Without making any adjustments, the decimal window appears to be best. It shows the main features more clearly (such as the graph crossing the x-axis twice).

3. $y = 2x^3 - 3x^2 + x - 4$

   ZDecimal

   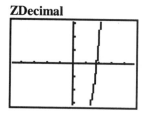

   ZStandard

   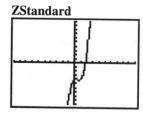

   ZInteger

   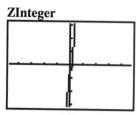

   Without making any adjustments, the standard window appears to be best. It shows the curvature of the graph more clearly.

Chapter 3: Relations, Functions, and Graphs

Part B

1. $y = 0.6x - 1.2$

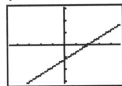

   Coordinate pairs:
   $(-2, -2.4), (-1, -1.8), (0, -1.2), (1, -0.6), (2, 0)$

2. $y = -0.5x + 2.2$

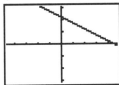

   Coordinate pairs:
   $(-2, 3.2), (-1, 2.7), (0, 2.2), (1, 1.7), (2, 1.2)$

3. $y = |x| - 2$

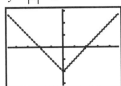

   Coordinate pairs:
   $(-2, 0), (-1, -1), (0, -2), (1, -1), (2, 0)$

4. $y = |x - 2|$

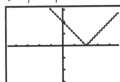

   Coordinate pairs:
   $(-2, 4), (-1, 3), (0, 2), (1, 1), (2, 0)$

5. $y = x - 2$

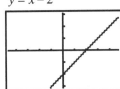

   Coordinate pairs:
   $(-2, -4), (-1, -3), (0, -2), (1, -1), (2, 0)$

6. $y = |x| - 2$

   | MATH | ▷ | 1 | X,T,Θ,n |) | − | 2 |

   $y = |x - 2|$

   | MATH | ▷ | 1 | X,T,Θ,n | − | 2 |) |

   $y = x - 2$

   | X,T,Θ,n | − | 2 |

   The important difference is the location of the absolute value sign. In #3, the absolute value is only on *x* while in #4 it is on the entire expression $x - 2$. Problem #5 has no absolute value sign at all.

3.3 Exercises

1. *A* is not a function because two elements in the domain, −2 and 2, correspond to more than one element in the range, 1 & 3 and −3 & −1, respectively.

3. *C* is a function because each element in the domain corresponds to exactly one element in the range.

5. *E* is not a function because the single element in the domain, 6, corresponds to 4 different elements in the range

7. *G* is a function because each element in the domain corresponds to exactly one element in the range.

9. This graph represents a function. All possible vertical lines cross the graph a maximum of one time.

11. This graph does not represent a function. It is possible to draw a vertical line (such as $x = 2$) that will cross the graph more than once.

SSM: Experiencing Introductory and Intermediate Algebra

13. This graph represents a function. All possible vertical lines cross the graph a maximum of one time.

15. $f(x) = 20x + 12$
$f(5) = 20(5) + 12$
$f(5) = 100 + 12$
$f(5) = 112$

17. $f(x) = 20x + 12$
$f(-7) = 20(-7) + 12$
$f(-7) = -140 + 12$
$f(-7) = -128$

19. $f(x) = 20x + 12$
$f(2.4) = 20(2.4) + 12$
$f(2.4) = 48 + 12$
$f(2.4) = 60$

21. $f(x) = 20x + 12$
$f\left(-\dfrac{1}{4}\right) = 20\left(-\dfrac{1}{4}\right) + 12$
$f\left(-\dfrac{1}{4}\right) = -5 + 12$
$f\left(-\dfrac{1}{4}\right) = 7$

23. $f(x) = 20x + 12$
$f(a) = 20(a) + 12$
$f(a) = 20a + 12$

25. $f(x) = 20x + 12$
$f(h+2) = 20(h+2) + 12$
$f(h+2) = 20h + 40 + 12$
$f(h+2) = 20h + 52$

27. $f(x) = 20x + 12$
$f(a-4) = 20(a-4) + 12$
$f(a-4) = 20a - 80 + 12$
$f(a-4) = 20a - 68$

29. $f(x) = 20x + 12$
$f(a+h) = 20(a+h) + 12$
$f(a+h) = 20a + 20h + 12$

31. $h(x) = 2x^2 - 4x + 5$
$h(7) = 2(7)^2 - 4(7) + 5$
$h(7) = 98 - 28 + 5$
$h(7) = 75$

33. $h(x) = 2x^2 - 4x + 5$
$h(-4) = 2(-4)^2 - 4(-4) + 5$
$h(-4) = 32 + 16 + 5$
$h(-4) = 53$

35. $h(x) = 2x^2 - 4x + 5$
$h(-1.1) = 2(-1.1)^2 - 4(-1.1) + 5$
$h(-1.1) = 2.42 + 4.4 + 5$
$h(-1.1) = 11.82$

37. $h(x) = 2x^2 - 4x + 5$
$h\left(-\dfrac{2}{5}\right) = 2\left(-\dfrac{2}{5}\right)^2 - 4\left(-\dfrac{2}{5}\right) + 5$
$h\left(-\dfrac{2}{5}\right) = \dfrac{8}{25} + \dfrac{8}{5} + 5$
$h\left(-\dfrac{2}{5}\right) = \dfrac{173}{25}$

39. $h(x) = 2x^2 - 4x + 5$
$h\left(2\dfrac{3}{5}\right) = 2\left(2\dfrac{3}{5}\right)^2 - 4\left(2\dfrac{3}{5}\right) + 5$
$h\left(2\dfrac{3}{5}\right) = \dfrac{338}{25} - \dfrac{52}{5} + 5$
$h\left(2\dfrac{3}{5}\right) = \dfrac{203}{25} = 8\dfrac{3}{25}$

41. $h(x) = 2x^2 - 4x + 5$
$h(b) = 2(b)^2 - 4(b) + 5$
$h(b) = 2b^2 - 4b + 5$

Chapter 3: Relations, Functions, and Graphs

43. $g(x) = |-3x+9|$
$g(5) = |-3(5)+9|$
$g(5) = |-15+9|$
$g(5) = 6$

45. $g(x) = |-3x+9|$
$g(-5) = |-3(-5)+9|$
$g(-5) = |15+9|$
$g(-5) = 24$

47. $g(x) = |-3x+9|$
$g(4.5) = |-3(4.5)+9|$
$g(4.5) = |-13.5+9|$
$g(4.5) = 4.5$

49. $g(x) = |-3x+9|$
$g(-4.5) = |-3(-4.5)+9|$
$g(-4.5) = |13.5+9|$
$g(-4.5) = 22.5$

51. $g(x) = |-3x+9|$
$g\left(\dfrac{2}{3}\right) = \left|-3\left(\dfrac{2}{3}\right)+9\right|$
$g\left(\dfrac{2}{3}\right) = |-2+9|$
$g\left(\dfrac{2}{3}\right) = 7$

53. $g(x) = |-3x+9|$
$g\left(-4\dfrac{2}{3}\right) = \left|-3\left(-4\dfrac{2}{3}\right)+9\right|$
$g\left(-4\dfrac{2}{3}\right) = |14+9|$
$g\left(-4\dfrac{2}{3}\right) = 23$

55. $F(x) = \sqrt{x+15} + 21$
$F(85) = \sqrt{(85)+15} + 21$
$F(85) = \sqrt{100} + 21$
$F(85) = 10 + 21$
$F(85) = 31$

57. $F(x) = \sqrt{x+15} + 21$
$F(-6) = \sqrt{(-6)+15} + 21$
$F(-6) = \sqrt{9} + 21$
$F(-6) = 3 + 21$
$F(-6) = 24$

59. $F(x) = \sqrt{x+15} + 21$
$F(-25) = \sqrt{(-25)+15} + 21$
$F(-25) = \sqrt{-10} + 21$,
which is a non-real number.

61. $F(x) = \sqrt{x+15} + 21$
$F(5.25) = \sqrt{(5.25)+15} + 21$
$F(5.25) = \sqrt{20.25} + 21$
$F(5.25) = 4.5 + 21$
$F(5.25) = 25.5$

63. $F(x) = \sqrt{x+15} + 21$
$F\left(-2\dfrac{3}{4}\right) = \sqrt{\left(-2\dfrac{3}{4}\right)+15} + 21$
$F\left(-2\dfrac{3}{4}\right) = \sqrt{\dfrac{49}{4}} + 21$
$F\left(-2\dfrac{3}{4}\right) = \dfrac{7}{2} + 21$
$F\left(-2\dfrac{3}{4}\right) = \dfrac{49}{2} = 24\dfrac{1}{2}$

SSM: Experiencing Introductory and Intermediate Algebra

65. Let x = the number of televisions in a production run.
$c(x) = 1500 + 35x$
$c(400) = 1500 + 35(400)$
$= 1500 + 14000$
$= 15,500$
The cost of the run would be $15,550.

67. Let x = the number of players.
$p(x) = 125x - 470$
$p(400) = 125(400) - 470$
$= 50,000 - 470$
$= 49,530$
The profit would be $49,530.

69. Let d = number of days.
$c(d) = 39 + 25d$
$c(3) = 39 + 25(3)$
$= 39 + 75$
$= 114$
The rental cost for three days would be $114.

71. Let x = number of half-hour increments.
$c(x) = 2.50 + 1.00x$
$3\frac{1}{2} = \frac{7}{2} = 7\left(\frac{1}{2}\right)$
$c(7) = 2.50 + 1.00(7)$
$= 2.50 + 7.00$
$= 9.50$
The total charge would be $9.50.

73. Let x = number of customers.
$R(x) = 175x - 3x^2$
$R(22) = 175(22) - 3(22)^2$
$= 3850 - 1452$
$= 2398$
The total revenue would be $2398.

75. Let x = number of CDs over 5.
$C(x) = 25 + 4x$
$12 - 5 = 7$
$C(7) = 25 + 4(7)$
$= 25 + 28$
$= 53$
The total cost for 12 CDs would be $53.

77. Let x = number of years since 1990.
$p(x) = -0.05x^2 + 1.21x + 60.08$
$2001 - 1990 = 11$
$p(11) = -0.05(11)^2 + 1.21(11) + 60.08$
$= -6.05 + 13.31 + 60.08$
$= 67.34$
The estimated percentage of households owning their own home in 2001 is 67.34%.

3.3 Calculator Exercises

Problems **1.** to **6.**

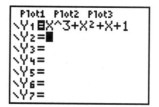

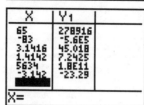

Chapter 3: Relations, Functions, and Graphs

Problems **7.** to **12.**

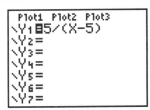

Problems **13.** to **18.**

3.4 Exercises

1. a. x-intercepts: $(-2,0)$ and $(6,0)$
    b. y-intercept: $(0,3)$
    c. Relative maxima: 4
    d. Relative minima: none
    e. Increasing: $x < 2$
    f. Decreasing: $x > 2$

3. a. x-intercepts: $(-5,0)$ and $(3,0)$
    b. y-intercept: $(0,-3)$
    c. Relative maxima: none
    d. Relative minima: -4
    e. Increasing: $x > -1$
    f. Decreasing: $x < -1$

5. $y = 3x - 6$

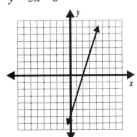

x-intercept: $(2,0)$

y-intercept: $(0,-6)$

7. $y = \dfrac{1}{2}x + 1$

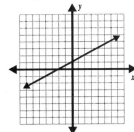

x-intercept: $(-2,0)$

y-intercept: $(0,1)$

85

9. $y = 1.2x - 6$

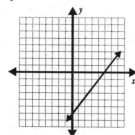

x-intercept: $(5, 0)$
y-intercept: $(0, -6)$

11. $f(x) = -12x + 24$

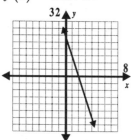

x-intercept: $(2, 0)$
y-intercept: $(0, 24)$

13. $f(x) = 9x + 15$

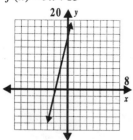

x-intercept: $\left(-\dfrac{5}{3}, 0\right)$
y-intercept: $(0, 15)$

15. $y = x^2 - 9$

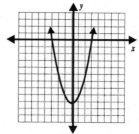

x-intercepts: $(-3, 0), (3, 0)$
y-intercept: $(0, -9)$

17. $y = x^2 + 6x + 9$

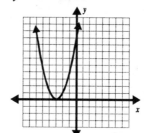

x-intercept: $(-3, 0)$
y-intercept: $(0, 9)$

19. $y = 4x^2 + 4x + 1$

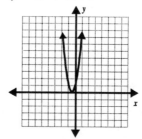

x-intercept: $\left(-\dfrac{1}{2}, 0\right)$
y-intercept: $(0, 1)$

Chapter 3: Relations, Functions, and Graphs

21. $g(x) = x^2 + 10x - 3$

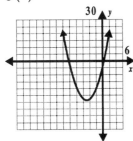

x-intercepts: $(-10.29, 0), (0.29, 0)$
y-intercept: $(0, -3)$

23. $H(x) = x^2 - 5x - 24$

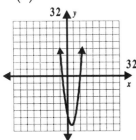

x-intercepts: $(-3, 0), (8, 0)$
y-intercept: $(0, -24)$

25. $y = x^3 + x^2 - 2x$

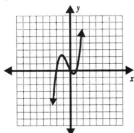

x-intercepts: $(-2, 0), (0, 0), (1, 0)$
y-intercept: $(0, 0)$

27. $f(x) = x^3 + 2x^2 - x - 2$

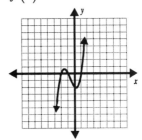

x-intercepts: $(-2, 0), (-1, 0), (1, 0)$
y-intercept: $(0, -2)$

29. $h(x) = |x| - 6$

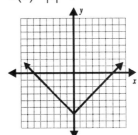

x-intercepts: $(-6, 0), (6, 0)$
y-intercept: $(0, -6)$

31. $y = |2x - 3| - 1$

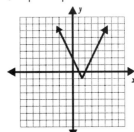

x-intercepts: $(1, 0), (2, 0)$
y-intercept: $(0, 2)$

33. $y = |x^2 - 2| - 1$

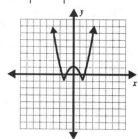

x-intercepts: $(-1.73, 0), (-1, 0), (1, 0), (1.73, 0)$

y-intercept: $(0, 1)$

35. $y = 2x + 8$

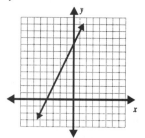

Always increasing

37. $f(x) = 3 - 2x$

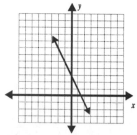

Always decreasing

39. $y = 1 - x^2$

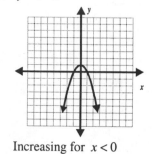

Increasing for $x < 0$

Decreasing for $x > 0$
Relative maximum is 1.

41. $g(x) = x^2 + 4x + 3$

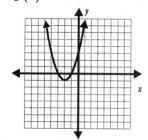

Increasing for $x > -2$
Decreasing for $x < -2$
Relative minimum is -1.

43. $y = |x + 3|$

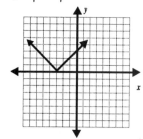

Increasing for $x > -3$
Decreasing for $x < -3$
Relative minimum is 0.

45. $f(x) = -|x + 3|$

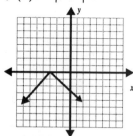

Increasing for $x < -3$
Decreasing for $x > -3$
Relative maximum is 0.

Chapter 3: Relations, Functions, and Graphs

47. $y = 3x - 5$ and $y = -2x + 15$

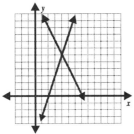

Intersect at $(4, 7)$

49. $f(x) = 2x + 7$ and $g(x) = -x + 1$

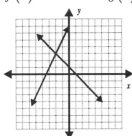

Intersect at $(-2, 3)$

51. $y = -5x + 2$ and $y = 3x + 8$

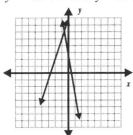

Intersect at $(-0.75, 5.75)$

53. $r(x) = 5x - 7$ and $c(x) = 12$

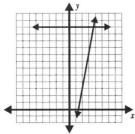

Intersect at $(3.8, 12)$

55. $y = 3$ and $y = -x^2 + 4$

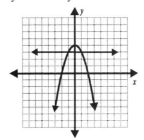

Intersect at $(-1, 3)$ and $(1, 3)$

57. $f(x) = 2x^2 - 4x + 5$ and $g(x) = 4x - 1$

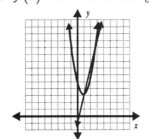

Intersect at $(1, 3)$ and $(3, 11)$

59. $y = \dfrac{1}{4}x^2 - 2$ and $y = \dfrac{1}{2}x$

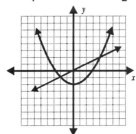

Intersect at $(-2, -1)$ and $(4, 2)$

61. $y = |x| - 5$ and $y = 2$

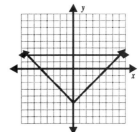

Intersect at $(-7, 2)$ and $(7, 2)$

SSM: Experiencing Introductory and Intermediate Algebra

63. Let x = number of pieces of equipment.
$Y_1 = 400x - 10x^2$ for $0 < x \le 25$

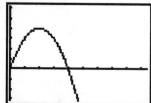

Increasing for $0 < x < 20$
Decreasing for $20 < x < 25$
(there was a maximum of 25 pieces)
Relative maximum is $c(20) = \$4000$.

65. Let x = number of craft items sold.
$p(x) = 5x - 50$ for $x > 0$

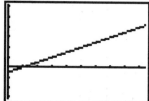

Always increasing for $x > 0$.

67. Let x = number of containers made in one run.

a. $c(x) = 4x + 50$

b. $r(x) = 10x$

c. $Y_1 = 4x + 50$;
$Y_2 = 10x$

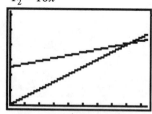

Intersect at $(8.\overline{3}, 83.\overline{3})$

d. This is the break-even point (where cost equals revenue). The company will need to make 9 or more baskets to make a profit.

69. Let x = number of credit hours with a passing grade.

a. $f(x) = 50x + 200$
$g(x) = 75x$

b. $Y_1 = 50x + 200$
$Y_2 = 75x$

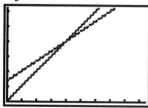

Intersect at $(8, 600)$

c. If Tatyana has 8 credit hours with a passing grade, she would get the same bonus (\$600) under either option.

71. The x-intercept is approximately $(2.5, 0)$. This means that Yahoo! Inc. broke even in 1997.

3.4 Calculator Exercises

Window settings may vary.

1. $g(x) = \frac{1}{3}x + 3$ and $f(x) = \frac{1}{4}x^2 - 4$

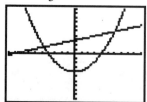

ZStandard
Intersect at $(-4.\overline{6}, 1.\overline{4})$ and $(6, 5)$

2. $y = |x| - 6$ and $y = -|x| + 4$

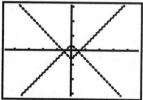

Zinteger
Intersect at $(-5, -1)$ and $(5, -1)$

3. $y = x^2 - 18$ and $y = -x^2 + 54$

ZStandard and Zinteger (scrolled up)

Intersect at $(-6, 18)$ and $(6, 18)$

Chapter 3 Section-By-Section Review

1. $b = -2a + 7$

a	$b = -2a + 7$	b
−3	$b = -2(-3) + 7$	13
−2	$b = -2(-2) + 7$	11
−1	$b = -2(-1) + 7$	9
0	$b = -2(0) + 7$	7
1	$b = -2(1) + 7$	5
2	$b = -2(2) + 7$	3
3	$b = -2(3) + 7$	1

2. $y = \dfrac{2}{3}x + 4$

x	$y = \dfrac{2}{3}x + 4$	y
9	$y = \dfrac{2}{3}(9) + 4$	10
6	$y = \dfrac{2}{3}(6) + 4$	8
3	$y = \dfrac{2}{3}(3) + 4$	6
0	$y = \dfrac{2}{3}(0) + 4$	4
−3	$y = \dfrac{2}{3}(-3) + 4$	2
−6	$y = \dfrac{2}{3}(-6) + 4$	0
−9	$y = \dfrac{2}{3}(-9) + 4$	−2

3. $y = 0.4x - 1.2$

x	$y = 0.4x - 1.2$	y
−3	$y = 0.4(-3) - 1.2$	−2.4
−2	$y = 0.4(-2) - 1.2$	−2.0
−1	$y = 0.4(-1) - 1.2$	−1.6
0	$y = 0.4(0) - 1.2$	−1.2
1	$y = 0.4(1) - 1.2$	−0.8
2	$y = 0.4(2) - 1.2$	−0.4
3	$y = 0.4(3) - 1.2$	0

SSM: Experiencing Introductory and Intermediate Algebra

4. $y = 5x^3 - 3x^2 + 2x - 22$

x	$y = 5x^3 - 3x^2 + 2x - 22$	y
-18	$y = 5(-18)^3 - 3(-18)^2 + 2(-18) - 22$	$-30{,}190$
-7	$y = 5(-7)^3 - 3(-7)^2 + 2(-7) - 22$	-1898
0	$y = 5(0)^3 - 3(0)^2 + 2(0) - 22$	-22
6	$y = 5(6)^3 - 3(6)^2 + 2(6) - 22$	962
21	$y = 5(21)^3 - 3(21)^2 + 2(21) - 22$	$45{,}002$
22.5	$y = 5(22.5)^3 - 3(22.5)^2 + 2(22.5) - 22$	$55{,}457$

5. $y = |x^2 - 6x + 5|$

x	$y = |x^2 - 6x + 5|$	y
-2	$y = |(-2)^2 - 6(-2) + 5|$	21
-1	$y = |(-1)^2 - 6(-1) + 5|$	12
0	$y = |(0)^2 - 6(0) + 5|$	5
1	$y = |(1)^2 - 6(1) + 5|$	0
2	$y = |(2)^2 - 6(2) + 5|$	3
3	$y = |(3)^2 - 6(3) + 5|$	4

6. $y = 3.6x^2 + 1.5x - 14.2$

x	$y = 3.6x^2 + 1.5x - 14.2$	y
-2.7	$y = 3.6(-2.7)^2 + 1.5(-2.7) - 14.2$	7.994
-1.9	$y = 3.6(-1.9)^2 + 1.5(-1.9) - 14.2$	-4.054
-0.6	$y = 3.6(-0.6)^2 + 1.5(-0.6) - 14.2$	-13.8
0	$y = 3.6(0)^2 + 1.5(0) - 14.2$	-14.2
0.8	$y = 3.6(0.8)^2 + 1.5(0.8) - 14.2$	-10.696
1.5	$y = 3.6(1.5)^2 + 1.5(1.5) - 14.2$	-3.85
2.4	$y = 3.6(2.4)^2 + 1.5(2.4) - 14.2$	10.136

7. $y = -1.6x + 4.5$; Answers will vary. Possible answer:

x	$y = -1.6x + 4.5$	y
-2	$y = -1.6(-2) + 4.5$	7.7
-1	$y = -1.6(-1) + 4.5$	6.1
0	$y = -1.6(0) + 4.5$	4.5
1	$y = -1.6(1) + 4.5$	2.9
2	$y = -1.6(2) + 4.5$	1.3

8. $y = \dfrac{3}{2}x - 6$; Answers will vary. Possible answer:

x	$y = \dfrac{3}{2}x - 6$	y
-4	$y = \dfrac{3}{2}(-4) - 6$	-12
-2	$y = \dfrac{3}{2}(-2) - 6$	-9
2	$y = \dfrac{3}{2}(2) - 6$	-3
4	$y = \dfrac{3}{2}(4) - 6$	0
6	$y = \dfrac{3}{2}(6) - 6$	3

9. $y = |2x - 9|$; Answers will vary. Possible answer:

x	$y = |2x-9|$	y
-3	$y = |2(-3)-9|$	15
-1	$y = |2(-1)-9|$	11
1	$y = |2(1)-9|$	7
3	$y = |2(3)-9|$	3
5	$y = |2(5)-9|$	1

10. $y = (5x+2)(x-4)$

x	$y = (5x+2)(x-4)$	y
-5	$y = (5(-5)+2)((-5)-4)$	207
-3	$y = (5(-3)+2)((-3)-4)$	91
-1	$y = (5(-1)+2)((-1)-4)$	15
1	$y = (5(1)+2)((1)-4)$	-21
3	$y = (5(3)+2)((3)-4)$	-17
5	$y = (5(5)+2)((5)-4)$	27

11. $y = \dfrac{3}{5}x + 8$

x	$y = \dfrac{3}{5}x+8$	y
-15	$y = \dfrac{3}{5}(-15)+8$	-1
-10	$y = \dfrac{3}{5}(-10)+8$	2
-5	$y = \dfrac{3}{5}(-5)+8$	5
0	$y = \dfrac{3}{5}(0)+8$	8
5	$y = \dfrac{3}{5}(5)+8$	11
10	$y = \dfrac{3}{5}(10)+8$	14
15	$y = \dfrac{3}{5}(15)+8$	17

12. $y = 17.1x - 12.9$

x	$y = 17.1x - 12.9$	y
-3	$y = 17.1(-3) - 12.9$	-64.2
-2	$y = 17.1(-2) - 12.9$	-47.1
-1	$y = 17.1(-1) - 12.9$	-30
0	$y = 17.1(0) - 12.9$	-12.9
1	$y = 17.1(1) - 12.9$	4.2
2	$y = 17.1(2) - 12.9$	21.3
3	$y = 17.1(3) - 12.9$	38.4

13. Let x = number of gallons of gasoline and c = the total cost.

   a. $c = 1.649x$

   b.

x	$c = 1.649x$	c
5	$c = 1.649(5)$	8.245
10	$c = 1.649(10)$	16.49
15	$c = 1.649(15)$	24.735
20	$c = 1.649(20)$	32.98

14. $A = \pi r^2$

r	$A = \pi r^2$	A
4	$A = \pi(4)^2$	50.265
6	$A = \pi(6)^2$	113.1
8	$A = \pi(8)^2$	201.06
10	$A = \pi(10)^2$	314.16

15. $b = 90 - a$

a	$b = 90 - a$	b
10	$b = 90 - 10$	80
20	$b = 90 - 20$	70
30	$b = 90 - 30$	60
40	$b = 90 - 40$	50
45	$b = 90 - 45$	45

SSM: Experiencing Introductory and Intermediate Algebra

16. $I = 2000(0.06)t = 120t$

t	$I = 120t$	I
2	$I = 120(2)$	240
3	$I = 120(3)$	360
4	$I = 120(4)$	480

17. $C = \dfrac{5}{9}(F - 32)$

F	$C = \dfrac{5}{9}(F-32)$	C
−23	$C = \dfrac{5}{9}(-23-32)$	−30.56
−14	$C = \dfrac{5}{9}(-14-32)$	−25.56
0	$C = \dfrac{5}{9}(0-32)$	−17.78
41	$C = \dfrac{5}{9}(41-32)$	5
50	$C = \dfrac{5}{9}(50-32)$	10
59	$C = \dfrac{5}{9}(59-32)$	15
100	$C = \dfrac{5}{9}(100-32)$	37.78

18. $y = 7 - 3x$

x	$y = 7-3x$	y
−10	$y = 7-3(-10)$	37
−5	$y = 7-3(-5)$	22
0	$y = 7-3(0)$	7
5	$y = 7-3(5)$	−8
10	$y = 7-3(10)$	−23

Ordered pairs: $(-10, 37)$, $(-5, 22)$, $(0, 7)$, $(5, -8)$, $(10, -23)$.

19. $y = \sqrt{x+8}$

x	$y = \sqrt{x+8}$	y
−8	$y = \sqrt{-8+8}$	0
−7	$y = \sqrt{-7+8}$	1
−4	$y = \sqrt{-4+8}$	2
1	$y = \sqrt{1+8}$	3
8	$y = \sqrt{8+8}$	4

Ordered pairs: $(-8, 0)$, $(-7, 1)$, $(-4, 2)$, $(1, 3)$, $(8, 4)$.

20. $d = \dfrac{2}{3}c - 1$

c	$d = \dfrac{2}{3}c-1$	d
−6	$d = \dfrac{2}{3}(-6)-1$	−5
−3	$d = \dfrac{2}{3}(-3)-1$	−3
0	$d = \dfrac{2}{3}(0)-1$	−1
3	$d = \dfrac{2}{3}(3)-1$	1
6	$d = \dfrac{2}{3}(6)-1$	3

Ordered pairs: $(-6, -5)$, $(-3, -3)$, $(0, -1)$, $(3, 1)$, $(6, 3)$.

Chapter 3: Relations, Functions, and Graphs

21. $r = \dfrac{d}{2}$

d	$r = \dfrac{d}{2}$	r
2	$r = \dfrac{2}{2}$	1
4	$r = \dfrac{4}{2}$	2
6	$r = \dfrac{6}{2}$	3
8	$r = \dfrac{8}{2}$	4
10	$r = \dfrac{10}{2}$	5

Ordered pairs: $(2,1)$, $(4,2)$, $(6,3)$, $(8,4)$, $(10,5)$.

22. Let *x* be the year and *y* be the number of trademarks issued (In thousands).
$(x, y) = \{(1993, 86.9), (1994, 70.1), (1995, 92.5),$
$(1996, 98.6), (1997, 119.9)\}$

23. *Domain*: $\{1, 3, 5, 7, 9\}$
Range: $\{2, 6, 10, 14, 18\}$

24. *Domain*: $\{..., -6, -4, -2, 0, 2, 4, 6, ...\}$
Range: $\{..., -6, -4, -2, 0, 2, 4, 6, ...\}$

25. $y = 4(x+5)-1$
$y = 4(-5+5)-1 = -1$
$y = 4(-4+5)-1 = 3$
$y = 4(-3+5)-1 = 7$
$y = 4(-2+5)-1 = 11$
$y = 4(-1+5)-1 = 15$
Domain: $\{-5, -4, -3, -2, -1\}$
Range: $\{-1, 3, 7, 11, 15\}$

26. $y = x^2 + 2.5$
$y = (0)^2 + 2.5 = 2.5$
$y = (0.5)^2 + 2.5 = 2.75$
$y = (1)^2 + 2.5 = 3.5$
$y = (1.5)^2 + 2.5 = 4.75$
$y = (2)^2 + 2.5 = 6.5$
Domain: $\{0, 0.5, 1, 1.5, 2\}$
Range: $\{2.5, 2.75, 3.5, 4.75, 6.5\}$

27. $y = \sqrt{5+x}$
$y = \sqrt{5+(-5)} = 0$
$y = \sqrt{5+(-4)} = 1$
$y = \sqrt{5+(-3)} = \sqrt{2}$
$y = \sqrt{5+(-2)} = \sqrt{3}$
$y = \sqrt{5+(-1)} = 2$
$y = \sqrt{5+(0)} = \sqrt{5}$
Domain: $\{-5, -4, -3, -2, -1, 0\}$
Range: $\{0, 1, \sqrt{2}, \sqrt{3}, 2, \sqrt{5}\}$

28. $y = \dfrac{12}{1-x}$
$y = \dfrac{12}{1-(-4)} = \dfrac{12}{5}$
$y = \dfrac{12}{1-(-2)} = \dfrac{12}{3} = 4$
$y = \dfrac{12}{1-(0)} = 12$
$y = \dfrac{12}{1-(2)} = -12$
$y = \dfrac{12}{1-(4)} = \dfrac{12}{-3} = -4$
Domain: $\{-4, -2, 0, 2, 4\}$
Range: $\left\{-12, -4, \dfrac{12}{5}, 4, 12\right\}$

29. $T = 2\pi\sqrt{\dfrac{L}{32}}$

L	$T = 2\pi\sqrt{\dfrac{L}{32}}$	T
1	$T = 2\pi\sqrt{\dfrac{1}{32}}$	1.111
8	$T = 2\pi\sqrt{\dfrac{8}{32}}$	3.142
16	$T = 2\pi\sqrt{\dfrac{16}{32}}$	4.443
24	$T = 2\pi\sqrt{\dfrac{24}{32}}$	5.441
32	$T = 2\pi\sqrt{\dfrac{32}{32}}$	6.283

30. $A(3,2)$

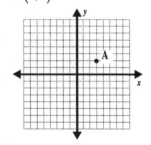

31. $B(4,-3)$

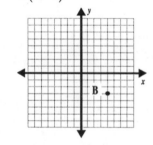

32. $C(-3,2)$

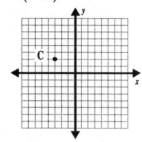

33. $D(-4,-3)$

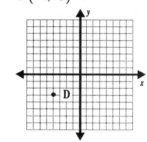

34. $E(0,5)$

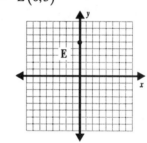

35. $F(-5,0)$

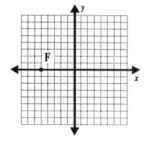

36. $A(5,3), B(-2,-5), C(2,-2), D(-3,5),$
    $E(5,0), F(0,-4), G(0,0)$

37. Quadrant I

Chapter 3: Relations, Functions, and Graphs

38. Quadrant III
39. Quadrant IV
40. Quadrant II
41. x-axis
42. y-axis
43. origin
44. $S = \{(0,-4),(1,-3),(2,-2),(3,-1),(4,0),(5,1)\}$

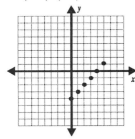

45. Ordered pairs: $(-1,-5)$, $(0,-3)$, $(1,-1)$, $(2,1)$, $(3,3)$, $(4,5)$

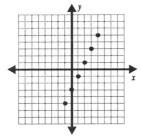

46. $y = 3 - 2x$

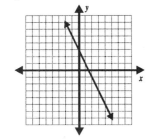

47. $T = \{(-5,3),(-3,3),(-1,3),(1,3),(1,3),(3,3),(5,3)\}$

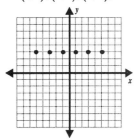

48. *Domain*: all real numbers
    Range: all real numbers ≤ 2

49. *Domain*: all real numbers
    Range: $\{-3\}$

50. *Domain*: the years 1980 to 1998
    Range: the range is roughly $71 \leq y \leq 90$

    The domain represents the time period between 1980 and 1998 when the data was collected. The range represents the U.S. energy consumption (in Quadrillion Btu) for each year.

51. Let x = number of children sent to daycare and y = weekly charge for daycare.

    a. $y = 40 + 10x$

    b. $Y_1 = 40 + 10x$

    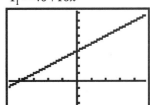

    c. *Domain*: integers > 0
       Range: $\{50, 60, 70, 80, ...\}$

52. Let x = the length of a side and y = area.

    a. $y = x^2$

b. $Y_1 = x^2$

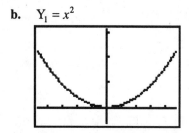

c. Domain: all real numbers > 0
Range: all real numbers > 0

53. S is not a function because two elements in the domain, -1 and -2, correspond to more than one element in the range, 5 and 9, and 3 and 11, respectively.

54. T is a function because each element in the domain corresponds to only one element in the range.

55. The graph is not of a function. A vertical line (such as $x = -1$) can cross the graph at more than one point (fails vertical line test).

56. This is a graph of a function. No vertical line can cross the graph at more than one point.

57. $f(x) = -4x + 13$
$f(13) = -4(13) + 13$
$\quad = -52 + 13$
$\quad = -39$

58. $f(x) = -4x + 13$
$f(-21) = -4(-21) + 13$
$\quad = 84 + 13$
$\quad = 97$

59. $f(x) = -4x + 13$
$f(2.5) = -4(2.5) + 13$
$\quad = -10 + 13$
$\quad = 3$

60. $f(x) = -4x + 13$
$f(-3.7) = -4(-3.7) + 13$
$\quad = 14.8 + 13$
$\quad = 27.8$

61. $f(x) = -4x + 13$
$f(3 + h) = -4(3 + h) + 13$
$\quad = -12 - 4h + 13$
$\quad = -4h + 1$

62. $f(x) = -4x + 13$
$f(-b) = -4(-b) + 13$
$\quad = 4b + 13$

63. $g(x) = 5x^2 + x - 4$
$g(3) = 5(3)^2 + (3) - 4$
$\quad = 45 + 3 - 4$
$\quad = 44$

64. $g(x) = 5x^2 + x - 4$
$g(-2) = 5(-2)^2 + (-2) - 4$
$\quad = 20 - 2 - 4$
$\quad = 14$

65. $g(x) = 5x^2 + x - 4$
$g(0.5) = 5(0.5)^2 + (0.5) - 4$
$\quad = 1.25 + 0.5 - 4$
$\quad = -2.25$

66. $g(x) = 5x^2 + x - 4$
$g(a) = 5(a)^2 + (a) - 4$
$\quad = 5a^2 + a - 4$

67. $g(x) = 5x^2 + x - 4$
$g(-a) = 5(-a)^2 + (-a) - 4$
$\quad = 5a^2 - a - 4$

68. $g(x) = 5x^2 + x - 4$
$g\left(-\frac{1}{4}\right) = 5\left(-\frac{1}{4}\right)^2 + \left(\frac{1}{4}\right) - 4$
$\quad = \frac{5}{16} - \frac{1}{4} - 4$
$\quad = \frac{5}{16} - \frac{4}{16} - \frac{64}{16}$
$\quad = -\frac{63}{16}$

69. $S(x) = \sqrt{2x+3} - 5$
$S(3) = \sqrt{2(3)+3} - 5$
$= \sqrt{9} - 5$
$= 3 - 5 = -2$

70. $S(x) = \sqrt{2x+3} - 5$
$S(11) = \sqrt{2(11)+3} - 5$
$= \sqrt{25} - 5$
$= 5 - 5 = 0$

71. $S(x) = \sqrt{2x+3} - 5$
$S(59) = \sqrt{2(59)+3} - 5$
$= \sqrt{121} - 5$
$= 11 - 5 = 6$

72. Let x = number of widgets in one run and c = cost of production run.
$c(x) = 4500 + 17x$
$c(1200) = 4500 + 17(1200)$
$= 4500 + 20,400$
$= 24,900$
The cost of producing 1200 widgets in one run is $24,900.

73. Let x = number of people attending and c = total charge for training.
$c(x) = 1500 + 125x$
$c(20) = 1500 + 125(20)$
$= 1500 + 2500$
$= 4000$
The cost of training 20 employees is $4000.

74. Let x = number of faces painted and p = total profit.
$p(x) = 1.5x - 15$
$p(135) = 1.50(135) - 15.00$
$= 202.50 - 15.00$
$= 187.50$
Dmitri will make a profit of $187.50 if he paints 135 faces at the church carnival.

75. Let x = years after 1995 and S = annual sales (in billions of dollars).
$S(x) = 2.1x^2 + 5.7x + 83.9$
$2000 - 1995 = 5$
$S(5) = 2.1(5)^2 + 5.7(5) + 83.9$
$= 52.5 + 28.5 + 83.9 = 164.90$
The estimated sales for Wal-Mart in the year 2000 is $164.9 billion.

76. a. x-intercepts: none
    b. y-intercept: $(0,7)$
    c. No relative maxima
    d. Relative minimum is 1
    e. Increasing for $x > 3$
    f. Decreasing for $x < 3$

77. $y = 3x + 9$

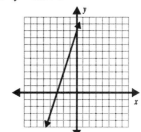

x-intercept: $(-3, 0)$
y-intercept: $(0, 9)$

78. $y = \dfrac{3}{4}x - 9$

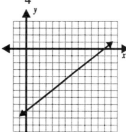

x-intercept: $(12, 0)$
y-intercept: $(0, -9)$

79. $y = x^2 - 0.36$

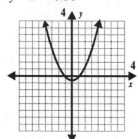

x-intercepts: $(-0.6, 0)$ and $(0.6, 0)$
y-intercept: $(0, -0.36)$

80. $y = |x| - 4$

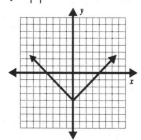

x-intercepts: $(-4, 0)$ and $(4, 0)$
y-intercept: $(0, -4)$

81. $h(x) = 6 - 2x$

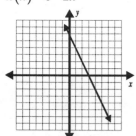

Always decreasing.

82. $y = 3 - x^2$

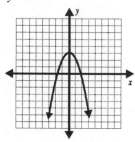

Increasing for $x < 0$
Decreasing for $x > 0$
Relative maximum is 3

83. $y = |x| + 2$

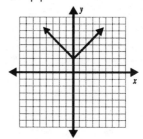

Increasing for $x > 0$
Decreasing for $x < 0$
Relative minimum is 2

84. $f(x) = |x^2 - 1|$

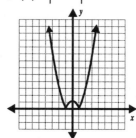

Increasing for $-1 < x < 0$ and $x > 1$
Decreasing for $x < -1$ and $0 < x < 1$
Relative minimum is $f(-1) = f(1) = 0$
Relative maximum is $f(0) = 1$

85. $y = 2x - 2$ and $y = -\frac{1}{3}x + 5$

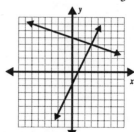

Intersect at $(3, 4)$

Chapter 3: Relations, Functions, and Graphs

86. $y = x^2 - 6$ and $y = x$

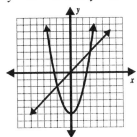

Intersect at $(-2,-2)$ and $(3,3)$

87. $f(x) = |x+5|$ and $g(x) = 2$

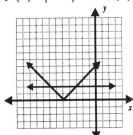

Intersect at $(-3,2)$ and $(-7,2)$

88. Let x = number of desks purchased and c = total cost to purchase the desks.

$c(x) = 325x - 15x^2$ for $0 < x \le 8$

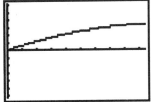

The function is always increasing for $0 < x < 8$. There is no relative maxima or minima.

89. Let x = number of harnesses sold and r = total revenue from selling harnesses.

$r(x) = 10.45x$ for $x \ge 0$

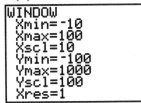

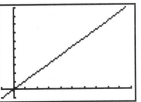

This function is always increasing for $x > 0$. There are no relative maxima or minima.

90. Let x = number of months and b = account balance.

$b(x) = 216 - 4.5x$ for $0 \le x \le 48$

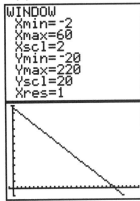

This function is always decreasing for $0 < x < 48$. There are no relative maxima or minima.

91. Let x = number of credit hours passed.

a. $f(x) = 400 + 65x$
   $g(x) = 100x$

b. Intersect at roughly $(11.43, 1142.86)$

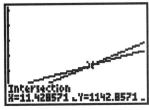

c. The point of intersection tells Hans the number of credit hours (with a passing grade) needed to get the same amount from either reimbursement plan. This is the x-coordinate. The y-coordinate shows how much he will be reimbursed.

92. Let x = the number of items sold.

a. $c(x) = 500 + 12x$
$r(x) = 25x$

b. Intersection point:

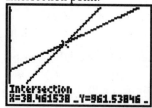

The point of intersection is roughly $(38.46, 961.54)$.

c. In business, an intersection point like this is called a break-even point. This point represents where (or when) the total costs equal the total revenue. You neither make money, nor lose any.

Chapter 3 Chapter Review

1. $f(x) = -x + 9$
$f(9) = -9 + 9$
$= 0$

2. $f(x) = -x + 9$
$f(-9) = -(-9) + 9 = 9 + 9$
$= 18$

3. $f(x) = -x + 9$
$f(1.8) = -(1.8) + 9$
$= 7.2$

4. $f(x) = -x + 9$
$f(-2.7) = -(-2.7) + 9$
$= 2.7 + 9$
$= 11.7$

5. $f(x) = -x + 9$
$f(-b) = -(-b) + 9$
$= b + 9$

6. $f(x) = -x + 9$
$f(1+h) = -(1+h) + 9$
$= -1 - h + 9$
$= -h + 8$

7. $g(x) = x^2 - 3x - 4$
$g(4) = (4)^2 - 3(4) - 4$
$= 16 - 12 - 4$
$= 0$

8. $g(x) = x^2 - 3x - 4$
$g(-1) = (-1)^2 - 3(-1) - 4$
$= 1 + 3 - 4$
$= 0$

9. $g(x) = x^2 - 3x - 4$
$g(1.5) = (1.5)^2 - 3(1.5) - 4$
$= 2.25 - 4.5 - 4$
$= -6.25$

10. $g(x) = x^2 - 3x - 4$
$g(v) = (v)^2 - 3(v) - 4$
$= v^2 - 3v - 4$

11. $g(x) = x^2 - 3x - 4$
$g(-v) = (-v)^2 - 3(-v) - 4$
$= v^2 + 3v - 4$

12. $g(x) = x^2 - 3x - 4$
$g\left(-\frac{2}{3}\right) = \left(-\frac{2}{3}\right)^2 - 3\left(-\frac{2}{3}\right) - 4$
$= \frac{4}{9} + 2 - 4$
$= \frac{4}{9} + \frac{18}{9} - \frac{36}{9} = -\frac{14}{9}$

13. $S(x) = \sqrt{6x - 8}$
$S(x) = \sqrt{6(4) - 8}$
$= \sqrt{24 - 8} = \sqrt{16}$
$= 4$

14. $S(x) = \sqrt{6x-8}$
    $S(12) = \sqrt{6(12)-8}$
    $ = \sqrt{72-8} = \sqrt{64} = 8$

15. $S(x) = \sqrt{6x-8}$
    $S(44) = \sqrt{6(44)-8}$
    $ = \sqrt{264-8} = \sqrt{256}$
    $ = 16$

16. P is a function. Each element in the domain corresponds to only one element in the range.

17. Q is not a function. An element in the domain, 2, corresponds to more than one element in the range.

18. $y = 4.8x - 1.2$

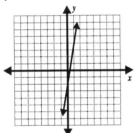

x-intercept: $(0.25, 0)$

y-intercept: $(0, -1.2)$

The function is always increasing.

19. $y = \dfrac{2}{5}x + 4$

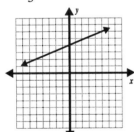

x-intercept: $(-10, 0)$

y-intercept: $(0, 4)$

The function is always increasing.

20. $y = x^2 - 1.21$

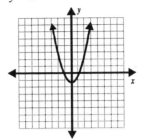

x-intercepts: $(-1.1, 0)$ and $(1.1, 0)$

y-intercept: $(0, -1.21)$

Increasing for $x > 0$

Decreasing for $x < 0$

Relative minimum is $y = -1.21$

21. $y = 2 - |x|$

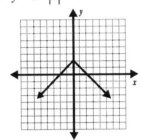

x-intercepts: $(-2, 0)$ and $(2, 0)$

y-intercept: $(0, 2)$

Increasing for $x < 0$

Decreasing for $x > 0$

Relative maximum is $y = 2$

22. $y = 2x + 2$ and $y = -2x - 10$

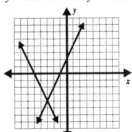

Intersect at $(-3, -4)$

23. $y = x^2$ and $y = 3x$

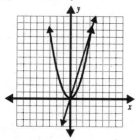

Intersect at $(0,0)$ and $(3,9)$

24. $f(x) = |2x|$ and $g(x) = x+3$

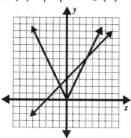

Intersect at $(-1,2)$ and $(3,6)$

25. Domain: $\{2,4,6,8,10\}$
    Range: $\{1,2,3,4,5\}$

26. Domain: $\{...,-6,-4,-2,0,2,4,6,...\}$
    Range: $\{3\}$

27. $y = x^2$

x	$y = x^2$
-5	$y = (-5)^2 = 25$
-4	$y = (-4)^2 = 16$
-3	$y = (-3)^2 = 9$
-2	$y = (-2)^2 = 4$
-1	$y = (-1)^2 = 1$

Domain: $\{-5,-4,-3,-2,-1\}$
Range: $\{1,4,9,16,25\}$

28. $y = x^2 - 1.5$

x	$y = x^2 - 1.5$
0	$y = (0)^2 - 1.5 = -1.5$
0.5	$y = (0.5)^2 - 1.5 = -1.25$
1	$y = (1)^2 - 1.5 = -0.5$
1.5	$y = (1.5)^2 - 1.5 = 0.75$
2	$y = (2)^2 - 1.5 = 2.5$

Domain: $\{0, 0.5, 1, 1.5, 2\}$
Range: $\{-1.5, -1.25, -0.5, 0.75, 2.5\}$

29. $y = 12 - 8x$

x	$y = 12 - 8x$	ordered pair
-6	$y = 12 - 8(-6) = 60$	$(-6, 60)$
-3	$y = 12 - 8(-3) = 36$	$(-3, 36)$
0	$y = 12 - 8(0) = 12$	$(0, 12)$
3	$y = 12 - 8(3) = -12$	$(3, -12)$
6	$y = 12 - 8(6) = -36$	$(6, -36)$

30. $y = \sqrt{10 - 3x}$

x	$y = \sqrt{10-3x}$	ordered pair
3	$y = \sqrt{10-3(3)} = 1$	$(3, 1)$
2	$y = \sqrt{10-3(2)} = 2$	$(2, 2)$
1	$y = \sqrt{10-3(1)} = \sqrt{7}$	$(1, \sqrt{7})$
0	$y = \sqrt{10-3(0)} = \sqrt{10}$	$(0, \sqrt{10})$
-1	$y = \sqrt{10-3(-1)} = \sqrt{13}$	$(-1, \sqrt{13})$
-2	$y = \sqrt{10-3(-2)} = 4$	$(-2, 4)$

31. $t = \dfrac{4}{7}s + 5$

s	$t = \dfrac{4}{7}s + 5$	ordered pair
-7	$t = \dfrac{4}{7}(-7) + 5 = 1$	$(-7, 1)$
0	$t = \dfrac{4}{7}(0) + 5 = 5$	$(0, 5)$
7	$t = \dfrac{4}{7}(7) + 5 = 9$	$(7, 9)$
14	$t = \dfrac{4}{7}(14) + 5 = 13$	$(14, 13)$
21	$t = \dfrac{4}{7}(21) + 5 = 17$	$(21, 17)$

32. $y = (3x - 5)(2x + 1)$

x	$y = (3x - 5)(2x + 1)$	y
-4	$y = (3(-4) - 5)(2(-4) + 1)$	119
-2	$y = (3(-2) - 5)(2(-2) + 1)$	33
0	$y = (3(0) - 5)(2(0) + 1)$	-5
2	$y = (3(2) - 5)(2(2) + 1)$	5
4	$y = (3(4) - 5)(2(4) + 1)$	63

33. $y = \dfrac{3}{4}x - 5$

x	$y = \dfrac{3}{4}x - 5$	y
-12	$y = \dfrac{3}{4}(-12) - 5$	-14
-8	$y = \dfrac{3}{4}(-8) - 5$	-11
-4	$y = \dfrac{3}{4}(-4) - 5$	-8
0	$y = \dfrac{3}{4}(0) - 5$	-5
4	$y = \dfrac{3}{4}(4) - 5$	-2
8	$y = \dfrac{3}{4}(8) - 5$	1
12	$y = \dfrac{3}{4}(12) - 5$	4

34. $y = 15.8 - 4.7x$

x	$y = 15.8 - 4.7x$	y
-2	$y = 15.8 - 4.7(-2)$	25.2
-1	$y = 15.8 - 4.7(-1)$	20.5
0	$y = 15.8 - 4.7(0)$	15.8
1	$y = 15.8 - 4.7(1)$	11.1
2	$y = 15.8 - 4.7(2)$	6.4

35. $y = 4x^2 - 17x - 15$

x	$y = 4x^2 - 17x - 15$	y
-2	$y = 4(-2)^2 - 17(-2) - 15$	35
$-\dfrac{3}{4}$	$y = 4\left(-\dfrac{3}{4}\right)^2 - 17\left(-\dfrac{3}{4}\right) - 15$	0
0	$y = 4(0)^2 - 17(0) - 15$	-15
$\dfrac{3}{4}$	$y = 4\left(\dfrac{3}{4}\right)^2 - 17\left(\dfrac{3}{4}\right) - 15$	$-\dfrac{51}{2}$
5	$y = 4(5)^2 - 17(5) - 15$	0

36. $y = \left|1 - 2x - 3x^2\right|$

x	$y = \left|1 - 2x - 3x^2\right|$	y
-6	$y = \left|1 - 2(-6) - 3(-6)^2\right|$	95
-3	$y = \left|1 - 2(-3) - 3(-3^2)\right|$	20
0	$y = \left|1 - 2(0) - 3(0)^2\right|$	1
3	$y = \left|1 - 2(3) - 3(3)^2\right|$	32
6	$y = \left|1 - 2(6) - 3(6)^2\right|$	119
9	$y = \left|1 - 2(9) - 3(9)^2\right|$	260

37. $y = 4.6x^2 + 2.8x + 10.4$

x	$y = 4.6x^2 + 2.8x + 10.4$	y
-3.7	$y = 4.6(-3.7)^2 + 2.8(-3.7) + 10.4$	63.014
-2.2	$y = 4.6(-2.2)^2 + 2.8(-2.2) + 10.4$	26.504
-0.7	$y = 4.6(-0.7)^2 + 2.8(-0.7) + 10.4$	10.694
0	$y = 4.6(0)^2 + 2.8(0) + 10.4$	10.4
0.8	$y = 4.6(0.8)^2 + 2.8(0.8) + 10.4$	15.584
2.3	$y = 4.6(2.3)^2 + 2.8(2.3) + 10.4$	41.174
3.8	$y = 4.6(3.8)^2 + 2.8(3.8) + 10.4$	87.464

38. Answers will vary; possible values:

x	$y = 4.5x - 1.6$	y
-2	$y = 4.5(-2) - 1.6$	-10.6
0	$y = 4.5(0) - 1.6$	-1.6
2	$y = 4.5(2) - 1.6$	7.4

39. Answers will vary; possible values:

x	$y = \dfrac{1}{4}x + 3$	y
-3	$y = \dfrac{1}{4}(-3) + 3$	$\dfrac{9}{4}$
2	$y = \dfrac{1}{4}(2) + 3$	$\dfrac{7}{2}$
8	$y = \dfrac{1}{4}(8) + 3$	5

40. Answers will vary; possible values:

x	$y = |3x - 10|$	y
-3	$y = |3(-3) - 10|$	19
1	$y = |3(1) - 10|$	7
5	$y = |3(5) - 10|$	5

41. $C = 2\pi r$

r	$C = 2\pi r$		ordered pair
$\dfrac{1}{4}$	$C = 2\pi\left(\dfrac{1}{4}\right)$	≈ 1.5708	$\left(\dfrac{1}{4}, 1.5708\right)$
$\dfrac{1}{2}$	$C = 2\pi\left(\dfrac{1}{2}\right)$	≈ 3.1416	$\left(\dfrac{1}{2}, 3.1416\right)$
1	$C = 2\pi(1)$	≈ 6.2832	$(1, 6.2832)$
$\dfrac{3}{2}$	$C = 2\pi\left(\dfrac{3}{2}\right)$	≈ 9.4248	$\left(\dfrac{3}{2}, 9.4248\right)$
2	$C = 2\pi(2)$	≈ 12.5664	$(2, 12.5664)$

42. $A = s^2$

s	$A = s^2$	A
3	$A = (3)^2$	9
5	$A = (5)^2$	25

43. $b = 180 - a$

a	$b = 180 - a$	b
30	$b = 180 - 30$	150
60	$b = 180 - 60$	120
90	$b = 180 - 90$	90
120	$b = 180 - 120$	60
150	$b = 180 - 150$	30

44. $I = (2000)(0.06)t = 120t$

t	$I = 120t$	I
2	$I = 120(2)$	240
3	$I = 120(3)$	360
4	$I = 120(4)$	480

45. $F = \dfrac{9}{5}C + 32$

C	$F = \dfrac{9}{5}C + 32$	F
-10	$F = \dfrac{9}{5}(-10) + 32$	14
-5	$F = \dfrac{9}{5}(-5) + 32$	23
0	$F = \dfrac{9}{5}(0) + 32$	32
5	$F = \dfrac{9}{5}(5) + 32$	41
10	$F = \dfrac{9}{5}(10) + 32$	50
15	$F = \dfrac{9}{5}(15) + 32$	59
20	$F = \dfrac{9}{5}(20) + 32$	68
25	$F = \dfrac{9}{5}(25) + 32$	77

46. Let x = number of zip disks purchased and c = total cost of purchase.

   a. $c(x) = 7.95x$

   b. For $c(x) = 7.95x$,

x	$c(x) = 7.95x$	$c(x)$
1	$c(1) = 7.95(1)$	7.95
2	$c(2) = 7.95(2)$	15.90
3	$c(3) = 7.95(3)$	23.85
4	$c(4) = 7.95(4)$	31.80
5	$c(5) = 7.95(5)$	39.75

47. Let x = number of items in run and c = the total cost of the production run.
$$c(x) = 2500 + 12x$$
$$c(1650) = 2500 + 12(1650)$$
$$= 2500 + 19,800$$
$$= 22,300$$
The production run of 1650 items will cost $22,300.

48. Let x = number of hours and c = total rental cost.
$$c(x) = 15 + 2x$$
$$c(10) = 15 + 2(10)$$
$$= 15 + 20$$
$$= 35$$
It will cost $35 to rent the grinder for 10 hours.

49. Let x = number of employees attending and c = total cost of luncheon.
$$c(x) = 275 + 9.5x$$
$$c(135) = 275 + 9.5(135)$$
$$= 275 + 1282.5$$
$$= 1557.5$$
The total charge for the luncheon will be $1557.50.

50. Let x = number of people attending and p = total profit.
$$p(x) = 4x - 185$$
$$p(310) = 4(310) - 185$$
$$= 1240 - 185$$
$$= 1055$$
The total profit for 310 admissions is $1055.

51. $f(x) = 250 - 3.5x$

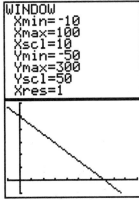

The function is decreasing for $0 < x < 71.429$.

52. $f(x) = 1000 + 50x$

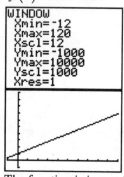

The function is increasing for $x > 0$

53. Let x = number of years.

   a. $f(x) = 25,000 + 5000x$

   $g(x) = 6000x$

   b. The two functions intersect at $(25, 150000)$.

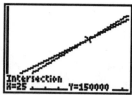

   c. The point of intersection represents the equilibrium point. That is, when $x = 25$ years, either option yields the same total payment of $150,000.

54. Let x = number of appliances sold.

   a. $c(x) = 22x + 600$

   $r(x) = 75x$

   b. The functions intersect at $(11.321, 849.057)$.

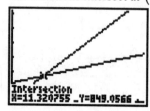

   c. The intersection point represents the break-even point. That is, when cost and revenue are the same. The retailer must sell at least 12 appliances to make a profit.

Chapter 3 Test

1. $y = 2x^2 + 17x - 9$

x	$y = 2x^2 + 17x - 9$	y
-9	$y = 2(-9)^2 + 17(9) - 9$	306
0	$y = 2(0)^2 + 17(0) - 9$	-9
3	$y = 2(3)^2 + 17(3) - 9$	60

2. $y = |2x - 3| - 1$

   a.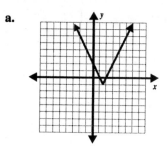

   b. No relative maxima

   c. Relative minimum is $y = -1$

   d. Increasing for $x > 1.5$

   e. Decreasing for $x < 1.5$

   f. x-intercepts: $(1, 0)$ and $(2, 0)$

   g. y-intercept: $(0, 2)$

3. a. This relation is not a function. The graph fails the vertical line test.

   b. The domain is all real numbers ≥ -4

   c. The range is all real numbers.

4. $A(1, 2)$, $B(-2, -4)$, $C(-5, 3)$, $D(2, -5)$, $E(0, -2)$

5. Point D lies in quadrant IV.

Chapter 3: Relations, Functions, and Graphs

6. $y = 2x^2 - 8x$

x	y
-2	24
-1	10
0	0
2	-8
4	0
5	10
6	24

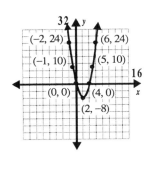

7. $y = \frac{3}{4}x - 2$

x	y
0	-2
4	1

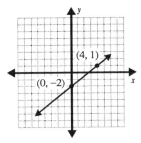

8. $y = \sqrt{7-x} + 1$

x	y
-9	5
-2	4
3	3
6	2
7	1

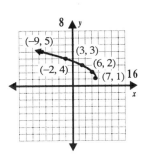

9. $f(x) = \frac{1}{2}x + 6$

$f(4) = \frac{1}{2}(4) + 6 = 2 + 6 = 8$

10. $f(x) = \frac{1}{2}x + 6$

$f(-6) = \frac{1}{2}(-6) + 6 = -3 + 6 = 3$

11. $f(x) = \frac{1}{2}x + 6$

$f(a) = \frac{1}{2}a + 6$

12. $f(x) = \frac{1}{2}x + 6$

$f(2+a) = \frac{1}{2}(2+a) + 6$

$= 1 + \frac{1}{2}a + 6$

$= \frac{1}{2}a + 7$

13. Let x = items in one production run and c = cost of production run.

$c(x) = 450 + 21.50x$

14. $c(250) = 450 + 21.5(250)$

$= 450 + 5375$

$= 5825$

The cost of the production run is $5825.

15. $y = 3x - 10$ and $y = -x - 2$

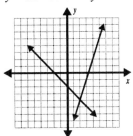

Intersect at $(2, -4)$

16. Answers will vary. One possible answer:

A point in a plane is an ordered pair. The first number (coordinate) is the distance and direction horizontally from the y-axis. The second number (coordinate) is the distance and direction vertically from the x-axis.

Chapters 1-3 Cumulative Review

1. 0 and 12 are whole numbers.

2. 0 and 12 are integers.

3. $-\frac{2}{3}$, 0, 12, $1\frac{4}{5}$, and -0.33 are rational numbers.

SSM: Experiencing Introductory and Intermediate Algebra

4. $\sqrt{7}$ is an irrational number.

5. $\dfrac{3}{8} > \dfrac{1}{3}$

6. $\dfrac{2}{3} > 0.66$

7. $-2.8 < -1.6$

8.

9. $-28 + 13 = -15$

10. $4.8 - 7.36 = -2.56$

11. $-87 \div (-29) = 3$

12. $-\dfrac{5}{8} - \dfrac{2}{3} = -\dfrac{15}{24} - \dfrac{16}{24} = -\dfrac{31}{24}$

13. $-2\dfrac{3}{4} \div 1\dfrac{3}{7}$
$= -\dfrac{11}{4} \div \dfrac{10}{7} = -\dfrac{11}{4} \cdot \dfrac{7}{10}$
$= -\dfrac{77}{40} = -1\dfrac{37}{40}$

14. $\left(-\dfrac{2}{3}\right)\left(\dfrac{3}{8}\right)\left(-\dfrac{7}{16}\right)\left(\dfrac{9}{10}\right)$
$= \dfrac{378}{3840} = \dfrac{63}{640}$

15. $(12.96)(-4.8) = -62.208$

16. $(14)(0)(5)(-6) = 0$

17. $(-12)(16) \div 4(-2)$
$= -192 \div 4(-2) = (-48)(-2) = 96$

18. $14 + (-7) + 22 - 16 - (-18)$
$= 14 - 7 + 22 - 16 + 18 = 31$

19. $-[3.8 - (-2.4)]$
$= -[3.8 + 2.4] = -[6.2] = -6.2$

20. $\dfrac{2(3^2 + 7) - 2^5}{3 \cdot 18}$
$= \dfrac{2(9+7) - 32}{3 \cdot 18} = \dfrac{2(16) - 32}{3 \cdot 18} = \dfrac{32 - 32}{3 \cdot 18} = 0$

21. $-|12 - 20| = -|-8| = -8$

22. $\sqrt{\dfrac{16}{25}} = \dfrac{\sqrt{16}}{\sqrt{25}} = \dfrac{4}{5}$

23. $-\sqrt{1.2} \approx -1.095$

24. $\sqrt{-16}$ is not a real number.

25. $\sqrt[3]{1\dfrac{13}{81}}$
$= \sqrt[3]{\dfrac{94}{81}} = \sqrt[3]{\dfrac{94 \cdot 9}{81 \cdot 9}} = \sqrt[3]{\dfrac{846}{729}} = \dfrac{\sqrt[3]{846}}{\sqrt[3]{729}} = \dfrac{\sqrt[3]{846}}{9}$
≈ 1.051

26. $14^0 = 1$

27. $1^{12} = 1$

28. 0^0 is indeterminant.

29. $-8^4 = -4096$

30. $(-8)^4 = 4096$

31. $\left(\dfrac{1}{10}\right)^{-2} = \left(\dfrac{10}{1}\right)^2 = 10^2 = 100$

32. $0.00000305 = 3.05 \times 10^{-6}$

33. $-4,235,600 = -4.2356 \times 10^6$

34. $3.56 \times 10^{-2} = 0.0356$

35. $6.78 \times 10^8 = 678,000,000$

36. $\sqrt{-x^2 + 5x - 2} + 5$
$\sqrt{-(3)^2 + 5(3) - 2} + 5 = \sqrt{-9 + 15 - 2} + 5$
$= \sqrt{4} + 5 = 2 + 5 = 7$

37. $a^3 - 2a^2 + a - 2a^3 + 7a - 5$
   a. There are 6 terms.

110

Chapter 3: Relations, Functions, and Graphs

b. a^3, $-2a^2$, a, $-2a^3$, $7a$

c. -5

d. $a^3 - 2a^2 + a - 2a^3 + 7a - 5$
$= a^3 - 2a^3 - 2a^2 + a + 7a - 5$
$= -a^3 - 2a^2 + 8a - 5$

38. $-(3y+2z)+(4y-2z)-(-3y-5z)$
$= -3y - 2z + 4y - 2z + 3y + 5z$
$= -3y + 4y + 3y - 2z - 2z + 5z$
$= 4y + z$

39. $\dfrac{3x}{4} + \dfrac{5y}{8} - \dfrac{1}{16} - \dfrac{3x}{4} + \dfrac{y}{8} - \dfrac{5}{6}$
$= \dfrac{3x}{4} - \dfrac{3x}{4} + \dfrac{5y}{8} + \dfrac{y}{8} - \dfrac{1}{16} - \dfrac{5}{6}$
$= \dfrac{6y}{8} - \dfrac{3}{48} - \dfrac{40}{48}$
$= \dfrac{3}{4}y - \dfrac{43}{48}$

40. $2[8+3(x-4)-2(3x+1)]$
$= 2[8+3x-12-6x-2]$
$= 2[3x-6x+8-12-2] = 2[-3x-6]$
$= -6x-12$

41. $-x^2 + 3x + 8 = -3x$
$-(-2)^2 + 3(-2) + 8 = -3(-2)$
$-4 - 6 + 8 = 6$
$-2 = 6$
Since $-2 \neq 6$, $x = -2$ is not a solution.

42. $y = 2x^2 + 3$

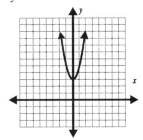

43. *Domain*: All real numbers.
Range: All real numbers ≥ 3.

44. This relation is a function. No vertical line can cross the graph at more than one point. The graph passes the vertical line test.

45. Relative minimum is $y = 3$.

46. Increasing for $x > 0$.

47. $f(x) = \dfrac{1}{3}x - 5$

a. $f(9) = \dfrac{1}{3}(9) - 5 = 3 - 5 = -2$

b. $f(3+h) = \dfrac{1}{3}(3+h) - 5 = 1 + \dfrac{h}{3} - 5$
$= \dfrac{1}{3}h - 4$

48. $V = L \cdot W \cdot H$
$V = (3.5)(2.25)(1.75)$
$V = 13.78125$
The volume of the solid is 13.7815 cubic feet.

49. $A = P(1+r)^t$
$= 500(1+.055)^4 = 500(1.055)^4$
$= 619.41$
$I = A - P$
$= 619.41 - 500 = 119.41$
Kelsey will receive $119.41 in interest. Her total investment amount will be $619.41.

50. Let x = number of ornaments in the production run and c = total cost of the production run.
$c(x) = 35 + 2.8x$
$c(150) = 35 + 2.8(150) = 455$
The total cost of the production run will be $455.

Chapter 4

4.1 Exercises

1. $6x-55 = x+72$ is linear because the expressions on both sides of the equation are in the form $ax+b$.

3. $4x^2 +5 = 2x-6$ is nonlinear because in the expression on the left side of the equals x has an exponent of 2.

5. $\frac{7}{9}z-\frac{2}{3}=0$ is linear because it is in standard form.

7. $\sqrt[3]{4x+16} = 27$ is nonlinear because the radical expression on the left side of the equation has a variable in its radicand.

9. $3(2x-5) = x+3(x-9)$ is linear because it simplifies to $6x-15 = 4x-27$, with expressions on both sides of the equation in the form $ax+b$.

11. The equation is $2x-7 = x+2$.
    The solution is 9.

13. The equation is $0.5x+1.25 = 0.5(x+2.5)$.
    The solution is the set of all real numbers.

15. The equation is $x-(4.5-0.5x)=1.5(x+2)$.
    The equation has no solution.

17. The equation is $\frac{1}{3}x+1 = \frac{3}{2}x-1$.

    The expression on the left is greater than the expression on the right for $x = 1$ and less than the expression on the right for $x = 2$. The solution is a non-integer number between 1 and 2.

19. Let $Y1 = 2x-7$ and $Y2 = 35-x$. A sample table is shown below.

X	Y1	Y2
11	15	24
12	17	23
13	19	22
14	21	21
15	23	20
16	25	19
17	27	18

    X=14

    Because $Y1 = Y2 = 21$ when $x = 14$, the solution is 14.

21. Let $Y1 = 3(2x+11)$ and $Y2 = 3(5+x)$. A sample table is shown below.

X	Y1	Y2
-9	-21	-12
-8	-15	-9
-7	-9	-6
-6	-3	-3
-5	3	0
-4	9	3
-3	15	6

    X=-6

    Because $Y1 = Y2 = -3$ when $x = -6$, the solution is -6.

23. Rewrite the equation in terms of x – that is, $6.8x+4.3 = 2.6x+33.7$. Let $Y1 = 6.8x+4.3$ and $Y2 = 2.6x+33.7$. A sample table is shown below.

X	Y1	Y2
4	31.5	44.1
5	38.3	46.7
6	45.1	49.3
7	51.9	51.9
8	58.7	54.5
9	65.5	57.1
10	72.3	59.7

    X=7

    Because $Y1 = Y2 = 51.9$ when $x = 7$, the solution is 7.

25. Let $Y1 = 7(x+10)+15$ and $Y2 = 6(x+15)+(x-5)$. A sample table is shown below.

X	Y1	Y2
-3	64	64
-2	71	71
-1	78	78
0	85	85
1	92	92
2	99	99
3	106	106

    X=0

    Because $Y1 = Y2$ for all values of x, the solution is the set of all real numbers. The equation is an identity.

27. Rewrite the equation in terms of x – that is, $(x-4)-(x+4) = (x+3)-(x-2)$. Let $Y1 = (x-4)-(x+4)$ and $Y2 = (x+3)-(x-2)$. A sample table is shown below.

Chapter 4: Linear Equations in One Variable

Because Y1 = –8 for all values of x and Y2 = 5 for all values of x, the equation has no solution. The equation is a contradiction.

29. Rewrite the equation in terms of x – that is, $3.5(x-1) = 7(0.5x+0.6)+2$. Let Y1 = $3.5(x-1)$ and Y2 = $7(0.5x+0.6)+2$. A sample table is shown below.

Because Y1 – Y2 = –9.7 for all values of x, the expressions will never be equal. The equation has no solution.

31. Let Y1 = $\frac{4}{5}(x-1)$ and Y2 = $6\left(\frac{1}{15}x - \frac{1}{10}\right)$. A sample table is shown below.

Because Y1 < Y2 when $x = 0$ and Y1 > Y2 when $x = 1$, the solution is a noninteger number between 0 and 1. The graphs of the functions are shown below.

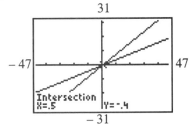

Because the intersection point is (0.5, – 0.4), the solution is 0.5.

33. The equation is $3x + 2 = 4 - x$. The solution is 0.5.

35. The equation is $\frac{1}{2}x + 5 = 4 - 0.5(6 - x)$. The equation has no solution.

37. Let Y1 = $x + 6$ and Y2 = $9 + 2x$. The graphs of the functions are shown below.

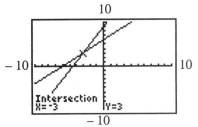

Because the intersection point is (– 3, 3), the solution is – 3.

39. Let Y1 = $(x+4)+(x+2)$ and Y2 = $(x-1)+(x-3)$. The graphs of the functions are shown below.

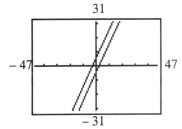

Because the lines are parallel, the equation has no solution.

41. Let Y1 = $2(x+3)$ and Y2 = $3(x-1)-(x-9)$. The graphs of the functions are shown below.

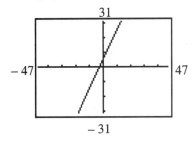

Because the lines are the same, the solution is the set of all real numbers. The equation is an identity.

43. Let $Y1 = 1.7x - 22.2$ and $Y2 = 13.8 - 0.7x$. The graphs of the functions are shown below.

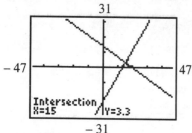

Because the intersection point is (15, 3.3), the solution is 15.

45. Let $Y1 = 2.2(x-1) + 1.7x$ and $Y2 = 3.5(x+1) + 0.4x$. The graphs of the functions are shown below.

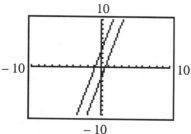

Because the lines are parallel, the equation has no solution.

47. Let $Y1 = \frac{4}{5}x + \frac{1}{5}$ and $Y2 = \frac{1}{5}x + 2$. The graphs of the functions are shown below.

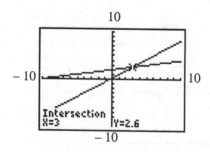

Because the intersection point is (3, 2.6), the solution is 3.

49. Let $Y1 = \frac{2}{3}(x+1) - \frac{1}{3}$ and $Y2 = \frac{1}{3}(x+1) + \frac{1}{3}x$. The graphs of the functions are shown below.

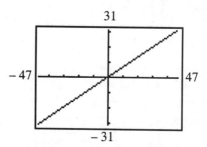

Because the lines are the same, the solution is the set of all real numbers. The equation is an identity.

51. Let $x =$ the number of miles. Then $49.95 = 29.95 + 0.25x$.
$Y1 = 49.95$; $Y2 = 29.95 + 0.25x$

X	Y1	Y2
77	49.95	49.2
78	49.95	49.45
79	49.95	49.7
80	49.95	49.95
81	49.95	50.2
82	49.95	50.45
83	49.95	50.7

X=80

$Y1 = Y2 = 49.95$ when $x = 80$
The cost will be the same under the two plans if the rental car is driven 80 miles.

53. Let $x =$ the number of pairs of shoes. Then $280 + 8x = 22x$.
$Y1 = 280 + 8x$; $Y2 = 22x$.

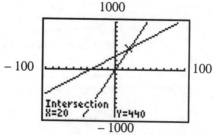

The intersection point is (20, 440). To break even, the shoe factory must produce and sell 20 pairs of shoes per day.

55. Let x = amount spent on the fifth day. Then
$$\frac{28+19+22+27+x}{5}=25.$$
$Y1 = \frac{28+19+22+27+x}{5}$; $Y2 = 25$.

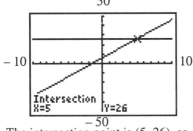

$Y1 = Y2 = 25$ when $x = 29$.
Ingrid can spend $29 on the fifth day.

57. Let x = the width of the rug, then
$x + 3$ = the length of the rug.
$2x + 2(x+3) = 26$.
$Y1 = 2x + 2(x+3)$; $Y2 = 26$.

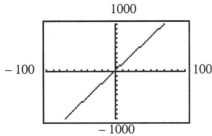

The intersection point is (5, 26), so $x = 5$ and $x + 3 = 8$.
The dimensions of the rug are 5 feet by 8 feet.

59. Let x = the number of rolls of paper. Then
$25 + 9x + 4x = 25 + 14x$.
$Y1 = 25 + 9x + 4x$; $Y2 = 25 + 14x$.

The lines are the same, so the solution is the set of all real numbers. Thus, Charlene's and Greta's charges will always be equal no matter how many rolls of paper are used.

61. Let x = the temperature on the seventy day, then
$$\frac{84+85+85+85+85+86+x}{7}=84.$$
$Y1 = \frac{84+85+85+85+85+86+x}{7}$; $Y2 = 84$.

$Y1 = Y2 = 84$ when $x = 78$.
The temperature on the seventh day must be $78°F$.

4.1 Calculator Exercises

A. 1. Let $Y1 = 10x - 156$ and $Y2 = 108 - 2x$.
Zooming out one time from the integer screen setting yields the graphs shown below.

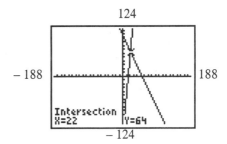

Because the intersection point is (22, 64), the solution is 22.

2. Let $Y1 = 9.2x + 55.8$ and $Y2 = 1.4x - 37.8$.
Zooming out one time from the integer screen setting yields the graphs shown below.

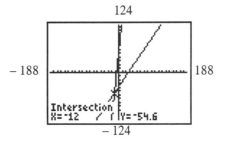

Because the intersection point is (−12, −54.6), the solution is −12.

3. Let $Y1 = 7x + 450$ and $Y2 = 2x + 1700$. Zooming out four times from the standard screen setting yields the graphs shown below.

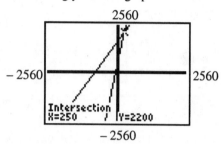

Because the intersection point is (250, 2200), the solution is 250.

4. Let $Y1 = 12x + 800$ and $Y2 = 8x - 1200$. Zooming out five times from the standard screen setting yields the graphs shown below.

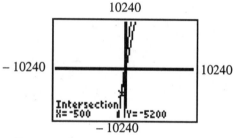

Because the intersection point is (−500, −5200), the solution is −500.

B. Students should experiment with some exercises.

4.2 Exercises

1. $x + 33 = 51$
   $x + 33 - 33 = 51 - 33$
   $x = 18$
   The solution is 18.

3. $75 = a - 41$
   $75 + 41 = a - 41 + 41$
   $116 = a$
   The solution is 116.

5. $-4.91 = y + 3.07$
   $-4.91 - 3.07 = y + 3.07 - 3.07$
   $-7.98 = y$
   The solution is -7.98.

7. $y - \dfrac{1}{6} = -\dfrac{1}{6}$
   $y - \dfrac{1}{6} + \dfrac{1}{6} = -\dfrac{1}{6} + \dfrac{1}{6}$
   $y = 0$
   The solution is 0.

9. $27 + (x - 13) = 11$
   $27 + x - 13 = 11$
   $x + 14 = 11$
   $x + 14 - 14 = 11 - 14$
   $x = -3$
   The solution is -3.

11. $(13.9 + x) + 0.88 = -2.07$
    $13.9 + x + 0.88 = -2.07$
    $x + 14.78 = -2.07$
    $x + 14.78 - 14.78 = -2.07 - 14.78$
    $x = -16.85$
    The solution is -16.85.

13. $\left(x - \dfrac{3}{10}\right) - \dfrac{2}{5} = -3\dfrac{1}{2}$
    $x - \dfrac{3}{10} - \dfrac{4}{10} = -\dfrac{7}{2}$
    $x - \dfrac{7}{10} = -\dfrac{35}{10}$
    $x - \dfrac{7}{10} + \dfrac{7}{10} = -\dfrac{35}{10} + \dfrac{7}{10}$
    $x = -\dfrac{28}{10} = -\dfrac{14}{5}$
    The solution is $-\dfrac{14}{5}$ or $-2\dfrac{4}{5}$.

15. $x - 9 = 27$
    $x - 9 + 9 = 27 + 9$
    $x = 36$
    The solution is 36.

17. $\left(\dfrac{1}{3}x+\dfrac{1}{8}\right)+\left(\dfrac{3}{4}+\dfrac{2}{3}x\right)=-\dfrac{3}{16}$

$\dfrac{1}{3}x+\dfrac{1}{8}+\dfrac{6}{8}+\dfrac{2}{3}x=-\dfrac{3}{16}$

$x+\dfrac{7}{8}=-\dfrac{3}{16}$

$x+\dfrac{14}{16}-\dfrac{14}{16}=-\dfrac{3}{16}-\dfrac{14}{16}$

$x=-\dfrac{17}{16}$

The solution is $-\dfrac{17}{16}$.

19. $(3x+76)-(2x-45)=31$

$3x+76-2x+45=31$

$x+121=31$

$x+121-121=31-121$

$x=-90$

The solution is -90.

21. $-324=-4y$

$\dfrac{-324}{-4}=\dfrac{-4y}{-4}$

$81=y$

The solution is 81.

23. $-5.1x=0.102$

$\dfrac{-5.1x}{-5.1}=\dfrac{0.102}{-5.1}$

$x=-0.02$

The solution is -0.02.

25. $-3\dfrac{1}{3}x=-1\dfrac{1}{3}$

$-\dfrac{10}{3}x=-\dfrac{4}{3}$

$-\dfrac{3}{10}\left(-\dfrac{10}{3}x\right)=-\dfrac{3}{10}\left(-\dfrac{4}{3}\right)$

$x=\dfrac{2}{5}$

The solution is $\dfrac{2}{5}$.

27. $\dfrac{x}{4}=1.22$

$4\left(\dfrac{x}{4}\right)=4(1.22)$

$x=4.88$

The solution is 4.88.

29. $-x=57$

$-1(-x)=-1(57)$

$x=-57$

The solution is -57.

31. $57=2x+17x$

$57=19x$

$\dfrac{57}{19}=\dfrac{19x}{19}$

$3=x$

The solution is 3.

33. $18.22x-12.9x=-12.76$

$5.32x=-12.76$

$\dfrac{5.32x}{5.32}=\dfrac{-12.76}{5.32}$

$x\approx -2.398$

The solution is approximately -2.398.

35. $\dfrac{5}{14}=\dfrac{9}{14}a+\dfrac{3}{7}a$

$\dfrac{5}{14}=\dfrac{9}{14}a+\dfrac{6}{14}a$

$\dfrac{5}{14}=\dfrac{15}{14}a$

$\dfrac{14}{15}\left(\dfrac{5}{14}\right)=\dfrac{14}{15}\left(\dfrac{15}{14}a\right)$

$\dfrac{1}{3}=a$

The solution is $\dfrac{1}{3}$.

SSM: Experiencing Introductory and Intermediate Algebra

37. $2(3x+6)+3(x-4)=126$
    $6x+12+3x-12=126$
    $9x=126$
    $\dfrac{9x}{9}=\dfrac{126}{9}$
    $x=14$
    The solution is 14.

39. $4.8(a+3)+2.4(a-6)=-7.2$
    $4.8a+14.4+2.4a-14.4=-7.2$
    $7.2a=-7.2$
    $\dfrac{7.2a}{7.2}=\dfrac{-7.2}{7.2}$
    $a=-1$
    The solution is -1.

41. $3\left(\dfrac{1}{2}x-\dfrac{3}{4}\right)-18\left(x-\dfrac{1}{8}\right)=0$
    $\dfrac{3}{2}x-\dfrac{9}{4}-18x+\dfrac{18}{8}=0$
    $\dfrac{3}{2}x-\dfrac{9}{4}-\dfrac{36}{2}x+\dfrac{9}{4}=0$
    $-\dfrac{33}{2}x=0$
    $-\dfrac{2}{33}\left(-\dfrac{33}{2}x\right)=-\dfrac{2}{33}(0)$
    $x=0$
    The solution is 0.

43. Let x = the number of remaining servings.
    $x+4=10$
    $x+4-4=10-4$
    $x=6$
    There are 6 servings remaining in the box.

45. Let x = Chuck's gross pay.
    $x-567.32=1784.26$
    $x-567.32+567.32=1784.26+567.32$
    $x=2351.58$
    Chuck's gross pay is $2351.58.

47. Let x = the amount of flour to borrow.
    $3\dfrac{3}{4}+x=5\dfrac{1}{2}$
    $\dfrac{15}{4}+x=\dfrac{11}{2}$

    $\dfrac{15}{4}+x-\dfrac{15}{4}=\dfrac{22}{4}-\dfrac{15}{4}$
    $x=\dfrac{7}{4}=1\dfrac{3}{4}$
    Glenda must borrow $1\dfrac{3}{4}$ cups of flour.

49. Let x = the amount of sales tax.
    $49.95+x=54.32$
    $49.95+x-49.95=54.32-49.95$
    $x=4.37$
    Tameka paid $4.37 in sales tax.

51. Let x = the amount of additional wallpaper needed.
    $35+x=18+22+18+22$
    $35+x=80$
    $35+x-35=80-35$
    $x=45$
    Karla must buy 45 more feet of boarder. If Karla buys two more rolls of boarder containing 20 feet each, then she will have $2\cdot 20=40$ feet, which is not enough.

53. Let x = the total number of paid admissions.
    $0.55x=264$
    $\dfrac{0.55x}{0.55}=\dfrac{264}{0.55}$
    $x=480$
    There were 480 paid admissions to the talent contest.

55. Let x = the number of packets to be sold.
    $2.50x=1450$
    $\dfrac{2.50x}{2.50}=\dfrac{1450}{2.50}$
    $x=580$
    Erika's class must sell 580 packets.

57. Let x = the value of the entire estate.
    $\dfrac{3}{5}x=45240$
    $\dfrac{5}{3}\left(\dfrac{3}{5}x\right)=\dfrac{5}{3}(45240)$
    $x=75400$
    The estate was worth $75,400.

59. Let h = the perpendicular distance across the library (i.e. the height of the parallelogram).
The area formula is $A = bh$.
$350 = 20h$
$\dfrac{350}{20} = \dfrac{20h}{20}$
$17.5 = h$
The perpendicular distance across the library (i.e. the height of the parallelogram) is 17.5 feet.

61. Let h = the height of the cylindrical tank.
The volume formula is $V = \pi r^2 h$.
$300 = \pi \cdot 4^2 \cdot h$
$300 = 16\pi h$
$\dfrac{300}{16\pi} = \dfrac{16\pi h}{16\pi}$
$5.968 \approx h$
The cylindrical tank must be about 6 feet high.

63. Let P = the amount of principal to be placed in savings.
The simple interest formula is $I = Prt$.
$864 = P(0.045)(3)$
$864 = 0.135P$
$\dfrac{864}{0.135} = \dfrac{0.135P}{0.135}$
$6400 = P$
You must place $6400 into savings.

65. Let x = the firm's quarterly profits.
$\dfrac{x}{5} = 12730$
$5\left(\dfrac{x}{5}\right) = 5(12730)$
$x = 63650$
The firm's quarterly profits were $63,650.

67. Let x = the sales for the day.
$0.03x = 40.50$
$\dfrac{0.03x}{0.03} = \dfrac{40.50}{0.03}$
$x = 1350$
Connie's sales for the day were $1350.

69. Let x = the maximum number of dogs allowed.
$0.75x = 12$
$\dfrac{0.75x}{0.75} = \dfrac{12}{0.75}$
$x = 16$
The maximum number of dogs allowed is 16.

71. Let x = the 1999 estimated U.S. population.
$x - 23300000 = 248700000$
$x = 248700000 + 23300000$
$x = 272000000$
The 1999 estimated U.S. population is 272,000,000.

73. Let x = the width of a single human hair.
$\dfrac{x}{2000} = 50$
$2000\left(\dfrac{x}{2000}\right) = 2000(50)$
$x = 100000$
A single human hair is about 100,000 nanometers wide.

75. Let x = the amount of 70% antifreeze needed.
$0.70x = 4$
$\dfrac{0.70x}{0.70} = \dfrac{4}{0.70}$
$x \approx 5.714$
The mechanic needs about 5.7 gallons of 70% antifreeze solution.

4.2 Calculator Exercises

1. Let x = the number of linear feet of fencing required.
The circumference formula is $C = 2\pi r$.
$x + 2(5) = 2\pi(75)$
$x + 2(5) - 2\pi(75) = 0$

```
solve(X-2π*75+2*
5,X,400)
        461.238898
```

SSM: Experiencing Introductory and Intermediate Algebra

Approximately 461.2 feet of fencing will be needed.

2. Let x = the difference between the circumference and the diameter of the circle.
   The circumference formula is $C = 2\pi r$.
   $$x + 25.8 = \pi(25.8)$$
   $$x + 25.8 - \pi(25.8) = 0$$

   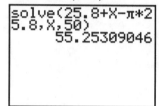

   The circumference is approximately 55.3 inches longer than the diameter.

3. Let x = the amount of increase per share.
   $$250x = 531\frac{1}{4}$$
   $$250x - 531\frac{1}{4} = 0$$

   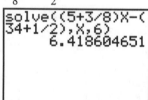

   The stock increased by $2\frac{1}{8}$ points per share.

4. Let x = the number of pieces.
   $$5\frac{3}{8}x = 34\frac{1}{2}$$
   $$5\frac{3}{8}x - 34\frac{1}{2} = 0$$

   6 pieces of the $5\frac{3}{8}$-inch tubing can be cut from the original piece.

5. Let x = the number of miles.
   $$18.5x = 220$$
   $$18.5x - 220 = 0$$

   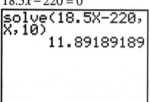

   The car will need approximately 12 gallons of gasoline for the 220-mile trip.

6. Let x = the number of square feet of wall.
   $$6.5x = 800$$
   $$6.5x - 800 = 0$$

   Approximately 123 square feet of wall can be constructed from the 800 bricks.

7. Let x = the daily recommended amount of fat.
   $$0.04x = 2.5$$
   $$0.04x - 2.5 = 0$$

   The daily recommended amount of fat is 62.5 grams.

8. Let x = the daily recommended amount of fat.
   $$0.05x = 3$$
   $$0.05x - 3 = 0$$

   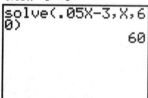

   The daily recommended amount of fat is 60 grams. This does not agree exactly with the answer to problem number 7. However, they are fairly close together. The difference is possibly due to round-off error.

9. Let r = the speed of the water flow.
$250 = r(15)$
$250 - r(15) = 0$

```
solve(250-R*15,R
,15)
        16.66666667
```

The speed of the water flow is approximately 16.7 feet per second.

4.3 Exercises

1. $4x + 8 = 0$
$4x + 8 - 8 = 0 - 8$
$4x = -8$
$\dfrac{4x}{4} = \dfrac{-8}{4}$
$x = -2$
The solution is -2.

3. $-x - 41 = 3$
$-x - 41 + 41 = 3 + 41$
$-x = 44$
$-1(-x) = -1(44)$
$x = -44$
The solution is -44.

5. $-3x + 7 = 7$
$-3x + 7 - 7 = 7 - 7$
$-3x = 0$
$\dfrac{-3x}{-3} = \dfrac{0}{-3}$
$x = 0$
The solution is 0.

7. $15.17 = 5.9x - 4.3$
$15.17 + 4.3 = 5.9x - 4.3 + 4.3$
$19.47 = 5.9x$
$\dfrac{19.47}{5.9} = \dfrac{5.9x}{5.9}$
$3.3 = x$
The solution is 3.3.

9. $-9.2x - 4.3 = -70.54$
$-9.2x - 4.3 + 4.3 = -70.54 + 4.3$
$-9.2x = -66.24$
$\dfrac{-9.2x}{-9.2} = \dfrac{-66.24}{-9.2}$
$x = 7.2$
The solution is 7.2.

11. $-\dfrac{5}{9}b + \dfrac{11}{12} = \dfrac{23}{36}$
$36\left(-\dfrac{5}{9}b + \dfrac{11}{12}\right) = 36\left(\dfrac{23}{36}\right)$
$36\left(-\dfrac{5}{9}b\right) + 36\left(\dfrac{11}{12}\right) = 36\left(\dfrac{23}{36}\right)$
$-20b + 33 = 23$
$-20b + 33 - 33 = 23 - 33$
$-20b = -10$
$\dfrac{-20b}{-20} = \dfrac{-10}{-20}$
$b = \dfrac{1}{2}$
The solution is $\dfrac{1}{2}$.

13. $-2\dfrac{2}{3}z - 3\dfrac{1}{2} = -8\dfrac{5}{6}$
$-\dfrac{8}{3}z - \dfrac{7}{2} = -\dfrac{53}{6}$
$6\left(-\dfrac{8}{3}z - \dfrac{7}{2}\right) = 6\left(-\dfrac{53}{6}\right)$
$6\left(-\dfrac{8}{3}z\right) - 6\left(\dfrac{7}{2}\right) = 6\left(-\dfrac{53}{6}\right)$
$-16z - 21 = -53$
$-16z - 21 + 21 = -53 + 21$

SSM: Experiencing Introductory and Intermediate Algebra

$-16z = -32$
$\dfrac{-16z}{-16} = \dfrac{-32}{-16}$
$z = 2$
The solution is 2.

15. $5x + 6 = x + 126$
$5x + 6 - x = x + 126 - x$
$4x + 6 = 126$
$4x + 6 - 6 = 126 - 6$
$4x = 120$
$\dfrac{4x}{4} = \dfrac{120}{4}$
$x = 30$
The solution is 30.

17. $27x - 49 = -12x - 10$
$27x - 49 + 12x = -12x - 10 + 12x$
$39x - 49 = -10$
$39x - 49 + 49 = -10 + 49$
$39x = 39$
$\dfrac{39x}{39} = \dfrac{39}{39}$
$x = 1$
The solution is 1.

19. $156z - 210 = 47z + 662$
$156z - 210 - 47z = 47z + 662 - 47z$
$109z - 210 = 662$
$109z - 210 + 210 = 662 + 210$
$109z = 872$
$z = 8$
The solution is 8.

21. $4x - (3x + 5) = x - 5$
$4x - 3x - 5 = x - 5$
$x - 5 = x - 5$
$x - 5 - x = x - 5 - x$
$-5 = -5$
This equation is always true (an identity). The solution is the set of all real numbers.

23. $6x - (x + 1) = 5x + 7$
$6x - x - 1 = 5x + 7$
$5x - 1 = 5x + 7$
$5x - 1 - 5x = 5x + 7 - 5x$
$-1 = 7$
This is a false equation (a contradiction). It has no solution.

25. $5(0.3x + 8.7) = 1.5x + 43.5$
$1.5x + 43.5 = 1.5x + 43.5$
$1.5x + 43.5 - 1.5x = 1.5x + 43.5 - 1.5x$
$43.5 = 43.5$
This equation is always true (an identity). The solution is the set of all real numbers.

27. $5.5x = 1.2x + 3.3(x - 2)$
$5.5x = 1.2x + 3.3x - 6.6$
$5.5x = 4.5x - 6.6$
$5.5x - 4.5x = 4.5x - 6.6 - 4.5x$
$x = -6.6$
The solution is -6.6.

29. $\dfrac{3}{4}x + 6 = \dfrac{1}{2}x + \dfrac{1}{4}x$
$4\left(\dfrac{3}{4}x + 6\right) = 4\left(\dfrac{1}{2}x + \dfrac{1}{4}x\right)$
$4\left(\dfrac{3}{4}x\right) + 4(6) = 4\left(\dfrac{1}{2}x\right) + 4\left(\dfrac{1}{4}x\right)$
$3x + 24 = 2x + x$
$3x + 24 = 3x$
$3x + 24 - 3x = 3x - 3x$
$24 = 0$
This is a false equation (a contradiction). It has no solution.

31. $3x - \dfrac{1}{4} = \left(x + \dfrac{1}{2}\right) + \left(2x + \dfrac{1}{3}\right)$

$12\left(3x - \dfrac{1}{4}\right) = 12\left[\left(x + \dfrac{1}{2}\right) + \left(2x + \dfrac{1}{3}\right)\right]$

$12(3x) - 12\left(\dfrac{1}{4}\right) = 12\left(x + \dfrac{1}{2}\right) + 12\left(2x + \dfrac{1}{3}\right)$

$36x - 3 = 12x + 12\left(\dfrac{1}{2}\right) + 12(2x) + 12\left(\dfrac{1}{3}\right)$

$36x - 3 = 12x + 6 + 24x + 4$

$36x - 3 = 36x + 10$

$36x - 3 - 36x = 36x + 10 - 36x$

$-3 = 10$

This is a false equation (a contradiction). It has no solution.

33. $11x - 12 = 7(3x - 6) - 2(x + 9)$

$11x - 12 = 21x - 42 - 2x - 18$

$11x - 12 = 19x - 60$

$11x - 12 - 19x = 19x - 60 - 19x$

$-8x - 12 = -60$

$-8x - 12 + 12 = -60 + 12$

$-8x = -48$

$\dfrac{-8x}{-8} = \dfrac{-48}{-8}$

$x = 6$

The solution is 6.

35. $7x - 5(3x + 9) = -2(4x + 35)$

$7x - 15x - 45 = -8x - 70$

$-8x - 45 = -8x - 70$

$-8x - 45 + 8x = -8x - 70 + 8x$

$-45 = -70$

This is a false equation (a contradiction). It has no solution.

37. $\dfrac{1}{4}x + \dfrac{5}{9} = \dfrac{5}{6}$

$36\left(\dfrac{1}{4}x + \dfrac{5}{9}\right) = 36\left(\dfrac{5}{6}\right)$

$36\left(\dfrac{1}{4}x\right) + 36\left(\dfrac{5}{9}\right) = 36\left(\dfrac{5}{6}\right)$

$9x + 20 = 30$

$9x + 20 - 20 = 30 - 20$

$9x = 10$

$\dfrac{9x}{9} = \dfrac{10}{9}$

$x = \dfrac{10}{9}$

The solution is $\dfrac{10}{9}$.

39. $3x + \dfrac{1}{4} = 2x + \dfrac{7}{36}$

$36\left(3x + \dfrac{1}{4}\right) = 36\left(2x + \dfrac{7}{36}\right)$

$36(3x) + 36\left(\dfrac{1}{4}\right) = 36(2x) + 36\left(\dfrac{7}{36}\right)$

$108x + 9 = 72x + 7$

$108x + 9 - 72x = 72x + 7 - 72x$

$36x + 9 = 7$

$36x + 9 - 9 = 7 - 9$

$36x = -2$

$\dfrac{36x}{36} = \dfrac{-2}{36}$

$x = -\dfrac{1}{18}$

The solution is $-\dfrac{1}{18}$.

SSM: Experiencing Introductory and Intermediate Algebra

41. $\frac{2}{5}b - 12 = \frac{2}{3}b + 20$

$15\left(\frac{2}{5}b - 12\right) = 15\left(\frac{2}{3}b + 20\right)$

$15\left(\frac{2}{5}b\right) - 15(12) = 15\left(\frac{2}{3}b\right) + 15(20)$

$6b - 180 = 10b + 300$

$6b - 180 - 10b = 10b + 300 - 10b$

$-4b - 180 = 300$

$-4b - 180 + 180 = 300 + 180$

$-4b = 480$

$\frac{-4b}{-4} = \frac{480}{-4}$

$b = -120$

The solution is -120.

43. $\frac{3}{4}\left(x + \frac{4}{5}\right) = -\frac{7}{8}x - \frac{2}{5}$

$\frac{3}{4}(x) + \frac{3}{4}\left(\frac{4}{5}\right) = -\frac{7}{8}x - \frac{2}{5}$

$\frac{3}{4}x + \frac{3}{5} = -\frac{7}{8}x - \frac{2}{5}$

$40\left(\frac{3}{4}x + \frac{3}{5}\right) = 40\left(-\frac{7}{8}x - \frac{2}{5}\right)$

$40\left(\frac{3}{4}x\right) + 40\left(\frac{3}{5}\right) = 40\left(-\frac{7}{8}x\right) - 40\left(\frac{2}{5}\right)$

$30x + 24 = -35x - 16$

$30x + 24 + 35x = -35x - 16 + 35x$

$65x + 24 = -16$

$65x + 24 - 24 = -16 - 24$

$65x = -40$

$\frac{65x}{65} = \frac{-40}{65}$

$x = -\frac{8}{13}$

The solution is $-\frac{8}{13}$.

45. $0.05x + 10.5 = 0.15x - 0.125$

$100(0.05x + 10.5) = 100(0.15x - 0.25)$

$5x + 1050 = 15x - 25$

$5x + 1050 - 15x = 15x - 25 - 15x$

$-10x + 1050 = -25$

$-10x + 1050 - 1050 = -25 - 1050$

$-10x = -1075$

$\frac{-10x}{-10} = \frac{-1075}{-10}$

$x = 107.5$

The solution is 107.5.

47. $21.1x + 0.46 = 10.9x + 0.46$

$100(21.1x + 0.46) = 100(10.9x + 0.46)$

$2110x + 46 = 1090x + 46$

$2110x + 46 - 1090x = 1090x + 46 - 1090x$

$1020x + 46 = 46$

$1020x + 46 - 46 = 46 - 46$

$1020x = 0$

$\frac{1020x}{1020} = \frac{0}{1020}$

$x = 0$

The solution is 0.

49. $15.2y - 175.43 = -2.4y - 176.31$

$100(15.2y - 175.43) = 100(-2.4y - 176.31)$

$1520y - 17543 = -240y - 17631$

$1520y - 17543 + 240y = -240y - 17631 + 240y$

$1760y - 17543 = -17631$

$1760y - 17543 + 17543 = -17631 + 17543$

$1760y = -88$

$1760y = -88$

$\frac{176y}{1760} = \frac{-88}{1760}$

$y = -0.05$

The solution is -0.05.

51. Let x = the maximum number miles possible.
$49.95 + 0.12x = 200$
$49.95 + 0.12x - 49.95 = 200 - 49.95$
$0.12x = 150.05$
$\dfrac{0.12x}{0.12} = \dfrac{150.05}{0.12}$
$x \approx 1250.417$
You can drive approximately 1250 miles.

53. Let x = the monthly payment.
$24x = 2252 - 200$
$24x = 2052$
$\dfrac{24x}{24} = \dfrac{2052}{24}$
$x = 85.5$
The monthly payment would be $85.50.

55. Let x = the number of liters of 30% solution.
$0.10(4) + 0.30x = 0.25(4 + x)$
$0.4 + 0.3x = 1 + 0.25x$
$0.4 + 0.3x - 0.25x = 1 + 0.25x - 0.25x$
$0.4 + 0.05x = 1$
$0.4 + 0.05x - 0.4 = 1 - 0.4$
$0.05x = 0.6$
$\dfrac{0.05x}{0.05} = \dfrac{0.6}{0.05}$
$x = 12$
12 liters of 30% vinegar solution were used.

57. Let x = the brother's average weekly earnings.
$25 + 2x = 730.10$
$25 + 2x - 25 = 730.1 - 25$
$2x = 705.1$
$\dfrac{2x}{2} = \dfrac{705.1}{2}$
$x = 352.55$
Gina's brother's average weekly earnings are $352.55.

59. Let x = the amount of sales.
$150 + 0.10x - 0.02x = 200 + 0.08x - 50$
$150 + 0.08x = 150 + 0.8x$

$150 + 0.08x - 0.08x = 150 + 0.8x - 0.08x$
$150 = 150$
This equation is always true (an identity). The solution is the set of all real numbers. The two companies always pay the same weekly amount to their salespeople for all values of sales.

61. Let x = the value of sales.
$300 + 0.04x = 700 + 0.04(x - 5000)$
$300 + 0.04x = 700 + 0.04x - 200$
$300 + 0.04x = 500 + 0.04x$
$300 + 0.04x - 0.04x = 500 + 0.04x - 0.04x$
$300 = 500$
This is a false equation (a contradiction). It has no solution. The two pay plans will never be equal for any value of sales.

63. Let x = the amount to be invested in the trust paying 8% simple interest.
$2000 = 600 + 0.045(20000) + 0.08x$
$2000 = 600 + 900 + 0.08x$
$2000 = 1500 + 0.08x$
$2000 - 1500 = 1500 + 0.08x - 1500$
$500 = 0.08x$
$\dfrac{500}{0.08} = \dfrac{0.08x}{0.08}$
$6250 = x$
An addition $6250 should be invested in the trust paying 8% simple interest.

65. Let x = the number of chirps.
$\dfrac{x}{8.6} + 46 = 81$
$\dfrac{x}{8.6} + 46 - 46 = 81 - 46$
$\dfrac{x}{8.6} = 35$
$8.6\left(\dfrac{x}{8.6}\right) = 8.6(35)$
$x = 301$
You will count about 301 chirps per minute when the temperature is $81°F$.

SSM: Experiencing Introductory and Intermediate Algebra

4.3 Calculator Exercises

1. Let x = the total sales for that month.
   $25 + 0.0375x = 49.48$
   Let $Y1 = 25 + 0.0375x$ and $Y2 = 49.48$.

   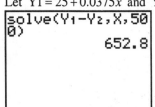

   The total sales for that month was $652.80.

2. Let x = the number of 120-grain tablets.
   $200 + 120x = 620$
   Let $Y1 = 200 + 120x$ and $Y2 = 620$.

   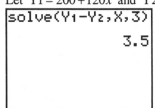

   The nurse must administer 3.5 of the 120-grain tablets.

3. Let x = the number of sheets of wood.
   $2 + \frac{3}{4}x = 4\frac{1}{4}$
   Let $Y1 = 2 + \frac{3}{4}x$ and $Y2 = 4\frac{1}{4}$.

   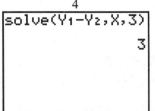

   The carpenter needs 3 sheets of wood.

4.4 Exercises

1. $P = 4s$
   $\frac{P}{4} = \frac{4s}{4}$
   $\frac{P}{4} = s$ or $s = \frac{P}{4}$

3. $C = \pi d$
   $\frac{C}{\pi} = \frac{\pi d}{\pi}$
   $\frac{C}{\pi} = d$ or $d = \frac{C}{\pi}$

5. $V = LWH$
   $\frac{V}{WH} = \frac{LWH}{WH}$
   $\frac{V}{WH} = L$ or $L = \frac{V}{WH}$

7. $S = 2LW + 2LH + 2WH$
   $S - 2WH = 2LW + 2LH + 2WH - 2WH$
   $S - 2WH = 2LW + 2LH$
   $S - 2WH = L(2W + 2H)$
   $\frac{S - 2WH}{2W + 2H} = \frac{L(2W + 2H)}{2W + 2H}$
   $\frac{S - 2WH}{2W + 2H} = L$ or $L = \frac{S - 2WH}{2W + 2H}$

9. $V = \pi r^2 h$
   $\frac{V}{\pi r^2} = \frac{\pi r^2 h}{\pi r^2}$
   $\frac{V}{\pi r^2} = h$ or $h = \frac{V}{\pi r^2}$

11. $I = Prt$
    $\frac{I}{rt} = \frac{Prt}{rt}$
    $\frac{I}{rt} = P$ or $P = \frac{I}{rt}$

13. $v = gt$
    $\frac{v}{t} = \frac{gt}{t}$
    $\frac{v}{t} = g$ or $g = \frac{v}{t}$

Chapter 4: Linear Equations in One Variable

15. $I = \dfrac{V}{R}$

$R(I) = R\left(\dfrac{V}{R}\right)$

$RI = V$

$\dfrac{RI}{I} = \dfrac{V}{I}$

$R = \dfrac{V}{I}$

17. $z = \dfrac{x-m}{s}$

$s(z) = s\left(\dfrac{x-m}{s}\right)$

$sz = x - m$

$sz + m = x - m + m$

$sz + m = x$

$sz + m - sz = x - sz$

$m = x - sz$

19. $4x + 3y = 0$

$4x + 3y - 4x = 0 - 4x$

$3y = -4x$

$\dfrac{3y}{3} = \dfrac{-4x}{3}$

$y = -\dfrac{4}{3}x$

21. $-5x + 10y = 0$

$-5x + 10y + 5x = 0 + 5x$

$10y = 5x$

$\dfrac{10y}{10} = \dfrac{5x}{10}$

$y = \dfrac{1}{2}x$

23. $-x - y = 0$

$-x - y + y = 0 + y$

$-x = y$ or $y = -x$

25. $5x + 4y = 20$

$5x + 4y - 5x = 20 - 5x$

$4y = 20 - 5x$

$\dfrac{4y}{4} = \dfrac{20 - 5x}{4}$

$\dfrac{4y}{4} = \dfrac{20}{4} - \dfrac{5x}{4}$

$y = 5 - \dfrac{5}{4}x$ or $y = -\dfrac{5}{4}x + 5$

27. $-x - y + y = 7 + y$

$-x = 7 + y$

$-x - 7 = 7 + y - 7$

$-x - 7 = y$ or $y = -x - 7$

29. $7x - 14y = -28$

$7x - 14y - 7x = -28 - 7x$

$-14y = -28 - 7x$

$\dfrac{-14y}{-14} = \dfrac{-28 - 7x}{-14}$

$y = \dfrac{-28}{-14} - \dfrac{7x}{-14}$

$y = 2 + \dfrac{1}{2}x$ or $y = \dfrac{1}{2}x + 2$

31. $-x + y = -1$

$-x + y + x = -1 + x$

$y = -1 + x$ or $y = x - 1$

33. $y - 5 = 4(x - 6)$

$y - 5 = 4x - 24$

$y - 5 + 5 = 4x - 24 + 5$

$y = 4x - 19$

35. $y + 6 = -2(x - 7)$

$y + 6 = -2x + 14$

$y + 6 - 6 = -2x + 14 - 6$

$y = -2x + 8$

37. $y + 2 = -1(x + 4)$

$y + 2 = -x - 4$

$y + 2 - 2 = -x - 4 - 2$

$y = -x - 6$

SSM: Experiencing Introductory and Intermediate Algebra

39. $y - 4 = \dfrac{2}{3}(x + 9)$

$y - 4 = \dfrac{2}{3}x + 6$

$y - 4 + 4 = \dfrac{2}{3}x + 6 + 4$

$y = \dfrac{2}{3}x + 10$

41. $y + \dfrac{5}{9} = -\dfrac{2}{3}\left(x - \dfrac{1}{3}\right)$

$y + \dfrac{5}{9} = -\dfrac{2}{3}x + \dfrac{2}{9}$

$y + \dfrac{5}{9} - \dfrac{5}{9} = -\dfrac{2}{3}x + \dfrac{2}{9} - \dfrac{5}{9}$

$y = -\dfrac{2}{3}x - \dfrac{3}{9}$

$y = -\dfrac{2}{3}x - \dfrac{1}{3}$

43. $P = 200 + 85m$

$P - 200 = 200 + 85m - 200$

$P - 200 = 85m$

$\dfrac{P - 200}{85} = \dfrac{85m}{85}$

$\dfrac{P}{85} - \dfrac{200}{85} = \dfrac{85m}{85}$

$\dfrac{1}{85}P - \dfrac{40}{17} = m$ or $m = \dfrac{1}{85}P - \dfrac{40}{17}$

$m = \dfrac{1}{85}(2240) - \dfrac{40}{17} = 24$

It will take 24 months to pay off $2240.

$m = \dfrac{1}{85}(1200) - \dfrac{40}{17} \approx 11.765$

It will take 12 months to pay off $1200.

45. $T = 12.5 + 22c$

$T - 12.5 = 12.5 + 22c - 12.5$

$T - 12.5 = 22c$

$\dfrac{T - 12.5}{22} = \dfrac{22c}{22}$

$\dfrac{T}{22} - \dfrac{12.5}{22} = c$

$\dfrac{1}{22}T - \dfrac{25}{44} = c$ or $c = \dfrac{1}{22}T - \dfrac{25}{44}$

$c = \dfrac{1}{22}(35) - \dfrac{25}{44} \approx 1.02$

Mildred can spend about $1.02 per child for a $35 party.

$c = \dfrac{1}{22}(50) - \dfrac{25}{44} \approx 1.70$

Mildred can spend about $1.70 per child for a $50 party.

47. $B = 75 + 3T$

$B - 75 = 75 + 3T - 75$

$B - 75 = 3T$

$\dfrac{B - 75}{3} = \dfrac{3T}{3}$

$\dfrac{B}{3} - \dfrac{75}{3} = T$

$\dfrac{1}{3}B - 25 = T$ or $T = \dfrac{1}{3}B - 25$

$T = \dfrac{1}{3}(725) - 25 \approx 216.67$

Ted's weekly earnings are about $216.67 if his boss's earnings average $725 per week.

$T = \dfrac{1}{3}(1275) - 25 = 400$

Ted's weekly earnings are $400 if his boss's earnings average $1275 per week.

49. $C = 85 + 185d$

$C - 85 = 85 + 185d - 85$

$C - 85 = 185d$

$\dfrac{C-85}{185} = \dfrac{185d}{185}$

$\dfrac{C}{185} - \dfrac{85}{185} = d$

$\dfrac{1}{185}C - \dfrac{17}{37} = d$ or $d = \dfrac{1}{185}C - \dfrac{17}{37}$

$d = \dfrac{1}{185}(270) - \dfrac{17}{37} = 1$

The equipment can be rented for 1 day if the total cost is limited to $270.

$d = \dfrac{1}{185}(825) - \dfrac{17}{37} = 4$

The equipment can be rented for 4 days if the total cost is limited to $825.

51. $V = 3 \cdot 5 \cdot h$

$V = 15h$

$\dfrac{V}{15} = \dfrac{15h}{15}$

$\dfrac{1}{15}V = h$ or $h = \dfrac{1}{15}V$

$h = \dfrac{1}{15}(60) = 4$

The height must be 4 feet for the volume to be 60 cubic feet.

$h = \dfrac{1}{15}(100) = \dfrac{20}{3} = 6\dfrac{2}{3}$

The height must be $6\dfrac{2}{3}$ feet for the volume to be 100 cubic feet.

4.4 Calculator Exercises

$10000 = Pe^{0.045t}$

$\dfrac{10000}{e^{0.045t}} = \dfrac{Pe^{0.045t}}{e^{0.045t}}$

$\dfrac{10000}{e^{0.045t}} = P$ or $P = \dfrac{10000}{e^{0.045t}}$

$Y1 = \dfrac{10000}{e^{0.045x}}$

X	Y1
5	7985.2
7	7297.9
10	6376.3
12	5827.5

X=

A	r	t	P
$10,000	4.5%	5	$7,985
$10,000	4.5%	7	$7,298
$10,000	4.5%	10	$6,376
$10,000	4.5%	12	$5,827

$25000 = Pe^{0.07t}$

$\dfrac{25000}{e^{0.07t}} = \dfrac{Pe^{0.07t}}{e^{0.07t}}$

$\dfrac{25000}{e^{0.07t}} = P$ or $P = \dfrac{25000}{e^{0.07t}}$

$Y1 = \dfrac{25000}{e^{0.07x}}$

X	Y1
5	17617
7	15316
10	12415
12	10793

X=

SSM: Experiencing Introductory and Intermediate Algebra

A	r	t	P
$25,000	7%	5	$17,617
$25,000	7%	7	$15,316
$25,000	7%	10	$12,415
$25,000	7%	12	$10,793

When investing, increasing the time period over which the investment accrues interest results in a smaller amount of initial principal required in order to accumulate a certain amount of money. In the same way, decreasing the time period over which the investment accrues interest results in a larger amount of initial principal required in order to accumulate a certain amount of money.

4.5 Exercises

1. Let x = the number of grains in the 1st dose.
   $x+2$ = the number of grains in the 2nd dose.
   $x+4$ = the number of grains in the 3rd dose.
   $x+(x+2)+(x+4)=24$
   $3x+6=24$
   $3x+6-6=24-6$
   $3x=18$
   $\frac{3x}{3}=\frac{18}{3}$
   $x=6$
   $x+2=8$
   $x+4=10$
   The doses for days 1 through 3 are 6 grains, 8 grains, and 10 grains, respectively.

3. Let x = the number of prizes in the 1st stage.
   $x+2$ = the number of prizes in the 2nd stage.
   $x+4$ = the number of prizes in the 3rd stage.
   $x+6$ = the number of prizes in the 4th stage.
   $x+(x+2)+(x+4)+(x+6)=24$
   $4x+12=24$
   $4x+12-12=24-12$
   $4x=12$
   $\frac{4x}{4}=\frac{12}{4}$
   $x=3$
   $x+2=5$
   $x+4=7$
   $x+6=9$
   The lottery will award 3 prizes, 5 prizes, 7 prizes, and 9 prizes at each stage, respectively.

5. Let x = the highest grade.
   $x-1$ = the 2nd highest grade.
   $x-2$ = the 3rd highest grade
   $\vdots$
   $x-7$ = the lowest grade.
   $x+(x-1)+(x-2)+(x-3)+(x-4)+(x-5)$
   $+(x-6)+(x-7)=676$
   $8x-28=676$
   $8x-28+28=676+28$
   $8x=704$
   $\frac{8x}{8}=\frac{704}{8}$
   $x=88$
   $x-7=81$
   The highest grade was 88 points and the lowest grade was 81 points.

7. Let x = the length of each equal side.
   $x+x+x=29\frac{1}{4}$
   $3x=29.25$
   $\frac{3x}{3}=\frac{29.25}{3}$
   $x=9.75$ or $9\frac{3}{4}$
   Each side of the triangle is $9\frac{3}{4}$ inches long.

9. Let x = the length of each of the equal sides.

$\frac{2}{3}x$ = the length of the third side.

$x + x + \frac{2}{3}x = 16$

$3\left(x + x + \frac{2}{3}x\right) = 3(16)$

$3x + 3x + 2x = 48$

$8x = 48$

$\frac{8x}{8} = \frac{48}{8}$

$x = 6$

$\frac{2}{3}x = 4$

The sides are 6 feet, 6 feet, and 4 feet long.

11. Let x = the width of the rectangle.

$x + 30$ = the length of the rectangle.

$2x + 2(x + 30) = 400$

$2x + 2x + 60 = 400$

$4x + 60 = 400$

$4x + 60 - 60 = 400 - 60$

$4x = 340$

$\frac{4x}{4} = \frac{340}{4}$

$x = 85$

$x + 30 = 115$

The dimensions are 85 yd by 115 yd.

13. Let x = the length of the rectangle.

$0.55x$ = the width of the rectangle.

$2x + 2(0.55x) = 294.5$

$2x + 1.1x = 294.5$

$3.1x = 294.5$

$\frac{3.1x}{3.1} = \frac{294.5}{3.1}$

$x = 95$

$0.55x = 52.25$

The dimensions are 95 cm by 52.25 cm.

15. Let x = the width of the dog run.

$5x$ = the length of the dog run.

$2x + 2(5x) = 96$

$2x + 10x = 96$

$12x = 96$

$\frac{12x}{12} = \frac{96}{12}$

$x = 8$

$5x = 40$

The dimensions are 8 ft by 40 ft.

$A = 8(40) = 320$

The dog run will cover 320 ft^2 of yard.

17. Let x = the measure of the supplementary angle.

$x + 138 = 180$

$x + 138 - 138 = 180 - 138$

$x = 42$

The measure of the supplementary angle is 42°.

Now, two of the angles of the garden measure 42° and 90°.

Let x = the measure of the third angle.

$x + 42 + 90 = 180$

$x + 132 = 180$

$x + 132 - 132 = 180 - 132$

$x = 48$

The measures of the three angles of the garden are 48°, 42°, and 90°.

19. Let P = the amount of money borrowed.

$I = P(0.125)(1)$

$I = 0.125P$

$P + 0.125P = 4500$

$1.125P = 4500$

$\frac{1.125P}{1.125} = \frac{4500}{1.125}$

$P = 4000$

$0.125P = 500$

The amount borrowed was $4000. The interest was $500.

21. Let P = the amount of money to be invested.
$I = P(0.09)(1)$
$I = 0.09P$

$P + 0.09P = 5000$
$1.09P = 5000$
$\dfrac{1.09P}{1.09} = \dfrac{5000}{1.09}$
$P \approx 4587.16$
You must invest about $4587.16.

23. Let P = the amount of money to be invested.
$I = P(0.10)(1)$
$I = 0.1P$

$P + 0.1P = 500000$
$1.1P = 500000$
$\dfrac{1.1P}{1.1} = \dfrac{500000}{1.1}$
$P \approx 454545.45$
The businessman should invest about $455,000.

25. Let x = the amount invested at 8%.
$15000 - x$ = the amount invested at 6.5%.
$0.08x + 0.065(15000 - x) = 1117.5$
$0.08x + 975 - 0.065x = 1117.5$
$0.015x + 975 = 1117.5$
$0.015x + 975 - 975 = 1117.5 - 975$
$0.015x = 142.5$
$\dfrac{0.015x}{0.015} = \dfrac{142.5}{0.015}$
$x = 9500$
$15000 - x = 5500$
Megan invested $9500 at 8% and $5500 at 6.5%.

27. $A = P + 0.07P$
$A = 1.07P$
$\dfrac{A}{1.07} = \dfrac{1.07P}{1.07}$
$\dfrac{A}{1.07} = P$ or $P = \dfrac{A}{1.07}$
$P = \dfrac{1350}{1.07} \approx 1261.68$

About $1261.68 should be invested in order to have $1350 at the end of one year.
$P = \dfrac{2500}{1.07} \approx 2336.45$
About $2336.45 should be invested in order to have $2500 at the end of one year.

29. Let x = the original price of the dress.
$x - 0.2x = 68$
$0.8x = 68$
$\dfrac{0.8x}{0.8} = \dfrac{68}{0.8}$
$x = 85$
The original price of the dress was $85.

31. Let x = the boutique's cost for the item.
$x + 0.6x = 19.95$
$1.6x = 19.95$
$\dfrac{1.6x}{1.6} = \dfrac{19.95}{1.6}$
$x \approx 12.47$
The boutique paid about $12.47 for the item.

33. Let x = the markup percentage.
$9.5 + 9.5x = 17.1$
$9.5 + 9.5x - 9.5 = 17.1 - 9.5$
$9.5x = 7.6$
$\dfrac{9.5x}{9.5} = \dfrac{7.6}{9.5}$
$x = 0.8$ or 80%
The markup percentage at this boutique is 80%.

35. Let x = the regular price for the TV.
$x - 0.25x = 195$
$0.75x = 195$
$\dfrac{0.75x}{0.75} = \dfrac{195}{0.75}$
$x = 260$
The regular price for the TV is $260.

37. Let x = the SRP for the car.
$x - 0.35x = 12500$
$0.65x = 12500$
$\dfrac{0.65x}{0.65} = \dfrac{12500}{0.65}$
$x \approx 19230.77$
The SRP for the car should be about $19,230.77.

39. $y = x - 0.1x$

$y = 0.9x$

$\dfrac{y}{0.9} = \dfrac{0.9x}{0.9}$

$\dfrac{y}{0.9} = x$ or $x = \dfrac{y}{0.9}$

$x = \dfrac{53.96}{0.9} \approx 59.96$

The original price of the coat is about $59.96 if the sale price is $53.96.

$x = \dfrac{98.95}{0.9} \approx 109.94$

The original price of the coat is about $109.94 if the sale price is $98.95.

41. Let x = Denzel's hourly wage before increasing.

$x + 0.022x = 13.75$

$1.022x = 13.75$

$\dfrac{1.022x}{1.022} = \dfrac{13.75}{1.022}$

$x \approx 13.45$

Denzel's made about $13.45 per hour before the cost-of-living increase.

43. Let x = the bill before adding the gratuity.

$x + 0.15x = 143.24$

$1.15x = 143.24$

$\dfrac{1.15x}{1.15} = \dfrac{143.24}{1.15}$

$x \approx 124.56$

The bill was about $124.56 before adding the gratuity.

45. $5(180 - 2x) = 180$

$900 - 10x = 180$

$900 - 10x - 900 = 180 - 900$

$-10x = -720$

$\dfrac{-10x}{-10} = \dfrac{-720}{-10}$

$x = 72$

Each angle x measures $72°$.

4.5 Calculator Exercises

1. $F = \dfrac{9}{5}C + 32$

$F = \dfrac{9}{5} \cdot \{50, 55, 60, 65\} + 32$

```
(9/5){50,55,60,6
5}+32
{122 131 140 14…
```

```
(9/5){50,55,60,6
5}+32
…22 131 140 149}
```

$50°C = 122°F$, $55°C = 131°F$, $60°C = 140°F$, and $65°C = 149°F$.

2. $C = \dfrac{5}{9}(F - 32)$

$C = \dfrac{5}{9}(\{0, 25, 50, 75, 100\} - 32)$

```
(5/9)({0,25,50,7
5,100}-32)
{-17.77777778  -…
```

```
(5/9)({0,25,50,7
5,100}-32)
… -3.888888889  1…
```

SSM: Experiencing Introductory and Intermediate Algebra

```
(5/9)((0,25,50,7
5,100)-32)
...10 23.88888889...
```

```
(5/9)((0,25,50,7
5,100)-32)
...89 37.77777778}
```

$0°F \approx -17.8°C$, $25°F \approx -3.9°C$, $50°F = 10°C$, $75°F \approx 23.9°C$, and $100°F \approx 37.8°C$.

3. $A + B = 90$
   $A + \{15, 30, 45, 60\} = 90$
   $A = 90 - \{15, 30, 45, 60\}$

```
90-{15,30,45,60}
    {75 60 45 30}
```

The complement of a 15° angle is a 75° angle.
The complement of a 30° angle is a 60° angle.
The complement of a 45° angle is a 45° angle.
The complement of a 60° angle is a 30° angle.

4. $d = rt$
   $d = 60 \cdot \{1, 1.5, 2, 2.50\}$

```
60*{1,1.5,2,2.5}
    {60 90 120 150}
```

At 60 mph, in 1 hour the distance traveled will be 60 miles. In 1.5 hours the distance will be 90 miles. In 2 hours the distance will be 120 miles. In 2.5 hours the distance will be 150 miles.

4.6 Exercises

1. Let $Y1 = |x|$ and $Y2 = 9$. The graphs of the functions are shown below.

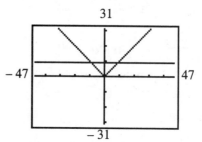

The graphs intersect at $(-9, 9)$ and $(9, 9)$. The solutions are -9 and 9.

3. Let $Y1 = |x+1|$ and $Y2 = 3$. The graphs of the functions are shown below.

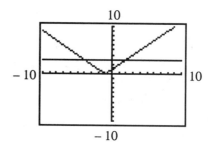

The graphs intersect at $(-4, 3)$ and $(2, 3)$. The solutions are -4 and 2.

5. Let $Y1 = |6x+6|$ and $Y2 = 12$. The graphs of the functions are shown below.

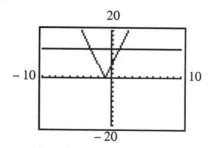

The graphs intersect at $(-3, 12)$ and $(1, 12)$. The solutions are -3 and 1.

Chapter 4: Linear Equations in One Variable

7. Let $Y1 = |-x - 1|$ and $Y2 = 1$. The graphs of the functions are shown below.

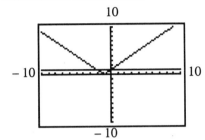

The graphs intersect at $(-2, 1)$ and $(0, 1)$. The solutions are -2 and 0.

9. Let $Y1 = |x + 21|$ and $Y2 = 26$. The graphs of the functions are shown below.

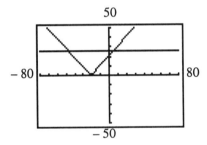

The graphs intersect at $(-47, 26)$ and $(5, 26)$. The solutions are -47 and 5.

The graphs intersect at $(-10, 29)$ and $(48, 29)$. The solutions are -10 and 48.

11. Let $Y1 = |-x - 19|$ and $Y2 = 22$. The graphs of the functions are shown below.

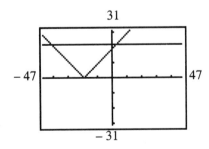

The graphs intersect at $(-41, 22)$ and $(3, 22)$. The solutions are -41 and 3.

13. Let $Y1 = |x - 2.17|$ and $Y2 = 6.09$. The graphs of the functions are shown below.

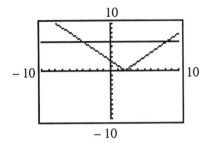

The graphs intersect at $(-3.92, 6.09)$ and $(8.26, 6.09)$. The solutions are -3.92 and 8.26.

15. Let $Y1 = \left|x + \dfrac{2}{5}\right|$ and $Y2 = \dfrac{11}{15}$. The graphs of the functions are shown below.

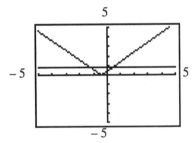

The graphs intersect approximately at $(-1.133, 0.733)$ and $(0.333, 0.733)$. The solutions are approximately -1.133 and 0.333.

17. Let $Y1 = 2|x - 15| + 25$ and $Y2 = 15$. The graphs of the functions are shown below.

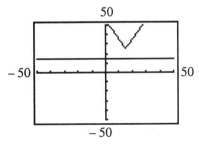

The graphs do not intersect. The equation has no solution.

19. Let $Y1 = 7|x-32|+21$ and $Y2 = 21$. The graphs of the functions are shown below.

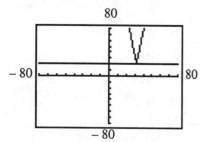

The graphs intersect at (32, 21). The solution is 32.

21. Let $Y1 = 3|x+5|-7$ and $Y2 = 4$. The graphs of the functions are shown below.

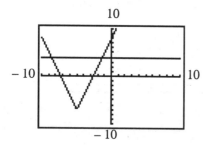

The graphs intersect approximately at $(-8.667, 4)$ and $(-1.333, 4)$. The solutions approximately -8.667 and -1.333.

23. $|x| = 138$

$x = -138$ or $x = 138$

The solutions are -138 and 138.

25. $|a| = -512$

Because the absolute value expression equals a negative number, the equation has no solution.

27. $|b| = 41.67$

$z = -41.67$ or $z = 41.67$

The solutions are -41.67 and 41.67.

29. $|c| = 14\frac{5}{9}$

$c = -14\frac{5}{9}$ or $c = 14\frac{5}{9}$

The solutions are $-14\frac{5}{9}$ and $14\frac{5}{9}$.

31. $|x+578| = 286$

$x + 578 = -286$

$x + 578 - 578 = -286 - 578$

$x = -846$

or

$x + 578 = 286$

$x + 578 - 578 = 286 - 578$

$x = -292$

The solutions are -846 and -292.

33. $|x-721| = 1942$

$x - 721 = -1942$

$x - 721 + 721 = -1942 + 721$

$x = -1221$

or

$x - 721 = 1942$

$x - 721 + 721 = 1942 + 721$

$x = 2663$

The solutions are -1221 and 2663.

35. $5|x-29| - 46 = 74$

$5|x-29| - 46 + 46 = 74 + 46$

$5|x-29| = 120$

$\dfrac{5|x-29|}{5} = \dfrac{120}{5}$

$|x-29| = 24$

$x - 29 = -24$ or $x - 29 = 24$

$x - 29 + 29 = -24 + 29$ $x - 29 + 29 = 24 + 29$

$x = 5$ $x = 53$

The solutions are 5 and 53.

37. $2|x+41|+96=48$

$2|x+41|+96-96=48-96$

$2|x+41|=-48$

$\dfrac{2|x+41|}{2}=\dfrac{-48}{2}$

$|x+41|=-24$

Because the absolute value expression equals a negative number, the equation has no solution.

39. $\left|\dfrac{x-14}{7}\right|=8$

$\dfrac{x-14}{7}=-8$ or $\dfrac{x-14}{7}=8$

$7\left(\dfrac{x-14}{7}\right)=7(-8)$ $7\left(\dfrac{x-14}{7}\right)=7(8)$

$x-14=-56$ $x-14=56$

$x-14+14=-56+14$ $x-14+14=56+14$

$x=-42$ $x=70$

The solutions are -42 and 70.

41. $|-3x-6|-18=-6$

$|-3x-6|-18+18=-6+18$

$|-3x-6|=12$

$-3x-6=-12$ or $-3x-6=12$

$-3x-6+6=-12+6$ $-3x-6+6=12+6$

$-3x=-6$ $-3x=18$

$\dfrac{-3x}{-3}=\dfrac{-6}{-3}$ $\dfrac{-3x}{-3}=\dfrac{18}{-3}$

$x=2$ $x=-6$

The solutions are -6 and 2.

43. $-4|x+12|=-16$

$\dfrac{-4|x+12|}{-4}=\dfrac{-16}{-4}$

$|x+12|=4$

$x+12=4$ or $x+12=-4$

$x+12-12=4-12$ $x+12-12=-4-12$

$x=-8$ $x=-16$

The solutions are -16 and -8.

45. Let $x=$ the maximum depth of Lake Superior.

$|x-923|=410$

$x-923=-410$

$x-923+923=-410+923$

$x=513$

or

$x-923=410$

$x-923+923=410+923$

$x=1333$

The maximum depth of Lake Superior is either 513 feet or 1333 feet. Given that Lake Superior is deeper than Lake Michigan, the maximum depth of Lake Superior is 1333 feet.

47. Let $x=$ the height in inches.

5 ft 9 in = 69 in

$|x-69|=6$

$x-69=-6$

$x=63$ in or 5 ft 3 in

or

$x-69=6$

$x=75$ in or 6 ft 3 in

The minimum and maximum heights that the clothes will fit are 5 ft 3 in and 6 ft 3 in.

49. Let $x=$ the percentage.

$|x-42|=3$

$x-42=-3$ or $x-42=3$

$x-42+42=-3+42$ $x-42+42=3+42$

$x=39$ $x=45$

The minimum and maximum percentages that may occur are 39% and 45%.

51. Let $x=$ Dee's actual weight.

$|x-132|=2$

$x-132=-2$

$x=130$

or

$x-132=2$

$x=134$

The range for Dee's actual weight is 130 pounds to 134 pounds.

53. Let x = the length of the pipe.

$\left|x - 5\frac{1}{2}\right| = \frac{1}{4}$

$x - 5\frac{1}{2} = -\frac{1}{4}$ or $x - 5\frac{1}{2} = \frac{1}{4}$

$x - 5\frac{1}{2} + 5\frac{1}{2} = -\frac{1}{4} + 5\frac{1}{2}$ $x - 5\frac{1}{2} + 5\frac{1}{2} = \frac{1}{4} + 5\frac{1}{2}$

$x = 5\frac{1}{4}$ $x = 5\frac{3}{4}$

The length can range from $5\frac{1}{4}$ to $5\frac{3}{4}$ inches.

55. Let x = the score.

$|x - 71| = 22$

$x - 71 = -22$ or $x - 71 = 22$

$x - 71 + 71 = -22 + 71$ $x - 71 + 71 = 22 + 71$

$x = 49$ $x = 93$

The possible extreme scores are 49 and 93 points.

57. Let x = the costar's age.

$|x - 47| = 12$

$x - 47 = -12$

$x - 47 + 47 = -12 + 47$

$x = 35$

or

$x - 47 = 12$

$x - 47 + 47 = 12 + 47$

$x = 59$

The pairs of ages can be 35 and 47 or 47 and 59.

4.6 Calculator Exercises

1. Let $Y1 = |x + 2|$ and $Y2 = |x - 3|$. The graphs of the functions are shown below.

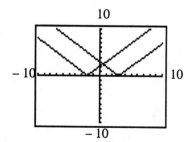

The graphs intersect at $(0.5, 2.5)$. The solution is 0.5.

2. Let $Y1 = 3|x + 2| - 5$ and $Y2 = 3|x + 2| + 5$. The graphs of the functions are shown below.

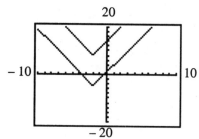

The graphs do not intersect. The equation has no solution.

3. Let $Y1 = |x + 6| - 9$ and $Y2 = -|x + 3| + 8$. The graphs of the functions are shown below.

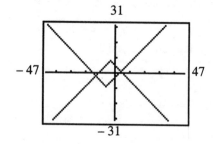

The graphs intersect at $(-13, -2)$ and $(4, 1)$. The solutions are -13 and 4.

4. Let $Y1 = |x+2|$ and $Y2 = -3x+8$. The graphs of the functions are shown below.

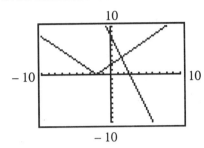

The graphs intersect at $(1.5, 3.5)$. The solution is 1.5.

5. Let $Y1 = |x+3| + |x-3|$ and $Y2 = 15$. The graphs of the functions are shown below.

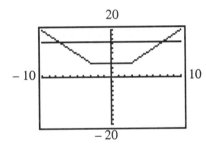

The graphs intersect at $(-7.5, 15)$ and $(7.5, 15)$. The solutions are -7.5 and 7.5.

Chapter 4 Section-By-Section Review

1. $4x^2 - 2x + 1 = 0$ is nonlinear because in the expression on the left side of the equals x has an exponent of 2.
2. $5x + 3 = x - 4$ is linear because the expressions on both sides of the equation are in the form $ax + b$.
3. $5.7x - 8.2(x + 4.6) = 0$ is linear because it simplifies to $-2.5x - 37.72 = 0$, which is in the form $ax + b = 0$.

4. $\sqrt{x} + 2 = 3x - 6$ is nonlinear because the radical expression on the left side of the equation has a variable in its radicand.

5. $\frac{1}{8}x + \frac{3}{4} = \frac{11}{16}$ is linear because the expressions on both sides of the equation are in the form $ax + b$.

6. $\frac{3}{x} + 5 = 12x$ is nonlinear because the expression on the left side of the equation has a variable in the denominator of a fraction.

7. The equation is $\frac{3}{4}(x+7) - 5 = \frac{1}{3}(x+12)$. The solution is 9.

8. Let $Y1 = 4x + 7$ and $Y2 = 2x - 5$. A sample table is shown below.

X	Y1	Y2
-9	-29	-23
-8	-25	-21
-7	-21	-19
-6	-17	-17
-5	-13	-15
-4	-9	-13
-3	-5	-11

X = -6

Because $Y1 = Y2 = -17$ when $x = -6$, the solution is -6.

9. Let $Y1 = 2.4x - 9.6$ and $Y2 = 4.8$. A sample table is shown below.

X	Y1	Y2
3	-2.4	4.8
4	0	4.8
5	2.4	4.8
6	4.8	4.8
7	7.2	4.8
8	9.6	4.8
9	12	4.8

X = 6

Because $Y1 = Y2 = 4.8$ when $x = 6$, the solution is 6.

10. Let $Y1 = \frac{3}{5}x - \frac{7}{10}$ and $Y2 = \frac{1}{5}x + \frac{1}{2}$. A sample table is shown next.

X	Y1	Y2
0	-.7	.5
1	-.1	.7
2	.5	.9
3	1.1	1.1
4	1.7	1.3
5	2.3	1.5
6	2.9	1.7

X = 3

Because $Y1 = Y2 = 1.1$ when $x = 3$, the solution is 3.

11. Let $Y1 = 14x + 12$ and $Y2 = 11(x-5) + 60$. A sample table is shown below.

X	Y1	Y2
-5	-58	-50
-4	-44	-39
-3	-30	-28
-2	-16	-17
-1	-2	-6
0	12	5
1	26	16

X = -3

Because $Y1 < Y2$ when $x = -3$ and $Y1 > Y2$ when $x = -2$, the solution is a noninteger number between -3 and -2.

12. Let $Y1 = 3(x-2) - 1$ and $Y2 = 4(x-1) - (x+3)$. A sample table is shown below.

X	Y1	Y2
-3	-16	-16
-2	-13	-13
-1	-10	-10
0	-7	-7
1	-4	-4
2	-1	-1
3	2	2

X = 0

Because $Y1 = Y2$ for all values of x, the solution is the set of all real numbers. The equation is an identity.

13. Let $Y1 = 2(2x+1) + x$ and $Y2 = 3(2x-1) - (x+1)$. A sample table is shown below.

X	Y1	Y2
-3	-13	-19
-2	-8	-14
-1	-3	-9
0	2	-4
1	7	1
2	12	6
3	17	11

X = -3

Because $Y1 - Y2 = 6$ for all values of x, the expressions will never be equal. The equation has no solution.

14. The equation is $2.3(x - 5.6) + 4 = 3x - 11.3$. The solution is 3.5.

15. Let $Y1 = 2x - 2$ and $Y2 = -x + 4$. The graphs of the functions are shown below.

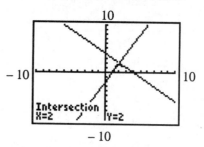

Because the intersection point is (2, 2), the solution is 2.

16. Let $Y1 = \frac{1}{2}x - 2$ and $Y2 = -\frac{1}{3}x - \frac{11}{3}$. The graphs of the functions are shown below.

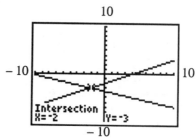

Because the intersection point is $(-2, -3)$, the solution is -2.

17. Let $Y1 = (x+3) + (x+1)$ and $Y2 = 3(x+1) - (x-1)$. The graphs of the functions are shown below.

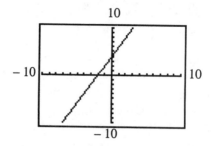

Because the lines are the same, the solution is the set of all real numbers. The equation is an identity.

18. Let $Y1 = (x+6) - 3(x+1)$ and $Y2 = (2x+5) - 2(2x+3)$. The graphs of the functions are shown below.

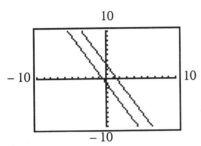

Because the lines are parallel, the equation has no solution.

19. Let $Y1 = 1.2x + 0.72$ and $Y2 = -2.1x + 8.64$. The graphs of the functions are shown below.

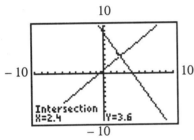

Because the intersection point is (2.4, 3.6), the solution is 2.4.

20. Let x = the number of hours. Then $90 = 25 + 6.5x$.
    $Y1 = 90$; $Y2 = 25 + 6.5x$

X	Y1	Y2
7	90	70.5
8	90	77
9	90	83.5
10	90	90
11	90	96.5
12	90	103
13	90	109.5

    X=10

    $Y1 = Y2 = 90$ when $x = 10$
    The two offers are equivalent at 10 hours.

21. Let x = the fourth week's donation amount.
    Then $\dfrac{2200 + 1750 + 1885 + x}{4} = 2000$.

    $Y1 = \dfrac{2200 + 1750 + 1885 + x}{4}$; $Y2 = 2000$

    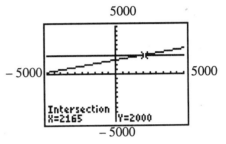

    The intersection point is (2165, 2000), so $x = -12$.

    The fourth week's donations must be $2165.

22. Let x = the length of each of the two equal sides.
    $x + 3$ = the length of the third side.
    $x + x + (x + 3) = 26.25$.
    $Y1 = x + x + (x + 3)$; $Y2 = 26.25$

    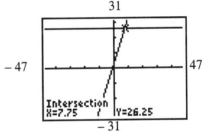

    The intersection point is (7.75, 26.25), so $x = 7.75$ and $x + 3 = 10.75$.

    The three sides of the triangle measure 7.75 feet, 7.75 feet, and 10.75 feet.

23. $41 + x = 67$
    $41 + x - 41 = 67 - 41$
    $x = 26$
    The solution is 26.

SSM: Experiencing Introductory and Intermediate Algebra

24. $y - \dfrac{7}{13} = \dfrac{11}{39}$

$y - \dfrac{7}{13} + \dfrac{7}{13} = \dfrac{11}{39} + \dfrac{7}{13}$

$y = \dfrac{32}{39}$

The solution is $\dfrac{32}{39}$.

25. $5 - (2 - x) = 1$

$5 - 2 + x = 1$

$3 - x = 1$

$3 - x - 3 = 1 - 3$

$x = -2$

The solution is -2.

26. $0.59(z - 1) + 0.41(z + 2) = 3.163$

$0.59z - 0.59 + 0.41z + 0.82 = 3.163$

$z + 0.23 = 3.163$

$z + 0.23 - 0.23 = 3.163 - 0.23$

$z = 2.933$

The solution is 2.933.

27. $-4x = 272$

$\dfrac{-4x}{-4} = \dfrac{272}{-4}$

$x = -68$

The solution is -68.

28. $45.86z = -1765.61$

$\dfrac{45.86z}{45.86} = \dfrac{-1765.61}{45.86}$

$z = -38.5$

The solution is -38.5.

29. $\dfrac{a}{7} = 15$

$7\left(\dfrac{a}{7}\right) = 7(15)$

$a = 105$

The solution is 105.

30. $\dfrac{4}{5}x = \dfrac{64}{125}$

$\dfrac{5}{4}\left(\dfrac{4}{5}x\right) = \dfrac{5}{4}\left(\dfrac{64}{125}\right)$

$x = \dfrac{16}{25}$

The solution is $\dfrac{16}{25}$.

31. $15x - 16x = -12$

$-x = -12$

$-1(-x) = -1(-12)$

$x = 12$

The solution is 12.

32. $-y = 2.98$

$-1(-y) = -1(2.98)$

$y = -2.98$

The solution is -2.98.

33. $4(2x - 6) + 6(x + 4) = 49$

$8x - 24 + 6x + 24 = 49$

$14x = 49$

$\dfrac{14x}{14} = \dfrac{49}{14}$

$x = \dfrac{7}{2}$

The solution is $\dfrac{7}{2}$.

34. Let x = the number of complementary passes.

$x + 189 = 247$

$x + 189 - 189 = 247 - 189$

$x = 58$

There were 58 complementary passes given.

35. Let x = the total number of graduates.

$0.55x = 154$

$\dfrac{0.55x}{0.55} = \dfrac{154}{0.55}$

$x = 280$

There were 280 graduates.

36. Let x = the number of books they must sell.
$15x = 80000$
$$\frac{15x}{15} = \frac{80000}{15}$$
$x = 5333\frac{1}{3}$
They must sell 5334 coupon books.

37. Let x = the total proceeds of the sale.
$$\frac{3}{7}x = 18270$$
$$\frac{7}{3}\left(\frac{3}{7}x\right) = \frac{7}{3}(18270)$$
$x = 42630$
The total proceeds of the sale were $42,630.

38. Let x = the current achieved by the scientists.
$0.0005x = 1000$
$$\frac{0.0005x}{0.0005} = \frac{1000}{0.0005}$$
$x = 2,000,000$
The current achieved by the scientists was 2,000,000 amperes per square centimeter.

39. Let x = the amount of 35% solution needed.
$0.35x = 4$
$$\frac{0.35x}{0.35} = \frac{4}{0.35}$$
$x \approx 11.429$
About 11.4 quarts of the solution are needed.

40. $3x + 7 = 4x + 21$
$3x + 7 - 7 = 4x + 21 - 7$
$3x = 4x + 14$
$3x - 4x = 4x - 4x + 14$
$-x = 14$
$-1(-x) = -1(14)$
$x = -14$
The solution is -14.

41. $14 - 2x = 5x$
$14 - 2x + 2x = 5x + 2x$
$14 = 7x$
$$\frac{14}{7} = \frac{7x}{7}$$
$2 = x$
The solution is 2.

42. $8.7x + 4.33 = -2.4x - 33.41$
$8.7x + 4.33 - 4.33 = -2.4x - 33.41 - 4.33$
$8.7x = -2.4x - 37.74$
$8.7x + 2.4x = -2.4x - 37.74 + 2.4x$
$11.1x = -37.74$
$$\frac{11.1x}{11.1} = \frac{-37.74}{11.1}$$
$x = -3.4$
The solution is -3.4.

43. $\frac{5}{7}a + \frac{11}{14} = \frac{2}{7}a$
$14\left(\frac{5}{7}a + \frac{11}{14}\right) = 14\left(\frac{2}{7}a\right)$
$10a + 11 = 4a$
$10a + 11 - 10a = 4a - 10a$
$11 = -6a$
$$\frac{11}{-6} = \frac{-6a}{-6}$$
$-\frac{11}{6} = a$
The solution is $-\frac{11}{6}$.

44. $2(x+5) - (x+6) = 2(x+2) - x$
$2x + 10 - x - 6 = 2x + 4 - x$
$x + 4 = x + 4$
$x + 4 - x = x + 4 - x$
$4 = 4$
This equation is always true. The solution is the set of all real numbers.

SSM: Experiencing Introductory and Intermediate Algebra

45. $3(x-4)+2(x+1)=5x+10$
$3x-12+2x+2=5x+10$
$5x-10=5x+10$
$5x-10-5x=5x+10-5x$
$-10=10$
This is a false equation. It has no solution.

46. Let $x=$ the maximum number of miles.
$49.95+0.22x=250$
$49.95+0.22x-49.95=250-49.95$
$0.22x=200.05$
$\dfrac{0.22x}{0.22}=\dfrac{200.05}{0.22}$
$x\approx 909.318$
You can drive a maximum of about 909 miles.

47. Let $x=$ the annual depreciation.
$7x+25000=150000$
$7x+25000-25000=150000-25000$
$7x=125000$
$\dfrac{7x}{7}=\dfrac{125000}{7}$
$x\approx 17857.14$
The annual depreciation is about $17857.

48. Let $x=$ the number of hours.
$175+35x=100+40x$
$175+35x-175=100+40x-175$
$35x=40x-75$
$35x-40x=40x-75-40x$
$-5x=-75$
$\dfrac{-5x}{-5}=\dfrac{-75}{-5}$
$x=15$
The two offers are the same at 15 hours.

49. Let $x=$ the amount to be invested at 8%.
$0.09(8000)+0.08x=1600$
$720+0.08x=1600$
$720+0.08x-720=1600-720$
$0.08x=880$
$\dfrac{0.08x}{0.08}=\dfrac{880}{0.08}$
$x=11000$
Erin should invest $11,000 at 8%.

50. $A=\dfrac{1}{2}h(b+B)$
$2\cdot A=2\cdot\dfrac{1}{2}h(b+B)$
$2A=h(b+B)$
$\dfrac{2A}{b+B}=\dfrac{h(b+B)}{b+B}$
$\dfrac{2A}{b+B}=h$ or $h=\dfrac{2A}{b+B}$

51. $S=2LW+2WH+2LH$
$S-2LH=2LW+2WH+2LH-2LH$
$S-2LH=2LW+2WH$
$S-2LH=W(2L+2H)$
$\dfrac{S-2LH}{2L+2H}=\dfrac{W(2L+2H)}{2L+2H}$
$\dfrac{S-2LH}{2L+2H}=W$ or $W=\dfrac{S-2LH}{2L+2H}$

52. $\dfrac{3}{4}x-\dfrac{5}{8}y=\dfrac{11}{12}$
$24\left(\dfrac{3}{4}x-\dfrac{5}{8}y\right)=24\left(\dfrac{11}{12}\right)$
$24\left(\dfrac{3}{4}x\right)-24\left(\dfrac{5}{8}y\right)=24\left(\dfrac{11}{12}\right)$
$18x-15y=22$
$18x-15y-18x=22-18x$
$-15y=-18x+22$

Chapter 4: Linear Equations in One Variable

$$\frac{-15y}{-15} = \frac{-18x+22}{-15}$$

$$y = \frac{-18x}{-15} + \frac{22}{-15}$$

$$y = \frac{6}{5}x - \frac{22}{15}$$

53. $A = 6000 + 8000n$

$A - 6000 = 6000 + 8000n - 6000$

$A - 6000 = 8000n$

$$\frac{A-6000}{8000} = \frac{8000n}{8000}$$

$$\frac{A-6000}{8000} = n \text{ or } n = \frac{A-6000}{8000}$$

$$n = \frac{78000 - 6000}{8000} = 9$$

The annuity will last 9 years if the amount is $78,000.

$$n = \frac{126000 - 6000}{8000} = 15$$

The annuity will last 15 years if the amount is $126,000.

54. If the lengths are consecutive even integers, then

Let x = the length of the 1st piece.

$x + 2$ = the length of the 2nd piece.

$x + 4$ = the length of the 3rd piece.

3 feet = 36 inches

$x + (x+2) + (x+4) = 36$

$3x + 6 = 36$

$3x + 6 - 6 = 36 - 6$

$3x = 30$

$$\frac{3x}{3} = \frac{30}{3}$$

$x = 10$

$x + 2 = 12$

$x + 4 = 14$

The lengths should be 10, 12 and 14 inches.

If the lengths are consecutive integers instead of consecutive even integers, then

Let x = the length of the 1st piece.

$x + 1$ = the length of the 2nd piece.

$x + 2$ = the length of the 3rd piece.

$x + (x+1) + (x+2) = 36$

$3x + 3 = 36$

$3x + 3 - 3 = 36 - 3$

$3x = 33$

$$\frac{3x}{3} = \frac{33}{3}$$

$x = 11$

$x + 1 = 12$

$x + 2 = 13$

The lengths should be 11, 12 and 13 inches.

55. Let x = the measure of the complementary angle.

$x + 48 = 90$

$x + 48 - 48 = 90 - 48$

$x = 42$

The measure of the complementary angle is 42°.

Let y = the measure of each of the two equal angles in the triangle.

$y + y + 42 = 180$

$2y + 42 = 180$

$2y + 42 - 42 = 180 - 42$

$2y = 138$

$$\frac{2y}{2} = \frac{138}{2}$$

$y = 69$

The measures of the angles in the triangle are 42°, 69°, and 69°.

145

SSM: Experiencing Introductory and Intermediate Algebra

56. Let x = the width of the pen.
$3x$ = the length of the pen.
$2x + 2(3x) - 4 - 6 = 230$
$2x + 6x - 10 = 230$
$8x - 10 = 230$
$8x - 10 + 10 = 230 + 10$
$8x = 240$
$\dfrac{8x}{8} = \dfrac{240}{8}$
$x = 30$
$3x = 90$
The dimensions are 30 ft by 90 ft.

57. Let t = the number of years.
$5000(0.075)t = 2250$
$375t = 2250$
$\dfrac{375t}{375} = \dfrac{2250}{375}$
$t = 6$
The money should be invested for 6 years.

58. Let P = the amount to be invested.
$P(0.075)(3) = 2250$
$0.225P = 2250$
$\dfrac{0.225P}{0.225} = \dfrac{2250}{0.225}$
$P = 10000$
You should invest $10,000.

59. Let x = the original price of the suit.
$x - 0.2x = 210$
$0.8x = 210$
$\dfrac{0.8x}{0.8} = \dfrac{210}{0.8}$
$x = 262.5$
The original price of the suit was $262.50.

60. Let x = the amount the artist was paid.
$x + 0.3x = 32.5$
$1.3x = 32.5$
$\dfrac{1.3x}{1.3} = \dfrac{32.5}{1.3}$
$x = 25$
The artist was paid $25.

61. $|a - 7| = 0$
$a - 7 = 0$
$a - 7 + 7 = 0 + 7$
$a = 7$
The solution is 7.

62. $|-b + 12| = 4$
$-b + 12 = -4$ or $-b + 12 = 4$
$-b + 12 - 12 = -4 - 12$ $-b + 12 - 12 = 4 - 12$
$-b = -16$ $-b = -8$
$-1(-b) = -1(-16)$ $-1(-b) = -1(-8)$
$b = 16$ $b = 8$
The solutions are 8 and 16.

63. $|c - 2| = -4$
Because the absolute value expression equals a negative number, the equation has no solution.

64. $|2x - 7| = 8$
$2x - 7 = -8$ or $2x - 7 = -8$
$2x - 7 + 7 = -8 + 7$ $2x - 7 + 7 = 8 + 7$
$2x = -1$ $2x = 15$
$\dfrac{2x}{2} = \dfrac{-1}{2}$ $\dfrac{2x}{2} = \dfrac{15}{2}$
$x = -\dfrac{1}{2}$ $x = \dfrac{15}{2}$
The solutions are $-\dfrac{1}{2}$ and $\dfrac{15}{2}$.

65. $2|x - 7| - 4 = 8$
$2|x - 7| - 4 + 4 = 8 + 4$
$2|x - 7| = 12$
$\dfrac{2|x - 7|}{2} = \dfrac{12}{2}$
$|x - 7| = 6$
$x - 7 = -6$ or $x - 7 = 6$
$x - 7 + 7 = -6 + 7$ $x - 7 + 7 = 6 + 7$
$x = 1$ $x = 13$
The solutions are 1 and 13.

66. $5|2x-7|+10=8$

$5|2x-7|+10-10=8-10$

$5|2x-7|=-2$

$\dfrac{5|2x-7|}{5}=\dfrac{-2}{5}$

$|2x-7|=-\dfrac{2}{5}$

Because the absolute value expression equals a negative number, the equation has no solution.

67. $2\left|\dfrac{x-1}{2}\right|-5=-2$

$2\left|\dfrac{x-1}{2}\right|-5+5=-2+5$

$2\left|\dfrac{x-1}{2}\right|=3$

$\dfrac{1}{2}\cdot 2\left|\dfrac{x-1}{2}\right|=\dfrac{1}{2}\cdot 3$

$\left|\dfrac{x-1}{2}\right|=\dfrac{3}{2}$

$\dfrac{x-1}{2}=-\dfrac{3}{2}$ or $\dfrac{x-1}{2}=\dfrac{3}{2}$

$2\left(\dfrac{x-1}{2}\right)=2\left(-\dfrac{3}{2}\right)$ $2\left(\dfrac{x-1}{2}\right)=2\left(\dfrac{3}{2}\right)$

$x-1=-3$ $x-1=3$

$x-1+1=-3+1$ $x-1+1=3+1$

$x=-2$ $x=4$

The solutions are -2 and 4.

68. Let $x =$ the permissable limit on the part.

$|x-62.79|=0.04$

$x-62.79=-0.04$

$x-62.79+62.79=-0.04+62.79$

$x=62.75$

or

$x-62.79=0.04$

$x-62.79+62.79=0.04+62.79$

$x=63.83$

The permissible limits on the part are 62.75 mm and 63.83 mm.

69. Let $x =$ the percentage of voters.

$|x-49|=4$

$x-49=-4$

$x-49+49=-4+49$

$x=45$

or

$x-49=4$

$x-49+49=4+49$

$x=53$

The limits on the percentage of voters are 45% and 53%.

Chapter 4 Chapter Review

1. Let $Y1=3(x+2)-2(x-1)$ and $Y2=(2x+5)-(x-3)$. The graphs of the functions are shown below.

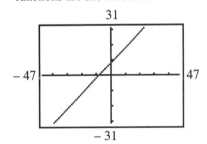

The lines are the same so the solution is the set of all real numbers. The equation is an identity.

2. Let $Y1=(2x+1)-(3x-7)$ and $Y2=(x+5)-2(x-2)$. The graphs of the functions are shown below.

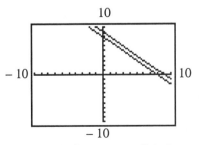

Because the lines are parallel, the equation has no solution.

3. Let $Y1 = x+1$ and $Y2 = 2x+5$. The graphs of the functions are shown below.

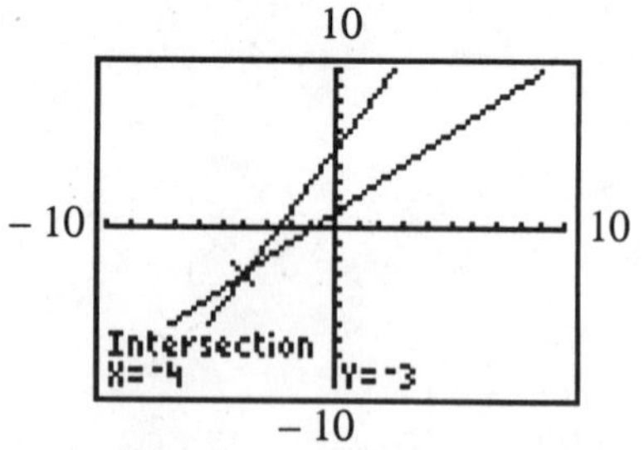

Because the intersection point is (– 4, – 3), the solution is – 4.

4. Let $Y1 = \frac{1}{3}x+3$ and $Y2 = 6-\frac{2}{3}x$. The graphs of the functions are shown below.

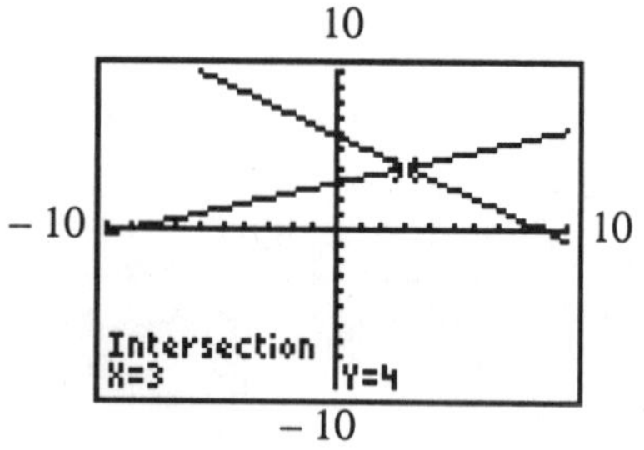

Because the intersection point is (3, 4), the solution is 3.

5. Let $Y1 = 1.2x-6.12$ and $Y2 = -2.2x+4.42$. The graphs of the functions are shown below.

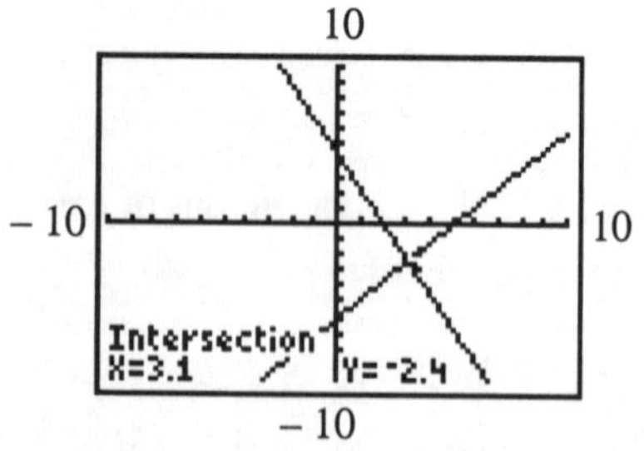

Because the intersection point is (3.1, – 2.4), the solution is 3.1.

6. Let $Y1 = 4x+7$ and $Y2 = 2x-5$. A sample table is shown below.

X	Y1	Y2
-1	-9	9
0	-5	7
1	-1	5
2	3	3
3	7	1
4	11	-1
5	15	-3

X=2

Because $Y1 = Y2 = 3$ when $x = 2$, the solution is 2.

7. Let $Y1 = 14(x+6)-17$ and $Y2 = 12(x+5)$. A sample table is shown below.

X	Y1	Y2
-7	-31	-24
-6	-17	-12
-5	-3	0
-4	11	12
-3	25	24
-2	39	36
-1	53	48

X= -4

Because $Y1 < Y2$ when $x = -4$ and $Y1 > Y2$ when $x = -3$, the solution is a noninteger number between – 4 and – 3.

8. Let $Y1 = 1.5x+5.5$ and $Y2 = -2.4x-6.2$. A sample table is shown below.

X	Y1	Y2
-6	-3.5	8.2
-5	-2	5.8
-4	-.5	3.4
-3	1	1
-2	2.5	-1.4
-1	4	-3.8
0	5.5	-6.2

X= -3

Because $Y1 = Y2 = 1$ when $x = -3$, the solution is – 3.

9. Let $Y1 = 2(2x-3)+3(x+1)$ and $Y2 = 6(x-1)+(x+3)$. A sample table is shown below.

X	Y1	Y2
-3	-24	-24
-2	-17	-17
-1	-10	-10
0	-3	-3
1	4	4
2	11	11
3	18	18

X= -3

Because $Y1 = Y2$ for all values of x, the solution is the set of all real numbers. The equation is an identity.

10. Let $Y1 = \frac{1}{3}x + \frac{14}{3}$ and $Y2 = \frac{5}{2} - \frac{3}{4}x$. A sample table is shown below.

X	Y1	Y2
-5	3	6.25
-4	3.3333	5.5
-3	3.6667	4.75
-2	4	4
-1	4.3333	3.25
0	4.6667	2.5
1	5	1.75

X = -2

Because $Y1 = Y2 = 4$ when $x = -2$, the solution is -2.

11. Let $Y1 = 4(x-2)-(x-1)$ and $Y2 = 3x+1$. A sample table is shown below.

X	Y1	Y2
-3	-16	-8
-2	-13	-5
-1	-10	-2
0	-7	1
1	-4	4
2	-1	7
3	2	10

X = -3

Because $Y1 - Y2 = -8$ for all values of x, the expressions will never be equal. The equation has no solution.

12. $5 - \frac{3}{7}x = \frac{2}{3}x + 2$ is linear because the expressions on both sides of the equation are of the form $ax + b$.

13. $2x - 4 = 1 + \frac{9}{x}$ is nonlinear because the expression on the right side of the equation has a variable in the denominator of a fraction.

14. $-2x^3 + 3x = x - 5$ is nonlinear because in the expression on the left side of the equals x has an exponent of 3.

15. $15 = 2(x-7)-(x-2)$ is linear because it simplifies to $15 = x - 12$, with expressions on both sides of the equation of the form $ax + b$.

16. $3.9(x-1.2)-6.7x = 0$ is linear because it simplifies to $-2.8x - 4.68 = 0$, which is of the form $ax + b = 0$.

17. $\sqrt[3]{x+1} - 5 = 4x$ is nonlinear because the radical expression on the left side of the equation has a variable in its radicand.

18. $\frac{z}{29} = -12$

$29\left(\frac{z}{29}\right) = 29(-12)$

$z = -348$

The solution is -348.

19. $\frac{5}{22} = \frac{25}{33}y$

$\frac{33}{25}\left(\frac{5}{22}\right) = \frac{33}{25}\left(\frac{25}{33}y\right)$

$\frac{3}{10} = y$

The solution is $\frac{3}{10}$.

20. $59a = 1888$

$\frac{59a}{59} = \frac{1888}{59}$

$a = 32$

The solution is 32.

21. $-174.243 = 2.41x$

$\frac{-174.243}{2.41} = \frac{2.41x}{2.41}$

$a = -72.3$

The solution is -72.3.

22. $2.3x - 3.3x = 14$

$-x = 14$

$-1(-x) = -1(14)$

$x = -14$

The solution is -14.

23. $-b = 14.59$

$-1(-b) = -1(14.59)$

$x = -14.59$

The solution is -14.59.

SSM: Experiencing Introductory and Intermediate Algebra

24. $3(x-8)+8(x+3)=77$
$3x-24+8x+24=77$
$11x=77$
$\dfrac{11x}{11}=\dfrac{77}{11}$
$x=7$
The solution is 7.

25. $z+193=-251$
$z+193-193=-251-193$
$z=-444$
The solution is -444.

26. $\dfrac{13}{17}+a=\dfrac{5}{51}$
$\dfrac{13}{17}+a-\dfrac{13}{17}=\dfrac{5}{51}-\dfrac{13}{17}$
$a=\dfrac{5}{51}-\dfrac{39}{51}$
$a=-\dfrac{34}{51}=-\dfrac{2}{3}$
The solution is $-\dfrac{2}{3}$.

27. $0.92(x-2)+0.08(x+1)=5.73$
$0.92x-1.84+0.08x+0.08=5.73$
$x-1.76=5.73$
$x-1.76+1.76=5.73+1.76$
$x=7.49$
The solution is 7.49.

28. $6.2x+5.67=4.9x+16.98$
$6.2x+5.67-4.9x=4.9x+16.98-4.9x$
$1.3x+5.67=16.98$
$1.3x+5.67-5.67=16.98-5.67$
$1.3x=11.31$
$\dfrac{1.3x}{1.3}=\dfrac{11.31}{1.3}$
$x=8.7$
The solution is 8.7.

29. $24.96-3.9a=0$
$24.96-3.9a+3.9a=0+3.9a$
$24.96=3.9a$
$\dfrac{24.96}{3.9}=\dfrac{3.9a}{3.9}$
$6.4=a$
The solution is 6.4.

30. $\dfrac{2}{3}x-\dfrac{3}{4}=-\dfrac{5}{6}$
$12\left(\dfrac{2}{3}x-\dfrac{3}{4}\right)=12\left(-\dfrac{5}{6}\right)$
$12\left(\dfrac{2}{3}x\right)-12\left(\dfrac{3}{4}\right)=12\left(-\dfrac{5}{6}\right)$
$8x-9=-10$
$8x-9+9=-10+9$
$8x=-1$
$\dfrac{8x}{8}=\dfrac{-1}{8}$
$x=-\dfrac{1}{8}$
The solution is $-\dfrac{1}{8}$.

31. $\dfrac{4}{9}y+\dfrac{11}{18}=\dfrac{5}{6}y$
$18\left(\dfrac{4}{9}y+\dfrac{11}{18}\right)=18\left(\dfrac{5}{6}y\right)$
$18\left(\dfrac{4}{9}y\right)+18\left(\dfrac{11}{18}\right)=18\left(\dfrac{5}{6}y\right)$
$8y+11=15y$
$8y+11-8y=15y-8y$
$11=7y$
$\dfrac{11}{7}=\dfrac{7y}{7}$
$\dfrac{11}{7}=y$
The solution is $\dfrac{11}{7}$.

32. $7x - 4 = 3x + 20$
    $7x - 4 - 3x = 3x + 20 - 3x$
    $4x - 4 = 20$
    $4x - 4 + 4 = 20 + 4$
    $4x = 24$
    $\dfrac{4x}{4} = \dfrac{24}{4}$
    $x = 6$
    The solution is 6.

33. $7x = 15 - 3x$
    $7x + 3x = 15 - 3x + 3x$
    $10x = 15$
    $\dfrac{10x}{10} = \dfrac{15}{10}$
    $x = \dfrac{3}{2}$
    The solution is $\dfrac{3}{2}$.

34. $2(x+2) + (x+1) = 5(x+1) - 2x$
    $2x + 4 + x + 1 = 5x + 5 - 2x$
    $3x + 5 = 3x + 5$
    $3x + 5 - 3x = 3x + 5 - 3x$
    $5 = 5$
    This equation is always true. The solution is the set of all real numbers.

35. $4(2x-1) = 3(2x+1) + 2(x+1)$
    $8x - 4 = 6x + 3 + 2x + 2$
    $8x - 4 = 8x + 5$
    $8x - 4 - 8x = 8x + 5 - 8x$
    $-4 = 5$
    This is a false equation. It has no solution.

36. $|2x + 6| = 0$
    $2x + 6 = 0$
    $2x + 6 - 6 = 0 - 6$
    $2x = -6$
    $\dfrac{2x}{2} = \dfrac{-6}{2}$
    $x = -3$
    The solution is -3.

37. $|4 - z| = 4$
    $4 - z = -4$ or $4 - z = 4$
    $4 - z - 4 = -4 - 4$ $4 - z - 4 = 4 - 4$
    $-z = -8$ $-z = 0$
    $\dfrac{-z}{-1} = \dfrac{-8}{-1}$ $\dfrac{-z}{-1} = \dfrac{0}{-1}$
    $z = 8$ $z = 0$
    The solutions are 0 and 8.

38. $|2x - 1| = -4$
    Because the absolute value expression equals a negative number, the equation has no solution.

39. $|5x + 2| = 12$
    $5x + 2 = -12$ or $5x + 2 = 12$
    $5x + 2 - 2 = -12 - 2$ $5x + 2 - 2 = 12 - 2$
    $5x = -14$ $5x = 10$
    $\dfrac{5x}{5} = \dfrac{-14}{5}$ $\dfrac{5x}{5} = \dfrac{10}{5}$
    $x = -\dfrac{14}{5}$ $x = 2$
    The solutions are $-\dfrac{14}{5}$ and 2.

40. $5|x - 7| = 15$
    $\dfrac{5|x-7|}{5} = \dfrac{15}{5}$
    $|x - 7| = 3$
    $x - 7 = -3$ or $x - 7 = 3$
    $x - 7 + 7 = -3 + 7$ $x - 7 + 7 = 3 + 7$
    $x = 4$ $x = 10$
    The solutions are 4 and 10.

SSM: Experiencing Introductory and Intermediate Algebra

41. $3\left|\dfrac{x+1}{4}\right| - 9 = -6$

$3\left|\dfrac{x+1}{4}\right| - 9 + 9 = -6 + 9$

$3\left|\dfrac{x+1}{4}\right| = 3$

$\dfrac{1}{3} \cdot 3\left|\dfrac{x+1}{4}\right| = \dfrac{1}{3} \cdot 3$

$\left|\dfrac{x+1}{4}\right| = 1$

$\dfrac{x+1}{4} = -1$ or $\dfrac{x+1}{4} = 1$

$4\left(\dfrac{x+1}{4}\right) = 4(-1)$ $\quad 2\left(\dfrac{x+1}{4}\right) = 2(1)$

$x + 1 = -4$ $\qquad\qquad \dfrac{x+1}{2} = 2$

$x + 1 - 1 = -4 - 1$ $\qquad x + 1 = 4$

$x = -5$ $\qquad\qquad\qquad x = 3$

The solutions are -5 and 3.

42. $y - 6 = \dfrac{2}{3}(x + 9)$

$y - 6 = \dfrac{2}{3}(x) + \dfrac{2}{3}(9)$

$y - 6 = \dfrac{2}{3}x + 6$

$y - 6 + 6 = \dfrac{2}{3}x + 6 + 6$

$y = \dfrac{2}{3}x + 12$

43. $\dfrac{1}{9}x + \dfrac{2}{3}y = \dfrac{1}{6}$

$18\left(\dfrac{1}{9}x + \dfrac{2}{3}y\right) = 18\left(\dfrac{1}{6}\right)$

$18\left(\dfrac{1}{9}x\right) + 18\left(\dfrac{2}{3}y\right) = 18\left(\dfrac{1}{6}\right)$

$2x + 12y = 3$

$2x + 12y - 2x = 3 - 2x$

$12y = -2x + 3$

$\dfrac{12y}{12} = \dfrac{-2x + 3}{12}$

$y = \dfrac{-2x}{12} + \dfrac{3}{12}$

$y = -\dfrac{1}{6}x + \dfrac{1}{4}$

44. $S = 2\pi r^2 + 2\pi rh$

$S - 2\pi r^2 = 2\pi r^2 + 2\pi rh - 2\pi r^2$

$S - 2\pi r^2 = 2\pi rh$

$\dfrac{S - 2\pi r^2}{2\pi r} = \dfrac{2\pi rh}{2\pi r}$

$\dfrac{S - 2\pi r^2}{2\pi r} = h$ or $h = \dfrac{S - 2\pi r^2}{2\pi r}$

45. $I = PRT$

$\dfrac{I}{RT} = \dfrac{PRT}{RT}$

$\dfrac{I}{RT} = P$ or $P = \dfrac{I}{RT}$

46. Let x = the height of the bank's tower.

$|x - 770| = 253$

$x - 770 = -253$

$x - 770 + 770 = -253 + 770$

$x = 517$

or

$x - 770 = 253$

$x - 770 + 770 = 253 + 770$

$x = 1023$

The two possible heights for the Bank of America Tower are 517 feet and 1023 feet. Given that the Peachtree Tower is shorter, the height of the bank's tower is 1023 feet.

47. Let x = the number of points.

$|x - 72| = 5$

$x - 72 = -5$ $\qquad$ or $\quad x - 72 = 5$

$x - 72 + 72 = -5 + 72$ $\quad x - 72 + 72 = 5 + 72$

$x = 67$ $\qquad\qquad\qquad x = 77$

The limits for a C grade are 67 and 77 points.

48. Let x = the subtotal of the purchases.

$0.0825x = 5.41$

$\dfrac{0.0825x}{0.0825} = \dfrac{5.41}{0.0825}$

$x \approx 65.58$

$65.58 + 5.41 = 70.99$

The subtotal was about $65.58 and the total bill was about $70.99.

49. Let x = the total amount Chuck expects to earn.

$\frac{2}{5}x = 5650$

$\frac{5}{2}\left(\frac{2}{5}x\right) = \frac{5}{2}(5650)$

$x = 14125$

Chuck expects to earn $14,125 for the project.

50. Let x = the number of hours she worked.

$13.25x = 16562.5$

$\frac{13.25x}{13.25} = \frac{16562.5}{13.25}$

$x = 1250$

She worked 1250 hours.

51. Let x = the number of houses.

$15 + 0.25x = 45$

$15 + 0.25x - 15 = 45 - 15$

$0.25x = 30$

$\frac{0.25x}{0.25} = \frac{30}{0.25}$

$x = 120$

The employee needs 120 houses.

52. Let x = the depreciation rate.

$20x + 15000 = 250000$

$20x + 15000 - 15000 = 250000 - 15000$

$20x = 235000$

$\frac{20x}{20} = \frac{235000}{20}$

$x = 11750$

The equipment will depreciate at an average rate of $11,750 per year.

53. Let x = the additional amount needed.

$985 + x = 1399$

$985 + x - 985 = 1399 - 985$

$x = 414$

Laurie needs to save an additional $414.

54. When each piece is 1 inch longer than the previous piece.

Let x = the length of the 1st piece.

$x + 1$ = the length of the 2nd piece.

$x + 2$ = the length of the 3rd piece.

$x + 3$ = the length of the 4th piece.

$x + (x+1) + (x+2) + (x+3) = 26$

$4x + 6 = 26$

$4x + 6 - 6 = 26 - 6$

$4x = 20$

$\frac{4x}{4} = \frac{20}{4}$

$x = 5$

$x + 1 = 6$

$x + 2 = 7$

$x + 3 = 8$

The pieces are 5, 6, 7, and 8 inches long.

When each piece is 2 inches longer than the previous piece.

Let x = the length of the 1st piece.

$x + 2$ = the length of the 2nd piece.

$x + 4$ = the length of the 3rd piece.

$x + 6$ = the length of the 4th piece.

$x + (x+2) + (x+4) + (x+6) = 26 - 2$

$4x + 12 = 24$

$4x + 12 - 12 = 24 - 12$

$4x = 12$

$\frac{4x}{4} = \frac{12}{4}$

$x = 3$

$x + 2 = 5$

$x + 4 = 7$

$x + 6 = 9$

The pieces are 3, 5, 7, and 9 inches long.

55. Let x = the measure of each of the equal angles.

$x + x + 62 = 180$

$2x + 62 = 180$

$2x + 62 - 62 = 180 - 62$

$2x = 118$

$\dfrac{2x}{2} = \dfrac{118}{2}$

$x = 59$

The measures of the angles in the triangle are 62°, 59°, and 59°.

Let y = the measure of the supplement.

$x + 59 = 180$

$x + 59 - 59 = 180 - 59$

$x = 121$

The measure of the supplement is 121°.

56. Let x = the original price before the markdown.

$x - 0.15x = 118.95$

$0.85x = 118.95$

$\dfrac{0.85x}{0.85} = \dfrac{118.95}{0.85}$

$x \approx 139.94$

The original price was about $139.94.

57. Let r = the simple interest rate.

$1560 = 8000(r)(3)$

$1560 = 24000r$

$\dfrac{1560}{24000} = \dfrac{24000r}{24000}$

$r = 0.065 = 6.5\%$

The simple interest rate needed is 6.5%.

58. Let P = the amount to be invested.

$562.50 = P(0.045)(1)$

$562.5 = 0.045P$

$\dfrac{562.5}{0.045} = \dfrac{0.045P}{0.045}$

$12500 = P$

You should invest $12,500.

59. Let x = the width of the room.

$1.5x$ = the length of the room.

$2x + 2(1.5x) = 84$

$2x + 3x = 84$

$5x = 84$

$\dfrac{5x}{5} = \dfrac{84}{5}$

$x = 16.8$

$1.5x = 25.2$

The dimensions are 16.8 feet by 25.2 feet.

60. Let x = the price before taxes.

$x + 0.0875x = 325.16$

$1.0875x = 325.16$

$\dfrac{1.0875x}{1.0875} = \dfrac{325.16}{1.0875}$

$x \approx 299.00$

$0.0875(299.00) \approx 26.16$

The price before taxes was about $299.00. The sales tax was about $26.16.

61. $A = 40 + 55.5h$

$A - 40 = 40 + 55.5h - 40$

$A - 40 = 55.5h$

$\dfrac{A - 40}{55.5} = \dfrac{55.5h}{55.5}$

$\dfrac{A - 40}{55.5} = h$ or $h = \dfrac{A - 40}{55.5}$

$h = \dfrac{178.75 - 40}{55.5} = 2.5$

A job that cost $178.75 takes 2.5 hours.

$h = \dfrac{95.50 - 40}{55.5} = 1$

A job that cost $95.50 takes 1 hour.

62. Let x = the angle made with the width of the rug.

$x + 35 + 90 = 180$

$x + 125 = 180$

$x + 125 - 125 = 180 - 125$

$x = 55$

The angle made with the width measures is 55°.

63. Let x = the amount of the 60% solution needed.

$0.6x = 2$

$\dfrac{0.6x}{0.6} = \dfrac{2}{0.6}$

$x = 3\dfrac{1}{3}$

The mechanic needs $3\dfrac{1}{3}$ gallons of the solution.

Chapter 4 Test

1. $3.14x + 9.07 = 5.72x$ is linear because the expressions on both sides of the equation are of the form $ax + b$.

2. $5x = 12 + \dfrac{19}{x}$ is nonlinear because the expression on the right side of the equation has a variable in the denominator of a fraction.

3. $4x + 21 = 5x^2$ is nonlinear because in the expression on the right side of the equals x has an exponent of 2.

4. $4(x - 6) = 3(5 - x) + 12$ is linear because it simplifies to $4x - 24 = -3x + 27$, with the expressions on both sides of the equation of the form $ax + b$.

5. $2(2x - 5) - 2(2 - x) = 6(x + 1) + 1$

$4x - 10 - 4 + 2x = 6x + 6 + 1$

$6x - 14 = 6x + 7$

$6x - 14 - 6x = 6x + 7 - 6x$

$-14 = 7$

This is a false equation. It has no solution.

6. $2(x + 5) = -3(x + 1) - 2$

$2x + 10 = -3x - 3 - 2$

$2x + 10 = -3x - 5$

$2x + 10 + 3x = -3x - 5 + 3x$

$5x + 10 = -5$

$5x + 10 - 10 = -5 - 10$

$5x = -15$

$\dfrac{5x}{5} = \dfrac{-15}{5}$

$x = -3$

The solution is -3.

7. $1.41(x + 5.08) + 1.17x + 0.00102 = -3.46x - 5.39334$

$1.41x + 7.1628 + 1.17x + 0.00102 = -3.46x - 5.39334$

$2.58x + 7.16382 = -3.46x - 5.39334$

$2.58x + 7.16382 + 3.46x = -3.46x - 5.39334 + 3.46x$

$6.04x + 7.16382 = -5.39334$

$6.04x + 7.16382 - 7.16382 = -5.39334 - 7.16382$

$6.04x = -12.55716$

$\dfrac{6.04x}{6.04} = \dfrac{-12.55716}{6.04}$

$x = -2.079$

The solution is -2.079.

8. $(x + 1) - 4(x - 1) = 3(2 - x) - 1$

$x + 1 - 4x + 4 = 6 - 3x - 1$

$-3x + 5 = -3x + 5$

$-3x + 3 + 3x = -3x + 5 + 3x$

$-5 = -5$

This equation is always true. The solution is the set of all real numbers.

9. $\dfrac{4}{5}x + \dfrac{31}{10} = \dfrac{-4}{3}x + \dfrac{41}{6}$

$30\left(\dfrac{4}{5}x + \dfrac{31}{10}\right) = 30\left(\dfrac{-4}{3}x + \dfrac{41}{6}\right)$

$30\left(\dfrac{4}{5}x\right) + 30\left(\dfrac{31}{10}\right) = 30\left(\dfrac{-4}{3}x\right) + 30\left(\dfrac{41}{6}\right)$

$24x + 93 = -40x + 205$

$24x + 93 + 40x = -40x + 205 + 40x$

$64x + 93 = 205$

$64x + 93 - 93 = 205 - 93$

$64x = 112$

$\dfrac{64x}{64} = \dfrac{112}{64}$

$x = \dfrac{7}{4}$

The solution is $\dfrac{7}{4}$.

10. $|x + 3| = 7$

$x + 3 = -7$ or $x + 3 = 7$

$x + 3 - 3 = -7 - 3$ $x + 3 - 3 = 7 - 3$

$x = -10$ $x = 4$

The solutions are -10 and 4.

11. $3|2x-9|+14=5$
    $3|2x-9|+14-14=5-14$
    $3|2x-9|=-9$
    $\dfrac{3|2x-9|}{3}=\dfrac{-9}{3}$
    $|2x-9|=-3$
    Because the absolute value expression equals a negative number, the equation has no solution.

12. Let $x=$ the length of the 1st piece.
    $x+1=$ the length of the 2nd piece.
    $x+2=$ the length of the 3rd piece.
    $x+(x+1)+(x+2)=45$
    $3x+3=45$
    $3x+3-3=45-3$
    $3x=42$
    $\dfrac{3x}{3}=\dfrac{42}{3}$
    $x=14$
    $x+1=15$
    $x+2=16$
    The pieces should be 14, 15, and 16 inches long.

13. Let $x=$ Ricardo's monthly payment to his dad.
    $12x+850=2470$
    $12x+850-850=2470-850$
    $12x=1620$
    $\dfrac{12x}{12}=\dfrac{1620}{12}$
    $x=135$
    Ricardo will pay his father $135 per month.

14. Let $x=$ the original price before the sale.
    $x-0.25x=179.95$
    $0.75x=179.95$
    $\dfrac{0.75x}{0.75}=\dfrac{179.95}{0.75}$
    $x\approx 239.93$
    The original price of the stereo before it went on sale was about $239.93.

15. $P=2L+2W$
    $P-2L=2L+2W-2L$
    $P-2L=2W$
    $\dfrac{P-2L}{2}=\dfrac{2W}{2}$
    $\dfrac{P-2L}{2}=W$ or $W=\dfrac{P-2L}{2}$
    $W=\dfrac{44.8-2(14.8)}{2}=7.6$
    The width of the rectangle is 7.6 inches.

16. Let $x=$ the measure of each of the equal angles.
    $x+x+42=180$
    $2x+42=180$
    $2x+42-42=180-42$
    $2x=138$
    $\dfrac{2x}{2}=\dfrac{138}{2}$
    $x=69$
    The measures of the angles in the triangle are 42°, 69°, and 69°.

    Let $y=$ the measure of the supplement.
    $x+69=180$
    $x+69-69=180-69$
    $x=111$
    The measure of the supplement is 111°.

17. Let $x=$ the number of liters of 60% apple juice.
    $0.6x+0.2(500)=0.5(x+500)$
    $0.6x+100=0.5x+250$
    $0.6x+100-0.5x=0.5x+250-0.5x$
    $0.1x+100=250$
    $0.1x+100-100=250-100$
    $0.1x=150$
    $\dfrac{0.1x}{0.1}=\dfrac{150}{0.1}$
    $x=1500$
    1500 liters of 60% apple juice must be added.

Chapter 4: Linear Equations in One Variable

18. Let x = the amount of sales.
$1500 = 1200 + 0.04x$
$1500 - 1200 = 1200 + 0.04x - 1200$
$300 = 0.04x$
$\dfrac{300}{0.04} = \dfrac{0.04x}{0.04}$
$7500 = x$
The two payment plans will be equal for $7500 worth of sales.

19. Let x = the number of hours to do the job.
$25x + 50 = 30x$
$25x + 50 - 25x = 30x - 25x$
$50 = 5x$
$\dfrac{50}{5} = \dfrac{5x}{5}$
$10 = x$
The two plans will cost the same if the job takes 10 hours.

20. Let x = the true percent of voters who support the new tax levy.
$|x - 52| = 3$
$x - 52 = -3$ or $x - 52 = 3$
$x - 52 + 52 = -3 + 52$ $x - 52 + 52 = 3 + 52$
$x = 49$ $x = 55$
The limits on the true percent of voters who support the levy are 49% and 55%.

21. To solve a linear equation in one variable graphically, graph the two linear functions defined by the expressions on the left and right sides of the equation. If the graphs intersect, the solution is the x-coordinate of the intersection point. If the graphs are parallel, the equation has no solution. If the graphs coincide, the solution is the set of all real numbers.

Chapter 5

5.1 Exercises

1. $5x + 7y = 35$ is linear and is written in standard form.
   $a = 5, b = 7, c = 35$

3. $-4\sqrt{x} + y = 8$ is not linear because the variable x is in the radicand.

5. $x^2 + y^2 = 1$ is not linear because the variables x and y are both squared.

7. $2x - 5 = 0$ is linear. It can be written in standard form as $2x = 5$.
   $a = 2, b = 0, c = 5$

9. Answers will vary.
   $x - 6y = 12$
   $x - 6y - x = 12 - x$
   $-6y = 12 - x$
   $\dfrac{-6y}{-6} = \dfrac{12-x}{-6}$
   $y = \dfrac{1}{6}x - 2$

x	$y = \dfrac{1}{6}x - 2$	y
-6	$y = \dfrac{1}{6}(-6) - 2$	-3
0	$y = \dfrac{1}{6}(0) - 2$	-2
6	$y = \dfrac{1}{6}(6) - 2$	-1

   $(-6, -3)$, $(0, -2)$, and $(6, -1)$ are three possible solutions.

   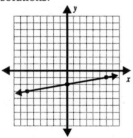

11. Answers will vary.
    $p(x) = \dfrac{4x+1}{3}$

x	$p(x) = \dfrac{4x+1}{3}$	$p(x)$
-1	$p(-1) = \dfrac{4(-1)+1}{3}$	-1
0	$p(0) = \dfrac{4(0)+1}{3}$	$\dfrac{1}{3}$
2	$p(1) = \dfrac{4(2)+1}{3}$	3

    $(-1, -1)$, $\left(0, \dfrac{1}{3}\right)$, and $(2, 3)$ are three possible solutions.

    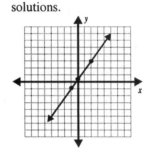

13. Answers will vary.
    $y = 8$

x	$y = 8$	y
-2	$y = 8$	8
0	$y = 8$	8
2	$y = 8$	8

    $(-2, 8)$, $(0, 8)$, and $(2, 8)$ are three possible solutions.

    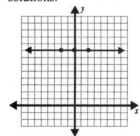

15. Answers will vary.

$r(x) = -\dfrac{3}{4}x + 4$

x	$r(x) = -\dfrac{3}{4}x + 4$	$r(x)$
-4	$r(-4) = -\dfrac{3}{4}(-4) + 4$	7
0	$r(0) = -\dfrac{3}{4}(0) + 4$	4
4	$r(4) = -\dfrac{3}{4}(4) + 4$	1

$(-4, 7)$, $(0, 4)$, and $(4, 1)$ are three possible solutions.

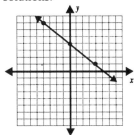

17. Answers will vary.
$y = 2.8x - 1.6$

x	$y = 2.8x - 1.6$	y
-1	$y = 2.8(-1) - 1.6$	-4.4
0	$y = 2.8(0) - 1.6$	-1.6
2	$y = 2.8(2) - 1.6$	4

$(-1, -4.4)$, $(0, -1.6)$, and $(2, 4)$ are three possible solutions.

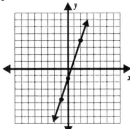

19. Answers will vary.
$y = \dfrac{3x - 5}{2}$

x	$y = \dfrac{3x - 5}{2}$	y
-3	$y = \dfrac{3(-3) - 5}{2}$	-7
1	$y = \dfrac{3(1) - 5}{2}$	-1
5	$y = \dfrac{3(5) - 5}{2}$	5

$(-3, -7)$, $(1, -1)$, and $(5, 5)$ are three possible solutions.

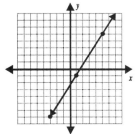

21. Answers will vary.
$3x + y - 4 = x + 2y - 3$
$3x + y - 4 + 4 = x + 2y - 3 + 4$
$3x + y = x + 2y + 1$
$3x + y - 3x = x + 2y + 1 - 3x$
$y = -2x + 2y + 1$
$y - 2y = -2x + 2y + 1 - 2y$
$-y = -2x + 1$
$y = 2x - 1$

x	$y = 2x - 1$	y
-3	$y = 2(-3) - 1$	-7
0	$y = 2(0) - 1$	-1
3	$y = 2(3) - 1$	5

$(-3, -7)$, $(0, -1)$, and $(3, 5)$ are three possible solutions.

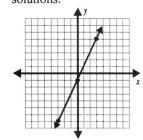

23. Answers will vary.
$$5y = -20$$
$$\frac{5y}{5} = \frac{-20}{5}$$
$$y = -4$$

x	$y = -4$	y
-3	$y = -4$	-4
0	$y = -4$	-4
3	$y = -4$	-4

$(-3,-4)$, $(0,-4)$, and $(3,-4)$ are three possible solutions.

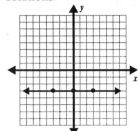

25. x-intercept: $(-2,0)$
y-intercept: $(0,4)$

27. x-intercept: $(0,0)$
y-intercept: $(0,0)$

29. x-intercept: $(3,0)$
y-intercept: none

31. $3x + 5y = 12$
Let $y = 0$.
$3x + 5(0) = 12$
$3x = 12$
$$\frac{3x}{3} = \frac{12}{3}$$
$x = 4$
The x-intercept is $(4,0)$.
Let $x = 0$.

$3(0) + 5y = 12$
$5y = 12$
$$\frac{5y}{5} = \frac{12}{5}$$
$$y = \frac{12}{5}$$
The y-intercept is $\left(0, \frac{12}{5}\right)$.

33. $4x - 7y = 14$
Let $y = 0$.
$4x - 7(0) = 14$
$4x = 14$
$$\frac{4x}{4} = \frac{14}{4}$$
$$x = \frac{7}{2}$$
The x-intercept is $\left(\frac{7}{2}, 0\right)$.

Let $x = 0$.
$4(0) - 7y = 14$
$-7y = 14$
$$\frac{-7y}{-7} = \frac{14}{-7}$$
$y = -2$
The y-intercept is $(0,-2)$.

35. $-2x - 9y = 27$
Let $y = 0$.
$-2x - 9(0) = 27$
$-2x = 27$
$$\frac{-2x}{-2} = \frac{27}{-2}$$
$$x = -\frac{27}{2}$$
The x-intercept is $\left(-\frac{27}{2}, 0\right)$.
Let $x = 0$.

$-2(0) - 9y = 27$
$-9y = 27$
$\dfrac{-9y}{-9} = \dfrac{27}{-9}$
$y = -3$
The y-intercept is $(0, -3)$.

37. $6x + 9y - 36 = 0$
$6x + 9y - 36 + 36 = 0 + 36$
$6x + 9y = 36$
Let $y = 0$.
$6x + 9(0) = 36$
$6x = 36$
$\dfrac{6x}{6} = \dfrac{36}{6}$
$x = 6$
The x-intercept is $(6, 0)$.
Let $x = 0$.
$6(0) + 9y = 36$
$9y = 36$
$\dfrac{9y}{9} = \dfrac{36}{9}$
$y = 4$
The y-intercept is $(0, 4)$.

39. $3x + 7y = 0$
Let $y = 0$.
$3x + 7(0) = 0$
$3x = 0$
$\dfrac{3x}{3} = \dfrac{0}{3}$
$x = 0$
The x-intercept is $(0, 0)$.
Let $x = 0$.
$3(0) + 7y = 0$
$7y = 0$
$\dfrac{7y}{7} = \dfrac{0}{7}$
$y = 0$
The y-intercept is $(0, 0)$.

41. $6x - 8 = 2x + 32$
$6x - 8 - 2x = 2x + 32 - 2x$
$4x - 8 = 32$
$4x - 8 + 8 = 32 + 8$
$4x = 40$
$\dfrac{4x}{4} = \dfrac{40}{4}$
$x = 10$
The x-intercept is $(10, 0)$.
There is no y-intercept since x is never 0.

43. $y = 3y - 22$
$y - 3y = 3y - 22 - 3y$
$-2y = -22$
$\dfrac{-2y}{-2} = \dfrac{-22}{-2}$
$y = 11$
The y-intercept is $(0, 11)$.
There is no x-intercept since y is never 0.

45. $12x - y = 24$
$12x - y - 12x = 24 - 12x$
$-y = 24 - 12x$
$\dfrac{-y}{-1} = \dfrac{24 - 12x}{-1}$
$y = 12x - 24$
The y-intercept is $(0, -24)$.

47. $y = 5(x - 3)$
$y = 5x - 15$
The y-intercept is $(0, -15)$.

49. $5x - 15y = 0$
$5x - 15y - 5x = 0 - 5x$
$-15y = -5x$
$\dfrac{-15y}{-15} = \dfrac{-5x}{-15}$
$y = \dfrac{1}{3}x$
The y-intercept is $(0, 0)$.

SSM: Experiencing Introductory and Intermediate Algebra

51. $3y = 12y + 18$
$3y - 12y = 12y + 18 - 12y$
$-9y = 18$
$\dfrac{-9y}{-9} = \dfrac{18}{-9}$
$y = -2$
The y-intercept is $(0, -2)$.

53. $y + 5 = 5$
$y + 5 - 5 = 5 - 5$
$y = 0$
The y-intercept is $(0, 0)$.

55. $-17.6x + 2.2y = 19.8$
$-17.6x + 2.2y + 17.6x = 19.8 + 17.6x$
$2.2y = 19.8 + 17.6x$
$\dfrac{2.2y}{2.2} = \dfrac{19.8 + 17.6x}{2.2}$
$y = 8x + 9$
The y-intercept is $(0, 9)$.

57. $x = 12y$
$12y = x$
$\dfrac{12y}{12} = \dfrac{x}{12}$
$y = \dfrac{1}{12}x$
The y-intercept is $(0, 0)$.

59. $3x + 5y = 30$
Let $y = 0$.
$3x + 5(0) = 30$
$3x = 30$
$\dfrac{3x}{3} = \dfrac{30}{3}$
$x = 10$
The x-intercept is $(10, 0)$.
Let $x = 0$.
$3(0) + 5y = 30$
$5y = 30$
$\dfrac{5y}{5} = \dfrac{30}{5}$
$y = 6$
The y-intercept is $(0, 6)$.

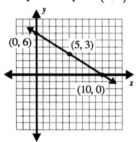

Check point: $(5, 3)$
$3(5) + 5(3) = 30$
$15 + 15 = 30$
$30 = 30$
Since $30 = 30$, $(5, 3)$ is a solution.

61. $4x - 3y = 24$
Let $y = 0$.
$4x - 3(0) = 24$
$4x = 24$
$\dfrac{4x}{4} = \dfrac{24}{4}$
$x = 6$
The x-intercept is $(6, 0)$.
Let $x = 0$.

$4(0) - 3y = 24$
$-3y = 24$
$\dfrac{-3y}{-3} = \dfrac{24}{-3}$
$y = -8$
The y-intercept is $(0, -8)$.

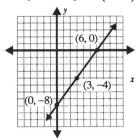

Check point: $(3, -4)$
$4(3) - 3(-4) = 24$
$12 + 12 = 24$
$24 = 24$
Since $24 = 24$, $(3, -4)$ is a solution.

63. $x - y = 9$
Let $y = 0$.
$x - (0) = 9$
$x = 9$
The x-intercept is $(9, 0)$.
Let $x = 0$.
$(0) - y = 9$
$-y = 9$
$\dfrac{-y}{-1} = \dfrac{9}{-1}$
$y = -9$
The y-intercept is $(0, -9)$.

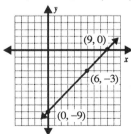

Check point: $(6, -3)$

$(6) - (-3) = 9$
$6 + 3 = 9$
$9 = 9$
Since $9 = 9$, $(6, -3)$ is a solution.

65. $-x - y = 9$
Let $y = 0$.
$-x - (0) = 9$
$-x = 9$
$\dfrac{-x}{-1} = \dfrac{9}{-1}$
$x = -9$
The x-intercept is $(-9, 0)$.
Let $x = 0$.
$-(0) - y = 9$
$-y = 9$
$\dfrac{-y}{-1} = \dfrac{9}{-1}$
$y = -9$
The y-intercept is $(0, -9)$.

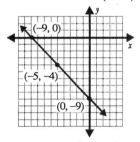

Check point: $(-5, -4)$
$-(-5) - (-4) = 9$
$5 + 4 = 9$
$9 = 9$
Since $9 = 9$, $(-5, -4)$ is a solution.

67. $2x - 7y = -14$
Let $y = 0$.
$2x - 7(0) = -14$
$2x = -14$
$\dfrac{2x}{2} = \dfrac{-14}{2}$
$x = -7$

The x-intercept is $(-7, 0)$.
Let $x = 0$.
$2(0) - 7y = -14$
$-7y = -14$
$\dfrac{-7y}{-7} = \dfrac{-14}{-7}$
$y = 2$
The y-intercept is $(0, 2)$.

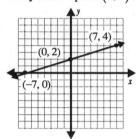

Check point: $(7, 4)$
$2(7) - 7(4) = -14$
$14 - 28 = -14$
$-14 = -14$
Since $-14 = -14$, $(7, 4)$ is a solution.

69. a. $(1, 0)$, $(2, 5)$, and $(4, 15)$

b.

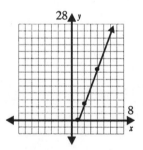

c. At $x = 5$, $y = 20$
When 5 people work on the assembly line, they can pack 20 boxes per minute.

d. Answers will vary.
The domain and range should both have a minimum of 0, though the domain may also start at 1. The maximum values, if any, would be determined by factors such as plant capacity, room on the production floor, etc.

71. a. $(2, 6)$, $(3.5, 10.5)$, and $(10.5, 31.5)$

b.

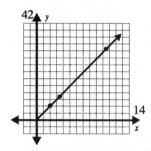

c. When $x = 4$, $y = 12$. An equilateral triangle with 4-inch sides will have a border of 12 inches.

d. Yes. All equilateral triangles have three sides that are the same length.

73. a. $D(20) = 3.75(20) + 250 = 325$
The cost of manufacturing 20 toasters per day would be $325 daily.

b. $D(30) = 3.75(30) + 250 = 362.5$
The cost of manufacturing 30 toasters per day would be $362.50 daily.

c. $D(25) = 3.75(25) + 250 = 343.75$
Yes, the daily cost for manufacturing 25 toasters per day would be $343.75.

d.

x	$D(x) = 3.75x + 250$	$D(x)$
0	$D(0) = 3.75(0) + 250$	250
5	$D(5) = 3.75(5) + 250$	268.75
10	$D(10) = 3.75(10) + 250$	287.50
15	$D(15) = 3.75(15) + 250$	306.25
20	$D(20) = 3.75(20) + 250$	325
25	$D(25) = 3.75(25) + 250$	343.75
30	$D(30) = 3.75(30) + 250$	362.50
35	$D(35) = 3.75(35) + 250$	381.25
40	$D(40) = 3.75(40) + 250$	400
45	$D(45) = 3.75(45) + 250$	418.75
50	$D(50) = 3.75(50) + 350$	437.50

75. Let x = the number of customers and p = daily profit of water park.
$$p(x) = 18.75x - (1000 + 4.5x)$$
$$= 18.75x - 1000 - 4.5x$$
$$= 14.25x - 1000$$
Let $p(x) = 0$.
$$0 = 14.25x - 1000$$
$$0 + 1000 = 14.25x - 1000 + 1000$$
$$1000 = 14.25x$$
$$\frac{1000}{14.25} = \frac{14.25x}{14.25}$$
$$70.18 \approx x$$
The x-intercept is roughly $(70.18, 0)$.
Let $x = 0$.
$$p(0) = 14.25(0) - 1000$$
$$p(0) = -1000$$
The y-intercept is $(0, -1000)$.

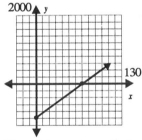

The x-intercept says that more than 70.18 customers must attend the park for the park to make a profit. The y-intercept says that if no customers attend the park, the park will lose $1000 due to fixed operating costs.

77. $d = 1944 - 425.7t$
$$0 = 1944 - 425.7t$$
$$0 + 425.7t = 1944 - 425.7t + 425.7t$$
$$425.7t = 1944$$
$$\frac{425.7t}{425.7} = \frac{1944}{425.7}$$
$$t \approx 4.57$$
The x-intercept is about $(4.57, 0)$.
$$d = 1944 - 425.7(0)$$
$$d = 1944$$
The y-intercept is $(0, 1944)$.

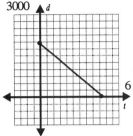

The x-intercept indicates how long it takes to fly from Atlanta to L.A. (that is, $t \approx 4.57$ hrs when $d = 0$). The y-intercept indicates how far apart the two cities are at the beginning of the trip (that is, when $t = 0$, $d = 1944$).
From the graph, it will take about 2.2 hours to fly 1000 miles from Atlanta.

5.1 Calculator Exercises

1. $5(x-1) = 2x - 4$
$$5x - 5 = 2x - 4$$
$$5x - 5 - 2x = 2x - 4 - 2x$$
$$3x - 5 = -4$$
$$3x - 5 + 5 = -4 + 5$$
$$3x = 1$$
$$\frac{3x}{3} = \frac{1}{3}$$
$$x = \frac{1}{3}$$

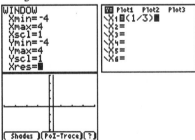

SSM: Experiencing Introductory and Intermediate Algebra

2. $\frac{3}{4}x+1=7$

$\frac{3}{4}x+1-1=7-1$

$\frac{3}{4}x=6$

$\frac{4}{3}\left(\frac{3}{4}\right)x=\frac{4}{3}(6)$

$x=8$

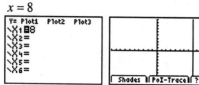

3. $4.5x+1.2=2.3x-5.4$

$4.5x+1.2-2.3x=2.3x-5.4-2.3x$

$2.2x+1.2=-5.4$

$2.2x+1.2-1.2=-5.4-1.2$

$2.2x=-6.6$

$\frac{2.2x}{2.2}=-\frac{6.6}{2.2}$

$x=-3$

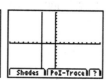

5.2 Exercises

1. The slope is $-\frac{3}{2}$. The graph is of a decreasing function.

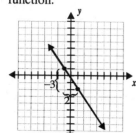

3. The slope is 0. The graph is of a constant function.

5. The slope is $\frac{1}{6}$. The graph is of an increasing function.

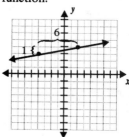

7. The slope is undefined. The graph is not of a function.

9. $(-7,-2)$ and $(5,6)$

$m=\frac{y_2-y_1}{x_2-x_1}$

$m=\frac{6-(-2)}{5-(-7)}=\frac{6+2}{5+7}=\frac{8}{12}=\frac{2}{3}$

The slope is $\frac{2}{3}$.

11. $(-12,-9)$ and $(4,-9)$

$m=\frac{y_2-y_1}{x_2-x_1}$

$m=\frac{-9-(-9)}{4-(-12)}=\frac{-9+9}{4+12}=\frac{0}{16}=0$

The slope is 0.

13. $(0,3)$ and $(0,8)$

$m=\frac{y_2-y_1}{x_2-x_1}$

$m=\frac{8-3}{0-0}=\frac{5}{0}=$ undefined

The slope is undefined.

15. $(6,-4)$ and $(7,-6)$

$m=\frac{y_2-y_1}{x_2-x_1}$

$m=\frac{-6-(-4)}{7-6}=\frac{-6+4}{1}=-2$

The slope is -2.

17. $(0,4)$ and $(5,0)$

$$m = \frac{y_2 - y_1}{x_2 - x_1}$$

$$m = \frac{0-4}{5-0} = -\frac{4}{5}$$

The slope is $-\frac{4}{5}$.

19. $(11.5, -9.2)$ and $(6.9, 18.4)$

$$m = \frac{y_2 - y_1}{x_2 - x_1}$$

$$m = \frac{18.4 - (-9.2)}{6.9 - 11.5} = \frac{18.4 + 9.2}{6.9 - 11.5}$$

$$= \frac{27.6}{-4.6} = -6$$

The slope is -6.

21. $\left(\frac{1}{2}, \frac{3}{4}\right)$ and $\left(-\frac{1}{2}, -\frac{5}{6}\right)$

$$m = \frac{y_2 - y_1}{x_2 - x_1}$$

$$m = \frac{-\frac{5}{6} - \frac{3}{4}}{-\frac{1}{2} - \frac{1}{2}} = \frac{-\frac{10}{12} - \frac{9}{12}}{-1} = \frac{19}{12}$$

The slope is $\frac{19}{12}$.

23. $y = 21x + 15$

$m = 21$ and $b = 15$

The slope is 21 and the y-intercept is $(0, 15)$.

25. $y = 5.95x - 2.01$

$m = 5.95$ and $b = -2.01$

The slope is 5.95 and the y-intercept is $(0, -2.01)$.

27. $y = 85{,}600 - 1255x$

$m = -1255$ and $b = 85{,}600$

The slope is -1255 and the y-intercept is $(0, 85600)$.

29. $16x - 4y = 64$

$16x - 4y - 16x = 64 - 16x$

$-4y = -16x + 64$

$\dfrac{-4y}{-4} = \dfrac{-16x + 64}{-4}$

$y = 4x - 16$

$m = 4$ and $b = -16$

The slope is 4 and the y-intercept is $(0, -16)$.

31. $7y + 18 = 2(y + 6) - 4$

$7y + 18 = 2y + 12 - 4$

$7y + 18 = 2y + 8$

$7y + 18 - 18 = 2y + 8 - 18$

$7y = 2y - 10$

$7y - 2y = 2y - 10 - 2y$

$5y = -10$

$\dfrac{5y}{5} = \dfrac{-10}{5}$

$y = -2$

$m = 0$ and $b = -2$

The slope is 0 and the y-intercept is $(0, -2)$.

33. $\dfrac{3}{2}x - \dfrac{3}{5}y = \dfrac{21}{10}$

$10\left(\dfrac{3}{2}x - \dfrac{3}{5}y\right) = 10\left(\dfrac{21}{10}\right)$

$15x - 6y = 21$

$15x - 6y - 15x = 21 - 15x$

$-6y = -15x + 21$

$\dfrac{-6y}{-6} = \dfrac{-15x + 21}{-6}$

$y = \dfrac{5}{2}x - \dfrac{7}{2}$

$m = \dfrac{5}{2}$ and $b = -\dfrac{7}{2}$

The slope is $\dfrac{5}{2}$ and the y-intercept is $\left(0, -\dfrac{7}{2}\right)$.

35. $x = -4\dfrac{7}{8}$

Undefined slope and no y-intercept. This is a vertical line.

SSM: Experiencing Introductory and Intermediate Algebra

37. $(8,3)$; $m = \dfrac{4}{7}$

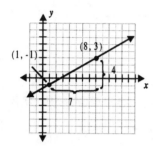

39. $(-10,4)$; $m = -\dfrac{5}{9}$

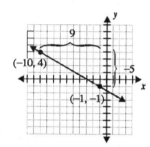

41. $(5,7)$; $m = 4$

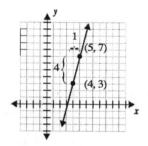

43. $(0,9)$; $m = 0$

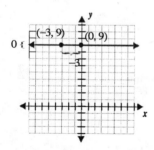

45. $(9,0)$; m is undefined

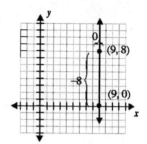

47. $y = \dfrac{5}{3}x - 4$

$m = \dfrac{5}{3}$; $b = -4$

The slope is $\dfrac{5}{3}$; the y-intercept is $(0,-4)$.

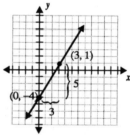

49. $16x - 8y = 40$

$16x - 8y - 16x = 40 - 16x$

$-8y = 40 - 16x$

$\dfrac{-8y}{-8} = \dfrac{40 - 16x}{-8}$

$y = 2x - 5$

$m = 2$; $b = -5$

The slope is 2; the y-intercept is $(0,-5)$.

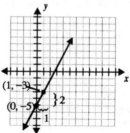

51. $7x + 2y = -16$

$7x + 2y - 7x = -16 - 7x$

$2y = -16 - 7x$

$\dfrac{2y}{2} = \dfrac{-16 - 7x}{2}$

$y = -\dfrac{7}{2}x - 8$

$m = -\dfrac{7}{2}; b = -8$

The slope is $-\dfrac{7}{2}$; the y-intercept is $(0, -8)$.

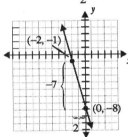

53. $14y + 21 = 6y + 5$

$14y + 21 - 6y = 6y + 5 - 6y$

$8y + 21 = 5$

$8y + 21 - 21 = 5 - 21$

$8y = -16$

$\dfrac{8y}{8} = \dfrac{-16}{8}$

$y = -2$

$m = 0; b = -2$

The slope is 0 and the y-intercept is $(0, -2)$.

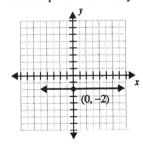

55. $5y = 150x + 350$

$\dfrac{5y}{5} = \dfrac{150x + 350}{5}$

$y = 30x + 70$

$m = 30; b = 70$

The slope is 30; the y-intercept is $(0, 70)$.

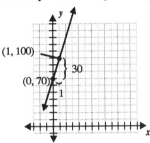

57. $f(x) = 0.3x - 1.2$

$m = 0.3; b = -1.2$

The slope is 0.3; the y-intercept is $(0, -1.2)$.

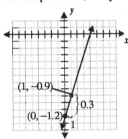

59. The grade is the slope written as a percent.

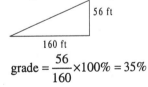

grade $= \dfrac{56}{160} \times 100\% = 35\%$

The grade of the advertised terrain is 35%.

61.

pitch $= \dfrac{3.3}{12} \times 100\% = 27.5\%$

The pitch of the roof is 27.5%.

63. Graph (b).

Assuming a constant speed of 2 mph (6 miles in 3 hours), the graph should have slope 2 and rise from $(0,0)$ to $(3,6)$. The troop worked at the site for an hour so their distance did not change during this time. The graph should be horizontal from $(3,6)$ to $(4,6)$. They returned at the same rate but they were heading home. Thus, the graph should have a slope of -2 and fall from $(4,6)$ to $(5,4)$. The troop rested for an hour so the graph should be horizontal from $(5,4)$ to $(6,4)$. They finished the remainder of the trip, 4 miles, in two hours. Thus, the graph should go from $(6,4)$ to $(8,0)$ to finish up.

65. Graph (a).

The temperature is increasing for the first 10 hours, constant for the next four, and decreasing for the last 10 hours. Since the y-axis represents the temperature, we expect to see the graph reflect this behavior.

67. $d = \dfrac{\$19{,}800}{5 \text{ yrs}} = \$3960/\text{yr}$

The depreciation on the minivan will be $3960 per year.

69. The average rate of change is the slope of the line.
$m = \dfrac{80.190 - 51.243}{2001 - 1997} = \dfrac{28.95}{4} = 7.2375$
The average rate of change per year for the net assets of this fund was $7.2375 billion.

71. $D(p) = 80 - \dfrac{4}{5}p$

  a. $m = -\dfrac{4}{5}; b = 80$

  The slope is $-\dfrac{4}{5}$ and the y-intercept is $(0, 80)$.

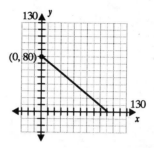

  b. At $p = 10$, the demand is $D = 72$.
  At $p = 20$, the demand is $D = 64$.
  At $p = 40$, the demand is $D = 48$.
  At $p = 64$, the demand is $D = 28.8$.

  c. Demand decreases as the price increases.

  d. We would assume that the price would be between 0 and 100 inclusive. A negative price does not make sense (you would be paying the customer to take the product off your hands) and for prices at $100 or more, the demand would be 0.

5.2 Calculator Exercises

The graphs in the following table use this window setting:

```
WINDOW
 Xmin=-12
 Xmax=12
 Xscl=1
 Ymin=-12
 Ymax=12
 Yscl=1
 Xres=1
```

Equation	$y = mx + b$		Conclusions		Graph
	Slope m	Constant b	Graph's Indication ↗ or ↘	Graph's y-intercept	
$y = 3x + 6$	3	6	↗	$(0, 6)$	
$y = -2x + 7$	-2	7	↘	$(0, 7)$	
$y = -x - 3$	-1	-3	↘	$(0, -3)$	
$y = 4x - 1$	4	-1	↗	$(0, -1)$	
$5x - 3y = 9$ $\left(y = \dfrac{5}{3}x - 3\right)$	$\dfrac{5}{3}$	-3	↗	$(0, -3)$	
$4x + 5y = 10$ $\left(y = -\dfrac{4}{5}x + 2\right)$	$-\dfrac{4}{5}$	2	↘	$(0, 2)$	

$y = \frac{7}{8}x - \frac{3}{4}$	$\frac{7}{8}$	$-\frac{3}{4}$	↗	$\left(0, -\frac{3}{4}\right)$	
$y = -1.7x + 3.2$	-1.7	3.2	↘	$(0, 3.2)$	

5.3 Exercises

1. $3x - 2y = 5(y+7)$
    $3x - 2y = 5y + 35$
    $3x - 2y - 5y = 5y + 35 - 5y$
    $3x - 7y = 35$
    $3x - 7y - 3x = 35 - 3x$
    $-7y = 35 - 3x$
    $\frac{-7y}{-7} = \frac{35 - 3x}{-7}$
    $y = \frac{3}{7}x - 5 \quad m = \frac{3}{7}, b = -5$
    $7x = 3(1 - y)$
    $7x = 3 - 3y$
    $7x + 3y = 3 - 3y + 3y$
    $7x + 3y = 3$
    $7x + 3y - 7x = 3 - 7x$
    $3y = 3 - 7x$
    $\frac{3y}{3} = \frac{3 - 7x}{3}$
    $y = -\frac{7}{3}x + 1 \quad m = -\frac{7}{3}, b = 1$
    The lines are intersecting and perpendicular because the product of their slopes is -1.

3. $x = 4(y - 3)$
    $x = 4y - 12$
    $x + 12 = 4y - 12 + 12$
    $x + 12 = 4y$
    $\frac{x + 12}{4} = \frac{4y}{4}$
    $y = \frac{1}{4}x + 3 \quad m = \frac{1}{4}, b = 3$
    $x = 4(y + 5)$
    $x = 4y + 20$
    $x - 20 = 4y + 20 - 20$
    $x - 20 = 4y$
    $\frac{x - 20}{4} = \frac{4y}{4}$
    $y = \frac{1}{4}x - 5 \quad m = \frac{1}{4}, b = -5$
    The lines are parallel because the have the same slope but different y-intercepts.

5. $4x - y = 6$
    $4x - y - 4x = 6 - 4x$
    $-y = 6 - 4x$
    $\frac{-y}{-1} = \frac{6 - 4x}{-1}$
    $y = 4x - 6 \quad m = 4, b = -6$
    $2x - y + 3 = 0$
    $2x - y + 3 + y = 0 + y$
    $y = 2x + 3 \quad m = 2, b = 3$
    The lines are only intersecting because they have different slopes whose product does not equal -1.

Chapter 5: Linear Equations and Functions

7. $x = 2(y-7)$
    $x = 2y - 14$
    $x + 14 = 2y - 14 + 14$
    $x + 14 = 2y$
    $\dfrac{x+14}{2} = \dfrac{2y}{2}$
    $y = \dfrac{1}{2}x + 7 \qquad m = \dfrac{1}{2}, b = 7$
    $y = \dfrac{1}{2}x + 7 \qquad m = \dfrac{1}{2}, b = 7$
    The two lines are coinciding because they have the same slope and the same y-intercept.

9. $5x + y = -6$
    $5x + y - 5x = -6 - 5x$
    $y = -5x - 6 \qquad m = -5, b = -6$
    $3x + y = 0$
    $3x + y - 3x = 0 - 3x$
    $y = -3x \qquad m = -3, b = 0$
    The lines are only intersecting because they have different slopes whose product does not equal -1.

11. $4x + y = 8$
    $4x + y - 4x = 8 - 4x$
    $y = -4x + 8 \qquad m = -4, b = 8$
    $4x + y + 2 = 0$
    $4x + y + 2 - 4x - 2 = 0 - 4x - 2$
    $y = -4x - 2 \qquad m = -4, b = -2$
    The lines are parallel because they have the same slope but different y-intercepts.

13. $y - 5 = 0$
    $y - 5 + 5 = 0 + 5$
    $y = 5 \qquad m = 0, b = 5$
    $2x + 6 = x + 9$
    $2x + 6 - x = x + 9 - x$
    $x + 6 = 9$
    $x + 6 - 6 = 9 - 6$
    $x = 3 \qquad m = \text{undefined}$
    The lines are intersecting and perpendicular because one is vertical and the other is horizontal.

15. $2y - 3 = 13$
    $2y - 3 + 3 = 13 + 3$
    $2y = 16$
    $\dfrac{2y}{2} = \dfrac{16}{2}$
    $y = 8 \qquad m = 0, b = 8$
    $y + 1 = 4$
    $y + 1 - 1 = 4 - 1$
    $y = 3 \qquad m = 0, b = 3$
    The lines are parallel because they are both

17. $x + 3 = 0$
    $x = -3 \qquad m = \text{undefined}$
    $x - 5 = 0$
    $x = 5 \qquad m = \text{undefined}$
    The lines are parallel because they are both vertical and have different x-intercepts.

19. $2x - 9 = 0$
    $2x - 9 + 9 = 0 + 9$
    $2x = 9$
    $\dfrac{2x}{2} = \dfrac{9}{2}$
    $x = \dfrac{9}{2} \qquad m = \text{undefined}$
    $x - 4 = 5 - x$
    $x - 4 + x = 5 - x + x$
    $2x - 4 = 5$
    $2x - 4 + 4 = 5 + 4$
    $2x = 9$
    $\dfrac{2x}{2} = \dfrac{9}{2}$
    $x = \dfrac{9}{2} \qquad m = \text{undefined}$
    The two lines are coinciding because they are both vertical and have the same x-intercept.

21. $3(y-3)=1$
$3y-9=1$
$3y-9+9=1+9$
$3y=10$
$\dfrac{3y}{3}=\dfrac{10}{3}$
$y=\dfrac{10}{3} \quad m=0, b=\dfrac{10}{3}$
$5y=10+2y$
$5y-2y=10+2y-2y$
$3y=10$
$\dfrac{3y}{3}=\dfrac{10}{3}$
$y=\dfrac{10}{3} \quad m=0, b=\dfrac{10}{3}$
The lines are coinciding because they have the same slope and the same y-intercept.

23. $x=2 \quad m=$ undefined
$y=2x-1 \quad m=2, b=-1$
The lines are only intersecting because they have different slopes whose product is not -1.

25. $y-3=0$
$y=3 \quad m=0, b=3$
$2x+3y=0$
$2x+3y-2x=0-2x$
$3y=-2x$
$y=-\dfrac{2}{3}x \quad m=-\dfrac{2}{3}, b=0$
The lines are only intersecting because they have different slopes whose product is not -1.

27. **a.** $y=0.15x+2.50+0.10x+1$
$y=0.25x+3.50$

   b. $y=0.25x$

   c. Because the slopes are the same and the y-intercepts are different, the graphs of the cost and revenue equations are parallel. There is no break-even point.

   d. $y=\dfrac{1}{3}x$

   e. $\dfrac{1}{3}x=0.25x+3.50$
$\dfrac{1}{3}x-0.25x=0.25x+3.50-0.25x$
$\dfrac{1}{12}x=3.50$
$12\left(\dfrac{1}{12}x\right)=12(3.50)$
$x=42$
At 42 candy bars, he will break even.

   f. Cost: $y=0.15x+1.00$
Revenue: $y=0.25x$

$0.15x+1.00=0.25x$
$0.15x+1.00-0.15x=0.25x-0.15x$
$1.00=0.10x$
$\dfrac{1.00}{0.10}=\dfrac{0.10x}{0.10}$
$x=10$
At 10 candy bars, Brook will break even.

   g. Cost: $y=0.15x+1.00$
Revenue: $y=\dfrac{1}{3}x$

$\dfrac{1}{3}x=0.15x+1.00$
$\dfrac{1}{3}x-0.15x=0.15x+1.00-0.15x$
$\dfrac{11}{60}x=1.00$
$\dfrac{60}{11}\left(\dfrac{11}{60}x\right)=\dfrac{60}{11}(1.00)$
$x=\dfrac{60}{11}\approx 5.45$
At about 5.45 candy bars, Brook will break even. He must sell 6 to make a profit.

   h. Answers will vary.
Based on the above information it seems as if Brook should sell his own candy bars at the rate of 3 for a dollar.

Chapter 5: Linear Equations and Functions

29. Krazy Kar: $y = 35 + 0.25x$ $m = 0.25, b = 35$
    Rational: $y = 60$ $m = 0, b = 60$
    Since the equations have different slopes, the graphs will intersect.

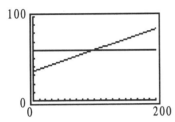

The intersection point is $(100, 60)$. At 100 miles per day, the daily rental costs are equal at $60.

31. Find expressions for the distance from the 50-yard line.
    Since Archie starts at the 50-yard line,
    $y = 10x$ $m = 10, b = 0$
    Since Speedie gets to the 50-yard line 4 seconds later,
    $y = 15(x - 4)$
    $y = 15x - 60$ $m = 15, b = -60$
    Since the slopes are different, the graphs of the equations will intersect.

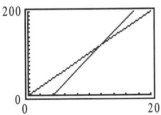

The intersection point is $(12, 120)$. After 12 seconds Speedie and Archie will be 120 feet from the 50-yard line. Since the endzone is 150 feet away, Speedie will catch Archie before he scores.

33. $y_1 = 21.33x + 1017$ $m = 21.33, b = 1017$
    $y_2 = 99.66x + 963$ $m = 99.66, b = 963$

Since the slopes of the two equations are different, their graphs will intersect.

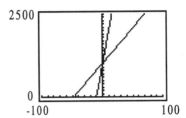

The intersection point is about $(0.689, 1031.705)$. In mid-1990 the number of returns by both types of corporations was the same at about 1031.705 thousand.

35. $y_1 = 50 + 2x$ $m = 2, b = 50$
    $y_2 = 75$ $m = 0, b = 75$
    $x = 8$ $m = $ undefined, $b = $ none
    Yes. The graph for her age is a vertical line. This line has a different slope than y_1 so the two graphs will intersect. The graph of y_2 is a horizontal line which means it will be perpendicular to the graph of her age.

The intersection of y_1 and $x = 8$ is $(8, 66)$. At 8 years, she will receive $66. The intersection of y_2 and $x = 8$ is $(8, 75)$. At 8 years, she will receive $75. The intersection of y_1 and y_2 is $(12.5, 75)$. At $12\frac{1}{2}$ years, she will receive the same either way, $75.

5.3 Calculator Exercises

1. ```
   WINDOW
   Xmin=-10
   Xmax=10
   Xscl=1
   Ymin=-10
   Ymax=10
   Yscl=1
   Xres=1
   ```
   ```

2. WINDOW
Xmin=-47
Xmax=47
Xscl=10
Ymin=-31
Ymax=31
Yscl=10
Xres=1

3. WINDOW
Xmin=0
Xmax=94
Xscl=10
Ymin=0
Ymax=62
Yscl=10
Xres=1

Yes.

4. WINDOW
Xmin=-470
Xmax=470
Xscl=100
Ymin=-310
Ymax=310
Yscl=100
Xres=1

Yes, they are still integer pairs.

5. The equations have slopes that are different but their product does not equal -1. These lines are only intersecting.

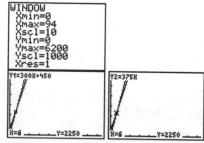

The intersection point is $(6, 2250)$.

6. Because the two slopes are the same and the y-intercepts are different, the two graphs will be parallel. There are no break-even points.

7. The equations have slopes that are different but their product does not equal -1. These lines are only intersecting.

The intersection point is $(14.4, 1800)$.

5.4 Exercises

1. $m = \dfrac{3}{2}, b = -1$

   $y = mx + b$

   $y = \dfrac{3}{2}x - 1$

3. $m =$ undefined, $b =$ none

   $x = c$

   $x = -5$

5. $m = -3, b = 0$

   $y = mx + b$

   $y = -3x + 0$

   $y = -3x$

7. $m = 0, b = -\dfrac{3}{2}$

   $y = mx + b$

   $y = 0x - \dfrac{3}{2}$

   $y = -\dfrac{3}{2}$

9. $m = -\dfrac{2}{5}, b = 4$

   $y = mx + b$

   $y = -\dfrac{2}{5}x + 4$

11. $m = \dfrac{5}{9}, b = 0$

$y = mx + b$

$y = \dfrac{5}{9}x$

13. $m = 4, b = -\dfrac{3}{4}$

$y = mx + b$

$y = 4x - \dfrac{3}{4}$

15. $m = -4.1, b = 0.5$

$y = mx + b$

$y = -4.1x + 0.5$

17. $m = 0, b = -33$

$y = mx + b$

$y = 0x - 33$

$y = -33$

19. $m = \dfrac{2}{3}, (3, -3)$

$y - y_1 = m(x - x_1)$

$y - (-3) = \dfrac{2}{3}(x - 3)$

$y + 3 = \dfrac{2}{3}x - 2$

$y = \dfrac{2}{3}x - 5$

21. $m = -3, (0, 4)$

$y - y_1 = m(x - x_1)$

$y - 4 = -3(x - 0)$

$y - 4 = -3x$

$y = -3x + 4$

23. $m = -1.7, (3, -1.5)$

$y - y_1 = m(x - x_1)$

$y - (-1.5) = -1.7(x - 3)$

$y + 1.5 = -1.7x + 5.1$

$y = -1.7x + 3.6$

25. $(-1, 1)$ and $(1, -2)$

$m = \dfrac{y_2 - y_1}{x_2 - x_1}$

$m = \dfrac{-2 - 1}{1 - (-1)} = -\dfrac{3}{2}$

$y - y_1 = m(x - x_1)$

$y - 1 = -\dfrac{3}{2}(x - (-1))$

$y - 1 = -\dfrac{3}{2}(x + 1)$

$y - 1 = -\dfrac{3}{2}x - \dfrac{3}{2}$

$y = -\dfrac{3}{2}x - \dfrac{1}{2}$

27. $(-1, -2)$ and $(-1, 5)$

$m = \dfrac{y_2 - y_1}{x_2 - x_1}$

$m = \dfrac{5 - (-2)}{-1 - (-1)} = \dfrac{5 + 2}{-1 + 1} = \dfrac{7}{0} =$ undefined

This is a vertical line.

$x = -1$

29. $(-1, 1)$ and $(-2, -1)$

$m = \dfrac{y_2 - y_1}{x_2 - x_1}$

$m = \dfrac{-1 - 1}{-2 - (-1)} = \dfrac{-2}{-2 + 1} = \dfrac{-2}{-1} = 2$

$y - y_1 = m(x - x_1)$

$y - 1 = 2(x - (-1))$

$y - 1 = 2(x + 1)$

$y - 1 = 2x + 2$

$y = 2x + 3$

31. $(-2, 2)$ and $(4, 2)$

$m = \dfrac{y_2 - y_1}{x_2 - x_1}$

$m = \dfrac{2 - 2}{4 - (-2)} = \dfrac{0}{4 + 2} = \dfrac{0}{6} = 0$

SSM: Experiencing Introductory and Intermediate Algebra

$y - y_1 = m(x - x_1)$
$y - 2 = 0(x - (-2))$
$y - 2 = 0$
$y = 2$

33. $\left(4\frac{1}{2}, 5\frac{1}{4}\right)$ and $(1, 4)$

$m = \dfrac{y_2 - y_1}{x_2 - x_1}$

$m = \dfrac{4 - 5\frac{1}{4}}{1 - 4\frac{1}{2}} = \dfrac{\frac{16}{4} - \frac{21}{4}}{\frac{2}{2} - \frac{9}{2}} = \dfrac{-\frac{5}{4}}{-\frac{7}{2}}$

$= \dfrac{5}{4} \cdot \dfrac{2}{7} = \dfrac{5}{14}$

$y - y_1 = m(x - x_1)$

$y - 4 = \dfrac{5}{14}(x - 1)$

$y - 4 = \dfrac{5}{14}x - \dfrac{5}{14}$

$y = \dfrac{5}{14}x - \dfrac{5}{14} + \dfrac{56}{14}$

$y = \dfrac{5}{14}x + \dfrac{51}{14}$

35. $\left(-1\frac{1}{3}, 2\right)$ and $(0, 0)$

$m = \dfrac{y_2 - y_1}{x_2 - x_1}$

$m = \dfrac{0 - 2}{0 - \left(-1\frac{1}{3}\right)} = \dfrac{-2}{1\frac{1}{3}} = -\dfrac{2}{1} \cdot \dfrac{3}{4} = -\dfrac{3}{2}$

$y - y_1 = m(x - x_1)$

$y - 0 = -\dfrac{3}{2}(x - 0)$

$y = -\dfrac{3}{2}x$

37. $(0.5, 0)$ and $(-0.8, 4.2)$

$m = \dfrac{y_2 - y_1}{x_2 - x_1}$

$m = \dfrac{4.2 - 0}{-0.8 - 0.5} = \dfrac{4.2}{-1.3} = -\dfrac{42}{13}$

$y - y_1 = m(x - x_1)$

$y - 0 = -\dfrac{42}{13}(x - 0.5)$

$y = -\dfrac{42}{13}x + \dfrac{21}{13}$

39. $(2.4, 2.8)$ and $(-2.6, -2.2)$

$m = \dfrac{y_2 - y_1}{x_2 - x_1}$

$m = \dfrac{-2.2 - 2.8}{-2.6 - 2.4} = \dfrac{-5}{-5} = 1$

$y - y_1 = m(x - x_1)$

$y - 2.8 = 1(x - 2.4)$

$y - 2.8 = x - 2.4$

$y = x - 2.4 + 2.8$

$y = x + 0.4$

41. $3x - 8y = 32$

$3x - 8y - 3x = 32 - 3x$

$-8y = 32 - 3x$

$\dfrac{-8y}{-8} = \dfrac{32 - 3x}{-8}$

$y = \dfrac{3}{8}x - 4 \qquad m_1 = \dfrac{3}{8}$

$(8, 7), m = \dfrac{3}{8}$

$y - y_1 = m(x - x_1)$

$y - 7 = \dfrac{3}{8}(x - 8)$

$y - 7 = \dfrac{3}{8}x - 3$

$y - 7 + 7 = \dfrac{3}{8}x - 3 + 7$

$y = \dfrac{3}{8}x + 4$

43. $x + 2y = 7$

$x + 2y - x = 7 - x$

$2y = 7 - x$

$\dfrac{2y}{2} = \dfrac{7-x}{2}$

$y = -\dfrac{1}{2}x + \dfrac{7}{2} \quad m_1 = -\dfrac{1}{2}$

$(4,0), m = -\dfrac{1}{2}$

$y - y_1 = m(x - x_1)$

$y - 0 = -\dfrac{1}{2}(x - 4)$

$y = -\dfrac{1}{2}x + 2$

45. $y = 3x + 4$

A parallel line has the same slope so $m = 3$. The y-coordinate of the y-intercept is $b = -5$.

$y = mx + b$

$y = 3x - 5$

47. $y = 3x - 10$

A parallel line has the same slope so $m = 3$.

Using the point $\left(\dfrac{4}{9}, -\dfrac{5}{9}\right)$, we have:

$y - y_1 = m(x - x_1)$

$y - \left(-\dfrac{5}{6}\right) = 3\left(x - \dfrac{4}{9}\right)$

$y + \dfrac{5}{6} = 3x - \dfrac{4}{3}$

$y + \dfrac{5}{6} - \dfrac{5}{6} = 3x - \dfrac{4}{3} - \dfrac{5}{6}$

$y = 3x - \dfrac{8}{6} - \dfrac{5}{6}$

$y = 3x - \dfrac{13}{6}$

49. $y = 1.2x + 3.5$

A parallel line has the same slope so $m = -1.2$.

Using the point $(4, -2)$, we have:

$y - y_1 = m(x - x_1)$

$y - (-2) = -1.2(x - 4)$

$y + 2 = -1.2x + 4.8$

$y + 2 - 2 = -1.2x + 4.8 - 2$

$y = -1.2x + 2.8$

51. $2x + 4y = 5$

$4y = 5 - 2x$

$\dfrac{4y}{4} = \dfrac{5 - 2x}{4}$

$y = -\dfrac{1}{2}x + \dfrac{5}{4}$

A parallel line has the same slope so $m = -\dfrac{1}{2}$.

Using the point $(0, 0)$, we have:

$y - y_1 = m(x - x_1)$

$y - 0 = -\dfrac{1}{2}(x - 0)$

$y = -\dfrac{1}{2}x$

53. $y = 3x + 12$

A perpendicular line has the opposite reciprocal slope. This slope is $m = -\dfrac{1}{3}$.

Using the point $(3.6, 5.8)$, we have:

$y - y_1 = m(x - x_1)$

$y - 5.8 = -\dfrac{1}{3}(x - 3.6)$

$y - 5.8 = -\dfrac{1}{3}x + 1.2$

$y - 5.8 + 5.8 = -\dfrac{1}{3}x + 1.2 + 5.8$

$y = -\dfrac{1}{3}x + 7$

55. $y = 5x - 1$

A perpendicular line has an opposite reciprocal slope. This slope is $m = -\dfrac{1}{5}$.

SSM: Experiencing Introductory and Intermediate Algebra

Using the point $(15, -30)$, we have:
$y - y_1 = m(x - x_1)$
$y - (-30) = -\frac{1}{5}(x - 15)$
$y + 30 = -\frac{1}{5}x + 3$
$y + 30 - 30 = -\frac{1}{5}x + 3 - 30$
$y = -\frac{1}{5}x - 27$

57. $y = -\frac{2}{3}x - 1$

A perpendicular line has an opposite reciprocal slope. This slope is $m = \frac{3}{2}$.

Using the origin $(0, 0)$, we have:
$y - y_1 = m(x - x_1)$
$y - 0 = \frac{3}{2}(x - 0)$
$y = \frac{3}{2}x$

59. $3x + 2y = 4$
$3x + 2y - 3x = 4 - 3x$
$2y = 4 - 3x$
$\frac{2y}{2} = \frac{4 - 3x}{2}$
$y = -\frac{3}{2}x + 2$

A perpendicular line has an opposite reciprocal slope. This slope is $m = \frac{2}{3}$.

Using an x-intercept of $(3, 0)$, we have:
$y - y_1 = m(x - x_1)$
$y - 0 = \frac{2}{3}(x - 3)$
$y = \frac{2}{3}x - 2$

61. a. $m = \frac{70864 - 77103}{1 - 0} = -6239$

b. $y - 77,103 = -6329(x - 0)$
$y - 77,103 + 77,103 = -6329x + 77,103$
$y = -6329x + 77,103$

c. $x = 1997 - 1994 = 3$
$y = -6329(3) + 77,103 = 58,116$
The predicted number of reported AIDS cases is 58,116. Yes, the estimate is pretty close.

d. $x = 2005 - 1994 = 11$
$y = -6329(11) + 77,103 = 7484$
The predicted number of reported AIDS cases is 7484.

e. $0 = -6329x + 77,103$
$6329x = 77,103$
$\frac{6329x}{6329} = \frac{77103}{6329}$
$x \approx 12.182$
The number of reported AIDS cases is predicted to be close to zero in 2007.

It is hard to predict since the trend may not continue. It is possible that the number of reported cases will start to decrease at a slower rate in the future.

63. $(0, 273)$ and $(-10, 263)$
$m = \frac{263 - 273}{-10 - 0} = \frac{-10}{-10} = 1$
$m = 1, b = 273$
$y = mx + b$
$y = x + 273$
When $x = 100$, $y = 100 + 273 = 373$
A Kelvin temperature of 373 corresponds to $100°$ C.

65. $m = \frac{16.96 - 15.92}{1.6 - 1.2} = \frac{1.04}{0.4} = 2.6$
$y - y_1 = m(x - x_1)$
$y - 15.92 = 2.6(x - 1.2)$

$y - 15.92 = 2.6x - 3.12$
$y - 15.92 + 15.92 = 2.6x - 3.12 + 15.92$
$y = 2.6x + 12.8$

x	$y = 2.6x + 12.8$	y
1.2	$y = 2.6(1.2) + 12.8$	15.92
1.6	$y = 2.6(1.6) + 12.8$	16.96
2.0	$y = 2.6(2.0) + 12.8$	18.00
2.4	$y = 2.6(2.4) + 12.8$	19.04
2.8	$y = 2.6(2.8) + 12.8$	20.08

67. Let x = weight in pounds and y = miles per gallon.
$(2500, 40)$ and $(3500, 35)$
$m = \dfrac{35 - 40}{3500 - 2500} = \dfrac{-5}{1000} = -0.005$
$y - y_1 = m(x - x_1)$
$y - 40 = -0.005(x - 2500)$
$y - 40 = -0.005x + 12.5$
$y = -0.005x + 52.5$

69. Let x = number of times ad runs each week and y = number of units sold.
$(6, 5000)$ and $(15, 14000)$
$m = \dfrac{14000 - 5000}{15 - 6} = \dfrac{9000}{9} = 1000$
$y - y_1 = m(x - x_1)$
$y - 5000 = 1000(x - 6)$
$y - 5000 = 1000x - 6000$
$y - 5000 + 5000 = 1000x - 6000 + 5000$
$y = 1000x - 1000$
When $x = 10$,
$y = 1000(10) - 1000 = 9000$
The predicted number of units sold is 9000 units.

71. Let x = number of years since 1990 and y = number of students enrolled.
$m = 120$, $(0, 5470)$, $b = 5470$
$y = mx + b$
$y = 120x + 5470$
When $x = 1996 - 1990 = 6$,

$y = 120(6) + 5470 = 6190$
The predicted number of enrolled students is 6190. Yes, the equation did a good job of predicting the number of students.
When $x = 2010 - 1990 = 20$,
$y = 120(20) + 5470 = 7870$
If the increase continues at the same rate, the equation predicts 7870 students to be enrolled.

73. Let x = number of years after 1980 and y = the economic loss from motor vehicle accidents in billions of dollars.

a. $(0, 57.1)$, $m = 3.78$
$y = mx + b$
$ = 3.78x + 57.1$

b. When $x = 1997 - 1980 = 17$,
$y = 3.78(17) + 57.1 = 121.36$
The predicted loss for 1997 is \$121.36 billion. This is close to the actual amount of \$123.7 billion.

c. When $x = 2005 - 1980 = 25$,
$y = 3.78(25) + 57.1 = 151.6$
The predicted loss for 2005 is \$151.6 billion.

5.4 Calculator Exercises

1. $(12, 925)$ and $(72, 4225)$

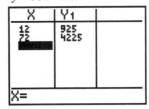

$y = 55x + 265$

2. $(16,-351)$ and $(40,417)$

```
LinReg
 y=ax+b
 a=32
 b=-863
```

$y = 32x - 863$

X	Y₁
16	-351
40	417

X=

3. $(0, 6.4)$ and $(36.4, 103.59)$

```
LinReg
 y=ax+b
 a=2.670054945
 b=6.4
```

$y = 2.670054945x + 6.4$

X	Y₁
0	6.4
36.4	103.59

X=

4. $\left(1, \dfrac{17}{4}\right)$ and $\left(3, \dfrac{57}{4}\right)$

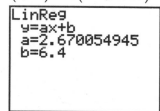

$y = 5x - \dfrac{3}{4}$

X	Y₁
1	4.25
3	14.25

X=

5. $(-5, 21)$ and $(-9, 7)$

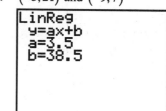

$y = 3.5x + 38.5$

X	Y₁
-5	21
-9	7

X=

6. $(16, -2)$ and $(7, 4)$

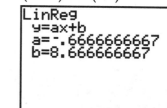

$y = -\dfrac{2}{3}x + \dfrac{26}{3}$

X	Y₁
16	-2
7	4

X=

Chapter 5 Section-By-Section Review

1. Linear;
   $y = 0.6x + 2.3$
   $0.6x - y = -2.3$

2. $y = 4x^2 - 2$ is non-linear because of the square on the x.

3. $5x - 3y + 7 = x - y + 9$ is linear.
   $5x - 3y = x - y + 2$
   $5x - 3y - x + y = x - y + 2 - x + y$
   $4x - 2y = 2$

4. $5y - 12 = 7 - y$ is linear.
   $5y = 19 - y$
   $5y + y = 19 - y + y$
   $6y = 19$

5. Answers will vary.
   $12x + 6y = 48$
   $6y = 48 - 12x$
   $y = -2x + 8$

x	$y = -2x + 8$	y
0	$y = -2(0) + 8$	8
1	$y = -2(1) + 8$	6
2	$y = -2(2) + 8$	4

   $(0,8)$, $(1,6)$, and $(2,4)$ are ordered pair solutions.

   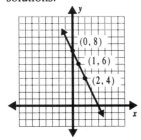

6. Answers will vary.

x	$y = \dfrac{8}{13}x - 7$	y
−13	$y = \dfrac{8}{13}(-13) - 7$	−15
0	$y = \dfrac{8}{13}(0) - 7$	−7
13	$y = \dfrac{8}{13}(13) - 7$	1

   $(-13,-15)$, $(0,-7)$, and $(13,1)$ are possible ordered pair solutions.

   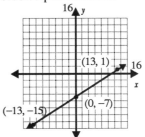

7. Answers will vary.

x	$y = -9$	y
−2	$y = -9$	−9
0	$y = -9$	−9
2	$y = -9$	−9

   $(-2,-9)$, $(0,-9)$, and $(2,-9)$ are possible ordered pair solutions.

   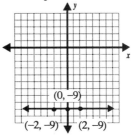

8. Answers will vary.
   $9x - y = 12$
   $9x - y + y = 12 + y$
   $9x = 12 + y$
   $9x - 12 = 12 + y - 12$
   $y = 9x - 12$

SSM: Experiencing Introductory and Intermediate Algebra

x	$y = 9x - 12$	y
0	$y = 9(0) - 12$	-12
1	$y = 9(1) - 12$	-3
2	$y = 9(2) - 12$	6

$(0, -12), (1, -3),$ and $(2, 6)$ are possible ordered pair solutions.

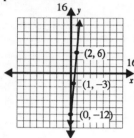

9. Answers will vary.

x	$y = -\dfrac{4}{3}x + 2$	y
-3	$y = -\dfrac{4}{3}(-3) + 2$	6
0	$y = -\dfrac{4}{3}(0) + 2$	2
3	$y = -\dfrac{4}{3}(3) + 2$	-2

$(-3, 6), (0, 2),$ and $(3, -2)$ are possible ordered pair solutions.

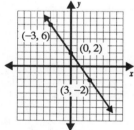

10. Answers will vary.
$5y - 2 = y - 10$
$5y - 2 - y = y - 10 - y$
$4y - 2 = -10$
$4y - 2 + 2 = -10 + 2$

$4y = -8$
$\dfrac{4y}{4} = \dfrac{-8}{4}$
$y = -2$

x	$y = -2$	y
-3	$y = -2$	-2
0	$y = -2$	-2
3	$y = -2$	-2

$(-3, -2), (0, -2),$ and $(3, -2)$ are possible ordered pair solutions.

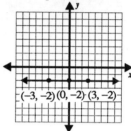

11. Answers will vary.
$2x + 16 = x + 18$
$2x + 16 - x = x + 18 - x$
$x + 16 = 18$
$x + 16 - 16 = 18 - 16$
$x = 2$

$(2, -3), (2, 0),$ and $(2, 3)$ are possible ordered pair solutions.

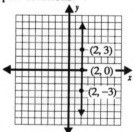

12. $8x + 12y = -24$
Let $y = 0$.
$8x + 12(0) = -24$
$8x = -24$
$\dfrac{8x}{8} = \dfrac{-24}{8}$
$x = -3$

Chapter 5: Linear Equations and Functions

The x-intercept is $(-3,0)$.
Let $x = 0$.
$8(0)+12y = -24$
$12y = -24$
$\dfrac{12y}{12} = \dfrac{-24}{12}$
$y = -2$
The y-intercept is $(0,-2)$.

13. $y = \dfrac{2}{5}x + 4$
Let $y = 0$.
$0 = \dfrac{2}{5}x + 4$
$0 - 4 = \dfrac{2}{5}x + 4 - 4$
$-4 = \dfrac{2}{5}x$
$\dfrac{5}{2}(-4) = \dfrac{5}{2}\left(\dfrac{2}{5}x\right)$
$-10 = x$
The x-intercept is $(-10,0)$.
Let $x = 0$.
$y = \dfrac{2}{5}(0) + 4$
$y = 4$
The y-intercept is $(0,4)$.

14. $2y - 4 = y - 7$
$2y - 4 - y = y - 7 - y$
$y - 4 = -7$
$y - 4 + 4 = -7 + 4$
$y = -3$
This is a horizontal line. There is no x-intercept but the y-intercept is $(0,-3)$.

15. $2(x-1)+5 = 9$
$2x - 2 + 5 = 9$
$2x + 3 = 9$
$2x + 3 - 3 = 9 - 3$
$2x = 6$
$\dfrac{2x}{2} = \dfrac{6}{2}$
$x = 3$
This is a vertical line. There is no y-intercept but the x-intercept is $(3,0)$.

16. $9x - 3y = -12$
$9x - 3y - 9x = -12 - 9x$
$-3y = -12 - 9x$
$\dfrac{-3y}{-3} = \dfrac{-12 - 9x}{-3}$
$y = 3x + 4$
The y-intercept is $(0,4)$.

17. $6x + 2y = x$
$6x + 2y - 6x = x - 6x$
$2y = -5x$
$\dfrac{2y}{2} = \dfrac{-5x}{2}$
$y = -\dfrac{5}{2}x$
The y-intercept is $(0,0)$.

18. $3y + 2 = 2(y - 4)$
$3y + 2 = 2y - 8$
$3y + 2 - 2y = 2y - 8 - 2y$
$y + 2 = -8$
$y + 2 - 2 = -8 - 2$
$y = -10$
The y-intercept is $(0,-10)$.

19. $7x + 11y = 77$
Let $x = 0$.
$7(0) + 11y = 77$
$11y = 77$
$y = 7$

The y-intercept is $(0,7)$.
Let $y = 0$.
$7x + 11(0) = 77$
$7x = 77$
$\dfrac{7x}{7} = \dfrac{77}{7}$
$x = 11$
The x-intercept is $(11,0)$.

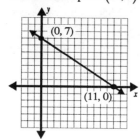

20. $-x - y = 2$
Let $x = 0$.
$-(0) - y = 2$
$-y = 2$
$\dfrac{-y}{-1} = \dfrac{2}{-1}$
$y = -2$
The y-intercept is $(0,-2)$.
Let $y = 0$.
$-x - (0) = 2$
$-x = 2$
$\dfrac{-x}{-1} = \dfrac{2}{-1}$
$x = -2$
The x-intercept is $(-2,0)$.

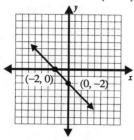

21. $7.40x + 14.80y = 29.60$
Let $x = 0$.
$7.40(0) + 14.80y = 29.60$
$14.80y = 29.60$
$\dfrac{14.80y}{14.80} = \dfrac{29.60}{14.80}$
$y = 2$
The y-intercept is $(0,2)$.
Let $y = 0$.
$7.40x + 14.80(0) = 29.60$
$7.40x = 29.60$
$\dfrac{7.40x}{7.40} = \dfrac{29.60}{7.40}$
$x = 4$
The x-intercept is $(4,0)$.

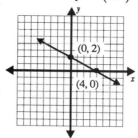

22. a. $(10, 85)$, $(20, 160)$

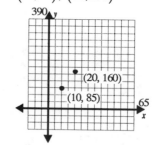

b.

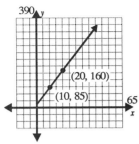

At $x = 30$, $y = 235$. Noriko will receive $235 for a job that is 30 pages long.

23. Let x = number of copies per week and y = weekly profit.

$y = 0.04x - (25 + 0.02x)$
$ = 0.04x - 25 - 0.02x$
$ = 0.02x - 25$

Let $x = 0$.
$y = 0.20(0) - 25$
$y = -25$

The y-intercept is $(0, -25)$.

Let $y = 0$.
$0 = 0.02x - 25$
$0 + 25 = 0.02x - 25 + 25$
$25 = 0.02x$
$\dfrac{25}{0.02} = \dfrac{0.02x}{0.02}$
$x = 1250$

The x-intercept is $(1250, 0)$.

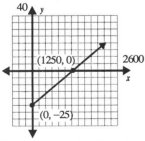

The y-intercept reflects fixed costs of operating the copy center. The x-intercept reflects the number of copies needed to break even.

24. $y = 1749 - 396x$

Let $x = 0$.
$y = 1749 - 396(0)$
$y = 1749$

The y-intercept is $(0, 1749)$.

Let $y = 0$.
$0 = 1749 - 396x$
$0 + 396x = 1749 - 396x + 396x$
$396x = 1749$
$\dfrac{396x}{396} = \dfrac{1749}{396}$
$x = \dfrac{53}{12}$

The x-intercept is $\left(\dfrac{53}{12}, 0\right)$.

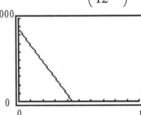

The y-intercept reflects the initial distance between Chicago and Los Angeles. The x-intercept reflects the total time it will take to make the trip.
When $y = 1000$, $x \approx 1.9$. It will take about 1.9 hours to travel 1000 miles from Chicago.

25. a. $y = 5200 - 100x$

b. Let $x = 0$.
$y = 5200 - 100(0)$
$y = 5200$

The y-intercept is $(0, 5200)$. Initially there is $5200 in the account.

Let $y = 0$.
$0 = 5200 - 100x$
$0 + 100x = 5200 - 100x + 100x$
$100x = 5200$
$\dfrac{100x}{100} = \dfrac{5200}{100}$
$x = 52$

The x-intercept is $(52,0)$. The account will be empty after 52 weeks.

c.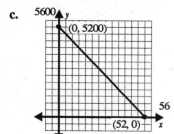

d. When $x = 32$, $y = 2000$. There will be $2000 left in the account after 32 weeks of payments.

26. The slope is 0. The graph is that of a constant function.

27. The slope is 5. The graph is that of an increasing function.

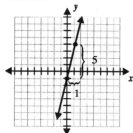

28. The slope is $-\dfrac{8}{5}$. The graph is that of a decreasing function.

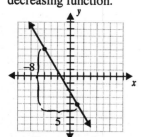

29. The slope is undefined. The graph is not of a function.

30. $m = \dfrac{y_2 - y_1}{x_2 - x_1}$

$= \dfrac{0-(-3)}{1-(-5)} = \dfrac{0+3}{1+5} = \dfrac{3}{6} = \dfrac{1}{2}$

31. $m = \dfrac{y_2 - y_1}{x_2 - x_1}$

$= \dfrac{2-7}{-4-(-7)} = \dfrac{-5}{-4+7} = -\dfrac{5}{3}$

32. $m = \dfrac{y_2 - y_1}{x_2 - x_1}$

$= \dfrac{-3-(-3)}{10-4} = \dfrac{-3+3}{6} = \dfrac{0}{6} = 0$

33. $m = \dfrac{y_2 - y_1}{x_2 - x_1}$

$= \dfrac{3-(-3)}{4-4} = \dfrac{3+3}{0} = \dfrac{6}{0} =$ undefined

34. $y = 23x - 51$

$m = 23, b = -51$

The slope is 23; the y-intercept is $(0,-51)$.

35. $6x + 5y = 12$

$6x + 5y - 6x = 12 - 6x$

$5y = 12 - 6x$

$\dfrac{5y}{5} = \dfrac{12-6x}{5}$

$y = -\dfrac{6}{5}x + \dfrac{12}{5}$

$m = -\dfrac{6}{5}, b = \dfrac{12}{5}$

The slope is $-\dfrac{6}{5}$; the y-intercept is $\left(0, \dfrac{12}{5}\right)$.

36. $4(y-2) = 3(2y-1)+4$

$4y-8 = 6y-3+4$

$4y-8 = 6y+1$

$4y-8-6y = 6y+1-6y$

$-2y-8 = 1$

$-2y-8+8 = 1+8$

$-2y = 9$

$\dfrac{-2y}{-2} = \dfrac{9}{-2}$

$y = -\dfrac{9}{2}$

$m = 0, b = -\dfrac{9}{2}$

The slope is 0 ; the y-intercept is $\left(0, -\dfrac{9}{2}\right)$.

37. $(-2, 3); m = \dfrac{5}{9}$

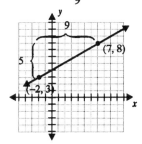

38. $(3, -2); m = -3$

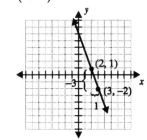

39. $(-2, -2); m = 0$

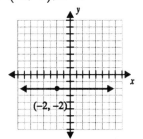

40. $(-2, -2); m = $ undefined

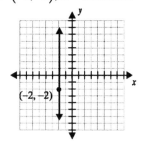

41. $y = -\dfrac{5}{3}x + 2$

$y = mx + b$

$m = -\dfrac{5}{3}, b = 2$

The slope is $-\dfrac{5}{3}$; the y-intercept is $(0, 2)$.

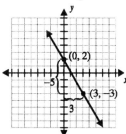

42. $3x + 2y = 12$

$3x + 2y - 3x = 12 - 3x$

$2y = 12 - 3x$

$\dfrac{2y}{2} = \dfrac{12 - 3x}{2}$

$y = -\dfrac{3}{2}x + 6$

$y = mx + b$

$m = -\dfrac{3}{2}, b = 6$

The slope is $-\dfrac{3}{2}$; the y-intercept is $(0, 6)$.

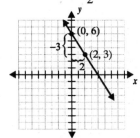

43. $m(x) = \dfrac{5}{8}x$

$y = mx + b$

$m = \dfrac{5}{8}, b = 0$

The slope is $\dfrac{5}{8}$; the y-intercept is $(0, 0)$.

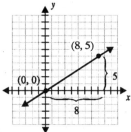

44. $y = -4$

$y = mx + b$

$m = 0, b = -4$

The slope is 0; the y-intercept is $(0, -4)$.

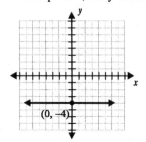

45. The grade is the slope written as a percent.

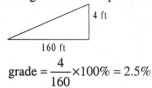

$\text{grade} = \dfrac{4}{160} \times 100\% = 2.5\%$

The grade of the road is 2.5%.

46. The depreciation per year is the slope of the linear relationship. It is the rate of change of value over time.

$m = \dfrac{485}{4} = 121.25$

The depreciation is $121.25 per year.

47. Graph (c).

The graph should begin decreasing at a constant rate. When $x = 40$ (that is, 1980), the graph should be constant.

48. Graph (a).

Assume $d(t)$ represents Janet's distance from home. The graph should be increasing as she heads toward the mall. The graph will be constant for one hour as she visits and then increase at the same rate. The graph is again constant as she shops, though only for one half-hour, after which it begins decreasing at a steeper rate than it increased. Graph (c) has her spending too much time shopping and graph (b) has her return rate slower than her initial rate.

49. $y = 2x + 6 \quad m = 2, b = 6$

$3y - x = 15$

$3y - x + x = 15 + x$

$3y = 15 + x$

$\dfrac{3y}{3} = \dfrac{15 + x}{3}$

$y = \dfrac{1}{3}x + 5 \quad m = \dfrac{1}{3}, b = 5$

The lines are only intersecting because they have different slopes whose product does not equal -1.

50. $2y - 2x = y + 3x + 2$
$2y - 2x + 2x = y + 3x + 2 + 2x$
$2y = y + 5x + 2$
$2y - y = y + 5x + 2 - y$
$y = 5x + 2 \qquad m = 5, b = 2$
$5x - y = -2$
$5x + 2 + y = -2 + 2 + y$
$y = 5x + 2 \qquad m = 5, b = 2$
These lines are coinciding because they have the same slope and same y-intercept.

51. $4x - 20 = 0$
$4x - 20 + 20 = 0 + 20$
$4x = 20$
$\dfrac{4x}{4} = \dfrac{20}{4}$
$x = 5 \qquad m = \text{undefined}, b = \text{none}$
$2x + 3 = x + 4$
$2x + 3 - x = x + 4 - x$
$x + 3 = 4$
$x + 3 - 3 = 4 - 3$
$x = 1 \qquad m = \text{undefined}, b = \text{none}$
These lines are parallel since they are both vertical and have different x-intercepts.

52. $2x + y = 4$
$2x + y - 2x = 4 - 2x$
$y = -2x + 4 \qquad m = -2, b = 4$
$y = -2(x + 2)$
$y = -2x - 4 \qquad m = -2, b = -4$
These lines are parallel because they have the same slope but different y-intercepts.

53. $3(y + 2) = 2(y + 4)$
$3y + 6 = 2y + 8$
$3y + 6 - 2y - 6 = 2y + 8 - 2y - 6$
$y = 2 \qquad m = 0, b = 2$
$2(x - 2) = 0$
$2x - 4 = 0$
$2x - 4 + 4 = 0 + 4$
$2x = 4$
$\dfrac{2x}{2} = \dfrac{4}{2}$
$x = 2 \qquad m = \text{undefined}, b = \text{none}$
These lines are intersecting and perpendicular because one is vertical and the other is horizontal.

54. $5y - 4(y + 3) = -10$
$5y - 4y - 12 = -10$
$y - 12 + 12 = -10 + 12$
$y = 2 \qquad m = 0, b = 2$
$y - 1 = -2(x - 1)$
$y - 1 = -2x + 2$
$y - 1 + 1 = -2x + 2 + 1$
$y = -2x + 3 \qquad m = -2, b = 3$
These lines are only intersecting because they have different slopes whose product does not equal -1.

55. $2x - 3y = -9$
$-3y = -9 - 2x$
$\dfrac{-3y}{-3} = \dfrac{-9 - 2x}{-3}$
$y = \dfrac{2}{3}x + 3 \qquad m = \dfrac{2}{3}, b = 3$
$3x + 2y = 6$
$2y = 6 - 3x$
$\dfrac{2y}{2} = \dfrac{6 - 3x}{2}$
$y = -\dfrac{3}{2}x + 3 \qquad m = -\dfrac{3}{2}, b = 3$
These lines are intersecting and perpendicular because they have different slopes whose product is -1.

SSM: Experiencing Introductory and Intermediate Algebra

56. $y = x+7 \quad m=1, b=7$
$y = -x+7 \quad m=-1, b=7$
These lines are intersecting and perpendicular because they have different slopes whose product is -1.

57. a. $y = 25+35+30x+25+5x$
$y = 35x+85$

b. $y = 35x$

c. Since the slopes are the same and the y-intercepts are different, the lines are parallel. There will be no break-even point.

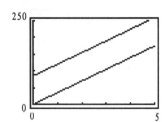

d. $y = 60x$

e.

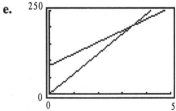

The break-even point (intersection point) is $(3.4, 204)$. He will start making a profit when the fourth calculator is sold.

f. Answers will vary. J.R. should charge more than $35 for the calculators. Given the two choices above, he should charge $60.

58. Males: $y = 0.9x+91 \quad m=0.9, b=91$
Females: $y = 0.9x+99 \quad m=0.9, b=99$

Because the slopes are the same and the y-intercepts are different, the graphs of these two equations will be parallel. The graph for the females will be above the graph for males since the y-intercept is larger for females.

59.

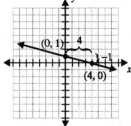

The slope is $-\dfrac{1}{4}$; the y-intercept is $(0,1)$
$y = mx+b$
$y = -\dfrac{1}{4}x+1$

60. The slope is -2; the y-intercept is $(0,3)$
$m = -2, b = 3$
$y = mx+b$
$y = -2x+3$

61. The slope is $\dfrac{3}{5}$; the y-intercept is $(0,-2)$
$m = \dfrac{3}{5}, b = -2$
$y = mx+b$
$y = \dfrac{3}{5}x-2$

62. Horizontal lines have a slope of 0.
The slope is 0; the y-intercept is $(0,-3.5)$
$y = mx+b$
$y = (0)x-3.5$
$y = -3.5$

63. Vertical lines have an undefined slope.
The x-intercept is $(2.6, 0)$
$x = 2.6$

64. $m = -5, (2,-3)$
$y - y_1 = m(x - x_1)$
$y - (-3) = -5(x-2)$
$y+3 = -5x+10$
$y+3-3 = -5x+10-3$
$y = -5x+7$

192

65. $(9,5)$ and $(-2,2)$

$$m = \frac{y_2 - y_1}{x_2 - x_1}$$
$$= \frac{2-5}{-2-9} = \frac{-3}{-11} = \frac{3}{11}$$
$$y - y_1 = m(x - x_1)$$
$$y - 5 = \frac{3}{11}(x - (9))$$
$$y - 5 = \frac{3}{11}(x - 9)$$
$$y - 5 = \frac{3}{11}x - \frac{27}{11}$$
$$y - 5 + 5 = \frac{3}{11}x - \frac{27}{11} + 5$$
$$y = \frac{3}{11}x - \frac{27}{11} + \frac{55}{11}$$
$$y = \frac{3}{11}x + \frac{28}{11}$$

66. $m = \frac{1}{3}, (4,6)$

$$y - y_1 = m(x - x_1)$$
$$y - 6 = \frac{1}{3}(x - 4)$$
$$y - 6 = \frac{1}{3}x - \frac{4}{3}$$
$$y - 6 + 6 = \frac{1}{3}x - \frac{4}{3} + 6$$
$$y = \frac{1}{3}x - \frac{4}{3} + \frac{18}{3}$$
$$y = \frac{1}{3}x + \frac{14}{3}$$

67. $y = 4x + 5$

A parallel line has the same slope so $m = 4$.
Using the given point, $(1,1)$, we have
$$y - y_1 = m(x - x_1)$$
$$y - 1 = 4(x - 1)$$
$$y - 1 = 4x - 4$$
$$y - 1 + 1 = 4x - 4 + 1$$
$$y = 4x - 3$$

68. $y = 2x - 1$

A perpendicular line has the opposite reciprocal slope. The given line has slope 2 so the new line will have slope $m = -\frac{1}{2}$.

Using the given point, $(2,4)$, we have
$$y - y_1 = m(x - x_1)$$
$$y - 4 = -\frac{1}{2}(x - 2)$$
$$y - 4 = -\frac{1}{2}x + 1$$
$$y - 4 + 4 = -\frac{1}{2}x + 1 + 4$$
$$y = -\frac{1}{2}x + 5$$

69. a. $(2,15)$ and $(8,30)$

$$m = \frac{y_2 - y_1}{x_2 - x_1}$$
$$= \frac{30-15}{8-2} = \frac{15}{6} = \frac{5}{2}$$
$$y - y_1 = m(x - x_1)$$
$$y - 15 = \frac{5}{2}(x - 2)$$
$$y - 15 = \frac{5}{2}x - 5$$
$$y - 15 + 15 = \frac{5}{2}x - 5 + 15$$
$$y = \frac{5}{2}x + 10$$

b. When $x = 5$,
$$y = \frac{5}{2}(5) + 10 = \frac{25}{2} + 10 = \frac{45}{2} = 22.5$$
Relatively, this is not very close to the actual value of 28.

70. a. $m = \frac{y_2 - y_1}{x_2 - x_1} = \frac{109.4 - 90.4}{10 - 0} = \frac{19}{10} = 1.9$

b. The slope is 1.9; the y-intercept is $(0, 90.4)$.
$$y = mx + b$$
$$y = 1.9x + 90.4$$

SSM: Experiencing Introductory and Intermediate Algebra

c. For 1998, $x = 18$.
$y = 1.9(18) + 90.4$
$ = 34.2 + 90.4 = 124.6$
The predicted number of nonfarm employees in 1998 is $124.6 million. This is close to the actual value of $125.8 million.

d. For 2005, $x = 25$.
$y = 1.9(25) + 90.4 = 47.5 + 90.4 = 137.9$
The predicted number of nonfarm employees in 2005 is 137.9 million.

Answers will vary.
This seems reasonable. The trend indicates that we should see growth. However, it would be important to check the size of the overall workforce to see if this number is feasible.

Chapter 5 Chapter Review

1. $3(y-5) = y+7$
$3y - 15 = y + 7$
$3y - 15 - y = y + 7 - y$
$2y - 15 = 7$
$2y - 15 + 15 = 7 + 15$
$2y = 22$
$y = 11$

x	$y = 11$	y
-3	$y = 11$	11
0	$y = 11$	11
3	$y = 11$	11

$(-3, 11), (0, 11)$, and $(3, 11)$ are three ordered pair solutions.

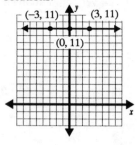

2. $14x + 7y = 7$
$14x + 7y - 14x = 7 - 14x$
$7y = 7 - 14x$
$\dfrac{7y}{7} = \dfrac{7 - 14x}{7}$
$y = -2x + 1$

x	$y = -2x + 1$	y
-3	$y = -2(-3) + 1$	7
0	$y = -2(0) + 1$	1
3	$y = -2(3) + 1$	-5

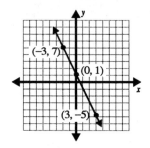

3. $y = -\dfrac{7}{5}x - 10$

x	$y = -\dfrac{7}{5}x - 10$	y
-10	$y = -\dfrac{7}{5}(-10) - 10$	4
-5	$y = -\dfrac{7}{5}(-5) - 10$	-3
0	$y = -\dfrac{7}{5}(0) - 10$	-10

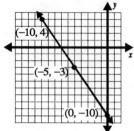

194

4. $(2,6); m = 0$

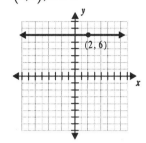

5. $(2,6); m$ is undefined

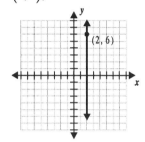

6. $(1,4); m = -2$

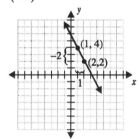

7. $(0,3); m = \dfrac{2}{7}$

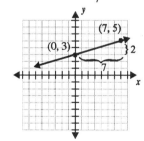

8. $y = -\dfrac{3}{8}x$

   $y = mx + b$

   $m = -\dfrac{3}{8}, b = 0$

   The slope is $-\dfrac{3}{8}$; the y-intercept is $(0,0)$.

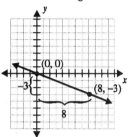

9. $y = 7$

   $y = mx + b$

   $m = 0, b = 7$

   The slope is 0; the y-intercept is $(0,7)$.

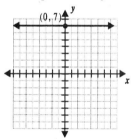

10. $5x - 4y = 12$

    $5x - 4y - 5x = 12 - 5x$

    $-4y = 12 - 5x$

    $\dfrac{-4y}{-4} = \dfrac{12 - 5x}{-4}$

    $y = \dfrac{5}{4}x - 3$

    $m = \dfrac{5}{4}, b = -3$

    The slope is $\dfrac{5}{4}$; the y-intercept is $(0,-3)$

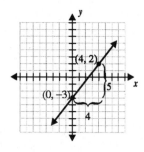

11. $15.50x + 21.70y = 108.50$

Let $x = 0$.

$15.50(0) + 21.70y = 108.50$

$21.70y = 108.50$

$\dfrac{21.70y}{21.70} = \dfrac{108.50}{21.70}$

$y = 5$

The y-intercept is $(0, 5)$.

Let $y = 0$.

$15.50x + 21.70(0) = 108.50$

$15.50x = 108.50$

$\dfrac{15.50x}{15.50} = \dfrac{108.50}{15.50}$

$x = 7$

The x-intercept is $(7, 0)$.

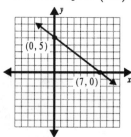

12. $12x - 24y = -48$

Let $x = 0$.

$12(0) - 24y = -48$

$-24y = -48$

$\dfrac{-24y}{-24} = \dfrac{-48}{-24}$

$y = 2$

The y-intercept is $(0, 2)$.

Let $y = 0$.

$12x - 24(0) = -48$

$12x = -48$

$\dfrac{12x}{12} = \dfrac{-48}{12}$

$x = -4$

The x-intercept is $(-4, 0)$.

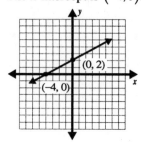

13. $-x + y = -6$

Let $x = 0$.

$-(0) + y = -6$

$y = -6$

The y-intercept is $(0, -6)$.

Let $y = 0$.

$-x + (0) = -6$

$-x = -6$

$\dfrac{-x}{-1} = \dfrac{-6}{-1}$

$x = 6$

The x-intercept is $(6, 0)$.

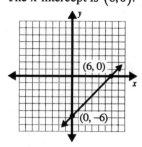

14. $2x - 3 = 1$

$2x = 4$

$\dfrac{2x}{2} = \dfrac{4}{2}$

$x = 2$ \quad $m = $ undefined, $b = $ none

$4x - 3y = 6$

$4x - 3y - 4x = 6 - 4x$

$-3y = 6 - 4x$

$\dfrac{-3y}{-3} = \dfrac{6-4x}{-3}$

$y = \dfrac{4}{3}x - 2 \qquad m = \dfrac{4}{3}, b = -2$

The lines are only intersecting because they have different slopes whose product does not equal -1.

15. $5x - 4y = 8$

$5x - 4y - 5x = 8 - 5x$

$-4y = 8 - 5x$

$\dfrac{-4y}{-4} = \dfrac{8-5x}{-4}$

$y = \dfrac{5}{4}x - 2 \qquad m = \dfrac{5}{4}, b = -2$

$4x - 5y = -15$

$4x - 5y - 4x = -15 - 4x$

$-5y = -15 - 4x$

$\dfrac{-5y}{-5} = \dfrac{-15-4x}{-5}$

$y = \dfrac{4}{5}x + 3 \qquad m = \dfrac{4}{5}, b = 3$

The lines are only intersecting because the slopes are different and their product does not equal -1.

16. $y = 5(x - 1)$

$y = 5x - 5 \qquad m = 5, b = -5$

$2(x - 1) = 7x - y + 3$

$2x - 2 = 7x - y + 3$

$2x - 2 + y = 7x - y + 3 + y$

$2x - 2 + y = 7x + 3$

$2x - 2 + y - 2x + 2 = 7x + 3 - 2x + 2$

$y = 5x + 5 \qquad m = 5, b = 5$

The lines are parallel because they have the same slope but different y-intercepts.

17. $5(x + 1) = 15$

$5x + 5 = 15$

$5x = 10$

$\dfrac{5x}{5} = \dfrac{10}{5}$

$x = 2 \qquad m = \text{undefined}, b = \text{none}$

$3y + 1 = 10$

$3y = 9$

$\dfrac{3y}{3} = \dfrac{9}{3}$

$y = 3 \qquad m = 0, b = 3$

The lines are intersecting and perpendicular because one is vertical and the other is horizontal.

18. $y = -2x + 3 \qquad m = -2, b = 3$

$2(x + y) = y + 3$

$2x + 2y = y + 3$

$2y = y + 3 - 2x$

$2y - y = y + 3 - 2x - y$

$y = -2x + 3 \qquad m = -2, b = 3$

The lines are coinciding because they have the same slope and the same y-intercept.

19. $y = x + 3 \qquad m = 1, b = 3$

$y = -x - 4 \qquad m = -1, b = -4$

The lines are intersecting and perpendicular because the product of their slopes is -1.

20. $5y = 20$

$\dfrac{5y}{5} = \dfrac{20}{5}$

$y = 4 \qquad m = 0, b = 4$

$2(y + 3) = -2$

$2y + 6 = -2$

$2y + 6 - 6 = -2 - 6$

$2y = -8$

$\dfrac{2y}{2} = \dfrac{-8}{2}$

$y = -4 \qquad m = 0, b = -4$

The lines are parallel because they have different slopes whose product does not equal -1.

SSM: Experiencing Introductory and Intermediate Algebra

21. Answers will vary.
$y = 8$

x	$y = 8$	y
-3	$y = 8$	8
0	$y = 8$	8
3	$y = 8$	8

$(-3, 8), (0, 8),$ and $(3, 8)$ are three ordered pair solutions.

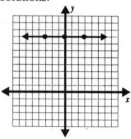

22. $8x - 9y = -72$
$8x - 9y - 8x = -72 - 8x$
$-9y = -72 - 8x$
$\dfrac{-9y}{-9} = \dfrac{-72 - 8x}{-9}$
$y = \dfrac{8}{9}x + 8$

x	$y = \dfrac{8}{9}x + 8$	y
-9	$y = \dfrac{8}{9}(-9) + 8$	0
0	$y = \dfrac{8}{9}(0) + 8$	8
9	$y = \dfrac{8}{9}(9) + 8$	16

$(-9, 0), (0, 8),$ and $(9, 16)$ are three ordered pair solutions.

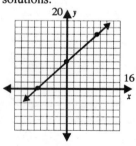

23. $t(x) = \dfrac{9}{11}x - 8$

x	$t(x) = \dfrac{9}{11}x - 8$	$t(x)$
-11	$t(-11) = \dfrac{9}{11}(-11) - 8$	-17
0	$t(0) = \dfrac{9}{11}(0) - 8$	-8
11	$t(11) = \dfrac{9}{11}(11) - 8$	1

$(-11, -17), (0, -8),$ and $(11, 1)$ are three ordered pair solutions.

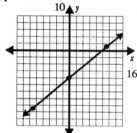

24. $y = x$ is linear.
$-x + y = 0$
$x - y = 0$ is in standard form.

25. $y = 2x^2 - 8$ is nonlinear because of the square on x.

26. $y = 1.3x - 0.5$ is linear.
$y - 1.3x = 1.3x - 0.5 - 1.3x$
$-1.3x + y = -0.5$
$1.3x - y = 0.5$ is in standard form.

27. $y = 15x$ is linear.
$y - y = 15x - y$
$0 = 15x - y$
$15x - y = 0$ is in standard form.

28. $8y - 5x = 21$ is linear.
$-8y + 5x = -21$
$5x - 8y = -21$ is in standard form.

29. $2x + 3y - 1 = y^2 + x$ is nonlinear because of the square on y.

198

Chapter 5: Linear Equations and Functions

30. $x + 4(x-2) = 2$

$x + 4x - 8 + 8 = 2 + 8$

$5x = 10$

$\dfrac{5x}{5} = \dfrac{10}{5}$

$x = 2 \qquad m = \text{undefined}, b = \text{none}$

This is a vertical line. The slope is undefined and there is no y-intercept.

31. $x - 3(y+2) = 4(x+1) - y$

$x - 3y - 6 = 4x + 4 - y$

$x - 3y - 6 - x + 6 = 4x + 4 - y - x + 6$

$-3y = 3x + 10 - y$

$-3y + y = 3x + 10 - y + y$

$-2y = 3x + 10$

$y = -\dfrac{3}{2}x - 5 \qquad m = -\dfrac{3}{2}, b = -5$

The slope is $-\dfrac{3}{2}$; the y-intercept is $(0, -5)$.

32. $12x - 4y = 8$

$12x - 4y - 12x = 8 - 12x$

$-4y = 8 - 12x$

$\dfrac{-4y}{-4} = \dfrac{8 - 12x}{-4}$

$y = 3x - 2 \qquad m = 3, b = -2$

The slope is 3; the y-intercept is $(0, -2)$.

33. $2y - 6 = 4(y-2) + 2$

$2y - 6 = 4y - 8 + 2$

$2y - 6 = 4y - 6$

$2y - 6 + 6 = 4y - 6 + 6$

$2y = 4y$

$2y - 4y = 4y - 4y$

$-2y = 0$

$\dfrac{-2y}{-2} = \dfrac{0}{-2}$

$y = 0 \qquad m = 0, b = 0$

The slope is 0; the y-intercept is $(0, 0)$.

34. $y = 13x - 15 \qquad m = 13, b = -15$

The slope is 13; the y-intercept is $(0, -15)$.

35. $y = -5.03x + 7.92 \qquad m = -5.03, b = 7.92$

The slope is -5.03; the y-intercept is $(0, 7.92)$.

36. $(5, 5)$ and $(8, 5)$

$m = \dfrac{y_2 - y_1}{x_2 - x_1} = \dfrac{5-5}{8-5} = \dfrac{0}{3} = 0$

The slope is 0.

37. $(-3, -3)$ and $(-3, 3)$

$m = \dfrac{y_2 - y_1}{x_2 - x_1} = \dfrac{3 - (-3)}{-3 - (-3)}$

$= \dfrac{3+3}{-3+3} = \dfrac{6}{0} = \text{undefined}$

The slope is undefined.

38. $(-4, 4)$ and $(-1, -5)$

$m = \dfrac{y_2 - y_1}{x_2 - x_1} = \dfrac{-5 - 4}{-1 - (-4)}$

$= \dfrac{-9}{-1 + 4} = \dfrac{-9}{3} = -3$

The slope is -3.

39. $\left(-\dfrac{2}{5}, \dfrac{4}{7}\right)$ and $\left(-\dfrac{4}{7}, \dfrac{1}{5}\right)$

$m = \dfrac{y_2 - y_1}{x_2 - x_1} = \dfrac{\dfrac{1}{5} - \dfrac{4}{7}}{-\dfrac{4}{7} - \left(-\dfrac{2}{5}\right)}$

$= \dfrac{\dfrac{7}{35} - \dfrac{20}{35}}{-\dfrac{20}{35} + \dfrac{14}{35}} = \dfrac{-\dfrac{13}{35}}{-\dfrac{6}{35}}$

$= \left(-\dfrac{13}{35}\right)\left(-\dfrac{35}{6}\right) = \dfrac{13}{6}$

The slope is $\dfrac{13}{6}$.

40. $(4,-2)$ and $(5,2)$

$m = \dfrac{y_2 - y_1}{x_2 - x_1} = \dfrac{2-(-2)}{5-4} = \dfrac{2+2}{1} = 4$

$y - y_1 = m(x - x_1)$

$y - (-2) = 4(x - 4)$

$y + 2 = 4x - 16$

$y + 2 - 2 = 4x - 16 - 2$

$y = 4x - 18$

41. $(-1,-1); m = -\dfrac{1}{4}$

$y - y_1 = m(x - x_1)$

$y - (-1) = -\dfrac{1}{4}(x - (-1))$

$y + 1 = -\dfrac{1}{4}(x + 1)$

$y + 1 = -\dfrac{1}{4}x - \dfrac{1}{4}$

$y = -\dfrac{1}{4}x - \dfrac{5}{4}$

42. $(0,5); m = -\dfrac{2}{3}$

The y-intercept is $(0,5)$ so $b = 5$.

$y = mx + b$

$y = -\dfrac{2}{3}x + 5$

43. Vertical line with $c = 4.1$.

$x = 4.1$

44. $y = -3x + 6 \qquad m = -3$

Parallel lines have the same slope.

$y - y_1 = m(x - x_1)$

$y - (-2) = -3(x - (-2))$

$y + 2 = -3(x + 2)$

$y + 2 = -3x - 6$

$y = -3x - 8$

45. Horizontal line with $b = 8$.

$y = 8$

46. $(1.2, 8); m = 4$

$y - y_1 = m(x - x_1)$

$y - 8 = 4(x - 1.2)$

$y - 8 = 4x - 4.8$

$y = 4x + 3.2$

47. $(0,-2); m = 3$

The y-intercept is $(0,-2)$ so $b = -2$.

$y = mx + b$

$y = 3x - 2$

48. $y = -\dfrac{1}{2}x - 1 \qquad m = -\dfrac{1}{2}$

Perpendicular lines have opposite reciprocal slopes. The slope is 2.

$(0,2); m = 2$

The y-intercept is $(0,2)$ so $b = 2$.

$y = mx + b$

$y = 2x + 2$

49. a. $(7, 24)$ and $(9, 28)$

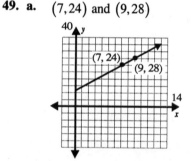

b. See the graph in part a.
When $x = 8$, $y = 26$.
Frank would earn \$26 for 8 innings.
When $x = 10$, $y = 30$.
Frank would earn \$30 for 10 innings.

50. a. $y = 180 + 10x - 15x$

$y = -5x + 180$

b. Let $x = 0$.

$y = -5(0) + 180 = 180$

The y-intercept is $(0, 180)$. The tank starts with 180 gallons.

Let $y = 0$.
$0 = -5x + 180$
$5x = 180$
$\dfrac{5x}{5} = \dfrac{180}{5}$
$x = 36$

The x-intercept is $(36, 0)$. After 36 hours, the tank will be empty.

c.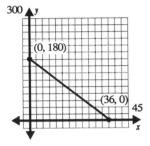

d. See the graph in part c.
When $x = 16$, $y = 100$.
After 16 hours, there will be 100 gallons in the tank.

51. a. $y = 50x + 20$

b. When $x = 2$,
$y = 50(2) + 20 = 100 + 20 = 120$
The trainer would earn $120 for 2 hours.

c. When $x = 1\dfrac{1}{2}$,
$y = 50\left(1\dfrac{1}{2}\right) + 20 = 75 + 20 = 95$
The trainer would earn $95 for $1\dfrac{1}{2}$ hours.

52. a. $c(x) = 75 + 150 + 0.75x$
$= 0.75x + 225$

b. $r(x) = 2x$

c.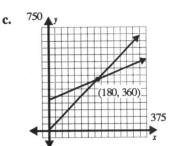

The intersection point is $(180, 360)$.

d. No. 150 is less than 180.
Yes. 200 is more than 180.
Shannon must sell at least 180 bows to avoid a loss.

53. grade $= \dfrac{60 \text{ ft}}{125 \text{ ft}} \times 100\% = 48\%$

54. pitch $= \dfrac{2 \text{ in}}{12 \text{ in}} \times 100\% = 16.67\%$

55. avg. rate of change $= \dfrac{275 - 15}{2000 - 1980}$
$= \dfrac{260}{20} = 13$

The average rate of change was $13 per year.

56. Graph (a) is best.

Jonathan's cumulative pay should be increasing at a constant rate for 6 months. Then the cumulative pay should stay constant for 2 months while he is on summer hiatus. After the hiatus, his cumulative pay begins to increase at a constant rate again. However, the rate is larger after summer due to the increase in his pay rate.

57. a. $m = \dfrac{y_2 - y_1}{x_2 - x_1} = \dfrac{1394 - 1361}{1 - 0} = 33$

b. The y-intercept is $(0, 1361)$ and the slope is $m = 33$.
$y = mx + b$
$y = 33x + 1361$

c. $y = 33(2) + 1361 = 66 + 1361 = 1427$
$y = 33(3) + 1361 = 99 + 1361 = 1460$
$y = 33(4) + 1361 = 132 + 1361 = 1493$
$y = 33(5) + 1361 = 165 + 1361 = 1526$

year	1994	1995	1996	1997
Actual	1414	1428	1479	1501
Predicted	1427	1460	1493	1526

The predicted amounts were always greater than the actual amount. The equation gave values that were fairly close.

d. In 2005, $x = 13$ and
$y = 33(13) + 1361 = 429 + 1361 = 1790$
The predicted emissions level for 2005 is 1790 million metric tons.

58. Let x = number of frames and p = profit.
$p = 40x - (35 + 6.5x)$
$= 40x - 35 - 6.5x$
$= 33.5x - 35$
Let $x = 0$.
$p = 33.5(0) - 35 = 0 - 35 = -35$
The y-intercept is $(0, -35)$.
Beckie's fixed costs total $35. She will pay this each week even if she does not make any frames.
Let $p = 0$.
$0 = 33.5x - 35$
$35 = 33.5x$
$x \approx 1.04$
The x-intercept is about $(1.04, 0)$.
At 1.04 frames, Beckie will break even. To make a profit, she needs to sell more than 1 frame.

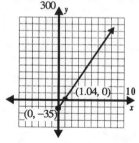

59. Graph (c).

Dennis' distance from home should be increasing initially, stay constant (flat) for a half-hour, and then decrease at a rate less than his initial rate (that is, the graph will fall slower than it initially rose).

Chapter 5 Test

1. $5x - 7y = 2(x+1) - 3$
$5x - 7y = 2x + 2 - 3$
$3x - 7y = -1$
is linear because it can be written in standard form.

2. $y = \frac{11}{12}x - 5$ is linear because it is in slope-intercept form.

3. $y^2 - 16 = 0$ is nonlinear because of the square on the y.

4. $7x + 2y = 3x^2 + 3$ is nonlinear because of the square on the x.

5. Answers will vary.
$2x - 8y = 0$
$-8y = -2x$
$y = \frac{1}{4}x$

x	$y = \frac{1}{4}x$	y
-4	$y = \frac{1}{4}(-4)$	-1
0	$y = \frac{1}{4}(0)$	0
4	$y = \frac{1}{4}(4)$	1

$(-4, -1), (0, 0)$, and $(4, 1)$ are three possible

solutions.

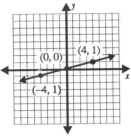

6. Answers may vary.
$3x + 2y = 7$
$2y = -3x + 7$
$\dfrac{2y}{2} = \dfrac{-3x+7}{2}$
$y = -\dfrac{3}{2}x + \dfrac{7}{2}$

x	$y = -\dfrac{3}{2}x + \dfrac{7}{2}$	y
-3	$y = -\dfrac{3}{2}(-3) + \dfrac{7}{2}$	8
-1	$y = -\dfrac{3}{2}(-1) + \dfrac{7}{2}$	5
5	$y = -\dfrac{3}{2}(5) + \dfrac{7}{2}$	-4

$(-3, 8), (-1, 5),$ and $(5, -4)$ are three possible solutions.

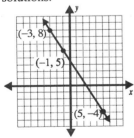

7. $12x + 15y = 60$
Let $x = 0$.
$12(0) + 15y = 60$
$15y = 60$
$\dfrac{15y}{15} = \dfrac{60}{15}$
$y = 4$
The y-intercept is $(0, 4)$.

Let $y = 0$.
$12x + 15(0) = 60$
$12x = 60$
$\dfrac{12x}{12} = \dfrac{60}{12}$
$x = 5$
The x-intercept is $(5, 0)$.

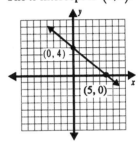

8. $2(x - 3) = x + 1$
$2x - 6 = x + 1$
$2x - 6 + 6 - x = x + 1 + 6 - x$
$x = 7$
This is a vertical line with x-intercept $(7, 0)$.

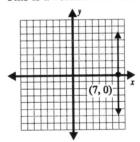

9. $4y - 3 = 2y + 9$
$4y - 3 - 2y + 3 = 2y + 9 - 2y + 3$
$2y = 12$
$\dfrac{2y}{2} = \dfrac{12}{2}$
$y = 6$

This is a horizontal line with y-intercept $(0,6)$.

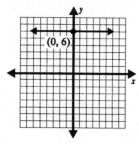

10. $g(x) = 3x - 5$

    $m = 3; \ b = -5$

    The slope is 3; the y-intercept is $(0, -5)$.

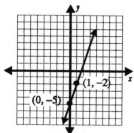

11. $3x - 4y = 4$

    $-4y = -3x + 4$

    $\dfrac{-4y}{-4} = \dfrac{-3x + 4}{-4}$

    $y = \dfrac{3}{4}x - 1$

    $m = \dfrac{3}{4}; \ b = -1$

    The slope is $\dfrac{3}{4}$; the y-intercept is $(0, -1)$.

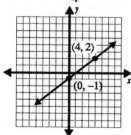

12. $m = \dfrac{y_2 - y_1}{x_2 - x_1} = \dfrac{-1 - (-1)}{-1 - 0}$

    $= \dfrac{-1 + 1}{-1} = \dfrac{0}{-1} = 0$

    The slope is 0.

13. $m = \dfrac{y_2 - y_1}{x_2 - x_1} = \dfrac{2.4 - 0.6}{-1.5 - 4.8} = \dfrac{1.8}{-6.3} = -\dfrac{2}{7}$

    The slope is $-\dfrac{2}{7}$.

14. $m = \dfrac{y_2 - y_1}{x_2 - x_1} = \dfrac{0 - (-4)}{1 - (-3)} = \dfrac{0 + 4}{1 + 3} = \dfrac{4}{4} = 1$

    The slope is 1.

15. $m = \dfrac{y_2 - y_1}{x_2 - x_1} = \dfrac{8 - 2}{-2 - (-2)} = \dfrac{6}{-2 + 2} = \dfrac{6}{0}$

    = undefined

    The slope is undefined. The line is vertical.

16. $y = -8x + 1 \qquad m = -8; \ b = 1$

    $y = x - 8 \qquad m = 1; \ b = -8$

    The lines are only intersecting since they have different slopes whose product does not equal -1.

17. $9(y - 2) + 2x = 7x$

    $9y - 18 + 2x = 7x$

    $9y - 18 = 5x$

    $9y = 5x + 18$

    $\dfrac{9y}{9} = \dfrac{5x + 18}{9}$

    $y = \dfrac{5}{9}x + 2 \qquad m = \dfrac{5}{9}; \ b = 2$

    $4x = 9(y + 4) - x$

    $4x = 9y + 36 - x$

    $5x = 9y + 36$

    $5x - 36 = 9y$

    $\dfrac{9y}{9} = \dfrac{5x - 36}{9}$

    $y = \dfrac{5}{9}x - 4 \qquad m = \dfrac{5}{9}; \ b = -4$

    The lines are parallel since they have the same slope but different y-intercepts.

Chapter 5: Linear Equations and Functions

18. $3x - y = 2$
$3x - 2 = y$
$y = 3x - 2 \quad m = 3; \, b = -2$
$y - x = 2(x - 1)$
$y - x = 2x - 2$
$y = 3x - 2 \quad m = 3; \, b = -2$
The lines are coinciding since they have the same slope and the same y-intercept.

19. $5y - 2 = 3y + 2$
$2y = 4$
$y = 2 \quad m = 0; b = 2$
$3x = 2(x + 2)$
$3x = 2x + 4$
$x = 4 \quad m = $ undefined; $b = $ none
The lines are intersecting and perpendicular because one is vertical and the other is horizontal.

20. $4y = 3x + 12$
$\dfrac{4y}{4} = \dfrac{3x + 12}{4}$
$y = \dfrac{3}{4}x + 3$
$m = \dfrac{3}{4}; \, b = 3$
The slope is $\dfrac{3}{4}$; the y-intercept is $(0, 3)$.

21. $7x + 7y = 21$
$7y = -7x + 21$
$\dfrac{7y}{7} = \dfrac{-7x + 21}{7}$
$y = -x + 3$
$m = -1; \, b = 3$
The slope is -1; the y-intercept is $(0, 3)$.

22. $y = \dfrac{3}{5}x \qquad m = \dfrac{3}{5}$
Perpendicular lines have opposite reciprocal slopes. The slope is $-\dfrac{5}{3}$.

$m = -\dfrac{5}{3}; \, (2, 1)$
$y - y_1 = m(x - x_1)$
$y - 1 = -\dfrac{5}{3}(x - 2)$
$y - 1 = -\dfrac{5}{3}x + \dfrac{10}{3}$
$y = -\dfrac{5}{3}x + \dfrac{13}{3}$

23. The y-intercept is $(0, 7)$ and the slope is $m = 9$.
$y = mx + b$
$y = 9x + 7$

24. $m = 6; \, (-1, 2)$
$y - y_1 = m(x - x_1)$
$y - 2 = 6(x - (-1))$
$y - 2 = 6(x + 1)$
$y - 2 = 6x + 6$
$y = 6x + 8$

25. $(2, 3)$ and $(-4, 1)$
$m = \dfrac{y_2 - y_1}{x_2 - x_1} = \dfrac{1 - 3}{-4 - 2} = \dfrac{-2}{-6} = \dfrac{1}{3}$
$y - y_1 = m(x - x_1)$
$y - 3 = \dfrac{1}{3}(x - 2)$
$y - 3 = \dfrac{1}{3}x - \dfrac{2}{3}$
$y = \dfrac{1}{3}x + \dfrac{7}{3}$

26. Answers will vary. Possible answer: $(-2, -9), (0, -5),$ and $(2, -1)$ are three ordered pair solutions.
$m = \dfrac{y_2 - y_1}{x_2 - x_1} = \dfrac{-5 - (-9)}{0 - (-2)}$
$= \dfrac{-5 + 9}{0 + 2} = \dfrac{4}{2} = 2$

SSM: Experiencing Introductory and Intermediate Algebra

The slope is 2.
The graph represents a function because it passes the vertical line test. Each value in the domain corresponds to exactly one value in the range. The function is increasing.

27. The amount of depreciation per year is the slope of the line.
$$m = \frac{y_2 - y_1}{x_2 - x_1} = \frac{0 - 4000}{6 - 0} = \frac{-4000}{6} = -666.67$$
The machine depreciates $666.67 each year.

28. The average rate of change is the slope of the line.
$$m = \frac{y_2 - y_1}{x_2 - x_1} = \frac{426 - 174}{1980 - 1940} = \frac{252}{40} = 6.3$$
The average rate of change was 6.3 acres per year (increase).

29. $y = -150x + 1450$

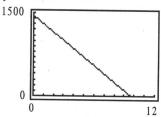

The y-intercept is $(0, 1450)$. The original amount of the loan is $1450.

The x-intercept is $\left(\frac{29}{3}, 0\right)$. After $\frac{29}{3}$ months, the loan will be paid off. Marty will actually pay the loan off in the 10th month but he will not make a full payment.

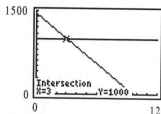

The loan balance will be below $1000 after the 4th month.

30. Answers will vary.

The intercept method uses the x-intercept and the y-intercept as the two points that determine the line. The slope-intercept method uses the y-intercept as one point, and then uses the slope to determine a second point.

Chapter 1-5 Cumulative Review

1. a. Since $\frac{2}{3} = \frac{2 \cdot 3}{3 \cdot 3} = \frac{6}{9}$, we have $\frac{5}{9} < \frac{2}{3}$.

    b. Since $1.6 = \frac{16}{10} = \frac{8}{5}$, we have $\frac{8}{5} = 1.6$.

    c. $-5.4 > -6.5$ because -5.4 is further to the right on a number line.

2.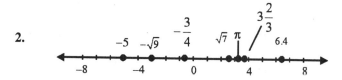

3. $5.7 - 4.68 = 1.02$

4. $-|17 - 29| = -|-12| = -(12) = -12$

5. $\sqrt{\frac{49}{81}} = \frac{\sqrt{49}}{\sqrt{81}} = \frac{7}{9}$

6. $-\sqrt{2.5} = -1.581$

7. $\sqrt{-4}$ is not a real number.

8. $-51 \div (-17) = \frac{-51}{-17} = 3$

9. $-\frac{3}{7} - \frac{4}{5} = -\frac{15}{35} - \frac{28}{35} = -\frac{43}{35}$

10. $(-8)(21) \div 3(-7) = -168 \div 3(-7)$
    $= -56(-7) = 392$

11. $\frac{4(5^2 - 9) - 14}{7 + 3} = \frac{4(25 - 9) - 14}{10} = \frac{4(16) - 14}{10}$
    $= \frac{64 - 14}{10} = \frac{50}{10} = 5$

206

Chapter 5: Linear Equations and Functions

12. $15+(-8)+13-12-(-7) = 15-8+13-12+7$
    $= 7+1+7 = 15$

13. $-[14.3-(-2.68)] = -[14.3+2.68]$
    $= -[16.98] = -16.98$

14. $\left(-\dfrac{3}{5}\right)\left(\dfrac{5}{9}\right)\left(-\dfrac{10}{21}\right)\left(\dfrac{7}{8}\right) = \left(-\dfrac{1}{3}\right)\left(-\dfrac{10}{21}\right)\left(\dfrac{7}{8}\right)$
    $= \left(\dfrac{10}{63}\right)\left(\dfrac{7}{8}\right) = \dfrac{70}{504} = \dfrac{5}{36}$

15. $25^0 = 1$

16. $-3^4 = -81$

17. $(-3)^4 = 81$

18. $\left(\dfrac{2}{3}\right)^{-2} = \left(\dfrac{3}{2}\right)^2 = \dfrac{3^2}{2^2} = \dfrac{9}{4}$

19. $-(4a-2b)+(7a-4b)-(-5a-b)$
    $= -4a+2b+7a-4b+5a+b$
    $= -4a+7a+5a+2b-4b+b$
    $= 8a-b$

20. $2[3(2x+3y)-(4x+y)]+7x$
    $= 2[6x+9y-4x-y]+7x$
    $= 12x+18y-8x-2y+7x$
    $= 12x-8x+7x+18y-2y$
    $= 11x+16y$

21. $x^3-2x^2+7x-5x^3+2x-4$

    a. There are 6 terms.

    b. $x^3, -2x^2, 7x, -5x^3$, and $2x$.

    c. -4

    d. $1, -2, 7, -5, 2, -4$

    e. x^3 and $-5x^3$, $7x$ and $2x$

22. $(-3)^2 + 5(-3) - 3 = 5(-3) + 6$
    $9 - 15 - 3 = -15 + 6$
    $-9 = -9$
    Since $-9 = -9$, $x = -3$ is a solution.

23. $y = 3x^2 - 1$

x	$y = 3x^2 - 1$	y
-2	$y = 3(-2)^2 - 1$	11
-1	$y = 3(-1)^2 - 1$	2
0	$y = 3(0)^2 - 1$	-1
1	$y = 3(1)^2 - 1$	2
2	$y = 3(2)^2 - 1$	11

    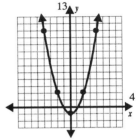

24. The domain is all real numbers.
    The range is all real numbers ≥ -1.

25. This relation is a function since it passes the vertical line test. Each element in the domain corresponds to exactly one element in the range.

26. The relation is increasing for $x > 0$.

27. The minimum of the relation is $y = -1$. There is no maximum.

28. $f(x) = 3x + 7$

    a. $f(-4) = 3(-4) + 7 = -12 + 7 = -5$

    b. $f(-4+h) = 3(-4+h) + 7$
       $= -12 + 3h + 7 = 3h - 5$

29. $5x - 3 = x - 7$
    $5x - 3 - x = x - 7 - x$
    $4x - 3 = -7$
    $4x - 3 + 3 = -7 + 3$
    $4x = -4$
    $\dfrac{4x}{4} = \dfrac{-4}{4}$
    $x = -1$

30. $2(x-2) + 1 = 3(x+6) + 2(x-7)$
    $2x - 4 + 1 = 3x + 18 + 2x - 14$
    $2x - 3 = 5x + 4$
    $2x - 5x = 4 + 3$
    $-3x = 7$
    $\dfrac{-3x}{-3} = \dfrac{7}{-3}$
    $x = -\dfrac{7}{3}$

31. $(7x+4) - (x+6) = 2(3x-1)$
    $7x + 4 - x - 6 = 6x - 2$
    $6x - 2 = 6x - 2$
    This is an identity. All real numbers are solutions.

32. $4(1.2x + 2) - 0.2(24x + 5) = 6$
    $4.8x + 8 - 4.8x - 1 = 6$
    $7 = 6$
    This is a contradiction. There is no solution.

33. $|5x - 1| = 6$
    $5x - 1 = 6$ or $5x - 1 = -6$
    $5x = 7$ or $5x = -5$
    $x = \dfrac{7}{5}$ or $x = -1$

34. $f(x) = -2x + 6$

x	$f(x) = -2x + 6$	ordered pair
-1	$f(-1) = -2(-1) + 6$	$(-1, 8)$
0	$f(0) = -2(0) + 6$	$(0, 6)$
3	$f(3) = -2(3) + 6$	$(3, 0)$

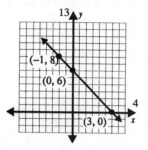

35. $8x + 2y = -4$
    $2y = -8x - 4$
    $y = -4x - 2$

x	$y = -4x - 2$	ordered pair
-2	$y = -4(-2) - 2$	$(-2, 6)$
0	$y = -4(0) - 2$	$(0, -2)$

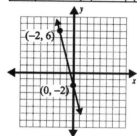

36. $3x - 4 = 5$
    $3x = 9$
    $x = 3$
    Vertical line; x-intercept is $(3, 0)$.

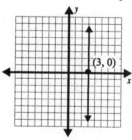

37. $2(y + 4) - 6 = 4$
    $2y + 8 - 6 = 4$
    $2y + 2 = 4$
    $2y = 2$
    $y = 1$

Horizontal line; y-intercept is $(0,1)$

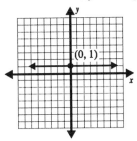

38. $4x+2y=8$
$2y=-4x+8$
$y=-2x+4 \quad m=-2, b=4$
$y=-2x+6 \quad m=-2, b=6$
The lines are parallel because they have the same slope but different y-intercepts.

39. $3x+2y=-2$
$2y=-3x-2$
$y=\dfrac{-3x-2}{2}$
$y=-\dfrac{3}{2}x-1 \quad m=-\dfrac{3}{2}, b=-1$
$-2x+3y=3$
$3y=2x+3$
$y=\dfrac{2x+3}{3}$
$y=\dfrac{2}{3}x+1 \quad m=\dfrac{2}{3}, b=1$
The lines are intersecting and perpendicular because the product of their slopes is -1.

40. $(-2,3)$ and $(-3,-1)$
$m=\dfrac{y_2-y_1}{x_2-x_1}=\dfrac{-1-3}{-3-(-2)}=\dfrac{-4}{-3+2}=\dfrac{-4}{-1}=4$

41. $y=3x-6$
The slope is $m=3$.

42. $x=4$
Vertical line; undefined slope.

43. $(1,-2)$ and $(-3,4)$
$m=\dfrac{y_2-y_1}{x_2-x_1}=\dfrac{4-(-2)}{-3-1}=\dfrac{6}{-4}=-\dfrac{3}{2}$

$y-y_1=m(x-x_1)$
$y-(-2)=-\dfrac{3}{2}(x-1)$
$y+2=-\dfrac{3}{2}x+\dfrac{3}{2}$
$y=-\dfrac{3}{2}x-\dfrac{1}{2}$

44. $4x-y=1$
$-y=-4x+1$
$y=4x-1 \quad m=4, b=-1$
Perpendicular lines have opposite reciprocal slopes. The slope is $-\dfrac{1}{4}$.
$y-y_1=m(x-x_1)$
$y-3=-\dfrac{1}{4}(x-2)$
$y-3=-\dfrac{1}{4}x+\dfrac{1}{2}$
$y=-\dfrac{1}{4}x+\dfrac{7}{2}$

45. $A=\dfrac{x+y+z}{3}$
$3A=x+y+z$
$3A-x-y=z$
$z=3A-x-y$

46. Let x = original price.
$87.49=x-0.3x$
$87.49=0.7x$
$\dfrac{87.49}{0.7}=x$
$124.99=x$
The dress originally cost $124.99.

47. Let x = the amount borrowed.
$1612.50=x+0.075x$
$1612.50=1.075x$
$\dfrac{1612.50}{1.075}=\dfrac{1.075x}{1.075}$
$1500=x$
Caroline borrowed $1500 for school.

48. Let x = number of CDs sold and p = profit.
$$p = 15x - (300 + 2.5x)$$
$$p = 15x - 300 - 2.5x$$
$$p = 12.5x - 300$$

The break-even point is where the profit is equal to 0.
$$0 = 12.5x - 300$$
$$12.5x = 300$$
$$x = \frac{300}{12.5} = 24$$

The producer will need to sell more than 24 CDs to make a profit.
No, 20 is less than 24. He will not make a profit.
Yes, 30 is greater than 24. He will make a profit.

49. a. $(0, 7500)$ and $(5, 9500)$
$$m = \frac{y_2 - y_1}{x_2 - x_1} = \frac{9500 - 7500}{5 - 0} = \frac{2000}{5} = 400$$

b. $y = mx + b$
$y = 400x + 7500$

c. For the year 2000, $x = 6$.
$y = 400(6) + 7500 = 2400 + 7500 = 9900$

The predicted enrollment for the year 2000 is 9900 students.

d. For the year 2010, $x = 16$.
$y = 400(16) + 7500 = 6400 + 7500 = 13900$

The predicted enrollment for the year 2010 is 13900 students. This could be possible for a rapidly growing school, but the constant growth rate for so long may not be realistic.

50. The amount of depreciation each year is the slope of the depreciation line.
$$m = \frac{12500 - 38500}{4 - 0} = \frac{-26000}{4} = -6500$$

The car depreciates $6500 each year for four years.

Chapter 6

6.1 Exercises

1. $y = x+1$
    $(4) = (3)+1$
    $4 = 4$ true

    $2y = x+5$
    $2(4) = (3)+5$
    $8 = 8$ true
    Since both equations are true, $(3,4)$ is a solution of the system.

3. $y = \dfrac{1}{4}x+1$
    $\left(\dfrac{6}{5}\right) = \dfrac{1}{4}\left(\dfrac{4}{5}\right)+1$
    $\dfrac{6}{5} = \dfrac{1}{5}+1$
    $\dfrac{6}{5} = \dfrac{6}{5}$ true

    $y = \dfrac{1}{2}x + \dfrac{4}{5}$
    $\left(\dfrac{6}{5}\right) = \dfrac{1}{2}\left(\dfrac{4}{5}\right)+\dfrac{4}{5}$
    $\dfrac{6}{5} = \dfrac{2}{5}+\dfrac{4}{5}$
    $\dfrac{6}{5} = \dfrac{6}{5}$ true

    Since both equations are true, $\left(\dfrac{4}{5}, \dfrac{6}{5}\right)$ is a solution of the system.

5. $3y = x+6$
    $3(2.1) = (0.3)+6$
    $6.3 = 6.3$ true

    $10y = 25$
    $10(2.1) = 25$
    $21 = 25$ false
    Since the second equation is false, $(0.3, 2.1)$ is not a solution of the system.

7. $7y+2 = 7$
    $7\left(\dfrac{5}{7}\right)+2 = 7$
    $5+2 = 7$
    $7 = 7$ true

    $5y+3 = 0$
    $5\left(\dfrac{5}{7}\right)+3 = 0$
    $\dfrac{25}{7}+3 = 0$
    $\dfrac{46}{7} = 0$ false
    Since the second equation is false, $\left(\dfrac{2}{3}, \dfrac{5}{7}\right)$ is not a solution of the system.

9. $2y = -x$
    $2(-2) = -(5)$
    $-4 = -5$ false

    $x = -4$
    $(5) = -4$ false
    Since both equations are false, $(5,-2)$ is not a solution to the system.

SSM: Experiencing Introductory and Intermediate Algebra

11. $2x+3y=-6$ $x-4y=8$
    $3y=-2x-6$ $-4y=-x+8$
    $y=-\dfrac{2}{3}x-2$ $y=\dfrac{1}{4}x-2$

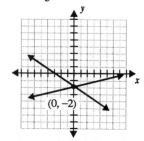

The solution is $(0,-2)$.

13. $a+b=4$ $a-2b=7$
    $b=-a+4$ $-2b=-a+7$
    $b=\dfrac{1}{2}a-\dfrac{7}{2}$

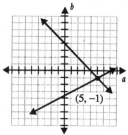

The solution is $(5,-1)$.

15. $x+2y=6$ $x+2y=2$
    $2y=-x+6$ $2y=-x+2$
    $y=-\dfrac{1}{2}x+3$ $y=-\dfrac{1}{2}x+1$

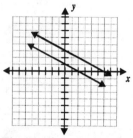

The lines are parallel. There is no solution to the system.

17. $x-y=-1$ $3x-3y=-3$
    $-y=-x-1$ $-3y=-3x-3$
    $y=x+1$ $y=x+1$

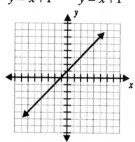

The lines are coinciding. There are an infinite number of solutions to the system. All ordered pairs (x,y) satisfying $y=x+1$ are solutions.

19. $y=\dfrac{1}{2}x+3$ $y=-\dfrac{3}{2}x+9$

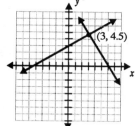

The solution is $(3,4.5)$.

21. $y=3x-1$ $y=-2x+4$

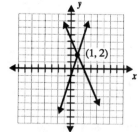

The solution is $(1,2)$.

23. $y = 2x+2 \qquad y = 2x-3$

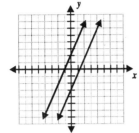

The lines are parallel. There is no solution to the system.

25. $y = \dfrac{1}{2}x+2 \qquad 6y = 3x+12$
$\qquad\qquad\qquad\qquad y = \dfrac{1}{2}x+2$

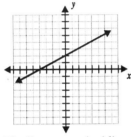

The lines are coinciding. There are an infinite number of solutions to the system. All ordered pairs (x,y) satisfying $y = \dfrac{1}{2}x+2$ are solutions.

27. $3x+2y=12 \qquad y-2=1$
$\quad\ 2y = -3x+12 \qquad y = 3$
$\quad\ y = -\dfrac{3}{2}x+6$

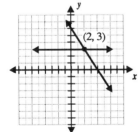

The solution is $(2,3)$.

29. $3y-2 = 2y+2 \qquad x-1=0$
$\quad\ y = 4 \qquad\qquad\qquad x = 1$

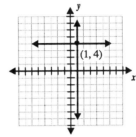

The solution is $(1,4)$.

31. $y-1=0 \qquad y+3=0$
$\quad\ y=1 \qquad\qquad y=-3$

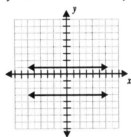

The lines are parallel. There is no solution to the system.

33. $y = 3x-1 \qquad 6x = -5-12y$
$\qquad\qquad\qquad 12y = -6x-5$
$\qquad\qquad\qquad y = -\dfrac{1}{2}x - \dfrac{5}{12}$

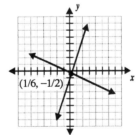

The solution is $\left(\dfrac{1}{6}, -\dfrac{1}{2}\right)$.

35. $y = x - 7$ $x = y + 7$
              $y = x - 7$

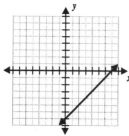

The lines are coinciding. There is no solution to the system.

37. $y = 1 + 4x$ $y = 4x - 2$
     $y = 4x + 1$

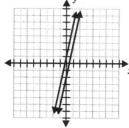

The lines are parallel. There is no solution to the system.

39. a. Case 1: Less than 200 miles

Turtle Rental: $y = 39.95$
Snail Rental: $y = 79.95$

Case 2: 200 or more miles
Let x = number of miles over 200
Turtle Rental: $y = 39.95 + 0.25x$
Snail Rental: $y = 79.95$

b. Case 1: Less than 200 miles

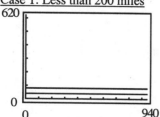

The lines are parallel. There is no solution.

Case 2: 200 or more miles

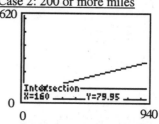

The solution is $(160, 79.95)$

c. If Ahmet drives less than 360 miles $(200 + 160)$, then Turtle Rental offers the better deal. If he drives more than 360 miles, Snail Rental has the better offer.

d. Turtle Rental is better for 300 miles.

e. Snail Rental is better for 600 miles.

41. a. Pre-sorted: $y = 125 + 0.23x$
     Regular: $y = 0.34x$

b.

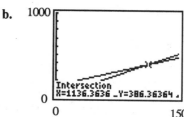

The solution is roughly $(1136.36, 386.36)$.

c. It is more cost effective to use the bulk-rate if there are at least 1137 pieces of mail. The graph representing the pre-sorted bulk-rate is lower than the graph representing first-class for values of x that are at least 1137.

d. It is more cost effective to use the first-class rate if there are no more than 1136 pieces of mail. The graph representing first-class is lower than the graph representing bulk-rate for values of x that are no greater than 1136.

43. Let x = car payment
    y = rent payment
    $x + y = 500$
    $y = 3x$

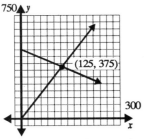

The solution is $(125, 375)$. Her car payment is $125 and her rent payment is $375.

45. Let x = hours at $7
y = hours at $8.20
$x + y = 40$
$7x + 8.2y = 298$

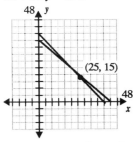

The solution is $(25, 15)$. Jenny should work 25 hours at $7 per hour and 15 hours at $8.20 per hour.

47. Let x = width
y = length
$y = 2x - 25$
$2x + 2y = 550$

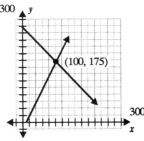

The solution is $(100, 175)$. The room is 100 feet wide and 175 feet long.

49. Let x = num. of hits for Reba
y = num. of hits for Shania
$x + y = 22$
$x = 2y + 1$

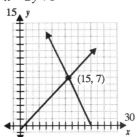

The solution is $(15, 7)$. Reba had 15 Top Ten hits and Shania had 7 Top Ten hits.

51. $m(x) = 0.056x + 3.354$
$f(x) = 0.103x + 1.842$

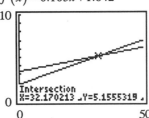

Equal participation by males and females will occur about 32.17 years after 1990 (that is, in early 2022).

6.1 Calculator Exercises

1. $5x + 2y = 5$ $x + 7 = 10$
$2y = -5x + 5$ $x = 3$
$y = -\dfrac{5}{2}x + \dfrac{5}{2}$

a. TRACE

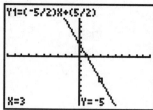

The solution is $(3,-5)$.

b. APPS

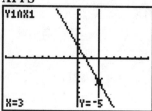

The solution is $(3,-5)$.

2. $5x - y = -19$ $5x + 20 = 13$
$y = 5x + 19$ $5x = -7$
$x = -\dfrac{7}{5}$

a. TRACE

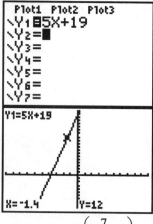

The solution is $\left(-\dfrac{7}{5}, 12\right)$.

b. APPS

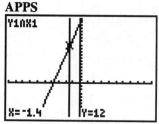

The solution is $\left(-\dfrac{7}{5}, 12\right)$.

3. $\dfrac{1}{2}x + \dfrac{1}{3}y = 7$ $2x - 7 = x + 1$
$y = -\dfrac{3}{2}x + 21$ $x = 8$

a. TRACE

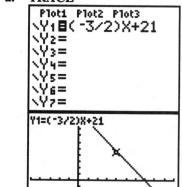

The solution is $(8, 9)$.

b. APPS

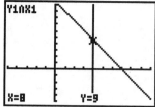

The solution is $(8, 9)$.

4. $x+5=5x-4$ $\quad$ $3y+5=y-2$

    $9=4x$ $\qquad\qquad$ $2y=-7$

    $x=\dfrac{9}{4}$ $\qquad\quad$ $y=-\dfrac{7}{2}$

    a. TRACE

    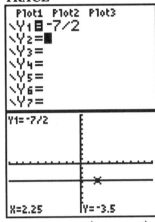

    The solution is $(2.25,-3.5)$.

    b. APPS

    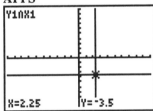

    The solution is $(2.25,-3.5)$.

6.2 Exercises

1. $x-2y=26$

    $5x+10y=-10$

    Solve the first equation for x.

    $x=2y+26$

    Substitute this result for x in the second equation.

    $5(2y+26)+10y=-10$

    $10y+130+10y=-10$

    $20y+130=-10$

    $20y=-140$

    $y=-7$

    Substitute this value for y in the first equation.

    $x-2(-7)=26$

    $x+14=26$

    $x=12$

    The solution is $(12,-7)$.

3. $y=3x-15$

    $x-5=0$

    Solve the second equation for x.

    $x=5$

    Substitute this value for x in the first equation.

    $y=3(5)-15$

    $y=15-15=0$

    The solution is $(5,0)$.

5. $x-5y=20$

    $2y+3=-7$

    Solve the second equation for y.

    $2y=-10$

    $y=-5$

    Substitute this value for y in the first equation.

    $x-5(-5)=20$

    $x+25=20$

    $x=-5$

    The solution is $(-5,-5)$.

7. $3y=x+6$

    $x-3y=9$

    Solve the second equation for x.

    $x=3y+9$

    Substitute this result for x in the first equation.

    $3y=(3y+9)+6$

    $3y=3y+15$

    $0=15$

    This is a contradiction. The system has no solution.

9. $y=x-3$

    $2x-y=19$

    Substitute $x-3$ for y in the second equation.

    $2x-(x-3)=19$

    $2x-x+3=19$

    $x+3=19$

    $x=16$

SSM: Experiencing Introductory and Intermediate Algebra

Substitute this value for x in the first equation.
$y = (16) - 3 = 13$
The solution is $(16, 13)$.

11. $y = x + 1$
$5x - 10y = 3$
Substitute $x + 1$ for y in the second equation.
$5x - 10(x + 1) = 3$
$5x - 10x - 10 = 3$
$-5x = 13$
$x = -\dfrac{13}{5}$
Substitute this value for x in the first equation.
$y = \left(-\dfrac{13}{5}\right) + 1 = -\dfrac{8}{5}$
The solution is $\left(-\dfrac{13}{5}, -\dfrac{8}{5}\right)$.

13. $4y = x + 4$
$5y = 2x + 11$
Solve the first equation for x.
$x = 4y - 4$
Substitute this result for x in the second equation.
$5y = 2(4y - 4) + 11$
$5y = 8y - 8 + 11$
$5y = 8y + 3$
$-3y = 3$
$y = -1$
Substitute this value for y in the first equation.
$4(-1) = x + 4$
$-4 = x + 4$
$x = -8$
The solution is $(-8, -1)$.

15. $3y = x + 6$
$10y = 25$
Solve the second equation for y.
$y = \dfrac{5}{2}$
Substitute this value for y in the first equation.

$3\left(\dfrac{5}{2}\right) = x + 6$
$\dfrac{15}{2} = x + 6$
$x = \dfrac{3}{2}$
The solution is $\left(\dfrac{3}{2}, \dfrac{5}{2}\right)$.

17. $y = -3x + 2$
$2y + 2x = 2 + y - x$
Substitute $-3x + 2$ for y in the second equation.
$2(-3x + 2) + 2x = 2 + (-3x + 2) - x$
$-6x + 4 + 2x = 2 - 3x + 2 - x$
$-4x + 4 = -4x + 4$
$4 = 4$
This is an identity. There are an infinite number of solutions to the system. All ordered pairs satisfying $y = -3x + 2$ are solutions.

19. $2x + y = -3$
$y = -0.5x + 3$
Substitute $-0.5x + 3$ for y in the first equation.
$2x + (-0.5x + 3) = -3$
$2x - 0.5x + 3 = -3$
$1.5x + 3 = -3$
$1.5x = -6$
$x = -4$
Substitute this value for x in the second equation.
$y = -0.5(-4) + 3$
$y = 2 + 3 = 5$
The solution is $(-4, 5)$.

21. $5x - 3y = -13$
$4x + y = 27$
Solve the second equation for y.
$y = -4x + 27$
Substitute this result for y in the first equation.

Chapter 6: Systems of Linear Equations

$5x - 3(-4x + 27) = -13$
$5x + 12x - 81 = -13$
$17x - 81 = -13$
$17x = 68$
$x = 4$
Substitute this value for x in the second equation.
$4(4) + y = 27$
$16 + y = 27$
$y = 11$
The solution is $(4, 11)$.

23. $x + 2y = -28$
$3x + y = -9$
Solve the first equation for x.
$x = -2y - 28$
Substitute this result for x in the second equation.
$3(-2y - 28) + y = -9$
$-6y - 84 + y = -9$
$-5y = 75$
$y = -15$
Substitute this value for y in the first equation.
$x + 2(-15) = -28$
$x - 30 = -28$
$x = 2$
The solution is $(2, -15)$.

25. $x - y = -1$
$3x - 3y = -3$
Solve the first equation for x.
$x = y - 1$
Substitute this result for x in the second equation.
$3(y - 1) - 3y = -3$
$3y - 3 - 3y = -3$
$-3 = -3$
This is an identity. There are an infinite number of solutions to the system. All ordered pairs satisfying $y = x + 1$ are solutions.

27. $y = 3x - 1$
$2x + y = 4$
Substitute $3x - 1$ for y in the second equation.
$2x + (3x - 1) = 4$
$5x - 1 = 4$
$5x = 5$
$x = 1$
Substitute this value for x in the first equation.
$y = 3(1) - 1 = 3 - 1 = 2$
The solution is $(1, 2)$.

29. $y = \frac{1}{2}x + 3$
$3x + 2y = -2$
Substitute $\frac{1}{2}x + 3$ for y in the second equation.
$3x + 2\left(\frac{1}{2}x + 3\right) = -2$
$3x + x + 6 = -2$
$4x + 6 = -2$
$4x = -8$
$x = -2$
Substitute this value for x in the first equation.
$y = \frac{1}{2}(-2) + 3 = -1 + 3 = 2$
The solution is $(-2, 2)$.

31. $y = \frac{1}{2}x + 2$
$6y = 3x + 12$
Substitute $\frac{1}{2}x + 2$ for y in the second equation.
$6\left(\frac{1}{2}x + 2\right) = 3x + 12$
$3x + 12 = 3x + 12$
$12 = 12$
This is an identity. There is an infinite number of solutions to the system. All ordered pairs satisfying $y = \frac{1}{2}x + 2$ are solutions.

SSM: Experiencing Introductory and Intermediate Algebra

33. $y = 2x + 2$
$2x - y = 3$
Substitute $2x + 2$ for y in the second equation.
$2x - (2x + 2) = 3$
$2x - 2x - 2 = 3$
$-2 = 3$
This is a contradiction. The system has no solution.

35. $3x + 2y = 12$
$y - 2 = 1$
Solve the second equation for y.
$y = 3$
Substitute this value in for y in the first equation.
$3x + 2(3) = 12$
$3x + 6 = 12$
$3x = 6$
$x = 2$
The solution is $(2, 3)$.

37. $3x - y = 5$
$x + 2y = 4$
Solve the second equation for x.
$x = -2y + 4$
Substitute this result for x in the first equation.
$3(-2y + 4) - y = 5$
$-6y + 12 - y = 5$
$-7y + 12 = 5$
$-7y = -7$
$y = 1$
Substitute this value for y in the second equation.
$x + 2(1) = 4$
$x + 2 = 4$
$x = 2$
The solution is $(2, 1)$.

39. $y = x - 7$
$x = y + 7$
Substitute $x - 7$ for y in the second equation.
$x = (x - 7) + 7$
$x = x - 7 + 7$
$x = x$

This is an identity. There are an infinite number of solutions to the system. All ordered pairs satisfying $y = x - 7$ are solutions.

41. $y = 1 + 4x$
$4x - y = 2$
Substitute $1 + 4x$ for y in the second equation.
$4x - (1 + 4x) = 2$
$4x - 1 - 4x = 2$
$-1 = 2$
This is a contradiction. The system has no solution.

43. $5x + 7y = 35$
$2x - 5y = 53$
Solve the second equation for x.
$2x = 5y + 53$
$x = \frac{5}{2}y + \frac{53}{2}$
Substitute this result for x in the first equation.
$5\left(\frac{5}{2}y + \frac{53}{2}\right) + 7y = 35$
$\frac{25}{2}y + \frac{265}{2} + 7y = 35$
$\frac{39}{2}y + \frac{265}{2} = 35$
$\frac{39}{2}y = -\frac{195}{2}$
$y = -5$
Substitute this value for y in the second equation.
$2x - 5(-5) = 53$
$2x + 25 = 53$
$2x = 28$
$x = 14$
The solution is $(14, -5)$.

45. Let x = smaller angle
y = larger angle
$x + y = 90$
$y = 4x - 10$
Substitute $4x - 10$ for y in the first equation.

$x+(4x-10)=90$
$x+4x-10=90$
$5x-10=90$
$5x=100$
$x=20$
Substitute this value for x in the second equation.
$y=4(20)-10=80-10=70$
The smaller angle measures $20°$ and the larger angle measures $70°$.

47. Let x = measure of equal angles
y = measure of other angle
$2x+y=180$
$y=2x+20$
Substitute $2x+20$ for y in the first equation.
$2x+(2x+20)=180$
$2x+2x+20=180$
$4x+20=180$
$4x=160$
$x=40$
Substitute this value for x in the second equation.
$y=2(40)+20=80+20=100$
The angles measure $40°, 40°,$ and $100°$.

49. Let r = radius of small circle
R = radius of large circle
$R=2r+5$
$2\pi R=283$
Solve the second equation for R.
$R=\dfrac{283}{2\pi}\approx 45$
Substitute this value for R in the first equation.
$\dfrac{283}{2\pi}=2r+5$
$2r=\dfrac{283}{2\pi}-5$
$r=\dfrac{283}{2(2\pi)}-\dfrac{5}{2}=\dfrac{283}{4\pi}-\dfrac{5}{2}\approx 20$
The smaller radius is about 20 inches and the larger radius is about 45 inches.

51. Let x = acres planted w/ corn
y = acres planted w/ soybean
$x+y=15$
$x=25-y$
Substitute $25-y$ for x in the first equation.
$(25-y)+y=15$
$25-y+y=15$
$25=15$
This is a contradiction. There is no solution to the system. No combination of corn and soybean will work for this problem.

53. a. Let x = number of ovens
$c(x)=22x+2500$

b. Let x = number of ovens
$r(x)=49x$

c. $y=22x+2500$
$y=49x$
Substitute $49x$ for y in the first equation.
$49x=22x+2500$
$27x=2500$
$x=\dfrac{2500}{27}\approx 92.59$
Substitute this value for x in the second equation.
$y=49(92.59)=\$4536.91$
The Homestore must sell about 92.59 ovens to break even. (The store will lose money if it sells 92 or fewer and it will make money if it sells 93 or more.)

55. $M(x)=1.09x+28.9$
$C(x)=0.78x+31.8$

$y=1.09x+28.9$
$y=0.78x+31.8$
Substitute $0.78x+31.8$ for y in the first equation.
$0.78x+31.8=1.09x+28.9$
$2.9=0.31x$
$x=\dfrac{2.9}{0.31}\approx 9.35$
The two net stock values will be equal 9.35 years

after 1990 (that is, in 1999). This does not match the data given. The data suggest that the values get closer but the coal-mining stock value is always higher.

57. Let x = amt. from individuals
y = amt. from other sources
$x + y = 1.9563 \times 10^{12}$
$y = x + 5.31 \times 10^{10}$
Substitute $x + 5.31 \times 10^{10}$ for y in the first equation.
$x + x + 5.31 \times 10^{10} = 1.9563 \times 10^{12}$
$2x + 5.31 \times 10^{10} = 1.9563 \times 10^{12}$
$2x = 1.9032 \times 10^{12}$
$x = 9.516 \times 10^{11}$
Substitute this value for x in the second equation.
$y = 9.516 \times 10^{11} + 5.31 \times 10^{10} = 1.0047 \times 10^{12}$
Revenue generated from individual income taxes amounted to about 9.516×10^{11} dollars while revenue generated from other sources amounted to about 1.0047×10^{12} dollars.

6.2 Calculator Exercises

1. $15.80x + y = 2655.10$
$18.40x + 73.20y = 19361.22$
Solve the first equation for y.
$y = 2655.10 - 15.80x$
Substitute this result for y in the second equation.
$18.40x + 73.20(2655.10 - 15.80x) = 19361.22$
$18.40x + 194353.32 - 1156.56x = 19361.22$
$-1138.16x + 194353.32 = 19361.22$
$-1138.16x = -174992.10$
$x = 153.75$
Substitute this value for x in the first equation.
$15.80(153.75) + y = 2655.10$
$2429.25 + y = 2655.10$
$y = 225.85$
The solution is $(153.75, 225.85)$.

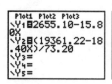

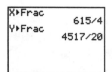

2. $4.055x - 8.752y = -42.949$
$x + 0.405y = 2.225$
Solve the second equation for x.
$x = 2.225 - 0.405y$
Substitute this result for x in the first equation.
$4.055(2.225 - 0.405y) - 8.752y = -42.949$
$9.022375 - 1.642275y - 8.752y = -42.949$
$9.022375 - 10.394275y = -42.949$
$-10.394275y = -51.971375$
$y = 5$
Substitute this value for y in the second equation.
$x + 0.405(5) = 2.225$
$x + 2.025 = 2.225$
$x = 0.2$
The solution is $(0.2, 5)$.

 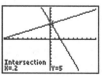

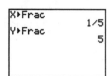

Chapter 6: Systems of Linear Equations

3. $\dfrac{7}{13}x + \dfrac{5}{17}y = \dfrac{2}{3}$

$x - \dfrac{5}{34}y = \dfrac{19}{42}$

Solve the second equation for x.

$x = \dfrac{5}{34}y + \dfrac{19}{42}$

Substitute this result for x in the first equation.

$\dfrac{7}{13}\left(\dfrac{5}{34}y + \dfrac{19}{42}\right) + \dfrac{5}{17}y = \dfrac{2}{3}$

$\dfrac{35}{442}y + \dfrac{19}{78} + \dfrac{5}{17}y = \dfrac{2}{3}$

$\dfrac{165}{442}y + \dfrac{19}{78} = \dfrac{2}{3}$

$\dfrac{165}{442}y = \dfrac{11}{26}$

$y = \dfrac{17}{15}$

Substitute this value for y in the second equation.

$x - \dfrac{5}{34}\left(\dfrac{17}{15}\right) = \dfrac{19}{42}$

$x - \dfrac{1}{6} = \dfrac{19}{42}$

$x = \dfrac{26}{42} = \dfrac{13}{21}$

The solution is $\left(\dfrac{13}{21}, \dfrac{17}{15}\right)$.

4. $\dfrac{28}{65}x + \dfrac{51}{56}y = \dfrac{43}{35}$

$\dfrac{35}{39}x - y = \dfrac{61}{51}$

Solve the second equation for y.

$y = \dfrac{35}{39}x - \dfrac{61}{51}$

Substitute this result for y in the first equation.

$\dfrac{28}{65}x + \dfrac{51}{56}\left(\dfrac{35}{39}x - \dfrac{61}{51}\right) = \dfrac{43}{35}$

$\dfrac{28}{65}x + \dfrac{85}{104}x - \dfrac{61}{56} = \dfrac{43}{35}$

$\dfrac{649}{520}x - \dfrac{61}{56} = \dfrac{43}{35}$

$\dfrac{649}{520}x = \dfrac{649}{280}$

$x = \dfrac{13}{7}$

Substitute this value for x in the second equation.

$\dfrac{35}{39}\left(\dfrac{13}{7}\right) - y = \dfrac{61}{51}$

$\dfrac{5}{3} - y = \dfrac{61}{51}$

$y = \dfrac{5}{3} - \dfrac{61}{51} = \dfrac{8}{17}$

The solution is $\left(\dfrac{13}{7}, \dfrac{8}{17}\right)$.

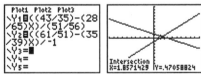

SSM: Experiencing Introductory and Intermediate Algebra

6.3 Exercises

1. $2x - y = -6$
   $5x + y = -8$
   $\overline{7x = -14}$
   $x = -2$
   Substitute this value for x in the second equation.
   $5(-2) + y = -8$
   $-10 + y = -8$
   $y = 2$
   The solution is $(-2, 2)$.

3. $x + 7y = 19$
   $2y = x - 1$
   Write the second equation in standard form and add the two equations.
   $x + 7y = 19$
   $\underline{-x + 2y = -1}$
   $ 9y = 18$
   $ y = 2$
   Substitute this value for y in the first equation.
   $x + 7(2) = 19$
   $x + 14 = 19$
   $x = 5$
   The solution is $(5, 2)$.

5. $5x + y = -24$
   $3x - 2y = 9$
   Multiply the first equation by 2 and add to the second equation.
   $10x + 2y = -48$
   $\underline{3x - 2y = 9}$
   $13x = -39$
   $ x = -3$
   Substitute this value for x in the first equation.
   $10(-3) + 2y = -48$
   $-30 + 2y = -48$
   $2y = -18$
   $y = -9$
   The solution is $(-3, -9)$.

7. $x + 3y = 2$
   $x + 5y = -2$
   Multiply the second equation by -1 and add to the first equation.
   $x + 3y = 2$
   $\underline{-x - 5y = 2}$
   $-2y = 4$
   $y = -2$
   Substitute this value for y in the first equation.
   $x + 3(-2) = 2$
   $x - 6 = 2$
   $x = 8$
   The solution is $(8, -2)$.

9. $2x + 7y = 29$
   $4x + 3y = 25$
   Multiply the first equation by -2 and add to the second equation.
   $-4x - 14y = -58$
   $\underline{4x + 3y = 25}$
   $-11y = -33$
   $y = 3$
   Substitute this value for y in the first equation.
   $2x + 7(3) = 29$
   $2x + 21 = 29$
   $2x = 8$
   $x = 4$
   The solution is $(4, 3)$.

11. $3x - 5y = 66$
    $4x + 3y = 1$
    Multiply the first equation by 4 and the second equation by -3. Add the new equations.
    $12x - 20y = 264$
    $\underline{-12x - 9y = -3}$
    $-29y = 261$
    $y = -9$
    Substitute this value for y in the second equation.

Chapter 6: Systems of Linear Equations

$4x+3(-9)=1$

$4x-27=1$

$4x=28$

$x=7$

The solution is $(7,-9)$.

13. $2x+9y=102$

    $5x-11y=-147$

    Multiply the first equation by -5 and the second equation by 2. Add the new equations.

    $-10x-45y=-510$

    $\underline{10x-22y=-294}$

    $-67y=-804$

    $y=12$

    Substitute this value for y in the first equation.

    $2x+9(12)=102$

    $2x+108=102$

    $2x=-6$

    $x=-3$

    The solution is $(-3,12)$.

15. $40x=23+10y$

    $50x+10y=94$

    Write the first equation in standard form and add to the second equation.

    $40x-10y=23$

    $\underline{50x+10y=94}$

    $90x=117$

    $x=\dfrac{13}{10}$

    Substitute this value for x in the second equation.

    $50\left(\dfrac{13}{10}\right)+10y=94$

    $65+10y=94$

    $10y=29$

    $y=\dfrac{29}{10}$

    The solution is $\left(\dfrac{13}{10},\dfrac{29}{10}\right)$.

17. $5x+40y=77$

    $5x+15y=17$

    Multiply the first equation by -1 and add to the second equation.

    $-5x-40y=-77$

    $\underline{5x+15y=17}$

    $-25y=-60$

    $y=\dfrac{12}{5}$

    Substitute this value for y in the second equation.

    $5x+15\left(\dfrac{12}{5}\right)=17$

    $5x+36=17$

    $5x=-19$

    $x=-\dfrac{19}{5}$

    The solution is $\left(-\dfrac{19}{5},\dfrac{12}{5}\right)$.

19. $10x-4y=28$

    $\underline{5x+4y=35}$

    $15x=63$

    $x=\dfrac{21}{5}$

    Substitute this value for x in the second equation.

    $5\left(\dfrac{21}{5}\right)+4y=35$

    $21+4y=35$

    $4y=14$

    $y=\dfrac{7}{2}$

    The solution is $\left(\dfrac{21}{5},\dfrac{7}{2}\right)$.

21. $2x+8y=29$

    $13y=3x+39$

    Write the second equation in standard form.

    $2x+8y=29$

    $-3x+13y=39$

    Multiply the first equation by 3 and the second equation by 2. Add the new equations.

SSM: Experiencing Introductory and Intermediate Algebra

$6x + 24y = 87$
$-6x + 26y = 78$
$\overline{50y = 165}$
$y = \dfrac{33}{10}$

Substitute this value for y in the first equation.
$2x + 8\left(\dfrac{33}{10}\right) = 29$
$2x + \dfrac{132}{5} = 29$
$2x = 29 - \dfrac{132}{5}$
$2x = \dfrac{13}{5}$
$x = \dfrac{13}{10}$

The solution is $\left(\dfrac{13}{10}, \dfrac{33}{10}\right)$.

23. $\dfrac{1}{4}x + \dfrac{1}{3}y = \dfrac{5}{12}$
$\dfrac{1}{4}x = \dfrac{1}{3}y - \dfrac{1}{12}$

Multiply both equations by 12 and write the second equation in standard form.
$3x + 4y = 5$
$3x - 4y = -1$

Add the equations.
$3x + 4y = 5$
$3x - 4y = -1$
$\overline{6x = 4}$
$x = \dfrac{2}{3}$

Substitute this value for x in the first equation.
$\dfrac{1}{4}\left(\dfrac{2}{3}\right) + \dfrac{1}{3}y = \dfrac{5}{12}$
$\dfrac{2}{12} + \dfrac{1}{3}y = \dfrac{5}{12}$

$\dfrac{1}{3}y = \dfrac{3}{12}$
$\dfrac{1}{3}y = \dfrac{1}{4}$
$y = \dfrac{3}{4}$

The solution is $\left(\dfrac{2}{3}, \dfrac{3}{4}\right)$.

25. $\dfrac{1}{3}x + \dfrac{1}{2}y = -\dfrac{1}{4}$
$\dfrac{1}{6}x - \dfrac{5}{6}y = \dfrac{11}{16}$

Multiply the first equation by -24 and the second equation by 48. Add the new equations.
$-8x - 12y = 6$
$8x - 40y = 33$
$\overline{-52y = 39}$
$y = -\dfrac{39}{52}$
$y = -\dfrac{3}{4}$

Substitute this value for y in the first equation.
$\dfrac{1}{3}x + \dfrac{1}{2}\left(-\dfrac{3}{4}\right) = -\dfrac{1}{4}$
$\dfrac{1}{3}x - \dfrac{3}{8} = -\dfrac{1}{4}$
$\dfrac{1}{3}x = \dfrac{1}{8}$
$x = \dfrac{3}{8}$

The solution is $\left(\dfrac{3}{8}, -\dfrac{3}{4}\right)$.

27. $y = \dfrac{3}{2}x - 9$
$\dfrac{1}{6}x - \dfrac{1}{9}y = 1$

Multiply the first equation by 2 and write in standard form. Multiply the second equation by 18. Add the new equations.

226

$-3x + 2y = -18$
$\underline{3x - 2y = 18}$
$0 = 0$
This is an identity. There are an infinite number of solutions. All ordered pairs (x, y) satisfying $y = \dfrac{3}{2}x - 9$ are solutions.

29. $\dfrac{4}{5}x - \dfrac{3}{5}y = -1$
$\dfrac{3}{2}x - \dfrac{9}{8}y = 1$
Multiply the first equation by -15 and the second equation by 8. Add the new equations.
$-12x + 9y = 15$
$\underline{12x - 9y = 8}$
$0 = 23$
This is a contradiction. There is no solution.

31. $x - 20y = 70$
$3x + 10y = 70$
Multiply the second equation by 2 and add to the first equation.
$x - 20y = 70$
$\underline{6x + 20y = 140}$
$7x \quad\quad = 210$
$x = 30$
Substitute this value for x in the second equation.
$6(30) + 20y = 140$
$180 + 20y = 140$
$20y = -40$
$y = -2$
The solution is $(30, -2)$.

33. $3x + y = 40$
$x + y = 20$
Multiply the second equation by -1 and add to the first equation.
$3x + y = 40$
$\underline{-x - y = -20}$
$2x \quad\quad = 20$
$x = 10$
Substitute this value for x in the second equation.

$10 + y = 20$
$y = 10$
The solution is $(10, 10)$.

35. $x - y = 300$
$2x - y = -100$
Multiply the first equation by -1 and add to the second.
$-x + y = -300$
$\underline{2x - y = -100}$
$x = -400$
Substitute this value for x in the first equation.
$(-400) - y = 300$
$-y = 700$
$y = -700$
The solution is $(-400, -700)$.

37. $10x - 10y = 22$
$y = 2x - 11$
Multiply the second equation by 5 and write it in standard form. Then add to the first equation.
$10x - 10y = 22$
$\underline{-10x + 5y = -55}$
$-5y = -33$
$y = \dfrac{33}{5}$
Substitute this value for y in the first equation.
$10x - 10\left(\dfrac{33}{5}\right) = 22$
$10x - 66 = 22$
$10x = 88$
$x = \dfrac{44}{5}$
The solution is $\left(\dfrac{44}{5}, \dfrac{33}{5}\right)$.

39. $3.2x + 4.2y = 368$
$4.4x - 2.1y = 128$
Multiply the second equation by 2 and add to the first equation.

$3.2x + 4.2y = 368$
$8.8x - 4.2y = 256$
_____
$12x \quad\quad = 624$
$\quad\quad x = 52$
Substitute this value for x in the first equation.
$3.2(52) + 4.2y = 368$
$166.4 + 4.2y = 368$
$4.2y = 201.6$
$y = 48$
The solution is $(52, 48)$.

41. $0.05x + 0.10y = 0.75$
$x + y = 11$
Multiply the first equation by -10 and add to the second equation.
$-0.5x - y = -7.5$
$x + y = 11$
_____
$0.5x \quad\quad = 3.5$
$\quad\quad x = 7$
Substitute this value for x in the second equation.
$(7) + y = 11$
$y = 4$
The solution is $(7, 4)$.

43. $0.3x + 0.45y = 43.5$
$x + y = 110$
Multiply the second equation by -0.3 and add to the first equation.
$0.3x + 0.45y = 43.5$
$-0.3x - 0.3y = -33$
_____
$0.15y = 10.5$
$y = 70$
Substitute this value for y in the second equation.
$x + (70) = 110$
$x = 40$
The solution is $(40, 70)$.

45. $12.50x + 6.50y = 1780$
$x + y = 200$
Multiply the second equation by -6.5 and add to the first equation.

$12.50x + 6.50y = 1780$
$-6.50x - 6.50y = -1300$
_____
$6x \quad\quad = 480$
$\quad x = 80$
Substitute this value for x in the second equation.
$(80) + y = 200$
$y = 120$
The solution is $(80, 120)$.

47. Let $x = $ num. of students
$\quad\quad y = $ num. of nonstudents
$x + y = 683$
$1.5x + 5y = 2645$
Multiply the first equation by -5 and add to the second equation.
$-5x - 5y = -3415$
$1.5x + 5y = 2645$
_____
$-3.5x \quad\quad = -770$
$\quad\quad x = 220$
There were 220 students at the game.

49. Let $x = $ num. of cookbooks
$\quad\quad y = $ num. of calendars
$x = y + 50$
$8.5x + 5y = 3462.5$
Multiply the first equation by 5 and write in standard form. Then add it to the second equation.
$5x - 5y = 250$
$8.5x + 5y = 3462.50$
_____
$13.5x \quad\quad = 3712.50$
$\quad\quad x = 275$
Substitute this value for x in the first equation.
$(275) = y + 50$
$y = 225$
The faculty and staff sold 275 cookbooks and 225 calendars.

51. Let $x = $ num. of adults
$\quad\quad y = $ num. of children
$x + y = 385$
$2x + 2y = 770$

Multiply the first equation by –2 and add to the second equation.
$-2x - 2y = -770$
$\underline{2x + 2y = 770}$
$0 = 0$
This is an identity. There is an infinite number of solutions. Any combination of adults, x, and children, y, satisfying $y = 385 - x$ is a solution.

53. Let x = length of smaller field
y = width of smaller field
$2x + 2y = 620$
$2(x + 90) + 2y = 800$
Multiply the first equation by –1 and add to the second equation written in standard form.
$-2x - 2y = -620$
$\underline{2x + 2y = 620}$
$0 = 0$
This is an identity. There is an infinite number of solutions. Any combination of length, x, and width, y, satisfying $y = 310 - x$ is a solution.

55. Let x = angle measure of same angle
y = angle measure of different angle
$2x + y = 180$
$y = 90 - 2x$
Multiply the second equation by –1 and write in standard form. Add the result to the first equation.
$2x + y = 180$
$\underline{-2x - y = -90}$
$0 = 90$
This is a contradiction. There is no solution.

57. Let x = distance from Sun to Mars
y = distance from Sun to Earth
$x - y = 4.864 \times 10^7$
$\dfrac{x+y}{2} = 1.1728 \times 10^8$
Multiply the second equation by 2 and add to the first equation.

$x - y = 4.864 \times 10^7$
$\underline{x + y = 2.3456 \times 10^8}$
$2x = 2.832 \times 10^8$
$x = 1.416 \times 10^8$
Substitute this value for x in the first equation.
$(1.416 \times 10^8) - y = 4.864 \times 10^7$
$y = 9.296 \times 10^7$
The Earth is about 9.296×10^7 miles from the Sun while Mars is about 1.416×10^8 miles from the Sun.

59. Let x = median weekly earnings for males
y = median weekly earnings for females
$x - y = 142$
$\underline{x + y = 1054}$
$2x = 1196$
$x = 598$
Substitute this value for x in the second equation.
$(598) + y = 1054$
$y = 456$
The median weekly earnings for males is $598 while the median weekly earnings for females is $456.

61. Let x = amt. earned by George Lucas
y = amt. earned by Oprah Winfrey
$x + y = 400$
$x = 2y - 50$
Multiply the first equation by 2 and add it to the second equation written in standard form.
$2x + 2y = 800$
$\underline{x - 2y = -50}$
$3x = 750$
$x = 250$
Substitute this value for x in the first equation.
$(250) + y = 400$
$y = 150$
In the year 2000, George Lucas made 250 million dollars and Oprah Winfrey made 150 million dollars.

SSM: Experiencing Introductory and Intermediate Algebra

6.3 Calculator Exercises

1. $6x + 35y = -52$
   $9x - 14y = 55$
   Multiply the first equation by 9 and the second equation by –6.
   $54x + 315y = -468$
   $-54x + 84y = -330$
   $399y = -798$
   $y = -2$
   Substituting into one of the original equations yields $x = 3$. The solution is $(3, -2)$.

2. $12x - 35y = 81$
   $15x - 28y = 54$
   Multiply the first equation by 15 and the second equation by –12. Add the new equations.
   $180x - 525y = 1215$
   $-180x + 336y = -648$
   $-189y = 567$
   $y = -3$
   Substituting into one of the original equations yields $x = -2$. The solution is $(-2, -3)$.

3. $21a + 10b = -88$
   $14a + 15b = 8$
   Multiply the first equation by 14 and the second equation by –21. Add the new equations.
   $294a + 140b = -1232$
   $-294a - 315b = -168$
   $-175b = -1400$
   $b = 8$
   Substituting into one of the original equations yields $a = -8$. The solution is $(-8, 8)$.

4. $33x + 7y = -3.1$
   $6x + 13y = -33.4$
   Multiply the first equation by 6 and the second equation by –33. Add the new equations.
   $198x + 42y = -18.6$
   $-198x - 429y = 1102.2$
   $-387y = 1083.6$
   $y = -2.8$
   Substituting into one of the original equations yields $x = 0.5$. The solution is $(0.5, -2.8)$.

5. $10x + 21y = 19$
   $14x - 9y = 1$
   Multiply the first equation by 14 and the second equation by –10. Add the new equations.
   $140x + 294y = 266$
   $-140x + 90y = -10$
   $384y = 256$
   $y = \dfrac{2}{3}$
   Substituting into one of the original equations yields $x = \dfrac{1}{2}$. The solution is $\left(\dfrac{1}{2}, \dfrac{2}{3}\right)$.

6. $8x - 9y = 8$
   $12x + 21y = -11$
   Multiply the first equation by 12 and the second equation by –8. Add the new equations.
   $96x - 108y = 96$
   $-96x - 168y = 88$
   $-276y = 184$
   $y = -\dfrac{2}{3}$
   Substituting into one of the original equations yields $x = \dfrac{1}{4}$. The solution is $\left(\dfrac{1}{4}, -\dfrac{2}{3}\right)$.

6.4 Exercises

1. Let d = distance Nolan travels
   t = time of travel
   $d = 60t$
   $2200 - d = 65t$
   Substitute $60t$ for d in the second equation.

$220 - 60t = 65t$
$220 = 125t$
$t = \dfrac{44}{25}$

Substitute this value for t in the first equation.
$d = 60\left(\dfrac{44}{25}\right) = \dfrac{528}{5} = 105.6$

They will meet 105.6 miles from Nashville.

3. Let x = hours spent walking
   y = hours spent riding
   $x = 2y$
   $3x + 60y = 11$

   Substitute $2y$ for x in the second equation.
   $3(2y) + 60y = 11$
   $6y + 60y = 11$
   $66y = 11$
   $y = \dfrac{1}{6}$

   Substitute this value for y in the first equation.
   $x = 2\left(\dfrac{1}{6}\right) = \dfrac{1}{3}$

   Kenny walked for 20 minutes (one-third of an hour) and rode for 10 minutes (one-sixth of an hour).

5. Let x = speed in still air
   y = speed of jet stream
   $6(x - y) = 2600$
   $5(x + y) = 2600$

   Multiply the first equation by 5 and the second equation by 6. Put both new equations in standard form and add the equations.
   $30x - 30y = 13000$
   $30x + 30y = 15600$
   $\overline{60x = 28600}$
   $x = \dfrac{1430}{3}$
   $x \approx 476.67$

   Substitute this value for x in the second equation.

$5\left(\dfrac{1430}{3} + y\right) = 2600$
$\dfrac{1430}{3} + y = 520$
$y = \dfrac{130}{3} \approx 43.33$

The planes fly at an average speed of roughly 476.67 miles per hour. The average speed due to the jet stream is roughly 43.33 miles per hour.

7. Let x = speed in still water
   y = speed of current
   $1.5(x - y) = 6$
   $0.75(x + y) = 6$

   Multiply the second equation by 2 and add to the first equation.
   $1.5x - 1.5y = 6$
   $1.5x + 1.5y = 12$
   $\overline{3x = 18}$
   $x = 6$

   Substitute this value for x in the second equation.
   $0.75(6 + y) = 6$
   $6 + y = 8$
   $y = 2$

   The girls can paddle an average of 6 miles per hour in still water. The average speed of the current is 2 miles per hour.

9. Let x = speed of motorboat
   y = speed of tidal current
   $0.25(x + y) = 13.4$
   $0.3(x - y) = 13.4$

   Multiply the first equation by 0.3 and the second equation by .25. Add the new equations.
   $0.075x + 0.075y = 4.02$
   $0.075x - 0.075y = 3.35$
   $\overline{0.15x = 7.37}$
   $x = \dfrac{737}{15}$
   $x \approx 49.13$

   Substitute this value for x in the first equation.

SSM: Experiencing Introductory and Intermediate Algebra

$$0.25\left(\frac{737}{15}+y\right)=13.4$$

$$\frac{737}{15}+y=53.6$$

$$y=\frac{67}{15}\approx 4.47$$

The average speed of Joel's motorboat is about 49.13 miles per hour. The average speed of the tidal current is about 4.47 miles per hour.

11. Let x = pounds of French vanilla coffee
    y = pounds of hazelnut coffee
    $x+y=20$
    $9.5x+7y=8.5(20)$
    Multiply the first equation by -7 and add to the second equation.
    $-7x-7y=-140$
    $\underline{9.5x+7y=170}$
    $2.5x=30$
    $x=12$
    Substitute this value for x in the first equation.
    $12+y=20$
    $y=8$
    Gary should mix 12 pounds of French vanilla coffee and 8 pounds of hazelnut coffee.

13. Let x = num. of azaleas
    y = num. of rhododendrons
    $x+y=30$
    $5x+12y=250$
    Multiply the first equation by -5 and add to the second equation.
    $-5x-5y=-150$
    $\underline{5x+12y=250}$
    $7y=100$
    $y=\frac{100}{7}\approx 14$
    Substitute this value for y in the first equation.
    $x+14=30$
    $x=16$
    Marian can buy 16 azaleas and 14 rhododendrons.

15. Let x = num. of adults
    y = num. of children
    $y=4x$
    $7.5x+4.5y=1938$
    Substitute $4x$ for y in the second equation.
    $7.5x+4.5(4x)=1938$
    $7.5x+18x=1938$
    $25.5x=1938$
    $x=76$
    Substitute this value for x in the first equation.
    $y=4(76)=304$
    There were 76 adults and 304 children at the first screening.

17. Let x = number of coins
    y = number of stamps
    $8x+4y=7(x+y)$
    $y=100$
    Substitute 100 for y in the first equation.
    $8x+4(100)=7(x+100)$
    $8x+400=7x+700$
    $x=300$
    The mixture should contain 100 stamps and 300 half-dollar coins.

19. Let x = hourly wage as waitress
    y = hourly wage as cook
    $15x+20y=320$
    $18x+24y=384$
    Divide the first equation by -5 and the second equation by 6. Add the new equations.
    $-3x-4y=-64$
    $\underline{3x+4y=64}$
    $0=0$
    This is an identity. There are an infinite number of solutions. Any combination of waitress wage, x, and cook wage, y, satisfying $15x+20y=320$ is a solution.

21. Let x = number of \$5 bills
y = number of \$10 bills
$x + y = 65$
$5x + 10y = 365$
Divide the second equation by -5 and add to the first equation.
$x + y = 65$
$-x - 2y = -73$
$\overline{}$
$-y = -8$
$y = 8$
Substitute this value for y in the first equation.
$x + 8 = 65$
$x = 57$
There are 57 \$5 bills and 8 \$10 bills.

23. Let x = gallons of grapefruit blend
y = gallons of orange beverage
$x + y = 200$
$0.45x + 0.75y = 0.55(200)$
Multiply the first equation by -0.45 and add to the second equation.
$-0.45x - 0.45y = -90$
$0.45x + 0.75y = 110$
$\overline{}$
$.3y = 20$
$y = \dfrac{200}{3}$
Substitute this value for y in the first equation.
$x + \dfrac{200}{3} = 200$
$x = \dfrac{400}{3}$
The blend should contain $133\dfrac{1}{3}$ gallons of the grapefruit beverage and $66\dfrac{2}{3}$ gallons of the orange beverage.

25. Let x = liters of 60% solution
y = liters of 35% solution
$x + y = 300$
$0.6x + 0.35y = 0.5(300)$
Multiply the first equation by -0.35 and add to the second equation.
$-0.35x - 0.35y = -105$
$0.6x + 0.35y = 150$
$\overline{}$
$0.25x = 45$
$x = 180$
Substitute this value for x in the first equation.
$180 + y = 300$
$y = 120$
The chemist should mix 180 liters of the 60% solution with 120 liters of the 35% solution to make 300 liters of a 50% solution.

27. Let x = gallons of 4.3% milk
y = gallons of skim milk
$x + y = 200$
$0.043x + 0y = 0.02(200)$
Solve the second equation for x.
$0.043x = 4$
$x \approx 93$
Substitute this value for x in the first equation.
$93 + y = 200$
$y = 107$
The farmer should mix 93 gallons of the 4.3% milk with 107 gallons of skim milk.

29. Let x = gallons drained and replaced
y = gallons remaining
$y = 4 - x$
$x + 0.45y = 0.6(4)$
Substitute $4 - x$ for y in the second equation.
$x + 0.45(4 - x) = 2.4$
$x + 1.8 - 0.45x = 2.4$
$0.55x = 0.6$
$x = \dfrac{12}{11}$
Substitute this for x in the first equation.
$y = 4 - \dfrac{12}{11} = \dfrac{32}{11}$
Mabel should drain $\dfrac{12}{11}$ gallons and replace it with pure antifreeze.

31. Let x = gallons of 12% alcohol wine
$\quad\quad y$ = total gallons of wine
$y = x + 5$
$0.12x + 0.20(5) = 0.15y$
Substitute $x+5$ for y in the second equation.
$0.12x + 1 = 0.15(x+5)$
$0.12x + 1 = 0.15x + 0.75$
$-0.03x = -0.25$
$x = \dfrac{25}{3}$

The wine maker should mix $8\dfrac{1}{3}$ gallons of the 12% alcohol wine with the 5 gallons of 20% alcohol wine.

33. Let x = amt. invested at 8.5%
$\quad\quad y$ = amt. invested at 7%
$x + y = 10,000$
$0.085x + 0.07y = 752.5$
Multiply the first equation by -0.07 and add to the second equation.
$-0.07x - 0.07y = -700$
$\underline{0.085x + 0.07y = 752.5}$
$0.015x \quad\quad\quad = 52.5$
$x = 3500$
Substitute this value for x in the first equation.
$3500 + y = 10,000$
$y = 6500$
Catherine invested $3500 at 8.5% and $6500 at 7%.

35. Let x = amt. in savings account
$\quad\quad y$ = amt. in certificate of deposit
$x + y = 16,500$
$0.05x + 0.0725y = 1000$
Multiply the first equation by -0.05 and add to the second equation.
$-0.05x - 0.05y = -825$
$\underline{0.05x + 0.0725y = 1000}$
$\quad\quad\quad .0225y = 175$
$\quad\quad\quad y \approx 7777.78$
Substitute this value for y in the first equation.

$x + 7777.78 = 16,500$
$x = 8722.22$
Zelda should invest $8722.22 in the savings account and $7777.78 in the certificate of deposit.

37. Let x = amt. in 7.25% account
$\quad\quad y$ = amt. in 6% account
$x = 2y$
$0.0725x + 0.06y = 1230$
Substitute $2y$ for x in the second equation.
$0.0725(2y) + 0.06y = 1230$
$0.145y + 0.06y = 1230$
$0.205y = 1230$
$y = 6000$
Substitute this value for y in the first equation.
$x = 2(6000) = 12000$
Zora should invest $12,000 in the 7.25% certificate account and $6000 in the 6% savings account.

39. Let x = interest rate from first company
$\quad\quad y$ = interest rate from second company
$x = 0.01 + y$
$45,000x + 55,000y = 6450$
Substitute $0.01 + y$ for x in the second equation.
$45,000(0.01 + y) + 55,000y = 6450$
$450 + 45,000y + 55,000y = 6450$
$450 + 100,000y = 6450$
$100,000y = 6000$
$y = 0.06$
Substitute this value for y in the first equation.
$x = 0.01 + 0.06 = 0.07$
The interest rate on the $45,000 loan was 7% and the interest rate on the $55,000 loan was 6%.

6.4 Calculator Exercises

1. Let x = her son's age
$\quad\quad y$ = Stella's age
Currently: $y = 3x - 4$.

5 years ago: $y - 5 =$ Stella's age

$x - 5 =$ her son's age

$y - 5 = 4(x - 5) - 1$

The last equation simplifies to $y = 4x - 16$.

The system is: $y = 3x - 4$

$y = 4x - 16$

```
Plot1 Plot2 Plot3
\Y1■3X-4
\Y2■4X-16
\Y3=
\Y4=
\Y5=
\Y6=
\Y7=
```

X	Y1	Y2
9	23	20
10	26	24
11	29	28
12	32	32
13	35	36
14	38	40
15	41	44

X=12

Stella is 32 years old and her son is 12 years old.

2. Let $x =$ Cecilia's age

$y =$ Gulen's age

In 6 years: $y + 6 = 5(x + 6)$

$y = 5x + 24$

In 13 years: $y + 13 = 3(x + 13)$

$y = 3x + 26$

The system is: $y = 5x + 24$

$y = 3x + 26$

```
Plot1 Plot2 Plot3
\Y1■5X+24
\Y2■3X+26
\Y3=
\Y4=
\Y5=
\Y6=
\Y7=
```

X	Y1	Y2
1	29	29
2	34	32
3	39	35
4	44	38
5	49	41
6	54	44
7	59	47

X=1

Gulen is 29 years old and Cecilia is 1 year old.

3. Let $x =$ Paul's age

$y =$ Joe's age

Currently: $y = x + 25$

In 10 years: $y + 10 = 2(x + 10)$

$y = 2x + 10$

The system is: $y = x + 25$

$y = 2x + 10$

```
Plot1 Plot2 Plot3
\Y1■X+25
\Y2■2X+10
\Y3=
\Y4=
\Y5=
\Y6=
\Y7=
```

X	Y1	Y2
11	36	32
12	37	34
13	38	36
14	39	38
15	40	40
16	41	42
17	42	44

X=15

Joe is 40 years old and Paul is 15 years old. Joe will be 50 years old when he retires.

4. Let $x =$ Jenny's age

$y =$ Katie's age

Currently: $y = x + 2$

In 5 years: $y + 5 = 1.1(x + 5)$

$y = 1.1x + .5$

The system is: $y = x + 2$

$y = 1.1x + .5$

```
Plot1 Plot2 Plot3
\Y1■X+2
\Y2■1.1X+.5
\Y3=
\Y4=
\Y5=
\Y6=
\Y7=
```

X	Y1	Y2
11	13	12.6
12	14	13.7
13	15	14.8
14	16	15.9
15	17	17
16	18	18.1
17	19	19.2

X=15

Jenny is 15 years old and Katie is 17 years old.

SSM: Experiencing Introductory and Intermediate Algebra

5. Let x = grandma's age in 1965
   y = mom's age in 1965

   In 1965: $y = \dfrac{1}{3}x$

   In 1977: $y + 12 = \dfrac{1}{2}(x + 12)$

   $y = \dfrac{1}{2}x - 6$

   The system is: $y = \dfrac{1}{3}x$

   $y = \dfrac{1}{2}x - 6$

   In 1965 her mom was 12 years old and her grandma was 36 years old. In 1977, her mom was 24 years old and her grandma was 48 years old.

6. Let x = age of bed in 1902
   y = age of dresser in 1902

   In 1902: $x + y = 148$

   $y = 148 - x$

   In 1906: $y + 4 = 2(x + 4)$

   $y = 2x + 4$

   The system is: $y = 148 - x$

   $y = 2x + 4$

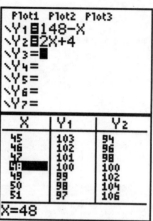

In 1902, the bed was 48 years old and the dresser was 100 years old.

7. Let x = Frac's age
   y = Fric's age

   Currently: $y = 3x$

   In 5 years: $y + 5 = 2(x + 5) + 2$

   $y = 2x + 7$

   The system is: $y = 3x$

   $y = 2x + 7$

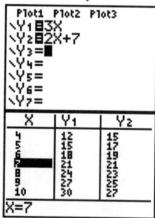

Fric is 21 years old and Frac is 7 years old.

6.5 Exercises

1. $3x + 4y - 5z = 46$

   $3(2) + 4(-3.5) - 5(-10.8) = 46$

   $6 - 14 + 54 = 46$

   $46 = 46$ true

This is a true statement so $(2,-3.5,-10.8)$ is a solution.

3. $3x+4y-5z=46$
$3(0)+4(8.5)-5(-2.4)=46$
$0+34+12=46$
$46=46$ true
This is a true statement so $(0,8.5,-2.4)$ is a solution.

5. $3x+4y-5z=46$
$3(5)+4(-1.5)-5(7.4)=46$
$15-6-37=46$
$-20=46$ false
This is not a true statement so $(5,-1.5,7.4)$ is not a solution.

7. $\frac{3}{4}a+\frac{1}{3}b=\frac{1}{2}c$
$\frac{3}{4}(8)+\frac{1}{3}(12)=\frac{1}{2}(10)$
$6+4=5$
$10=5$ false
This is not a true statement so $(8,12,10)$ is not a solution.

9. $\frac{3}{4}a+\frac{1}{3}b=\frac{1}{2}c$
$\frac{3}{4}(-16)+\frac{1}{3}(9)=\frac{1}{2}(-18)$
$-12+3=-9$
$-9=-9$ true
This is a true statement so $(-16,9,-18)$ is a solution.

11. $\frac{3}{4}a+\frac{1}{3}b=\frac{1}{2}c$
$\frac{3}{4}\left(\frac{2}{3}\right)+\frac{1}{3}\left(-\frac{3}{4}\right)=\frac{1}{2}\left(-\frac{1}{2}\right)$
$\frac{1}{2}-\frac{1}{4}=-\frac{1}{4}$
$\frac{2}{4}-\frac{1}{4}=-\frac{1}{4}$
$\frac{1}{4}=-\frac{1}{4}$ false
This is not a true statement so $\left(\frac{2}{3},-\frac{3}{4},-\frac{1}{2}\right)$ is not a solution.

13. $5x-3y+7z=26$
$5(-3)-3(5)+7(8)=26$
$-15-15+56=26$
$26=26$ true

$4x-5z=-37$
$4(-3)-5(8)=-37$
$-12-40=-37$
$-52=-37$ false

$4y+9z=92$
$4(5)+9(8)=92$
$20+72=92$
$92=92$ true
Since the second equation is false, $(-3,5,8)$ is not a solution of the system.

15. $x-\frac{1}{2}y=z-46$
$(14)-\frac{1}{2}(78)=(21)-46$
$14-39=21-46$
$-25=-25$ true

$x+y+z=113$
$(14)+(78)+(21)=113$
$113=113$ true

$2x-3y+4z=-122$
$2(14)-3(78)+4(21)=-122$
$28-234+84=-122$
$-122=-122$ true
Since all three equations are true, $(14,78,21)$ is a solution of the system.

SSM: Experiencing Introductory and Intermediate Algebra

17. $x - y + z = 14$
$x + y + z = 8$
$x + 2y - 5z = -37$
Eliminate y.
Add the first equation to the third equation.
$x - y + z = 14$
$\underline{x + y + z = 8}$
$2x + 2z = 22$ (1)
Multiply the first equation by 2 and add to the third equation.
$2x - 2y + 2z = 28$
$\underline{x + 2y - 5z = -37}$
$3x - 3z = -9$ (2)
Make a new system of two equations using equations (1) and (2).
$2x + 2z = 22$
$3x - 3z = -9$
Eliminate z.
Divide the first equation by 2 and the second equation by 3. Add the new equations.
$x + z = 11$
$\underline{x - z = -3}$
$2x = 8$
$x = 4$
Substitute this value for x in the first equation of the new system.
$2(4) + 2z = 22$
$8 + 2z = 22$
$2z = 14$
$z = 7$
Substitute $x = 4$ and $z = 7$ into one of the original equations to find y.
$4 + y + 7 = 8$
$y + 11 = 8$
$y = -3$
The solution is $(4, -3, 7)$.
Check:
```
4→X: -3→Y: 7→Z: X-Y
+Z=14
                    1
X+Y+Z=8
                    1
X+2Y-5Z=-37
                    1
■
```

19. $\frac{1}{3}a + \frac{2}{3}b - \frac{3}{5}c = -\frac{5}{8}$
$2a - \frac{1}{3}b - \frac{4}{5}c = \frac{5}{4}$
$8a + 8b + 8c = 5$
Eliminate a.
Multiply the first equation by -6 and add to the second equation.
$-2a - 4b + \frac{18}{5}c = \frac{15}{4}$
$\underline{2a - \frac{1}{3}b - \frac{4}{5}c = \frac{5}{4}}$
$-\frac{13}{3}b + \frac{14}{5}c = 5$ (1)
Multiply the second equation by -4 and add to the third equation.
$-8a + \frac{4}{3}b + \frac{16}{5}c = -5$
$\underline{8a + 8b + 8c = 5}$
$\frac{28}{3}b + \frac{56}{5}c = 0$ (2)
Make a new system of two equations using equations (1) and (2).
$-\frac{13}{3}b + \frac{14}{5}c = 5$
$\frac{28}{3}b + \frac{56}{5}c = 0$
Eliminate b.
Multiply the first equation by -4 and add to the second equation.
$\frac{52}{3}b - \frac{56}{5}c = -20$
$\underline{\frac{28}{3}b + \frac{56}{5}c = 0}$
$\frac{80}{3}b = -20$
$b = -\frac{3}{4}$
Substitute this value for b in the first equation.

$\frac{28}{3}\left(-\frac{3}{4}\right)+\frac{56}{5}c=0$

$-7+\frac{56}{5}c=0$

$\frac{56}{5}c=7$

$c=\frac{5}{8}$

Substitute $b=-\frac{3}{4}$ and $c=\frac{5}{8}$ in one of the original equations to solve for a.

$8a+8\left(-\frac{3}{4}\right)+8\left(\frac{5}{8}\right)=5$

$8a-6+5=5$

$8a=6$

$a=\frac{3}{4}$

The solution is $\left(\frac{3}{4},-\frac{3}{4},\frac{5}{8}\right)$.

Check:

21. $4x-5y-10z=63$

$2x+15z=-108$ (1)

$x-5y-15z=99$

Eliminate y.
Multiply the third equation by -1 and add to the first equation.

$4x-5y-10z=63$

$\underline{-x+5y+15z=-99}$

$3x+5z=-36$ (2)

Make a new system of two equations using equations (1) and (2).

$2x+15z=-108$

$3x+5z=-36$

Eliminate z.
Multiply the second equation by -3 and add to the first equation.

$2x+15z=-108$

$\underline{-9x-15z=108}$

$-7x=0$

$x=0$

Substitute this value for x in the second equation.

$3(0)+5z=-36$

$5z=-36$

$z=-\frac{36}{5}$

Substitute $x=0$ and $z=-\frac{36}{5}$ into one of the original equations to solve for y.

$4(0)-5y-10\left(-\frac{36}{5}\right)=63$

$-5y+72=63$

$-5y=-9$

$y=\frac{9}{5}$

The solution is $\left(0,\frac{9}{5},-\frac{36}{5}\right)$.

23. $a-b+c=385$

$2a-3b+c=16$

$4a-5b+2c=503$

Eliminate c.
Multiply the first equation by -1 and add to the second equation.

$-a+b-c=-385$

$\underline{2a-3b+c=16}$

$a-2b=-369$ (1)

Multiply the second equation by -2 and add to the third equation.

$-2a+2b-2c=-770$

$\underline{4a-5b+2c=503}$

$2a-3b=-267$ (2)

Make a new system of two equations using equations (1) and (2).

$a-2b=-369$

$2a-3b=-267$

Eliminate a.
Multiply the first equation by -2 and add to the second equation.

SSM: Experiencing Introductory and Intermediate Algebra

$-2a + 4b = 738$
$\underline{2a - 3b = -267}$
$b = 471$
Substitute this value for b in the first equation.
$a - 2(471) = -369$
$a - 942 = -369$
$a = 573$
Substitute $a = 573$ and $b = 471$ into one of the original equations to solve for c.
$(573) - (471) + c = 385$
$102 + c = 385$
$c = 283$
The solution is $(573, 471, 283)$.

25. $5x - 3y + 7z = 8$
$x + 4y - 3z = 11$
$6x + y + 4z = 19$
Eliminate y.
Multiply the third equation by 3 and add to the first equation.
$5x - 3y + 7z = 8$
$\underline{18x + 3y + 12z = 57}$
$23x + 19z = 65$ (1)
Multiply the third equation by -4 and add to the second equation.
$x + 4y - 3z = 11$
$\underline{-24x - 4y - 16z = -76}$
$-23x - 19z = -65$ (2)
Make a new system of two equations using equations (1) and (2).
$23x + 19z = 65$
$-23x - 19z = -65$
Eliminate x.
Add the equations.
$23x + 19z = 65$
$\underline{-23x - 19z = -65}$
$0 = 0$
This is an identity. There are an infinite number of solutions.

Notice that the third original equation is the sum of the first two original equations so it provides no additional information. We have:

$23x + 19z = 65$
$19z = 65 - 23x$
$z = -\dfrac{23}{19}x + \dfrac{65}{19}$
Going back to the original equations, we eliminate z instead of y.
Multiply the first equation by 3 and the second equation by 7. Add the new equations.
$15x - 9y + 21z = 24$
$\underline{7x + 28y - 21z = 77}$
$22x + 21y = 101$
$21y = -22x + 101$
$y = -\dfrac{22}{21}x + \dfrac{101}{21}$
We have both y and z in terms of x. A general solution can be written as
$\left(x, -\dfrac{22}{21}x + \dfrac{101}{21}, -\dfrac{23}{19}x + \dfrac{65}{19}\right)$.

27. $8a - 3b + 2c = -29$
$5a + b + 7c = 2$
$3a - 4b - 5c = 30$
Eliminate b.
Multiply the second equation by 3 and add to the first equation.
$8a - 3b + 2c = -29$
$\underline{15a + 3b + 21c = 6}$
$23a + 23c = -23$ (1)
Multiply the second equation by 4 and add to the third equation.
$20a + 4b + 28c = 8$
$\underline{3a - 4b - 5c = 30}$
$23a + 23c = 38$ (2)
Make a new system of two equations using equations (1) and (2).
$23a + 23c = -23$
$23a + 23c = 38$
Eliminate c.
Multiply the second equation by -1 and add to the first equation.

240

$23a + 23b = -23$
$-23a - 23b = -38$
_____
$0 = -61$
This is a contradiction. There is no solution.

29. Let x = num. of adults
y = num. of children
z = num. of resort guests
$x + y + z = 1257$
$12x + 5y + 3.5z = 9547$
$-x + y - z = 87$
Eliminate x.
Add the first equation to the third equation.
$x + y + z = 1257$
$-x + y - z = 87$
_____
$2y = 1344$
$y = 672$ (1)
Multiply the first equation by -12 and add to the second equation.
$-12x - 12y - 12z = -15084$
$12x + 5y + 3.5z = 9547$
_____
$-7y - 8.5z = -5537$ (2)
Substitute $y = 672$ into equation (2).
$-7(672) - 8.5z = -5537$
$-4704 - 8.5z = -5537$
$-8.5z = -833$
$z = 98$
Substitute $y = 672$ and $z = 98$ into one of the original equations to solve for x.
$x + (672) + (98) = 1257$
$x + 770 = 1257$
$x = 487$
There were 487 adult tickets, 672 child tickets, and 98 guest tickets sold.

31. Let x = amt. invested at 5%
y = amt. invested at 5.5%
z = amt. invested at 6%
$x + y + z = 11,200$
$x - z = -1400$ (1)
$0.05x + 0.055y + 0.06z = 623$
Eliminate y.

Multiply the first equation by -0.055 and add to the third equation.
$-0.055x - 0.055y - 0.055z = -616$
$0.05x + 0.055y + 0.06z = 623$
_____
$-0.005x + 0.005z = 7$ (2)
Make a new system of two equations using equations (1) and (2).
$x - z = -1400$
$-0.005x + 0.005z = 7$
Eliminate x.
Multiply the first equation by 0.005 and add to the second equation.
$0.005x - 0.005z = -7$
$-0.005x + 0.005z = 7$
_____
$0 = 0$
This is an identity. There is an infinite number of solutions.

A general solution can be written as
$(z - 1400, 12600 - 2z, z)$
where $1400 \le z \le 6300$.

33. Let x = num. of \$1
y = num. of \$5
z = num. of \$10
$x + y + z = 60$
$x + 5y + 10z = 184$
$x - 3y = 0$ (1)
Eliminate z.
Multiply the first equation by -10 and add to the second equation.
$-10x - 10y - 10z = -600$
$x + 5y + 10z = 184$
_____
$-9x - 5y = -416$ (2)
Make a new system of two equations using equations (1) and (2).
$x - 3y = 0$
$-9x - 5y = -416$
Eliminate x.
Multiply the first equation by 9 and add to the second equation.

SSM: Experiencing Introductory and Intermediate Algebra

$9x - 27y = 0$
$\underline{-9x - 5y = -416}$
$-32y = -416$
$y = 13$

Substitute this value for y in the first equation.
$9x - 27(13) = 0$
$9x = 351$
$x = 39$

Substitute $x = 39$ and $y = 13$ into one of the original equations to solve for z.
$(39) + (13) + z = 60$
$52 + z = 60$
$z = 8$

Ron collected 39 $1 bills, 13 $5 bills and 8 $10 bills.

35. Let $x =$ num. of small boxes
$y =$ num. of medium boxes
$z =$ num. of large boxes
$x + y + z = 145$
$2.39x + 2.79y + 3.59z = 446.55$
$10.8x + 13.5y + 20.4z = 2346$

Eliminate z.
Multiply the first equation by -3.59 and add to the second equation.
$-3.59x - 3.59y - 3.59z = -520.55$
$\underline{2.39x + 2.79y + 3.59z = 446.55}$
$-1.2x - 0.8y = -74 \quad (1)$

Multiply the first equation by -20.4 and add to the third equation.
$-20.4x - 20.4y - 20.4z = -2958$
$\underline{10.8x + 13.5y + 20.4z = 2346}$
$-9.6x - 6.9y = -612 \quad (2)$

Make a new system of two equations using equations (1) and (2).
$-1.2x - 0.8y = -74$
$-9.6x - 6.9y = -612$

Eliminate x.
Multiply the first equation by -8 and add to the second equation.

$9.6x + 6.4y = 592$
$\underline{-9.6x - 6.9y = -612}$
$-0.5y = -20$
$y = 40$

Substitute this value for y into the first equation.
$-1.2x - 0.8(40) = -74$
$-1.2x - 32 = -74$
$-1.2x = -42$
$x = 35$

Substitute $x = 35$ and $y = 40$ into one of the original equations to solve for z.
$(35) + (40) + z = 145$
$75 + z = 145$
$z = 70$

Henry sold 35 small boxes, 40 medium boxes, and 70 large boxes.

37. Let $x =$ num. of gold investment stocks
$y =$ num. of oil related investment stocks
$z =$ num. of money market fund stocks
$x + y + z = 600$
$19.75x + 9.5y + 12z = 7225$
$0.5x + 0.1y + 0.9z = 260$

Eliminate z.
Multiply the first equation by -12 and add to the second equation.
$-12x - 12y - 12z = -7200$
$\underline{19.75x + 9.5y + 12z = 7225}$
$7.75x - 2.5y = 25 \quad (1)$

Multiply the first equation by -0.9 and add to the third equation.
$-0.9x - 0.9y - 0.9z = -540$
$\underline{0.5x + 0.1y + 0.9z = 260}$
$-0.4x - 0.8y = -280 \quad (2)$

Make a new system of two equations using equations (1) and (2).
$7.75x - 2.5y = 25$
$-0.4x - 0.8y = -280$

Eliminate y.
Divide the first equation by 2.5 and the second equation by -0.8. Add the new equations.

242

$3.1x - y = 10$
$0.5x + y = 350$
———————
$3.6x = 360$
$x = 100$
Substitute this value for x in the first equation.
$7.75(100) - 2.5y = 25$
$775 - 2.5y = 25$
$-2.5y = -750$
$y = 300$
Substitute $x = 100$ and $y = 300$ into one of the original equations to solve for z.
$(100) + (300) + z = 600$
$400 + z = 600$
$z = 200$

$100(19.75) = \$1975$
$300(9.50) = \$2850$
$200(12.00) = \$2400$
Tenisha invested $1975 in gold stocks, $2850 in oil stocks, and $2400 in a money market fund.

39. Let $x = $ num. of pies
$\quad y = $ num. of cookie dozens
$\quad z = $ num. of cakes
$1.25x + y + 1.5z = 69.5$
$3.5x + 2.5y + 4.5z = 200$
$3x - z = 0 \quad\quad (1)$
Eliminate y.
Multiply the first equation by -2.5 and add to the second equation.
$-3.125x - 2.5y - 3.75z = -173.75$
$\underline{3.5x + 2.5y + 4.5z = 200}$
$\quad 0.375x + 0.75z = 26.25 \quad (2)$
Make a new system of two equations using equations (1) and (2).
$\quad 3x - z = 0$
$0.375x + 0.75z = 26.25$
Eliminate z.
Multiply the first equation by 0.75 and add to the second equation.

$2.25x - 0.75z = 0$
$\underline{0.375x + 0.75z = 26.25}$
$2.625x = 26.25$
$x = 10$
Substitute this value for x in the first equation.
$3(10) - z = 0$
$30 - z = 0$
$z = 30$
Substitute $x = 10$ and $z = 30$ into one of the original equations to solve for y.
$1.25(10) + y + 1.5(30) = 69.5$
$12.50 + y + 45 = 69.50$
$57.50 + y = 69.50$
$y = 12$
Concetta bakes 10 pies, 12 dozen cookies, and 30 cakes.

41. Let $x = $ servings of cereal
$\quad y = $ servings of orange juice
$\quad z = $ servings of milk
$225x + 6.8y + 69.8z = 736$
$4.5x + 31.2y + 116.2z = 505$
$0.6x + 35.4y + 0.6z = 39$
Eliminate x.
Multiply the third equation by -375 and add to the first equation.
$\quad 225x + 6.8y + 69.8z = 736$
$\underline{-225x - 13275y - 225z = -14625}$
$\quad\quad -13268.2y - 155.2z = -13889 \quad (1)$
Multiply the third equation by -7.5 and add to the second equation.
$\quad 4.5x + 31.2y + 116.2z = 505$
$\underline{-4.5x - 265.5y - 4.5z = -292.5}$
$\quad\quad -234.3y + 111.7z = 212.5 \quad (2)$
Make a new system of two equations using equations (1) and (2).
$-13268.2y - 155.2z = -13889$
$-234.3y + 111.7z = 212.5$
Eliminate z.
Multiply the first equation by 111.7 and the second equation by 155.2. Add the new equations.

$-1482057.94y - 17335.84z = -1551401.3$
$-36363.36y + 17335.84z = 32980$
$-1518421.3y = -1518421.3$
$y = 1$

Substitute this value for y in the second equation.
$-234.3(1) + 111.7z = 212.5$
$111.7z = 446.8$
$z = 4$

Substitute $y = 1$ and $z = 4$ into one of the original equations to solve for x.
$0.6x + 35.4(1) + 0.6(4) = 39$
$0.6x + 37.8 = 39$
$0.6x = 1.2$
$x = 2$

The baby should receive 2 servings of cereal, 1 serving of orange juice, and 4 servings of milk.

6.5 Calculator Exercises

1. $x - y - z = 0$
    $x + y + z = 4$
    $x + y - z = 6$
    Coefficient table:
    1 −1 −1 0
    1   1   1 4
    1   1 −1 6

    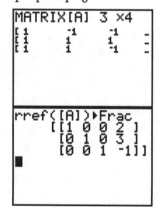

    The solution is $(2, 3, -1)$.

2. $x + y - z = -4$
    $x + 2y + z = -3$
    $x - y + z = 6$
    Coefficient table:
    1   1 −1 −4
    1   2   1 −3
    1 −1   1   6

    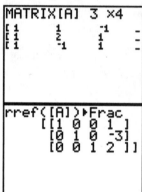

    The solution is $(1, -3, 2)$.

3. $2x + 8z = 4$
    $4x - y + 8z = 3$
    $2x + y = 3$
    Coefficient table:
    2   0 8 4
    4 −1 8 3
    2   1 0 3

    The solution is $\left(\dfrac{1}{2}, 2, \dfrac{3}{8}\right)$.

4. $4x + y - 3z = 2$

$8x - y + 6z = 3$

$y + 9z = 9$

Coefficient table:

4   1   −3   2
8   −1   6   3
0   1   9   9

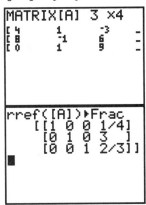

The solution is $\left(\dfrac{1}{4}, 3, \dfrac{2}{3}\right)$.

5. $x + 2y + 3z = 6$

$2x - y + z = 2$

$x + 2y - z = 2$

Coefficient table:

1   2   3   6
2   −1   1   2
1   2   −1   2

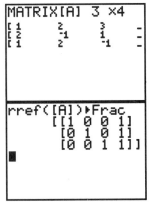

The solution is $(1, 1, 1)$.

6. $x - 2y + z = 0$

$x - 3y - z = 3$

$2x + y - 4z = 1$

Coefficient table:

1   −2   1   0
1   −3   −1   3
2   1   −4   1

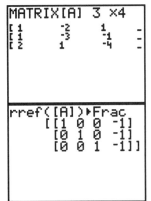

The solution is $(-1, -1, -1)$.

7. $4x + y - z = 30$

$-x + 2y - z = 7$

$9x - 8y + 4z = -3$

Coefficient table:

4   1   −1   30
−1   2   −1   7
9   −8   4   −3

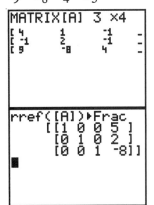

The solution is $(5, 2, -8)$.

SSM: Experiencing Introductory and Intermediate Algebra

8. $3x - y + 3z = 5$
    $x + 7y - 2z = 24$
    $5x - 6y + 4z = 9$
    Coefficient table:
    3 −1 3   5
    1   7 −2 24
    5 −6 4   9

    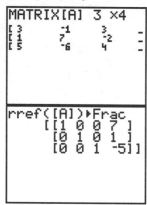

    The solution is $(7, 1, -5)$.

9. $x - y + z = -2$
    $2x + 3y + z = 8$
    $3x + 2y + 2z = 10$
    Coefficient table:
    1 −1 1   −2
    2   3 1   8
    3   2 2   10

    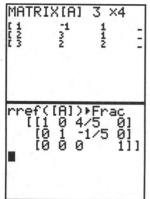

    The last row is $0x + 0y + 0z = 1$
    $0 = 1$
    This is a contradiction. There is no solution.

10. $3x - 2y + z = -3$
    $x - y - 4z = 7$
    $4x - 3y - 3z = 4$
    Coefficient table:
    3 −2 1   −3
    1 −1 −4   7
    4 −3 −3   4

    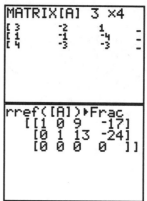

    The last row is $0x + 0y + 0z = 0$
    $0 = 0$
    This is an identity. There are an infinite number of solutions to the system. All ordered triples satisfying $x - y - 4z = 7$ are solutions.

Chapter 6 Section-By-Section Review

1. $3x + 2y = -2$
    $3(2) + 2(-4) = -2$
    $6 - 8 = -2$
    $-2 = -2$ true

    $4x - 3y = 20$
    $4(2) - 3(-4) = 20$
    $8 + 12 = 20$
    $20 = 20$ true

    Since both equations are true, $(2, -4)$ is a solution.

Chapter 6: Systems of Linear Equations

2. $2x + y = 6$
$2\left(\dfrac{17}{7}\right) + \left(\dfrac{8}{7}\right) = 6$
$\dfrac{34}{7} + \dfrac{8}{7} = 6$
$\dfrac{42}{7} = 6$
$6 = 6$ true

$-x + 3y = 1$
$-\left(\dfrac{17}{7}\right) + 3\left(\dfrac{8}{7}\right) = 1$
$-\dfrac{17}{7} + \dfrac{24}{7} = 1$
$\dfrac{7}{7} = 1$
$1 = 1$

Since both equations are true, $\left(\dfrac{17}{7}, \dfrac{8}{7}\right)$ is a solution.

3. $4x - 5y = 3$
$4(0.25) - 5(-0.45) = 3$
$1 + 2.25 = 3$
$3.25 = 3$ false

$8x + 5y = 0$
$8(0.25) + 5(-0.45) = 0$
$2 - 2.25 = 0$
$-0.25 = 0$ false

Neither equation is true so $(0.25, -0.45)$ is not a solution.

4. $x + 3y = 13$
$(7) + 3(-2) = 13$
$7 - 6 = 13$
$1 = 13$ false

$x - y = 5$
$(7) - (-2) = 5$
$7 + 2 = 5$
$9 = 5$ false

Neither equation is true so $(7, -2)$ is not a solution.

5. $3x + 6y = -2$
$3\left(\dfrac{2}{3}\right) + 6\left(\dfrac{2}{3}\right) = -2$
$2 + 12 = -2$
$14 = -2$ false

$6x - 3y = 6$
$6\left(\dfrac{2}{3}\right) - 3\left(\dfrac{2}{3}\right) = 6$
$4 - 2 = 6$
$2 = 6$ false

Neither equation is true so $\left(\dfrac{2}{3}, \dfrac{2}{3}\right)$ is not a solution.

6. $6x + 5y = -3$
$6(1.5) + 5(-2.4) = -3$
$9 - 12 = -3$
$-3 = -3$ true

$2x - 10y = 27$
$2(1.5) - 10(-2.4) = 27$
$3 + 24 = 27$
$27 = 27$ true

Since both equations are true, $(1.5, -2.4)$ is a solution.

7. $2(x - 3) = 2$
$2((4) - 3) = 2$
$2(1) = 2$
$2 = 2$ true

$3(y + 1) = -3$
$3((2) + 1) = -3$
$3(3) = -3$
$9 = -3$ false

Since the second equation is false, $(4, 2)$ is not a solution.

SSM: Experiencing Introductory and Intermediate Algebra

8. $x + 5 = 2$
    $(-3) + 5 = 2$
    $2 = 2$ true

    $2y - 3 = 7$
    $2(5) - 3 = 7$
    $10 - 3 = 7$
    $7 = 7$ true
    Since both equations are true, $(-3, 5)$ is a solution.

9. $2x + y = 17 \qquad y = 3x - 18$
    $y = -2x + 17$

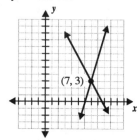

    The solution is $(7, 3)$.

10. $3(x + 2) + 1 = -5 \qquad 2x - y = -10$
    $3x + 6 + 1 = -5 \qquad -y = -2x - 10$
    $3x + 7 = -5 \qquad y = 2x + 10$
    $3x = -12$
    $x = -4$

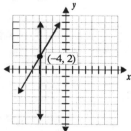

    The solution is $(-4, 2)$.

11. $x + 6 = 4 \qquad 2(y + 2) = y + 3$
    $x = -2 \qquad 2y + 4 = y + 3$
    $\qquad\qquad y + 4 = 3$
    $\qquad\qquad y = -1$

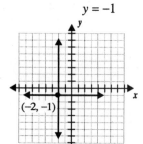

    The solution is $(-2, -1)$.

12. $x - 2y = -3 \qquad 2x + 4y = 8$
    $-2y = -x - 3 \qquad 4y = -2x + 8$
    $y = \dfrac{1}{2}x + \dfrac{3}{2} \qquad y = -\dfrac{1}{2}x + 2$

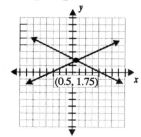

    The solution is $\left(\dfrac{1}{2}, \dfrac{7}{4}\right)$.

13. $y = 2(x + 2) \qquad 2x - y = 5$
    $y = 2x + 4 \qquad -y = -2x + 5$
    $\qquad\qquad y = 2x - 5$

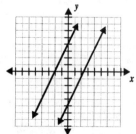

    The lines are parallel. There is no solution.

248

14. $y = \dfrac{3}{2}x - 6$ $3x - 2y = 12$
   $-2y = -3x + 12$
   $y = \dfrac{3}{2}x - 6$

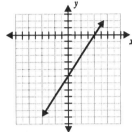

The lines are coinciding. There are an infinite number of solutions. All ordered pairs, (x, y) satisfying $y = \dfrac{3}{2}x - 6$ are solutions.

15. $y = 3x - 6$ $y = -3$

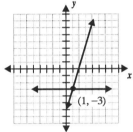
$(1, -3)$

The solution is $(1, -3)$.

16. $2y - 3 = 1$ $5(y - 4) = 10$
    $2y = 4$ $y - 4 = 2$
    $y = 2$ $y = 6$

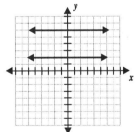

The lines are parallel. There is no solution.

17. $2x - 11 = 3$ $14 - 2x = x - 7$
    $2x = 14$ $14 - 3x = -7$
    $x = 7$ $-3x = -21$
                     $x = 7$

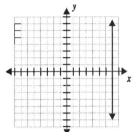

The lines are coinciding. There are an infinite number of solutions. All ordered pairs, (x, y), with $x = 7$ are solutions.

18. $y - 1 = 4$ $3y - 2 = 13$
    $y = 5$ $3y = 15$
                   $y = 5$

The lines are coinciding. All ordered pairs, (x, y), with $y = 5$ are solutions.

19. $2x + 3y = 6$ $x + y = 1$
    $3y = -2x + 6$ $y = -x + 1$
    $y - \dfrac{2}{3}x + 2$

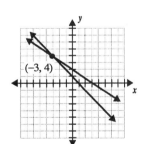
$(-3, 4)$

The solution is $(-3, 4)$.

20. $y = 2x - 15 \qquad y = -3x + 10$

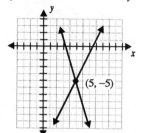

The solution is $(5, -5)$.

21. $y = 2x + 1 \qquad x = 2y + 7$
$\qquad\qquad\qquad\quad 2y = x - 7$
$\qquad\qquad\qquad\quad y = \dfrac{1}{2}x - \dfrac{7}{2}$

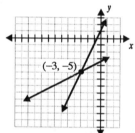

The solution is $(-3, -5)$.

22. $5y + 6 = 3y + 5 \qquad 2(x-3) + 1 = 4$
$2y + 6 = 5 \qquad\qquad 2x - 6 + 1 = 4$
$2y = -1 \qquad\qquad\quad 2x - 5 = 4$
$y = -\dfrac{1}{2} \qquad\qquad\quad 2x = 9$
$\qquad\qquad\qquad\qquad\quad x = \dfrac{9}{2}$

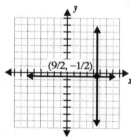

The solution is $\left(\dfrac{9}{2}, -\dfrac{1}{2}\right)$.

23. Let x = num. of Democrats
$\qquad y$ = num. of Republicans

$x + y = 380 \qquad y = x + 40$
$y = -x + 380$

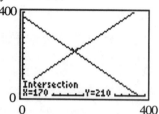

There are 170 Democrats; 50 more than the number of independent voters.

24. $F(x) = 0.507x + 9.92$
$M(x) = 0.330x + 10.80$

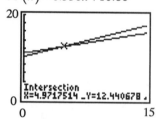

The average hourly earnings will be the same for both industries about 4.97 years after 1990 (that is, in 1995).

25. $8y = 5$
$4x + 8y = 2$
Solve the first equation for y.
$y = \dfrac{5}{8}$
Substitute this value for y in the second equation.
$4x + 8\left(\dfrac{5}{8}\right) = 2$
$4x + 5 = 2$
$4x = -3$
$x = -\dfrac{3}{4}$

The solution is $\left(-\dfrac{3}{4}, \dfrac{5}{8}\right)$.

Chapter 6: Systems of Linear Equations

26. $10x - 5y = -14$
$5x - 1 = 0$
Solve the second equation for x.
$5x = 1$
$x = \dfrac{1}{5}$
Substitute this value for x in the first equation.
$10\left(\dfrac{1}{5}\right) - 5y = -14$
$2 - 5y = -14$
$-5y = -16$
$y = \dfrac{16}{5}$
The solution is $\left(\dfrac{1}{5}, \dfrac{16}{5}\right)$.

27. $4x - y = 5$
$y = 4x + 3$
Substitute $4x + 3$ for y in the first equation.
$4x - (4x + 3) = 5$
$4x - 4x - 3 = 5$
$-3 = 5$
This is a contradiction. The system has no solution.

28. $x - 2y = 71$
$3x - 7y = 275$
Solve the first equation for x.
$x = 2y + 71$
Substitute this result for x in the second equation.
$3(2y + 71) - 7y = 275$
$6y + 213 - 7y = 275$
$-y + 213 = 275$
$y = -62$
Substitute this value for y in the first equation.
$x - 2(-62) = 71$
$x + 124 = 71$
$x = -53$
The solution is $(-53, -62)$.

29. $2x + 3y = -69$
$2x - 4y = 218$
Solve the first equation for x.
$2x = -3y - 69$
$x = -\dfrac{3}{2}y - \dfrac{69}{2}$
Substitute this result for x in the second equation.
$2\left(-\dfrac{3}{2}y - \dfrac{69}{2}\right) - 4y = 218$
$-3y - 69 - 4y = 218$
$-7y - 69 = 218$
$-7y = 287$
$y = -41$
Substitute this value for y in the first equation.
$2x + 3(-41) = -69$
$2x - 123 = -69$
$2x = 54$
$x = 27$
The solution is $(27, -41)$.

30. $3(x + 2) - y = -1$
$y = 3x + 7$
Substitute $3x + 7$ for y in the first equation.
$3(x + 2) - (3x + 7) = -1$
$3x + 6 - 3x - 7 = -1$
$-1 = -1$
This is an identity. There are an infinite number of solutions. All ordered pairs, (x, y), satisfying $y = 3x + 7$ are solutions.

31. $x + 8y = 5$
$12x + y = 10$
Solve the first equation for x.
$x = -8y + 5$
Substitute this result for x in the second equation.
$12(-8y + 5) + y = 10$
$-96y + 60 + y = 10$
$-95y + 60 = 10$
$-95y = -50$
$y = \dfrac{10}{19}$

SSM: Experiencing Introductory and Intermediate Algebra

Substitute this value for y in the first equation.
$$x+8\left(\frac{10}{19}\right)=5$$
$$x+\frac{80}{19}=5$$
$$x=\frac{15}{19}$$
The solution is $\left(\frac{15}{19},\frac{10}{19}\right)$.

32. $5x-3y=406$
$2x-y=327$
Solve the second equation for y.
$y=2x-327$
Substitute this result for y in the first equation.
$5x-3(2x-327)=406$
$5x-6x+981=406$
$-x+981=406$
$-x=-575$
$x=575$
Substitute this value for x in the second equation.
$2(575)-y=327$
$1150-y=327$
$y=823$
The solution is $(575,823)$.

33. $y=\frac{1}{3}x-4$
$x-5y=0$
Substitute $\frac{1}{3}x-4$ for y in the second equation.
$x-5\left(\frac{1}{3}x-4\right)=0$
$x-\frac{5}{3}x+20=0$
$-\frac{2}{3}x+20=0$
$\frac{2}{3}x=20$
$x=30$
Substitute this value for x in the first equation.

$y=\frac{1}{3}(30)-4$
$y=10-4=6$
The solution is $(30,6)$.

34. $y=\frac{1}{2}x+7$
$y=-\frac{3}{5}x-4$
Substitute $\frac{1}{2}x+7$ for y in the second equation.
$\frac{1}{2}x+7=-\frac{3}{5}x-4$
$\frac{1}{2}x+\frac{3}{5}x=-7-4$
$\frac{11}{10}x=-11$
$x=-10$
Substitute this value for x in the first equation.
$y=\frac{1}{2}(-10)+7$
$y=-5+7=2$
The solution is $(-10,2)$.

35. $3x-y=10$
$2x-6y=26$
Solve the first equation for y.
$y=3x-10$
Substitute this result for y in the second equation.
$2x-6(3x-10)=26$
$2x-18x+60=26$
$-16x+60=26$
$-16x=-34$
$x=\frac{17}{8}$
Substitute this value for x in the first equation.

$3\left(\dfrac{17}{8}\right) - y = 10$

$\dfrac{51}{8} - y = 10$

$y = \dfrac{51}{8} - 10 = \dfrac{51}{8} - \dfrac{80}{8} = -\dfrac{29}{8}$

The solution is $\left(\dfrac{17}{8}, -\dfrac{29}{8}\right)$.

36. $x = 160y$

    $3x - 440y = 30$

    Substitute $160y$ for x in the second equation.

    $3(160y) - 440y = 30$

    $480y - 440y = 30$

    $40y = 30$

    $y = \dfrac{3}{4}$

    Substitute this value for y in the first equation.

    $x = 160\left(\dfrac{3}{4}\right) = 120$

    The solution is $\left(120, \dfrac{3}{4}\right)$.

37. Let $x =$ first angle

    $y =$ second angle

    $x + y = 90$

    $y = 12 + 2x$

    Substitute $12 + 2x$ for y in the first equation.

    $x + (12 + 2x) = 90$

    $3x + 12 = 90$

    $3x = 78$

    $x = 26$

    Substitute this value for x in the second equation.

    $y = 12 + 2(26) = 12 + 52 = 64$

    The angles measure $64°$ and $26°$ so the difference between them is $38°$.

38. **a.** $c(x) = 450 + 29x$

    where x is the number of scooters.

    b. $r(x) = 89.95x$

    where x is the number of scooters.

    c. $y = 450 + 29x$

    $y = 89.95x$

    Substitute $89.95x$ for y in the first equation.

    $89.95x = 450 + 29x$

    $60.95x = 450$

    $x \approx 7.383$

    Kidsports must sell approximately 7.383 scooters to break even. The store will lose money if it sells 7 or less, and make money if it sells 8 or more.

39. $W(x) = 78.1x + 237.8$

    $S(x) = 40.4x + 406.5$

    $y = 78.1x + 237.8$

    $y = 40.4x + 406.5$

    Substitute $78.1x + 237.8$ for y in the second equation.

    $78.1x + 237.8 = 40.4x + 406.5$

    $78.1x - 40.4x = 406.5 - 237.8$

    $37.7x = 168.7$

    $x \approx 4.475$

    The two companies spent approximately the same amount for advertising about 4.475 years after 1990 (that is, in 1995).

40. $5x + 3y = -10$

    $\underline{5x - 3y = 80}$

    $10x = 70$

    $x = 7$

    Substitute this value for x in the first equation.

    $5(7) + 3y = -10$

    $35 + 3y = -10$

    $3y = -45$

    $y = -15$

    The solution is $(7, -15)$.

41. $x = 18 - 2y$

$3x + 2y = 30$

Multiply the second equation by -1 and add to the first equation written in standard form.

$x + 2y = 18$
$\underline{-3x - 2y = -30}$
$-2x = -12$
$x = 6$

Substitute this value for x in the second equation.

$3(6) + 2y = 30$

$18 + 2y = 30$

$2y = 12$

$y = 6$

The solution is $(6, 6)$.

42. $\frac{1}{2}x + \frac{1}{3}y = 3$

$\frac{1}{4}x - \frac{2}{5}y = -7$

Multiply the second equation by -2 and add to the first equation.

$\frac{1}{2}x + \frac{1}{3}y = 3$
$\underline{-\frac{1}{2}x + \frac{4}{5}y = 14}$
$\frac{17}{15}y = 17$
$y = 15$

Substitute this value for y in the first equation.

$\frac{1}{2}x + \frac{1}{3}(15) = 3$

$\frac{1}{2}x + 5 = 3$

$\frac{1}{2}x = -2$

$x = -4$

The solution is $(-4, 15)$.

43. $5x + 10y = -55$

$2x - 3y = 6$

Multiply the first equation by 2 and the second equation by -5. Add the new equations.

$10x + 20y = -110$
$\underline{-10x + 15y = -30}$
$35y = -140$
$y = -4$

Substitute this value for y in the second equation.

$2x - 3(-4) = 6$

$2x + 12 = 6$

$2x = -6$

$x = -3$

The solution is $(-3, -4)$.

44. $2x = 3y + 1$

$15y = 10x + 5$

Divide the second equation by 5. Write both equations in standard form and add.

$2x - 3y = 1$
$\underline{-2x + 3y = 1}$
$0 = 2$

This is a contradiction. The system has no solution.

45. $5x + 7y = 21$

$3x - 2y = 13$

Multiply the first equation by 3 and the second equation by -5. Add the new equations.

$15x + 21y = 63$
$\underline{-15x + 10y = -65}$
$31y = -2$
$y = -\frac{2}{31}$

Substitute this value for y in the second equation.

$3x - 2\left(-\frac{2}{31}\right) = 13$

$3x + \frac{4}{31} = 13$

$3x = \frac{399}{31}$

$x = \frac{133}{31}$

The solution is $\left(\frac{133}{31}, -\frac{2}{31}\right)$.

46. $3x + y = 20$

$y = \dfrac{2}{3}x + \dfrac{16}{3}$

Multiply the first equation by -1 and add to the second equation written in standard form.
$-3x - y = -20$

$-\dfrac{2}{3}x + y = \dfrac{16}{3}$

$-\dfrac{11}{3}x = -\dfrac{44}{3}$

$x = 4$

Substitute this value for x in the first equation.
$3(4) + y = 20$
$12 + y = 20$
$y = 8$

The solution is $(4, 8)$.

47. $7(y - 1) = 5x$

$y = \dfrac{5}{7}x + 1$

Write both equations in standard form.
$-5x + 7y = 7$

$-\dfrac{5}{7}x + y = 1$

Multiply the second equation by -7 and add to the first equation.
$-5x + 7y = 7$
$5x - 7y = -7$

$0 = 0$

This is an identity. The system has an infinite number of solutions. Any ordered pair, (x, y), satisfying $y = \dfrac{5}{7}x + 1$ is a solution.

48. $3x - 3y = 4$

$9x + 9y = -2$

Multiply the first equation by 3 and add to the second equation.
$9x - 9y = 12$
$9x + 9y = -2$

$18x = 10$

$x = \dfrac{5}{9}$

Substitute this value for x in the second equation.
$9\left(\dfrac{5}{9}\right) + 9y = -2$
$5 + 9y = -2$
$9y = -7$
$y = -\dfrac{7}{9}$

The solution is $\left(\dfrac{5}{9}, -\dfrac{7}{9}\right)$.

49. $3x - y = 0$

$2x + 2y = 7$

Multiply the first equation by 2 and add to the second equation.
$6x - 2y = 0$
$2x + 2y = 7$

$8x = 7$

$x = \dfrac{7}{8}$

Substitute this value for x in the second equation.
$2\left(\dfrac{7}{8}\right) + 2y = 7$
$\dfrac{7}{4} + 2y = 7$
$2y = \dfrac{21}{4}$
$y = \dfrac{21}{8}$

The solution is $\left(\dfrac{7}{8}, \dfrac{21}{8}\right)$.

50. $0.25x + 0.3y = 4$

$0.5x - 0.2y = 4$

Multiply the first equation by 4 and the second equation by -2. Add the new equations.
$x + 1.2y = 16$
$-x + 0.4y = -8$

$1.6y = 8$
$y = 5$

Substitute this value for y in the second equation.

SSM: Experiencing Introductory and Intermediate Algebra

$0.5x - 0.2(5) = 4$
$0.5x - 1 = 4$
$0.5x = 5$
$x = 10$
The solution is $(10, 5)$.

51. $0.2x + 0.1y = 5$
$0.02x - 0.01y = 13.5$
Multiply the second equation by 10 and add to the first equation.
$0.2x + 0.1y = 5$
$\underline{0.2x - 0.1y = 135}$
$0.4x = 140$
$x = 350$
Substitute this value for x in the first equation.
$0.2(350) + 0.1y = 5$
$70 + 0.1y = 5$
$0.1y = -65$
$y = -650$
The solution is $(350, -650)$.

52. Let x = J. K. Rowling's earnings (millions)
y = Stephen King's earnings (millions)
$x + y = 80$
$y = x + 8$
Write the second equation in standard form and add to the first equation.
$x + y = 80$
$\underline{-x + y = 8}$
$2y = 88$
$y = 44$
Substitute this value for y in the first equation.
$x + (44) = 80$
$x = 36$
In the year 2000, J.K. Rowling earned $36 million and Stephen King earned $44 million.

53. Let x = num. of small offices
y = num. of large offices
$x + y = 40$
$500x + 1800y = 35,600$
Multiply the first equation by –500 and add to the second equation.
$-500x - 500y = -20,000$
$\underline{500x + 1800y = 35,600}$
$1300y = 15,600$
$y = 12$
Substitute this value for y in the first equation.
$x + (12) = 40$
$x = 28$
The realtor should have 28 small offices and 12 large offices.

54. Let x = mass of Mars
y = mass of Earth
$x + y = 6.612 \times 10^{24}$
$y = x + 5.328 \times 10^{24}$
Write the second equation in standard form and add to the first equation.
$x + y = 6.612 \times 10^{24}$
$\underline{-x + y = 5.328 \times 10^{24}}$
$2y = 1.194 \times 10^{25}$
$y = 5.97 \times 10^{24}$
Substitute this value for y in the first equation.
$x + 5.97 \times 10^{24} = 6.612 \times 10^{24}$
$x = 6.42 \times 10^{23}$
The mass of Mars is approximately 6.42×10^{23} kg and the mass of Earth is approximately 5.97×10^{24} kg.

55. Let x = num. of components from A
y = num. of components from B
$x + y = 200$
$25x + 35y = 5500$
Multiply the first equation by –25 and add to the second equation.
$-25x - 25y = -5000$
$\underline{25x + 35y = 5500}$
$10y = 500$
$y = 50$
Substitute this value for y in the first equation.
$x + (50) = 200$
$x = 150$

Chapter 6: Systems of Linear Equations

Each month the plant should order 150 components from supplier A and 50 components from supplier B.

56. Let x = pounds of 10%
y = pounds of 5%
$x + y = 150$
$.1x + 0.05y = .08(150)$
Multiply the second equation by -10 and add to the first equation.
$$\begin{array}{r} x + y = 150 \\ -x - 0.5y = -120 \\ \hline 0.5y = 30 \\ y = 60 \end{array}$$
Substitute this value for y in the first equation.
$x + (60) = 150$
$x = 90$
The manufacturer should mix 90 pounds of the 10% nitrogen fertilizer with 60 pounds of the 5% nitrogen fertilizer.

57. Let x = pounds of broccoli
y = pounds of cauliflower
$x + y = 200$
$0.49x + 0.99y = 0.69(200)$
Multiply the first equation by -0.49 and add to the second equation.
$$\begin{array}{r} -0.49x - 0.49y = -98 \\ 0.49x + 0.99y = 138 \\ \hline 0.5y = 40 \\ y = 80 \end{array}$$
Substitute this value in for y in the first equation.
$x + (80) = 200$
$x = 120$
The plant should mix 120 pounds of broccoli with 80 pounds of cauliflower.

58. Let x = hours of interstate driving
y = hours of highway driving
$x + y = 10$
$65x + 45y = 600$
Multiply the first equation by -45 and add to the second equation.

$$\begin{array}{r} -45x - 45y = -450 \\ 65x + 45y = 600 \\ \hline 20x = 150 \end{array}$$
$x = \dfrac{15}{2} = 7.5$
The driver was on the interstate for 7.5 hours. At 65mph, he traveled $65(7.5) = 487.5$ miles on the interstate.

59. Let x = speed in still water
y = speed of current
$\dfrac{1}{3}(x + y) = 20.5$
$\dfrac{7}{20}(x - y) = 20.5$
Solve the first equation for x.
$x + y = 61.5$
$x = 61.5 - y$
Substitute this result for x in the second equation.
$\dfrac{7}{20}((61.5 - y) - y) = 20.5$
$61.5 - 2y = \dfrac{410}{7}$
$-2y = -\dfrac{41}{14}$
$y = \dfrac{41}{28} \approx 1.5$
Substitute this value for y in the first equation.
$\dfrac{1}{3}\left(x + \dfrac{41}{28}\right) = 20.5$
$x + \dfrac{41}{28} = 61.5$
$x = \dfrac{1681}{28} \approx 60.0$
The boat's average speed in still water is about 60.0 mph. The average speed of the current is about 1.5 mph.

60. Let x = pounds of 25% copper alloy
y = total pounds
$y = x + 40$
$0.25x + 0.30(40) = 0.27y$
Substitute $x + 40$ for y in the second equation.

SSM: Experiencing Introductory and Intermediate Algebra

$0.25x+12 = 0.27(x+40)$
$0.25x+12 = 0.27x+10.8$
$1.2 = 0.02x$
$x = 60$
Substitute this value for x in the first equation.
$y = (60)+40 = 100$
60 pounds of the 25% copper alloy must be combined with 40 pounds of the 30% copper alloy.

61. Let x = miles driven
$\quad y$ = cost of rental
$y = 150+0.25x$
$y = 175+0.20x$
Substitute $150+0.25x$ for y in the second equation.
$150+0.25x = 175+0.20x$
$0.05x = 25$
$x = 500$
He must drive more than 500 miles during the week for company B to be more economical.

62. Let x = amt. invested at 4.5%
$\quad y$ = amt. invested at 6%
$x+y = 10,000,000$
$0.045x+0.06y = 487,500$
Multiply the first equation by -0.06 and add to the second equation.
$\underline{\begin{array}{l}-0.06x-0.06y = -600,000\\ 0.045x+0.06y = 487,500\end{array}}$
$\quad\quad\quad -0.015x = -112,500$
$\quad\quad\quad\quad\quad\quad x = 7,500,000$
Substitute this value for x in the first equation.
$(7,500,000)+y = 10,000,000$
$y = 2,500,000$
Julia invested \$7,500,000 at 4.5% and \$2,500,000 at 6%.

63. $2x-3y+4z = -14$
$2(3)-3(4)+4(-2) = -14$
$6-12-8 = -14$
$-14 = -14$ true

$5x-2z = 7$
$5(3)-2(-2) = 7$
$15+4 = 7$
$19 = 7$ false

$3x+4y = 25$
$3(3)+4(4) = 25$
$9+16 = 25$
$25 = 25$ true
Since the second equation is false, $(3,4,-2)$ is not a solution.

64. $8x+y+2z = 5$
$8(0.5)+(3.4)+2(-1.2) = 5$
$4+3.4-2.4 = 5$
$5 = 5$ true

$2x-5y-5z = -10$
$2(0.5)-5(3.4)-5(-1.2) = -10$
$1-17+6 = -10$
$-10 = -10$ true

$4x+2y+4z = 4$
$4(0.5)+2(3.4)+4(-1.2) = 4$
$2+6.8-4.8 = 4$
$4 = 4$ true
Since all three equations are true, $(0.5, 3.4, -1.2)$ is a solution to the system.

65. $4x+y+z = 7$
$x-y+z = 4$
$2x-y+z = 15$
Eliminate y.
Add the first equation to the second equation.
$4x+y+z = 7$
$\underline{x-y+z = 4}$
$5x+2z = 11$ (1)
Add the first equation to the third equation.
$4x+y+z = 7$
$\underline{2x-y+z = 15}$
$6x+2z = 22$ (2)

258

Make a new system of two equations using equations (1) and (2).
$5x + 2z = 11$
$6x + 2z = 22$
Eliminate z.
Multiply the first equation by -1 and add to the second equation.
$-5x - 2z = -11$
$\underline{6x + 2z = 22}$
$x = 11$
Substitute this value for x in the second equation.
$6(11) + 2z = 22$
$66 + 2z = 22$
$2z = -44$
$z = -22$
Substitute $x = 11$ and $z = -22$ into one of the original equations.
$(11) - y + (-22) = 4$
$-11 - y = 4$
$-y = 15$
$y = -15$
The solution is $(11, -15, -22)$.

66. $x + y + z = -1$
$x + y - z = 3$
$3y - z = 4$ (1)
Eliminate x.
Multiply the first equation by -1 and add to the second equation.
$-x - y - z = 1$
$\underline{x + y - z = 3}$
$ -2z = 4$
$z = -2$
Substitute $z = -2$ in equation (1).
$3y - (-2) = 4$
$3y + 2 = 4$
$3y = 2$
$y = \dfrac{2}{3}$
Substitute $y = \dfrac{2}{3}$ and $z = -2$ into one of the original equations to solve for x.

$x + \left(\dfrac{2}{3}\right) + (-2) = -1$
$x - \dfrac{4}{3} = -1$
$x = \dfrac{1}{3}$
The solution is $\left(\dfrac{1}{3}, \dfrac{2}{3}, -2\right)$.

67. $x + 2y - z = 3$
$3x - y + z = 6$
$2x - 3y + 2z = 3$
Eliminate z.
Add the first equation to the second equation.
$x + 2y - z = 3$
$\underline{3x - y + z = 6}$
$4x + y = 9$ (1)
Multiply the first equation by 2 and add to the third equation.
$2x + 4y - 2z = 6$
$\underline{2x - 3y + 2z = 3}$
$4x + y = 9$ (2)
Make a new system of two equations using equations (1) and (2).
$4x + y = 9$
$4x + y = 9$
These two equations are exactly the same. The system has an infinite number of solutions.

68. $x - 3y + 4z = 3$
$4x + y - z = 8$
$3x + 4y - 5z = 5$
Eliminate x.
Multiply the first equation by -4 and add to the second equation.
$-4x + 12y - 16z = -12$
$\underline{4x + y - z = 8}$
$13y - 17z = -4$ (1)
Multiply the first equation by -3 and add to the third equation.

$-3x+9y-12z=-9$
$3x+4y-5z=5$
$13y-17z=-4$ (2)

Make a new system of equations using equations (1) and (2).
$13y-17z=-4$
$13y-17z=-4$
These equations are identical. The system will have an infinite number of solutions.

69. Let x = number of students
y = number of teachers
z = number of visitors

$x+y+z=140$
$x-5y=0$ (1)
$2.5x+5y+7.5z=637.50$

Eliminate z.
Multiply the first equation by -7.5 and add to the third equation.
$-7.5x-7.5y-7.5z=-1050$
$2.5x+5y+7.5z=637.50$
$-5x-2.5y=-412.50$ (2)

Make a new system of two equations using equations (1) and (2).
$x-5y=0$
$-5x-2.5y=-412.50$

Eliminate x.
Multiply the first equation by 5 and add to the second equation.
$5x-25y=0$
$-5x-2.5y=-412.50$
$-27.5y=-412.50$
$y=15$

Substitute this value for y in the first equation.
$x-5(15)=0$
$x-75=0$
$x=75$

Substitute $x=75$ and $y=15$ into one of the original equations to solve for z.

$(75)+(15)+z=140$
$90+z=140$
$z=50$

There were 75 students, 15 teachers, and 50 visitors at the game.

70. Let x = number of clubs
y = number of veggies
z = number of chicken

$5x+3y+6z=22$
$26x+48z=100$ (1)
$312x+237y+348z=1446$

Eliminate y.
Multiply the first equation by -79 and add to the third equation.
$-395x-237y-474z=-1738$
$312x+237y+348z=1446$
$-83x-126z=-292$ (2)

Make a new system of two equations using equations (1) and (2).
$26x+48z=100$
$-83x-126z=-292$

Eliminate x.
Multiply the first equation by 83 and the second equation by 26. Add the new equations.
$2158x+3984z=8300$
$-2158x-3276z=-7592$
$708z=708$
$z=1$

Substitute this value for z in the first equation.
$26x+48(1)=100$
$26x=52$
$x=2$

Substitute $x=2$ and $z=1$ into one of the original equations.
$5(2)+3y+6(1)=22$
$10+3y+6=22$
$3y+16=22$
$3y=6$
$y=2$

Robin ate 2 club sandwiches, 2 veggie sandwiches, and 1 roast chicken sandwich.

71. $I_1 - I_2 + I_3 = 0$
$2I_1 + 3I_2 = 11$
$3I_2 + 4I_3 = 17$ (1)

Eliminate I_1.
Multiply the first equation by -2 and add to the second equation.
$-2I_1 + 2I_2 - 2I_3 = 0$
$\underline{2I_1 + 3I_2 \quad\quad = 11}$
$5I_2 - 2I_3 = 11$ (2)

Make a new system of two equations using equations (1) and (2).
$3I_2 + 4I_3 = 17$
$5I_2 - 2I_3 = 11$

Multiply the second equation by 2 and add to the first equation.
$3I_2 + 4I_3 = 17$
$\underline{10I_2 - 4I_3 = 22}$
$13I_2 = 39$
$I_2 = 3$

Substitute this value for I_2 in the first equation.
$3(3) + 4I_3 = 17$
$9 + 4I_3 = 17$
$4I_3 = 8$
$I_3 = 2$

Substitute $I_2 = 3$ and $I_3 = 2$ into one of the original equations to solve for I_1.
$I_1 - (3) + (2) = 0$
$I_1 - 1 = 0$
$I_1 = 1$

The three currents are $I_1 = 1, I_2 = 3,$ and $I_3 = 2$.

Chapter 6 Chapter Review

1. $y = \dfrac{4}{7}x + 2$
$x - 3y = 4$

Multiply the first equation, in standard form, and add to the second equation.

$-\dfrac{12}{7}x + 3y = 6$
$\underline{x - 3y = 4}$
$-\dfrac{5}{7}x = 10$
$x = -14$

Substitute this value for x in the first equation.
$y = \dfrac{4}{7}(-14) + 2$
$y = -8 + 2 = -6$

The solution is $(-14, -6)$.

2. $y = \dfrac{4}{5}x - 6$
$5y + 2 = 4(x - 7)$

Write both equations in standard form.
$\dfrac{4}{5}x - y = 6$
$4x - 5y = 30$

Multiply the first equation by -5 and add to the second equation.
$-4x + 5y = -30$
$\underline{4x - 5y = 30}$
$0 = 0$

This is an identity. There are an infinite number of solutions. Any ordered pair, (x, y), satisfying $y = \dfrac{4}{5}x - 6$ is a solution.

3. $1.25x + 3.5y = -25$
$4.5x + 2.8y = 8$

Multiply the first equation by 4 and the second equation by -5. Add the new equations.
$5x + 14y = -100$
$\underline{-22.5x - 14y = -40}$
$-17.5x = -140$
$x = 8$

Substitute this value for x in the first equation.

$1.25(8)+3.5y=-25$
$10+3.5y=-25$
$3.5y=-35$
$y=-10$
The solution is $(8,-10)$.

4. $0.55x-0.68y=48$
$\underline{-0.51x+0.68y=0}$
$0.04x=48$
$x=1200$
Substitute this value for x in the first equation.
$0.55(1200)-0.68y=48$
$660-0.68y=48$
$-0.68y=-612$
$y=900$
The solution is $(1200,900)$.

5. $8x+2y=-6$
$\underline{6x-2y=-78}$
$14x=-84$
$x=-6$
Substitute this value for x in the first equation.
$8(-6)+2y=-6$
$-48+2y=-6$
$2y=42$
$y=21$
The solution is $(-6,21)$.

6. $2x+8y=-12$
$y=2x+3$
Write the second equation in standard form and add to the first equation.
$2x+8y=-12$
$\underline{-2x+\ y=3}$
$9y=-9$
$y=-1$
Substitute this value for y in the first equation.

$2x+8(-1)=-12$
$2x-8=-12$
$2x=-4$
$x=-2$
The solution is $(-2,-1)$.

7. $3x+y=4$
$12x+4y=9$
Multiply the first equation by -4 and add to the second equation.
$-12x-4y=-16$
$\underline{12x+4y=9}$
$0=-7$
This is a contradiction. There is no solution to the system.

8. $3x-5y=11$
$4x+2y=13$
Multiply the first equation by 4 and the second equation by -3. Add the new equations.
$12x-20y=44$
$\underline{-12x-\ 6y=-39}$
$-26y=5$
$y=-\dfrac{5}{26}$
Substitute this value for y in the first equation.
$3x-5\left(-\dfrac{5}{26}\right)=11$
$3x+\dfrac{25}{26}=11$
$3x=\dfrac{261}{26}$
$x=\dfrac{87}{26}$
The solution is $\left(\dfrac{87}{26},-\dfrac{5}{26}\right)$.

9. $\dfrac{1}{3}x-\dfrac{2}{5}y=10$
$\dfrac{1}{2}x+\dfrac{4}{5}y=-13$
Multiply the first equation by 2 and add to the second equation.

$\frac{2}{3}x - \frac{4}{5}y = 20$

$\frac{1}{2}x + \frac{4}{5}y = -13$

$\frac{7}{6}x = 7$

$x = 6$

Substitute this value for x in the first equation.

$\frac{1}{3}(6) - \frac{2}{5}y = 10$

$2 - \frac{2}{5}y = 10$

$-\frac{2}{5}y = 8$

$y = -20$

The solution is $(6, -20)$.

10. $4x + 8y = 68$

$5x - 3y = -6$

Multiply the first equation by 5 and the second equation by -4. Add the new equations.

$20x + 40y = 340$

$-20x + 12y = 24$

$52y = 364$

$y = 7$

Substitute this value for y in the first equation.

$4x + 8(7) = 68$

$4x + 56 = 68$

$4x = 12$

$x = 3$

The solution is $(3, 7)$.

11. $2x + y = 0$

$x + 4y = 3$

Multiply the second equation by -2 and add to the first equation.

$2x + y = 0$

$-2x - 8y = -6$

$-7y = -6$

$y = \frac{6}{7}$

Substitute this value for y in the first equation.

$2x + \left(\frac{6}{7}\right) = 0$

$2x = -\frac{6}{7}$

$x = -\frac{3}{7}$

The solution is $\left(-\frac{3}{7}, \frac{6}{7}\right)$.

12. $5x + 3y = 9$

$x - y = 6$

Multiply the second equation by 3 and add to the first equation.

$5x + 3y = 9$

$3x - 3y = 18$

$8x = 27$

$x = \frac{27}{8}$

Substitute this value for x in the second equation.

$\frac{27}{8} - y = 6$

$-y = \frac{21}{8}$

$y = -\frac{21}{8}$

The solution is $\left(\frac{27}{8}, -\frac{21}{8}\right)$.

13. $2y + 2 = y$ $y = -x + 1$

$y + 2 = 0$

$y = -2$

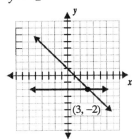

The solution is $(3, -2)$.

14. $2x+3=x$ $\qquad$ $2x=5$
    $x+3=0$ $\qquad$ $x=\dfrac{5}{2}$
    $x=-3$

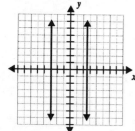

The lines are parallel. There is no solution to the system.

15. $y=2x+10$ $\qquad$ $3x+y=-10$
    $\qquad\qquad\qquad\quad$ $y=-3x-10$

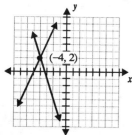

The solution is $(-4, 2)$.

16. $2(x-1)-4=0$ $\qquad$ $2x+y=3$
    $2x-2-4=0$ $\qquad\quad$ $y=-2x+3$
    $2x=6$
    $x=3$

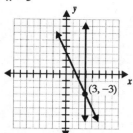

The solution is $(3, -3)$.

17. $x+y=9-x$ $\qquad$ $2x+y=5$
    $y=-2x+9$ $\qquad\;\,$ $y=-2x+5$

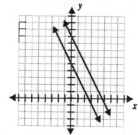

The lines are parallel. The system has no solution.

18. $y=-4x-3$ $\qquad$ $4x+2y=y-3$
    $\qquad\qquad\qquad\;\;$ $4x+y=-3$
    $\qquad\qquad\qquad\;\;$ $y=-4x-3$

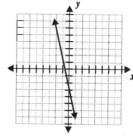

The lines are coinciding. There are an infinite number of solutions to the system. All ordered pairs (x, y) satisfying $y=-4x-3$ are solutions.

19. $3x+4=2x$ $\qquad$ $x+7=3$
    $x+4=0$ $\qquad\quad\;$ $x=-4$
    $x=-4$

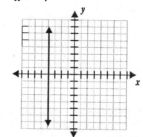

The lines are coinciding. There are an infinite number of solutions to the system. All ordered pairs (x, y) satisfying $x=-4$ are solutions.

20. $3y+7=2y+5 \qquad y+9=7$
$y+7=5 \qquad\qquad y=-2$
$y=-2$

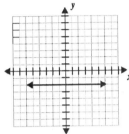

The lines are coinciding. There are an infinite number of solutions to the system. All ordered pairs (x,y) satisfying $y=-2$ are solutions.

21. $3(x-3)-1=8 \qquad 3y-10=15-2y$
$3x-9-1=8 \qquad\quad 5y-10=15$
$3x=18 \qquad\qquad\quad 5y=25$
$x=6 \qquad\qquad\qquad y=5$

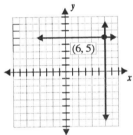

The solution is $(6,5)$.

22. $y=3x+7 \qquad x-y=-1$
$\qquad\qquad\qquad y=x+1$

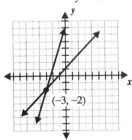

The solution is $(-3,-2)$.

23. $x-5y=15 \qquad x+5y=-5$
$5y=x-15 \qquad\; 5y=-x-5$
$y=\dfrac{1}{5}x-3 \qquad y=-\dfrac{1}{5}x-1$

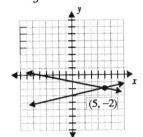

The solution is $(5,-2)$.

24. $y=-4x-6 \qquad y=3x+8$

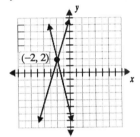

The solution is $(-2,2)$.

25. $y=x \qquad x=2-y$
$\qquad\qquad y=-x+2$

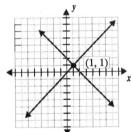

The solution is $(1,1)$.

26. $x+9=3 \quad 3y=2y-1$
$\quad x=-6 \quad y=-1$

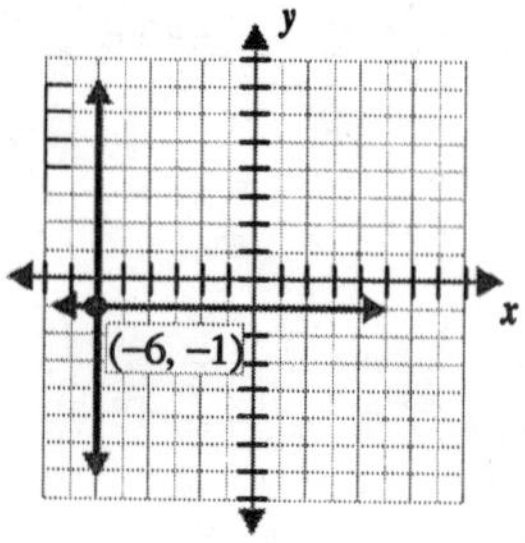

[Graph showing horizontal and vertical lines intersecting at $(-6,-1)$]

The solution is $(-6,-1)$.

27. $y=\dfrac{5}{11}x-43$
$x+2y=-2$

Substitute $\dfrac{5}{11}x-43$ for y in the second equation.

$x+2\left(\dfrac{5}{11}x-43\right)=-2$

$x+\dfrac{10}{11}x-86=-2$

$\dfrac{21}{11}x-86=-2$

$\dfrac{21}{11}x=84$

$x=44$

Substitute this value for x in the first equation.

$y=\dfrac{5}{11}(44)-43=20-43=-23$

The solution is $(44,-23)$.

28. $y=\dfrac{3}{4}x+14$

$y=-\dfrac{2}{3}x-3$

Substitute $\dfrac{3}{4}x+14$ for y in the second equation.

$\dfrac{3}{4}x+14=-\dfrac{2}{3}x-3$

$\dfrac{17}{12}x+14=-3$

$\dfrac{17}{12}x=-17$

$x=-12$

Substitute this value for x in the first equation.

$y=\dfrac{3}{4}(-12)+14=-9+14=5$

The solution is $(-12,5)$.

29. $2x+3y=53$
$2x-4y=-248$

Solve the second equation for x.

$2x=4y-248$

$x=2y-124$

Substitute this result for x in the first equation.

$2(2y-124)+3y=53$

$4y-248+3y=53$

$7y-248=53$

$7y=301$

$y=43$

Substitute this value for y in the second equation.

$2x-4(43)=-248$

$2x-172=-248$

$2x=-76$

$x=-38$

The solution is $(-38,43)$.

30. $y=4x-7$
$2y+12=4x+y+5$

Substitute $4x-7$ for y in the second equation.

$2(4x-7)+12=4x+(4x-7)+5$

$8x-14+12=8x-2$

$8x-2=8x-2$

$0=0$

This is an identity. There are an infinite number of solutions to the system. All ordered pairs (x,y) satisfying $y=4x-7$ are solutions.

31. $y = -5x$
$x - y = 1$
Substitute $-5x$ for y in the second equation.
$x - (-5x) = 1$
$6x = 1$
$x = \dfrac{1}{6}$
Substitute this value for x in the first equation.
$y = -5\left(\dfrac{1}{6}\right) = -\dfrac{5}{6}$
The solution is $\left(\dfrac{1}{6}, -\dfrac{5}{6}\right)$.

32. $5x - 10y = 21$
$5x = -3$
Solve the second equation for x.
$x = -\dfrac{3}{5}$
Substitute this value for x in the first equation.
$5\left(-\dfrac{3}{5}\right) - 10y = 21$
$-3 - 10y = 21$
$-10y = 24$
$y = -\dfrac{12}{5}$
The solution is $\left(-\dfrac{3}{5}, -\dfrac{12}{5}\right)$.

33. $x + 7y = 3$
$4x + y = 9$
Solve the second equation for y.
$y = -4x + 9$
Substitute this expression for y in the first equation.
$x + 7(-4x + 9) = 3$
$x - 28x + 63 = 3$
$-27x + 63 = 3$
$-27x = -60$
$x = \dfrac{20}{9}$
Substitute this value for x in the second equation.

$4\left(\dfrac{20}{9}\right) + y = 9$
$\dfrac{80}{9} + y = 9$
$y = \dfrac{1}{9}$
The solution is $\left(\dfrac{20}{9}, \dfrac{1}{9}\right)$.

34. $5x + 6y = 669$
$x - 2y = 1705$
Solve the second equation for x.
$x = 2y + 1705$
Substitute this result for x in the first equation.
$5(2y + 1705) + 6y = 669$
$10y + 8525 + 6y = 669$
$16y = -7856$
$y = -491$
Substitute this value for y in the second equation.
$x - 2(-491) = 1705$
$x + 982 = 1705$
$x = 723$
The solution is $(723, -491)$.

35. $x - y = 5$
$6x + 2y = 9$
Solve the first equation for x.
$x = y + 5$
Substitute this value for x in the second equation.
$6(y + 5) + 2y = 9$
$6y + 30 + 2y = 9$
$8y + 30 = 9$
$8y = -21$
$y = -\dfrac{21}{8}$
Substitute this value for y in the first equation.

$x - \left(-\dfrac{21}{8}\right) = 5$

$x + \dfrac{21}{8} = 5$

$x = \dfrac{19}{8}$

The solution is $\left(\dfrac{19}{8}, -\dfrac{21}{8}\right)$.

36. $x = 324y$

$3x - 810y = 90$

Substitute $324y$ for x in the second equation.

$3(324y) - 810y = 90$

$972y - 810y = 90$

$162y = 90$

$y = \dfrac{5}{9}$

Substitute this value for x in the first equation.

$x = 324\left(\dfrac{5}{9}\right) = 180$

The solution is $\left(180, \dfrac{5}{9}\right)$.

37. $y = -2x + 7$

$2x + y = 0$

Substitute $-2x + 7$ for y in the second equation.

$2x + (-2x + 7) = 0$

$7 = 0$

This is a contradiction. There is no solution to the system.

38. $2x - y = 60$

$5x - 3y = 60$

Solve the first equation for y.

$y = 2x - 60$

Substitute this result for y in the second equation.

$5x - 3(2x - 60) = 60$

$5x - 6x + 180 = 60$

$-x + 180 = 60$

$x = 120$

Substitute this value for x in the first equation.

$2(120) - y = 60$

$240 - y = 60$

$y = 180$

The solution is $(120, 180)$.

39. $4x + 2y = 3$

$4\left(\dfrac{5}{6}\right) + 2\left(-\dfrac{1}{6}\right) = 3$

$\dfrac{10}{3} - \dfrac{1}{3} = 3$

$\dfrac{9}{3} = 3$

$3 = 3$ true

$x - y = -1$

$\left(\dfrac{5}{6}\right) - \left(-\dfrac{1}{6}\right) = -1$

$\dfrac{5}{6} + \dfrac{1}{6} = -1$

$\dfrac{6}{6} = -1$

$1 = -1$ false

Since the second equation is false, $\left(\dfrac{5}{6}, -\dfrac{1}{6}\right)$ is not a solution.

40. $3x + y = -1$

$3(-1.6) + (3.8) = -1$

$-4.8 + 3.8 = -1$

$-1 = -1$ true

$5x + 5y = 11$

$5(-1.6) + 5(3.8) = 11$

$-8 + 19 = 11$

$11 = 11$ true

Since both equations are true, $(-1.6, 3.8)$ is a solution.

Chapter 6: Systems of Linear Equations

41. $5x + 7y = -6$
$5(-4) + 7(2) = -6$
$-20 + 14 = -6$
$-6 = -6$ true

$2x - 7y = -22$
$2(-4) - 7(2) = -22$
$-8 - 14 = -22$
$-22 = -22$ true
Since both equations are true, $(-4, 2)$ is a solution.

42. $x + y = 1$
$\left(\frac{13}{9}\right) + \left(-\frac{4}{9}\right) = 1$
$\frac{9}{9} = 1$
$1 = 1$ true

$8x - y = 12$
$8\left(\frac{13}{9}\right) - \left(-\frac{4}{9}\right) = 12$
$\frac{104}{9} + \frac{4}{9} = 12$
$\frac{108}{9} = 12$
$12 = 12$ true
Both equations are true so $\left(\frac{13}{9}, -\frac{4}{9}\right)$. is a solution.

43. $2x - 1 = 5$
$2(3) - 1 = 5$
$6 - 1 = 5$
$5 = 5$ true

$3y + 5 = 2$
$3(-2) + 5 = 2$
$-6 + 5 = 2$
$-1 = 2$ false
Since the second equation is false, $(3, -2)$ is not a solution.

44. $x + 10 = 5$
$(-5) + 10 = 5$
$5 = 5$ true

$3y + 11 = 2$
$3(-3) + 11 = 2$
$-9 + 11 = 2$
$2 = 2$ true
Since both equations are true, $(-5, -3)$ is a solution.

45. $x + 2y = 2$
$(-0.44) + 2(1.22) = 2$
$-0.44 + 2.44 = 2$
$2 = 2$ true

$2x + 4y = 5$
$2(-0.44) + 4(1.22) = 5$
$-0.88 + 4.88 = 5$
$4 = 5$ false
Since the second equation is false, $(-0.44, 1.22)$ is not a solution.

46. $2x + 5y = -19$
$2(-2) + 5(-3) = -19$
$-4 - 15 = -19$
$-19 = -19$ true

$3x - 7y = 10$
$3(-2) - 7(-3) = 10$
$-6 + 21 = 10$
$15 = 10$ false
Since the second equation is false, $(-2, -3)$ is not a solution.

47. Let x = pounds of 15% alloy
y = pounds of 35% alloy
$x + y = 200$
$0.15x + 0.35y = 0.27(200)$
Multiply the first equation by -0.15 and add to the second equation.
$-0.15x - 0.15y = -30$
$\underline{0.15x + 0.35y = 54}$
$0.2y = 24$
$y = 120$
Substitute this value for y in the first equation.
$x + (120) = 200$
$x = 80$
The mixture should contain 80 pounds of the 15% alloy and 120 pounds of the 35% alloy. There should be 40 more pounds of the 35% alloy than the 15% alloy.

48. Let x = rent for 1-bedroom apartment
y = rent for 2-bedroom apartment
$8x + 12y = 7300$
$y = 150 + x$
Substitute $150 + x$ for y in the first equation.
$8x + 12(150 + x) = 7300$
$8x + 1800 + 12x = 7300$
$20x = 5500$
$x = 275$
Substitute this value for x in the second equation.
$y = 150 + 275 = 425$
$8x = 8(275) = 2200$
$12y = 12(425) = 5100$
The landlord will receive $2200 from the 1-bedroom apartments and $5100 from the 2-bedroom apartments.

49. Let x = measure of first angle
y = measure of second angle
$x + y = 180$
$y = x + 10$
Substitute $x + 10$ for y in the first equation.

$x + (x + 10) = 180$
$2x = 170$
$x = 85$
Substitute this value for x in the second equation.
$y = (85) + 10 = 95$
The smaller angle measures 85°.

50. Let x = number of men surveyed
y = number of women surveyed
$x + y = 700$
$0.4x + 0.7y = 400$
Multiply the first equation by -0.4 and add to the second equation.
$-0.4x - 0.4y = -280$
$\underline{0.4x + 0.7y = 400}$
$0.3y = 120$
$y = 400$
Substitute this value for y in the first equation.
$x + (400) = 700$
$x = 300$
$400 - 300 = 100$
There were 100 more women than men surveyed.

51. Let x = measure of first angle
y = measure of second angle
$x + y = 90$
$y = 12 + 3x$
Substitute $12 + 3x$ for y in the first equation.
$x + (12 + 3x) = 90$
$4x + 12 = 90$
$4x = 78$
$x = 19.5$
Substitute this value for x in the second equation.
$y = 12 + 3(19.5) = 12 + 58.5 = 70.5$
The angles measure 19.5° and 70.5°.

52. Let x = cc of 15% sulfuric acid
y = total cc of mixture
$y = x + 18$
$0.15x + 0.25(18) = 0.21y$
Substitute $x + 18$ for y in the second equation.

$0.15x + 4.5 = 0.21(x + 18)$
$0.15x + 4.5 = 0.21x + 3.78$
$-0.06x = -0.72$
$x = 12$
Substitute this value for x in the first equation.
$y = (12) + 18 = 30$
The chemist should use 12 cc of the 15% sulfuric acid to obtain 30 cc of a 21% sulfuric acid mixture.

53. Let x = rate of walking
$\quad\quad y$ = rate of jogging
$\frac{1}{2}x + \frac{1}{3}y = 3.5$
$\frac{2}{3}x + \frac{1}{3}y = 4$
Multiply the second equation by -1 and add to the first equation.
$\frac{1}{2}x + \frac{1}{3}y = 3.5$
$-\frac{2}{3}x - \frac{1}{3}y = -4$
$\overline{\quad -\frac{1}{6}x = -.5 \quad}$
$x = 3$
Substitute this value for x in the second equation.
$\frac{2}{3}(3) + \frac{1}{3}y = 4$
$2 + \frac{1}{3}y = 4$
$\frac{1}{3}y = 2$
$y = 6$
The patient walked at a rate of 3 miles per hour and jogged at a rate of 6 miles per hour.

54. Let x = pounds of gourmet coffee
$\quad\quad y$ = pounds of Dutch chocolate
$x + y = 50$
$8.5x + 12.5y = 9.5(50)$
Multiply the first equation by -8.5 and add to the second equation.

$-8.5x - 8.5y = -425$
$8.5x + 12.5y = 475$
$\overline{\quad 4y = 50 \quad}$
$y = 12.5$
Substitute this value for y in the first equation.
$x + (12.5) = 50$
$x = 37.5$
The shop should mix 37.5 pounds of gourmet chocolate with 12.5 pounds of Dutch chocolate.

55. Let x = hours on production line
$\quad\quad y$ = hours as clerk
$x + y = 20$
$10.75x + 6.5y = 181$
Multiply the first equation by -6.5 and add to the second equation.
$-6.5x - 6.5y = -130$
$10.75x + 6.5y = 181$
$\overline{\quad 4.25x = 51 \quad}$
$x = 12$
Substitute this value for x in the first equation.
$(12) + y = 20$
$y = 8$
Nikki should work 12 hours on the production line and 8 hours as a night clerk.

56. Let x = avg. speed in still air
$\quad\quad y$ = avg. speed of jet stream
$5.75(x - y) = 2600$
$5(x + y) = 2600$
Solve the second equation for x.
$x + y = 520$
$x = -y + 520$
Substitute this result for x in the first equation.
$5.75(-y + 520 - y) = 2600$
$5.75(-2y + 520) = 2600$
$-11.5y + 2990 = 2600$
$-11.5y = -390$
$y = \frac{780}{23} \approx 33.91$
Substitute this value for y in the second equation.

$5\left(x+\dfrac{780}{23}\right)=2600$

$x+\dfrac{780}{23}=520$

$x=\dfrac{11180}{23}\approx 486.09$

The planes average speed is approximately 486.09 miles per hour. The jet stream's average speed is approximately 33.91 miles per hour.

57. Let x = number of miles driven

$U(x)=39.95+0.15x$

$B(x)=19.95+0.22x$

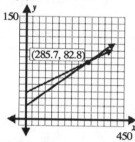

You must drive at least 286 miles to make U Rent less costly.

58. Let x = number of items produced and sold

$r(x)=6.5x$

$c(x)=75+2.5x$

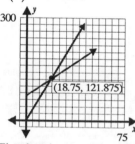

The shop keeper must sell at least 19 items to make a profit.

59. $x+y+z=23$

$2x-y+z=-11$

$x+3y-z=5$

Eliminate z.
Add the first equation to the second equation.

$x+y+z=23$

$x+3y-z=5$

$\overline{2x+4y=28}$ (1)

Multiply the second equation by -1 and add to the first equation.

$x+y+z=23$

$-2x+y-z=11$

$\overline{-x+2y=34}$ (2)

Make a new system of two equations using equations (1) and (2).

$2x+4y=28$

$-x+2y=34$

Eliminate x.
Multiply the second equation by 2 and add to the first equation.

$2x+4y=28$

$-2x+4y=68$

$\overline{8y=96}$

$y=12$

Substitute this value for y in the first equation.

$2x+4(12)=28$

$2x+48=28$

$2x=-20$

$x=-10$

Substitute $x=-10$ and $y=12$ into one of the original equations to solve for z.

$(-10)+(12)+z=23$

$z=21$

The solution is $(-10,12,21)$.

60. $x+2y+z=-1$

$x-2y-2z=5$

$x+6y+z=2$

Eliminate y.
Add the first equation and the second equation.

$x+2y+z=-1$

$x-2y-2z=5$

$\overline{2x-z=4}$ (1)

Multiply the second equation by 3 and add to the third equation.

$3x - 6y - 6z = 15$

$\underline{x + 6y + z = 2}$

$4x - 5z = 17$ (2)

Make a new system of two equations using equations (1) and (2).

$2x - z = 4$

$4x - 5z = 17$

Multiply the first equation by -2 and add to the second equation.

$-4x + 2z = -8$

$\underline{4x - 5z = 17}$

$-3z = 9$

$z = -3$

Substitute this value for z in the first equation.

$2x - (-3) = 4$

$2x + 3 = 4$

$2x = 1$

$x = \dfrac{1}{2}$

Substitute $x = \dfrac{1}{2}$ and $z = -3$ into one of the original equations to solve for y.

$\left(\dfrac{1}{2}\right) + 2y + (-3) = -1$

$2y - \dfrac{5}{2} = -1$

$2y = \dfrac{3}{2}$

$y = \dfrac{3}{4}$

The solution is $\left(\dfrac{1}{2}, \dfrac{3}{4}, -3\right)$.

61. $I_1 - I_2 - I_3 = 0$

$3I_1 + 4I_2 = 18.5$ (1)

$ 4I_2 - 6I_3 = -1$

Eliminate I_3.

Multiply the first equation by -6 and add to the third equation.

$-6I_1 + 6I_2 + 6I_3 = 0$

$\underline{ 4I_2 - 6I_3 = -1}$

$-6I_1 + 10I_2 = -1$ (2)

Make a new system of two equations using equations (1) and (2).

$3I_1 + 4I_2 = 18.5$

$-6I_1 + 10I_2 = -1$

Multiply the first equation by 2 and add to the second equation.

$6I_1 + 8I_2 = 37$

$\underline{-6I_1 + 10I_2 = -1}$

$18I_2 = 36$

$I_2 = 2$

Substitute this value for I_2 in the first equation.

$3I_1 + 4(2) = 18.5$

$3I_1 + 8 = 18.5$

$3I_1 = 10.5$

$I_1 = \dfrac{7}{2}$

Substitute $I_1 = \dfrac{7}{2}$ and $I_2 = 2$ into one of the original equations to solve for I_3.

$\left(\dfrac{7}{2}\right) - (2) - I_3 = 0$

$\dfrac{3}{2} - I_3 = 0$

$I_3 = \dfrac{3}{2}$

The three currents (in amperes) are

$(I_1, I_2, I_3) = \left(\dfrac{7}{2}, 2, \dfrac{3}{2}\right)$.

62. Let x = number of turkey breast sandwiches

y = number of ham sandwiches

z = number of hamburgers

$4x + 5y + 39z = 92$

$19x + 28y + 90z = 255$

$289x + 302y + 640z = 2173$

Eliminate x.

Multiply the first equation by 19 and the second equation by -4. Add the new equations.

$76x + 95y + 741z = 1748$

$\underline{-76x - 112y - 360z = -1020}$

$-17y + 381z = 728$ (1)

Multiply the first equation by 289 and the third equation by –4. Add the new equations.
$1156x + 1445y + 11271z = 26588$
$-1156x - 1208y - 2560z = -8692$
$\overline{}$
$\qquad 237y + 8711z = 17896 \quad (2)$

Make a new system of two equations using equations (1) and (2).
$-17y + 381z = 728$
$237y + 8711z = 17896$

Eliminate y.
Multiply the first equation by 237 and the second equation by 17. Add the new equations.
$-4029y + 90297z = 172{,}536$
$4029y + 148087z = 304{,}232$
$\overline{}$
$\qquad 238384z = 476768$
$\qquad\qquad z = 2$

Substitute this value for z in the first equation.
$-17y + 381(2) = 728$
$-17y + 762 = 728$
$-17y = -34$
$y = 2$

Substitute $y = 2$ and $z = 2$ into one of the original equations to solve for x.
$4x + 5(2) + 39(2) = 92$
$4x + 10 + 78 = 92$
$4x = 4$
$x = 1$

Matthew ate 1 turkey breast sandwich, 2 ham sandwiches, and 2 hamburgers.

Chapter 6 Test

1. $y = -3.5x + 15.5$
$(-64.3) = -3.5(22.8) + 15.5$
$-64.3 = -79.8 + 15.5$
$-64.3 = -64.3$ true

$15x + 10y = -301$
$15(22.8) + 10(-64.3) = -301$
$342 - 643 = -301$
$-301 = -301$ true

Since both equations are true, $(22.8, -64.3)$ is a solution of the system.

2. $2y - 8 = 0 \qquad y = 4$
$2y = 8$
$y = 4$

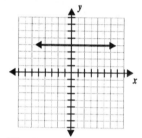

The lines are coinciding. There are an infinite number of solutions to the system. All ordered pairs (x, y), such that $y = 4$, are solutions to the system.

3. $y = \dfrac{1}{2}x + 3 \qquad y = \dfrac{1}{2}x - 5$

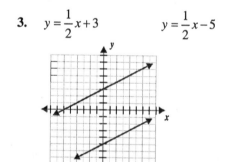

The lines are parallel. There is no solution to the system.

4. $y = 5x - 9 \qquad 4x + 8y = 16$
$\qquad\qquad\qquad\; 8y = -4x + 16$
$\qquad\qquad\qquad\; y = -\dfrac{1}{2}x + 2$

Chapter 6: Systems of Linear Equations

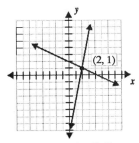

The solution is $(2,1)$.

5. $x + 2y = 6$
$5x - 11y = -54$
Solve the first equation for x.
$x = -2y + 6$
Substitute this result for x in the second equation.
$5(-2y + 6) - 11y = -54$
$-10y + 30 - 11y = -54$
$-21y + 30 = -54$
$-21y = -84$
$y = 4$
Substitute this value for y in the first equation.
$x + 2(4) = 6$
$x + 8 = 6$
$x = -2$
The solution is $(-2, 4)$.

6. $3x + 6y = 12$
$x + 2y = -5$
Solve the second equation for x.
$x = -2y - 5$
Substitute this result for x in the first equation.
$3(-2y - 5) + 6y = 12$
$-6y - 15 + 6y = 12$
$-15 = 12$
This is a contradiction. There is no solution to the system.

7. $x + 6y = -2$
$x - 2y = 1$
Solve the second equation for x.
$x = 2y + 1$
Substitute this result for x in the first equation.
$(2y + 1) + 6y = -2$
$8y + 1 = -2$
$8y = -3$
$y = -\dfrac{3}{8}$
Substitute this value for y in the second equation.
$x - 2\left(-\dfrac{3}{8}\right) = 1$
$x + \dfrac{3}{4} = 1$
$x = \dfrac{1}{4}$
The solution is $\left(\dfrac{1}{4}, -\dfrac{3}{8}\right)$.

8. $3x + 7y = 11$
$6x + 14y = 2$
Multiply the first equation by -2 and add to the second equation.
$-6x - 14y = -22$
$\underline{6x + 14y = 2}$
$0 = -20$
This is a contradiction. There is no solution to the system.

9. $2x + 9y = 16$
$5y = 8 - x$
Multiply the second equation by -2 and write in standard form. Add to the first equation.
$2x + 9y = 16$
$\underline{-2x - 10y = -16}$
$-y = 0$
$y = 0$
Substitute this value for y in the first equation.
$2x + 9(0) = 16$
$2x = 16$
$x = 8$
The solution is $(8, 0)$.

10. $5x + 3y = 11$
$2x + 5y = 7$
Multiply the first equation by 2 and the second equation by -5. Add the new equations.

SSM: Experiencing Introductory and Intermediate Algebra

$$10x + 6y = 22$$
$$-10x - 25y = -35$$
$$-19y = -13$$
$$y = \frac{13}{19}$$

Substitute this value for y in the first equation.

$$5x + 3\left(\frac{13}{19}\right) = 11$$

$$5x + \frac{39}{19} = 11$$

$$5x = \frac{170}{19}$$

$$x = \frac{34}{19}$$

The solution is $\left(\frac{34}{19}, \frac{13}{19}\right)$.

11. Let $x =$ the number of items.

   a. $c(x) = 5500 + 65x$

   b. $r(x) = 125x$

   c. $y = 5500 + 65x$
      $y = 125x$
      Substitute $125x$ for y in the first equation.
      $125x = 5500 + 65x$
      $60x = 5500$
      $x = \frac{275}{3} = 91\frac{2}{3}$

      The factory must sell at least 92 items to make a profit. The factory will lose money if it sells 91 or fewer items.

12. Let $x =$ Kenny's speed
    $y =$ miles to Dolly
    $y = 3x$
    $y = 2(x + 4)$
    Substitute $3x$ for y in the second equation.
    $3x = 2x + 8$
    $x = 8$
    Substitute this value for x in the first equation.
    $y = 3(8) = 24$

    Kenny can pedal at a rate of 8 miles per hour. Dolly was 24 miles away.

13. Let $x =$ nanoseconds for 1 addition
    $y =$ nanoseconds for 1 multiplication
    $6x + 8y = 58$
    $10x + 5y = 55$
    Multiply the first equation by 5 and the second equation by -3. Add the new equations.
    $$30x + 40y = 290$$
    $$-30x - 15y = -165$$
    $$25y = 125$$
    $$y = 5$$
    Substitute this value for y in the first equation.
    $6x + 8(5) = 58$
    $6x + 40 = 58$
    $6x = 18$
    $x = 3$
    It takes the computer 3 nanoseconds for one addition and 5 nanoseconds for one multiplication.

14. Let $x =$ pounds of raisins
    $y =$ pounds of peanuts
    $1.25x + 2y = 1.5(30)$
    $x + y = 30$
    Multiply the second equation by -2 and add to the first equation.
    $$1.25x + 2y = 45$$
    $$-2x - 2y = -60$$
    $$-0.75x = -15$$
    $$x = 20$$
    Substitute this value for x in the second equation.
    $(20) + y = 30$
    $y = 10$
    The candy shop owner should mix 20 pounds of the chocolate-covered raisins with 10 pounds of the chocolate-covered peanuts.

Chapter 6: Systems of Linear Equations

15. Let x = measure of smaller angle
y = measure of larger angle
$x + y = 90$
$y = 2x$
Substitute $2x$ for y in the first equation.
$x + (2x) = 90$
$3x = 90$
$x = 30$
Substitute this value for x in the second equation.
$y = 2(30) = 60$
The angle measures should be $30°$ and $60°$.

16. Let x = cc of 50% solution
y = cc of 10% solution
$x + y = 100$
$.5x + .1y = .3(100)$
Multiply the second equation by -2 and add to the first equation.
$x + y = 100$
$-x - 0.2y = -60$
$\overline{0.8y = 40}$
$y = 50$
Substitute this value for y in the first equation.
$x + (50) = 100$
$x = 50$
The chemist should mix 50 cc of both the 50% and 10% solutions.

17. Let x = measure of smaller angle
y = measure of larger angle
$x + y = 180$
$y = 15 + 2x$
Substitute $15 + 2x$ for y in the first equation.
$x + (15 + 2x) = 180$
$3x + 15 = 180$
$3x = 165$
$x = 55$
Substitute this value for x in the second equation.
$y = 15 + 2(55)$
$y = 15 + 110$
$y = 125$
The angles measure $55°$ and $125°$.

18. Let x = amount invested in fund A
y = amount invested in fund B
$0.095x + 0.07y = 780$
$x = 2y$
Substitute $2y$ for x in the first equation.
$0.095(2y) + 0.07y = 780$
$0.19y + 0.07y = 780$
$0.26y = 780$
$y = 3000$
Substitute this value for y in the second equation.
$x = 2(3000) = 6000$
Caitlin should invest $6000 in Mutual Fund A and $3000 in Mutual Fund B.

19. Let x = boat's average speed in still water
y = average speed of current
$\dfrac{3}{5}(x - y) = 36$
$\dfrac{1}{2}(x + y) = 36$
Multiply the first equation by 5 and the second equation by 6. Add the new equations.
$3x - 3y = 180$
$3x + 3y = 216$
$\overline{6x = 396}$
$x = 66$
Substitute this value for x in the second equation.
$\dfrac{1}{2}(66 + y) = 36$
$66 + y = 72$
$y = 6$
The boat has an average speed of 66 miles per hour. The river current averages 6 miles per hour.

20. $5x - y + 4z = -1$
$3y + 4z = 0$ (1)
$10x - 7y = 0$
Eliminate x.
Multiply the first equation by -2 and add to the third equation.

277

SSM: Experiencing Introductory and Intermediate Algebra

$-10x + 2y - 8z = 2$
$\underline{10x - 7y = 0}$
$ -5y - 8z = 2$ (2)

Make new system of two equations using equations (1) and (2).
$3y + 4z = 0$
$-5y - 8z = 2$

Eliminate z.
Multiply the first equation by 2 and add to the second equation.
$6y + 8z = 0$
$\underline{-5y - 8z = 2}$
$y = 2$

Substitute this value for y in the first equation.
$3(2) + 4z = 0$
$4z = -6$
$z = -\dfrac{3}{2}$

Substitute $y = 2$ and $z = -\dfrac{3}{2}$ into one of the original equations to solve for x.
$5x - (2) + 4\left(-\dfrac{3}{2}\right) = -1$
$5x - 2 - 6 = -1$
$5x = 7$
$x = \dfrac{7}{5}$

The solution is $\left(\dfrac{7}{5}, 2, -\dfrac{3}{2}\right)$.

21. $3x - 2y + z = -4$
$x - y - z = 3$
$2x - y + 2z = -7$

Eliminate x.
Multiply the second equation by -3 and add to the first equation.
$3x - 2y + z = -4$
$\underline{-3x + 3y + 3z = -9}$
$y + 4z = -13$ (1)

Multiply the second equation by -2 and add to the third equation.

$-2x + 2y + 2z = -6$
$\underline{2x - y + 2z = -7}$
$y + 4z = -13$ (2)

Make a new system of two equations using equations (1) and (2).
$y + 4z = -13$
$y + 4z = -13$

The two equations are identical. There are an infinite number of solutions to the system. All ordered triples satisfying $x - y - z = 3$ are solutions.

22. Let x = number of bird feeders
y = number of birdhouses
z = number of snack tables
$2x + 1.5y + 3z = 29$
$3x + 2y + 4z = 40$
$x - z = 1$

Eliminate y.
Multiply the first equation by 4 and the second equation by -3. Add the new equations.
$8x + 6y + 12z = 116$
$\underline{-9x - 6y - 12z = -120}$
$-x = -4$
$x = 4$

Substitute this value for x in the third equation.
$(4) - z = 1$
$-z = -3$
$z = 3$

Substitute $x = 4$ and $z = 3$ into the second equation to solve for y.
$3(4) + 2y + 4(3) = 40$
$12 + 2y + 12 = 40$
$2y + 24 = 40$
$2y = 16$
$y = 8$

Mike makes 4 bird feeders, 8 birdhouses, and 3 snack tables each week.

23. One solution exists.
No solution exists.
An infinite number of solutions will exist.
Answers will vary.

Chapter 7

7.1 Exercises

1. $3(x+1) > -(5x-4)$ is linear because it simplifies to $3x+3 > -5x+4$, with expressions on both sides of the inequality symbol in the form $ax+b$.

3. $4x^2 + 1 < 3x + 2$ is nonlinear because in the expression on the left side of the inequality symbol x has an exponent of 2.

5. $\frac{5}{7}(a+2) \geq \frac{3}{7}a + \frac{2}{3}$ is linear because it simplifies to $\frac{5}{7}a + \frac{10}{7} \geq \frac{3}{7}a + \frac{2}{3}$, with expressions on both sides of the inequality symbol in the form $ax+b$.

7. $0.6x + 2.7 \leq 5.2 - 1.9x$ is linear because the expressions on both sides of the inequality symbol are in the form $ax+b$.

9. $x + 4(x-8) > 2(3x+1)$ is linear because it simplifies to $5 - 32 > 6x + 2$, with expressions on both sides of the inequality symbol in the form $ax+b$.

11. $\sqrt{2x-7} \leq x + 3$ is nonlinear because the radical expression on the left side of the inequality symbol has a variable in its radicand.

13. $\frac{5}{x} + 2x > 0$ is nonlinear because the expression on the left side of the inequality symbol has a variable in the denominator of a fraction.

15. $2p + 6 < 0$ is linear because it is in the form $ax + b < 0$.

17. $x \geq 6$

    Interval notation: $[6, \infty)$.

19. $z < 12$

    Interval notation: $(-\infty, 12)$.

21. $b > -\frac{13}{5}$

    Interval notation: $\left(-\frac{13}{5}, \infty\right)$.

23. $x \leq 4\frac{2}{3}$

    Interval notation: $\left(-\infty, 4\frac{2}{3}\right]$.

25. $p \leq 12.59$

    Interval notation: $(-\infty, 12.59]$.

27. $q > -6.7$

    Interval notation: $(-6.7, \infty)$.

29. $5 < x < 13$

    Interval notation: $(5, 13)$.

31. $2 < x < 8$

    Interval notation: $(2, 8)$.

SSM: Experiencing Introductory and Intermediate Algebra

33. $-4 < d \leq 0$

Interval notation: $(-4, 0]$.

35. $-1 \leq m < 6$

Interval notation: $[-1, 6)$.

37. $2 \leq t \leq 7$

Interval notation: $[2, 7]$.

39. $\frac{2}{5} < s \leq 3\frac{1}{3}$

Interval notation: $\left(\frac{2}{5}, 3\frac{1}{3}\right]$.

41. $-2.5 \leq q < 3.5$

Interval notation: $[-2.5, 3.5)$.

43. $\frac{4}{5} \leq x \leq 4.5$

Interval notation: $\left[\frac{4}{5}, 4.5\right]$.

45.

	Inequality	Number line	Interval notation
a.	$x < 4.5$		$(-\infty, 4.5)$
b.	$x \leq 3$		$(-\infty, 3]$
c.	$x < -2$		$(-\infty, -2)$
d.	$z \geq 5.7$		$[5.7, \infty)$
e.	$z > -7$		$(-7, \infty)$
f.	$z > 2$		$(2, \infty)$
g.	$2 < y < 8$		$(2, 8)$
h.	$-3 \leq y < 2$		$[-3, 2)$
i.	$0 \leq y \leq 9$		$[0, 9]$

280

47.

Classification	Richter magnitudes
Generally not felt, but recorded.	$r < 3.5$
Often felt, but rarely causes damage.	$3.5 \leq r \leq 5.4$
At most slight damage to well-designed buildings and major damage to poorly constructed buildings over small regions.	$r < 6.0$

49. Let $x =$ the number of miles to be driven.
$39.95 + 0.2x \leq 150$

51. Let $x =$ the amount of weekly sales.
$450 + 0.05x > 800$

53. Let $x =$ the number of people to be invited.
$150 \leq 25 + 12.5x \leq 200$

7.1 Calculator Exercises

1. $x < -2$

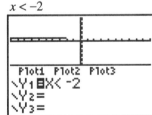

2. $x > -2$

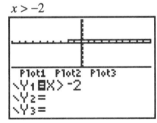

3. $x \leq -2$

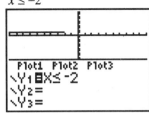

4. $x \geq -2$

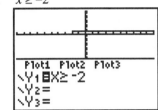

5. $-2.5 \leq x \leq 2.5$

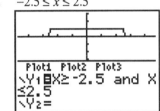

6. $-2.5 \leq x < 2.5$

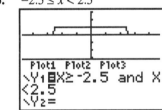

7. $-2.5 < x < 2.5$

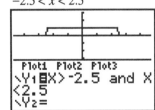

7.2 Exercises

1. Let $Y1 = 6(x-12)$ and $Y2 = 3x$. A sample table is shown below.

SSM: Experiencing Introductory and Intermediate Algebra

X	Y1	Y2
20	48	60
21	54	63
22	60	66
23	66	69
24	72	72
25	78	75
26	84	78
X=24		

Because $Y1 < Y2$ for values $x < 24$, the solutions are the integers less than 24.

3. Let $Y1 = 3(x+5)+3$ and $Y2 = 3x+18$. A sample table is shown below.

X	Y1	Y2
-3	9	9
-2	12	12
-1	15	15
0	18	18
1	21	21
2	24	24
3	27	27
X=-3		

Because $Y1 = Y2$ for all values of x, the inequality is never true. This inequality has no solution.

5. Let $Y1 = 3x+3$ and $Y2 = x+2$. A sample table is shown below.

X	Y1	Y2
-3	-6	-1
-2	-3	0
-1	0	1
0	3	2
1	6	3
2	9	4
3	12	5
X=0		

Because $Y1 > Y2$ for integer values $x \geq 0$, the solutions are the integers greater than or equal to 0.

7. Let $Y1 = 2(4x+2)+2x-7$ and $Y2 = 2(5x+4)$. A sample table is shown below.

X	Y1	Y2
-3	-33	-22
-2	-23	-12
-1	-13	-2
0	-3	8
1	7	18
2	17	28
3	27	38
X=-3		

Because $Y1 - Y2 = -11$ for all values of x, the inequality is always true. The solutions are all integers.

9. Let $Y1 = \frac{2}{3}x - \frac{2}{3}$ and $Y2 = -\frac{3}{4}x - \frac{7}{2}$. The graphs of the functions are shown below.

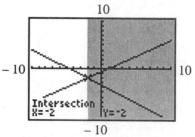

The intersection point is $(-2, -2)$. Because $Y1 \geq Y2$ for $x \geq -2$, the solution set is all x values that satisfy $x \geq -2$, or $[-2, \infty)$.

11. Let $Y1 = 2(x+3)-(x-1)$ and $Y2 = 8-(x+9)$. The graphs of the functions are shown below.

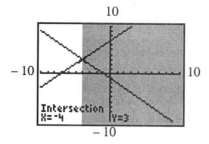

The intersection point is $(-4, 3)$. Because $Y1 > Y2$ for $x > -4$, the solution set is all x values that satisfy $x > -4$, or $(-4, \infty)$.

13. Let $Y1 = 0.4x - 3.2$ and $Y2 = -0.6x - 0.2$. The graphs of the functions are shown below.

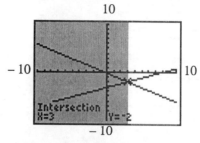

The intersection point is $(3, -2)$. Because $Y1 \leq Y2$ for $x \leq 3$, the solution set is all x values that satisfy $x \leq 3$, or $(-\infty, 3]$.

15. Let $Y1 = 2(x+1)$ and $Y2 = 3(x-1) - x$. The graphs of the functions are shown below.

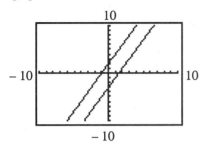

The two lines are parallel with the graph of $Y1$ is always above that of $Y2$. The solution is the set of all real numbers, or $(-\infty, \infty)$.

17. Let $Y1 = x - (3x+1)$ and $Y2 = -x - (x+1)$. The graphs of the functions are shown below.

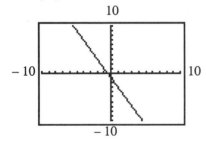

The two lines coincide. Because $Y1 = Y2$ for all values of x and because the inequality does not contain an equals, the inequality is never true. This inequality has no solution.

19. $4x + 12 > 0$
$4x + 12 - 12 > 0 - 12$
$4x > -12$
$\dfrac{4x}{4} > \dfrac{-12}{4}$
$x > -3$
The solution set is $(-3, \infty)$.

21. $-3x + 12 \geq 12$
$-3x + 12 - 12 \geq 12 - 12$
$-3x \geq 0$
$\dfrac{-3x}{-3} \leq \dfrac{0}{-3}$
$x \leq 0$
The solution set is $(-\infty, 0]$.

23. $-7x - 12 < -26$
$-7x - 12 + 12 < -26 + 12$
$-7x < -14$
$\dfrac{-7x}{-7} > \dfrac{-14}{-7}$
$x > 2$
The solution set is $(2, \infty)$.

25. $15.17 < 5.9x - 4.3$
$15.17 + 4.3 < 5.9x - 4.3 + 4.3$
$19.47 < 5.9x$
$\dfrac{19.47}{5.9} < \dfrac{5.9x}{5.9}$
$3.3 < x$ or $x > 3.3$
The solution set is $(3.3, \infty)$.

27. $6.1 > -0.55a + 6.1$
$6.1 - 6.1 > -0.55a + 6.1 - 6.1$
$0 > -0.55a$
$\dfrac{0}{-0.55} < \dfrac{-0.55a}{-0.55}$
$0 < a$ or $a > 0$
The solution set is $(0, \infty)$.

29. $2.07z + 4.12 \geq 16.54$
$2.07z + 4.12 - 4.12 \geq 16.54 - 4.12$
$2.07z \geq 12.42$
$\dfrac{2.07z}{2.07} \geq \dfrac{12.42}{2.07}$
$z \geq 6$
The solution set is $[6, \infty)$.

31. $-\dfrac{5}{9}b+\dfrac{11}{12}<\dfrac{23}{36}$

$36\left(-\dfrac{5}{9}b+\dfrac{11}{12}\right)<36\left(\dfrac{23}{36}\right)$

$-20b+33<23$

$-20b+33-33<23-33$

$-20b<-10$

$\dfrac{-20b}{-20}>\dfrac{-10}{-20}$

$b>\dfrac{1}{2}$

The solution set is $\left(\dfrac{1}{2},\infty\right)$.

33. $156z-210>47z+662$

$156z-210-47z>47z+662-47z$

$109z-210>662$

$109z-210+210>662+210$

$109z>872$

$\dfrac{109z}{109}>\dfrac{872}{109}$

$z>8$

The solution set is $(8,\infty)$.

35. $-1.05x-15.41<2.55x-47.09$

$-1.05x-15.41-2.55x<2.55x-47.09-2.55x$

$-3.6x-15.41<-47.09$

$-3.6x-15.41+15.41<-47.09+15.41$

$-3.6x<-31.68$

$\dfrac{-3.6x}{-3.6}>\dfrac{-31.68}{-3.6}$

$x>8.8$

The solution set is $(8.8,\infty)$.

37. $11x+\dfrac{1}{4}\le 2x+\dfrac{7}{36}$

$36\left(11x+\dfrac{1}{4}\right)\le 36\left(2x+\dfrac{7}{36}\right)$

$396x+9\le 72x+7$

$396x+9-72x\le 72x+7-72x$

$324x+9\le 7$

$324x+9-9\le 7-9$

$324x\le -2$

$\dfrac{324x}{324}\le \dfrac{-2}{324}$

$x\le -\dfrac{1}{162}$

The solution set is $\left(-\infty,-\dfrac{1}{162}\right]$.

39. $\dfrac{2}{5}b-12<\dfrac{2}{3}b+20$

$15\left(\dfrac{2}{5}b-12\right)<15\left(\dfrac{2}{3}b+20\right)$

$6b-180<10b+300$

$6b-180-10b<10b+300-10b$

$-4b-180<300$

$-4b-180+180<300+180$

$-4b<480$

$\dfrac{-4b}{-4}>\dfrac{480}{-4}$

$b>-120$

The solution set is $(-120,\infty)$.

41. $4x-(3x+5)<x-5$

$4x-3x-5<x-5$

$x-5<x-5$

$x-5-x<x-5-x$

$-5<-5$

Because this inequality must always be false, it has no solution.

43. $4x-(3x+5) \leq x-5$
$4x-3x-5 \leq x-5$
$x-5 \leq x-5$
$x-5-x \leq x-5-x$
$-5 \leq -5$
Because this inequality must always be true, its solution is the set of all real numbers.

45. $0.05x+10.5 < 0.15x - 0.25$
$0.05x + 10.5 - 0.15x < 0.15x - 0.25 - 0.15x$
$-0.1x + 10.5 < -0.25$
$-0.1x + 10.5 - 10.5 < -0.25 - 10.5$
$-0.1x < -10.75$
$\dfrac{-0.1x}{-0.1} > \dfrac{-10.75}{-0.1}$
$x > 107.5$
The solution set is $(107.5, \infty)$.

47. $5 < 4 - 3x \leq 11$
$5 - 4 < 4 - 3x - 4 \leq 11 - 4$
$1 < -3x \leq 7$
$\dfrac{1}{-3} > \dfrac{-3x}{-3} \geq \dfrac{7}{-3}$
$-\dfrac{1}{3} > x \geq -\dfrac{7}{3}$ or $-\dfrac{7}{3} \leq x < -\dfrac{1}{3}$
The solution set is $\left[-\dfrac{7}{3}, -\dfrac{1}{3}\right)$.

49. $-4 \leq 3(x+1) - 5 \leq 7$
$-4 \leq 3x + 3 - 5 \leq 7$
$-4 \leq 3x - 2 \leq 7$
$-4 + 2 \leq 3x - 2 + 2 \leq 7 + 2$
$-2 \leq 3x \leq 9$
$\dfrac{-2}{3} \leq \dfrac{3x}{3} \leq \dfrac{9}{3}$
$-\dfrac{2}{3} \leq x \leq 3$
The solution set is $\left[-\dfrac{2}{3}, 3\right]$.

51. Let x = Lee's score on the sixth test.
$\dfrac{93 + 97 + 92 + 89 + 95 + x}{6} \geq 93$
$\dfrac{466 + x}{6} \geq 93$
$6\left(\dfrac{466 + x}{6}\right) \geq 6(93)$
$466 + x \geq 558$
$466 + x - 466 \geq 558 - 466$
$x \geq 92$
Lee's score on the test six must be 92 or higher.

53. Let x = the number of hours Luigi can work.
$9.75x + 5200 \leq 7500$
$9.75x + 5200 - 5200 \leq 7500 - 5200$
$9.75x \leq 2300$
$\dfrac{9.75x}{9.75} \leq \dfrac{2300}{9.75}$
$x \leq 235.897$ (rounded to the nearest thousandth)
Luigi can work 235 hours or less.

55. Let x = the number of people.
$165x \leq 2000$
$\dfrac{165x}{165} \leq \dfrac{2000}{165}$
$x \leq 12.121$ (rounded to the nearest thousandth)
12 or fewer people can ride the elevator safely.

57. Let x = the number of meat entrée plates.
$120 - x$ = the number of vegetarian plates.
$35x + 30(120 - x) \leq 4000$
$35x + 3600 - 30x \leq 4000$
$5x + 3600 \leq 4000$
$5x + 3600 - 3600 \leq 4000 - 3600$
$5x \leq 400$
$\dfrac{5x}{5} \leq \dfrac{400}{5}$
$x \leq 80$
Angie can order 80 or fewer meat entrée plates.

SSM: Experiencing Introductory and Intermediate Algebra

59. Let x = the width of the garden.
$\frac{3}{4}x + 30$ = the length of the garden.
$2x + \left(\frac{3}{4}x + 30\right) - 4 \leq 185$
$4\left[2x + \left(\frac{3}{4}x + 30\right) - 4\right] \leq 4(185)$
$8x + 3x + 120 - 16 \leq 740$
$11x + 104 \leq 740$
$11x + 104 - 104 \leq 740 - 104$
$11x \leq 636$
$\frac{11x}{11} \leq \frac{636}{11}$
$x \leq 57.818$ (rounded to the nearest thousandth)
The width of the garden must be 57 feet or less.

61. Let x = the number of prints.
$0.25x - (0.08x + 45) \geq 50$
$0.25x - 0.08x - 45 \geq 50$
$0.17x - 45 \geq 50$
$0.17x - 45 + 45 \geq 50 + 45$
$0.17x \geq 95$
$\frac{0.17x}{0.17} \geq \frac{95}{0.17}$
$x \geq 558.824$ (rounded to the nearest thousandth)
559 or more prints must be produced each day.

63. Let x = high temperature on fifth day.
$82 < \frac{82 + 83 + 88 + 92 + x}{5} < 88$
$82 < \frac{345 + x}{5} < 88$
$5(82) < 5\left(\frac{345 + x}{5}\right) < 5(88)$
$410 < 345 + x < 440$
$410 - 345 < 345 + x - 345 < 440 - 345$
$65 < x < 95$
The temperature on the fifth day must be between $65°F$ and $95°F$.

65. $0.146x + 1 < 30$
$0.146x + 1 - 1 < 30 - 1$
$0.146x < 29$
$\frac{0.146x}{0.146} < \frac{29}{0.146}$
$x < 198.630$ (rounded to the nearest thousandth)
The doctor may not prescribe medication for weights of 198 pounds or less.

67. $(6.95 \times 10^{10})n + (4.36 \times 10^{11}) > 1.5 \times 10^{12}$
$(6.95 \times 10^{10})n + (4.36 \times 10^{11}) > 1.5 \times 10^{12}$
$\qquad -(4.36 \times 10^{11}) \qquad -(4.36 \times 10^{11})$
$(6.95 \times 10^{10})n > 1.064 \times 10^{12}$
$\frac{(6.95 \times 10^{10})n}{6.95 \times 10^{10}} > \frac{1.064 \times 10^{12}}{6.95 \times 10^{10}}$
$x > 15.309$ (rounded to the nearest thousandth)
Total expenditures exceeded 1.5 trillion dollars for years 16 and later.

7.2 Calculator Exercises

Students should experiment with some exercises.

7.3 Exercises

1. $2x + 1.7y > x - 4.6$ is linear.
In standard form,
$2x + 1.7y - x > x - 4.6 - x$
$x + 1.7y > -4.6$

3. $y < x^2 + 2x - 3$ is nonlinear because the x term is squared.

5. $y > \sqrt{x} + 9$ is nonlinear because the radical expression contains a variable.

7. $\dfrac{x}{2} - \dfrac{y}{6} > \dfrac{1}{12}$ is linear.
In standard form,
$12\left(\dfrac{x}{2} - \dfrac{y}{6}\right) > 12\left(\dfrac{1}{12}\right)$
$6x - 2y > 1$

9. $4x + 16 \leq y$ is linear.
In standard form,
$4x + 16 - 16 \leq y - 16$
$4x \leq y - 16$
$4x - y \leq y - 16 - y$
$4x - y \leq -16$

11. $y \leq -\dfrac{2}{5}x + \dfrac{7}{15}$ is linear.
In standard form,
$15(y) \leq 15\left(-\dfrac{2}{5}x + \dfrac{7}{15}\right)$
$15y \leq -6x + 7$
$15y + 6x \leq -6x + 7 + 6x$
$6x + 15y \leq 7$

13. $2x + y < 3$
$2x + y - 2x < 3 - 2x$
$y < -2x + 3$

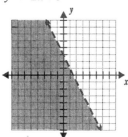

15. $5x - 3y \geq 6$
$5x - 3y - 5x \geq 6 - 5x$
$-3y \geq -5x + 6$
$\dfrac{-3y}{-3} \leq \dfrac{-5x + 6}{-3}$
$y \leq \dfrac{5}{3}x - 2$

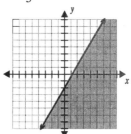

17. $y < -\dfrac{3}{4}x + 4$

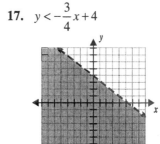

19. $y \geq 2.8x - 1.6$

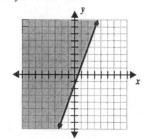

21. $2y > 3x - 5$

$\dfrac{2y}{2} > \dfrac{3x-5}{2}$

$y > \dfrac{3}{2}x - \dfrac{5}{2}$

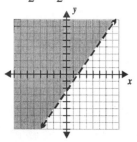

23. $3x + y - 4 \leq x + 2y - 3$

$3x + y - 4 - 3x \leq x + 2y - 3 - 3x$

$y - 4 \leq -2x + 2y - 3$

$y - 4 - 2y \leq -2x + 2y - 3 - 2y$

$-y - 4 \leq -2x - 3$

$-y - 4 + 4 \leq -2x - 3 + 4$

$-y \leq -2x + 1$

$-1(-y) \geq -1(-2x + 1)$

$y \geq 2x - 1$

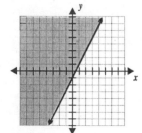

25. $5y > -20$

$\dfrac{5y}{5} > \dfrac{-20}{5}$

$y > -4$

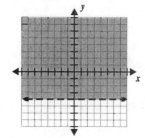

27. $3x + 6 \leq 9$

$3x + 6 - 6 \leq 9 - 6$

$3x \leq 3$

$\dfrac{3x}{3} \leq \dfrac{3}{3}$

$x \leq 1$

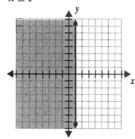

29. $3x + 5y \geq 12$

$3x + 5y - 3x \geq 12 - 3x$

$5y \geq -3x + 12$

$\dfrac{5y}{5} \geq \dfrac{-3x + 12}{5}$

$y \geq -\dfrac{3}{5}x + \dfrac{12}{5}$

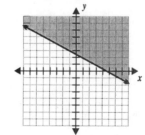

31. $-x - y < 7$

$-x - y + x < 7 + x$

$-y < x + 7$

$-1(-y) > -1(x+7)$

$y > -x - 7$

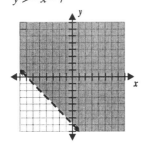

33. $-3x - 9y > 27$

$-3x - 9y + 3x > 27 + 3x$

$-9y > 3x + 27$

$\dfrac{-9y}{-9} < \dfrac{3x + 27}{-9}$

$y < -\dfrac{1}{3}x - 3$

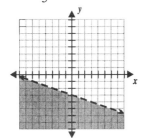

35. $\dfrac{x}{8} + \dfrac{y}{3} \leq 1$

$\dfrac{x}{8} + \dfrac{y}{3} - \dfrac{x}{8} \leq 1 - \dfrac{x}{8}$

$\dfrac{y}{3} \leq -\dfrac{x}{8} + 1$

$3\left(\dfrac{y}{3}\right) \leq 3\left(-\dfrac{x}{8} + 1\right)$

$y \leq -\dfrac{3}{8}x + 3$

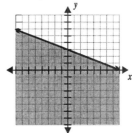

37. $-\dfrac{4}{7}x + \dfrac{2}{3}y \geq \dfrac{10}{21}$

$-\dfrac{4}{7}x + \dfrac{2}{3}y + \dfrac{4}{7}x \geq \dfrac{10}{21} + \dfrac{4}{7}x$

$\dfrac{2}{3}y \geq \dfrac{4}{7}x + \dfrac{10}{21}$

$\dfrac{3}{2}\left(\dfrac{2}{3}y\right) \geq \dfrac{3}{2}\left(\dfrac{4}{7}x + \dfrac{10}{21}\right)$

$y \geq \dfrac{6}{7}x + \dfrac{5}{7}$

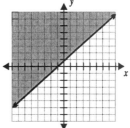

39. $1.8x - 3.2y > 0$
$1.8x - 3.2y - 1.8x > 0 - 1.8x$
$-3.2y > -1.8x$
$\dfrac{-3.2y}{-3.2} < \dfrac{-1.8x}{-3.2}$
$y < 0.5625x$

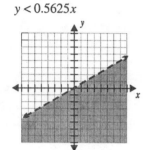

41. $4.6y < 3.5y + 5.94$
$4.6y - 3.5y < 3.5y + 5.94 - 3.5y$
$1.1y < 5.94$
$\dfrac{1.1y}{1.1} < \dfrac{5.94}{1.1}$
$y < 5.4$

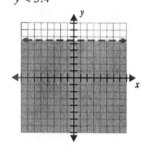

43. $x - y > 9$
$x - y - x > 9 - x$
$-y > -x + 9$
$-1(-y) < -1(-x + 9)$
$y < x - 9$

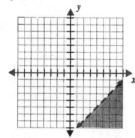

45. $-x - y > 9$
$-x - y + x > 9 + x$
$-y > x + 9$
$-1(-y) < -1(x + 9)$
$y < -x - 9$

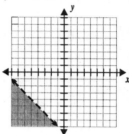

47. $y \leq x$

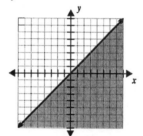

49. Let $x =$ the number of village pieces.
$y =$ the number of angel ornaments.
$25x + 12y \leq 225$
Solve for y.
$25x + 12y - 25x \leq 225 - 25x$
$12y \leq -25x + 225$
$\dfrac{12y}{12} \leq \dfrac{-25x + 225}{12}$
$y \leq -\dfrac{25}{12}x + \dfrac{75}{4}$

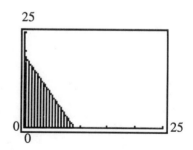

The ordered pairs in the shaded region represent the possible combinations of village pieces and angel ornaments.

Yes, (4, 7) is in the shaded region. She can buy 4 village pieces and 7 angel ornaments.

No, (7, 9) is not in the shaded region. She cannot buy 7 village pieces and 9 angel ornaments.

51. Let $x =$ the width of the room.

    $y =$ the length of the room.

    $2x + 2y \leq 220$

    Solve for y.

    $2x + 2y - 2x \leq 220 - 2x$

    $2y \leq -2x + 220$

    $\dfrac{2y}{2} \leq \dfrac{-2x + 220}{2}$

    $y \leq -x + 110$

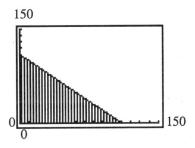

The ordered pairs in the shaded region represent the possible combinations of widths and lengths.
No, (70, 60) is not in the shaded region. Barbara does not have enough border for a 70-by-60 foot room.
Yes, (40, 60) is in the shaded region. Barbara has enough border for a 40-by-60 foot room.

53. Let $x =$ the number of days.

    $y =$ the number of items sold.

    $15x + 12y \geq 400$

    Solve for y.

    $15x + 12y - 15x \geq 400 - 15x$

    $15x + 12y - 15x \geq 400 - 15x$

    $12y \geq -15x + 400$

    $\dfrac{12y}{12} \geq \dfrac{-15x + 400}{12}$

    $y \geq -\dfrac{5}{4}x + \dfrac{100}{3}$

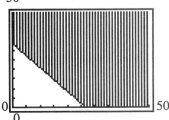

The ordered pairs in the shaded region represent the possible combinations of days and items.
No, (3, 20) is not in the shaded region. Pablo will not earn at least $400 working 3 days and selling 20 items.
Yes, (5, 30) is in the shaded region. Pablo will earn at least $400 working 5 days and selling 30 items.

55. Let $x =$ the number of adults.

    $y =$ the number of children.

    $4.5x + 2y \geq 250$

    Solve for y.

    $4.5x + 2y - 4.5x \geq 250 - 4.5x$

    $2y \geq -4.5x + 250$

    $\dfrac{2y}{2} \geq \dfrac{-4.5x + 250}{2}$

    $y \geq -2.25x + 125$

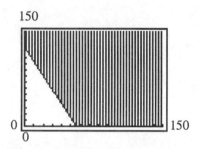

The ordered pairs in the shaded region represent the possible combinations of adults and children.
No, (40, 25) is not in the shaded region. They will not meet their goal with 40 adults and 25 children
Yes, (42, 45) is in the shaded region. They will meet their goal with 42 adults and 45 children.

57. Let $x =$ the number of 2.25-pound packages.

$y =$ the number of 4-pound packages.

According to the table, the cost of shipping a 2.25-pound package to Zone 5 is $2.11. The cost of shipping a 4-pound package to Zone 5 is $2.37.
$2.11x + 2.37y \leq 150$
Solve for y.
$2.11x + 2.37y - 2.11x \leq 150 - 2.11x$
$2.37y \leq -2.11x + 150$
$\dfrac{2.37y}{2.37} \leq \dfrac{-2.11x + 150}{2.37}$
$y \leq -\dfrac{211}{237}x + \dfrac{5000}{79}$

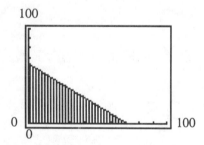

The ordered pairs in the shaded region represent the possible combinations of 2.25-pound packages and 4-pound packages.
Yes, (25, 35) is in the shaded region. Twenty-five 2.25-pound packages and thirty-five 4-pound packages can be mailed to Zone 5 without exceeding the budget.
No, (40, 30) is not in the shaded region. Forty 2.25-pound packages and thirty-five 4-pound packages cannot be mailed to Zone 5 without exceeding the budget.

7.4 Exercises

1. Solve the first inequality for y.
$5x - 3y > 15$
$5x - 3y - 5x > 15 - 5x$
$-3y > -5x + 15$
$\dfrac{-3y}{-3} < \dfrac{-5x + 15}{-3}$
$y < \dfrac{5}{3}x - 5$
Solve the second inequality for y.
$4x + y > 12$
$4x + y - 4x > 12 - 4x$
$y > -4x + 12$
Find the intersection point (using elimination).
$5x - 3y = 15$
$\underline{12x + 3y = 36}$
$17x = 51$
$\dfrac{17x}{17} = \dfrac{51}{17}$
$x = 3$
$y = \dfrac{5}{3}x - 5$
$y = \dfrac{5}{3}(3) - 5 = 0$
The intersection point is $(3, 0)$.

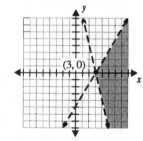

3. Solve the first inequality for *y*.
$$2x + 4y \geq 8$$
$$2x + 4y - 2x \geq 8 - 2x$$
$$4y \geq -2x + 8$$
$$\frac{4y}{4} \geq \frac{-2x + 8}{4}$$
$$y \geq -\frac{1}{2}x + 2$$
Solve the second inequality for *y*.
$$3x + y \leq -8$$
$$3x + y - 3x \leq -8 - 3x$$
$$y \leq -3x - 8$$
Find the intersection point (using elimination).
$$2x + 4y = 8$$
$$\underline{-12x - 4y = 32}$$
$$-10x = 40$$
$$\frac{-10x}{-10} = \frac{40}{-10}$$
$$x = -4$$
$$y = -3x - 8$$
$$y = -3(-4) - 8 = 4$$
The intersection point is $(-4, 4)$.

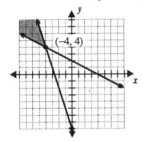

5. $y > 3x - 1$
$y > -2x + 4$
Find the intersection point (using substitution).
$$3x - 1 = -2x + 4$$
$$3x - 1 + 2x = -2x + 4 + 2x$$
$$5x - 1 = 4$$
$$5x - 1 + 1 = 4 + 1$$
$$5x = 5$$
$$5x = 5$$
$$\frac{5x}{5} = \frac{5}{5}$$
$$x = 1$$
$$y = 3x - 1$$
$$y = 3(1) - 1 = 2$$
The intersection point is $(1, 2)$.

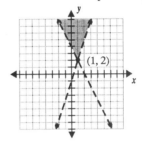

7. $y < \frac{3}{4}x + 1$

$y < -\frac{2}{3}x + 1$

Find the intersection point (using substitution).
$$\frac{3}{4}x + 1 = -\frac{2}{3}x + 1$$
$$12\left(\frac{3}{4}x + 1\right) = 12\left(-\frac{2}{3}x + 1\right)$$
$$9x + 12 = -8x + 12$$
$$9x + 12 + 8x = -8x + 12 + 8x$$
$$17x + 12 = 12$$
$$17x + 12 - 12 = 12 - 12$$
$$17x = 0$$
$$\frac{17x}{17} = \frac{0}{17}$$
$$x = 0$$
$$y = \frac{3}{4}x + 1$$

$y = \frac{3}{4}(0) + 1 = 1$

The intersection point is $(0, 1)$.

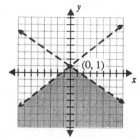

9. Solve the first inequality for y.
$3x + 2y < 12$
$3x + 2y - 3x < 12 - 3x$
$2y < -3x + 12$
$\frac{2y}{2} < \frac{-3x + 12}{2}$
$y < -\frac{3}{2}x + 6$

Solve the second inequality for y.
$y - 2 < 1$
$y - 2 + 2 < 1 + 2$
$y < 3$

Find the intersection point (using substitution).
$y = 3$
$3x + 2y = 12$
$3x + 2(3) = 12$
$3x + 6 = 12$
$3x + 6 - 6 = 12 - 6$
$3x = 6$
$\frac{3x}{3} = \frac{6}{3}$
$x = 2$

The intersection point is $(2, 3)$.

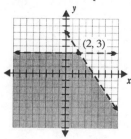

11. Solve the first inequality for y.
$3y - 2 > 2y + 2$
$3y - 2 - 2y > 2y + 2 - 2y$
$y - 2 > 2$
$y - 2 + 2 > 2 + 2$
$y > 4$

Solve the second inequality for x.
$x - 1 < 0$
$x - 1 + 1 < 0 + 1$
$x < 1$

Find the intersection point.
$y = 4$
$x = 1$

The intersection point is $(1, 4)$.

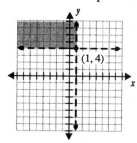

13. $y \le 3x - 5$

Solve the second inequality for y.
$x > 4 - 2y$
$x - 4 > 4 - 2y - 4$
$x - 4 > -2y$
$\frac{x - 4}{-2} < \frac{-2y}{2}$
$-\frac{1}{2}x + 2 < y$ or $y > -\frac{1}{2}x + 2$

Find the intersection point (using substitution).
$x = 4 - 2y$
$x = 4 - 2(3x - 5)$
$x = 4 - 6x + 10$
$x = -6x + 14$

294

$x + 6x = -6x + 14 + 6x$
$7x = 14$
$\dfrac{7x}{7} = \dfrac{14}{7}$
$x = 2$
$y = 3(2) - 5 = 1$
The intersection point is $(2, 1)$.

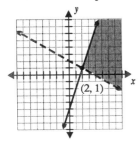

15. $y \leq \dfrac{1}{2}x + 3$

Solve the second inequality for y.
$x + 2y < 6$
$x + 2y - x < 6 - x$
$2y < -x + 6$
$\dfrac{2y}{2} < \dfrac{-x + 6}{2}$
$y < -\dfrac{1}{2}x + 3$

Find the intersection point (using substitution).
$\dfrac{1}{2}x + 3 = -\dfrac{1}{2}x + 3$
$2\left(\dfrac{1}{2}x + 3\right) = 2\left(-\dfrac{1}{2}x + 3\right)$
$x + 6 = -x + 6$
$x + 6 + x = -x + 6 + x$
$2x + 6 = 6$
$2x + 6 - 6 = 6 - 6$
$2x = 0$
$\dfrac{2x}{2} = \dfrac{0}{2}$

$x = 0$
$y = \dfrac{1}{2}(0) + 3 = 3$
The intersection point is $(0, 3)$.

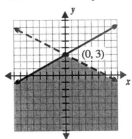

17. $y \leq 10 - x$

Solving the second inequality for y.
$2y > x + 6$
$\dfrac{2y}{2} > \dfrac{x + 6}{2}$
$y > \dfrac{1}{2}x + 3$
$y > 0$
$x < 3$

The intersection points of the bounded area are $(-6, 0)$, $\left(3, \dfrac{9}{2}\right)$, and $(3, 7)$.

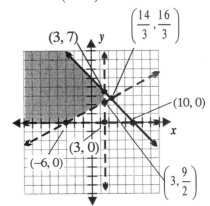

19. Solve the first inequality for y.
$x - 2y > 7$
$x - 2y - x > 7 - x$
$-2y > -x + 7$
$\dfrac{-2y}{-2} < \dfrac{-x+7}{-2}$
$y < \dfrac{1}{2}x - \dfrac{7}{2}$

Solve the second inequality for y.
$5x + 10y < -3$
$5x + 10y - 5x < -3 - 5x$
$10y < -5x - 3$
$\dfrac{10y}{10} < \dfrac{-5x-3}{10}$
$y < -\dfrac{1}{2}x - \dfrac{3}{10}$

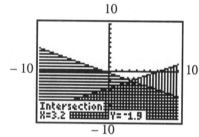

The intersection point is $(3.2, -1.9)$.

21. Solve the first inequality for y.
$3x - 5y \geq 5$
$3x - 5y - 3x \geq 5 - 3x$
$-5y \geq -3x + 5$
$\dfrac{-5y}{-5} \leq \dfrac{-3x+5}{-5}$
$y \leq \dfrac{3}{5}x - 1$

Solve the second inequality for y.
$10y + 15 > 2$
$10y + 15 - 15 > 2 - 15$
$10y > -13$
$\dfrac{10y}{10} > -\dfrac{13}{10}$
$y > -\dfrac{13}{10}$

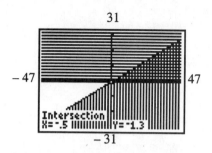

The intersection point is $(-0.5, -1.3)$.

23. $y \geq x - 3$

Solve the second inequality for y.
$10x - 5y \geq 14$
$10x - 5y - 10x \geq 14 - 10x$
$-5y \geq -10x + 14$
$\dfrac{-5y}{-5} \leq \dfrac{-10x}{-5} + \dfrac{14}{-5}$
$y \leq 2x - \dfrac{14}{5}$

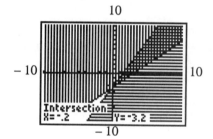

The intersection point is $(-0.2, -3.2)$.

25. Solve the first inequality for y.
$2x + 9y < 102$
$2x + 9y - 2x < 102 - 2x$
$9y < -2x + 102$
$\dfrac{9y}{9} < \dfrac{-2x + 102}{9}$
$y < -\dfrac{2}{9}x + \dfrac{34}{3}$

Solve the second inequality for y.
$5x - 11y < -147$
$-11y < -5x - 147$
$\dfrac{-11y}{-11} > \dfrac{-5x}{-11} - \dfrac{147}{-11}$
$y > \dfrac{5}{11}x + \dfrac{147}{11}$

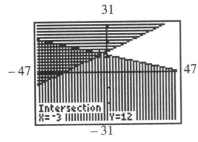

The intersection point is $(-3, 12)$.

27. Solve the first inequality for y.
$3.2x + 4.2y > 368$
$3.2x + 4.2y - 3.2x > 368 - 3.2x$
$4.2y > -3.2x + 368$
$\dfrac{4.2y}{4.2} > \dfrac{-3.2x + 368}{4.2}$
$y > -\dfrac{16}{21}x + \dfrac{1840}{21}$

Solve the second inequality for y.
$4.4x - 2.1y > 128$
$4.4x - 2.1y - 4.4x > 128 - 4.4x$
$-2.1y > -4.4x + 128$
$\dfrac{-2.1y}{-2.1} < \dfrac{-4.4x}{-2.1} + \dfrac{128}{-2.1}$
$y < \dfrac{44}{21}x - \dfrac{1280}{21}$

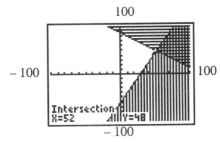

The intersection point is $(52, 48)$.

29. Solve the first inequality for y.
$0.05x + 0.10y < 0.75$
$0.05x + 0.10y - 0.05x < 0.75 - 0.05x$
$0.10y < -0.05x + 0.75$
$\dfrac{0.10y}{0.10} < \dfrac{-0.05x + 0.75}{0.10}$
$y < -\dfrac{1}{2}x + \dfrac{15}{2}$

Solve the second inequality for y.
$x + y > 11$
$x + y - x > 11 - x$
$y > -x + 11$

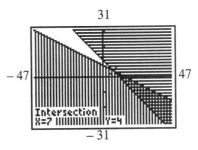

The intersection point is $(7, 4)$.

31. Let $x =$ the width of the patio.
$y =$ the length of the patio.
$2x + 2y \le 100$
$y \ge x + 10$
$x \ge 0$
$y \ge 0$

Solve the first inequality for y.
$y \le -x + 50$

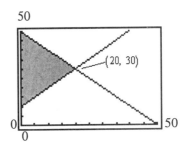

The ordered pairs in the shaded region represent the possible combinations of widths and lengths of the patio. One possible set of dimensions are 10 feet by 30 feet.

33. Let $x =$ the amount invested at 6%.

$y =$ the amount invested at 8%.

$x + y \leq 3000$

$0.06x + 0.08y \geq 200$

$x \geq 0$

$y \geq 0$

Solve the first two inequalities for y.

$y \leq -x + 3000$

$y \geq -\dfrac{3}{4}x + 2500$

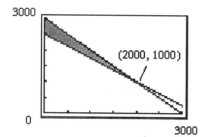

The ordered pairs in the shaded region represent the possible combinations of investment amounts. One possible pair of investment amounts is $1000 at 6% and $1800 at 8%.

35. Let $x =$ the number of hours on the first job.

$y =$ the number of hours at on second job.

$x + y \leq 20$

$6.5x + 8.25y \geq 150$

$x \geq 0$

$y \geq 0$

Solve the first two inequalities for y.

$y \leq -x + 20$

$y \geq -\dfrac{26}{33}x + \dfrac{200}{11}$

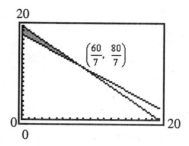

The ordered pairs in the shaded region represent the possible combinations of work times. One possible pair of times is 2 hours on the first job and 17 hours on the second job.

No, (7, 10) is not in the shaded area. He could not work 7 hours on the first job and 10 hours on the second job.

Yes, (5, 15) is in the shaded area. He can work 5 hours on the first job and 15 hours on the second. job

37. Let $x =$ the number servings of lasagna.

$y =$ the number servings of veal.

$1.75x + 2.25y \leq 200$

$x \geq 50$

$y \geq 25$

Solve the first inequality for y.

$y \leq -\dfrac{7}{9}x + \dfrac{800}{9}$

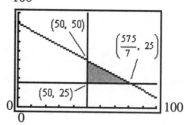

The ordered pairs in the shaded region represent the possible combinations of lasagna and veal entrées. One possible combination is 70 servings of lasagna and 30 servings of veal parmigiana.
Yes, (60, 35) is in the shaded area. Rosie can serve 60 lasagna and 35 veal entrées.

No, (60, 50) is not in the shaded area. Rosie cannot serve 60 lasagna and 50 veal entrées.

39. Let $x = $ the number of 2.25-pound packages.

$y = $ the number of 4-pound packages.

The cost of shipping a 2.25-pound package to Zone 5 is \$2.11. The cost of shipping a 4-pound package to Zone 5 is \$2.37.

$2.11x + 2.37y \leq 150$

$y < 2x$

$x \geq 0$

$y \geq 0$

Solve the first inequality for y.

$y \leq -\dfrac{211}{237}x + \dfrac{5000}{79}$

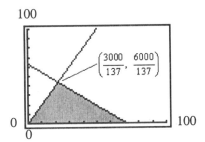

The ordered pairs in the shaded region represent the possible combinations of 2.25-pound packages and 4-pound packages. One possible combination is twenty-five 2.25-pound packages and forty 4-pound packages.

No, (50, 50) is not in the shaded region. Fifty 2.25-pound packages and fifty 4-pound packages cannot be mailed to Zone 5 without exceeding the restrictions.

Yes, (20, 20) is in the shaded region. Twenty 2.25-pound packages and twenty 4-pound packages can be mailed to Zone 5 without exceeding the restrictions.

7.4 Calculator Exercises

1. $2x - 3y > -6$

$-2x + 3y > -6$

Solve both inequalities for y.

$y < \dfrac{2}{3}x + 2$

$y > \dfrac{2}{3}x - 2$

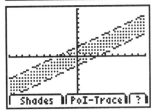

The solution set of this system is the set of all points that lie between the two parallel lines $y = \dfrac{2}{3}x + 2$ and $y = \dfrac{2}{3}x - 2$, not including the points on either of the lines.

2. $x + 2y < 6$

$x + 2y > 2$

Solve both inequalities for y.

$y < -\dfrac{1}{2}x + 3$

$y > -\dfrac{1}{2}x + 1$

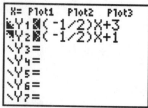

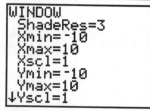

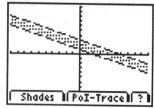

The solution set of this system is the set of all points that lie between the two parallel lines $y = -\frac{1}{2}x+3$ and $y = -\frac{1}{2}x+1$, not including the points on either of the lines.

3. $y > -3x+2$

    $y \le -3x$

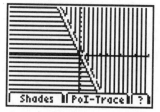

This system has no solution. The two inequalities have no points in common.

4. $y > 2x+2$

    $y \le 2x-3$

This system has no solution. The two inequalities have no points in common.

5. $x+4 > 0$

    $3x+2 < -4$

    Solve both inequalities for x.

    $x > -4$

    $x < -2$

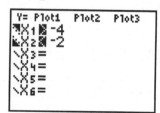

Chapter 7: Linear Inequalities

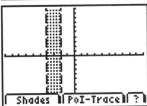

The solution set of this system is the set of all points that lie between the two vertical lines $x = -4$ and $x = -2$, not including the points on either of the lines.

6. $y - 1 < 0$

   $y + 3 > 0$

   Solve both inequalities for y.

   $y < 1$

   $y > -3$

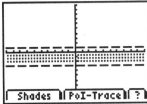

The solution set of this system is the set of all points that lie between the two horizontal lines $y = -3$ and $y = 1$, not including the points on either of the lines.

7. $y < 4 - 2x$

   $2x + y \le -1$

   Solve the second inequality for y.

   $y \le -2x - 1$

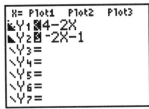

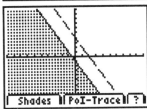

The solution set of this system is the set of all points that lie on and below the line $y = -2x - 1$.

8. $y < x - 7$

   $x > y + 9$

   Solve the second inequality for y.

   $y < x - 9$

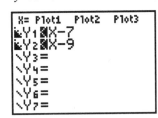

301

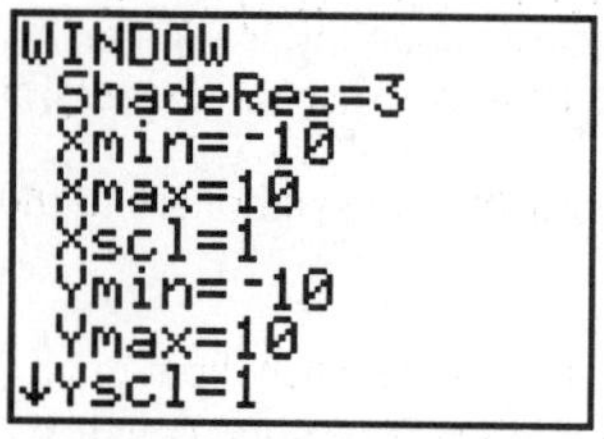

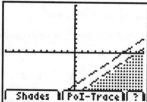

The solution set of this system is the set of all points that lie below the line $y = x - 9$, not including the points on the line.

9. $3y \leq x + 6$

   $3y - x \geq 6$

   Solve both inequalities for y.

   $y \leq \dfrac{1}{3}x + 2$

   $y \geq \dfrac{1}{3}x + 2$

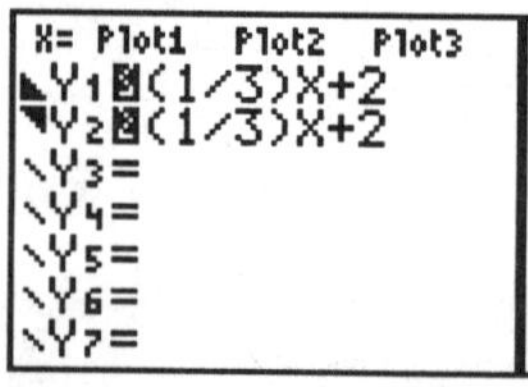

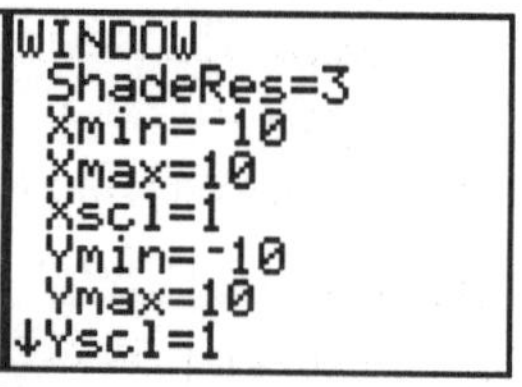

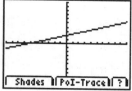

The solution set of this system is the set of all points that lie on the line $y = \dfrac{1}{3}x + 2$.

10. $y < 2x + 3$

    $y - x + 4 > x + 7$

    Solve the second inequality for y.

    $y > 2x + 3$

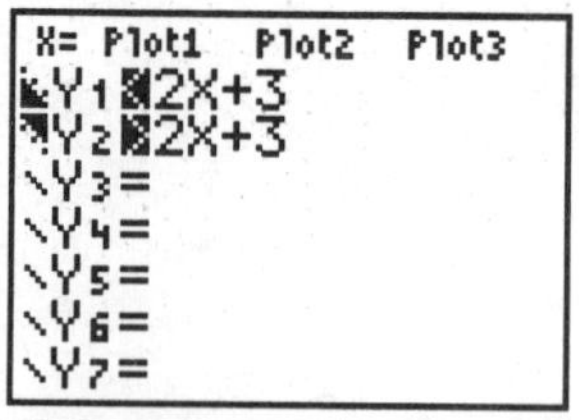

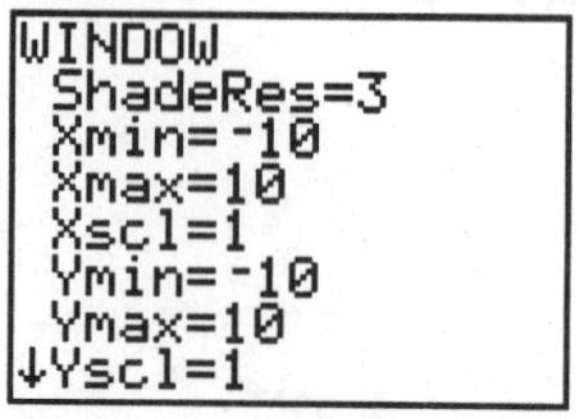

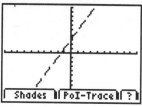

This system has no solution. The two inequalities have no points in common.

11. $y < 7$

    $2y - 3 \leq y + 4$

    Solve the second inequality for y.

    $y \leq 7$

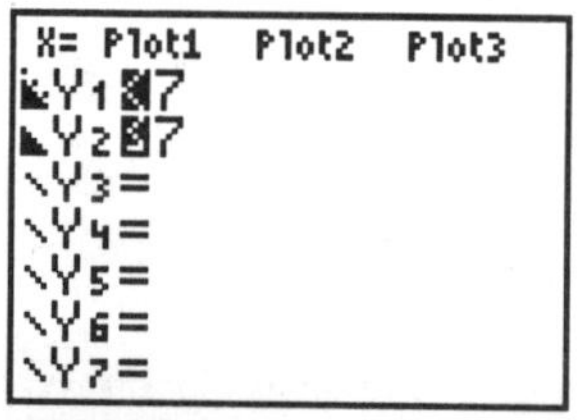

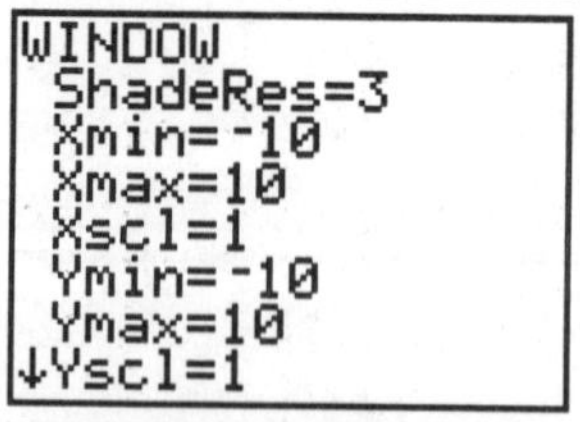

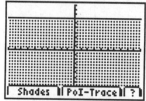

The solution set of this system is the set of all points that lie below the horizontal line $y = 7$, not including the points on the line. (Note: Because the solid line and dashed line overlap, the calculator display incorrectly makes the points on the line appear to be part of the solution set.)

12. $y \geq 3x + 4$

    $3x + 5 \leq y + 1$

    Solve the second inequality for y.

    $y \geq 3x + 4$

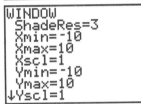

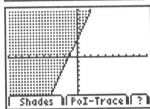

The solution set of this system is the set of all points that lie on and above the line $y = 3x + 4$.

Chapter 7 Section-By-Section Review

1. $3x^2 - 2x + 1 < 0$ is nonlinear because in the expression on the left side of the inequality symbol x has an exponent of 2.

2. $5x - 4 > x + 7$ is linear because the expressions on both sides of the inequality symbol are in the form $ax + b$.

3. $\dfrac{2}{3}x + \dfrac{4}{5} \leq \dfrac{7}{15}$ is linear because the expressions on both sides of the inequality symbol are in the form $ax + b$.

4. $\sqrt{x} - 3 \geq 6$ is nonlinear because the radical expression on the left side of the inequality symbol has a variable in its radicand.

5. $x - 3 \geq 6$ is linear because the expressions on both sides of the inequality symbol are in the form $ax + b$.

6. $\dfrac{1}{x} - 3x \geq 6$ is nonlinear because the expression on the left side of the inequality symbol has a variable in the denominator of a fraction.

7. $1.5z - 12.6 < 14.7z$ is linear because the expressions on both sides of the inequality symbol are in the form $ax + b$.

8. $3(a + 2) < 15a - (2a + 1)$ is linear because it simplifies to $3a + 6 < 13a - 1$, with expressions on both sides of the inequality symbol in the form $ax + b$.

9. $x < 3$

    Interval notation: $(-\infty, 3)$.

10. $x > -2$

    Interval notation: $(-2, \infty)$.

11. $x \leq -5$

    Interval notation: $(-\infty, -5]$.

12. $x \geq -3.5$

    Interval notation: $[-3.5, \infty)$.

13. $-2 < a < 4$

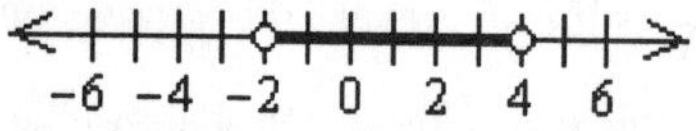

Interval notation: $(-2, 4)$.

14. $-1 < b \le 0$

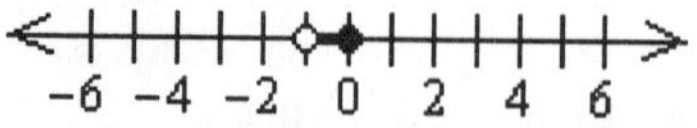

Interval notation: $(-1, 0]$.

15. $3 \le c \le 5.5$

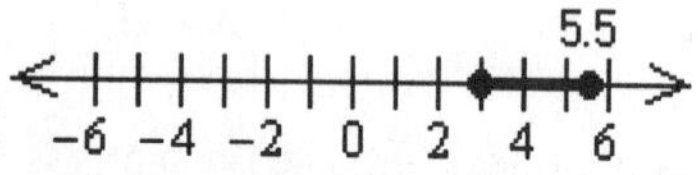

Interval notation: $[3, 5.5]$.

16. $2\frac{1}{2} < d < 8$

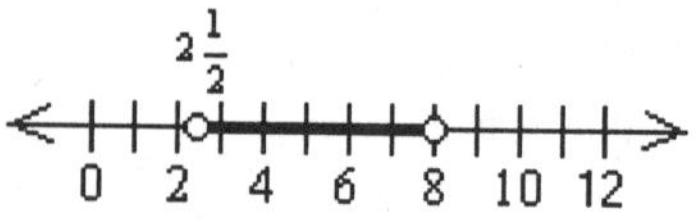

Interval notation: $\left(2\frac{1}{2}, 8\right)$.

17. $-2.3 \le f \le -1\frac{1}{3}$

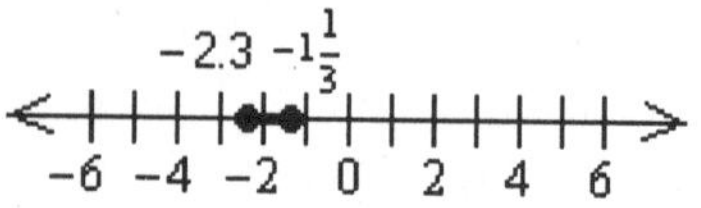

Interval notation: $\left[-2.3, -1\frac{1}{3}\right]$.

18. Let x = the number of minutes.
$30x + 25x \ge 300$

19. Let x = the cost of the base material.
$2x$ = the cost of the pavement.
$\frac{1}{4}(2x)$ = the cost of the sidewalk.

$x + 2x + \frac{1}{4}(2x) < 200{,}000$

20. Let x = the number of pieces of mail.
$125 + 0.28x < 0.34x$

21.

F-Scale number	Wind speed (in mph)
F0	$40 \le x \le 72$
F1	$73 \le x \le 112$
F2	$113 \le x \le 157$

22. Let $Y1 = 4x + 7$ and $Y2 = 2x - 5$. A sample table is shown below.

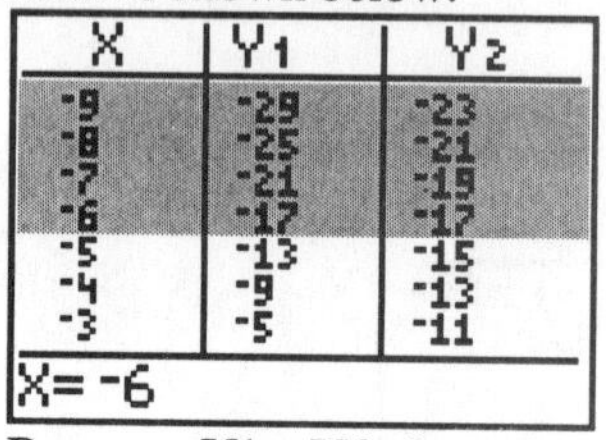

X	Y1	Y2
-9	-29	-23
-8	-25	-21
-7	-21	-19
-6	-17	-17
-5	-13	-15
-4	-9	-13
-3	-5	-11

X= -6

Because $Y1 < Y2$ for values $x < -6$, the solutions are the integers less than -6.

23. Let $Y1 = 2.4x - 9.6$ and $Y2 = 4.8$. A sample table is shown below.

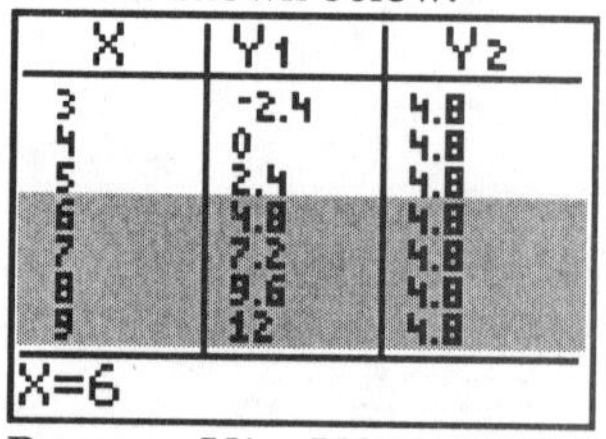

X	Y1	Y2
3	-2.4	4.8
4	0	4.8
5	2.4	4.8
6	4.8	4.8
7	7.2	4.8
8	9.6	4.8
9	12	4.8

X=6

Because $Y1 > Y2$ for values $x > 6$, the solutions are the integers greater than 6.

24. Let $Y1 = \frac{3}{5}x - \frac{7}{10}$ and $Y2 = \frac{1}{5}x + \frac{1}{2}$. A sample table is shown below.

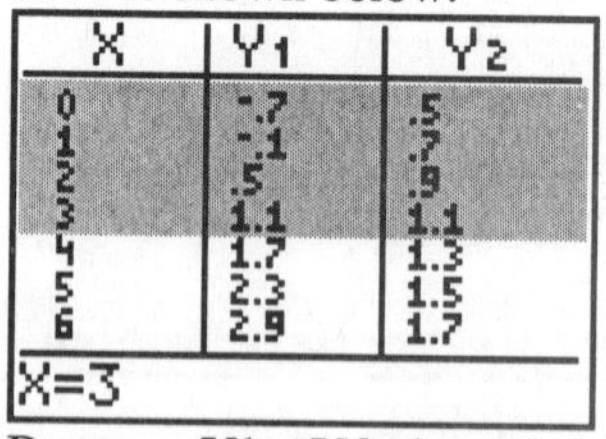

X	Y1	Y2
0	-.7	.5
1	-.1	.7
2	.5	.9
3	1.1	1.1
4	1.7	1.3
5	2.3	1.5
6	2.9	1.7

X=3

Because $Y1 \le Y2$ for values $x \le 3$, the solutions are the integers less than or equal to 3.

Chapter 7: Linear Inequalities

25. Let $Y1 = \frac{1}{2}x - 2$ and $Y2 = -\frac{1}{3}x - \frac{11}{3}$. A sample table is shown below.

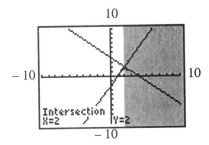

Because $Y1 \geq Y2$ for values $x \geq -2$, the solutions are the integers greater than or equal to -2.

26. Let $Y1 = 2x - 2$ and $Y2 = -x + 4$. The graphs of the functions are shown below.

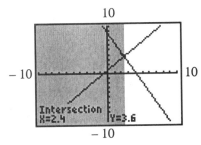

The intersection point is (2, 2). Because $Y1 > Y2$ for $x > 2$, the solution set is all x values that satisfy $x > 2$, or $(2, \infty)$.

27. Let $Y1 = 1.2x + 0.72$ and $Y2 = -2.1x + 8.64$. The graphs of the functions are shown below.

The intersection point is (2.4, 3.6). Because $Y1 \leq Y2$ for $x \leq 2.4$, the solution set is all x values that satisfy $x \leq 2.4$, or $(-\infty, 2.4]$.

28. Let $Y1 = (x+6) - 3(x+1)$ and $Y2 = (2x+5) - 2(2x+3)$. The graphs of the functions are shown below.

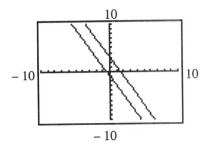

The two lines are parallel with the graph of Y1 is always above that of Y2. This inequality is never true; it has no solution.

29. Let $Y1 = (x+3) + (x+1)$ and $Y2 = 3(x+1) - (x-1)$. The graphs of the functions are shown below.

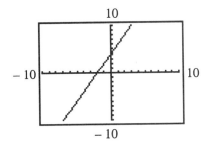

The two lines coincide. Because $Y1 = Y2$ for all values of x and because the inequality contains an equals, the inequality is always true. The solution is the set of all real numbers, or $(-\infty, \infty)$.

30. $412 + x > 671$

$412 + x - 412 > 671 - 412$

$x > 259$

The solution set is $(259, \infty)$.

31. $y - \frac{7}{13} < \frac{11}{39}$

$y - \frac{7}{13} + \frac{7}{13} < \frac{11}{39} + \frac{7}{13}$

$y < \frac{32}{39}$

The solution set is $\left(-\infty, \frac{32}{39}\right)$.

32. $3x+7 < 4x+21$
$3x+7-3x < 4x+21-3x$
$7 < x+21$
$7-21 < x+21-21$
$-14 < x$ or $x > -14$
The solution set is $(-14, \infty)$.

33. $14+2x < 2x$
$14+2x-2x < 2x-2x$
$14 < 0$
Because this inequality must always be false, it has no solution.

34. $8.7x+4.33 \le -2.4x-33.41$
$8.7x+4.33+2.4x \le -2.4x-33.41+2.4x$
$11.1x+4.33 \le -33.41$
$11.1x+4.33-4.33 \le -33.41-4.33$
$11.1x \le -37.74$
$\dfrac{11.1x}{11.1} \le \dfrac{-37.74}{11.1}$
$x \le -3.4$
The solution set is $(-\infty, -3.4]$.

35. $6.8z-9.52 \ge 0$
$6.8z-9.52+9.52 \ge 0+9.52$
$6.8z \ge 9.52$
$\dfrac{6.8z}{6.8} \ge \dfrac{9.52}{6.8}$
$x \ge 1.4$
The solution set is $[1.4, \infty)$.

36. $2(x+5)-(x+6) < 2(x+2)$
$2x+10-x-6 < 2x+4$
$x+4 < 2x+4$
$x+4-x < 2x+4-x$
$4 < x+4$
$4-4 < x+4-4$
$0 < x$ or $x > 0$
The solution set is $(0, \infty)$.

37. $3(x-4)+2(x+1) > 5x+10$
$3x-12+2x+2 > 5x+10$
$5x-10 > 5x+10$
$5x-10-5x > 5x+10-5x$
$-10 > 10$
Because this inequality must always be false, it has no solution.

38. $-3 < 4-2x \le 0$
$-3-4 < 4-2x-4 \le 0-4$
$-7 < -2x \le -4$
$\dfrac{-7}{-2} > \dfrac{-2x}{-2} \ge \dfrac{-4}{-2}$
$\dfrac{7}{2} > x \ge 2$ or $2 \le x < \dfrac{7}{2}$
The solution set is $\left[2, \dfrac{7}{2}\right)$.

39. $8 \le 3(x-7)+5 < 20$
$8 \le 3x-21+5 < 20$
$8 \le 3x-16 < 20$
$8+16 \le 3x-16+16 < 20+16$
$24 \le 3x < 36$
$\dfrac{24}{3} \le \dfrac{3x}{3} < \dfrac{36}{3}$
$8 \le x < 12$
The solution set is $[8, 12)$.

40. Let x = the number of miles.
$49.95+0.18x \le 150$
$49.95+0.18x-49.95 \le 150-49.95$
$0.18x \le 100.05$
$\dfrac{0.18x}{0.18} \le \dfrac{100.05}{0.18}$
$x \le 555.833$ (rounded to the nearest thousandth)
You can drive 555 or fewer miles.

41. Let x = Ali's sales in the sixth month.
$$\frac{2100+1300+1650+1250+1725+x}{6} > 1500$$
$$\frac{8025+x}{6} > 1500$$
$$6\left(\frac{8025+x}{6}\right) > 6(1500)$$
$$8025 + x > 9000$$
$$8025 + x - 8025 > 9000 - 8025$$
$$x > 975$$
Ali's sales must be greater than $975.

42. Let x = the width of the flowerbed.
$x + 4$ = the length of the flowerbed.
$$2x + 2(x+4) \le 40$$
$$2x + 2x + 8 \le 40$$
$$4x + 8 \le 40$$
$$4x + 8 - 8 \le 40 - 8$$
$$4x \le 32$$
$$\frac{4x}{4} \le \frac{32}{4}$$
$$x \le 8$$
The width of the flowerbed can be 8 feet or less.

43. Let x = the number of students needed.
$$35x - 20x - 225 \ge 500$$
$$15x - 225 \ge 500$$
$$15x - 225 + 225 \ge 500 + 225$$
$$15x \ge 725$$
$$\frac{15x}{15} \ge \frac{725}{15}$$
$x \ge 48.333$ (rounded to the nearest thousandth)
The testing center must test 49 or more students daily in order to realize the $500 daily profit.

44. Let x = Paul's sixth test score.
$$88 \le \frac{89+96+89+80+100+x}{6} \le 92$$
$$88 \le \frac{454+x}{6} \le 92$$
$$6(88) \le 6\left(\frac{454+x}{6}\right) \le 6(92)$$
$$528 \le 454 + x \le 552$$
$$528 - 454 \le 454 + x - 454 \le 552 - 454$$
$$74 \le x \le 98$$
To earn a B+, Paul's sixth test score must be from 74 to 98 points, inclusive.

45. $0.15x + 2 \ge 27$
$$0.15x + 2 - 2 \ge 27 - 2$$
$$0.15x \ge 25$$
$$\frac{0.15x}{0.15} \ge \frac{25}{0.15}$$
$x \ge 166.667$ (rounded to the nearest thousandth)
A doctor will prescribe medication if the man 166 or more pounds.

46. $x + 2y < 12$ is linear. It is in standard form.

47. $y < \frac{2}{3}x + \frac{5}{9}$ is linear.
In standard form,
$$9(y) < 9\left(\frac{2}{3}x + \frac{5}{9}\right)$$
$$9y < 6x + 5$$
$$9y - 6x < 6x + 5 - 6x$$
$$-6x + 9y < 5$$

48. $x^2 + y^2 \ge 1$ is nonlinear because the x term and y term are squared.

49. $0.3x + 2.9 > 1.4y$ is linear.
In standard form,
$$0.3x + 2.9 - 1.4y > 1.4y - 1.4y$$
$$0.3x - 1.4y + 2.9 > 0$$
$$0.3x - 1.4y + 2.9 - 2.9 > 0 - 2.9$$
$$0.3x - 1.4y > -2.9$$

50. $y \ge \sqrt{x} - 1.44$ is nonlinear because the radical expression contains a variable.

51. $y < x^2 + 9$ is nonlinear because the x term is squared.

SSM: Experiencing Introductory and Intermediate Algebra

52. $12x + 6y < 48$
$12x + 6y - 12x < 48 - 12x$
$6y < -12x + 48$
$\dfrac{6y}{6} < \dfrac{-12x + 48}{6}$
$y < -2x + 8$

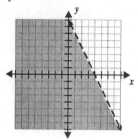

53. $y > \dfrac{3}{5}x - 6$

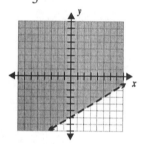

54. $4y \le x + 12$
$\dfrac{4y}{4} \le \dfrac{x + 12}{4}$
$y \le \dfrac{1}{4}x + 3$

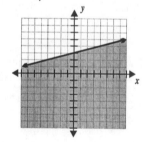

55. $y + 9 \ge 12$
$y + 9 - 9 \ge 12 - 9$
$y \ge 3$

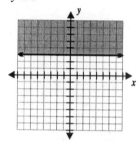

56. $5y - 2 < y - 10$
$5y - 2 - y < y - 10 - y$
$4y - 2 < -10$
$4y - 2 + 2 < -10 + 2$
$4y < -8$
$\dfrac{4y}{4} < \dfrac{-8}{4}$
$y < -2$

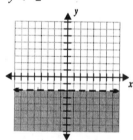

57. $2x + 16 > x + 18$
$2x + 16 - x > x + 18 - x$
$x + 16 > 18$
$x + 16 - 16 > 18 - 16$
$x > 2$

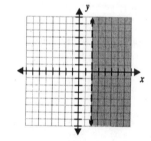

58. $8x - 12y > 24$

$8x - 12y - 8x > 24 - 8x$

$-12y > -8x + 24$

$\dfrac{-12y}{-12} < \dfrac{-8x + 24}{-12}$

$y < \dfrac{2}{3}x - 2$

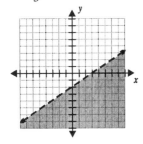

59. $4.4x + 1.1y \geq 12.1$

$4.4x + 1.1y - 4.4x \geq 12.1 - 4.4x$

$1.1y \geq -4.4x + 12.1$

$\dfrac{1.1y}{1.1} \geq \dfrac{-4.4x + 12.1}{1.1}$

$y \geq -4x + 11$

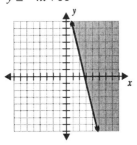

60. $7.4x - 14.8y \leq 29.6$

$7.4x - 14.8y - 7.4x \leq 29.6 - 7.4x$

$-14.8y \leq -7.4x + 29.6$

$\dfrac{-14.8y}{-14.8} \geq \dfrac{-7.4x + 29.6}{-14.8}$

$y \geq \dfrac{1}{2}x - 2$

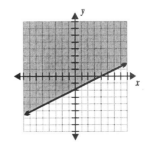

61. $\dfrac{x}{12} + \dfrac{y}{8} > -\dfrac{5}{4}$

$24\left(\dfrac{x}{12} + \dfrac{y}{8}\right) > 24\left(-\dfrac{5}{4}\right)$

$2x + 3y > -30$

$2x + 3y - 2x > -30 - 2x$

$3y > -2x - 30$

$\dfrac{3y}{3} > \dfrac{-2x - 30}{3}$

$y > -\dfrac{2}{3}x - 10$

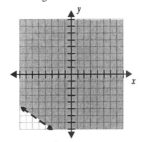

62. $-x - y < 2$

$-x - y + x < 2 + x$

$-y < x + 2$

$-1(-y) > -1(x + 2)$

$y > -x - 2$

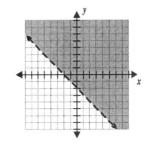

SSM: Experiencing Introductory and Intermediate Algebra

63. $y > -9x + 6$

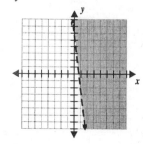

64. Let x = the number of rhododendrons.
y = the number of azaleas.
$4x + 6y \leq 85$
Solve for y.
$4x + 6y - 4x \leq 85 - 4x$
$6y \leq -4x + 85$
$\dfrac{6y}{6} \leq \dfrac{-4x + 85}{6}$
$y \leq -\dfrac{2}{3}x + \dfrac{85}{6}$

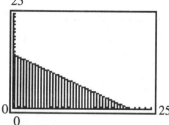

The ordered pairs in the shaded region represent the possible combinations of rhododendrons and azaleas.
Two possible solutions: Oksana could buy 10 rhododendrons and 7 azaleas or 5 rhododendrons and 10 azaleas.

65. Let x = the number solved correctly.
y = the number not solved correctly.
$5x - 3y \geq 80$
Solve for y.
$5x - 3y - 5x \geq 80 - 5x$
$-3y \geq -5x + 80$
$\dfrac{-3y}{-3} \leq \dfrac{-5x + 80}{-3}$
$y \leq \dfrac{5}{3}x - \dfrac{80}{3}$

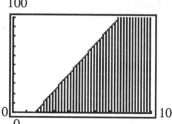

The ordered pairs in the shaded region represent the possible combinations of problems solved correctly and not solved correctly.
Two possible solutions: Rosa could solve 50 problems correctly and 50 problems incorrectly or 30 problems correctly and 20 problems incorrectly.

66. Let x = the number of 2.5 pound packages.
y = the number of 5 pound packages.
According to the table, the cost of shipping a 2.5-pound package $5.20. The cost of shipping a 5-pound package is $7.70.
$5.2x + 7.7y + 10.25 \leq 120$
Solve for y.
$5.2x + 7.7y + 10.25 - 10.25 \leq 120 - 10.25$
$5.2x + 7.7y \leq 109.75$
$5.2x + 7.7y - 5.2x \leq 109.75 - 5.2x$
$7.7y \leq -5.2x + 109.75$
$\dfrac{7.7y}{7.7} \leq \dfrac{-5.2x + 109.75}{7.7}$
$y \leq -\dfrac{52}{77}x + \dfrac{2195}{154}$

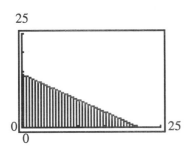

The ordered pairs in the shaded region represent the possible combinations of 2.5-pound and 5-pound packages that can be shipped without exceeding the budget.
Two possible solutions: Five 2.5-pound and ten 5-pound packages can be shipped or fifteen 2.5-pound and four 5-pound packages can be shipped without exceeding the budget.
Yes, (10, 6) is in the shaded region. Ten 2.5-pound and six 5-pound packages can be shipped without exceeding the budget.
No, (12, 8) is not in the shaded region. Twelve 2.5-pound and eight 5-pound packages cannot be shipped without exceeding the budget.

67. Solve the first inequality for y.
$2x + y > 10$
$2x + y - 2x > 10 - 2x$
$y > -2x + 10$
The second inequality is already solved for y.
$y < 3x - 5$
Find the intersection point (using substitution).
$-2x + 10 = 3x - 5$
$-2x + 10 - 3x = 3x - 5 - 3x$
$-5x + 10 = -5$
$-5x + 10 - 10 = -5 - 10$
$-5x = -15$
$\dfrac{-5x}{-5} = \dfrac{-15}{-5}$
$x = 3$

$y = 3x - 5$
$y = 3(3) - 5 = 4$
The intersection point is $(3, 4)$.

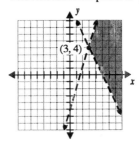

68. Solve the first inequality for x.
$3(x + 2) + 1 > -5$
$3x + 6 + 1 > -5$
$3x + 7 > -5$
$3x + 7 - 7 > -5 - 7$
$3x > -12$
$\dfrac{3x}{3} > \dfrac{-12}{3}$
$x > -4$
Solve the second inequality for y.
$2x - y < -8$
$2x - y + y < -8 + y$
$2x < y - 8$
$2x + 8 < y - 8 + 8$
$2x + 8 < y$ or $y > 2x + 8$
Find the intersection point (using substitution).
$y = 2x + 8$
$y = 2(-4) + 8 = 0$
The intersection point is $(-4, 0)$.

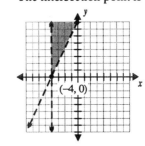

69. Solve the first inequality for x.
$x+6<4$
$x+6-6<4-6$
$x<-2$
Solve the second inequality for y.
$2(y+2)>y+3$
$2y+4>y+3$
$2y+4-y>y+3-y$
$y+4>3$
$y+4-4>3-4$
$y<-1$
The intersection point is $(-2,-1)$.

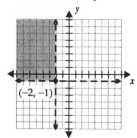

70. Solve the first inequality for y.
$x-2y\geq 12$
$x-2y-x\geq 12-x$
$-2y\geq -x+12$
$\dfrac{-2y}{-2}\leq \dfrac{-x+12}{-2}$
$y\leq \dfrac{1}{2}x-6$

Solve the second inequality for y.
$2x+3y<-6$
$2x+3y-2x<-6-2x$
$3y<-2x-6$
$\dfrac{3y}{3}<\dfrac{-2x-6}{3}$
$y<-\dfrac{2}{3}x-2$

Find the intersection point (using elimination).
$-2x+4y=-24$
$\underline{2x+3y=-6}$
$7y=-30$
$\dfrac{7y}{7}=\dfrac{-30}{7}$
$y=-\dfrac{30}{7}$
$x-2\left(-\dfrac{30}{7}\right)=12$
$x+\dfrac{60}{7}=12$
$x+\dfrac{60}{7}-\dfrac{60}{7}=12-\dfrac{60}{7}$
$x=\dfrac{24}{7}$

The intersection point is $\left(\dfrac{24}{7},-\dfrac{30}{7}\right)$.

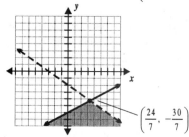

71. $y<3x-6$
$y<-3$
Find the intersection point (using substitution).
$-3=3x-6$
$-3+6=3x-6+6$
$3=3x$
$\dfrac{3}{3}=\dfrac{3x}{3}$
$1=x$
$x=1$
$y=-3$
The intersection point is $(1,-3)$.

Chapter 7: Linear Inequalities

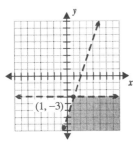

72. Solve the first inequality for y.
$2x+3y \leq 6$
$2x+3y-2x \leq 6-2x$
$3y \leq -2x+6$
$\dfrac{3y}{3} \leq \dfrac{-2x+6}{3}$
$y \leq -\dfrac{2}{3}x+2$

Solve the second inequality for y.
$x+y \leq 1$
$x+y-x \leq 1-x$
$y \leq -x+1$

Find the intersection point (using elimination).
$2x+3y=6$
$\underline{-3x-3y=-3}$
$-x=3$
$\dfrac{-x}{-1}=\dfrac{3}{-1}$
$x=-3$
$y=-x+1$
$y=-(-3)+1=4$

The intersection point is $(-3, 4)$.

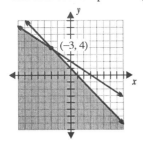

73. $y < 2x-15$
$y > -3x+10$

Find the intersection point (using substitution).
$2x-15=-3x+10$
$2x-15+3x=-3x+10+3x$
$5x-15=10$
$5x-15+15=10+15$
$5x=25$
$\dfrac{5x}{5}=\dfrac{25}{5}$
$x=5$
$y=2x-15$
$y=2(5)-15=-5$

The intersection point is $(5, -5)$.

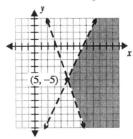

74. $y > 2x+1$

Solve the second inequality for y.
$x > 2y+7$
$x-7 > 2y+7-7$
$x-7 > 2y$
$\dfrac{x-7}{2} > \dfrac{2y}{2}$
$\dfrac{1}{2}x-\dfrac{7}{2} > y$ or $y < \dfrac{1}{2}x-\dfrac{7}{2}$

Find the intersection point (using substitution).

$x = 2(2x+1)+7$
$x = 4x+2+7$
$x = 4x+9$
$x-4x = 4x+9-4x$
$-3x = 9$
$\dfrac{-3x}{-3} = \dfrac{9}{-3}$
$x = -3$
$y = 2x+1$
$y = 2(-3)+1 = -5$
The intersection point is $(-3, -5)$.

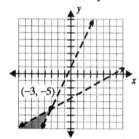

75. Solve the first inequality for y.
$2y+6 \le 3y+5$
$2y+6-2y \le 3y+5-2y$
$6 \le y+5$
$6-5 \le y+5-6$
$-1 \le y$ or $y \ge -1$
Solve the second inequality for x.
$2(x-3)+1 > 5$
$2x-6+1 > 5$
$2x-5 > 5$
$2x-5+5 > 5+5$
$2x > 10$
$\dfrac{2x}{2} > \dfrac{10}{2}$
$x > 5$

The intersection point is $(5, -1)$

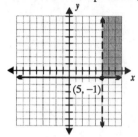

76. $y < 6-x$
$y < 2x+1$
$x \ge 0$
$y \ge 0$
The intersection points of the bounded area are
$(0, 0)$, $(6, 0)$, $\left(\dfrac{5}{3}, \dfrac{13}{3}\right)$, and $(0, 1)$.

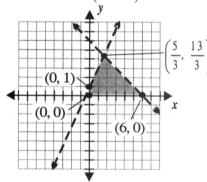

77. $y < 25 - \dfrac{1}{4}x$
$y < \dfrac{2}{3}x+5$
$x \ge 0$
$y \ge 0$
The intersection points of the bounded area are

$(0, 0)$, $(100, 0)$, $\left(\dfrac{240}{11}, \dfrac{215}{11}\right)$, and $(0, 5)$.

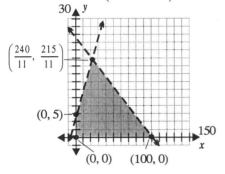

78. Let x = the number of rhododendrons.

    y = the number of azaleas.

    $4x + 6y \leq 85$

    $y + 4 \leq x$

    $x \geq 0$

    $y \geq 0$

    Solve the first two inequalities for y.

    $y \leq -\dfrac{2}{3}x + \dfrac{85}{6}$

    $y \leq x - 4$

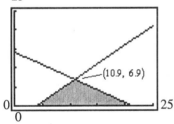

The ordered pairs in the shaded region represent the possible combinations of rhododendrons and azaleas.
One possible combination is 10 rhododendrons and 6 azaleas.

79. Let x = the amount invested at 5%.

    y = the amount invested at 6%.

    $x + y \leq 4000$

    $0.05x + 0.06y \geq 225$

    $x \geq 0$

    $y \geq 0$

    Solve the first two inequalities for y.

    $y \leq -x + 4000$

    $y \geq -\dfrac{5}{6}x + 3750$

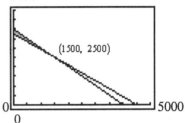

The ordered pairs in the shaded region represent the possible combinations of investment.
One possible combination is $500 at 5% and $3400 at 6%.

80. Let x = the number of 2.5 pound packages.

    y = the number of 5 pound packages.

    According to the table, the cost of shipping a 2.5-pound package $5.20. The cost of shipping a 5-pound package is $7.70.

    $5.2x + 7.7y + 10.25 \leq 120$

    $x \leq 2y$

    $x \geq 0$

    $y \geq 0$

    Solve the first two equations for y.

    $y \leq -\dfrac{52}{77}x + \dfrac{2195}{154}$

    $y \geq \dfrac{1}{2}x$

SSM: Experiencing Introductory and Intermediate Algebra

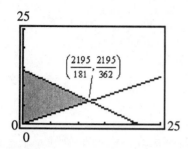

The ordered pairs in the shaded region represent the possible combinations of 2.5-pound and 5-pound packages that can be shipped without exceeding the budget.
One possible combination is five 2.5-pound and nine 5-pound packages can be shipped without exceeding the budget.
No, (10, 12) is not in the shaded region. Ten 2.5-pound and twelve 5-pound packages cannot be shipped without exceeding the budget.
Yes, (7, 6) is in the shaded region. Seven 2.5-pound and six 5-pound packages can be shipped without exceeding the budget.

Chapter 7 Chapter Review

1. $x < -2$

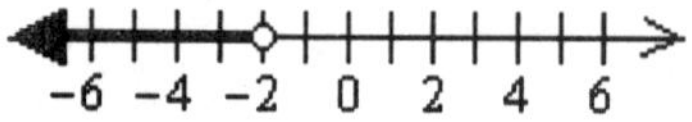

   Interval notation: $(-\infty, -2)$.

2. $x > 7$

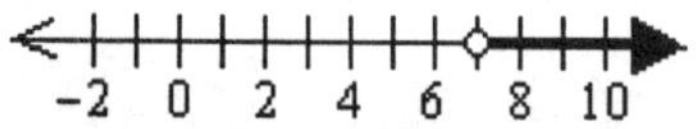

   Interval notation: $(7, \infty)$.

3. $-1 < x < 3$

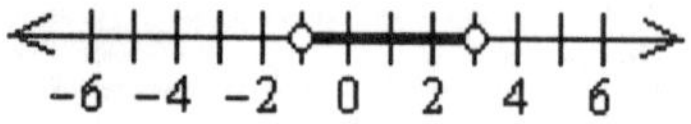

   Interval notation: $(-1, 3)$.

4. $x \geq 2.6$

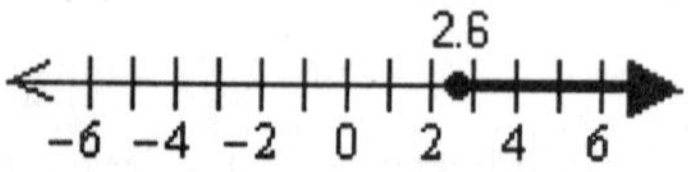

   Interval notation: $[2.6, \infty)$.

5. $x \leq 3\frac{2}{3}$

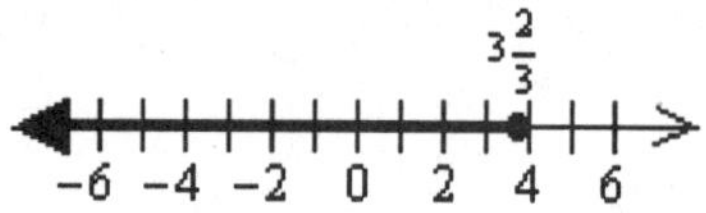

   Interval notation: $\left(-\infty, 3\frac{2}{3}\right]$.

6. $-2.4 < x \leq 4\frac{1}{3}$

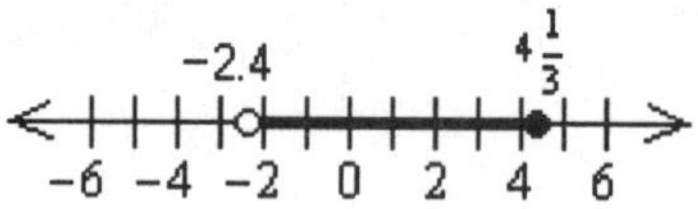

   Interval notation: $\left(-2.4, 4\frac{1}{3}\right]$.

7. Let $Y1 = 5x + 3$ and $Y2 = 2x - 9$. A sample table is shown below.

   Because $Y1 < Y2$ for values $x < -4$, the solutions are the integers less than -4.

8. Let $Y1 = 5.6x - 15.3$ and $Y2 = 1.3x + 19.1$. A sample table is shown below.

   Because $Y1 > Y2$ for values $x > 8$, the solutions are the integers greater than 8.

9. Let $Y1 = \frac{1}{6}x + \frac{23}{3}$ and $Y2 = \frac{13}{6} - \frac{5}{3}x$. A sample table is shown below.

Because $Y1 \leq Y2$ for values $x \leq -3$, the solutions are the integers less than or equal to -3.

10. Let $Y1 = \dfrac{3}{7}x + \dfrac{9}{5}$ and $Y2 = \dfrac{4}{5}x - \dfrac{17}{5}$. A sample table is shown below.

X	Y1	Y2
11	6.5143	5.4
12	6.9429	6.2
13	7.3714	7
14	7.8	7.8
15	8.2286	8.6
16	8.6571	9.4
17	9.0857	10.2

X=14

Because $Y1 \geq Y2$ for values $x \leq 14$, the solutions are the integers less than or equal to 14.

11. Let $Y1 = 5x - 13$ and $Y2 = 3(1-x)$. The graphs of the functions are shown below.

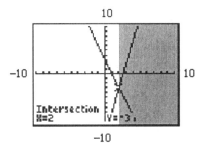

The intersection point is $(2, -3)$. Because $Y1 > Y2$ for $x > 2$, the solution set is all x values that satisfy $x > 2$, or $(2, \infty)$.

12. Let $Y1 = 2.1x + 31.71$ and $Y2 = 8.19 - 3.5x$. The graphs of the functions are shown below.

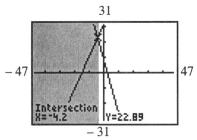

The intersection point is $(-4.2, 22.89)$. Because $Y1 < Y2$ for $x \leq -4.2$, the solution set is all x values that satisfy $x \leq -4.2$, or $(-\infty, 4.2]$.

13. Let $Y1 = 3(x-1) + (x-1)$ and $Y2 = 2(x-1) + 2x - 1$. The graphs of the functions are shown below.

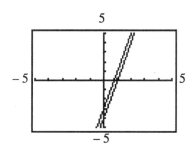

The two lines are parallel with the graph of $Y1$ is always below that of $Y2$. This inequality is always true. The solution is the set of all real numbers, or $(-\infty, \infty)$.

14. Let $Y1 = 4(x+2) - 3(x-5)$ and $Y2 = 5x + 8 - 4(x+3)$. The graphs of the functions are shown below.

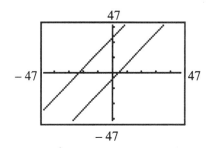

The two lines are parallel. Because $Y1 > Y2$ for all values of x, the inequality is never true. This inequality has no solution.

15. $173 - x < 359$
$173 - x - 173 < 359 - 173$
$-x < 186$
$-1(-x) > -1(186)$
$x > -186$
The solution set is $(-186, \infty)$.

SSM: Experiencing Introductory and Intermediate Algebra

16. $z - \dfrac{4}{17} > \dfrac{15}{34}$

    $z - \dfrac{4}{17} + \dfrac{4}{17} > \dfrac{15}{34} + \dfrac{4}{17}$

    $z > \dfrac{23}{34}$

    The solution set is $\left(\dfrac{23}{34}, \infty\right)$.

17. $5x + 4 > 3x + 18$

    $5x + 4 - 3x > 3x + 18 - 3x$

    $2x + 4 > 18$

    $2x + 4 - 4 > 18 - 4$

    $2x > 14$

    $\dfrac{2x}{2} > \dfrac{14}{2}$

    $x > 7$

    The solution set is $(7, \infty)$.

18. $8x < 8x - 16$

    $8x - 8x < 8x - 16 - 8x$

    $0 < -16$

    Because this inequality must always be false, it has no solution.

19. $3.5x + 19.88 \geq -1.9x + 4.76$

    $3.5x + 19.88 + 1.9x \geq -1.9x + 4.76 + 1.9x$

    $5.4x + 19.88 \geq 4.76$

    $5.4x + 19.88 - 19.88 \geq 4.76 - 19.88$

    $5.4x \geq -15.12$

    $\dfrac{5.4x}{5.4} \geq \dfrac{-15.12}{5.4}$

    $x \geq -2.8$

    The solution set is $[-2.8, \infty)$.

20. $2.6y + 9.62 \leq 0$

    $2.6y + 9.62 - 9.62 \leq 0 - 9.62$

    $2.6y \leq -9.62$

    $\dfrac{2.6y}{2.6} \leq \dfrac{-9.62}{2.6}$

    $x \leq -3.7$

    The solution set is $(-\infty, -3.7]$.

21. $3(x+3) < 4(x+1) - 2(x-2)$

    $3x + 9 < 4x + 4 - 2x + 4$

    $3x + 9 < 2x + 8$

    $3x + 9 - 2x < 2x + 8 - 2x$

    $x + 9 < 8$

    $x + 9 - 9 < 8 - 9$

    $x < -1$

    The solution set is $(-\infty, -1)$.

22. $3(x-3) + 2(x+2) < 5x + 7$

    $3x - 9 + 2x + 4 < 5x + 7$

    $5x - 5 < 5x + 7$

    $5x - 5 - 5x < 5x + 7 - 5x$

    $-5 < 7$

    Because this inequality must always be true, its solution is the set of all real numbers, or $(-\infty, \infty)$.

23. $15 < 4(2x - 1) + 5 \leq 25$

    $15 < 8x - 4 + 5 \leq 25$

    $15 < 8x + 1 \leq 25$

    $15 - 1 < 8x + 1 - 1 \leq 25 - 1$

    $14 < 8x \leq 24$

    $\dfrac{14}{8} < \dfrac{8x}{8} \leq \dfrac{24}{8}$

    $\dfrac{7}{4} < x \leq 3$

    The solution set is $\left(\dfrac{7}{4}, 3\right]$.

24. Solve the first inequality for y.

    $2y + 2 > y$

    $2y + 2 - y > y - y$

    $y + 2 > 0$

    $y + 2 - 2 > 0 - 2$

    $y > -2$

    The second inequality is already solved for y.

    $y < -x + 1$

    Find the intersection point (using substitution).

$y = -x + 1$
$-2 = -x + 1$
$-2 - 1 = -x + 1 - 1$
$-3 = -x$
$-1(-3) = -1(-x)$
$3 = x$
The intersection point is $(3, -2)$.

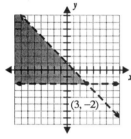

25. Solve the first inequality for x.
$2x + 3 > x$
$2x + 3 - x > x - x$
$x + 3 > 0$
$x + 3 - 3 > 0 - 3$
$x > -3$
Solve the second inequality for x.
$2x < 6$
$\dfrac{2x}{2} < \dfrac{6}{2}$
$x < 3$
The two vertical lines are parallel. There is no intersection point.

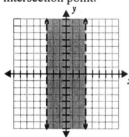

26. $y \geq 2x - 8$
Solve the second inequality for y.
$3x + y < 8$
$3x + y - 3x < 8 - 3x$
$y < -3x + 8$
Find the intersection point (using substitution).

$2x - 8 = -3x + 8$
$2x - 8 + 3x = -3x + 8 + 3x$
$5x - 8 = 8$
$5x - 8 + 8 = 8 + 8$
$5x = 16$
$\dfrac{5x}{5} = \dfrac{16}{5}$
$x = \dfrac{16}{5}$
$y = -3x + 8$
$y = -3\left(\dfrac{16}{5}\right) + 8 = -\dfrac{8}{5}$
The intersection point is $\left(\dfrac{16}{5}, -\dfrac{8}{5}\right)$.

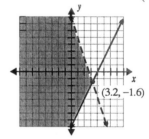

27. Solve the first inequality for x.
$2(x - 1) - 4 > 0$
$2x - 2 - 4 > 0$
$2x - 6 > 0$
$2x - 6 + 6 > 0 + 6$
$2x > 6$
$\dfrac{2x}{2} > \dfrac{6}{2}$
$x > 3$
Solve the second inequality for y.
$2x + y \leq 3$
$2x + y - 2x \leq 3 - 2x$
$y \leq -2x + 3$

Find the intersection point (using substitution).
$y = -2x + 3$
$y = -2(3) + 3 = -3$
The intersection point is $(3, -3)$.

SSM: Experiencing Introductory and Intermediate Algebra

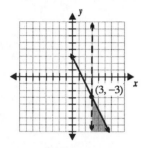

28. Solve the first inequality for *y*.
$x + y < 9 - x$
$x + y - x < 9 - x - x$
$y < -2x + 9$
Solve the second inequality for *y*.
$2x + y > 5$
$2x + y - 2x > 5 - 2x$
$y > -2x + 5$
The two lines are parallel. There is no intersection point.

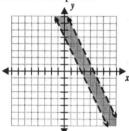

29. Solve the first inequality for *x*.
$3(x-3) - 1 > 8$
$3x - 9 - 1 > 8$
$3x - 10 > 8$
$3x - 10 + 10 > 8 + 10$
$3x > 18$
$\frac{3x}{3} > \frac{18}{3}$
$x > 6$
Solve the second inequality for *y*.
$3y - 10 < 15 - 2y$
$3y - 10 + 2y < 15 - 2y + 2y$
$5y - 10 < 15$
$5y - 10 + 10 < 15 + 10$
$5y < 25$
$\frac{5y}{5} < \frac{25}{5}$
$y < 5$
The intersection point is (6, 5).

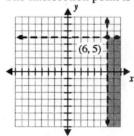

30. $y < 3x + 7$
Solve the second inequality for *y*.
$x - y > -1$
$x - y - x > -1 - x$
$-y > -x - 1$
$-1(-y) > -1(-x - 1)$
$y < x + 1$
Find the intersection point (using substitution).
$3x + 7 = x + 1$
$2x + 7 - x = x + 1 - x$
$2x + 7 = 1$
$2x + 7 - 7 = 1 - 7$
$2x = -6$
$\frac{2x}{2} = \frac{-6}{2}$
$x = -3$
$y = x + 1$
$y = -3 + 1 = -2$
The intersection point is $(-3, -2)$.

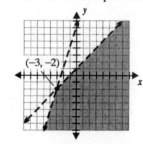

31. Solve the first inequality for y.
$$x - 5y \leq 15$$
$$x - 5y - x \leq 15 - x$$
$$-5y \leq -x + 15$$
$$\frac{-5y}{-5} \geq \frac{-x + 15}{-5}$$
$$y \geq \frac{1}{5}x - 3$$

Solve the second inequality for y.
$$x + 5y \leq -5$$
$$x + 5y - x \leq -5 - x$$
$$5y \leq -x - 5$$
$$\frac{5y}{5} \leq \frac{-x - 5}{5}$$
$$y \leq -\frac{1}{5}x - 1$$

Find the intersection point (using elimination).
$$x - 5y = 15$$
$$x + 5y = -5$$
$$\overline{2x = 10}$$
$$\frac{2x}{2} = \frac{10}{2}$$
$$x = 5$$
$$y = \frac{1}{5}x - 3$$
$$y = \frac{1}{5}(5) - 3 = -2$$

The intersection point is $(5, -2)$.

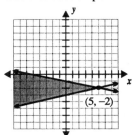

32. $y < -4x + 6$
$y > 3x - 8$

Find the intersection point (using substitution).

$$-4x + 6 = 3x - 8$$
$$-4x + 6 - 3x = 3x - 8 - 3x$$
$$-7x + 6 = -8$$
$$-7x + 6 - 6 = -8 - 6$$
$$-7x = -14$$
$$\frac{-7x}{-7} = \frac{-14}{-7}$$
$$x = 2$$
$$y = -4x + 6$$
$$y = -4(2) + 6 = -2$$

The intersection point is $(2, -2)$.

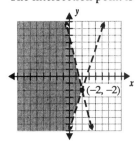

33. $y > x$

Solve the second inequality for y.
$$x > 2 - y$$
$$x + y > 2 - y + y$$
$$x + y > 2$$
$$x + y - x > 2 - x$$
$$y > -x + 2$$

Find the intersection point (using substitution).
$$x = -x + 2$$
$$x + x = -x + 2 + x$$
$$2x = 2$$
$$x = 1$$
$$y = x \text{ so } y = 1.$$

The intersection point is $(1, 1)$.

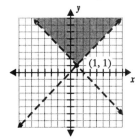

SSM: Experiencing Introductory and Intermediate Algebra

34. Solve the first inequality for x.
$x + 9 \leq 3$
$x + 9 - 9 \leq 3 - 9$
$x \leq -6$
Solve the second inequality for y.
$3y > 2y - 1$
$3y - 2y > 2y - 1 - 2y$
$y > -1$
The intersection point is $(-6, -1)$.

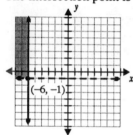

35. $y < -2x + 7$
$y < 2x$
$x \geq 0$
$y \geq 0$
The intersection points of the bounded area are $(0, 0)$, $\left(\frac{7}{4}, \frac{7}{2}\right)$, and $\left(\frac{7}{2}, 0\right)$.

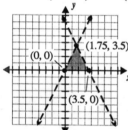

36. $y > \frac{3}{4}x - 4$
$y > -\frac{2}{3}x + 3$
$x \geq 0$
$y \geq 0$
The intersection points for the bounded area are $(0, 3)$, $\left(\frac{9}{2}, 0\right)$, and $\left(\frac{16}{3}, 0\right)$.

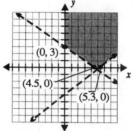

37. $9x - 5y < 45$
$9x - 5y - 9x < 45 - 9x$
$-5y < -9x + 45$
$\frac{-5y}{-5} > \frac{-9x + 45}{-5}$
$y > \frac{9}{5}x - 9$

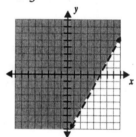

38. $y > \frac{4}{3}x - 5$

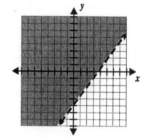

39. $3y \leq 2x+9$

$\dfrac{3y}{3} \leq \dfrac{2x+9}{3}$

$y \leq \dfrac{2}{3}x+3$

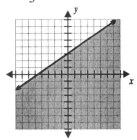

40. $y+13 \geq 8$

$y+13-13 \geq 8-13$

$y \geq -5$

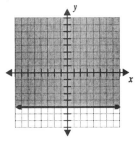

41. $-2y-6 < y-11$

$-2y-6-y < y-11-y$

$-3y-6 < -11$

$-3y-6+6 < -11+6$

$-3y < -5$

$\dfrac{-3y}{-3} > \dfrac{-5}{-3}$

$y > \dfrac{5}{3}$

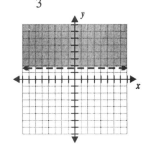

42. $x-13 > 3x+19$

$x-13-x > 3x+19-x$

$-13 > 2x+19$

$-13-19 > 2x+19-19$

$-32 > 2x$

$\dfrac{-32}{2} > \dfrac{2x}{2}$

$-16 > x$ or $x < -16$

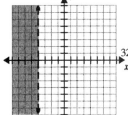

43. $x-7y > 21$

$x-7y-x > 21-x$

$-7y > -x+21$

$\dfrac{-7y}{-7} < \dfrac{-x+21}{-7}$

$y < \dfrac{1}{7}x-3$

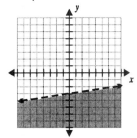

44. $2.7x+5.4y \geq 16.2$

$2.7x+5.4y-2.7x \geq 16.2-2.7x$

$5.4y \geq -2.7x+16.2$

SSM: Experiencing Introductory and Intermediate Algebra

$$\frac{5.4y}{5.4} \geq \frac{-2.7x+16.2}{5.4}$$

$$y \geq -\frac{1}{2}x+3$$

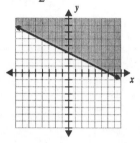

45. $4.8x - 1.8y \leq 14.4$

$4.8x - 1.8y - 4.8x \leq 14.4 - 4.8x$

$-1.8y \leq -4.8x + 14.4$

$\frac{-1.8y}{-1.8} \geq \frac{-4.8x + 14.4}{-1.8}$

$y \geq \frac{8}{3}x - 8$

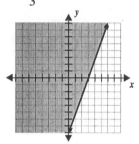

46. $\frac{x}{12} + \frac{y}{9} > -\frac{1}{3}$

$36\left(\frac{x}{12} + \frac{y}{9}\right) > 36\left(-\frac{1}{3}\right)$

$3x + 4y > -12$

$3x + 4y - 3x > -12 - 3x$

$4y > -3x - 12$

$\frac{4y}{4} > \frac{-3x - 12}{4}$

$y > -\frac{3}{4}x - 3$

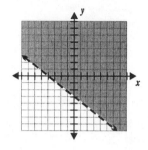

47. $x - y < 7$

$x - y - x < 7 - x$

$-y < -x + 7$

$-1(-y) > -1(-x + 7)$

$y > x - 7$

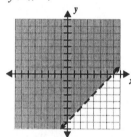

48. $y > -5x + 7$

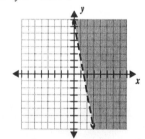

49.
Classification	Weight, lb
Healthy weight	$130 \leq x \leq 165$
Moderately overweight	$166 \leq x \leq 190$
Severely overweight	$x > 190$

50. Let x = the number of packs to be sold.

$12.5x - 2.5x - 255 \geq 1200$

$10x - 255 \geq 1200$

324

$10x - 255 + 255 \geq 1200 + 255$

$10x \geq 1455$

$\dfrac{10x}{10} \geq \dfrac{1455}{10}$

$x \geq 145.5$

At least 146 packs must be sold in order to make a profit of $1200 or more.

One possible answer: 150 packs sold.

51. Let x = Catherine's phone bill in the sixth month.

$\dfrac{45 + 36 + 52 + 48 + 31 + x}{6} < 42$

$\dfrac{212 + x}{6} < 42$

$6\left(\dfrac{212 + x}{6}\right) < 6(42)$

$212 + x < 252$

$212 + x - 212 < 252 - 212$

$x < 40$

Catherine's sixth phone bill must be below $40.

One possible answer: A $35 phone bill.

52. Let x = the length of the street sign.

$18x > 600$

$\dfrac{18x}{18} > \dfrac{600}{18}$

$x > 33\dfrac{1}{3}$

The sign must be over $33\dfrac{1}{3}$ inches long.

One possible answer: A 35-inch long street sign.

53. Let x = the number of packages.

$19.95x - 1.45x - 17 \geq 600$

$18.5x - 17 \geq 600$

$18.5x - 17 + 17 \geq 600 + 17$

$18.55x \geq 617$

$\dfrac{18.55x}{18.55} \geq \dfrac{617}{18.55}$

$x \geq 33.261$ (rounded to the nearest thousandth)

Tony must sell at least 34 packages.

One possible answer: Tony sells 40 packages.

54. Let x = Tony's profit for the sixth month.

$500 < \dfrac{344 + 434 + 254 + 705 + 723 + x}{6} < 600$

$500 < \dfrac{2460 + x}{6} < 600$

$6(500) < 6\left(\dfrac{2460 + x}{6}\right) < 6(600)$

$3000 < 2460 + x < 3600$

$3000 - 2460 < 2460 + x - 2460 < 3600 - 2460$

$540 < x < 1140$

Tony's profit for the sixth month must be between $540 and $1140.

One possible answer: A $1000 profit.

55. Let x = the number of 4-pound packages.

$2.37x \leq 100$

$\dfrac{2.37x}{2.37} \leq \dfrac{100}{2.37}$

$x \leq 42.194$ (rounded to the nearest thousandth)

You can send 42 or fewer 4-pound packages to Zone 5 and remain within the $100 budget.

One possible answer: Mail 35 packages.

56. Let x = the number of wins.

y = the number of ties.

$3x + y \geq 25$

Solve for y.

$3x + y - 3x \geq 25 - 3x$

$y \geq -3x + 25$

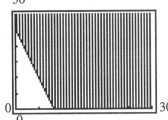

The ordered pairs in the shaded region represent the possible combinations of wins and ties.

One possible solution: The team could win 7 games and tie 5 games.

57. Let x = the number of 4-pound packages.

   y = the number of 3-pound packages.

$2.37x + 2.20y \leq 100$

$y \leq x$

$x \geq 0$

$y \geq 0$

Solve the first equation for y.

$y \leq -\dfrac{237}{220}x + \dfrac{500}{11}$

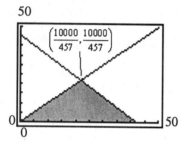

The ordered pairs in the shaded region represent the possible combinations of 4-pound and 3-pound packages that can be shipped without exceeding the budget.

One possible combination is fifteen 4-pound and ten 3-pound packages can be shipped without exceeding the budget.

Yes, (25, 12) is in the shaded region. Twenty-five 4-pound and twelve 3-pound packages can be shipped without exceeding the budget.

No, (32, 20) is not in the shaded region. Thirty-two 4-pound and twenty 3-pound packages cannot be shipped without exceeding the budget.

58. Let x = the number of acres of oats.

   y = the number of acres of wheat.

$y \geq 2x$

$x + y \leq 540$

$x \geq 0$

$y \geq 0$

Solve the second equation for y.

$y \leq -x + 540$

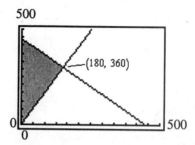

The ordered pairs in the shaded region represent the possible combinations of acres planted in oats and wheat.

One possible combination is 150 acres in oats and 350 acres in wheat.

59. Let x = the rent for the efficiency apartments.

   y = the rent for the regular apartments.

$y \geq x + 75$

$3x + 5y \geq 6000$

$x \geq 0$

$y \geq 0$

Solve the second equation for y.

$y \leq -\dfrac{3}{5}x + 1200$

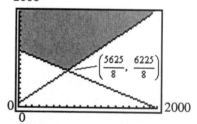

The ordered pairs in the shaded region represent the possible combinations of rent charged for the efficiency apartments and regular apartments.

One possible combination is $600 for efficiency apartments and $900 for regular apartments.

Chapter 7 Test

1. $2x - 3y > x + 8$ is linear.
   In standard form,
   $2x - 3y - 2x > x + 8 - 2x$
   $-3y > -x + 8$
   $\dfrac{-3y}{-3} < \dfrac{-x+8}{-3}$
   $y < \dfrac{1}{3}x - \dfrac{8}{3}$

2. $4x^2 + 2x < x - 9$ is nonlinear because the x term is squared.

3. $5(x-3) \geq 4 - (x+1)$ is linear.
   In standard form,
   $5(x-3) \geq 4 - (x+1)$
   $5x - 15 \geq 4 - x - 1$
   $5x - 15 \geq -x + 3$
   $5x - 15 + x \geq -x + 3 + x$
   $6x - 15 \geq 3$
   $6x - 15 + 15 \geq 3 + 15$
   $6x \geq 18$
   $\dfrac{6x}{6} \geq \dfrac{18}{6}$
   $x \geq 3$

4. $\dfrac{1}{2}x - 4 \geq y + \dfrac{3}{8}$ is linear.
   In standard form,
   $\dfrac{1}{2}x - 4 - \dfrac{3}{8} \geq y + \dfrac{3}{8} - \dfrac{3}{8}$
   $\dfrac{1}{2}x - \dfrac{35}{8} \geq y$ or $y \leq \dfrac{1}{2}x - \dfrac{35}{8}$

5. $7(x+2) - 3(x+1) > 4(x+8)$
   $7x + 14 - 3x - 3 > 4x + 32$
   $4x + 11 > 4x + 32$
   $4x + 11 - 4x > 4x + 32 - 4x$
   $11 > 32$
   Because this inequality must always be false, it has no solution.

6. $5x + 9 < 2x - 3$
   $5x + 9 - 2x < 2x - 3 - 2x$
   $3x + 9 < -3$
   $3x + 9 - 9 < -3 - 9$
   $3x < -12$
   $\dfrac{3x}{3} < \dfrac{-12}{3}$
   $x < -4$
   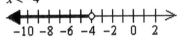
   The solution set is $(-\infty, -4)$.

7. $\dfrac{4}{5}(x-10) < \dfrac{1}{5}(4x+5) + 1$
   $\dfrac{4}{5}x - 8 < \dfrac{4}{5}x + 1 + 1$
   $\dfrac{4}{5}x - 8 < \dfrac{4}{5}x + 2$
   $\dfrac{4}{5}x - 8 - \dfrac{4}{5}x < \dfrac{4}{5}x + 2 - \dfrac{4}{5}x$
   $-8 < 2$
   Because this inequality must always be true, its solution is the set of all real numbers.
   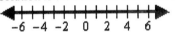
   The solutions set is $(-\infty, \infty)$.

8. $5a - 7 \geq 8a + 1$
   $5a - 7 - 8a \geq 8a + 1 - 8a$
   $-3a - 7 \geq 1$
   $-3a - 7 + 7 \geq 1 + 7$
   $-3a \geq 8$
   $\dfrac{-3a}{-3} \leq \dfrac{8}{-3}$
   $a \leq -\dfrac{8}{3}$
   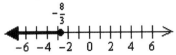
   The solution set is $\left(-\infty, -\dfrac{8}{3}\right]$.

9. $-2 < 3(x-4) - 2 \leq 1$
    $-2 < 3x - 12 - 2 \leq 1$
    $-2 < 3x - 14 \leq 1$
    $-2 + 14 < 3x - 14 + 14 \leq 1 + 14$
    $12 < 3x \leq 15$
    $\dfrac{12}{3} < \dfrac{3x}{3} \leq \dfrac{15}{3}$
    $4 < x \leq 5$

    The solution set is $(4, 5]$.

10. $y < -2x + 5$

    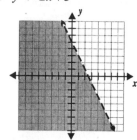

11. Solve the inequality for x.
    $x + 3 > 2x - 1$
    $x + 3 - x > 2x - 1 - x$
    $3 > x - 1$
    $3 + 1 > x - 1 + 1$
    $4 > x$ or $x < 4$

    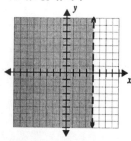

12. Solve the first inequality for y.
    $x - y > 4$
    $x - y - x > 4 - x$
    $-y > -x + 4$
    $-1(-y) < -1(-x + 4)$
    $y < x - 4$
    The second inequality is already solved for y.

    $x + 2y > 4$
    $x + 2y - x > 4 - x$
    $2y > -x + 4$
    $\dfrac{2y}{2} > \dfrac{-x + 4}{2}$
    $y > -\dfrac{1}{2}x + 2$

    Find the intersection point (using elimination).
    $2x - 2y = 8$
    $x + 2y = 4$
    $3x = 12$
    $\dfrac{3x}{3} = \dfrac{12}{3}$
    $x = 4$
    $y = x - 4$
    $y = 4 - 4 = 0$
    The intersection point is $(4, 0)$.

    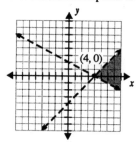

13. $y \geq 4x - 5$
    $y > -3x + 4$

    Find the intersection point (using substitution).
    $4x - 5 = -3x + 4$
    $4x - 5 + 3x = -3x + 4 + 3x$
    $7x - 5 = 4$
    $7x - 5 + 5 = 4 + 5$
    $7x = 9$
    $\dfrac{7x}{7} = \dfrac{9}{7}$
    $x = \dfrac{9}{7}$
    $y = -3x + 4$

$y = -3\left(\dfrac{9}{7}\right) + 4 = \dfrac{1}{7}$

The intersection point is $\left(\dfrac{9}{7}, \dfrac{1}{7}\right)$.

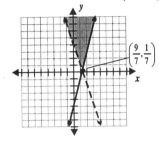

14. Solve the first inequality for y.
$3y - 1 > 2$
$3y - 1 + 1 > 2 + 1$
$3y > 3$
$\dfrac{3y}{3} > \dfrac{3}{3}$
$y > 1$

Solve the second inequality for x.
$x - 3 \leq -5$
$x - 3 + 3 \leq -5 + 3$
$x \leq -2$

The intersection point is $(-2, 1)$.

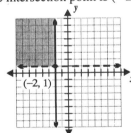

15.

F-Scale number	Wind speed (mph)
F3	$158 \leq x \leq 206$
F4	$207 \leq x \leq 260$
F5	$261 \leq x \leq 318$
F6	$x \geq 319$

16. Let x = the score on the third exam.
$\dfrac{83 + 72 + x}{3} \geq 80$
$\dfrac{155 + x}{3} \geq 80$
$3\left(\dfrac{155 + x}{3}\right) \geq 3(80)$
$155 + x \geq 240$
$155 + x - 155 \geq 240 - 155$
$x \geq 85$

The score on the third exam must be at least 85.
One possible solution: A score of 90.

17. Let x = the wind speed of the fourth tornado.
$158 \leq \dfrac{168 + 172 + 225 + x}{4} \leq 206$
$158 \leq \dfrac{565 + x}{4} \leq 206$
$4(158) \leq 4\left(\dfrac{565 + x}{4}\right) \leq 4(206)$
$632 \leq 565 + x \leq 824$
$632 - 565 \leq 565 + x - 565 \leq 824 - 565$
$67 \leq x \leq 259$

The fourth tornado's wind speed can be from 67 mph to 259 mph.
One possible solution: A wind speed of 100 mph.

18. Let x = the number baskets.
$49x - 23.5x - 250 \geq 300$
$25.5x - 250 \geq 300$
$25.5x - 250 + 250 \geq 300 + 250$
$25.5x \geq 550$
$\dfrac{25.5x}{25.5} \geq \dfrac{550}{25.5}$
$x \geq 21.569$ (rounded to the nearest thousandth)

Alexandra must sell at least 22 baskets per week.
One possible answer: She sells 25 baskets.

19. Let x = the number of good deeds.
    y = the number of activity sheets.
    $5x + 2y \geq 20$
    Solve for y.
    $5x + 2y - 5x \geq 20 - 5x$
    $2y \geq -5x + 20$
    $\dfrac{2y}{2} \geq \dfrac{-5x + 20}{2}$
    $y \geq -\dfrac{5}{2}x + 10$

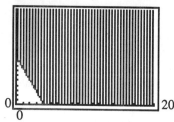

The ordered pairs in the shaded region represent the possible combinations of good deeds and activity sheets.
One possible solution: A Cub Scout could have 3 good deeds and 3 activity sheets.

20. Let x = the number shipped to Zone 102.
    y = the number shipped to Zone 103.
    $22.5x + 26.25y \leq 500$
    $y \geq x$
    $x \geq 0$
    $y \geq 0$
    Solve the first equation for y.
    $y \leq -\dfrac{6}{7}x + \dfrac{400}{21}$

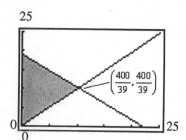

The ordered pairs in the shaded region represent the possible combinations of 10-pound packages that can be shipped to Zones 102 and 103 without exceeding the budget.
One possible combination is 5 packages to Zone 102 and 10 packages to Zone 103.
No, (10, 15) is not in the shaded region. Alexandria cannot ship 10 baskets to Zone 102 and 15 baskets to Zone 103 without exceeding the budget.
Yes, (5, 10) is in the shaded region. Alexandria can ship 5 baskets to Zone 102 and 10 baskets to Zone 103 without exceeding the budget.

21. Let x = the amount invested at 4%.
    y = the amount invested at 8%.
    $x + y \leq 6000$
    $0.04x + 0.08y \geq 400$
    $x \geq 0$
    $y \geq 0$
    Solve the first two equations for y.
    $y \leq -x + 6000$
    $y \geq -\dfrac{1}{2}x + 5000$

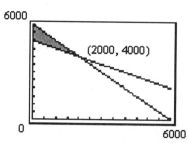

The ordered pairs in the shaded region represent the possible combinations of investments at 4% and 8%.
One possible combination is $1000 at 4% interest and $4750 at 8% interest.

22. A solution of a system of linear inequalities in two variables is an ordered pair that makes all of the inequalities comprising the system true.

Chapters 1 – 7 Cumulative Review

1. $-(-9) = 9$

2. $-|-9| = -(9) = -9$

3. $\sqrt[3]{-\dfrac{27}{64}} = -\dfrac{3}{4}$ because $\left(-\dfrac{3}{4}\right)^3 = -\dfrac{27}{64}$

4. $\sqrt{10} \approx 3.162$

5. $\left(-\dfrac{9}{14}\right)\left(\dfrac{7}{3}\right) = -\dfrac{63}{42} = -\dfrac{3 \cdot 21}{2 \cdot 21} = -\dfrac{3}{2}$

6. $-\dfrac{3}{8} \div \left(-1\dfrac{2}{3}\right) = -\dfrac{3}{8} \div \left(-\dfrac{5}{3}\right) = -\dfrac{3}{8} \cdot \left(-\dfrac{3}{5}\right) = \dfrac{9}{40}$

7. $12(-3) \div (-6)(2) \div (-2) = -36 \div (-6)(2) \div (-2)$
$= 6(2) \div (-2)$
$= 12 \div (-2)$
$= -6$

8. $40 + 16 \div 8 - \sqrt{3^2 + 7 \cdot 5} + 5$
$= 40 + 16 \div 8 - \sqrt{9 + 7 \cdot 5} + 5$
$= 40 + 16 \div 8 - \sqrt{9 + 35} + 5$
$= 40 + 16 \div 8 - \sqrt{49}$
$= 40 + 16 \div 8 - 7$
$= 40 + 2 - 7$
$= 42 - 7$
$= 35$

9. $\dfrac{2(5^2 - 10) + 4^2 - 1}{8^2 - 2(32)} = \dfrac{2(25 - 10) + 4^2 - 1}{8^2 - 2(32)}$
$= \dfrac{2(15) + 4^2 - 1}{8^2 - 2(32)}$
$= \dfrac{2(15) + 16 - 1}{64 - 2(32)}$
$= \dfrac{30 + 16 - 1}{64 - 64}$
$= \dfrac{45}{0}$
$= \text{Undefined}$

10. $6x - 2(4x - 1) = 6x - 8x + 2$
$= -2x + 2$

11. $4[2(x-3) + 1] - [5(2x - 4) - 6]$
$= 4[2x - 6 + 1] - [10x - 20 - 6]$
$= 4[2x - 5] - [10x - 26]$
$= 8x - 20 - 10x + 26$
$= -2x + 6$

12. $y = 2x^2 - 3$

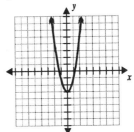

13. The domain is the set of all real numbers, or $(-\infty, \infty)$.
The range is the set of all real numbers greater than or equal to -3, or $[-3, \infty)$.

14. The relation is a function because all possible vertical lines cross the graph a maximum of one time.

15. The relative minimum value is -3 at $x = 0$. There is no relative maximum.

16. The relation is decreasing for all $x < 0$ and increasing for all $x > 0$.

17. For the x-intercept, solve $0 = 2x^2 - 3$.
$2x^2 - 3 = 0$
$2x^2 = 3$
$x^2 = \dfrac{3}{2}$
$x = \pm\sqrt{\dfrac{3}{2}} \approx \pm 1.225$

The x-intercepts are $(-1.225, 0)$ and $(1.225, 0)$.

For the y-intercept, solve $y = 2(0)^2 - 3$.
$y = 2(0)^2 - 3 = -3$
The y-intercept is $(0, -3)$.

18. $7x - 3 = 5$
$7x = 8$
$x = \dfrac{8}{7}$
The solution is $\dfrac{8}{7}$.

19. $(x+3) - 2(3x+4) = 5$
$x + 3 - 6x - 8 = 5$
$-5x - 5 = 5$
$-5x = 10$
$x = -2$
The solution is -2.

20. $1.2(x+3) - 4(0.3x + 0.15) = 3$
$1.2x + 3.6 - 1.2x - 0.6 = 3$
$3 = 3$
Because this equation must always be true (an identity), the solution is the set of all real numbers.

21. $|2x + 6| = 10$
$2x + 6 = -10$ or $2x + 6 = 10$
$2x = -16$ $\qquad 2x = 4$
$x = -8$ $\qquad\quad x = 2$
The solutions are -8 and 2.

22. $5|x| + 8 = 8$
$5|x| = 0$
$|x| = 0$
$x = 0$
The solution is 0.

23. $-12x + 4 \geq -2x + 8$
$4 \geq 10x + 8$
$-4 \geq 10x$
$-\dfrac{2}{5} \geq x$
$x \leq -\dfrac{2}{5}$

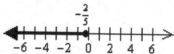

Interval notation: $\left(-\infty, -\dfrac{2}{5}\right]$.

24. $4(x+3) - 3(x-2) \leq x - 5$
$4x + 12 - 3x + 6 \leq x - 5$
$x + 18 \leq x - 5$
$18 \leq -5$
Because this inequality can never be true, it has no solution.

25. $30 < 2x + 10 \leq 70$
$20 < 2x \leq 60$
$10 < x \leq 30$

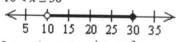

Interval notation $(10, 30]$

26. $f(x) = -3x + 4$

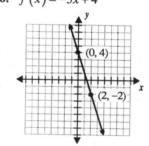

27. $6x + 3y = 9$
$3y = -6x + 9$
$y = -2x + 3$

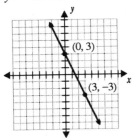

28. $2x + 3 = 11$
$2x = 8$
$x = 4$

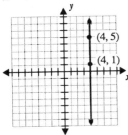

29. $3y - 2 = 5$
$3y = 7$
$y = \dfrac{7}{3}$

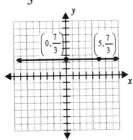

30. $y < 4x - 3$
Graph $y = 4x - 3$ using a dashed line.
Use $(0,0)$ as a test point.
$0 < 4(0) - 3$
$0 < -3$ false

Shade the half-plane not containing $(0,0)$.

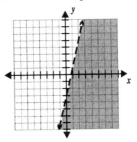

31. Solve the first equation for y.
$2x + 3y = 21$
$3y = -2x + 21$
$y = -\dfrac{2}{3}x + 7$

Solve the second equation for y.
$3x - 2y = 2$
$-2y = -3x + 2$
$y = \dfrac{3}{2}x - 1$

Because the products of the slopes of the two lines is $\left(-\dfrac{2}{3}\right)\left(\dfrac{3}{2}\right) = -1$, the graphs of the two lines are both intersecting and perpendicular.

32. Using substitution,
$3x + 4 = 2x - 5$
$x + 4 = -5$
$x = -9$
$y = 3(-9) + 4 = -23$
The solution is $(-9, -23)$.

33. Using substitution,
$4x + 2(-2x + 4) = 8$
$4x - 4x + 8 = 8$
$8 = 8$
This is an identity. Thus, the solution is the set of all ordered pairs that satisfy the equation $y = -2x + 4$

34. $y \geq -x$
$y < 2x + 4$
Graph $y = -x$ with a solid line and $y = 2x + 4$ with a dashed line. Shade above the line $y = -x$ and below the line $y = 2x + 4$.

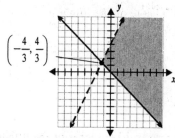
$\left(-\frac{4}{3}, \frac{4}{3}\right)$

One possible solution is $(1,0)$.

35. $m = \dfrac{y_2 - y_1}{x_2 - x_1} = \dfrac{-2-3}{-1-(-2)} = \dfrac{-5}{1} = -5$

36. $m = \dfrac{y_2 - y_1}{x_2 - x_1} = \dfrac{3-(-5)}{6-6} = \dfrac{8}{0} =$ undefined

37. $y = mx + b$
    $y = 2x + 3$
    The slope is 2.

38. $y = mx + b$
    $y = 0x + 1.4$
    The slope is 0.

39. $m = \dfrac{y_2 - y_1}{x_2 - x_1} = \dfrac{-2-3}{-1-(-2)} = \dfrac{-5}{1} = -5$
    $y - y_1 = m(x - x_1)$
    $y - 3 = -5[x - (-2)]$
    $y - 3 = -5(x + 2)$
    $y - 3 = -5x - 10$
    $y = -5x - 7$

40. Write $2x + 3y = 1$ in slop-intercept form.
    $2x + 3y = 1$
    $3y = -2x + 1$
    $y = -\dfrac{2}{3}x + \dfrac{1}{3}$

The slope is $-\dfrac{2}{3}$.
$y - y_1 = m(x - x_1)$
$y - 4 = -\dfrac{2}{3}(x - 5)$
$y - 4 = -\dfrac{2}{3}x + \dfrac{10}{3}$
$y = -\dfrac{2}{3}x + \dfrac{22}{3}$

41. $5,340,000 = 5.34 \times 10^6$

42. $1.2 \times 10^{-4} = 0.00012$

43. $-4.783\text{E} - 5 = -0.00004783$

44. $P = 2L + 2W$
    $P - 2W = 2L$
    $\dfrac{P - 2W}{2} = \dfrac{2L}{2}$
    $L = \dfrac{P - 2W}{2}$

45. $f(x) = x^2 + 2x - 1$
    $f(-3) = (-3)^2 + 2(-3) - 1$
    $ = 9 - 6 - 1$
    $ = 2$

46. Let $x =$ the amount Lance invests.
    $x + 0.11x = 2775$
    $1.11x = 2775$
    $x = 2500$
    Lance should invest $2500.

47. Let $x =$ the measure of the smaller angle.
$2x + 25 =$ the measure of the larger angle.
$x + (2x + 25) = 90$
$3x + 25 = 90$
$3x = 65$
$x = \dfrac{65}{3} = 21\dfrac{2}{3}$
$2x + 25 = 2\left(\dfrac{65}{3}\right) + 25 = \dfrac{205}{3} = 68\dfrac{1}{3}$

The two angles measure $\left(21\dfrac{2}{3}\right)^\circ$ and $\left(68\dfrac{1}{3}\right)^\circ$.

48. Let $x =$ the amount of Hazelnut Coffee used.
$10 - x =$ the amount of Cinnamon Coffee used.
$7.5x + 6.75(10 - x) = 7(10)$
$7.5x + 67.5 - 6.75x = 70$
$0.75x + 67.5 = 70$
$0.75x = 2.5$
$x = \dfrac{2.5}{0.75} = \dfrac{10}{3} = 3\dfrac{1}{3}$
$10 - x = 6\dfrac{2}{3}$

Michael should use $3\dfrac{1}{3}$ pounds of Hazelnut Coffee and $6\dfrac{2}{3}$ pounds of Cinnamon Coffee.

49. Let $x =$ April's score on the fifth test.
$\dfrac{82 + 88 + 80 + 95 + x}{5} \geq 85$
$\dfrac{345 + x}{5} \geq 85$
$345 + x \geq 425$
$x \geq 80$
April must score an 80 or higher on the fifth test in order to earn a B in her Algebra class.

50. Cost function: $C(x) = 35 + 1.5x$
$C(12) = 35 + 1.5(12) = 53$
It costs $53 to rent the chainsaw for 12 hours.

Chapter 8

8.1 Exercises

1. Polynomial
3. Polynomial
5. Not a polynomial. The term $x^{1/2}$ is not a monomial.
7. Polynomial
9. Polynomial.
11. Not a polynomial. The terms $5\sqrt{x}$ and $3\sqrt{y}$ are not monomials. The radicands contain variables.
13. Polynomial
15. Not a polynomial. The terms $\dfrac{4}{x^2}$ and $\dfrac{1}{x}$ are not monomials.
17. Not a polynomial. The terms a^{-2} and $17a^{-1}$ are not monomials.
19. Polynomial
21. $3a - 4b - 5c$ has three terms. Therefore, it is a trinomial.
23. $2z^2$ has 1 term. Therefore it is a monomial.
25. $x - y$ has 2 terms. Therefore, it is a binomial.
27. $4p^4 - 2p^3 + 11p - 57$ has 4 terms. Therefore, it is a polynomial.
29. $6x^2 - 12 + 8x - 5x^2 + x - 17$ reduces to $x^2 + 9x - 29$ which has 3 terms. Therefore, it is a trinomial.
31. $3b - 4 + 7b$ reduces to $10b - 4$ which has 2 terms. Therefore, it is a binomial.
33. $x + 2x + 3x + 4x + 5x$ reduces to $15x$ which has 1 term. Therefore, it is a monomial.

35. $\dfrac{1}{2}x + \dfrac{2}{3}y - \dfrac{3}{4}z$ has 3 terms. Therefore, it is a trinomial.

37. $2 - 15c$
    The term 2 has degree 0 and the term $-15c$ has degree 1. The degree of the polynomial is 1.

39. 123
    The term 123 has degree 0. The degree of the polynomial is 0.

41. $5 + 5x - 4x - x$
    $= 5$
    There is one term, 5, with degree 0. The degree of the polynomial is 0.

43. $7x^5 + 2x^2y^2 - 12$
    The terms are $7x^5$, with degree 5, $2x^2y^2$, with degree 4, and -12, with degree 0. The degree of the polynomial is 5.

45. $\pi r^2 + 2\pi rh$
    The terms are πr^2, with degree 2, and $2\pi rh$, with degree 2. The degree of the polynomial is 2.

47. $4x^2yz^{12} - 8xy^2z^9 + 3x^3y^3z$
    The terms are $4x^2yz^{12}$, with degree 15, $-8xy^2z^9$, with degree 12, and $3x^3y^3z$ with degree 7. The degree of the polynomial is 15.

49. $a^3 + 3a^2 - 2a + 5$

51. $\dfrac{3}{5}x + \dfrac{4}{5}x^3 - \dfrac{7}{15}x - \dfrac{8}{15}x^4$
    $= -\dfrac{8}{15}x^4 + \dfrac{4}{5}x^3 + \left(\dfrac{3}{5} - \dfrac{7}{15}\right)x$
    $= -\dfrac{8}{15}x^4 + \dfrac{4}{5}x^3 + \dfrac{2}{15}x$

53. $3.06x^4 + 0.1x^3 + 4.6x^2 - 1.72$

55. $11 - 2b + 7b^2 - 6b^3$

Chapter 8: Polynomial Functions

57. $\dfrac{2}{7}x + \dfrac{1}{7}x^5 - \dfrac{3}{14}x - \dfrac{5}{14}x^3$

$= \left(\dfrac{2}{7} - \dfrac{3}{14}\right)x - \dfrac{5}{14}x^3 + \dfrac{1}{7}x^5$

$= \dfrac{1}{14}x - \dfrac{5}{14}x^3 + \dfrac{1}{7}x^5$

59. $-2.77 + 3.2x^3 + 9.76x^5 + 0.5x^7$

61. $3x^2 - 4x + 1$

   a. $3(4)^2 - 4(4) + 1$
$= 3(16) - 16 + 1 = 48 - 15$
$= 33$

   b. $3(-2)^2 - 4(-2) + 1$
$= 3(4) + 8 + 1 = 12 + 9$
$= 21$

   c. $3(0)^2 - 4(0) + 1$
$= 3(0) - 0 + 1 = 0 + 1$
$= 1$

63. $-x^2 + 4xy + y^2$

   a. $-(2)^2 + 4(2)(3) + (3)^2$
$= -4 + 24 + 9$
$= 29$

   b. $-(-2)^2 + 4(-2)(-3) + (-3)^2$
$= -4 + 24 + 9$
$= 29$

   c. $-(0)^2 + 4(0)(0) + (0)^2$
$= 0$

65. $1.3m^3 - 2.5m^2 + 3.7m - 4.9$

   a. $1.3(1)^3 - 2.5(1)^2 + 3.7(1) - 4.9$
$= 1.3 - 2.5 + 3.7 - 4.9$
$= -2.4$

   b. $1.3(2.5)^3 - 2.5(2.5)^2 + 3.7(2.5) - 4.9$
$= 20.3125 - 15.625 + 9.25 - 4.9$
$= 9.0375$

   c. $1.3(-1.5)^3 - 2.5(-1.5)^2 + 3.7(-1.5) - 4.9$
$= -4.3875 - 5.625 - 5.55 - 4.9$
$= -20.4625$

67. $\dfrac{2}{3}x^2 - x - 3$

   a. $\dfrac{2}{3}\left(-\dfrac{3}{2}\right)^2 - \left(-\dfrac{3}{2}\right) - 3$

$= \dfrac{2}{3}\left(\dfrac{9}{4}\right) + \dfrac{3}{2} - 3 = 0$

   b. $\dfrac{2}{3}\left(\dfrac{1}{4}\right)^2 - \left(\dfrac{1}{4}\right) - 3$

$= \dfrac{2}{3}\left(\dfrac{1}{16}\right) - \dfrac{1}{4} - 3$

$= \dfrac{1}{24} - \dfrac{13}{4} = -\dfrac{77}{24}$

   c. $\dfrac{2}{3}(3)^2 - (3) - 3$

$= \dfrac{2}{3}(9) - 6 = 6 - 6$

$= 0$

69. Let x = length of first side.
Perimeter:
$A + B + C$
$= x + 2x + (1 + x)$
$= 3x + 1 + x$
$= 4x + 1$ units
$4(8) + 1 = 32 + 1 = 33$

If the first side measures 8 inches, the perimeter would be 33 inches.

71. Let $x =$ width of rectangular solid.
Surface area:
$2LW + 2WH + 2LH$
$= 2(x)(x) + 2(x)(3) + 2(x)(3)$
$= 2x^2 + 6x + 6x$
$= 2x^2 + 12x$ square units

$2(1.5)^2 + 12(1.5)$
$= 2(2.25) + 18$
$= 22.5$

If the width of the rectangular solid is 1.5 meters, the surface area is 22.5 square meters.

73. Area of A = x^2
Area of B = $12x$
Area of C = $3(7) = 21$
total area = area of A + area of B + area of C
$= x^2 + 12x + 21$ m^2

75. Let $x =$ yard width
$y =$ yard length
area not covered = area of yard − area of patio
$xy - 15(20)$
$= xy - 300$ ft^2
$(75)(120) - 300 = 9000 - 300 = 8700$
There are about 8700 square inches of yard not covered by the patio.

77. Let $x =$ length of one side.
Cost:
$12x^2$ dollars
$12(12)^2 = 1728$
The cost to replace a square patio that is 12 feet on each side is roughly $1728.

79. Let $x =$ length of one side of square deck

a. Revenue: $200 + 12x^2$

b. Cost: $75 + 2.75x^2$

c. Revenue:
$200 + 12(15)^2 = 200 + 2700 = 2900$
Cost:

$75 + 2.75(15)^2 = 75 + 618.75 = 693.75$
Profit:
$2900 - 693.75 = 2206.25$
For a deck that is 15 feet on all sides, the revenue would be $2900, the cost will be $693.75, and the profit will be $2206.25.

81. Let $x =$ score on IQ test.

a. $0.006(65)^2 + 0.824(65) + 42.706$
$= 25.35 + 53.56 + 42.706$
$= 121.616$
A score of 65 indicates an IQ score of 122.

b. $0.006(40)^2 + 0.824(40) + 42.706$
$= 9.6 + 32.96 + 42.706$
$= 85.266$
A score of 40 indicates an IQ score of 85.

c. $0.006(50)^2 + 0.824(50) + 42.706$
$= 15 + 41.20 + 42.706 = 98.906$
A score of 50 indicates an IQ score of 99.

83. Let $x =$ tire pressure (psi)
$-1.143(32)^2 + 75.214(32) - 1200$
$= -1170.432 + 2406.848 - 1200$
$= 36.416$
If the tire pressure is 32 psi, the expected mileage is 36,416 miles.

8.1 Calculator Exercises

1. ```
{-3,-1,1,3}→L1:3
L1^3-8L1²+9L1-12
         {-192 -32 -8 24}
```
```

2.
```
{-5,-3,-1}→L₁:{2
,0,-2}→L₂:4L₁^3+
2L₁²*L₂+2L₁*L₂²+
L₂^3
        {-432 -108 -24}
```

3.
```
{1/2,1/4,-1/2,0}
→L₁:{2/5,3/5,5,2
/5}→L₂:(1/2)L₁²+
(3/5)L₁*L₂-(1/4)
L₂^3
   {.229 .06725 -3…
```

```
{1/2,1/4,-1/2,0}
→L₁:{2/5,3/5,5,2
/5}→L₂:(1/2)L₁²+
(3/5)L₁*L₂-(1/4)
L₂^3
   …06725 -32.625…
```

```
{1/2,1/4,-1/2,0}
→L₁:{2/5,3/5,5,2
/5}→L₂:(1/2)L₁²+
(3/5)L₁*L₂-(1/4)
L₂^3
   …  -32.625 -.016}
```

8.2 Exercises

1. $y = x^3 + 2x^2 - 5x - 6$

```
 X  | Y₁
-4  | -18
-2  |  4
-1  |  0
 0  | -6
 2  |  0
X=
```

Some possible solutions:
$\{(-4,-18),(-2,4),(-1,0),(0,-6),(2,0)\}$

3. $y = x^2 + 4x + 1$

```
 X  | Y₁
-3  | -2
-2  | -3
-1  | -2
 0  |  1
 2  | 13
X=
```

Some possible solutions:
$\{(-3,-2),(-2,-3),(-1,-2),(0,1),(2,13)\}$

5. $y = 2x - 5$

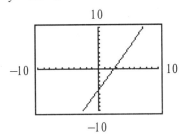

The range is all real numbers.

7. $y = -x^2 + 6x - 4$

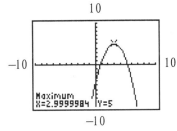

The range is all real numbers less than or equal to 5.

9. $y = 2x^2 - 8x + 3$

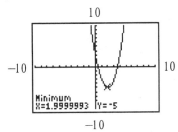

The range is all real numbers greater than or equal to -5.

11. $y = \frac{1}{2}x^2 + 6$

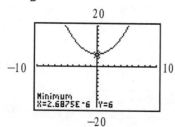

The range is all real numbers greater than or equal to 6.

13. $y = -\frac{1}{2}x^2 - 3x$

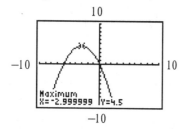

The range is all real numbers less than or equal to $\frac{9}{2}$.

15. $y = x^3 - 8$

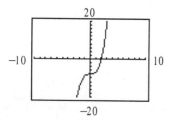

There is no absolute maximum or absolute minimum value. The range is all real numbers.

17. $y = -\frac{1}{8}x^3 + 1$

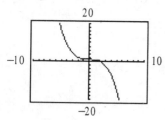

There is no absolute maximum or absolute minimum value. The range is all real numbers.

19. $y = x^3 - 4x$

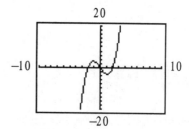

There is no absolute maximum or absolute minimum value. The range is all real numbers.

21. $y = x^3 + 3x^2 - 10x - 24$

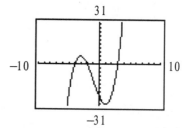

There is no absolute maximum or absolute minimum value. The range is all real numbers.

23. $y = \frac{1}{16}x^4 - x^2$

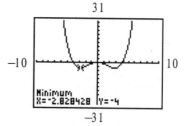

There is an absolute minimum of -4. The range is all real numbers greater than or equal to -4.

25. $y = -\dfrac{16}{81}x^4 + x^2$

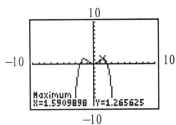

There is an absolute maximum of approximately 1.27. The range is all real numbers less than or equal to 1.27.

27. $y = -x^4 - x^3 + 11x^2 + 9x - 18$

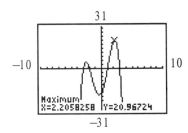

There is an absolute maximum of approximately 20.97. The range is all real numbers less than or equal to 20.97.

29. $f(x) = x^2 + 16x + 64$
$f(2) = (2)^2 + 16(2) + 64$
$= 4 + 32 + 64$
$= 100$

31. $f(x) = x^2 + 16x + 64$
$f(-2) = (-2)^2 + 16(-2) + 64$
$= 4 - 32 + 64$
$= 36$

33. $g(x) = 4x^2 - 4x + 1$
$g(3) = 4(3)^2 - 4(3) + 1$
$= 36 - 12 + 1$
$= 25$

35. $g(x) = 4x^2 - 4x + 1$
$g\left(\dfrac{3}{2}\right) = 4\left(\dfrac{3}{2}\right)^2 - 4\left(\dfrac{3}{2}\right) + 1$
$= 9 - 6 + 1$
$= 4$

37. $h(x) = 9x^2 + 12x + 4$
$h(-2) = 9(-2)^2 + 12(-2) + 4$
$= 36 - 24 + 4$
$= 16$

39. $h(x) = 9x^2 + 12x + 4$
$h(-1.5) = 9(-1.5)^2 + 12(-1.5) + 4$
$= 20.25 - 18 + 4$
$= 6.25$

41. $f(x) = -2x^3 + x^2 - 5x + 8$
$f(-2) = -2(-2)^3 + (-2)^2 - 5(-2) + 8$
$= 16 + 4 + 10 + 8$
$= 38$

43. $f(x) = -2x^3 + x^2 - 5x + 8$
$f(2) = -2(2)^3 + (2)^2 - 5(2) + 8$
$= -16 + 4 - 10 + 8$
$= -14$

45. $g(x) = 2.7x^3 - 1.5x^2 + 3.5x - 6.7$
$g(2) = 2.7(2)^3 - 1.5(2)^2 + 3.5(2) - 6.7$
$= 21.6 - 6 + 7 - 6.7$
$= 15.9$

47. $g(x) = 2.7x^3 - 1.5x^2 + 3.5x - 6.7$
$g(-4) = 2.7(-4)^3 - 1.5(-4)^2 + 3.5(-4) - 6.7$
$= -172.8 - 24 - 14 - 6.7$
$= -217.5$

SSM: Experiencing Introductory and Intermediate Algebra

49. $g(x) = 2.7x^3 - 1.5x^2 + 3.5x - 6.7$

$g(10) = 2.7(10)^3 - 1.5(10)^2 + 3.5(10) - 6.7$
$= 2700 - 150 + 35 - 6.7$
$= 2578.3$

51. $g(x) = 2.7x^3 - 1.5x^2 + 3.5x - 6.7$

$g(1) = 2.7(1)^3 - 1.5(1)^2 + 3.5(1) - 6.7$
$= 2.7 - 1.5 + 3.5 - 6.7$
$= -2$

53. $h(x) = \frac{1}{2}x^3 - \frac{3}{4}x^2 + \frac{3}{8}x - \frac{5}{8}$

$h(2) = \frac{1}{2}(2)^3 - \frac{3}{4}(2)^2 + \frac{3}{8}(2) - \frac{5}{8}$
$= 4 - 3 + \frac{3}{4} - \frac{5}{8}$
$= \frac{9}{8}$

55. $h(x) = \frac{1}{2}x^3 - \frac{3}{4}x^2 + \frac{3}{8}x - \frac{5}{8}$

$h(-4) = \frac{1}{2}(-4)^3 - \frac{3}{4}(-4)^2 + \frac{3}{8}(-4) - \frac{5}{8}$
$= -32 - 12 - \frac{3}{2} - \frac{5}{8}$
$= -\frac{369}{8}$

57. a. $R(x) = 150x - 5x^2$

$R(5) = 150(5) - 5(5)^2 = 625$
$R(10) = 150(10) - 5(10)^2 = 1000$
$R(15) = 150(15) - 5(15)^2 = 1125$
$R(20) = 150(20) - 5(20)^2 = 1000$
$R(25) = 150(25) - 5(25)^2 = 625$
$R(30) = 150(30) - 5(30)^2 = 0$

The revenue if 5, 10, 15, 20, 25, or 30 watches are ordered is $625, $1000, $1125, $1000, $625, and $0, respectively.

b. The range can be determined by graphing the function.

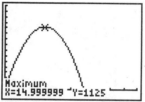

The range is all non-negative real numbers less than or equal to 1125.

c. Dave's Wholesale Jewelers will have an increase in revenue with the discount until they reach a maximum revenue of $1125 (15 watches ordered). The revenue then begins to decrease, reaching a value of $0 when 30 watches are ordered

59. $S(x) = -0.011x^2 + 1.91x - 72.54$

Age, x	63	65	70	75	80	85
Days, $S(x)$	4.13	5.14	7.26	8.84	9.86	10.34

a. The function predicted a stay of a little over 10 days for an 85-year-old woman. The stay was not predicted very well relative to the actual value of 8 days.. The prediction was over 2 days off (25% of the actual length).

b. The function predicted a stay of a little over 7 days for an 85-year-old woman. The predictions were somewhat close, relative to the data. The prediction was better for the 6-day stay than for the 9-day stay.

c. Answers will vary.

61. $s(t) = -16t^2 + 270t$

The domain is all real numbers between 0 and 8 (not inclusive). The rocket will be above the ground for only 8 seconds before it explodes.

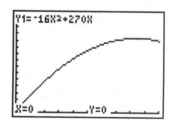

Chapter 8: Polynomial Functions

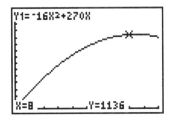

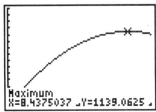

The range is $0 \le s \le 1136$. The rocket will explode when it is 1136 feet above the ground.

63. $s(t) = -16t^2 + 1242$

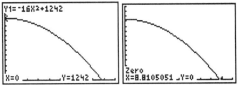

The range is all real numbers between 0 and 1242. The top floor of the Petronas Towers is 1242 feet above the ground. The parachutists will hit the ground in about 8.81 seconds.

8.2 Calculator Exercises

Part 1

1. $y = x$

x-intercept: $(0,0)$

y-intercept: $(0,0)$

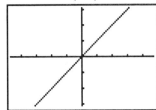

2. $y = x^2 + x$

x-intercept: $(0,0)$, $(-1,0)$

y-intercept: $(0,0)$

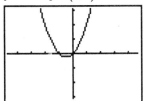

3. $y = x^3 - x$

x-intercepts: $(0,0)$, $(-1,0), (0,-1)$

y-intercept: $(0,0)$

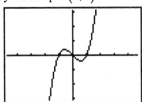

4. $y = x^4 + 2x^3 - x^2 - 2x$

x-intercepts: $(-2,0), (-1,0), (0,0), (1,0)$

y-intercept: $(0,0)$

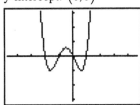

5. $y = x^5 - 5x^3 + 4x$

x-intercepts: $(-2,0), (-1,0), (0,0), (1,0), (2,0)$

y-intercept: $(0,0)$

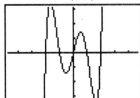

SSM: Experiencing Introductory and Intermediate Algebra

Polynomial	Degree of Polynomial	Number of changes in direction
$y = x$	1	0
$y = x^2 + x$	2	1
$y = x^3 - x$	3	2
$y = x^4 + 2x^3 - x^2 - 2x$	4	3
$y = x^5 - 5x^3 + 4x$	5	4

Part 2

Problems 1. – 3.

```
Plot1 Plot2 Plot3
\Y1■2.5X^3-1.6X2
+4.2X-9.3
\Y2=
\Y3=
\Y4=
\Y5=
\Y6=
```

```
Y1(-2.4)
           -63.156
Y1(155.8)
        9416390.016
Y1(0.944)
         -4.65793664
```

Problems 4. – 6.

```
Plot1 Plot2 Plot3
\Y1■(2/3)X^3+(1/
2)X2-2X+(1/3)
\Y2=
\Y3=
\Y4=
\Y5=
\Y6=
```

```
Y1(6)▶Frac
            451/3
Y1(3/4)▶Frac
            -29/48
Y1(-9.)▶Frac
            -2563/6
```

8.3 Exercises

1. $y = 1 - x - x^2 - x^3$ is not quadratic. This is a third degree polynomial.

3. $f(x) = 8x + 11$ is not quadratic since there is no quadratic term. This is a linear function.

5. $g(x) = 0.5x^2 + 2.6x - 8.4$ is quadratic. It is already in standard form.

7. $y = -2x^2 - 5x - 7$ is quadratic. It is already in standard form.

9. $a = \pi r^2$ is quadratic. It is already in standard form: $a = \pi r^2 + 0r + 0$.

11. $S = 6e^2$ is quadratic. It is already in standard form: $S = 6e^2 + 0e + 0$.

13. $y = \dfrac{7}{x^2} + 3x - 12$ is not quadratic. It is not a polynomial function because of the variable in the denominator.

15. $y = 2x^2 - 2x + 5$ is quadratic. It is already in standard form.

17. $y = -5x^2 + 10x + 1$
    The coefficients are $a = -5, b = 10,$ and $c = 1$.
    The graph is concave down because $a < 0$.
    The graph is narrow compared to the graph of $y = x^2$ because $|a| > 1$.
    The y-intercept is $(0,1)$.

The x-coordinate of the vertex is

$-\dfrac{b}{2a} = -\left(\dfrac{10}{2(-5)}\right) = 1$

The y-coordinate of the vertex is

$y = -5(1)^2 + 10(1) + 1$
$= -5 + 10 + 1$
$= 6$

The vertex is: $(1, 6)$

The axis of symmetry is $x = 1$.

	Function	Coefficients			Properties of graph				
		a	b	c	Wide or narrow	Concave Up/down	Vertex	Axis of Symmetry	y-intercept
19.	$y = 0.6x^2 + 6x - 2$	0.6	6	-2	Wide	Upward	$(-5, -17)$	$x = -5$	$(0, -2)$
21.	$y = 2x^2 + 3x + 5$	2	3	5	Narrow	Upward	$\left(-\dfrac{3}{4}, \dfrac{31}{8}\right)$	$x = -\dfrac{3}{4}$	$(0, 5)$
23.	$y = -\dfrac{1}{4}x^2 + x - 3$	$-\dfrac{1}{4}$	1	-3	Wide	Downward	$(2, -2)$	$x = 2$	$(0, -3)$
25.	$f(x) = x^2 + 8x + 1$	1	8	1	Same	Upward	$(-4, -15)$	$x = -4$	$(0, 1)$

27. $f(x) = 2x^2 + 5x - 7$

x-coordinate of vertex:

$-\dfrac{b}{2a} = -\dfrac{5}{2(2)} = -\dfrac{5}{4}$

y-coordinate of vertex:

$f\left(-\dfrac{5}{4}\right) = 2\left(-\dfrac{5}{4}\right)^2 + 5\left(-\dfrac{5}{4}\right) - 7 = -\dfrac{81}{8}$

Vertex: $\left(-\dfrac{5}{4}, -\dfrac{81}{8}\right)$

Axis of symmetry: $x = -\dfrac{5}{4}$

x	$f(x) = 2x^2 + 5x - 7$	$f(x)$
-3	$f(-3) = 2(-3)^2 + 5(-3) - 7 = -4$	-4
-2	$f(-2) = 2(-2)^2 + 5(-2) - 7 = -9$	-9
0	$f(0) = 2(0)^2 + 5(0) - 7 = -7$	-7
1	$f(1) = 2(1)^2 + 5(1) - 7 = 0$	0

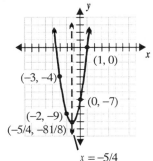

29. $y = \frac{1}{6}x^2 + 3x + 12$

x-coordinate of vertex:

$-\frac{b}{2a} = -\frac{3}{2(1/6)} = -\frac{3}{1/3} = -9$

y-coordinate of vertex:

$y = \frac{1}{6}(-9)^2 + 3(-9) + 12 = -\frac{3}{2}$

Vertex: $\left(-9, -\frac{3}{2}\right)$

Axis of symmetry: $x = -9$

x	$y = \frac{1}{6}x^2 + 3x + 12$	y
-14	$y = \frac{1}{6}(-14)^2 + 3(-14) + 12 = \frac{8}{3}$	$\frac{8}{3}$
-12	$y = \frac{1}{6}(-12)^2 + 3(-12) + 12 = 0$	0
-6	$y = \frac{1}{6}(-6)^2 + 3(-6) + 12 = 0$	0
-3	$y = \frac{1}{6}(-3)^2 + 3(-3) + 12 = \frac{9}{2}$	$\frac{9}{2}$

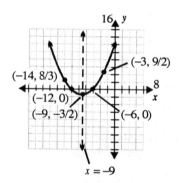

31. $h(x) = 14 + 5x - x^2$

x-coordinate of vertex:

$-\frac{b}{2a} = -\frac{5}{2(-1)} = \frac{5}{2}$

y-coordinate of vertex:

$h\left(\frac{5}{2}\right) = 14 + 5\left(\frac{5}{2}\right) - \left(\frac{5}{2}\right)^2 = \frac{81}{4}$

Vertex: $\left(\frac{5}{2}, \frac{81}{4}\right)$

Axis of symmetry: $x = \frac{5}{2}$

x	$h(x) = 14 + 5x - x^2$	$h(x)$
0	$h(0) = 14 + 5(0) - (0)^2 = 14$	14
1	$h(1) = 14 + 5(1) - (1)^2 = 18$	18
4	$h(4) = 14 + 5(4) - (4)^2 = 18$	18
5	$h(5) = 14 + 5(5) - (5)^2 = 14$	14

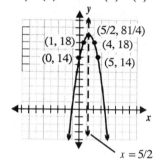

33. $y = -2x^2 + 8x - 3$

x-coordinate of vertex:

$-\frac{b}{2a} = -\frac{8}{2(-2)} = 2$

y-coordinate of vertex:

$y = -2(2)^2 + 8(2) - 3 = 5$

Vertex: $(2, 5)$

Axis of symmetry: $x = 2$

x	$y = -2x^2 + 8x - 3$	y
0	$y = -2(0)^2 + 8(0) - 3 = -3$	-3
1	$y = -2(1)^2 + 8(1) - 3 = 3$	3
3	$y = -2(3)^2 + 8(3) - 3 = 3$	3
4	$y = -2(4)^2 + 8(4) - 3 = -3$	-3

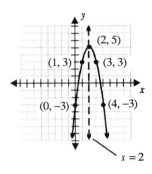

35. $g(x) = 0.8x^2 - 1.2x$

x-coordinate of vertex:

$$-\frac{b}{2a} = -\frac{(-1.2)}{2(0.8)} = \frac{1.2}{1.6} = 0.75$$

y-coordinate of vertex:

$$g(0.75) = 0.8(0.75)^2 - 1.2(0.75) = -0.45$$

Vertex: $(0.75, -0.45)$

Axis of symmetry: $x = 0.75$

x	$g(x) = 0.8x^2 - 1.2x$	$g(x)$
-1	$g(-1) = 0.8(-1)^2 - 1.2(-1) = 2$	2
0	$g(0) = 0.8(0)^2 - 1.2(0) = 0$	0
2	$g(2) = 0.8(2)^2 - 1.2(2) = 0.8$	0.8
4	$g(4) = 0.8(4)^2 - 1.2(4) = 8$	8

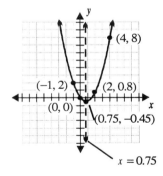

37. $y = 0.4x^2$

x-coordinate of vertex:

$$-\frac{b}{2a} = -\frac{0}{2(0.4)} = 0$$

y-coordinate of vertex:

$y = -0.4(0)^2 = 0$

Vertex: $(0, 0)$

Axis of symmetry: $x = 0$

x	$y = 0.4x^2$	y
-5	$y = 0.4(-5)^2 = 10$	10
-2	$y = 0.4(-2)^2 = 1.6$	1.6
2	$y = 0.4(2)^2 = 1.6$	1.6
5	$y = 0.4(5)^2 = 10$	10

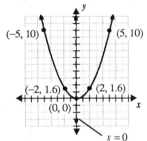

39. $y = 3x^2 - 3$

x-coordinate of vertex:

$$-\frac{b}{2a} = -\frac{0}{2(3)} = 0$$

y-coordinate of vertex:

$y = 3(0)^2 - 3 = -3$

Vertex: $(0, -3)$

Axis of symmetry: $x = 0$

x	$y = 3x^2 - 3$	y
-2	$y = 3(-2)^2 - 3 = 9$	9
-1	$y = 3(-1)^2 - 3 = 0$	0
1	$y = 3(1)^2 - 3 = 0$	0
2	$y = 3(2)^2 - 3 = 9$	9

SSM: Experiencing Introductory and Intermediate Algebra

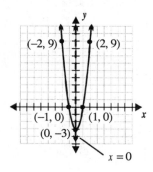

41. $f(x) = -0.5x^2 + 3x$

x-coordinate of vertex:

$$-\frac{b}{2a} = -\frac{3}{2(-0.5)} = 3$$

y-coordinate of vertex:

$$f(3) = -0.5(3)^2 + 3(3) = -4.5$$

Vertex: $(3, -4.5)$

Axis of symmetry: $x = 3$

x	$f(x) = -0.5x^2 + 3x$	$f(x)$
-2	$f(-2) = -0.5(-2)^2 + 3(-2) = -8$	-8
0	$f(0) = -0.5(0)^2 + 3(0) = 0$	0
6	$f(6) = -0.5(6)^2 + 3(6) = 0$	0
8	$f(8) = -0.5(8)^2 + 3(8) = -8$	-8

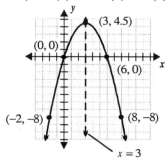

43. $s(t) = -16t^2 + 12t + 24$

The coefficients are $a = -16, b = 12,$ and $c = 24$.

x-coordinate of vertex:

$$-\frac{b}{2a} = -\frac{12}{2(-16)} = \frac{3}{8}$$

y-coordinate of vertex:

$$s\left(\frac{3}{8}\right) = -16\left(\frac{3}{8}\right)^2 + 12\left(\frac{3}{8}\right) + 24$$

$$= -\frac{9}{4} + \frac{9}{2} + 24$$

$$= \frac{105}{4} = 26.25$$

Vertex: $\left(\frac{3}{8}, \frac{105}{4}\right)$

The maximum height the apple will reach is 26.25 feet.

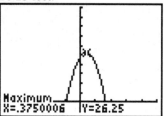

45. $A(x) = 140x - x^2$

The coefficients are $a = -1, b = 140,$ and $c = 0$.

x-coordinate of vertex:

$$-\frac{b}{2a} = -\frac{140}{2(-1)} = 70$$

y-coordinate of vertex:

$$A(70) = 140(70) - (70)^2 = 4900$$

Vertex: $(70, 4900)$

The cottage will have a maximum area of 4900 ft^2 when the width is 70 feet.

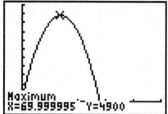

348

47. $A(x) = 10x - x^2$

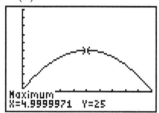

Vertex: $(5, 25)$.

The maximum area of the triangle is 25 in^2 when the height is 5 inches.

49. $R(x) = 46x - x^2$

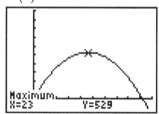

Vertex: $(23, 529)$

The maximum revenue is $529 when 23 dolls are sold. The seller should limit the number of dolls sold. The revenue increases until 23 dolls are sold and then it starts to decrease.

51. $P(x) = 6x - 0.05x^2 - 5$

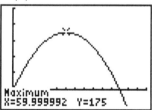

Vertex: $(60, 175)$

The producer will achieve a maximum profit of $175 when 60 compact discs are sold.

8.3 Calculator Exercises

1. $y_1 = x^2 - 2x \quad y_2 = x^2 \quad y_3 = x^2 + 2x$

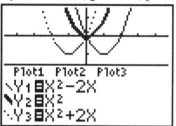

2. $y_1 = x^2 - 1 \quad y_2 = x^2 \quad y_3 = x^2 + 1$

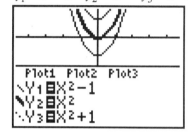

3. $y_1 = 0.3x^2 \quad y_2 = x^2 \quad y_3 = 3x^2$

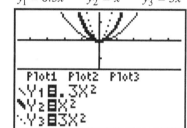

If the absolute value of the leading coefficient is between 0 and 1, the graph will get narrower. If the absolute value of the leading coefficient is greater than 1, the graph will get wider.

4. $y_1 = -3x^2 \quad y_2 = x^2 \quad y_3 = 3x^2$

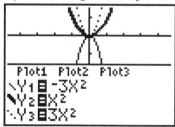

If the leading coefficient is negative, the graph will open down. If the leading coefficient is positive, the graph will open up.

5. $y_1 = 3x^2 \quad y_2 = 3x^2 + 1 \quad y_3 = 3x^2 - 2$

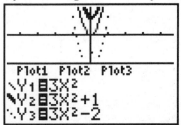

Adding a constant will move the graph up and change the y-intercept. Subtracting a constant will move the graph down and also change the y-intercept.

6. $y_1 = x^2 + 1 \quad y_2 = x^2 + 2x + 1 \quad y_3 = x^2 - 2x + 1$

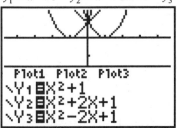

If the coefficient on x is negative, the graph will shift to the right and down. If the coefficient on x is positive, the graph will shift to the left and down.

8.4 Exercises

1. $(2, 0), (-4, 0), (0, -8)$

    The coefficient c is the y-coordinate of the y-intercept. Thus, $c = -8$.
    Use this result and the other two points to form a system of two equations.
    $$0 = a(2)^2 + b(2) - 8$$
    $$0 = a(-4)^2 + b(-4) - 8$$
    Write the equations in standard form.
    $4a + 2b = 8$
    $16a - 4b = 8$
    Multiply the first equation by -4 and add to the second equation.

    $-16a - 8b = -32$
    $\underline{16a - 4b = 8}$
    $-12b = -24$
    $b = 2$
    Substitute this value for b in one of the two equations.
    $4a + 2(2) = 8$
    $4a + 4 = 8$
    $4a = 4$
    $a = 1$
    The quadratic function is $y = x^2 + 2x - 8$.

    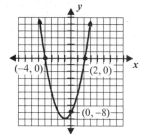

    The graph passes through the given points.

3. $(-4, 0), \left(\dfrac{1}{2}, 0\right), (0, -4)$

    The coefficient c is the y-coordinate of the y-intercept. Thus, $c = -4$.
    Use this result and the other two points to form a system of two equations.
    $$0 = a(-4)^2 + b(-4) - 4$$
    $$0 = a\left(\dfrac{1}{2}\right)^2 + b\left(\dfrac{1}{2}\right) - 4$$
    Write the equations in standard form.
    $16a - 4b = 4$
    $\dfrac{1}{4}a + \dfrac{1}{2}b = 4$
    Multiply the second equation by 8 and add to the first equation.
    $16a - 4b = 4$
    $\underline{2a + 4b = 32}$
    $18a = 36$
    $a = 2$
    Substitute this value for a in one of the two equations.

$16(2) - 4b = 4$

$32 - 4b = 4$

$-4b = -28$

$b = 7$

The quadratic function is $y = 2x^2 + 7x - 4$.

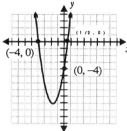

The graph passes through the given points.

5. $(0, 3), (3, 0), (-3, 0)$

The coefficient c is the y-coordinate of the y-intercept. Thus, $c = 3$.

Use this result and the other two points to form a system of two equations.

$0 = a(3)^2 + b(3) + 3$

$0 = a(-3)^2 + b(-3) + 3$

Write the equations in standard form.

$9a + 3b = -3$

$9a - 3b = -3$

Add the two equations.

$9a + 3b = -3$

$\underline{9a - 3b = -3}$

$18a = -6$

$a = -\dfrac{1}{3}$

Substitute this value for a in one of the two equations.

$9\left(-\dfrac{1}{3}\right) + 3b = -3$

$-3 + 3b = -3$

$3b = 0$

$b = 0$

The quadratic function is $y = -\dfrac{1}{3}x^2 + 3$.

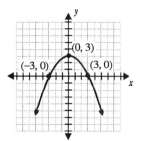

The graph passes through the given points.

7. $(0, -4), (2, 0), (-4, 0)$

The coefficient c is the y-coordinate of the y-intercept. Thus, $c = -4$.

Use this result and the other two points to form a system of two equations.

$0 = a(2)^2 + b(2) - 4$

$0 = a(-4)^2 + b(-4) - 4$

Write the equations in standard form.

$4a + 2b = 4$

$16a - 4b = 4$

Multiply the first equation by 2 and add to the second equation.

$8a + 4b = 8$

$\underline{16a - 4b = 4}$

$24a = 12$

$a = \dfrac{1}{2}$

Substitute this value for a in one of the two equations.

$4\left(\dfrac{1}{2}\right) + 2b = 4$

$2 + 2b = 4$

$2b = 2$

$b = 1$

The quadratic function is $y = \dfrac{1}{2}x^2 + x - 4$.

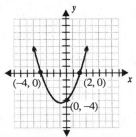

The graph passes through the given points.

9. $\left(0, \frac{2}{3}\right), (-2, 0), (2, 0)$

The coefficient c is the y-coordinate of the y-intercept. Thus, $c = \frac{2}{3}$.

Use this result and the other two points to form a system of two equations.

$0 = a(-2)^2 + b(-2) + \frac{2}{3}$

$0 = a(2)^2 + b(2) + \frac{2}{3}$

Write the equations in standard form.

$4a - 2b = -\frac{2}{3}$

$4a + 2b = -\frac{2}{3}$

Add the two equations.

$4a - 2b = -\frac{2}{3}$
$4a + 2b = -\frac{2}{3}$
$\overline{8a = -\frac{4}{3}}$

$a = -\frac{1}{6}$

Substitute this value for a in one of the two equations.

$4\left(-\frac{1}{6}\right) - 2b = -\frac{2}{3}$

$-\frac{2}{3} - 2b = -\frac{2}{3}$

$-2b = 0$

$b = 0$

The quadratic function is $y = -\frac{1}{6}x^2 + \frac{2}{3}$.

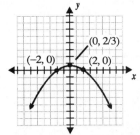

The graph passes through the given points.

11. $(0, -2), \left(\frac{1}{2}, 0\right), \left(\frac{9}{2}, 0\right)$

The coefficient c is the y-coordinate of the y-intercept. Thus, $c = -2$.

Use this result and the other two points to form a system of two equations.

$0 = a\left(\frac{1}{2}\right)^2 + b\left(\frac{1}{2}\right) - 2$

$0 = a\left(\frac{9}{2}\right)^2 + b\left(\frac{9}{2}\right) - 2$

Write the equations in standard form.

$\frac{1}{4}a + \frac{1}{2}b = 2$

$\frac{81}{4}a + \frac{9}{2}b = 2$

Multiply the first equation by -9 and add to the second equation.

$-\frac{9}{4}a - \frac{9}{2}b = -18$

$\frac{81}{4}a + \frac{9}{2}b = 2$
$\overline{18a = -16}$

$a = -\frac{8}{9}$

Substitute this value for a in one of the two equations.

$\frac{1}{4}\left(-\frac{8}{9}\right)+\frac{1}{2}b = 2$

$-\frac{2}{9}+\frac{1}{2}b = 2$

$\frac{1}{2}b = \frac{20}{9}$

$b = \frac{40}{9}$

The quadratic function is $y = -\frac{8}{9}x^2 + \frac{40}{9}x - 2$.

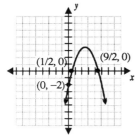

The graph passes through the given points.

13. $(2,-15),(-2,-7),(-4,9)$

Write a system of three equations.
$-15 = a(2)^2 + b(2) + c$
$-7 = a(-2)^2 + b(-2) + c$
$9 = a(-4)^2 + b(-4) + c$
Simplify the equations.
$-15 = 4a + 2b + c$
$-7 = 4a - 2b + c$
$9 = 16a - 4b + c$
Solve for a, b, and c.
$-4a - 2b - c = 15$
$\underline{4a - 2b + c = -7}$
$-4b = 8$
$b = -2$
$4a - 2(-2) + c = -7$
$16a - 4(-2) + c = 9$

$-4a - c = 11$
$\underline{16a + c = 1}$
$12a = 12$
$a = 1$

$4(1) + 2(-2) + c = -15$
$c = -15$
$a = 1, b = -2, c = -15$
$y = x^2 - 2x - 15$

The graph passes through the given points.

15. $(1,-10),(-2,5),(2,-7)$

Write a system of three equations.
$a(1)^2 + b(1) + c = -10$
$a(-2)^2 + b(-2) + c = 5$
$a(2)^2 + b(2) + c = 7$
Simplify the equations.
$a + b + c = -10$
$4a - 2b + c = 5$
$4a + 2b + c = 7$
Solve for a, b, and c.
$-4a + 2b - c = -5$
$\underline{4a + 2b + c = -7}$
$4b = -12$
$b = -3$
$a + (-3) + c = -10$
$4a - 2(-3) + c = 5$

$-a - c = 7$
$\underline{4a + c = -1}$
$3a = 6$
$a = 2$
$2 - 3 + c = -10$
$c = -9$
$a = 2, b = -3, c = -9$
$y = 2x^2 - 3x - 9$

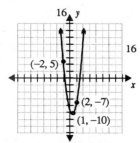

The graph passes through the given points.

17. $(3,2),(-3,14),(6,5)$

Write a system of three equations.

$a(3)^2 + b(3) + c = 2$

$a(-3)^2 + b(-3) + c = 14$

$a(6)^2 + b(6) + c = 5$

Simplify the equations.

$9a + 3b + c = 2$

$9a - 3b + c = 14$

$36a + 6b + c = 5$

Solve for a, b, and c.

$-9a - 3b - c = -2$

$\underline{9a - 3b + c = 14}$

$-6b = 12$

$b = -2$

$9a - 3(-2) + c = 14$

$36a + 6(-2) + c = 5$

$-9a - c = -8$

$\underline{36a + c = 17}$

$27a = 9$

$a = \dfrac{1}{3}$

$9\left(\dfrac{1}{3}\right) + 3(-2) + c = 2$

$3 - 6 + c = 2$

$c = 5$

$a = \dfrac{1}{3}, b = -2, c = 5$

$y = \dfrac{1}{3}x^2 - 2x + 5$

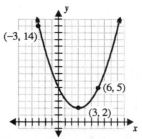

The graph passes through the given points.

19. $(-1,-2),(1,6),(4,3)$

Write a system of three equations.

$a(-1)^2 + b(-1) + c = -2$

$a(1)^2 + b(1) + c = 6$

$a(4)^2 + b(4) + c = 3$

Simplify the equations.

$a - b + c = -2$

$a + b + c = 6$

$16a + 4b + c = 3$

Solve for a, b, and c.

$-a + b - c = 2$

$\underline{a + b + c = 6}$

$2b = 8$

$b = 4$

$a + (4) + c = 6$

$16a + 4(4) + c = 3$

$-a - c = -2$

$\underline{16a + c = -13}$

$15a = -15$

$a = -1$

$(-1) - (4) + c = -2$

$c = 3$

$a = -1, b = 4, c = 3$

$y = -x^2 + 4x + 3$

Chapter 8: Polynomial Functions

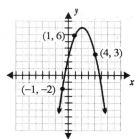

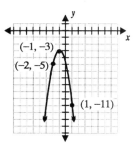

The graph passes through the given points.

21. $(-1,-3),(-2,-5),(1,-11)$

Write a system of three equations.
$a(-1)^2 + b(-1) + c = -3$
$a(-2)^2 + b(-2) + c = -5$
$a(1)^2 + b(1) + c = -11$
Simplify the equations.
$a - b + c = -3$
$4a - 2b + c = -5$
$a + b + c = -11$
Solve for a, b, c.
$-a + b - c = 3$
$\underline{a + b + c = -11}$
$2b = -8$
$b = -4$
$a - (-4) + c = -3$
$4a - 2(-4) + c = -5$

$-a - c = 7$
$\underline{4a + c = -13}$
$3a = -6$
$a = -2$
$(-2) - (-4) + c = -3$
$-2 + 4 + c = -3$
$c = -5$
$a = -2, b = -4, c = -5$
$y = -2x^2 - 4x - 5$

The graph passes through the given points.

23. $(-2,4), \left(-3, \dfrac{5}{2}\right), \left(-1, \dfrac{13}{2}\right)$

Write a system of three equations.
$a(-2)^2 + b(-2) + c = 4$
$a(-3)^2 + b(-3) + c = \dfrac{5}{2}$
$a(-1)^2 + b(-1) + c = \dfrac{13}{2}$
Simplify the equations.
$4a - 2b + c = 4$
$9a - 3b + c = \dfrac{5}{2}$
$a - b + c = \dfrac{13}{2}$
Solve for $a, b,$ and c.
$-4a + 2b - c = -4$
$\underline{9a - 3b + c = \dfrac{5}{2}}$
$5a - b = -\dfrac{3}{2}$

$-9a + 3b - c = -\dfrac{5}{2}$
$\underline{a - b + c = \dfrac{13}{2}}$
$-8a + 2b = 4$

$10a - 2b = -3$
$\underline{-8a + 2b = 4}$
$2a = 1$
$a = \dfrac{1}{2}$

355

$10\left(\dfrac{1}{2}\right) - 2b = -3$

$5 - 2b = -3$

$-2b = -8$

$b = 4$

$4\left(\dfrac{1}{2}\right) - 2(4) + c = 4$

$2 - 8 + c = 4$

$c = 10$

$a = \dfrac{1}{2}, b = 4, c = 10$

$y = \dfrac{1}{2}x^2 + 4x + 10$

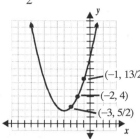

The graph passes through the given points.

25. Write three ordered pairs, (t, s).

$(0, 150), (1, 146), (2, 110)$

Use a graphing utility to fit a quadratic model to the data.

```
QuadReg
y=ax²+bx+c
a=-16
b=12
c=150
```

$s(t) = -16t^2 + 12t + 150$

Window: $(-1, 8, 1, -20, 200, 20)$

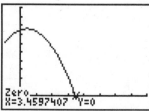

The ball will hit the ground in approximately 3.46 seconds.

27. Write three ordered pairs, (t, s).

$(0, 400), (2, 336), (4, 144)$

Use a graphing utility to fit a quadratic model to the data.

```
QuadReg
y=ax²+bx+c
a=-16
b=0
c=400
```

$s(t) = -16t^2 + 400$

Window: $(-1, 8, 1, -50, 500, 50)$

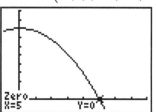

The dummy will hit the ground in 5 seconds.

29. Write three ordered pairs, (t, s).

$(0, 2000), (10, 1730), (20, 920)$

Use a graphing utility to fit a quadratic model to the data.

```
QuadReg
y=ax²+bx+c
a=-2.7
b=0
c=2000
```

$s(t) = -2.7t^2 + 2000$

Window: $(-10, 40, 5, -100, 2500, 100)$

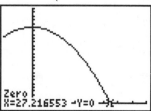

The cylinder will touch the surface in about 27.22 seconds.

31. From the problem statement, write three ordered pairs.

$(0, 20.45), (5, 22.60), (10, 25.18)$

Use a graphing utility to fit a quadratic model to the data.

```
QuadReg
 y=ax²+bx+c
 a=.0086
 b=.387
 c=20.45
```

$y = 0.0086x^2 + 0.387x + 20.45$

In 2005, $x = 20$.

$y = 0.0086(20)^2 + 0.387(20) + 20.45$

$= 3.44 + 7.74 + 20.45$

$= 31.63$

The function predicts that natural gas production in the U.S. will be up to 31.63 trillion cubic feet. Answers will vary.

33. Let x = number of years after 1975

y = number of franchises

From the problem statement, write three ordered pairs.

$(0, 3352), (10, 6972), (20, 18299)$

Use a graphing utility to fit a quadratic model to the data.

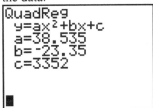

$y = 38.535x^2 - 23.35x + 3352$

In 2005, $x = 30$.

$y = 38.535(30)^2 - 23.35(30) + 3352 = 37,333$

The number of McDonald's franchises in 2005 is predicted to be 37,333. Explanations will vary.

35. Let x = distance from the middle of towers

y = height above roadway

Superimpose a coordinate grid on the center span placing the origin in the middle on the roadway.

From the problem statement, we can write the following ordered pairs:

$(-797.75, 130), (0, 0), (797.75, 130)$

```
QuadReg
 y=ax²+bx+c
 a=2.0427242E-4
 b=0
 c=0
```

$y = 0.0002043x^2$

37. Let x = distance from the center of the bridge

y = height of arch

Superimpose a coordinate grid on the arch bridge with the origin at the center of the bridge on the roadway. From the problem statement, we can write the following ordered pairs:

$(-850, 0), (0, 360), (850, 0)$

Use a graphing utility to fit a quadratic model to the data.

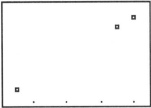

$y = -0.0005x^2 + 360$

8.4 Calculator Exercises

1. $(5, 155), (35, 335), (40, 365)$

The data appears to be linear.

SSM: Experiencing Introductory and Intermediate Algebra

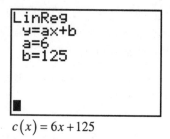

$c(x) = 6x + 125$

2. $(5, 450), (30, 1200), (35, 1050)$

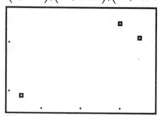

There is some curvature in the data. A quadratic model would seem appropriate.

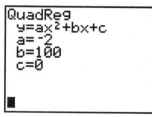

$r(x) = -2x^2 + 100x$

3. $c(30) = 6(30) + 125 = 305$
$r(40) = -2(40)^2 + 100(40) = 800$
$450 - 155 = 295$
$1200 - 305 = 895$
$1050 - 335 = 715$
$800 - 365 = 435$

Number	$5	$30	$35	$40
Cost	$155	**$305**	$335	$365
Revenue	$450	$1200	$1050	**$800**
Profit	**$295**	**$895**	**$715**	**$435**

4. $(5, 295), (30, 895), (40, 435)$

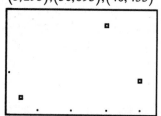

There is some curvature in the data. A quadratic model would seem appropriate.

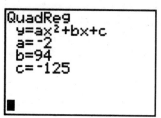

$p(x) = -2x^2 + 94x - 125$

$p(35) = -2(35)^2 + 94(35) - 125 = 715$

The function works with the fourth ordered pair.

5. $p(x) = -2x^2 + 94x - 125$

Window: $(-5, 60, 5, -100, 1200, 100)$

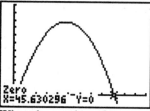

When the number of items produced and sold is greater than or equal to 46, the profit will be negative.

6. $p(x) = -2x^2 + 94x - 125$
Window: $(-5, 60, 5, -100, 1200, 100)$

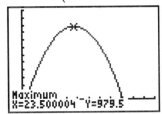

The vertex is $(23.5, 979.5)$. This indicates that the maximum profit is \$979.50 when about 23 items are sold. (This profit will not be attainable if fractional values are not allowed. In that case, the maximum profit would be \$979 when 23 or 24 items are sold.)

Chapter 8 Section-By-Section Review

1. x; monomial

2. $5x - 3$; binomial

3. $\sqrt{x} + 2$; not a polynomial because the variable is in the radicand.

4. $3x^3 - 4x^2 + x - 1$; polynomial

5. $\dfrac{3}{a} - 2a + 1$; not a polynomial because $\dfrac{3}{a}$ is not a monomial.

6. $3a^4 - 2a^2 + 5$; trinomial

7. $5x + 3x^3 - 2$
The term $5x$ has degree 1, the term $3x^3$ has degree 3, and the term -2 has degree 0. The degree of the polynomial is 3.

8. $2x^2y + 3xy - 5$
The term $2x^2y$ has degree 3, the term $3xy$ has degree 2, and the term -5 has degree 0. The degree of the polynomial is 3.

9. $x + 9$
The term x has degree 1 and the term 9 has degree 0. The degree of the polynomial is 1.

10. $0.5a - 3.1a^2 + 9.6a + 3.1a^2$ reduces to $10.1a$. The term $10.1a$ has degree 1. The degree of the polynomial is 1.

11. $12x^2 + 30x + 3$
The term $12x^2$ has degree 2, the term $30x$ has degree 1, and the term 3 has degree 0. The degree of the polynomial is 2.

12. $11y^4 + 9y^3 + 5y^2 - 6y + 12$

13. $-p + 5$

14. $\dfrac{1}{4}z^4 + \dfrac{1}{3}z^3 + \dfrac{1}{2}z^2 + z + 1$

15. $-2.3b^5 - 9.1b^3 + 0.6b + 1.8$

16. $2x^3 + 11x^2 - 21x - 90$
$2(3)^3 + 11(3)^2 - 21(3) - 90$
$= 54 + 99 - 63 - 90$
$= 0$

17. $2x^3 + 11x^2 - 21x - 90$
$2(0)^3 + 11(0)^2 - 21(0) - 90$
$= -90$

18. $2x^3 + 11x^2 - 21x - 90$
$2(1)^3 + 11(1)^2 - 21(1) - 90$
$= 2 + 11 - 21 - 90$
$= -98$

19. $2x^3 + 11x^2 - 21x - 90$
$2(-6)^3 + 11(-6)^2 - 21(-6) - 90$
$= -432 + 396 + 126 - 90$
$= 0$

20. $2x^3 + 11x^2 - 21x - 90$

$2\left(-\dfrac{5}{2}\right)^3 + 11\left(-\dfrac{5}{2}\right)^2 - 21\left(-\dfrac{5}{2}\right) - 90$

$= -\dfrac{125}{4} + \dfrac{275}{4} + \dfrac{105}{2} - 90$

$= 0$

21. $a^3 + 2a^2b - 3ab^2 - b^3$

$(0)^3 + 2(0)^2(1) - 3(0)(1)^2 - (1)^3$

$= 0 + 0 + 0 - 1$

$= -1$

22. $a^3 + 2a^2b - 3ab^2 - b^3$

$(-1)^3 + 2(-1)^2(0) - 3(-1)(0)^2 - (0)^3$

$= -1 + 0 + 0 - 0$

$= -1$

23. $a^3 + 2a^2b - 3ab^2 - b^3$

$(0)^3 + 2(0)^2(0) - 3(0)(0)^2 - (0)^3$

$= 0 + 0 - 0 - 0$

$= 0$

24. $a^3 + 2a^2b - 3ab^2 - b^3$

$(1)^3 + 2(1)^2(1) - 3(1)(1)^2 - (1)^3$

$= 1 + 2 - 3 - 1$

$= -1$

25. $a^3 + 2a^2b - 3ab^2 - b^3$

$(-1)^3 + 2(-1)^2(1) - 3(-1)(1)^2 - (1)^3$

$= -1 + 2 + 3 - 1$

$= 3$

26. $a^3 + 2a^2b - 3ab^2 - b^3$

$(-1)^3 + 2(-1)^2(-1) - 3(-1)(-1)^2 - (-1)^3$

$= -1 - 2 + 3 + 1$

$= 1$

27. Let x = width
Perimeter:
$2L + 2W$
$2(x^2) + 2(x)$
$2x^2 + 2x$ units
$2(7)^2 + 2(7) = 98 + 14 = 112$
If the width is 7 yards, the perimeter of the rectangle would be 112 yards.

28. Let x = length of first side
Perimeter:
$A + B + C$
$(x) + (3x) + (x^2 + 1)$
$x^2 + 4x + 1$ units
$(4)^2 + 4(4) + 1 = 16 + 16 + 1 = 33$
If the first side of the triangle is 4 inches, the perimeter of the triangle would be 33 inches.

29. Total area:

$a^2 + 17a + \dfrac{1}{2}(6)a$

$= a^2 + 17a + 3a$

$= a^2 + 20a$ in^2

30. Shaded area:

$10x - \dfrac{1}{2}(x)(x)$

$= 10x - \dfrac{1}{2}x^2$ in^2

31. Area of garden: $\dfrac{1}{2}xy$ ft^2

Area of lawn: z^2 ft^2

Area of lawn not covered by garden:

$z^2 - \dfrac{1}{2}xy$ ft^2

$(80)^2 - \dfrac{1}{2}(6)(10) = 6400 - 30 = 6370$

If the triangle has a height of 6 feet and a base of 10 feet, and the lawn measures 80 feet on a side, the area of the lawn not covered by the garden would be 6370 square feet.

32. a. Total charge:
$500 + 40xy$ dollars

b. Total cost:
$275 + 12xy$ dollars

c. $500 + 40(25)(15) = 500 + 15000 = 15500$
$275 + 12(25)(15) = 275 + 4500 = 4775$

For a room that is 25 feet long and 15 feet wide, the contractor's revenue and cost would be $15,500 and $4,775, respectively.

33. $0.01x^2 - 0.67x + 20.09$ minutes

a. $0.01(3)^2 - 0.67(3) + 20.09 = 18.17$

With 3 months experience, the estimated time to complete the task is 18.17 minutes.

b. $0.01(6)^2 - 0.67(6) + 20.09 = 16.43$

With 6 months experience, the estimated time to complete the task is 16.43 minutes.

c. $0.01(12)^2 - 0.67(12) + 20.09 = 13.49$

With 12 months experience, the estimated time to complete the task is 13.49 minutes.

34.

x	$y = x^3 - x^2 - 6x$
-4	$y = (-4)^3 - (-4)^2 - 6(-4) = -56$
-3	$y = (-3)^3 - (-3)^2 - 6(-3) = -18$
-2	$y = (-2)^3 - (-2)^2 - 6(-2) = 0$
-1	$y = (-1)^3 - (-1)^2 - 6(-1) = 4$
0	$y = (0)^3 - (0)^2 - 6(0) = 0$
1	$y = (1)^3 - (1)^2 - 6(1) = -6$
2	$y = (2)^3 - (2)^2 - 6(2) = -8$
3	$y = (3)^3 - (3)^2 - 6(3) = 0$
4	$y = (4)^3 - (4)^2 - 6(4) = 24$

35. $y = -3x + 5$

Window: $(-10, 10, 1, -10, 10, 1)$

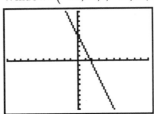

This polynomial has no absolute maximum or absolute minimum. The range is all real numbers.

36. $y = 2x^2 - 2x - 12$

Window: $(-5, 5, 1, -25, 15, 5)$

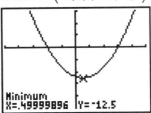

The polynomial has an absolute minimum at $x = 0.5$. The range is all real numbers greater than or equal to -12.5.

37. $y = x^3 + 2x^2 - 5x - 6$

Window: $(-5, 5, 1, -25, 15, 5)$

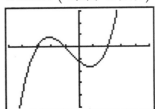

The polynomial has no absolute maximum or absolute minimum. The range is all real numbers.

38. $y = x^4 + 2x^3 - 5x^2 - 6x$
Window: $(-5, 5, 1, -25, 15, 5)$

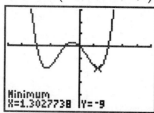

The polynomial has an absolute minimum of -9. The range is all real numbers greater than or equal to -9.

39. $f(x) = 3x^3 - x^2 + 2x - 4$
$f(-2) = 3(-2)^3 - (-2)^2 + 2(-2) - 4$
$= -24 - 4 - 4 - 4$
$= -36$

40. $f(x) = 3x^3 - x^2 + 2x - 4$
$f(0) = 3(0)^3 - (0)^2 + 2(0) - 4$
$= 0 - 0 + 0 - 4$
$= -4$

41. $f(x) = 3x^3 - x^2 + 2x - 4$
$f(2) = 3(2)^3 - (2)^2 + 2(2) - 4$
$= 24 - 4 + 4 - 4$
$= 20$

42. $f(x) = 3x^3 - x^2 + 2x - 4$
$f\left(-\frac{1}{2}\right) = 3\left(-\frac{1}{2}\right)^3 - \left(-\frac{1}{2}\right)^2 + 2\left(-\frac{1}{2}\right) - 4$
$= -\frac{3}{8} - \frac{1}{4} - 1 - 4$
$= -\frac{45}{8}$

43. $f(x) = 3x^3 - x^2 + 2x - 4$
$f(1.7) = 3(1.7)^3 - (1.7)^2 + 2(1.7) - 4$
$= 14.739 - 2.89 + 3.4 - 4$
$= 11.249$

44. a. $R(x) = 12x - 0.50x^2$
$R(5) = 12(5) - 0.50(5)^2 = 47.5$
$R(10) = 12(10) - 0.50(10)^2 = 70$
$R(15) = 12(15) - 0.50(15)^2 = 67.5$
$R(20) = 12(20) - 0.50(20)^2 = 40$
$R(25) = 12(25) - 0.50(25)^2 = -12.5$
The revenue for selling 5, 10, 15, 20, or 25 shirts is $47.50, $70, $67.50, $40, or $-$12.50, respectively. Tony's Tees should limit the number of shirts that can be bought to 10 or revenue will be lost.
$12 - 0.50(10) = 12 - 5 = 7$
The price per shirt should be no less than $7.

b. $C(x) = 4x$
$C(5) = 4(5) = 20$
$C(10) = 4(10) = 40$
$C(15) = 4(15) = 60$
$C(20) = 4(20) = 80$
The cost of selling 5, 10, 15, or 20 shirts is $20, $40, $60, or $80, respectively.

c. $P(5) = 47.5 - 20 = 27.5$
$P(10) = 70 - 40 = 30$
$P(15) = 67.5 - 60 = 7.5$
$P(20) = 40 - 80 = -40$
The profit for selling 5, 10, 15, or 20 shirts is $27.50, $30, $7.5, or $-$40, respectively.

d. Profit seems to be maximized at 10 shirts. The price per shirt should be $7.

45. $y = 15{,}800 + 2.2x - 0.001x^2$

Hundreds of gallons, x	900	1000	1100
Cost, y	16,970	17,000	17,010

Hundreds of gallons, x	1200	1300	1400
Cost, y	17,000	16,970	16,920

The cost of production starts increasing but reaches a maximum and then starts to decrease. An owner of a dairy farm should produce at least 110,00 gallons to start taking advantage of the decreasing costs. Assuming the selling price is the same, the profit should begin increasing after the maximum cost is reached.

46. $s(t) = -16t^2 + 40t + 6000$

Window: $(-25, 50, 5, -500, 7500, 500)$

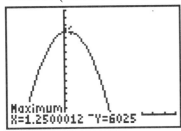

The range of the function is all real numbers between 0 and 6025 (inclusive). The rock will reach a maximum height of 6025 feet above the river. It has a minimum height of 0 (when it hits the river).

47. $y = x^2 + x + 1$ is quadratic. It is written in standard form.

48. $y = x^3 - x - 1$ is not quadratic. It is a third degree polynomial.

49. $y = \dfrac{5}{x^2} + x + 1$ is not quadratic. It is not a polynomial because the variable is in the denominator.

50. $y = x^2 + 4x + 4$ is quadratic. It is written in standard form.

	Function	Coefficients			Properties of graph				
		a	b	c	Wide or narrow	Concave Up/down	Vertex	Axis of Symmetry	y-intercept
51.	$y = -\dfrac{1}{4}x^2 + \dfrac{1}{2}x + 1$	$-\dfrac{1}{4}$	$\dfrac{1}{2}$	1	Wider	Downward	$\left(1, \dfrac{5}{4}\right)$	$x = 1$	$(0, 1)$
52.	$f(x) = -2x^2 + 4x$	-2	4	0	Narrower	Downward	$(1, 2)$	$x = 1$	$(0, 0)$
53.	$g(x) = \dfrac{1}{3}x^2 + x$	$\dfrac{1}{3}$	1	0	Wider	Upward	$\left(-\dfrac{3}{2}, -\dfrac{3}{4}\right)$	$x = -\dfrac{3}{2}$	$(0, 0)$
54.	$y = 3x^2 - 3x + 1$	3	-3	1	Narrower	Upward	$\left(\dfrac{1}{2}, \dfrac{1}{4}\right)$	$x = \dfrac{1}{2}$	$(0, 1)$

55. $f(x) = x^2 + 2x - 8$

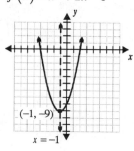

56. $y = -\dfrac{1}{2}x^2 + x - 2$

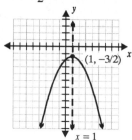

57. $h(x) = 2x^2 - 8$

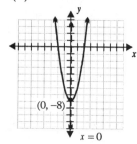

58. $A(w) = w^2 + 8w$

Window: $(-16, 16, 2, -32, 32, 2)$

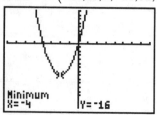

The vertex is $(-4, -16)$. It does not have any physical meaning; we can't have negative length or negative area.

59. $R(x) = 30x - 0.50x^2$

Window: $(0, 80, 5, 0, 600, 50)$

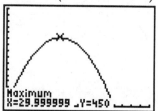

The vertex is $(30, 450)$. The photographer's revenue is maximized at $450 when 30 photos are ordered.

60. $s(t) = -16t^2 + 60t + 120$

Window: $(0, 8, 1, 0, 225, 25)$

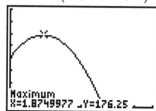

The vertex is $(1.875, 176.25)$. The egg reaches a maximum height of 176.25 feet after 1.875 seconds.

61. $(8, 0), (-3, 0), (0, -24)$

Write a system of three equations.

$a(8)^2 + b(8) + c = 0$

$a(-3)^2 + b(-3) + c = 0$

$a(0)^2 + b(0) + c = -24$

Simplify the equations.

$64a + 8b + c = 0$

$9a - 3b + c = 0$

$c = -24$

$64a + 8b = 24$

$9a - 3b = 24$

Solve for a and b.

$192a + 24b = 72$

$\underline{72a - 24b = 192}$

$264a = 264$

$a = 1$

$9(1) - 3b - 24 = 0$
$-3b = 15$
$b = -5$
$a = 1, b = -5, c = -24$
$y = x^2 - 5x - 24$

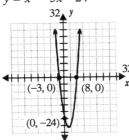

The graph passes through the given points.

62. $(0,5), (5,0), \left(-\dfrac{1}{2}, 0\right)$

Write a system of three equations.
$a(0)^2 + b(0) + c = 5$
$a(5)^2 + b(5) + c = 0$
$a\left(-\dfrac{1}{2}\right)^2 + b\left(-\dfrac{1}{2}\right) + c = 0$

Simplify the equations.
$c = 5$
$25a + 5b + c = 0$
$\dfrac{1}{4}a - \dfrac{1}{2}b + c = 0$

$25a + 5b = -5$
$\dfrac{1}{4}a - \dfrac{1}{2}b = -5$

Solve for a and b.
$5a + b = -1$
$\dfrac{1}{2}a - b = -10$
$\overline{\qquad\qquad}$
$\dfrac{11}{2}a = -11$
$a = -2$
$25(-2) + 5b + 5 = 0$
$5b = 45$
$b = 9$

$a = -2, b = 9, c = 5$
$y = -2x^2 + 9x + 5$

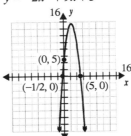

The graph passes through the given points.

63. $(2,-4), (4,-6), (7,6)$

Write a system of three equations.
$a(2)^2 + b(2) + c = -4$
$a(4)^2 + b(4) + c = -6$
$a(7)^2 + b(7) + c = 6$

Simplify the equations.
$4a + 2b + c = -4$
$16a + 4b + c = -6$
$49a + 7b + c = 6$

Solve for a, b, and c.
$-4a - 2b - c = 4$
$\underline{16a + 4b + c = -6}$
$12a + 2b = -2$

$-4a - 2b - c = 4$
$\underline{49a + 7b + c = 6}$
$45a + 5b = 10$

$-60a - 10b = 10$
$\underline{90a + 10b = 20}$
$30a = 30$
$a = 1$
$12(1) + 2b = -2$
$2b = -14$
$b = -7$
$4(1) + 2(-7) + c = -4$
$4 - 14 + c = -4$
$c = 6$

$a = 1, b = -7, c = 6$

$y = x^2 - 7x + 6$

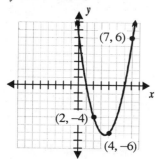

The graph passes through the given points.

64. $s(t) = at^2 + bt + c$

    Write three ordered pairs, (t, s).

    $(0, 160), (1, 176), (2, 160)$

    Use a graphing utility to fit a quadratic model.

    ```
    QuadReg
    y=ax²+bx+c
    a=-16
    b=32
    c=160
    ```

    $s(t) = -16t^2 + 32t + 160$

    The hammer will hit the ground when $s(t) = 0$.

    Window: $(0, 6, 1, -50, 250, 50)$

    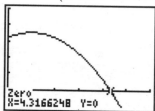

    The hammer will hit the ground in about 4.3 seconds.

65. Let x = number of years since 1900

    y = number over age 65 in U.S. (millions)

    $y = ax^2 + bx + c$

    Write three ordered pairs, (x, y), from the data.

    $(10, 4), (50, 12.4), (90, 31.2)$

    Use a graphing utility to fit a quadratic model.

    ```
    QuadReg
    y=ax²+bx+c
    a=.00325
    b=.015
    c=3.525
    ```

    $y = 0.00325x^2 + 0.015x + 3.525$

    In 1980, $x = 1980 - 1900 = 80$.

    $y = 0.00325(80)^2 + 0.015(80) + 3.525 = 25.525$

    In 1970, $x = 1970 - 1900 = 70$.

    $y = 0.00325(70)^2 + 0.015(70) + 3.525 = 20.5$

    In 1960, $x = 1960 - 1900 = 60$.

    $y = 0.00325(60)^2 + 0.015(60) + 3.525 = 16.125$

    In 1980, 1970, and 1960, the predictions are 25.525 million, 20.5 million, and 16.125 million, respectively. These are close to the actual values. In 2010, $x = 2010 - 1900 = 110$.

    $y = 0.00325(110)^2 + 0.015(110) + 3.525 = 44.5$

    The predicted 2010 population in the U.S. over age 65 is 44.5 million.

66. Let x = distance from the center of the bridge

    y = height of arch

    Superimpose a coordinate grid on the bridge with the origin in the center and at the bottom of the main span. From the problem statement, we can write the following ordered pairs:

    $(-160, 0), (0, 280), (160, 0)$

    Use a graphing utility to fit a quadratic model to the data.

    ```
    QuadReg
    y=ax²+bx+c
    a=-.0109375
    b=0
    c=280
    ```

    The curve of the arch can be approximated by $y = -0.0109375x^2 + 280$.

Chapter 8 Chapter Review

1. $2x^3 - 3x^2 - 29x - 30$
 $= 2(5)^3 - 3(5)^2 - 29(5) - 30$
 $= 250 - 75 - 145 - 30$
 $= 0$

2. $2x^3 - 3x^2 - 29x - 30$
 $= 2(0)^3 - 3(0)^2 - 29(0) - 30$
 $= 0 - 0 - 0 - 30$
 $= -30$

3. $2x^3 - 3x^2 - 29x - 30$
 $= 2(1)^3 - 3(1)^2 - 29(1) - 30$
 $= 2 - 3 - 29 - 30$
 $= -60$

4. $2x^3 - 3x^2 - 29x - 30$
 $= 2(-2)^3 - 3(-2)^2 - 29(-2) - 30$
 $= -16 - 12 + 58 - 30$
 $= 0$

5. $2x^3 - 3x^2 - 29x - 30$
 $= 2\left(-\dfrac{3}{2}\right)^3 - 3\left(-\dfrac{3}{2}\right)^2 - 29\left(-\dfrac{3}{2}\right) - 30$
 $= -\dfrac{27}{4} - \dfrac{27}{4} + \dfrac{87}{2} - 30$
 $= 0$

6. $2a^3 + 4a^2b - 2ab^2 + b^3$
 $= 2(-1)^3 + 4(-1)^2(0) - 2(-1)(0)^2 + (0)^3$
 $= -2 + 0 - 0 + 0$
 $= -2$

7. $2a^3 + 4a^2b - 2ab^2 + b^3$
 $= 2(1)^3 + 4(1)^2(1) - 2(1)(1)^2 + (1)^3$
 $= 2 + 4 - 2 + 1$
 $= 5$

8. $2a^3 + 4a^2b - 2ab^2 + b^3$
 $= 2(-1)^3 + 4(-1)^2(1) - 2(-1)(1)^2 + (1)^3$
 $= -2 + 4 + 2 + 1$
 $= 5$

9. $2a^3 + 4a^2b - 2ab^2 + b^3$
 $= 2(-1)^3 + 4(-1)^2(-1) - 2(-1)(-1)^2 + (-1)^3$
 $= -2 - 4 + 2 - 1$
 $= -5$

10. a. binomial; degree of each term is 0 and 2; degree of the polynomial is 2;
 $3x^2 + 5$

 b. polynomial; degree of each term is 2, 3, 0, and 1; degree of the polynomial is 3;
 $-5a^3 + 15a^2 + a + 4$

 c. polynomial; degree of each term is 4, 1, 0, 5, and 2; degree of the polynomial is 5;
 $x^5 + 5x^4 - 3x^2 + x - 2$

11. a. $3b^2 + 13b - 4$; trinomial; degree of each term is 2, 1, and 0; degree of the polynomial is 2.

 b. $3x^2y - 4xy + 3xy^2 + 5 - 4x^2y^2$; polynomial; degree of each term is 3, 2, 3, 0, and 4; degree of the polynomial is 4.

 c. $6xyz$; monomial; degree of the term is 3; degree of the polynomial is 3.

12. $f(x) = 2x^3 - 3x^2 - 23x + 12$
 $f(-3) = 2(-3)^3 - 3(-3)^2 - 23(-3) + 12$
 $= -54 - 27 + 69 + 12$
 $= 0$

13. $f(x) = 2x^3 - 3x^2 - 23x + 12$
 $f(0) = 2(0)^3 - 3(0)^2 - 23(0) + 12$
 $= 0 - 0 - 0 + 12$
 $= 12$

14. $f(x) = 2x^3 - 3x^2 - 23x + 12$

$f(4) = 2(4)^3 - 3(4)^2 - 23(4) + 12$
$= 128 - 48 - 92 + 12$
$= 0$

15. $f(x) = 2x^3 - 3x^2 - 23x + 12$

$f\left(\dfrac{1}{2}\right) = 2\left(\dfrac{1}{2}\right)^3 - 3\left(\dfrac{1}{2}\right)^2 - 23\left(\dfrac{1}{2}\right) + 12$
$= \dfrac{1}{4} - \dfrac{3}{4} - \dfrac{23}{2} + 12$
$= 0$

16. $f(x) = 2x^3 - 3x^2 - 23x + 12$

$f(2.2) = 2(2.2)^3 - 3(2.2)^2 - 23(2.2) + 12$
$= 21.296 - 14.52 - 50.6 + 12$
$= -31.824$

17.

x	$y = 2x^3 + 2x^2 - 12x$	y
-3	$y = 2(-3)^3 + 2(-3)^2 - 12(-3) = 0$	0
-2	$y = 2(-2)^3 + 2(-2)^2 - 12(-2) = 16$	16
-1	$y = 2(-1)^3 + 2(-1)^2 - 12(-1) = 12$	12
0	$y = 2(0)^3 + 2(0)^2 - 12(0) = 0$	0
1	$y = 2(1)^3 + 2(1)^2 - 12(1) = -8$	-8
2	$y = 2(2)^3 + 2(2)^2 - 12(2) = 0$	0
3	$y = 2(3)^3 + 2(3)^2 - 12(3) = 36$	36

18. $y = 4x - 2$

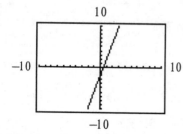

The range is all real numbers.

19. $y = 3x^2 + 3x - 6$

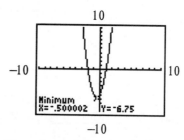

The polynomial has an absolute minimum. The range is all real numbers greater than or equal to -6.75.

20. $y = x^3 - 3x^2 - 13x + 15$

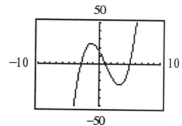

The polynomial has no absolute maximum or absolute minimum. The range is all real numbers.

21. $y = x^4 - 3x^3 - 13x^2 + 15x$

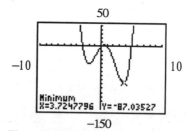

The polynomial has an absolute minimum. The range is all real numbers greater than or equal to -87.04.

22. $f(x) = 2x^2 + 7x - 4$

x-coordinate of vertex:

$-\dfrac{b}{2a} = -\dfrac{7}{2(2)} = -\dfrac{7}{4}$

y-coordinate of vertex:

$$f\left(-\frac{7}{4}\right) = 2\left(-\frac{7}{4}\right)^2 + 7\left(-\frac{7}{4}\right) - 4$$
$$= \frac{49}{8} - \frac{49}{4} - 4$$
$$= -\frac{81}{8}$$

Vertex: $\left(-\frac{7}{4}, -\frac{81}{8}\right)$

Axis of symmetry: $x = -\frac{7}{4}$

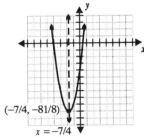

23. $y = \frac{1}{4}x^2 - x + 3$

x-coordinate of vertex:
$$-\frac{b}{2a} = -\frac{(-1)}{2\left(\frac{1}{4}\right)} = 2$$

y-coordinate of vertex:
$$y = \frac{1}{4}(2)^2 - (2) + 3$$
$$= 1 - 2 + 3$$
$$= 2$$

Vertex: $(2, 2)$

Axis of symmetry: $x = 2$

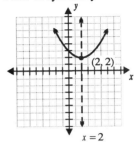

24. $h(x) = -x^2 + 9$

x-coordinate of vertex:
$$-\frac{b}{2a} = -\frac{0}{2(-1)} = 0$$

y-coordinate of vertex:
$$h(0) = -(0)^2 + 9 = 9$$

Vertex: $(0, 9)$

Axis of symmetry: $x = 0$

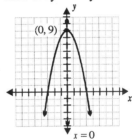

		Coefficients			Properties of graph				
	Function	a	b	c	Wide or Narrow	Concave Up/down	Vertex	Axis of Symmetry	y-intercept
25.	$y = \frac{1}{3}x^2 + \frac{2}{3}x + 1$	$\frac{1}{3}$	$\frac{2}{3}$	1	Wider	Upward	$\left(-1, \frac{2}{3}\right)$	$x = -1$	$(0, 1)$
26.	$f(x) = -3x^2 + 6x$	-3	6	0	Narrower	Downward	$(1, 3)$	$x = 1$	$(0, 0)$
27.	$y = -\frac{1}{4}x^2 + x + 3$	$-\frac{1}{4}$	1	3	Wider	Downward	$(2, 4)$	$x = 2$	$(0, 3)$
28.	$g(x) = 2x^2 + 4x - 6$	2	4	-6	Narrower	Upward	$(-1, -8)$	$x = -1$	$(0, -6)$

29. $(-2,12), (-4,8), (-7,17)$
Write a system of three equations.
$a(-2)^2 + b(-2) + c = 12$
$a(-4)^2 + b(-4) + c = 8$
$a(-7)^2 + b(-7) + c = 17$
Simplify the equations.
$4a - 2b + c = 12$
$16a - 4b + c = 8$
$49a - 7b + c = 17$
Solve for a, b, and c.
$-4a + 2b - c = -12$
$\underline{16a - 4b + c = 8}$
$12a - 2b = -4$

$-4a + 2b - c = -12$
$\underline{49a - 7b + c = 17}$
$45a - 5b = 5$

$-60a + 10b = 20$
$\underline{90a - 10b = 10}$
$30a = 30$
$a = 1$
$12(1) - 2b = -4$
$-2b = -16$
$b = 8$
$4(1) - 2(8) + c = 12$
$4 - 16 + c = 12$
$c = 24$
$a = 1, b = 8, c = 24$
$y = x^2 + 8x + 24$

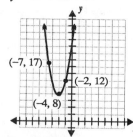

The graph passes through the given points.

30. $(9,0), (-4,0), (0,-36)$
Write a system of three equations.
$a(9)^2 + b(9) + c = 0$
$a(-4)^2 + b(-4) + c = 0$
$a(0)^2 + b(0) + c = -36$
Simplify the equations.
$81a + 9b + c = 0$
$16a - 4b + c = 0$
$c = -36$

$81a + 9b = 36$
$16a - 4b = 36$
Solve for a and b.
$9a + b = 4$
$\underline{4a - b = 9}$
$13a = 13$
$a = 1$
$81(1) + 9b - 36 = 0$
$9b = -45$
$b = -5$
$a = 1, b = -5, c = -36$
$y = x^2 - 5x - 36$

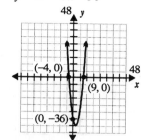

The graph passes through the given points.

31. $(0,-12), (3,0), \left(-\dfrac{4}{3}, 0\right)$
Write a system of three equations.
$a(0)^2 + b(0) + c = -12$
$a(3)^2 + b(3) + c = 0$
$a\left(-\dfrac{4}{3}\right)^2 + b\left(-\dfrac{4}{3}\right) + c = 0$

Simplify the equations.
$c = -12$
$9a + 3b + c = 0$
$\frac{16}{9}a - \frac{4}{3}b + c = 0$

$9a + 3b = 12$
$\frac{16}{9}a - \frac{4}{3}b = 12$

Solve for a and b.
$36a + 12b = 48$
$\underline{16a - 12b = 108}$
$52a = 156$
$a = 3$
$9(3) + 3b = 12$
$3b = -15$
$b = -5$
$a = 3, b = -5, c = -12$
$y = 3x^2 - 5x - 12$

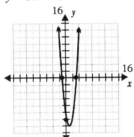

The graph passes through the given points.

32. Let x = width of rectangle
Perimeter:
$2x + 2(5 + x^3)$
$= 2x + 10 + 2x^3$
$= 2x^3 + 2x + 10$ units
$2(3)^3 + 2(3) + 10 = 70$
When the width is 3 feet, the perimeter is 70 feet.

33. Let x = first side of triangle
Perimeter:
$x + (2x - 3) + (x^2 - 7)$
$= x + 2x - 3 + x^2 - 7$
$= x^2 + 3x - 10$ units
$(10)^2 + 3(10) - 10 = 120$
When the length of the first side is 10 cm, the perimeter of the triangle is 120 cm.

34. Area of first shape: $\frac{1}{2}(5)x = \frac{5}{2}x$
Area of second shape: $(x+2)x = x^2 + 2x$
Area of third shape: $\frac{1}{2}x(5+8) = \frac{13}{2}x$
Total area:
$\frac{5}{2}x + x^2 + 2x + \frac{13}{2}x = x^2 + 11x$
The total area is $x^2 + 11x$ square units.

35. Area of rectangle: $3x(x) = 3x^2$
Area of semicircle: $\frac{1}{2}\pi\left(\frac{x}{2}\right)^2 = \frac{\pi x^2}{8}$
Total area:
$3x^2 + \frac{\pi x^2}{8} = \frac{24x^2 + \pi x^2}{8} = \frac{24 + \pi}{8}x^2$
The total area is $\frac{24 + \pi}{8}x^2$ square units.

36. Area not covered by pool
= area of yard − area of pool
= area of rectangle − area of circle
= $xy - \pi z^2$ square units
The area not covered is $xy - \pi z^2$.
$(80)(50) - \pi(8)^2 = 4000 - 64\pi \approx 3798.94$
The area not covered would be about 3798.94 square feet.

SSM: Experiencing Introductory and Intermediate Algebra

37. Let x = length of kitchen
    y = width of kitchen
   a. Cost: $1.5xy$
   b. Profit: $800 - 1.5xy$
   c. $800 - 1.5(15)(10) = 575$
      The profit will be $575.

38. $P(x) = 10x + 2x^2$

   a.

x	$P(x) = 10x + 2x^2$	$P(x)$
0	$P(0) = 10(0) + 2(0)^2 = 0$	0
1	$P(1) = 10(1) + 2(1)^2 = 12$	12
2	$P(2) = 10(2) + 2(2)^2 = 28$	28
3	$P(3) = 10(3) + 2(3)^2 = 48$	48
4	$P(4) = 10(4) + 2(4)^2 = 72$	72
5	$P(5) = 10(5) + 2(5)^2 = 100$	100
6	$P(6) = 10(6) + 2(6)^2 = 132$	132
7	$P(7) = 10(7) + 2(7)^2 = 168$	168

   b. Answers will vary.
      The payments increase with each A. The amount of the increase is greater each time.

39. a. $y = 1.5 + 0.0082x - 0.0000081x^2$
       $y = 1.5 + 0.0082(750) - 0.0000081(750)^2$
       $y = 3.09375$
       The predicted GPA is 3.09375.

   b.
GMAT	500	550	600
GPA	3.575	3.55975	3.504

GMAT	650	700	750
GPA	3.40775	3.271	3.09375

   c. As the entrance exam score goes up, the grade point average goes down.

40. $s(t) = -0.8t^2 + 1500$

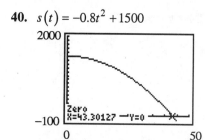

It will take about 43.3 seconds for the supply package to hit the ground.

41. Let x = number of years since 1900
    y = divorces per 1000 population
    Write three ordered pairs from the problem statement.
    $(20, 1.6), (40, 2), (70, 3.5)$

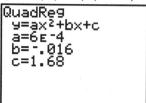

The quadratic function is:
$y = 0.0006x^2 - 0.016x + 1.68$

$y = 0.0006(30)^2 - 0.016(30) + 1.68 = 1.74$

$y = 0.0006(90)^2 - 0.016(90) + 1.68 = 5.1$

In 1930 and 1990, the predicted divorce rate per 1000 population is 1.74 and 5.1, respectively. The model works fairly well, particularly for 1930.

$y = 0.0006(110)^2 - 0.016(110) + 1.68 = 7.18$

The model predicts the divorce rate in 2010 to be 7.18 per 1000 population.

42. $0.046x^2 - 10.98x + 729.87$

$0.046(99)^2 - 10.98(99) + 729.87 = 93.696$

$0.046(119)^2 - 10.98(119) + 729.87 = 74.656$

They can sell about 94 cards if the price is $0.99 and about 75 cards if the price is $1.19.

Chapter 8: Polynomial Functions

43. $s(t) = -16t^2 - 60t + 19{,}500$

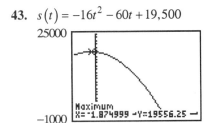

The range is all real numbers between 0 and 19,500. The rock starts at 19,500 feet (the highest it will ever be) and falls until it hits the ground ($s = 0$). The rock will not be able to reach the maximum of the function.

44. $R(x) = 15x - x^2$

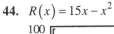

The vertex is $(7.5, 56.25)$, but the number of books sold must be a whole number.

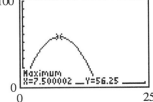

The maximum revenue is $56 when 7 or 8 books are sold.

45. Let x = distance from the center of the bridge
y = height of arch

Superimpose a coordinate grid on the bridge with the origin in the center and on the tracks. From the problem statement, we can write the following ordered pairs:
$(-488.75, 0), (0, 170), (488.75, 0)$
Use a graphing utility to fit a quadratic model.

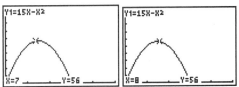

The model would be: $y = -0.000712x^2 + 170$

Chapter 8 Test

1. $123x^2y^3z$ has 1 term; monomial.

2. $3a^3 + 5a^2b + 7ab^2 + 9b^3$ has 4 terms; polynomial.

3. $2 - c$ has 2 terms; binomial.

4. The term with the largest degree is $-0.5x^5$. Therefore, the degree of the polynomial is 5.

5. The term with the largest degree is $5x^2y^3$. Therefore, the degree of the polynomial is 5.

6. $15 + 3x^4 - 7x + x^5 + 9x^2 + 21x$
$= x^5 + 3x^4 + 9x^2 + 14x + 15$

7. $a - \dfrac{2}{3}a^3 - \dfrac{5}{6}a^2 - \dfrac{4}{9}$
$= -\dfrac{4}{9} + a - \dfrac{5}{6}a^2 - \dfrac{2}{3}a^3$

8. $a^2 + 3ab^3 - 7b^2 - b - 6$
$= (0)^2 + 3(0)(3)^3 - 7(3)^2 - (3) - 6$
$= 0 + 0 - 63 - 3 - 6$
$= -72$

9. $a^2 + 3ab^3 - 7b^2 - b - 6$
$= (-2)^2 + 3(-2)(0)^3 - 7(0)^2 - (0) - 6$
$= 4 + 0 - 0 - 0 - 6$
$= -2$

SSM: Experiencing Introductory and Intermediate Algebra

10. $a^2 + 3ab^3 - 7b^2 - b - 6$
$= (2)^2 + 3(2)(-3)^3 - 7(-3)^2 - (-3) - 6$
$= 4 - 162 - 63 + 3 - 6$
$= -224$

11. Let x = length of room
y = width of room
Cost:
$16.5xy + 75$ dollars
$16.5\left(7\dfrac{1}{3}\right)(5) + 75 = 605 + 75 = 680$

If the room measures 5 yards by $7\dfrac{1}{3}$ yards, the cost for carpeting would be $680.

12. $g(x) = 3x^2 + 7x - 6$
$g(-3) = 3(-3)^2 + 7(-3) - 6$
$= 27 - 21 - 6$
$= 0$

13. $g(x) = 3x^2 + 7x - 6$
$g(0) = 3(0)^2 + 7(0) - 6$
$= -6$

14. $g(x) = 3x^2 + 7x - 6$
$g(1) = 3(1)^2 + 7(1) - 6$
$= 3 + 7 - 6$
$= 4$

15. $y = \dfrac{1}{2}x^2 - 2x - 6$
x-coordinate of vertex:
$x = -\dfrac{b}{2a} = -\dfrac{(-2)}{2\left(\dfrac{1}{2}\right)} = 2$

y-coordinate of vertex:
$y = \dfrac{1}{2}(2)^2 - 2(2) - 6$
$= 2 - 4 - 6 = -8$
The vertex is $(2, -8)$.

16. $y = \dfrac{1}{2}x^2 - 2x - 6$

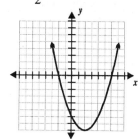

17. $y = \dfrac{1}{2}x^2 - 2x - 6$

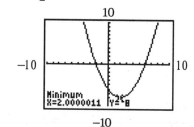

The relation has an absolute minimum of -8. The range is all real numbers greater than or equal to -8.

18. The relation is a function. It passes the vertical line test.

	Coefficients			Properties of graph				
Function	a	b	c	Wide or Narrow	Concave Up/down	Vertex	Axis of Symmetry	y-intercept
19. $y = \dfrac{1}{2}x^2 + 2x + 3$	$\dfrac{1}{2}$	2	3	Wider	Upward	$(-2, 1)$	$x = -2$	$(0, 3)$
20. $y = 3x^2 - 3x + \dfrac{1}{4}$	3	-3	$\dfrac{1}{4}$	Narrower	Upward	$\left(\dfrac{1}{2}, -\dfrac{1}{2}\right)$	$x = \dfrac{1}{2}$	$\left(0, \dfrac{1}{4}\right)$

21. $(0,8),(2,0),(4,0)$

Write a system of three equations.

$a(0)^2 + b(0) + c = 8$

$a(2)^2 + b(2) + c = 0$

$a(4)^2 + b(4) + c = 0$

Simplify the equations.

$c = 8$

$4a + 2b + c = 0$

$16a + 4b + c = 0$

$4a + 2b = -8$

$16a + 4b = -8$

Solve for a and b.

$-8a - 4b = 16$

$\underline{16a + 4b = -8}$

$8a = 8$

$a = 1$

$4(1) + 2b = -8$

$2b = -12$

$b = -6$

$a = 1, b = -6, c = 8$

$y = x^2 - 6x + 8$

22. $(-1,6),(1,4),(2,9)$

Write a system of three equations.

$a(-1)^2 + b(-1) + c = 6$

$a(1)^2 + b(1) + c = 4$

$a(2)^2 + b(2) + c = 9$

Simplify the equations.

$a - b + c = 6$

$a + b + c = 4$

$4a + 2b + c = 9$

Solve for a, b, and c.

$-a + b - c = -6$

$\underline{a + b + c = 4}$

$2b = -2$

$b = -1$

$a - (-1) + c = 6$

$4a + 2(-1) + c = 9$

$-a - c = -5$

$\underline{4a + c = 11}$

$3a = 6$

$a = 2$

$4(2) + c = 11$

$c = 3$

$a = 2, b = -1, c = 3$

$y = 2x^2 - x + 3$

23. $1020x^2 - 19,195x + 114,446$

Associate's degree:

$1020(14)^2 - 19,195(14) + 114,446 = 45,636$

High school:

$1020(12)^2 - 19,195(12) + 114,446 = 30,986$

The expected average annual earnings for a male who completes an associate's degree or just has a high school diploma is $45,636 or $30,986, respectively.

$45,636 - 30,986 = 14,650$

A male who earns an associate's degree can expect to earn an average of $14,650 more per yearn than a male with just a high school diploma.

24. $s(t) - 16t^2 + 100$

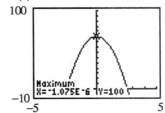

The range is all real numbers between 0 and 100 (inclusive). The note starts at a maximum height of 100 feet (the top) and falls until it hits the ground ($s = 0$).

25. $P(x) = 75x - 2x^2$

x-coordinate of vertex:

$$-\frac{b}{2a} = -\frac{(75)}{2(-2)} = \frac{75}{4} = 18.75$$

y-coordinate of vertex:

$$P(18.75) = 75(18.75) - 2(18.75)^2 = 703.125$$

The vertex is $(18.75, 703.125)$.

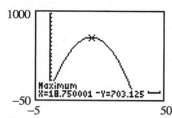

Since the supplier will only sell a whole number of keyboards, we need to check the following:

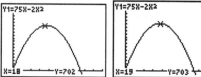

The maximum revenue is $703 when 19 keyboards are sold.

26. Let x = distance from the center of the bridge

   y = height of arch

   Superimpose a coordinate grid on the bridge with the origin in the center and at the bottom of the arch. From the problem statement, we can write the following ordered pairs:

   $(-837.5, 0), (0, 325), (837.5, 0)$

   Use a graphing utility to fit a quadratic model.

   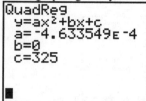

   The quadratic model is $y = -0.000463x^2 + 325$.

27. Answers will vary. One possible answer:

   A quadratic function is a polynomial function, in one variable, of degree 2. The graph of a quadratic function is always parabolic and either opens up (if the leading coefficient is positive) or down (if the leading coefficient is negative). The graph is symmetric about a vertical line passing through the vertex (the high or low point). Like all polynomials, quadratic functions have a y-intercept that is determined by its constant term.

Chapter 9

9.1 Exercises

1. $-3x^4 = -3 \cdot x \cdot x \cdot x \cdot x$

3. $(-3x)^4 = (-3x)(-3x)(-3x)(-3x)$

5. $a^3 b^0 c^5 = a^3 \cdot 1 \cdot c^5 = a \cdot a \cdot a \cdot c \cdot c \cdot c \cdot c \cdot c$

7. $\left(\dfrac{3}{4}\right)^3 \cdot x^2 = \dfrac{3}{4} \cdot \dfrac{3}{4} \cdot \dfrac{3}{4} \cdot x \cdot x$

9. $5(x+y)^2 = 5(x+y)(x+y)$

11. $x^5 \cdot x^8 = x^{5+8} = x^{13}$

13. $y \cdot y^{13} = y^{1+13} = y^{14}$

15. $\dfrac{1}{2}x^2 \cdot x \cdot x^5 = \dfrac{1}{2}x^{2+1+5} = \dfrac{1}{2}x^8$

17. $(x+y)^4 (x+y)^2 = (x+y)^{4+2} = (x+y)^6$

19. $(x+3)^2 (x+3) = (x+3)^{2+1} = (x+3)^3$

21. $\dfrac{p^{11}}{p^6} = p^{11-6} = p^5$

23. $\dfrac{54q^7}{18q} = \left(\dfrac{54}{18}\right)\left(\dfrac{q^7}{q}\right) = \dfrac{3}{1} q^{7-1} = 3q^6$

25. $\dfrac{(2x-3)^8}{(2x-3)^3} = (2x-3)^{8-3} = (2x-3)^5$

27. $\dfrac{-3(p+q)^2}{9(p+q)} = \left(-\dfrac{3}{9}\right)\dfrac{(p+q)^2}{(p+q)}$
$= -\dfrac{1}{3}(p+q)^{2-1} = -\dfrac{1}{3}(p+q)$

29. $\left(a^5\right)^6 = a^{5 \cdot 6} = a^{30}$

31. $(-3x)^4 = (-3)^4 x^4 = 81x^4$

33. $(abc)^{21} = a^{21} b^{21} c^{21}$

35. $\left(5m^3\right)^3 = 5^3 \left(m^3\right)^3 = 125 m^{3 \cdot 3} = 125 m^9$

37. $\left[(x+y)^3\right]^2 = (x+y)^{3 \cdot 2} = (x+y)^6$

39. $\left[(a-b)^4\right]^1 = (a-b)^{4 \cdot 1} = (a-b)^4$

41. $\left(x^2\right)^0 = x^{2 \cdot 0} x = x^0 = 1$

43. $\left(\dfrac{b}{d}\right)^4 = \dfrac{b^4}{d^4}$

45. $\left(\dfrac{3b}{c}\right)^4 = \dfrac{3^4 b^4}{c^4} = \dfrac{81 b^4}{c^4}$

47. $\left(\dfrac{-d}{2c}\right)^6 = \dfrac{(-1)^6 d^6}{2^6 c^6} = \dfrac{d^6}{64 c^6}$

49. $-7a^3 b \cdot 5a^2 b^4 = -35 a^{3+2} b^{1+4} = -35 a^5 b^5$

51. $\dfrac{15 x^3 y^2}{3xy^2} = \dfrac{15}{3} \cdot \dfrac{x^3}{x} \cdot \dfrac{y^2}{y^2} = 5 x^{3-1} y^{2-2}$
$= 5x^2 y^0 = 5x^2 \cdot 1 = 5x^2$

53. $\dfrac{-27 ab^2 c^3}{15bc} = \dfrac{-27}{15} \cdot a \cdot \dfrac{b^2}{b} \cdot \dfrac{c^3}{c}$
$= -\dfrac{9}{5} ab^{2-1} c^{3-1} = -\dfrac{9}{5} abc^2$

55. $\dfrac{-4x(4-x)^4}{2(4-x)} = \dfrac{-4}{2} \cdot x \cdot \dfrac{(4-x)^4}{(4-x)}$
$= -2x(4-x)^{4-1} = -2x(4-x)^3$

SSM: Experiencing Introductory and Intermediate Algebra

57. $\dfrac{\left(p^5 q^7\right)^3}{p^6 q} = \dfrac{\left(p^5\right)^3 \left(q^7\right)^3}{p^6 q} = \dfrac{p^{5 \cdot 3} q^{7 \cdot 3}}{p^6 q}$

$= \dfrac{p^{15} q^{21}}{p^6 q} = p^{15-6} q^{21-1} = p^9 q^{20}$

59. $\left(\dfrac{4x^3}{y^2}\right)^2 = \dfrac{\left(4x^3\right)^2}{\left(y^2\right)^2} = \dfrac{4^2 (x^3)^2}{y^{2 \cdot 2}}$

$= \dfrac{16 x^{3 \cdot 2}}{y^4} = \dfrac{16 x^6}{y^4}$

61. $\left(\dfrac{-3 p^4 q^2}{p^2 q}\right)^3 = \left(-3 p^{4-2} q^{2-1}\right)^3 = \left(-3 p^2 q\right)^3$

$= (-3)^3 \left(p^2\right)^3 q^3 = -27 p^{2 \cdot 3} q^3 = -27 p^6 q^3$

63. $\left[(2a)^2\right]^5 = (2a)^{2 \cdot 5} = (2a)^{10}$

$= 2^{10} a^{10} = 1024 a^{10}$

65. $\left[\left(\dfrac{x}{2y}\right)^2\right]^3 = \left(\dfrac{x}{2y}\right)^{2 \cdot 3} = \left(\dfrac{x}{2y}\right)^6 = \dfrac{x^6}{2^6 y^6} = \dfrac{x^6}{64 y^6}$

67. $x^3 \cdot x^4 = x^{12}$

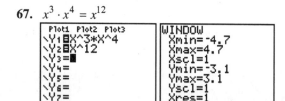

The values in the table are not always the same and the graphs do not coincide. The simplification is incorrect.

69. $x^3 \cdot x^4 = x^7$

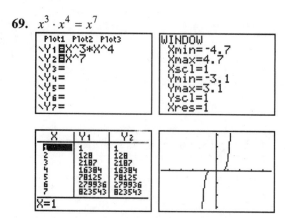

The values in the table match and the graphs are coinciding. The simplification is correct.

71. $\left(3x^4\right)^2 = 9x^8$

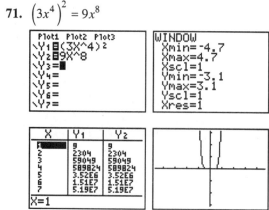

The values in the table match and the graphs are coinciding. The simplification is correct.

73. $\left(3x^4\right)^2 = 9x^6$

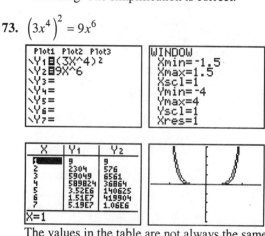

The values in the table are not always the same and the graphs do not coincide. The simplification is incorrect.

75. $\left(\dfrac{8x^3}{2x^6}\right)^2 = \dfrac{16}{x^6}$

The values in the table match and the graphs are coinciding. The simplification is correct.

77. $\left(\dfrac{8x^3}{2x^6}\right)^2 = \dfrac{16}{x^5}$

The values in the table are not always the same and the graphs do not coincide. The simplification is incorrect.

79. Let x = length of original side

$5x$ = length of enlarged side

The original area is x^2 square units.

The enlarged area is $(5x)^2 = 5^2 x^2 = 25x^2$ square units.

The enlarged area is 25 times bigger.

$x^2 = (6)^2 = 36$

$25x^2 = 25(6)^2 = 25(36) = 900$

If the original side is 6 feet, the original area and enlarged area are 36 square feet and 900 square feet, respectively.

81. Let x = length of original side

$4x$ = length of enlarged side

The original volume is x^3 cubic units.

The enlarged volume is $(4x)^3 = 4^3 x^3 = 64x^3$ cubic units.

The enlarged volume is 64 times bigger.

$x^3 = (1.5)^3 = 3.375$

$64x^3 = 64(1.5)^3 = 64(3.375) = 216$

If the original side is 1.5 feet, the original volume and enlarged volume are 3.375 ft^3 and 216 ft^3, respectively.

83. Let x = original radius

$2x$ = larger radius

The original volume is $\dfrac{4}{3}\pi x^3$ cubic units.

The larger sphere has a volume of

$\dfrac{4}{3}\pi (2x)^3 = \dfrac{4}{3}\pi 2^3 x^3 = \dfrac{32}{3}\pi x^3$.

The larger sphere has a volume that is 8 times bigger than the original volume.

$\dfrac{4}{3}\pi x^3 = \dfrac{4}{3}\pi(4)^3 = \dfrac{4}{3}\pi \cdot 64 = \dfrac{256}{3}\pi$

$\dfrac{32}{3}\pi x^3 = \dfrac{32}{3}\pi(4)^3 = \dfrac{32}{3}\pi \cdot 64 = \dfrac{2048}{3}\pi$

If the original radius is 4 feet, the volumes of the original sphere and the larger sphere are $\dfrac{256}{3}\pi$ ft^3 and $\dfrac{2048}{3}\pi$ ft^3, respectively.

9.1 Calculator Exercises

1. $x^4 \cdot x^5 = x^{20}$

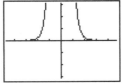

SSM: Experiencing Introductory and Intermediate Algebra

The simplification is incorrect.

2. $x^4 \cdot x^5 = x^9$

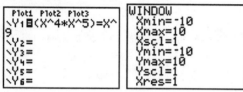

The simplification is correct.

3. $\left(x^4\right)^5 = x^{20}$

The simplification is correct.

4. $\left(x^4\right)^5 = x^9$

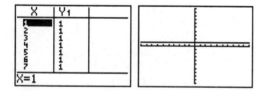

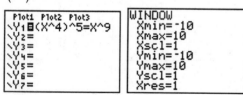

The simplification is not correct.

5. $\left(3x^4\right)^2 = 9x^8$

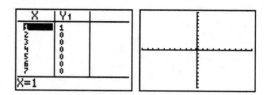

The simplification is correct.

6. $\left(3x^4\right)^2 = 9x^6$

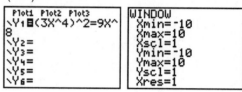

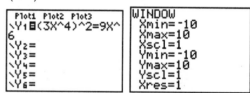

The simplification is not correct.

9.2 Exercises

1. $p^{-3} = \dfrac{1}{p^3}$

Chapter 9: Exponents and Polynomials

3. $\dfrac{1}{q^{-5}} = q^5$

5. $\dfrac{p^{-3}}{q^{-5}} = \dfrac{q^5}{p^3}$

7. $\dfrac{d^4}{c^{-3}} = c^3 d^4$

9. $\dfrac{5^{-2} h^3}{2^{-4} k^{-4}} = \dfrac{2^4 h^3 k^4}{5^2} = \dfrac{16 h^3 k^4}{25}$

11. $p^{-3} q^5 = \dfrac{q^5}{p^3}$

13. $(5a^3)(-4a^{-1}) = [5(-4)]a^{3+(-1)} = -20a^2$

15. $\dfrac{4m^{-2}}{16m^{-3}} = \dfrac{4}{16}\left(\dfrac{m^{-2}}{m^{-3}}\right) = \dfrac{1}{4} m^{-2-(-3)} = \dfrac{1}{4} m$

17. $\dfrac{4^{-3} n^{-2}}{3^{-4} n^{-3}} = \dfrac{4^{-3}}{3^{-4}}\left(\dfrac{n^{-2}}{n^{-3}}\right) = \dfrac{3^4}{4^3} n^{-2-(-3)} = \dfrac{81}{64} n$

19. $\left(c^{-4}\right)^{-2} = c^{(-4)(-2)} = c^8$

21. $\left(5a^{-2} b^2\right)^{-3} = 5^{-3} a^{(-2)(-3)} b^{2(-3)} = \dfrac{1}{5^3} a^6 b^{-6}$
$= \dfrac{a^6}{125 b^6}$

23. $(-6p^3 q)(7p^{-3} q^{-4}) = (-6 \cdot 7) p^{3+(-3)} q^{1+(-4)}$
$= -42 p^0 q^{-3} = -\dfrac{42}{q^3}$

25. $(1.4x^2)(4.3x^3 y^{-2}) = (1.4)(4.3) x^{2+3} y^{-2}$
$= \dfrac{6.02 x^5}{y^2}$

27. $\left(\dfrac{3}{7} x^2 y^{-1}\right)\left(\dfrac{14}{15} x^{-1} y^4\right) = \left(\dfrac{3}{7} \cdot \dfrac{14}{15}\right) x^{2+(-1)} y^{-1+4}$
$= \dfrac{2}{5} xy^3$

29. $(3x^5 yz)(-7xy^{-4} z^{-1}) = 3(-7) x^{5+1} y^{1+(-4)} z^{1+(-1)}$
$= -21 x^6 y^{-3} z^0 = -\dfrac{21 x^6}{y^3}$

31. $\dfrac{m^{-2}}{m^5} = m^{-2-5} = m^{-7} = \dfrac{1}{m^7}$

33. $\dfrac{-7c^{-5}}{21c^{-7}} = \dfrac{-7}{21}\left(\dfrac{c^{-5}}{c^{-7}}\right) = -\dfrac{1}{3} c^{-5-(-7)} = -\dfrac{1}{3} c^2$

35. $\dfrac{8x^2 y}{2x^{-1} y^{-3}} = \dfrac{8}{2} \cdot \dfrac{x^2}{x^{-1}} \cdot \dfrac{y}{y^{-3}} = 4 x^{2-(-1)} y^{1-(-3)}$
$= 4x^3 y^4$

37. $\dfrac{5^{-3} h^{-1} k}{5^{-2} h k^{-6}} = \dfrac{5^{-3}}{5^{-2}} \cdot \dfrac{h^{-1}}{h} \cdot \dfrac{k}{k^{-6}} = 5^{-3-(-2)} h^{-1-1} k^{1-(-6)}$
$= 5^{-1} h^{-2} k^7 = \dfrac{k^7}{5h^2}$

39. $\dfrac{121 a^{-3} b^3}{11 a^2 b^{-2}} = \dfrac{121}{11} \cdot \dfrac{a^{-3}}{a^2} \cdot \dfrac{b^3}{b^{-2}} = 11 a^{-3-2} b^{3-(-2)}$
$= 11 a^{-5} b^5 = \dfrac{11 b^5}{a^5}$

41. $\left(\dfrac{a^{-3} b}{2ab^{-3}}\right)^3 = \left(\dfrac{1}{2} \cdot \dfrac{a^{-3}}{a} \cdot \dfrac{b}{b^{-3}}\right)^3 = \left(\dfrac{b^4}{2a^4}\right)^3$
$= \dfrac{\left(b^4\right)^3}{2^3 \left(a^4\right)^3} = \dfrac{b^{12}}{8a^{12}}$

43. $\left(\dfrac{-4p^3 q^2}{2pq^{-3}}\right)^5 = \left(\dfrac{-4}{2} \cdot \dfrac{p^3}{p} \cdot \dfrac{q^2}{q^{-3}}\right)^5 = \left(-2p^2 q^5\right)^5$
$= (-2)^5 \left(p^2\right)^5 \left(q^5\right)^5 = -32 p^{10} q^{25}$

45. $\left(\dfrac{a}{b}\right)^{-3} = \left(\dfrac{b}{a}\right)^3 = \dfrac{b^3}{a^3}$

47. $\left(\dfrac{4x}{y}\right)^{-3} = \left(\dfrac{y}{4x}\right)^3 = \dfrac{y^3}{4^3 x^3} = \dfrac{y^3}{64 x^3}$

49. $\left(\dfrac{-3p^2}{q^{-2}}\right)^{-4} = \left(\dfrac{q^{-2}}{-3p^2}\right)^4 = \dfrac{\left(q^{-2}\right)^4}{(-3)^4\left(p^2\right)^4}$

$= \dfrac{q^{-8}}{81p^8} = \dfrac{1}{81p^8 q^8}$

51. $\left(\dfrac{y^{-1}}{z^{-2}}\right)^{-3} = \dfrac{\left(y^{-1}\right)^{-3}}{\left(z^{-2}\right)^{-3}} = \dfrac{y^3}{z^6}$

53. $\left(\dfrac{5a^{-2}b}{25ab^{-3}}\right)^{-3} = \left(\dfrac{5b \cdot b^3}{25a \cdot a^2}\right)^{-3} = \left(\dfrac{b^4}{5a^3}\right)^{-3}$

$= \left(\dfrac{5a^3}{b^4}\right)^3 = \dfrac{5^3\left(a^3\right)^3}{\left(b^4\right)^3} = \dfrac{125a^9}{b^{12}}$

55. $\left[2.8\left(10^4\right)\right]\left[5.9\left(10^{12}\right)\right] = (2.8)(5.9)\left(10^{4+12}\right)$

$= 16.52\left(10^{16}\right) = 1.652(10)\left(10^{16}\right) = 1.652\left(10^{17}\right)$

The Sun is approximately 1.652×10^{17} miles from the center of the Milky Way.

57. $\dfrac{2.998\left(10^8\right)}{107.7\left(10^6\right)} = \dfrac{2.998}{107.7}\left(\dfrac{10^8}{10^6}\right)$

$\approx 0.02784\left(10^{8-6}\right) = 0.02784\left(10^2\right)$

$= 0.02784(100) = 2.784$ meters

The wavelength is approximately 2.784 meters.

9.2 Calculator Exercises

1. Step 1: $\left(\dfrac{3}{5}x^3\right)\left(\dfrac{5}{7}x^{-2}\right) = \left(\dfrac{3}{5} \cdot \dfrac{5}{7}\right)\left(x^3 \cdot x^{-2}\right)$

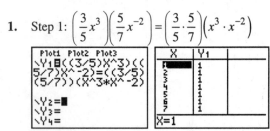

This step is correct.

Step 2: $\left(\dfrac{3}{5} \cdot \dfrac{5}{7}\right)\left(x^3 \cdot x^{-2}\right) = \left(\dfrac{3}{7}\right)\left(x^{-6}\right)$

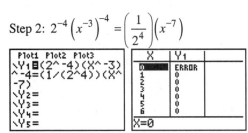

This step is incorrect.

The simplification contains an error.

2. Step 1: $\left(2x^{-3}\right)^{-4} = 2^{-4}\left(x^{-3}\right)^{-4}$

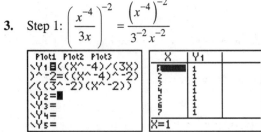

This step is correct.

Step 2: $2^{-4}\left(x^{-3}\right)^{-4} = \left(\dfrac{1}{2^4}\right)\left(x^{-7}\right)$

This step is incorrect.

The simplification contains an error.

3. Step 1: $\left(\dfrac{x^{-4}}{3x}\right)^{-2} = \dfrac{\left(x^{-4}\right)^{-2}}{3^{-2}x^{-2}}$

This step is correct.

Step 2: $\dfrac{\left(x^{-4}\right)^{-2}}{3^{-2}x^{-2}} = \dfrac{x^8}{-6x^{-2}}$

Chapter 9: Exponents and Polynomials

```
Plot1 Plot2 Plot3
\Y1■((X^-4)^-2)/
((3^-2)(X^-2))=(
X^8)/((-6)(X^-2)
)
\Y2=
\Y3=
\Y4=
```

X	Y1
0	ERROR
1	0
2	0
3	0
4	0
5	0
6	0

X=0

This step is incorrect.

The simplification contains an error.

9.3 Exercises

1. $(9x^2 - 17x + 31) + (2x^4 + 3x^2 + 12)$
    $9x^2 - 17x + 31 + 2x^4 + 3x^2 + 12$
    $2x^4 + 12x^2 - 17x + 43$

3. $(5x^4 + 6x + 3x^3 - 2x^2 - 12) + (4x^4 + 21 - 8x^2 - 9x)$
    $5x^4 + 6x + 3x^3 - 2x^2 - 12 + 4x^4 + 21 - 8x^2 - 9x$
    $9x^4 + 3x^3 - 10x^2 - 3x + 9$

5. $(5x^2y - 3xy^2 + 6y^3) + (15x^3 - 8x^2y + 3xy^2)$
    $5x^2y - 3xy^2 + 6y^3 + 15x^3 - 8x^2y + 3xy^2$
    $15x^3 - 3x^2y + 6y^3$

7. $(6 - 7a^3 + 3a^2 - 5a) + (6a + 8a^3 + 2) + (5a^2 - 8a - 9)$
    $6 - 7a^3 + 3a^2 - 5a + 6a + 8a^3 + 2 + 5a^2 - 8a - 9$
    $a^3 + 8a^2 - 7a - 1$

9. $\left(\dfrac{2}{3}y^4 + \dfrac{1}{6}y^3 + 3y^2 - \dfrac{1}{3}y + \dfrac{5}{9}\right) + \left(\dfrac{7}{3}y^3 - \dfrac{8}{9}y^2 + \dfrac{5}{6}y - 3\right)$
    $\dfrac{2}{3}y^4 + \dfrac{1}{6}y^3 + 3y^2 - \dfrac{1}{3}y + \dfrac{5}{9} + \dfrac{7}{3}y^3 - \dfrac{8}{9}y^2 + \dfrac{5}{6}y - 3$
    $\dfrac{2}{3}y^4 + \left(\dfrac{1}{6} + \dfrac{7}{3}\right)y^3 + \left(3 - \dfrac{8}{9}\right)y^2 + \left(-\dfrac{1}{3} + \dfrac{5}{6}\right)y + \left(\dfrac{5}{9} - 3\right)$
    $\dfrac{2}{3}y^4 + \dfrac{5}{2}y^3 + \dfrac{19}{9}y^2 + \dfrac{1}{2}y - \dfrac{22}{9}$

11. $(12.07x^3 + 8.6x^2 - 3.19x + 14) + (6.7x^3 - 9.83x^2 + 7x - 4.265)$
    $12.07x^3 + 8.6x^2 - 3.19x + 14 + 6.7x^3 - 9.83x^2 + 7x - 4.265$
    $18.77x^3 - 1.23x^2 + 3.81x + 9.735$

SSM: Experiencing Introductory and Intermediate Algebra

13. $(4756a^3 - 3219a^2 - 1816a + 2083) + (361a^3 + 54217a^2 + 12)$
    $4756a^3 - 3219a^2 - 1816a + 2083 + 361a^3 + 54217a^2 + 12$
    $5117a^3 + 50998a^2 - 1816a + 2095$

15. $(3a + 4b) + (5b + 6c)$
    $3a + 4b + 5b + 6c$
    $3a + 9b + 6c$

17. $(5z^3 + 27z^2 - 35z + 42) - (3z^3 + 16z - 72)$
    $5z^3 + 27z^2 - 35z + 42 - 3z^3 - 16z + 72$
    $2z^3 + 27z^2 - 51z + 114$

19. $(a^5 - 9) - (a^5 - a^4 + a^3 - a^2 + a - 9)$
    $a^5 - 9 - a^5 + a^4 - a^3 + a^2 - a + 9$
    $a^4 - a^3 + a^2 - a$

21. $(16x^2 - 32 + 9x) - (12x + 7 + 9x^2)$
    $16x^2 - 32 + 9x - 12x - 7 - 9x^2$
    $7x^2 - 3x - 39$

23. $(13a^3 - 6a^2 + 11) - (12a - 3 + 18a^3)$
    $13a^3 - 6a^2 + 11 - 12a + 3 - 18a^3$
    $-5a^3 - 6a^2 - 12a + 14$

25. $(42x^3 + 17x^2y + 3xy^2 + 23y^3) - (47x^2y + 12y^3)$
    $42x^3 + 17x^2y + 3xy^2 + 23y^3 - 47x^2y - 12y^3$
    $42x^3 - 30x^2y + 3xy^2 + 11y^3$

27. $(4a + 7c) - (2b + 6d)$
    $4a + 7c - 2b - 6d$

29. $\left(\dfrac{5}{7}x^2 + \dfrac{8}{21}x - \dfrac{11}{14}\right) - \left(\dfrac{1}{2}x^2 + \dfrac{5}{6}x + \dfrac{19}{42}\right)$
    $\dfrac{5}{7}x^2 + \dfrac{8}{21}x - \dfrac{11}{14} - \dfrac{1}{2}x^2 - \dfrac{5}{6}x - \dfrac{19}{42}$
    $\left(\dfrac{5}{7} - \dfrac{1}{2}\right)x^2 + \left(\dfrac{8}{21} - \dfrac{5}{6}\right)x + \left(-\dfrac{11}{14} - \dfrac{19}{42}\right)$
    $\dfrac{3}{14}x^2 - \dfrac{19}{42}x - \dfrac{26}{21}$

31. $(21.2x^3 + 0.9x^2y - 13.22xy^2 + 81.07y^3) - (12.2x^3 - 0.1x^2y + 0.78xy^2 + 13.07y^3)$
    $21.2x^3 + 0.9x^2y - 13.22xy^2 + 81.07y^3 - 12.2x^3 + 0.1x^2y - 0.78xy^2 - 13.07y^3$
    $9x^3 + x^2y - 14xy^2 + 68y^3$

33. $(5062z^2 - 106z + 8295) - (379z^2 + 4297z + 1108)$
    $5062z^2 - 106z + 8295 - 379z^2 - 4297z - 1108$
    $4683z^2 - 4403z + 7187$

35. Let x = number of pots

$$C(x) = 200 + (2.50 + 2.00)x$$
$$C(x) = 200 + 4.50x$$
$$R(x) = 13.50x$$
$$P(x) = R(x) - C(x)$$
$$= (13.50x) - (200 + 4.50x)$$
$$= 13.50x - 200 - 4.50x$$
$$= 9.00x - 200$$
$$P(20) = 9.00(20) - 200$$
$$= 180 - 200$$
$$= -20$$
$$P(30) = 9.00(30) - 200$$
$$= 270 - 200$$
$$= 70$$

The profit for selling 20 and 30 pots is $-\$20$ and $\$70$, respectively.

37. Let x = number of cakes baked and sold

$$c(x) = b(x) + d(x)$$
$$= (3.50x + 25) + (1.50x + 5)$$
$$= 3.50x + 25 + 1.50x + 5$$
$$= 5.00x + 30$$
$$r(x) = 28.50x$$
$$p(x) = r(x) - c(x)$$
$$= (28.50x) - (5.00x + 30)$$
$$= 28.50x - 5.00x - 30$$
$$= 23.50x - 30$$

39. $d(t) = \left(-16t^2 + 240t + 100\right) - \left(-16t^2 + 250t\right)$
$$= -16t^2 + 240t + 100 + 16t^2 - 250t$$
$$= -10t + 100$$

9.3 Calculator Exercises

1. $\left(2x^4 + 4x^2 - 3x + 12\right) + \left(3x^3 - 5x^2 + 7x - 9\right)$

$2x^4 + 4x^2 - 3x + 12 + 3x^3 - 5x^2 + 7x - 9$

$2x^4 + 3x^3 - x^2 + 4x + 3$

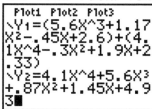 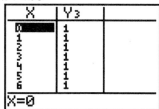 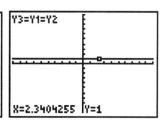

2. $\left(5.6x^3 + 1.17x^2 - 0.45x + 2.6\right) + \left(4.1x^4 - 0.3x^2 + 1.9x + 2.33\right)$

$5.6x^3 + 1.17x^2 - 0.45x + 2.6 + 4.1x^4 - 0.3x^2 + 1.9x + 2.33$

$4.1x^4 + 5.6x^3 + 0.87x^2 + 1.45x + 4.93$

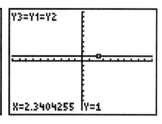

SSM: Experiencing Introductory and Intermediate Algebra

3. $\left(3x^4 + 2x^3 - 3x^2 - 2x + 8\right) - \left(5x^5 - 3x^2 + 7x - 4\right)$

$3x^4 + 2x^3 - 3x^2 - 2x + 8 - 5x^5 + 3x^2 - 7x + 4$

$-5x^5 + 3x^4 + 2x^3 - 9x + 12$

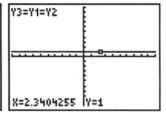

4. $\left(\dfrac{5}{6}x^3 + \dfrac{1}{2}x^2 - \dfrac{7}{9}x + 4\right) - \left(\dfrac{2}{3}x^2 + \dfrac{5}{6}x - \dfrac{4}{9}\right)$

$\dfrac{5}{6}x^3 + \dfrac{1}{2}x^2 - \dfrac{7}{9}x + 4 - \dfrac{2}{3}x^2 - \dfrac{5}{6}x + \dfrac{4}{9}$

$\dfrac{5}{6}x^3 + \left(\dfrac{1}{2} - \dfrac{2}{3}\right)x^2 + \left(-\dfrac{7}{9} - \dfrac{5}{6}\right)x + \left(4 + \dfrac{4}{9}\right)$

$\dfrac{5}{6}x^3 - \dfrac{1}{6}x^2 - \dfrac{29}{18}x + \dfrac{40}{9}$

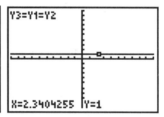

9.4 Exercises

1. $\left(8ab^2\right)\left(-2a^3b\right)$

   $= -16a^{1+3}b^{2+1}$

   $= -16a^4b^3$

3. $-2x(3x - y + 2z)$

   $= -2x(3x) - 2x(y) - 2x(2z)$

   $= -6x^2 - 2xy - 4xz$

5. $2a^3(3a + 2b - c)$

   $= 2a^3(3a) + 2a^3(2b) + 2a^3(-c)$

   $= 6a^4 + 4a^3b - 2a^3c$

7. $(x + 4)(x + 2)$

   $= x(x + 2) + 4(x + 2)$

   $= x^2 + 2x + 4x + 8$

   $= x^2 + 6x + 8$

9. $(5 + x)(3 + 2x)$

   $= 5(3 + 2x) + x(3 + 2x)$

   $= 15 + 10x + 3x + 2x^2$

   $= 15 + 13x + 2x^2$

11. $(2x + 5)(3y - 2)$

    $= 2x(3y - 2) + 5(3y - 2)$

    $= 6xy - 4x + 15y - 10$

13. $(3x+4y)(x-2y)$
$= 3x(x-2y)+4y(x-2y)$
$= 3x^2 - 6xy + 4xy - 8y^2$
$= 3x^2 - 2xy - 8y^2$

15. $(a-2.4)(5a+3.8)$
$= a(5a+3.8) - 2.4(5a+3.8)$
$= 5a^2 + 3.8a - 12a - 9.12$
$= 5a^2 - 8.2a - 9.12$

17. $(2x+1.1)(3y+3.2)$
$= 2x(3y+3.2) + 1.1(3y+3.2)$
$= 6xy + 6.4x + 3.3y + 3.52$

19. $\left(a+\dfrac{2}{3}\right)\left(a+\dfrac{1}{3}\right)$
$= a\left(a+\dfrac{1}{3}\right) + \dfrac{2}{3}\left(a+\dfrac{1}{3}\right)$
$= a^2 + \dfrac{1}{3}a + \dfrac{2}{3}a + \dfrac{2}{9}$
$= a^2 + a + \dfrac{2}{9}$

21. $(2x^2 - 3)(x^2 + 4)$
$= 2x^2(x^2+4) - 3(x^2+4)$
$= 2x^4 + 8x^2 - 3x^2 - 12$
$= 2x^4 + 5x^2 - 12$

23. $(4x^2+3)(2x+1)$
$= 4x^2(2x+1) + 3(2x+1)$
$= 8x^3 + 4x^2 + 6x + 3$

25. $(2x+3y^2)(3x^2 - 5y)$
$= 2x(3x^2-5y) + 3y^2(3x^2-5y)$
$= 6x^3 - 10xy + 9x^2y^2 - 15y^3$

27. $(3a^2 + 5b^3)(a^2 + b^3)$
$= 3a^2(a^2+b^3) + 5b^3(a^2+b^3)$
$= 3a^4 + 3a^2b^3 + 5a^2b^3 + 5b^6$
$= 3a^4 + 8a^2b^3 + 5b^6$

29. $(x+4)(x^2 - 4x + 16)$
$= x(x^2 - 4x + 16) + 4(x^2 - 4x + 16)$
$= x^3 - 4x^2 + 16x + 4x^2 - 16x + 64$
$= x^3 + 64$

31. $(3x-2)(2x^2 - 5x - 3)$
$= 3x(2x^2 - 5x - 3) - 2(2x^2 - 5x - 3)$
$= 6x^3 - 15x^2 - 9x - 4x^2 + 10x + 6$
$= 6x^3 - 19x^2 + x + 6$

33. $(x^2 + x + 1)(x^2 + 2x + 3)$
$= x^2(x^2+2x+3) + x(x^2+2x+3) + (x^2+2x+3)$
$= x^4 + 2x^3 + 3x^2 + x^3 + 2x^2 + 3x + x^2 + 2x + 3$
$= x^4 + 3x^3 + 6x^2 + 5x + 3$

35. $(a+b+c)^2$
$= (a+b+c)(a+b+c)$
$= a(a+b+c) + b(a+b+c) + c(a+b+c)$
$= a^2 + ab + ac + ab + b^2 + bc + ac + bc + c^2$
$= a^2 + b^2 + c^2 + 2ab + 2ac + 2bc$

37. $(z+3)^3 = (z+3)(z+3)(z+3)$
$= [z(z+3) + 3(z+3)](z+3)$
$= (z^2 + 3z + 3z + 9)(z+3)$
$= (z^2 + 6z + 9)(z+3)$
$= (z^2 + 6z + 9)z + (z^2 + 6z + 9)3$
$= z^3 + 6z^2 + 9z + 3z^2 + 18z + 27$
$= z^3 + 9z^2 + 27z + 27$

39. $(3a-2b)^3 = (3a-2b)(3a-2b)(3a-2b)$
$= [3a(3a-2b)-2b(3a-2b)](3a-2b)$
$= (9a^2 -6ab-6ab+4b^2)(3a-2b)$
$= (9a^2 -12ab+4b^2)(3a-2b)$
$= (9a^2 -12ab+4b^2)(3a)+(9a^2 -12ab+4b^2)(-2b)$
$= 27a^3 -36a^2b+12ab^2 -18a^2b+24ab^2 -8b^3$
$= 27a^3 -54a^2b+36ab^2 -8b^3$

41. $(x-5)(x+5)$
$= x^2 -5^2$
$= x^2 -25$

43. $(3m+7)(3m-7)$
$= (3m)^2 -7^2$
$= 9m^2 -49$

45. $(2a+3b)(2a-3b)$
$= (2a)^2 -(3b)^2$
$= 4a^2 -9b^2$

47. $(4x-1.5)(4x+1.5)$
$= (4x)^2 -(1.5)^2$
$= 16x^2 -2.25$

49. $\left(\frac{2}{5}x-1\right)\left(\frac{2}{5}x+1\right)$
$= \left(\frac{2}{5}x\right)^2 -1^2$
$= \frac{4}{25}x^2 -1$

51. $\left(\frac{1}{3}x+\frac{4}{5}\right)\left(\frac{1}{3}x-\frac{4}{5}\right)$
$= \left(\frac{1}{3}x\right)^2 -\left(\frac{4}{5}\right)^2$
$= \frac{1}{9}x^2 -\frac{16}{25}$

53. $(x^2 +7)(x^2 -7)$
$= (x^2)^2 -7^2$
$= x^4 -49$

55. $(2x^3 +5y)(2x^3 -5y)$
$= (2x^3)^2 -(5y)^2$
$= 4x^6 -25y^2$

57. $(m+7)^2$
$= m^2 +2(m)(7)+7^2$
$= m^2 +14m+49$

59. $(x-y)^2$
$= x^2 -2xy+y^2$

61. $(2p+9q)^2$
$= (2p)^2 +2(2p)(9q)+(9q)^2$
$= 4p^2 +36pq+81q^2$

63. $(6c-5)^2$
$= (6c)^2 -2(6c)(5)+5^2$
$= 36c^2 -60c+25$

65. $(3x^3 + 2)^2$
$= (3x^3)^2 + 2(3x^3)(2) + 2^2$
$= 9x^6 + 12x^3 + 4$

67. $(2x^2 - 3y^3)^2$
$= (2x^2)^2 - 2(2x^2)(3y^3) + (3y^3)^2$
$= 4x^4 - 12x^2y^3 + 9y^6$

69. a. Length: $18 - 2x$ inches
Width: $12 - 2x$ inches
Height: x inches

b. $V = L \cdot W \cdot H$
$= (18 - 2x)(12 - 2x)(x)$
$= [18(12 - 2x) - 2x(12 - 2x)]x$
$= (216 - 36x - 24x + 4x^2)x$
$= (216 - 60x + 4x^2)x$
$= 4x^3 - 60x^2 + 216x$
The volume is $4x^3 - 60x^2 + 216x$ in^3.

c. $S = L \cdot W + 2W \cdot H + 2L \cdot H$
$= (18 - 2x)(12 - 2x) + 2(12 - 2x)x + 2(18 - 2x)x$
$= 18(12 - 2x) - 2x(12 - 2x) + 24x - 4x^2 + 36x - 4x^2$
$= 216 - 36x - 24x + 4x^2 + 24x - 4x^2 + 36x - 4x^2$
$= 216 - 4x^2$
The surface area is $216 - 4x^2$ in^2.

71. a. Area of outer circle:
(πx^2) ft^2

b. Area of inner circle:
$[\pi(x - 5)^2]$ ft^2

c. Difference of the areas:
$\pi x^2 - \pi(x - 5)^2$
$= \pi x^2 - \pi(x^2 - 2(x)(5) + 5^2)$
$= \pi x^2 - \pi(x^2 - 10x + 25)$
$= \pi x^2 - \pi x^2 + 10\pi x - 25\pi$
$= 10\pi x - 25\pi$
The area of the deck is $(10\pi x - 25\pi)$ ft^2.

SSM: Experiencing Introductory and Intermediate Algebra

9.4 Calculator Exercises

1. $(2x-1)(2x+1) = 2x^2 - 1$

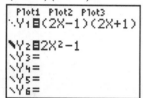 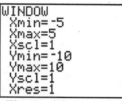 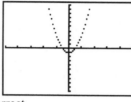 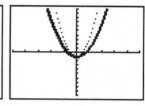

The graphs do not coincide. The simplification is incorrect.

2. $(x+1)(x-1) = x^2 + x + 2$

 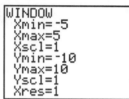 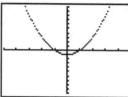 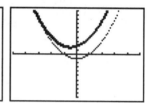

The graphs do not coincide. The simplification is incorrect.

3. $(x-2)(x-1) = x^2 - 3x + 2$

 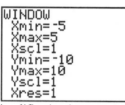 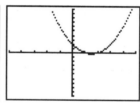 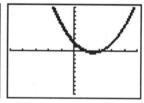

The graphs coincide. The simplification is correct.

4. $(0.5x+1)(4x-0.8) = 2x^2 + 3.6x - 0.8$

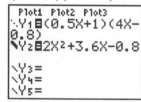 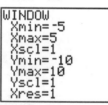 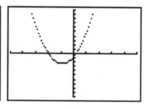 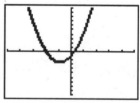

The graphs coincide. The simplification is correct.

5. $\left(\frac{1}{2}x - 3\right)\left(2x + \frac{1}{3}\right) = x^2 - \frac{35}{6}x + 1$

 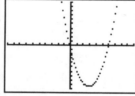 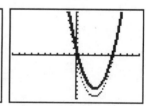

The graphs do not coincide. The simplification is incorrect.

390

6. $(x^2+4)(x^2-1) = x^4 + 3x^2 - 4$

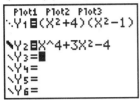 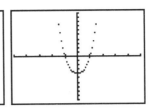 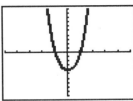

The graphs coincide. The simplification is correct.

7. $(x^2+1)(2x-1) = 2x^3 - x^2 + 2x - 1$

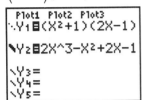 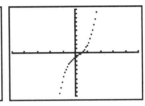 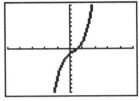

The graphs coincide. The simplification is correct.

8. $(x^2-1)(0.3x-1.4) = 0.3x^3 - 1.4x^2 - 0.3x - 1.4$

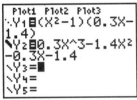 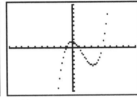 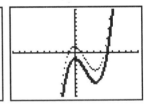

The graphs do not coincide. The simplification is incorrect.

9.5 Exercises

1. $\dfrac{20a^4b^2c}{-5a^2bc} = \dfrac{20}{-5} \cdot \dfrac{a^4}{a^2} \cdot \dfrac{b^2}{b} \cdot \dfrac{c}{c}$

   $= -4a^{4-2}b^{2-1}c^{1-1} = -4a^2b^1c^0 = -4a^2b$

3. $\dfrac{6x^4y^6z^2 + 18x^2y^3z}{3x^2y^3z} = \dfrac{6x^4y^6z^2}{3x^2y^3z} + \dfrac{18x^2y^3z}{3x^2y^3z}$

   $= 2x^{4-2}y^{6-3}z^{2-1} + 6x^{2-2}y^{3-3}z^{1-1}$

   $= 2x^2y^3z + 6$

5. $\dfrac{3p^3 - 9p^2q + 6pq^2 - 12q^3}{3pq}$

   $= \dfrac{3p^3}{3pq} - \dfrac{9p^2q}{3pq} + \dfrac{6pq^2}{3pq} - \dfrac{12q^3}{3pq}$

   $= p^{3-1}q^{-1} - 3p^{2-1}q^{1-1} + 2p^{1-1}q^{2-1} - 4p^{-1}q^{3-1}$

   $= \dfrac{p^2}{q} - 3p + 2q - \dfrac{4q^2}{p}$

7. $\dfrac{6x^3 + 12x^2 - 18x}{3x} = \dfrac{6x^3}{3x} + \dfrac{12x^2}{3x} - \dfrac{18x}{3x}$

   $= 2x^{3-1} + 4x^{2-1} - 6x^{1-1}$

   $= 2x^2 + 4x - 6$

9. $\dfrac{4x^2 - 2x + 12}{8x} = \dfrac{4x^2}{8x} - \dfrac{2x}{8x} + \dfrac{12}{8x}$

$= \dfrac{1}{2}x^{2-1} - \dfrac{1}{4}x^{1-1} + \dfrac{3}{2x}$

$= \dfrac{1}{2}x - \dfrac{1}{4} + \dfrac{3}{2x}$

11. $\dfrac{3.72x^4 - 6.96x^2 + 1.08}{1.2x^2}$

$= \dfrac{3.72x^4}{1.2x^2} - \dfrac{6.96x^2}{1.2x^2} + \dfrac{1.08}{1.2x^2}$

$= 3.1x^{4-2} - 5.8x^{2-2} + \dfrac{0.9}{x^2}$

$= 3.1x^2 - 5.8 + \dfrac{0.9}{x^2}$

13. $\dfrac{9x^4 + 6x^3y - 18x^2y^2 - 24xy^3 + 72y^4}{-3xy}$

$= \dfrac{9x^4}{-3xy} + \dfrac{6x^3y}{-3xy} - \dfrac{18x^2y^2}{-3xy} - \dfrac{24xy^3}{-3xy} + \dfrac{72y^4}{-3xy}$

$= -\dfrac{3x^3}{y} - 2x^2 + 6xy + 8y^2 - \dfrac{24y^3}{x}$

15. $\quad\quad\quad\quad\;\; 5x\; - 6$
$3x+2 \overline{\smash{\big)}\, 15x^2 - 8x - 12}$
$\quad\quad\; \underline{-\left(15x^2 + 10x\right)}$
$\quad\quad\quad\quad\quad -18x - 12$
$\quad\quad\quad\quad\; \underline{-(-18x - 12)}$
$\quad\quad\quad\quad\quad\quad\quad\quad 0$
The quotient is $5x - 6$.

17. $\quad\quad\quad\quad\; 2x - 5$
$x+7 \overline{\smash{\big)}\, 2x^2 + 9x - 35}$
$\quad\quad\; \underline{-\left(2x^2 + 14x\right)}$
$\quad\quad\quad\quad\quad -5x - 35$
$\quad\quad\quad\quad\; \underline{-(-5x - 35)}$
$\quad\quad\quad\quad\quad\quad\quad\; 0$
The quotient is $2x - 5$.

19. $\quad\quad\quad\quad\; 5x + 6$
$x-6 \overline{\smash{\big)}\, 5x^2 - 24x - 36}$
$\quad\quad\; \underline{-\left(5x^2 - 30x\right)}$
$\quad\quad\quad\quad\quad 6x - 36$
$\quad\quad\quad\quad\; \underline{-(6x - 36)}$
$\quad\quad\quad\quad\quad\quad\quad 0$
The quotient is $5x + 6$.

21. $\quad\quad\quad\quad\; 3y - 5$
$y+8 \overline{\smash{\big)}\, 3y^2 + 19y - 20}$
$\quad\quad\; \underline{-\left(3y^2 + 24y\right)}$
$\quad\quad\quad\quad\quad -5y - 20$
$\quad\quad\quad\quad\; \underline{-(-5y - 40)}$
$\quad\quad\quad\quad\quad\quad\quad 20$
The quotient is $3y - 5 + \dfrac{20}{y+8}$.

23. $\quad\quad\quad\quad\; 3a + 5$
$2a-5 \overline{\smash{\big)}\, 6a^2 - 5a - 30}$
$\quad\quad\; \underline{-\left(6a^2 - 15a\right)}$
$\quad\quad\quad\quad\quad 10a - 30$
$\quad\quad\quad\quad\; \underline{-(10a - 25)}$
$\quad\quad\quad\quad\quad\quad\quad -5$
The quotient is $3a + 5 - \dfrac{5}{2a-5}$.

25.

$$\begin{array}{r}x^2-3.2x+.68\\5x+2\overline{\smash{\big)}5x^3-14x^2-3x+2}\\\underline{-(5x^3+2x^2)}\\-16x^2-3x\\\underline{-(-16x^2-6.4x)}\\3.4x+2\\\underline{-(3.4x+1.36)}\\0.64\end{array}$$

The quotient is $x^2-3.2x+0.68+\dfrac{0.64}{5x+2}$.

27.

$$\begin{array}{r}4x^2+2x+5\\2x-1\overline{\smash{\big)}8x^3+8x-5}\\\underline{-(8x^3-4x^2)}\\4x^2+8x\\\underline{-(4x^2-2x)}\\10x-5\\\underline{-(10x-5)}\\0\end{array}$$

The quotient is $4x^2+2x+5$.

29.

$$\begin{array}{r}3a-7\\3a+7\overline{\smash{\big)}9a^2-49}\\\underline{-(9a^2+21a)}\\-21a-49\\\underline{-(-21a-49)}\\0\end{array}$$

The quotient is $3a-7$.

31.

$$\begin{array}{r}x+5.4\\5x+3\overline{\smash{\big)}5x^2+30x+9}\\\underline{-(5x^2+3x)}\\27x+9\\\underline{-(27x+16.2)}\\-7.2\end{array}$$

The quotient is $x+5.4-\dfrac{7.2}{5x+3}$.

33.

$$\begin{array}{r}4z-11\\4z-11\overline{\smash{\big)}16z^2-88z+121}\\\underline{-(16z^2-44z)}\\-44z+121\\\underline{-(-44z+121)}\\0\end{array}$$

The quotient is $4z-11$.

35.

$$\begin{array}{r}x^2-3x+9\\x+3\overline{\smash{\big)}x^3+27}\\\underline{-(x^3+3x^2)}\\-3x^2\\\underline{-(-3x^2-9x)}\\9x+27\\\underline{-(9x+27)}\\0\end{array}$$

The quotient is x^2-3x+9.

SSM: Experiencing Introductory and Intermediate Algebra

37.
$$\begin{array}{r}a^2+5a+25\\a-5\overline{)a^3-125}\\\underline{-(a^3-5a^2)}\\5a^2\\\underline{-(5a^2-25a)}\\25a-125\\\underline{-(25a-125)}\\0\end{array}$$

The quotient is $a^2+5a+25$.

39.
$$\begin{array}{r}16x^2-12x+9\\4x+3\overline{)64x^3+27}\\\underline{-(64x^3+48x^2)}\\-48x^2\\\underline{-(-48x^2-36x)}\\36x+27\\\underline{-(36x+27)}\\0\end{array}$$

The quotient is $16x^2-12x+9$.

41. Let x = number of calculators

  a. $c(x)=45+35x$

  b. $r(x)=50x$

  c. $p(x)=r(x)-c(x)$
  $=50x-(45+35x)$
  $=50x-45-35x$
  $=15x-45$

  d. $a(x)=\dfrac{p(x)}{x}=\dfrac{15x-45}{x}$
  $=\dfrac{15x}{x}-\dfrac{45}{x}=15-\dfrac{45}{x}$

  e. $a(30)=15-\dfrac{45}{30}=15-1.5=13.5$

  Jennifer's average profit would be $13.50.

43. $m(t)=\dfrac{6000(1-t^5)}{1-t}$

$m(1.06)=\dfrac{6000(1-1.1^5)}{1-1.1}=36630.6$

$$\begin{array}{r}t^4+t^3+t^2+t+1\\-t+1\overline{)-t^5+1}\\\underline{-(-t^5+t^4)}\\-t^4\\\underline{-(-t^4+t^3)}\\-t^3\\\underline{-(-t^3+t^2)}\\-t^2\\\underline{-(-t^2+t)}\\-t+1\\\underline{-(-t+1)}\\0\end{array}$$

$m(t)=6000(t^4+t^3+t^2+t+1)$

$m(1.1)=6000(1.1^4+1.1^3+1.1^2+1.1+1)$
$=36630.6$

The total amount received would be roughly $36630.60.

$m(1.15)=\dfrac{6000(1-1.15^5)}{1-1.15}\approx 40454.29$

If the percent increase were raised to 15%, the total amount received would be roughly $40,454.29.

45. $c(n)=\dfrac{250(1-n^5)}{1-n}$

$c(0.9)=\dfrac{250(1-0.9^5)}{1-0.9}=1023.775$

394

Chapter 9: Exponents and Polynomials

$$\begin{array}{r}n^4+n^3+n^2+n+1\\-n+1\overline{)-n^5+1}\end{array}$$

$$\underline{-\left(-n^5+n^4\right)}$$
$$-n^4$$
$$\underline{-\left(-n^4+n^3\right)}$$
$$-n^3$$
$$\underline{-\left(-n^3+n^2\right)}$$
$$-n^2$$
$$\underline{-\left(-n^2+n\right)}$$
$$-n+1$$
$$\underline{-\left(-n+1\right)}$$
$$0$$

$$c(n) = 250\left(n^4+n^3+n^2+n+1\right)$$
$$c(0.9) = 250\left(.9^4+.9^3+.9^2+.9+1\right)$$
$$= 250(4.0951) = 1023.775$$

The total rental cost for 5 weeks would be roughly $1023.78.

9.5 Calculator Exercises

1. Pressure from woman: $\dfrac{110}{2\pi r^2}$

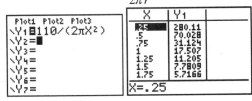

The woman exerts roughly the same pressure as the elephant if the radius of her heel is 1 inch.

2. Pressure from man: $\dfrac{175}{2\pi r^2}$

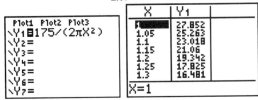

The man exerts roughly the same pressure as the elephant if the radius of his heel is about 1.25 inches.

Chapter 9 Section-By-Section Review

1. $-5c^2 = -5 \cdot c \cdot c$

2. $(-5c)^2 = (-5c)(-5c)$

3. $4(x+y)^2 z^0 = 4(x+y)^2(1) = 4(x+y)(x+y)$

4. $\left(\dfrac{2}{3}\right)^4 x^2 = \dfrac{2}{3} \cdot \dfrac{2}{3} \cdot \dfrac{2}{3} \cdot \dfrac{2}{3} \cdot x \cdot x$

5. $a^2 \cdot a^7 = a^{2+7} = a^9$

6. $(p+q)^3 (p+q) = (p+q)^{3+1} = (p+q)^4$

7. $\dfrac{2}{3}b^3 \cdot b \cdot b^4 = \dfrac{2}{3}b^{3+1+4} = \dfrac{2}{3}b^8$

8. $\dfrac{t^{12}}{t^9} = t^{12-9} = t^3$

SSM: Experiencing Introductory and Intermediate Algebra

9. $\dfrac{24a^4}{3a^2} = \dfrac{24}{3} \cdot \dfrac{a^4}{a^2} = 8a^{4-2} = 8a^2$

10. $\dfrac{(x+3y)^5}{(x+3y)^4} = (x+3y)^{5-4} = (x+3y)$

11. $(cd)^{22} = c^{22}d^{22}$

12. $(-2a)^4 = (-2)^4 a^4 = 16a^4$

13. $(-2a)^5 = (-2)^5 a^5 = -32a^5$

14. $(3x^3)^2 = 3^2(x^3)^2 = 9x^{3\cdot 2} = 9x^6$

15. $\left[(a+b)^2\right]^5 = (a+b)^{2\cdot 5} = (a+b)^{10}$

16. $(a^3)^0 = 1$

17. $\left(\dfrac{-2x}{3z}\right)^3 = \dfrac{(-2)^3 x^3}{3^3 z^3} = -\dfrac{8x^3}{27z^3}$

18. $\left(\dfrac{-4d}{e}\right)^3 = \dfrac{(-4)^3 d^3}{e^3} = -\dfrac{64d^3}{e^3}$

19. $-2xy^2 \cdot 4x^3y^5 = -8x^{1+3}y^{2+5} = -8x^4y^7$

20. $\dfrac{72x^4y^5z^3}{-24x^2y^4z^3} = \dfrac{72}{-24}x^{4-2}y^{5-4}z^{3-3}$
    $= -3x^2yz^0 = -3x^2y$

21. $\dfrac{(x^2y^3)^4}{x^3y^5} = \dfrac{(x^2)^4(y^3)^4}{x^3y^5} = \dfrac{x^{2\cdot 4}y^{3\cdot 4}}{x^3y^5}$
    $= \dfrac{x^8y^{12}}{x^3y^5} = x^{8-3}y^{12-5} = x^5y^7$

22. $\left(\dfrac{2p^2q}{pq^2}\right)^3 = \left(2p^{2-1}q^{1-2}\right)^3 = \left(2pq^{-1}\right)^3$
    $= 2^3 p^3 q^{-1\cdot 3} = \dfrac{4p^3}{q^3}$

23. $\left[\left(\dfrac{m}{4n}\right)^2\right]^2 = \left(\dfrac{m}{4n}\right)^{2\cdot 2} = \left(\dfrac{m}{4n}\right)^4$
    $= \dfrac{m^4}{(4)^4 n^4} = \dfrac{m^4}{256n^4}$

24. Let x = length of original size

    original area: x^2

    reduced area: $\left(\dfrac{1}{4}x\right)^2$

    $\left(\dfrac{1}{4}x\right)^2 = \left(\dfrac{1}{4}\right)^2 x^2 = \dfrac{1}{16}x^2$

    The reduced area is one-sixteenth the size of the original area.

25. Let r = original radius

    original area: πr^2

    enlarged area: $\pi(2r)^2$

    $\pi(2r)^2 = \pi \cdot 2^2 \cdot r^2 = 4\pi r^2$

    The enlarged garden has 4 times the area of the original garden.

    $4\pi(6)^2 = 144\pi \approx 452.39$

    If the original garden had a radius of 6 feet, the enlarged garden would have an area of about 452.39 ft^2.

26. $\dfrac{3^{-2}h^{-4}}{k^{-2}} = \dfrac{k^2}{3^2 h^4} = \dfrac{k^2}{9h^4}$

27. $c^8 d^{-5} = \dfrac{c^8}{d^5}$

28. $(-b^2)(-3b^{-5}) = 3b^{2+(-5)} = 3b^{-3} = \dfrac{3}{b^3}$

29. $\dfrac{5z^{-3}}{15z^{-7}} = \dfrac{5}{15}z^{-3-(-7)} = \dfrac{1}{3}z^4$

30. $(5a^{-4}b^2)^{-3} = 5^{-3}a^{(-4)(-3)}b^{2(-3)}$
    $= 5^{-3}a^{12}b^{-6} = \dfrac{a^{12}}{5^3 b^6} = \dfrac{a^{12}}{125b^6}$

31. $\left(\dfrac{2}{3}p^3q^{-2}\right)\left(\dfrac{3}{5}p^{-5}q^{-3}\right) = \dfrac{2}{5}p^{3+(-5)}q^{-2+(-3)}$
$= \dfrac{2}{5}p^{-2}q^{-5} = \dfrac{2}{5p^2q^5}$

32. $\dfrac{144x^{-4}y^{-3}}{12x^2y^{-4}} = \dfrac{144}{12}x^{-4-2}y^{-3-(-4)}$
$= 12x^{-6}y = \dfrac{12y}{x^6}$

33. $\left(\dfrac{27a^5b^{-3}}{9a^2b^7}\right)^2 = \left(3a^{5-2}b^{-3-7}\right)^2 = \left(3a^3b^{-10}\right)^2$
$= 3^2\left(a^3\right)^2\left(b^{-10}\right)^2 = 9a^6b^{-20}$
$= \dfrac{9a^6}{b^{20}}$

34. $\left(\dfrac{a^{-3}}{b^{-4}}\right)^{-2} = \left(\dfrac{b^{-4}}{a^{-3}}\right)^2 = \left(\dfrac{a^3}{b^4}\right)^2 = \dfrac{\left(a^3\right)^2}{\left(b^4\right)^2} = \dfrac{a^6}{b^8}$

35. $\left(\dfrac{2a^{-2}b}{24a^5b^{-2}}\right)^{-2} = \left(\dfrac{24a^5b^{-2}}{2a^{-2}b}\right)^2$
$= \left(\dfrac{24}{2}a^{5-(-2)}b^{-2-1}\right)^2 = \left(12a^7b^{-3}\right)^2$
$= 12^2\left(a^7\right)^2\left(b^{-3}\right)^2 = 144a^{14}b^{-6}$
$= \dfrac{144a^{14}}{b^6}$

36. $(3.37\times 10^1)(5.9\times 10^{12}) = 3.37\cdot 10^1\cdot 5.9\cdot 10^{12}$
$= 19.883\cdot 10^{13} = 1.9883\cdot 10\cdot 10^{13} = 1.9883\cdot 10^{14}$
$= 1.9883\times 10^{14}$

The star Pollux is about 1.9883×10^{14} miles from Earth.

37. $(5x^4+3x^3+6x-3)+(4x^3+5x^2+7)$
$= 5x^4+3x^3+6x-3+4x^3+5x^2+7$
$= 5x^4+7x^3+5x^2+6x+4$

38. $(3.57z^3-2.08z^2+8.77z-1.99)+(4.73-2.98z+5.64z^2)$
$= 3.57z^3-2.08z^2+8.77z-1.99+4.73-2.98z+5.64z^2$
$= 3.57z^3+3.56z^2+5.79z+2.74$

39. $(5a^4+a^3+a^2+a+1)-(a^4+2a^3+3a^2+4a+5)$
$= 5a^4+a^3+a^2+a+1-a^4-2a^3-3a^2-4a-5$
$= 4a^4-a^3-2a^2-3a-4$

40. $(65z^4+27z^2+36)-(16z^3+8z+12)$
$= 65z^4+27z^2+36-16z^3-8z-12$
$= 65z^4-16z^3+27z^2-8z+24$

SSM: Experiencing Introductory and Intermediate Algebra

41. $\left(\dfrac{5}{8}b^4 + \dfrac{7}{8}b^3 - \dfrac{3}{4}b^2 + \dfrac{1}{2}b - \dfrac{1}{4}\right) - \left(\dfrac{1}{2}b^4 + \dfrac{3}{8}b^2 + \dfrac{1}{8}b\right)$

$= \dfrac{5}{8}b^4 + \dfrac{7}{8}b^3 - \dfrac{3}{4}b^2 + \dfrac{1}{2}b - \dfrac{1}{4} - \dfrac{1}{2}b^4 - \dfrac{3}{8}b^2 - \dfrac{1}{8}b$

$= \left(\dfrac{5}{8} - \dfrac{1}{2}\right)b^4 + \dfrac{7}{8}b^3 + \left(-\dfrac{3}{4} - \dfrac{3}{8}\right)b^2 + \left(\dfrac{1}{2} - \dfrac{1}{8}\right)b - \dfrac{1}{4}$

$= \dfrac{1}{8}b^4 + \dfrac{7}{8}b^3 - \dfrac{9}{8}b^2 + \dfrac{3}{8}b - \dfrac{1}{4}$

42. Let x = number of items

  a. $C(x) = 10 + 3.5x$

  b. $R(x) = 10x$

  c. $P(x) = R(x) - C(x)$
  $= (10x) - (10 + 3.5x)$
  $= 10x - 10 - 3.5x$
  $= 6.5x - 10$

  d. $P(10) = 6.5(10) - 10 = 65 - 10 = 55$
  $P(25) = 6.5(25) - 10 = 162.5 - 10 = 152.5$
  The profits for 10 items and 25 items are $55 and $152.50, respectively.

43. $c(x) = f(x) + g(x)$
$= (150 + 25x) + (25 + 4.5x)$
$= 150 + 25x + 25 + 4.5x$
$= 175 + 29.5x$
$r(x) = 125x$
$p(x) = r(x) - c(x)$
$= (125x) - (175 + 29.5x)$
$= 125x - 175 - 29.5x$
$= 95.5x - 175$

44. $-3a^3b(5ab^3) = -15a^{3+1}b^{1+3} = -15a^4b^4$

45. $(-6.9x^3z^4)(3.4xz^3) = -23.46x^{3+1}z^{4+3}$
$= -23.46x^4z^7$

46. $6x^3(3x^2 + 2x - 7)$
$= 6x^3(3x^2) + 6x^3(2x) + 6x^3(-7)$
$= 18x^{3+2} + 12x^{3+1} - 42x^3$
$= 18x^5 + 12x^4 - 42x^3$

47. $-7a(3a^6 + a^4 - 2a^2)$
$= -7a(3a^6) - 7a(a^4) - 7a(-2a^2)$
$= -21a^{1+6} - 7a^{1+4} + 14a^{1+2}$
$= -21a^7 - 7a^5 + 14a^3$

48. $4a^2(2a - 3b + c)$
$= 4a^2(2a) + 4a^2(-3b) + 4a^2(c)$
$= 8a^{2+1} - 12a^2b + 4a^2c$
$= 8a^3 - 12a^2b + 4a^2c$

49. $(p+6)(p-9)$
$= p(p-9) + 6(p-9)$
$= p^2 - 9p + 6p - 54$
$= p^2 - 3p - 54$

50. $(5x-2)(x+11)$
$= 5x(x+11) - 2(x+11)$
$= 5x^2 + 55x - 2x - 22$
$= 5x^2 + 53x - 22$

51. $(y-1.8)(y+3.4)$
$= y(y+3.4)-1.8(y+3.4)$
$= y^2 +3.4y-1.8y-6.12$
$= y^2 +1.6y-6.12$

52. $(2x+1)(3x^2+5x-4)$
$= 2x(3x^2+5x-4)+1(3x^2+5x-4)$
$= 6x^3 +10x^2 -8x+3x^2 +5x-4$
$= 6x^3 +13x^2 -3x-4$

53. $(a-3)(a^2+3a+9)$
$= a(a^2+3a+9)-3(a^2+3a+9)$
$= a^3 +3a^2 +9a-3a^2 -9a-27$
$= a^3 -27$

54. $(x^2+2x+3)(x^2-x+5)$
$= x^2(x^2-x+5)+2x(x^2-x+5)+3(x^2-x+5)$
$= x^4 -x^3 +5x^2 +2x^3 -2x^2 +10x+3x^2 -3x+15$
$= x^4 +x^3 +6x^2 +7x+15$

55. $(z^2+2z-3)^2$
$= (z^2+2z-3)(z^2+2z-3)$
$= z^2(z^2+2z-3)+2z(z^2+2z-3)-3(z^2+2z-3)$
$= z^4 +2z^3 -3z^2 +2z^3 +4z^2 -6z-3z^2 -6z+9$
$= z^4 +4z^3 -2z^2 -12z+9$

56. $(b-4)^3$
$= (b-4)(b-4)(b-4)$
$= [b(b-4)-4(b-4)](b-4)$
$= (b^2 -4b-4b+16)(b-4)$
$= (b^2 -8b+16)(b-4)$
$= (b^2 -8b+16)b+(b^2 -8b+16)(-4)$
$= b^3 -8b^2 +16b-4b^2 +32b-64$
$= b^3 -12b^2 +48b-64$

57. $(2x-5)(2x+5) = (2x)^2 -(5)^2 = 4x^2 -25$

58. $\left(\frac{4}{5}x-\frac{1}{2}\right)\left(\frac{4}{5}x+\frac{1}{2}\right) = \left(\frac{4}{5}x\right)^2 -\left(\frac{1}{2}\right)^2$
$= \frac{16}{25}x^2 -\frac{1}{4}$

59. $(z^2-10)(z^2+10) = (z^2)^2 -(10)^2 = z^4 -100$

60. $(y+9)^2 = y^2 +2(y)(9)+9^2 = y^2 +18y+81$

61. $(3x-5)^2 = (3x)^2 -2(3x)(5)+(5)^2$
$= 9x^2 -30x+25$

62. $(x^3+3)^2 = (x^3)^2 + 2(x^3)(3) + 3^2$
$= x^6 + 6x^3 + 9$

63. Let x = height of the box

   a. length: $x+3$
      width: $x-3$
      height: x

   b. Volume:
      $x(x+3)(x-3)$
      $= x(x^2 - 3^2) = x(x^2 - 9) = x(x^2) - x(9)$
      $= x^3 - 9x$
      The volume is $x^3 - 9x$ in^3.

   c. Surface Area:
      $2x(x+3) + 2x(x-3) + 2(x+3)(x-3)$
      $= 2x^2 + 6x + 2x^2 - 6x + 2(x^2 - 3^2)$
      $= 4x^2 + 2(x^2 - 9)$
      $= 4x^2 + 2x^2 - 18$
      $= 6x^2 - 18$
      The surface area is $6x^2 - 18$ in^2.

64. Let x = original width

   a. Current area:
      original length: $3x$
      $x(3x) = 3x^2$
      The current area is $3x^2$ ft^2.

   b. Enlarged area:
      New width: $2x$
      New length: $3x+9$
      $(2x)(3x+9)$
      $= 2x(3x) + 2x(9) = 6x^2 + 18x$
      The enlarged area is $6x^2 + 18x$ ft^2.

   c. Difference:
      $(6x^2 + 18x) - (3x^2)$
      $= 6x^2 + 18x - 3x^2$
      $= 3x^2 + 18x$
      The difference in the areas is $3x^2 + 18x$ ft^2.

65. $\dfrac{144x^6 y^3 z^4}{2x^5 y^2 z^6} = \dfrac{144}{2} x^{6-5} y^{3-2} z^{4-6}$
    $= 72xyz^{-2} = \dfrac{72xy}{z^2}$

66. $\dfrac{6ab^2 + 18a^2 b}{3a^2 b^2} = \dfrac{6ab^2}{3a^2 b^2} + \dfrac{18a^2 b}{3a^2 b^2}$
    $= 2a^{1-2} b^{2-2} + 6a^{2-2} b^{1-2} = 2a^{-1} b^0 + 6a^0 b^{-1}$
    $= \dfrac{2}{a} + \dfrac{6}{b}$

67. $\dfrac{15b^4 - 10b^3 - 25b^2 + 5b}{-5b}$
    $= \dfrac{15b^4}{-5b} + \dfrac{-10b^3}{-5b} + \dfrac{-25b^2}{-5b} + \dfrac{5b}{-5b}$
    $= -3b^{4-1} + 2b^{3-1} + 5b^{2-1} - b^{1-1}$
    $= -3b^3 + 2b^2 + 5b - 1$

68. $\ \ \ \ \ \ \ 3x + 8$
    $5x - 7 \overline{)15x^2 + 19x - 56}$
    $\underline{-(15x^2 - 21x)}$
    $\ 40x - 56$
    $\underline{-(40x - 56)}$
    $\ 0$
    The quotient is $3x + 8$.

Chapter 9: Exponents and Polynomials

69.
$$\begin{array}{r} 4x-5 \\ 2x+3\overline{)8x^2+2x-19} \\ \underline{-(8x^2+12x)} \\ -10x-19 \\ \underline{-(-10x-15)} \\ -4 \end{array}$$

The quotient is $4x-5-\dfrac{4}{2x+3}$.

70.
$$\begin{array}{r} x^2-7x+5 \\ x+7\overline{)x^3-44x+35} \\ \underline{-(x^3+7x^2)} \\ -7x^2-44x \\ \underline{-(-7x^2-49x)} \\ 5x+35 \\ \underline{-(5x+35)} \\ 0 \end{array}$$

The quotient is x^2-7x+5.

71.
$$\begin{array}{r} z^2+2z+4 \\ z-2\overline{)z^3-8} \\ \underline{-(z^3-2z^2)} \\ 2z^2 \\ \underline{-(2z^2-4z)} \\ 4z-8 \\ \underline{-(4z-8)} \\ 0 \end{array}$$

The quotient is z^2+2z+4.

72.
$$\begin{array}{r} 4a-5 \\ 4a+5\overline{)16a^2-25} \\ \underline{-(16a^2+20a)} \\ -20a-25 \\ \underline{-(-20a-25)} \\ 0 \end{array}$$

The quotient is $4a-5$.

73.
$$\begin{array}{r} 2x+9 \\ 2x+9\overline{)4x^2+36x+81} \\ \underline{-(4x^2+18x)} \\ 18x+81 \\ \underline{-(18x+81)} \\ 0 \end{array}$$

The quotient is $2x+9$.

74. **a.** $c(x) = 35 + 1.25x$

   b. $r(x) = 3.75x$

   c. $p(x) = r(x) - c(x)$
   $= 3.75x - (35 + 1.25x)$
   $= 3.75x - 35 - 1.25x$
   $= 2.5x - 35$

   d. $a(x) = \dfrac{p(x)}{x} = \dfrac{2.5x-35}{x}$
   $= \dfrac{2.5x}{x} - \dfrac{35}{x} = 2.5 - \dfrac{35}{x}$

   e. $a(25) = 2.5 - \dfrac{35}{25} = 1.1$

   The average profit for the produce stand owner is $1.10.

75. $c(t) = \dfrac{2000(1-t^4)}{1-t}$

   $c(0.85) = \dfrac{2000(1-0.85^4)}{1-0.85} = 6373.25$

SSM: Experiencing Introductory and Intermediate Algebra

$$\begin{array}{r}t^3+t^2+t+1\\-t+1{\overline{\smash{\big)}\,-t^4+1}}\\\underline{-\left(-t^4+t^3\right)}\\-t^3\\\underline{-\left(-t^3+t^2\right)}\\-t^2\\\underline{-\left(-t^2+t\right)}\\-t+1\\\underline{-\left(-t+1\right)}\\0\end{array}$$

$c(t) = 2000\left(t^3 + t^2 + t + 1\right)$

$c(0.85) = 2000\left(.85^3 + .85^2 + .85 + 1\right)$

$ = 2000(3.186625)$

$ = 6373.25$

The total charge for four years was $6373.25.

Chapter 9 Chapter Review

1. $m^{-7}n^5 = \dfrac{n^5}{m^7}$

2. $\left(3a^{-5}b^4\right)^{-2} = 3^{-2}a^{(-5)(-2)}b^{4(-2)}$

   $= \dfrac{1}{3^2}a^{10}b^{-8} = \dfrac{a^{10}}{9b^8}$

3. $\left(\dfrac{21x^4y^{-3}}{7x^3y^2}\right)^2 = \left(3x^{4-3}y^{-3-2}\right)^2 = \left(3xy^{-5}\right)^2$

   $= \left(\dfrac{3x}{y^5}\right)^2 = \dfrac{3^2x^2}{y^{5(2)}} = \dfrac{9x^2}{y^{10}}$

4. $\dfrac{5^{-3}t^3}{4^{-3}s^4} = \dfrac{4^3t^3}{5^3s^4} = \dfrac{64t^3}{125s^4}$

5. $\dfrac{128a^{-3}b^{-2}}{32a^2b^{-5}} = 4a^{-3-2}b^{-2-(-5)} = 4a^{-5}b^3 = \dfrac{4b^3}{a^5}$

6. $\left(\dfrac{2}{5}x^2y^{-3}\right)\left(\dfrac{5}{7}x^{-5}y^{-1}\right) = \dfrac{2}{7}x^{2+(-5)}y^{-3+(-1)}$

   $= \dfrac{2}{7}x^{-3}y^{-4} = \dfrac{2}{7x^3y^4}$

7. $\left(\dfrac{15h^2k^5}{3h^{-1}k^7}\right)^{-2} = \left(5h^{2-(-1)}k^{5-7}\right)^{-2} = \left(5h^3k^{-2}\right)^{-2}$

   $= \left(\dfrac{5h^3}{k^2}\right)^{-2} = \left(\dfrac{k^2}{5h^3}\right)^2 = \dfrac{k^{2(2)}}{5^2h^{3(2)}} = \dfrac{k^4}{25h^6}$

8. $s^2 \cdot s^8 = s^{2+8} = s^{10}$

9. $(m+n)^5(m+n) = (m+n)^{5+1} = (m+n)^6$

10. $\left(3y^2\right)^3 = 3^3y^{2(3)} = 27y^6$

11. $\dfrac{c^9}{c^6} = c^{9-6} = c^3$

12. $\left(x^0\right)^8 = (1)^8 = 1$

13. $-5a^2b^3 \cdot 3ab^4 = -15a^{2+1}b^{3+4} = -15a^3b^7$

14. $\dfrac{1}{2}z^4 \cdot z \cdot z^2 = \dfrac{1}{2}z^{4+1+2} = \dfrac{1}{2}z^7$

15. $\left(b^5\right)^7 = b^{5\cdot7} = b^{35}$

16. $(-3d)^6 = (-3)^6d^6 = 729d^6$

17. $(-3d)^3 = (-3)^3d^3 = -27d^3$

18. $\left(x^4y^5z\right)^4 = \left(x^4\right)^4\left(y^5\right)^4z^4 = x^{16}y^{20}z^4$

19. $\dfrac{25b^5}{5b^3} = 5b^{5-3} = 5b^2$

20. $\dfrac{(a+2b)^9}{(a+2b)^2} = (a+2b)^{9-2} = (a+2b)^7$

21. $\dfrac{64p^5q^7}{-4p^2q} = -16p^{5-2}q^{7-1} = -16p^3q^6$

22. $\left[(4x)^2\right]^3 = (4x)^{2(3)} = (4x)^6 = 4^6 x^6 = 4096 x^6$

23. $\left[\left(\dfrac{c}{2d}\right)^3\right]^2 = \left(\dfrac{c}{2d}\right)^{3(2)} = \left(\dfrac{c}{2d}\right)^6$

$= \dfrac{c^6}{2^6 d^6} = \dfrac{c^6}{64 d^6}$

24. $\left(\dfrac{-3a^2}{4b}\right)^3 = \dfrac{(-3)^3 (a^2)^3}{4^3 b^3} = -\dfrac{27 a^6}{64 b^3}$

25. $\dfrac{(3a^3)^3}{54 a^7} = \dfrac{3^3 a^9}{54 a^7} = \dfrac{27}{54} a^{9-7} = \dfrac{1}{2} a^2$

26. $\left(\dfrac{21 a^6 b^2}{7 a^3 b^6}\right)^2 = (3 a^{6-3} b^{2-6})^2 = (3 a^3 b^{-4})^2$

$= \left(\dfrac{3 a^3}{b^4}\right)^2 = \dfrac{3^2 a^{3 \cdot 2}}{b^{4 \cdot 2}} = \dfrac{9 a^6}{b^8}$

27. $(2y^2 + 4y - 3) + (8y^2 - 12y + 15)$

$= 2y^2 + 4y - 3 + 8y^2 - 12y + 15$

$= 10y^2 - 8y + 12$

28. $\left(\dfrac{3}{8}a^3 + \dfrac{3}{4}a^2 - \dfrac{5}{8}a + \dfrac{1}{4}\right) + \left(\dfrac{3}{4}a^3 + \dfrac{7}{16}a - \dfrac{5}{8}a^2 + \dfrac{11}{16}\right)$

$= \dfrac{3}{8}a^3 + \dfrac{3}{4}a^2 - \dfrac{5}{8}a + \dfrac{1}{4} + \dfrac{3}{4}a^3 + \dfrac{7}{16}a - \dfrac{5}{8}a^2 + \dfrac{11}{16}$

$= \left(\dfrac{3}{8} + \dfrac{3}{4}\right)a^3 + \left(\dfrac{3}{4} - \dfrac{5}{8}\right)a^2 + \left(-\dfrac{5}{8} + \dfrac{7}{16}\right)a + \left(\dfrac{1}{4} + \dfrac{11}{16}\right)$

$= \dfrac{9}{8}a^3 + \dfrac{1}{8}a^2 - \dfrac{3}{16}a + \dfrac{15}{16}$

29. $(4.9z^3 - 6.82z^2 + 12z - 11.07) + (4.6 - 1.83z + 4.9z^2)$

$= 4.9z^3 - 6.82z^2 + 12z - 11.07 + 4.6 - 1.83z + 4.9z^2$

$= 4.9z^3 - 1.92z^2 + 10.17z - 6.47$

30. $(2x^3 + 6x^2 - 9x + 13) - (4x^3 + 17x^2 - x + 6)$

$= 2x^3 + 6x^2 - 9x + 13 - 4x^3 - 17x^2 + x - 6$

$= -2x^3 - 11x^2 - 8x + 7$

31. $(6a^4 + 3a^2 + 4a + 5) - (5a^4 + 4a^3 + 3a^2 + 2a + 1)$

$= 6a^4 + 3a^2 + 4a + 5 - 5a^4 - 4a^3 - 3a^2 - 2a - 1$

$= a^4 - 4a^3 + 2a + 4$

32. $(117z^4 + 43z^2 + 88) - (18z^3 + 50z + 32)$

$= 117z^4 + 43z^2 + 88 - 18z^3 - 50z - 32$

$= 117z^4 - 18z^3 + 43z^2 - 50z + 56$

33. $11x^5y^6(13x^2y^6) = 143x^{5+2}y^{6+6} = 143x^7y^{12}$

34. $-6x(2x^4 - 4x^3 + 6x^2 + 8x - 10)$
    $= -12x^{1+4} + 24x^{1+3} - 36x^{1+2} - 48x^{1+1} + 60x$
    $= -12x^5 + 24x^4 - 36x^3 - 48x^2 + 60x$

35. $9a(4a^5 + 2a^3 - 3a)$
    $= 36a^{1+5} + 18a^{1+3} - 27a^{1+1}$
    $= 36a^6 + 18a^4 - 27a^2$

36. $(m-11)(m+11) = m(m+11) - 11(m+11)$
    $= m^2 + 11m - 11m - 121$
    $= m^2 - 121$

37. $(z-8)^2 = z^2 - 2(z)(8) + 8^2$
    $= z^2 - 16z + 64$

38. $(4a-7)(a+13) = 4a(a+13) - 7(a+13)$
    $= 4a^2 + 52a - 7a - 91$
    $= 4a^2 + 45a - 91$

39. $(13-x)(13+x) = (13)^2 - x^2 = 169 - x^2$

40. $(b^3+4)^2 = (b^3)^2 + 2(b^3)(4) + 4^2$
    $= b^6 + 8b^3 + 16$

41. $(t+2)^3$
    $= (t+2)(t+2)(t+2)$
    $= [t(t+2) + 2(t+2)](t+2)$
    $= (t^2 + 2t + 2t + 4)(t+2)$
    $= (t^2 + 4t + 4)(t+2)$
    $= (t^2 + 4t + 4)t + (t^2 + 4t + 4)2$
    $= t^3 + 4t^2 + 4t + 2t^2 + 8t + 8$
    $= t^3 + 6t^2 + 12t + 8$

42. $(7x-2)(x^2 + 4x - 3)$
    $= 7x(x^2 + 4x - 3) - 2(x^2 + 4x - 3)$
    $= 7x^3 + 28x^2 - 21x - 2x^2 - 8x + 6$
    $= 7x^3 + 26x^2 - 29x + 6$

43. $(p+q+r)^2 = (p+q+r)(p+q+r)$
    $= p(p+q+r) + q(p+q+r) + r(p+q+r)$
    $= p^2 + pq + pr + pq + q^2 + qr + pr + qr + r^2$
    $= p^2 + q^2 + r^2 + 2pq + 2pr + 2qr$

44. $(b-4)(b^2 + 4b + 16)$
    $= b(b^2 + 4b + 16) - 4(b^2 + 4b + 16)$
    $= b^3 + 4b^2 + 16b - 4b^2 - 16b - 64$
    $= b^3 - 64$

45. $(y+5)(y-9)$
    $= y(y-9) + 5(y-9)$
    $= y^2 - 9y + 5y - 45$
    $= y^2 - 4y - 45$

46. $(x^2 + 3x + 1)(x^2 - x + 4)$
    $= x^2(x^2 - x + 4) + 3x(x^2 - x + 4) + 1(x^2 - x + 4)$
    $= x^4 - x^3 + 4x^2 + 3x^3 - 3x^2 + 12x + x^2 - x + 4$
    $= x^4 + 2x^3 + 2x^2 + 11x + 4$

47. $\dfrac{81x^5y^7z}{27x^2y^6z^2} = \dfrac{81}{27}x^{5-2}y^{7-6}z^{1-2}$
    $= 3x^3y^1z^{-1} = \dfrac{3x^3y}{z}$

48. $\dfrac{14b^4 - 21b^3 - 35b^2 + 28b}{-7b}$

$= \dfrac{14b^4}{-7b} + \dfrac{-21b^3}{-7b} + \dfrac{-35b^2}{-7b} + \dfrac{28b}{-7b}$

$= -2b^{4-1} + 3b^{3-1} + 5b^{2-1} - 4b^{1-1}$

$= -2b^3 + 3b^2 + 5b - 4$

49. $\dfrac{16cd^2 + 8c^2d}{4c^2d^2} = \dfrac{16cd^2}{4c^2d^2} + \dfrac{8c^2d}{4c^2d^2}$

$= 4c^{1-2}d^{2-2} + 2c^{2-2}d^{1-2}$

$= 4c^{-1}d^0 + 2c^0d^{-1}$

$= \dfrac{4}{c} + \dfrac{2}{d}$

50. $3x - 5 \overline{\smash{\big)}\, 24x^2 - 37x - 5}$ with quotient $8x + 1$

$\underline{-(24x^2 - 40x)}$

$3x - 5$

$\underline{-(3x - 5)}$

0

The quotient is $8x + 1$.

51. $2x + 5 \overline{\smash{\big)}\, 2x^2 - 13x - 50}$ with quotient $x - 9$

$\underline{-(2x^2 + 5x)}$

$-18x - 50$

$\underline{-(-18x - 45)}$

-5

The quotient is $x - 9 - \dfrac{5}{2x + 5}$.

52. $3a - 2 \overline{\smash{\big)}\, 27a^3 - 8}$ with quotient $9a^2 + 6a + 4$

$\underline{-(27a^3 - 18a^2)}$

$18a^2$

$\underline{-(18a^2 - 12a)}$

$12a - 8$

$\underline{-(12a - 8)}$

0

The quotient is $9a^2 + 6a + 4$.

53. $5x - 1 \overline{\smash{\big)}\, 25x^3 + 39x - 8}$ with quotient $5x^2 + x + 8$

$\underline{-(25x^3 - 5x^2)}$

$5x^2 + 39x$

$\underline{-(5x^2 - x)}$

$40x - 8$

$\underline{-(40x - 8)}$

0

The quotient is $5x^2 + x + 8$.

54. $-3d^4 = -3 \cdot d \cdot d \cdot d \cdot d$

55. $(-3d)^4 = (-3d)(-3d)(-3d)(-3d)$

56. a. $c(x) = 35 + 2.75x$

b. $r(x) = 10x$

c. $p(x) = r(x) - c(x)$

$= 10x - (35 + 2.75x)$

$= 10x - 35 - 2.75x$

$= 7.25x - 35$

d. $a(x) = \dfrac{p(x)}{x} = \dfrac{7.25x - 35}{x} = \dfrac{7.25x}{x} + \dfrac{-35}{x}$

$= 7.25 - \dfrac{35}{x}$

SSM: Experiencing Introductory and Intermediate Algebra

e. $a(10) = 7.25 - \dfrac{35}{10} = 7.25 - 3.5 = 3.75$

The average profit per child would be $3.75.

57. $m(s) = \dfrac{25(1-s^5)}{1-s}$

$m(0.95) = \dfrac{25(1-0.95^5)}{1-0.95} \approx 113.11$

$$
\begin{array}{r}
s^4 + s^3 + s^2 + s + 1 \\
-s+1 \overline{) -s^5 + 1} \\
\underline{-(-s^5 + s^4)} \\
-s^4 \\
\underline{-(-s^4 + s^3)} \\
-s^3 \\
\underline{-(-s^3 + s^2)} \\
-s^2 \\
\underline{-(-s^2 + s)} \\
-s + 1 \\
\underline{-(-s+1)} \\
0
\end{array}
$$

$m(s) = 25(s^4 + s^3 + s^2 + s + 1)$

$m(0.95) = 25(.95^4 + .95^3 + .95^2 + .95 + 1)$

$\approx 25(4.52438) = 113.1095 \approx 113.11$

Reggie received a total of $113.11 for the tutoring session.

58. $1600(5.9 \times 10^{12}) = (1600)(5.9)(10^{12})$
$= (9440)(10^{12}) = (9.44)(10^3)(10^{12})$
$= 9.44(10^{15}) = 9.44 \times 10^{15}$

1,600 light-years is roughly 9.44×10^{15} miles.

59. Let x = original height.

a. Length: $2x + 3$
Width: x
Height: x

b. Volume:
$(2x+3)(x)(x) = (2x+3)x^2 = 2x^3 + 3x^2$

c. Surface Area:
$2(2x+3)(x) + 2(2x+3)(x) + 2(x)(x)$
$2x(2x+3) + 2x(2x+3) + 2x^2$
$4x^2 + 6x + 4x^2 + 6x + 2x^2$
$10x^2 + 12x$

d. $2(8)^3 + 3(8)^2 = 1024 + 3(64) = 1216$

The volume of the box is 1216 in^3.

e. $10(8)^2 + 12(8) = 640 + 96 = 736$

The surface area of the box is 736 in^2.

60. a. Area:
$\dfrac{1}{2}(2x+5)(x) = \left(x + \dfrac{5}{2}\right)(x) = x^2 + \dfrac{5}{2}x$

b. $(12)^2 + \dfrac{5}{2}(12) = 144 + 30 = 174$

The area of the triangle is 174 cm^2.

61. a. Original area:
x^2

b. Enlarged area:
$(x+5)(2x) = 2x^2 + 10x$

c. Difference:
$(2x^2 + 10x) - x^2 = x^2 + 10x$

Chapter 9 Test

1. $(2x-1)(2x-1)^8 = (2x-1)^9$

2. $3x^2 y^0 z^{-3} = \dfrac{3x^2}{z^3}$

3. $m^{-9} n^4 = \dfrac{n^4}{m^9}$

4. $(-2x^2 y^{-1})^{-3} = \left(\dfrac{-2x^2}{y}\right)^{-3} = \left(\dfrac{y}{-2x^2}\right)^3$

   $= \dfrac{y^3}{(-2)^3 (x^2)^3} = \dfrac{y^3}{-8x^6}$

5. $\dfrac{a^{12}}{a^5} = a^{12-5} = a^7$

6. $\left[\dfrac{(2p)^3}{3q}\right]^{-2} = \left[\dfrac{3q}{(2p)^3}\right]^2 = \dfrac{3^2 q^2}{(2p)^6}$

   $= \dfrac{9q^2}{2^6 p^6} = \dfrac{9q^2}{64 p^6}$

7. $\dfrac{24 a^2 b^{-3} c}{3ab^2 c^{-2}} = 8 a^{2-1} b^{-3-2} c^{1-(-2)}$

   $= 8 a^1 b^{-5} c^3 = \dfrac{8ac^3}{b^5}$

8. $\dfrac{(a^2 b^4)^3}{a^2 b^5 c} = \dfrac{(a^2)^3 (b^4)^3}{a^2 b^5 c} = \dfrac{a^6 b^{12}}{a^2 b^5 c} = \dfrac{a^4 b^7}{c}$

9. $(3y^2 + 16 - 7y + 5y^3) + (7y + 6y^2 + 4y^4 - 11)$

   $= 3y^2 + 16 - 7y + 5y^3 + 7y + 6y^2 + 4y^4 - 11$

   $= 4y^4 + 5y^3 + 9y^2 + 5$

10. $(7x^5 + 23x^3 + 17x^2 - 39) - (2x^4 - 4x^2 - 2x - 9)$

    $= 7x^5 + 23x^3 + 17x^2 - 39 - 2x^4 + 4x^2 + 2x + 9$

    $= 7x^5 - 2x^4 + 23x^3 + 21x^2 + 2x - 30$

11. $(-2p^3 q^{-5} r^2)(5.7 p^6 q^7 r)$

    $= -11.4 p^{3+6} q^{-5+7} r^{2+1}$

    $= -11.4 p^9 q^2 r^3$

12. $-4t(2t^3 - 3t^2 - 8t + 6)$

    $= -4t(2t^3) - 4t(-3t^2) - 4t(-8t) - 4t(6)$

    $= -8t^4 + 12t^3 + 32t^2 - 24t$

13. $(9 - 5d)(9 + 5d)$

    $= 9^2 - (5d)^2 = 81 - 5^2 d^2$

    $= 81 - 25d^2$

14. $(3x + 4)(5x - 7)$

    $= 3x(5x - 7) + 4(5x - 7)$

    $= 15x^2 - 21x + 20x - 28$

    $= 15x^2 - x - 28$

15. $(4z - 3)(2z^2 - z + 5)$

    $= 4z(2z^2 - z + 5) - 3(2z^2 - z + 5)$

    $= 8z^3 - 4z^2 + 20z - 6z^2 + 3z - 15$

    $= 8z^3 - 10z^2 + 23z - 15$

16. $(x + 3)^2 = (x + 3)(x + 3)$

    $= x(x + 3) + 3(x + 3)$

    $= x^2 + 3x + 3x + 9$

    $= x^2 + 6x + 9$

17. $\dfrac{15x^6 + 25x^5 - 5x^3}{5x^2} = \dfrac{15x^6}{5x^2} + \dfrac{25x^5}{5x^2} + \dfrac{-5x^3}{5x^2}$

    $= 3x^{6-2} + 5x^{5-2} - x^{3-2}$

    $= 3x^4 + 5x^3 - x$

18.
$$\begin{array}{r} 5x+7 \\ 3x-4\overline{\smash{\big)}15x^2+x-28} \\ \underline{-\left(15x^2-20x\right)} \\ 21x-28 \\ \underline{-(21x-28)} \\ 0 \end{array}$$

The quotient is $5x+7$.

19. Let x = height of the box.

   a. Volume:
   $$(2x)(x+4)(x) = \left(2x^2\right)(x+4) = 2x^3 + 8x^2$$

   b. Surface Area:
   $$2(2x)(x+4) + 2(2x)(x) + 2(x+4)(x)$$
   $$= 4x(x+4) + 4x(x) + 2x(x+4)$$
   $$= 4x^2 + 16x + 4x^2 + 2x^2 + 8x$$
   $$= 10x^2 + 24x$$

   c. $2(5)^3 + 8(5)^2 = 250 + 200 = 450$

   $10(5)^2 + 24(5) = 250 + 120 = 370$

   The box would have a volume of 450 in^3 and a surface area of 370 in^2.

20. a. $c(x) = 235 + 5.75x$

   b. $r(x) = 25x$

   c. $p(x) = r(x) - c(x)$
   $= 25x - (235 + 5.75x)$
   $= 25x - 235 - 5.75x$
   $= 19.25x - 235$

   d. $a(x) = \dfrac{p(x)}{x} = \dfrac{19.25x - 235}{x}$
   $= \dfrac{19.25x}{x} - \dfrac{235}{x}$
   $= 19.25 - \dfrac{235}{x}$

   e. $a(45) = 19.25 - \dfrac{235}{45} \approx 14.03$

   Dallie's average profit is roughly $14.03 per basket.

21. $x^5 \cdot x^8 = x^{5+8} = x^{13}$ is simplified using the product rule for exponents while $\left(x^5\right)^8 = x^{5 \cdot 8} = x^{40}$ is simplified using the power-to-a-power rule. Recall that $\left(x^5\right)^8$ in expanded form is $x^5 \cdot x^5 \cdot x^5 \cdot x^5 \cdot x^5 \cdot x^5 \cdot x^5 \cdot x^5$. We could apply the product rule several times and obtain the same result, x^{40}.

Chapter 10

10.1 Exercises

1. $60 = 2^2 \cdot 3 \cdot 5$
   $50 = 2 \cdot 5^2$
   The GCF of the coefficients is $2 \cdot 5 = 10$.
   The GCF of the variable factors is a^2bc^2.
   The GCF of $60a^2b^4c^3$ and $50a^3bc^2$ is $10a^2bc^2$.

3. $252 = 2^2 \cdot 3^2 \cdot 7$
   $180 = 2^2 \cdot 3^2 \cdot 5$
   The GCF of the coefficients is $2^2 \cdot 3^2 = 36$.
   The GCF of the variable factors is x^2.
   The GCF of $252x^3y^4$ and $180x^2z$ is $36x^2$.

5. $45 = 3^2 \cdot 5$
   $135 = 3^3 \cdot 5$
   The GCF of the coefficients is $3^2 \cdot 5 = 45$.
   There are no common variable factors.
   The GCF of $45pq^4$ and $135r^4s$ is 45.

7. $63 = 3^2 \cdot 7$
   $98 = 2 \cdot 7^2$
   The GCF of the coefficients is 7.
   The GCF of the variable factors is xyz.
   The GCF of $63xyz$ and $98xyz$ is $7xyz$.

9. $60 = 2^2 \cdot 3 \cdot 5$
   $90 = 2 \cdot 3^2 \cdot 5$
   $150 = 2 \cdot 3 \cdot 5^2$
   The GCF of the coefficients is $2 \cdot 3 \cdot 5 = 30$.
   The GCF of the variable factors is ab^2c.
   The GCF of $60a^2b^3c^3$, $90ab^2c$, and $150a^2b^4c^2$ is $30ab^2c$.

11. $4x + 12y = 4 \cdot x + 4 \cdot 3y$
    $= 4(x + 3y)$

13. $8x^3 - 4x^2 + 12x - 24$
    $= 4 \cdot 2x^3 - 4 \cdot x^2 + 4 \cdot 3x - 4 \cdot 6$
    $= 4(2x^3 - x^2 + 3x - 6)$

15. $3a^4 - 5a^3 + 7a^2 = a^2 \cdot 3a^2 - a^2 \cdot 5a + a^2 \cdot 7$
    $= a^2(3a^2 - 5a + 7)$

17. $-3x^5 - 9x^4 - 12x^3$
    $= -3x^3 \cdot x^2 + (-3x^3) \cdot 3x + (-3x^3) \cdot 4$
    $= -3x^3(x^2 + 3x + 4)$

19. $7x^4y^2 - 3x^2y^2 + 9x^2y^4$
    $= x^2y^2 \cdot 7x^2 - x^2y^2 \cdot 3 + x^2y^2 \cdot 9y^2$
    $= x^2y^2(7x^2 - 3 + 9y^2)$

21. $8a^5b^3c + 4a^4b^2 + 16a^3c$
    $= 4a^3 \cdot 2a^2b^3c + 4a^3 \cdot ab^2 + 4a^3 \cdot 4c$
    $= 4a^3(2a^2b^3c + ab^2 + 4c)$

23. $66u^3v^4 - 88u^4v^3 = 22u^3v^3 \cdot 3v - 22u^3v^3 \cdot 4u$
    $= 22u^3v^3(3v - 4u)$

25. $3x^3 + 5y^4$
    There is no GCF. This binomial does not factor.

27. $5x(x+3) - 4(x+3) = (x+3)(5x-4)$

29. $x(2x+y) + 2y(2x+y) = (2x+y)(x+2y)$

31. $6x^2 + 10x + 21x + 35 = (6x^2 + 10x) + (21x + 35)$
    $= 2x(3x+5) + 7(3x+5)$
    $= (3x+5)(2x+7)$

33. $x^2 + 8x + x + 8 = (x^2 + 8x) + (x + 8)$
$= x(x+8) + 1(x+8)$
$= (x+8)(x+1)$

35. $2a^2 + 3a - 2a - 3 = (2a^2 + 3a) + (-2a - 3)$
$= a(2a+3) - 1(2a+3)$
$= (2a+3)(a-1)$

37. $2x^2 + xy + 4xy + 2y^2 = (2x^2 + xy) + (4xy + 2y^2)$
$= x(2x+y) + 2y(2x+y)$
$= (2x+y)(x+2y)$

39. $x^2 + xy - xy - y^2 = (x^2 + xy) + (-xy - y^2)$
$= x(x+y) - y(x+y)$
$= (x+y)(x-y)$

41. $10xy - 55y + 24x - 132$
$= (10xy - 55y) + (24x - 132)$
$= 5y(2x-11) + 12(2x-11)$
$= (2x-11)(5y+12)$

43. $12ac + 3bc + 4ad + bd$
$= (12ac + 3bc) + (4ad + bd)$
$= 3c(4a+b) + d(4a+b)$
$= (4a+b)(3c+d)$

45. $2x^2y^2 + 3xy - 8xy - 12$
$= (2x^2y^2 + 3xy) + (-8xy - 12)$
$= xy(2xy+3) - 4(2xy+3)$
$= (2xy+3)(xy-4)$

47. $-x^2 - 3x - xy - 3y = -1(x^2 + 3x + xy + 3y)$
$= -1\left[(x^2 + 3x) + (xy + 3y)\right]$
$= -1\left[x(x+3) + y(x+3)\right]$
$= -1(x+3)(x+y)$

49. $x^4 + x^2y^2 + 2x^2y^2 + 2y^4$
$= (x^4 + x^2y^2) + (2x^2y^2 + 2y^4)$
$= x^2(x^2 + y^2) + 2y^2(x^2 + y^2)$
$= (x^2 + y^2)(x^2 + 2y^2)$

51. $ac + bc + ad + bd = (ac + bc) + (ad + bd)$
$= c(a+b) + d(a+b)$
$= (a+b)(c+d)$

53. $8x^2 + 4x + 24x + 12 = 4(2x^2 + x + 6x + 3)$
$= 4\left[x(2x+1) + 3(2x+1)\right]$
$= 4(2x+1)(x+3)$

55. $u^4 + u^3v - 2u^3v - 2u^2v^2$
$= u^2(u^2 + uv - 2uv - 2v^2)$
$= u^2\left[(u^2 + uv) + (-2uv - 2v^2)\right]$
$= u^2\left[u(u+v) - 2v(u+v)\right]$
$= u^2(u+v)(u-2v)$

57. $6a^4 + 6a^3b^2 + 6a^3b + 6a^2b^3$
$= 6a^2(a^2 + ab^2 + ab + b^3)$
$= 6a^2\left[(a^2 + ab^2) + (ab + b^3)\right]$
$= 6a^2\left[a(a+b^2) + b(a+b^2)\right]$
$= 6a^2(a+b^2)(a+b)$

59. $-4x^3 - 12x^2 - 2x^2 - 6x$
$= -2x(2x^2 + 6x + x + 3)$
$= -2x\left[(2x^2 + 6x) + (x+3)\right]$
$= -2x\left[2x(x+3) + 1(x+3)\right]$
$= -2x(x+3)(2x+1)$

61. $4x^2 + 6xz + 6xz + 9z^2$
$= (4x^2 + 6xz) + (6xz + 9z^2)$
$= 2x(2x + 3z) + 3z(2x + 3z)$
$= (2x + 3z)(2x + 3z)$
$= (2x + 3z)^2$

63. $36a^2 - 30ab - 30ab + 25b^2$
$= (36a^2 - 30ab) + (-30ab + 25b^2)$
$= 6a(6a - 5b) - 5b(6a - 5b)$
$= (6a - 5b)(6a - 5b)$
$= (6a - 5b)^2$

65. $5m^3 + 5m^2n + 5m^2n + 5mn^2$
$= 5m(m^2 + mn + mn + n^2)$
$= 5m[(m^2 + mn) + (mn + n^2)]$
$= 5m[m(m + n) + n(m + n)]$
$= 5m(m + n)(m + n)$
$= 5m(m + n)^2$

67. $2x^3 + 3x + 8x^2 + 12 = (2x^3 + 3x) + (8x^2 + 12)$
$= x(2x^2 + 3) + 4(2x^2 + 3)$
$= (2x^2 + 3)(x + 4)$

69. $5a^2x + 2b^2x + 15a^2y + 6b^2y$
$= (5a^2x + 2b^2x) + (15a^2y + 6b^2y)$
$= x(5a^2 + 2b^2) + 3y(5a^2 + 2b^2)$
$= (5a^2 + 2b^2)(x + 3y)$

71. a. $x + (x+1) + (x+2) + (x+3) + (x+4)$
$\quad + (x+5) + (x+6) + (x+7) + (x+8)$
$= 9x + 36$

Amy will receive $(9x + 36)$ dollars.

b. $9x + 36 = 9(x + 4)$

c. The binomial $(x + 4)$ is the average amount in dollars that Amy will receive for each of her 9 goals.

d. $9(x + 4) = 9(10 + 4) = 9(14) = 126$.
Amy will receive $126.
Check:
9 days $\cdot$ $14/day = $126

73. a. $n^2 - n = n(n - 1)$

b. $n^2 - n = 21^2 - 21 = 441 - 21 = 420$
$n(n - 1) = 21(21 - 1) = 21(20) = 420$

c. The factored expression was easier to evaluate. Reasons provided will vary.

75. $x^2 + 7x - 3x - 21 = (x^2 + 7x) + (-3x - 21)$
$= x(x + 7) - 3(x + 7)$
$= (x + 7)(x - 3)$

The rectangle's length is $(x + 7)$ units and its width is $(x - 3)$ units.

77. $(50 + 2x)(200 + 2x) - (50)(200)$
$= 10{,}000 + 100x + 400x + 4x^2 - 10{,}000$
$= 4x^2 + 500x$

The area of the green strip is $(4x^2 + 500x)$ ft^2.

Factoring:
$4x^2 + 500x = 4x(x + 125)$

Meaning:
The area of the green strip is equivalent to a rectangle measuring $(4x)$ ft by $(x + 125)$ ft.

If $x = 10$:
$4x(x + 125) = 4 \cdot 10(10 + 125)$
$\quad\quad\quad\quad\quad = 4 \cdot 10 \cdot 135$
$\quad\quad\quad\quad\quad = 5400$

If the green strip is 10 feet wide, then there are 5400 ft^2 of greenery.

SSM: Experiencing Introductory and Intermediate Algebra

10.1 Calculator Exercises

Part 1

1. ```
   gcd(105,147)
                21
   ```

   The GCF of the coefficients is 21.
   The GCF of the variable factors is xyz^2.
   The GCF of $105x^2yz^3$ and $147xy^2z^2$ is $21xyz^2$.

2. ```
   gcd(104,65)
                13
   gcd(104,143)
                13
   gcd(65,143)
                13
   ```

   The GCF of the coefficients is 13.
   The GCF of the variable factors is c.
   The GCF of $104ab^2c^2$, $65b^3c$, and $143a^2c^2$ is $13c$.

3. ```
   gcd(108,96)
                12
   gcd(108,72)
                36
   gcd(96,72)
                24
   ```

   The GCF of the coefficients is 12.
   The GCF of the variable factors is x^2.
   The GCF of $108x^3$, $96x^2$, and $72x^5$ is $12x^2$.

4. ```
   gcd(64,128)
                64
   gcd(64,192)
                64
   gcd(64,224)
                32
   ```

   ```
   gcd(128,192)
                32
   gcd(128,224)
                64
   gcd(192,224)
                32
   ```

   The GCF of the coefficients is 32.
   The GCF of the variable factors is abc.
   The GCF of $64abc$, $128abc^2$, $192ab^2c$, and $224a^2bc$ is $32abc$.

Part 2

1. $30 = 2 \cdot 3 \cdot 5$

   ```
   prgmPRIME
   ?30
                2
                3
                5
             Done
   ```

2. $108 = 2^2 \cdot 3^3$

   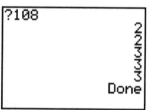

3. $525 = 3 \cdot 5^2 \cdot 7$

   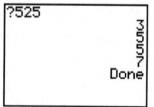

4. $1287 = 3^2 \cdot 11 \cdot 13$

   ```
   ?1287
                3
                3
               11
               13
             Done
   ```

5. $1547 = 7 \cdot 13 \cdot 17$

```
?1547
        7
       13
       17
     Done
```

6. $4500 = 2^2 \cdot 3^2 \cdot 5^3$

```
?4500
     ...
```

```
     2
     2
     3
     3
     5
     5
     5
   Done
```

10.2 Exercises

1. $x^2 + 4x + 4 = x^2 + 2 \cdot x \cdot 2 + 2^2$
$= (x+2)^2$

3. $16z^2 + 40z + 25 = (4z)^2 + 2(4z)(5) + 5^2$
$= (4z+5)^2$

5. $x^2 + 13x + 169$ does not factor. The first and last terms are both perfect squares, x^2 and 13^2, but the middle term is not $2 \cdot x \cdot 13$.

7. $x^2 - 10x + 25 = x^2 - 2 \cdot x \cdot 5 + 5^2 = (x-5)^2$

9. $36z^2 - 60z + 25 = (6z)^2 - 2(6z)(5) + 5^2$
$= (6z-5)^2$

11. $c^2 - 16d + 16d^2$ does not factor. The first and last terms are both perfect squares, c^2 and $(4d)^2$, but the middle term is not $2(c)(4d)$.

13. $a^4 - 32a^2 + 256 = (a^2)^2 - 2(a^2)(16) + 16^2$
$= (a^2 - 16)^2$
$= (a^2 - 16)(a^2 - 16)$
$= (a+4)(a-4)(a+4)(a-4)$
$= (a+4)^2 (a-4)^2$

15. $16x^4 - 72x^2 + 81$
$= (4x^2)^2 - 2(4x^2)(9) + 9^2$
$= (4x^2 - 9)^2$
$= (4x^2 - 9)(4x^2 - 9)$
$= \left[(2x)^2 - 3^2\right]\left[(2x)^2 - 3^2\right]$
$= (2x+3)(2x-3)(2x+3)(2x-3)$
$= (2x+3)^2 (2x-3)^2$

17. $3x^2 + 24x + 48 = 3(x^2 + 8x + 16)$
$= 3(x^2 + 2 \cdot x \cdot 4 + 4^2)$
$= 3(x+4)^2$

19. $2a^2 - 12a + 18 = 2(a^2 - 6a + 9)$
$= 2(a^2 - 2 \cdot x \cdot 3 + 3^2)$
$= 2(a-3)^2$

21. $p^3 + 2p^2q + pq^2 = p(p^2 + 2pq + q^2)$
$= p(p+q)^2$

SSM: Experiencing Introductory and Intermediate Algebra

23. $2p^5 - 4p^3q^2 + 2pq^4$
$= 2p(p^4 - 2p^2q^2 + q^4)$
$= 2p\left[(p^2)^2 - 2p^2q^2 + (q^2)^2\right]$
$= 2p(p^2 - q^2)^2$
$= 2p(p^2 - q^2)(p^2 - q^2)$
$= 2p(p+q)(p-q)(p+q)(p-q)$
$= 2p(p+q)^2(p-q)^2$

25. $m^5 + 2m^3n^2 + mn^4$
$= m(m^4 + 2m^2n^2 + n^4)$
$= m\left[(m^2)^2 + 2(m^2)(n^2) + (n^2)^2\right]$
$= m(m^2 + n^2)^2$

27. $x^2 - 100 = x^2 - 10^2 = (x+10)(x-10)$

29. $121 - c^2 = 11^2 - c^2 = (11+c)(11-c)$

31. $49a^2 - 4 = (7a)^2 - 2^2 = (7a+2)(7a-2)$

33. $25 - 4y^2 = 5^2 - (2y)^2 = (5+2y)(5-2y)$

35. $16u^2 - 9v^2 = (4u)^2 - (3v)^2 = (4u+3v)(4u-3v)$

37. $7z^2 - 28 = 7(z^2 - 4)$
$= 7(z^2 - 2^2)$
$= 7(z+2)(z-2)$

39. $25 + 4p^2$ will not factor. This is a sum of two squares, not a difference of two squares.

41. $x^4 - 625 = (x^2)^2 - 25^2$
$= (x^2 + 25)(x^2 - 25)$
$= (x^2 + 25)(x^2 - 5^2)$
$= (x^2 + 25)(x+5)(x-5)$

43. $256 - z^4 = 16^2 - (z^2)^2$
$= (16 + z^2)(16 - z^2)$
$= (16 + z^2)(4^2 - z^2)$
$= (16 + z^2)(4+z)(4-z)$

45. $x^8 - 1 = (x^4)^2 - 1^2$
$= (x^4 + 1)(x^4 - 1)$
$= (x^4 + 1)\left[(x^2)^2 - 1^2\right]$
$= (x^4 + 1)(x^2 + 1)(x^2 - 1)$
$= (x^4 + 1)(x^2 + 1)(x^2 - 1^2)$
$= (x^4 + 1)(x^2 + 1)(x+1)(x-1)$

47. $x^3 - 27 = x^3 - 3^3$
$= (x-3)(x^2 + x \cdot 3 + 3^2)$
$= (x-3)(x^2 + 3x + 9)$

49. $a^3 + 64 = a^3 + 4^3$
$= (a+4)(a^2 - a \cdot 4 + 4^2)$
$= (a+4)(a^2 - 4a + 16)$

51. $27x^3 + 64y^3$
$= (3x)^3 + (4y)^3$
$= (3x + 4y)\left[(3x)^2 - (3x)(4y) + (4y)^2\right]$
$= (3x + 4y)(9x^2 - 12xy + 16y^2)$

53. $8p^3 - 125q^3$
$= (2p)^3 - (5q)^3$
$= (2p - 5q)\left[(2p)^2 + (2p)(5q) + (5q)^2\right]$
$= (2p - 5q)(4p^2 + 10pq + 25q^2)$

55. $p^4 + 64pq^3 = p(p^3 + 64q^3)$
$= p\left[p^3 + (4q)^3\right]$
$= p(p + 4q)\left[p^2 - p(4q) + (4q)^2\right]$
$= p(p + 4q)(p^2 - 4pq + 16q^2)$

57. $81u^4 - 3uv^3 = 3u(27u^3 - v^3)$
$= 3u\left[(3u)^3 - v^3\right]$
$= 3u(3u - v)\left[(3u)^2 + (3u)v + v^2\right]$
$= 3u(3u - v)(9u^2 + 3uv + v^2)$

59. a. Large area − small area
$= x^2 - 15^2$
$= x^2 - 225$
The garden area is $(x^2 - 225)$ ft^2.

b. $x^2 - 225 = x^2 - 15^2 = (x + 15)(x - 15)$
A rectangle with an equivalent area will have dimensions of $(x + 15)$ ft by $(x - 15)$ ft.

c. $x^2 - 225 = (100)^2 - 225 = 9775$
$85(100) = 8500$
Yes, the square plot that is 100 ft on each side has a larger garden area than the 85 ft by 100 ft plot.

61. $1225 + 140x + 4x^2 = 35^2 + 2(35)(2x) + (2x)^2$
$= (35 + 2x)^2$
Length of each side: $(35 + 2x) - 2x = 35$ ft.
Total area of stage and mosh pit:
$(35 + 2x)^2 = (35 + 2 \cdot 25)^2 = (85)^2 = 7225$ ft^2.
Area of stage: $35^2 = 1225$ ft^2.
Area of the mosh pit: $7225 - 1225 = 6000$ ft^2.

10.2 Calculator Exercises

1.
```
√(841)
           29
³√(841)
     9.439130677
```
841 is a perfect square, 29^2. 841 is not a perfect cube.

2.
```
√(42875)
     207.0627924
³√(42875)
           35
```
42,875 is not a perfect square. 42,875 is a perfect cube, 35^3.

3.
```
√(361)
           19
³√(361)
     7.120367359
```
361 is a perfect square, 19^2. 361 is not a perfect cube.

4.  ```
    √(729)
                27
    ³√(729)
                 9
    ```

    729 is a perfect square, 27^2. 729 is also a perfect cube, 9^3.

5. $x^2 - 1225 = x^2 - 35^2$
    $\quad\quad\quad\quad = (x+35)(x-35)$

    ```
    √(1225)
                35
    ```

6. $a^3 - 6895 = a^3 - 19^3$
    $\quad\quad\quad\quad = (a-19)(a^2 + a \cdot 19 + 19^2)$
    $\quad\quad\quad\quad = (a-19)(a^2 + 19a + 361)$

    ```
    ³√(6859)
                19
    19²
               361
    ```

7. $256y^2 - 441 = (16y)^2 - 21^2$
    $\quad\quad\quad\quad\quad = (16y+21)(16y-21)$

    ```
    √(256)
                16
    √(441)
                21
    ```

8. $343b^3 - 1728$
    $= (7b)^3 - 12^3$
    $= (7b-12)\left[(7b)^2 + (7b)(12) + 12^2\right]$
    $= (7b-12)(49b^2 + 84b + 144)$

    ```
    ³√(343)
                 7
    ³√(1728)
                12
    7²
                49
    ```

    ```
    7²
                49
    7*12
                84
    12²
               144
    ```

9. $z^2 - 729 = z^2 - 27^2$
    $\quad\quad\quad\quad = (x+27)(x-27)$

    ```
    √(729)
                27
    ```

10. $z^3 - 729 = z^3 - 9^3$
    $\quad\quad\quad\quad = (z-9)(z^2 + z \cdot 9 + 9^2)$
    $\quad\quad\quad\quad = (z-9)(z^2 + 9z + 81)$

    ```
    ³√(729)
                 9
    9²
                81
    ```
    ```

10.3 Exercises

1. $x^2 + 14x + 45$

$a = 1, b = 14, c = 45$

Factor	Factor	Sum of factors
1	45	46
3	15	18
5	9	14	← b

$x^2 + 14x + 45 = (x+5)(x+9)$

3. $y^2 - 15y + 56$

$a = 1, b = -15, c = 56$

Factor	Factor	Sum of factors
−1	−56	−57
−2	−28	−30
−4	−14	−18
−7	−8	−15	← b

$y^2 - 15y + 58 = (y-7)(y-8)$

5. $p^2 - 9p - 36$

$a = 1, b = -9, c = -36$

Factor	Factor	Sum of factors
−1	36	35
1	−36	−35
−2	18	16
2	−18	−16
−3	12	9
3	−12	−9	← b
−4	9	5
4	−9	−5
−6	6	0

$p^2 - 9p - 36 = (p+3)(p-12)$

7. $z^2 + 6z + 12$

$a = 1, b = 6, c = 12$

Factor	Factor	Sum of factors
1	12	13
2	6	8
3	4	4

Because none of the possible factors of 12 add up to 6, $z^2 + 6z + 12$ does not factor.

9. $x^4 + 25x^2 + 144$

$a = 1, b = 25, c = 144$

Factor	Factor	Sum of factors
1	144	145
2	72	74
3	48	51
4	36	40
6	24	30
8	18	26
9	16	25	← b
12	12	24

$x^4 + 25x^2 + 144 = (x^2+9)(x^2+16)$

11. $x^4 - 2x^2 - 3$

$a = 1, b = -2, c = -3$

Factor	Factor	Sum of factors
−1	3	2
1	−3	−2	← b

$x^4 - 2x^2 - 3 = (x^2+1)(x^2-3)$

SSM: Experiencing Introductory and Intermediate Algebra

13. $3a^2 + 48a + 165 = 3(a^2 + 16a + 55)$

$a = 1,\ b = 16,\ c = 55$

Factor	Factor	Sum of factors	
1	55	56	
5	11	16	←b

$3a^2 + 48a + 165 = 3(a+5)(a+11)$

15. $4c^2 + 44c - 104 = 4(c^2 + 11c - 26)$

$a = 1,\ b = 11,\ c = -26$

Factor	Factor	Sum of factors	
−1	26	25	
1	−26	−25	
−2	13	11	←b
2	−13	−11	

$4c^2 + 44c - 104 = 4(c-2)(c+13)$

17. $x^2 - 11xy + 24y^2$

$a = 1,\ b = -11,\ c = 24$

Factor	Factor	Sum of factors	
−1	−24	−25	
−2	−12	−24	
−3	−8	−11	←b
−4	−6	−10	

$x^2 - 11xy + 24y^2 = (x-3y)(x-8y)$

19. $x^2 + 11xy - 12y^2$

$a = 1,\ b = 11,\ c = -12$

Factor	Factor	Sum of factors	
−1	12	11	←b
1	−12	−11	
−2	6	4	
2	−6	−4	

−3	4	1
3	−4	−1

$x^2 + 11xy - 12y^2 = (x-y)(x+12y)$

21. $-3a^2 - 15ab - 18b^2 = -3(a^2 + 5ab + 6b^2)$

$a = 1,\ b = 5,\ c = 6$

Factor	Factor	Sum of factors	
1	6	7	
2	3	5	←b

$-3a^2 - 15ab - 18b^2 = -3(a+2b)(a+3b)$

23. $-2x^2 + 14xy + 36y^2 = -2(x^2 - 7xy - 18y^2)$

$a = 1,\ b = -7,\ c = -18$

Factor	Factor	Sum of factors	
−1	18	17	
1	−18	−17	
−2	9	7	
2	−9	−7	←b
−3	6	3	
3	−6	−3	

$-2x^2 + 14xy + 36y^2 = -2(x+2y)(x-9y)$

25. $3x^2 + 10x + 3 = (\_\_\ \_\_)(\_\_\ \_\_)$

Possible factors	Middle term	
$(x+1)(3x+3)$	$3x + 3x = 6x$	
$(x+3)(3x+1)$	$x + 9x = 10x$	←bx

$3x^2 + 10x + 3 = (x+3)(3x+1)$

27. $2x^2 - 15x + 7 = (\_\_\ \_\_)(\_\_\ \_\_)$

Possible factors	Middle term
$(x-1)(2x-7)$	$-7x - 2x = -9x$
$(x-7)(2x-1)$	$-x - 14x = -15x$ ← bx

$2x^2 - 15x + 7 = (x-7)(2x-1)$

29. $3x^2 - x - 2 = (\_\_\ \_\_)(\_\_\ \_\_)$

Possible factors	Middle term
$(x-1)(3x+2)$	$2x - 3x = -x$ ← bx
$(x+1)(3x-2)$	$-2x + 3x = x$
$(x-2)(3x+1)$	$x - 6x = -5x$
$(x+2)(3x-1)$	$-x + 6x = 5x$

$3x^2 - x - 2 = (x-1)(3x+2)$

31. $5m^2 + 9m - 2 = (\_\_\ \_\_)(\_\_\ \_\_)$

Possible factors	Middle term
$(m-1)(5m+2)$	$2m - 5m = -3m$
$(m+1)(5m-2)$	$-2m + 5m = 3m$
$(m-2)(5m+1)$	$m - 10m = -9m$
$(m+2)(5m-1)$	$-m + 10m = 9m$ ← bx

$5m^2 + 9m - 2 = (m+2)(5m-1)$

33. $2m^2 + 7m - 3 = (\_\_\ \_\_)(\_\_\ \_\_)$

Possible factors	Middle term
$(m-1)(2m+3)$	$2m - 5m = -3m$
$(m+1)(2m-3)$	$-2m + 5m = 3m$
$(m-3)(2m+1)$	$m - 10m = -9m$
$(m+3)(2m-1)$	$-m + 10m = 9m$

Because none of the possible factors result in the middle term $7m$, $2m^2 + 7m - 3$ does not factor.

35. $4a^2 + 25a + 6 = (\_\_\ \_\_)(\_\_\ \_\_)$

Possible factors	Middle term
$(a+1)(4a+6)$	$6a + 4a = 10a$
$(a+6)(4a+1)$	$a + 24a = 25a$ ← bx
$(a+2)(4a+3)$	$3a + 8a = 11a$
$(a+3)(4a+2)$	$2a + 12a = 14a$
$(2a+1)(2a+6)$	$12a + 2a = 14a$
$(2a+2)(2a+3)$	$6a + 4a = 10a$

$4a^2 + 25a + 6 = (a+6)(4a+1)$

37. $9d^2 - 13d + 4 = (\_\_\ \_\_)(\_\_\ \_\_)$

Possible factors	Middle term
$(d-1)(9d-4)$	$-4d - 9d = -13d$ ← bx
$(d-4)(9d-1)$	$-d - 36d = -37d$
$(d-2)(9d-2)$	$-2d - 18d = -20d$
$(3d-1)(3d-4)$	$-12d - 3d = -15d$
$(3d-2)(3d-2)$	$-6d - 6d = -12d$

$9d^2 - 13d + 4 = (d-1)(9d-4)$

39. $6x^2 - 23x - 4 = (\_\_\ \_\_)(\_\_\ \_\_)$

Possible factors	Middle term
$(x-1)(6x+4)$	$4x - 6x = -2x$
$(x+1)(6x-4)$	$-4x + 6x = 2x$
$(x-4)(6x+1)$	$x - 24x = -23x$ ← bx
$(x+4)(6x-1)$	$-x + 24x = 23x$
$(x-2)(6x+2)$	$2x - 12x = -10x$
$(x+2)(6x-2)$	$-2x + 12x = 10x$
$(2x-1)(3x+4)$	$8x - 3x = 5x$
$(2x+1)(3x-4)$	$-8x + 3x = -5x$
$(2x-4)(3x+1)$	$2x - 12x = -10x$
$(2x+4)(3x-1)$	$-2x + 12x = 10x$
$(2x-2)(3x+2)$	$4x - 6x = -2x$

SSM: Experiencing Introductory and Intermediate Algebra

$(2x+2)(3x-2)$	$-4x+6x = 2x$

$6x^2 - 23x - 4 = (x-4)(6x+1)$

41. $8y^2 + 7y - 18 = (\underline{\quad}\underline{\quad})(\underline{\quad}\underline{\quad})$

Possible factors	Middle term	
$(y-1)(8y+18)$	$18y - 8y = 10y$	
$(y+1)(8y-18)$	$-18y + 8y = -10y$	
$(y-18)(8y+1)$	$y - 144y = -143y$	
$(y+18)(8y-1)$	$-y + 144y = 143y$	
$(y-2)(8y+9)$	$9y - 16y = -7y$	
$(y+2)(8y-9)$	$-9y + 16y = 7y$	← bx
$(y-9)(8y+2)$	$2y - 72y = -70y$	
$(y+9)(8y-2)$	$-2y + 72y = 70y$	
$(y-3)(8y+6)$	$6y - 24y = -18y$	
$(y+3)(8y-6)$	$-6y + 24y = 18y$	
$(y-6)(8y+3)$	$3y - 48y = -45y$	
$(y+6)(8y-3)$	$-3y + 48y = 45y$	
$(2y-1)(4y+18)$	$36y - 4y = 32y$	
$(2y+1)(4y-18)$	$-36y + 4y = -32y$	
$(2y-18)(4y+1)$	$2y - 72y = -70y$	
$(2y+18)(4y-1)$	$-2y + 72y = 70y$	
$(2y-2)(4y+9)$	$18y - 8y = 10y$	
$(2y+2)(4y-9)$	$-18y + 8y = -10y$	
$(2y-9)(4y+2)$	$4y - 36y = -32y$	
$(2y+9)(4y-2)$	$-4y + 36y = 32y$	
$(2y-3)(4y+6)$	$12y - 12y = 0$	
$(2y+3)(4y-6)$	$-12y + 12y = 0$	
$(2y-6)(4y+3)$	$6y - 24y = -18y$	
$(2y+6)(4y-3)$	$-6y + 24y = 18y$	

$8y^2 + 7y - 18 = (y+2)(8y-9)$

43. $6b^2 + 17b + 12 = (\underline{\quad}\underline{\quad})(\underline{\quad}\underline{\quad})$

Possible factors	Middle term	
$(b+1)(6b+12)$	$12b + 6b = 18b$	
$(b+12)(6b+1)$	$b + 72b = 72b$	
$(b+2)(6b+6)$	$6b + 12b = 18b$	
$(b+6)(6b+2)$	$2b + 36b = 38b$	
$(b+3)(6b+4)$	$4b + 18b = 22b$	
$(b+4)(6b+3)$	$3b + 24b = 27b$	
$(2b+1)(3b+12)$	$24b + 3b = 27b$	
$(2b+12)(3b+1)$	$2b + 36b = 38b$	
$(2b+2)(3b+6)$	$12b + 6b = 18b$	
$(2b+6)(3b+2)$	$4b + 18b = 22b$	
$(2b+3)(3b+4)$	$8b + 9b = 17b$	← bx
$(2b+4)(3b+3)$	$6b + 12b = 18b$	

$6b^2 + 17b + 12 = (2b+3)(3b+4)$

45. $20x^2 - 31x + 12 = (\underline{\quad}\underline{\quad})(\underline{\quad}\underline{\quad})$

Possible factors	Middle term
$(x-1)(20x-12)$	$-12x - 20x = -32x$
$(x-12)(20x-1)$	$-1x - 240x = -242x$
$(x-2)(20x-6)$	$-6x - 40x = -46x$
$(x-6)(20x-2)$	$-2x - 120x = -122x$
$(x-3)(20x-4)$	$-4x - 60x = -64x$
$(x-4)(20x-3)$	$-3x - 80x = -83x$
$(2x-1)(10x-12)$	$-12x - 20x = -32x$
$(2x-12)(10x-1)$	$-2x - 120x = -122x$
$(2x-2)(10x-6)$	$-12x - 20x = -32x$
$(2x-6)(10x-2)$	$-4x - 60x = -64x$
$(2x-3)(10x-4)$	$-8x - 30x = -38x$
$(2x-4)(10x-3)$	$-6x - 40x = -46x$
$(4x-1)(5x-12)$	$-48x - 5x = -53x$
$(4x-12)(5x-1)$	$-4x - 60x = -64x$

Possible factors	Middle term	
$(4x-2)(5x-6)$	$-24x-10x=-34x$	
$(4x-6)(5x-2)$	$-8x-30x=-38x$	
$(4x-3)(5x-4)$	$-16x-15x=-31x$	← bx
$(4x-4)(5x-3)$	$-12x-20x=-32x$	

$$20x^2-31x+12=(4x-3)(5x-4)$$

47. $18x^2-9x-20=(\underline{}\ \underline{})(\underline{}\ \underline{})$

Possible factors	Middle term
$(x-1)(18x+20)$	$20x-18x=2x$
$(x+1)(18x-20)$	$-20x+18x=-2x$
$(x-20)(18x+1)$	$x-360x=-359x$
$(x+20)(18x-1)$	$-x+360x=359x$
$(x-2)(18x+10)$	$10x-36x=-26x$
$(x+2)(18x-10)$	$-10x+36x=26x$
$(x-10)(18x+2)$	$2x-180x=-178x$
$(x+10)(18x-2)$	$-2x+180x=178x$
$(x-4)(18x+5)$	$5x-72x=-67x$
$(x+4)(18x-5)$	$-5x+72x=67x$
$(x-5)(18x+4)$	$4x-90x=-86x$
$(x+5)(18x-4)$	$-4x+90x=86x$
$(2x-1)(9x+20)$	$40x-9x=31x$
$(2x+1)(9x-20)$	$-40x+9x=-31x$
$(2x-20)(9x+1)$	$2x-180x=-178x$
$(2x+20)(9x-1)$	$-2x+180x=178x$
$(2x-2)(9x+10)$	$20x-18x=2x$
$(2x+2)(9x-10)$	$-20x+18x=-2x$
$(2x-10)(9x+2)$	$4x-90x=-86x$
$(2x+10)(9x-2)$	$-4x+90x=86x$
$(2x-4)(9x+5)$	$10x-36x=-26x$
$(2x+4)(9x-5)$	$-10x+36x=26x$
$(2x-5)(9x+4)$	$8x-45x=-37x$
$(2x+5)(9x-4)$	$-8x+45x=37x$

Possible factors	Middle term	
$(3x-1)(6x+20)$	$60x-6x=54x$	
$(3x+1)(6x-20)$	$-60x+6x=-54x$	
$(3x-20)(6x+1)$	$3x-120x=-117x$	
$(3x+20)(6x-1)$	$-3x+120x=117x$	
$(3x-2)(6x+10)$	$30x-12x=18x$	
$(3x+2)(6x-10)$	$-20x+12x=-18x$	
$(3x-10)(6x+2)$	$6x-60x=-54x$	
$(3x+10)(6x-2)$	$-6x+60x=54x$	
$(3x-4)(6x+5)$	$15x-24x=-9x$	← bx
$(3x+4)(6x-5)$	$-15x+24x=9x$	
$(3x-5)(6x+4)$	$12x-30x=-18x$	
$(3x+5)(6x-4)$	$-12x+30x=18x$	

$$18x^2-9x-20=(3x-4)(6x+5)$$

49. $18p^2-57p-21=3(6p^2-19p-7)$
$=3(\underline{}\ \underline{})(\underline{}\ \underline{})$

Possible factors	Middle term	
$(p-1)(6p+7)$	$7p-6p=p$	
$(p+1)(6p-7)$	$-7p+6p=-p$	
$(p-7)(6p+1)$	$p-42p=-41p$	
$(p+7)(6p-1)$	$-p+42p=41p$	
$(2p-1)(3p+7)$	$14p-3p=11p$	
$(2p+1)(3p-7)$	$-14p+3p=-11p$	
$(2p-7)(3p+1)$	$2p-21p=-19p$	← bx
$(2p+7)(3p-1)$	$-2p+21p=19p$	

$$18p^2-57p-21=3(2p-7)(3p+1)$$

51. $2x^4 + 11x^2 + 9 = (\_\_ \ \_\_)(\_\_ \ \_\_)$

Possible factors	Middle term
$(x^2 + 1)(2x^2 + 9)$	$9x^2 + 2x^2 = 11x^2$ ← bx
$(x^2 + 9)(2x^2 + 1)$	$x^2 + 18x^2 = 19x^2$
$(x^2 + 3)(2x^2 + 3)$	$3x^2 + 6x^2 = 9x^2$

$2x^4 + 11x^2 + 9 = (x^2 + 1)(2x^2 + 9)$

53. $4m^4 + 13m^2 - 12 = (\_\_ \ \_\_)(\_\_ \ \_\_)$

Possible factors	Middle term
$(m^2 - 1)(4m^2 + 12)$	$12m^2 - 4m^2 = 8m^2$
$(m^2 + 1)(4m^2 - 12)$	$-12m^2 + 4m^2 = -8m^2$
$(m^2 - 12)(4m^2 + 1)$	$m^2 - 48m^2 = -47m^2$
$(m^2 + 12)(4m^2 - 1)$	$-m^2 + 48m^2 = 47m^2$
$(m^2 - 2)(4m^2 + 6)$	$6m^2 - 8m^2 = -2m^2$
$(m^2 + 2)(4m^2 - 6)$	$-6m^2 + 8m^2 = 2m^2$
$(m^2 - 6)(4m^2 + 2)$	$2m^2 - 24m^2 = -22m^2$
$(m^2 + 6)(4m^2 - 2)$	$-2m^2 + 24m^2 = 22m^2$
$(m^2 - 3)(4m^2 + 4)$	$4m^2 - 12m^2 = -8m^2$
$(m^2 + 3)(4m^2 - 4)$	$-4m^2 + 12m^2 = 8m^2$
$(m^2 - 4)(4m^2 + 3)$	$3m^2 - 16m^2 = -13m^2$
$(m^2 + 4)(4m^2 - 3)$	$-3m^2 + 16m^2 = 13m^2$ ← bx
$(2m^2 - 1)(2m^2 + 12)$	$24m^2 - 2m^2 = 22m^2$
$(2m^2 + 1)(2m^2 - 12)$	$-24m^2 + 2m^2 = -22m^2$
$(2m^2 - 2)(2m^2 + 6)$	$12m^2 - 4m^2 = 8m^2$
$(2m^2 + 2)(2m^2 - 6)$	$-12m^2 + 4m^2 = -8m^2$
$(2m^2 - 3)(2m^2 + 4)$	$8m^2 - 6m^2 = 2m^2$
$(2m^2 + 3)(2m^2 - 4)$	$-8m^2 + 6m^2 = -2m^2$

$4m^4 + 13m^2 - 12 = (m^2 + 4)(4m^2 - 3)$

55. $5x - 6x^2 + 56 = -6x^2 + 5x + 56$
$= -1(6x^2 - 5x - 56)$
$= -1(\_\_ \ \_\_)(\_\_ \ \_\_)$

Possible factors	Middle term
$(x - 1)(6x + 56)$	$56x - 6x = 50x$
$(x + 1)(6x - 56)$	$-56x + 6x = -50x$
$(x - 56)(6x + 1)$	$x - 336x = -335x$
$(x + 56)(6x - 1)$	$-x + 336x = 335x$
$(x - 2)(6x + 28)$	$28x - 12x = 14x$
$(x + 2)(6x - 28)$	$-28x + 12x = -14x$
$(x - 28)(6x + 2)$	$2x - 168x = -166x$
$(x + 28)(6x - 2)$	$-2x + 168x = 166x$
$(x - 4)(6x + 14)$	$14x - 24x = -10x$
$(x + 4)(6x - 14)$	$-14x + 24x = 10x$
$(x - 14)(6x + 4)$	$4x - 84x = -80x$
$(x + 14)(6x - 4)$	$-4x + 84x = 80x$
$(x - 7)(6x + 8)$	$8x - 42x = -34x$
$(x + 7)(6x - 8)$	$-8x + 42x = 34x$
$(x - 8)(6x + 7)$	$7x - 48x = -41x$
$(x + 8)(6x - 7)$	$-7x + 48x = 41x$
$(2x - 1)(3x + 56)$	$112x - 3x = 109x$
$(2x + 1)(3x - 56)$	$-112x + 3x = -109x$
$(2x - 56)(3x + 1)$	$2x - 168x = -166x$
$(2x + 56)(3x - 1)$	$-2x + 168x = 166x$
$(2x - 2)(3x + 28)$	$56x - 6x = 50x$
$(2x + 2)(3x - 28)$	$-56x + 6x = -50x$
$(2x - 28)(3x + 2)$	$4x - 84x = -80x$
$(2x + 28)(3x - 2)$	$-4x + 84x = 80x$
$(2x - 4)(3x + 14)$	$28x - 12x = 16x$
$(2x + 4)(3x - 14)$	$-28x + 12x = -16x$
$(2x - 14)(3x + 4)$	$8x - 42x = -34x$
$(2x + 14)(3x - 4)$	$-8x + 42x = 34x$

Chapter 10: Factoring

$(2x-7)(3x+8)$	$16x - 21x = -5x$ ← bx
$(2x+7)(3x-8)$	$-16x + 21x = 5x$
$(2x-8)(3x+7)$	$14x - 24x = -10x$
$(2x+8)(3x-7)$	$-14x + 24x = 10x$

$$5x - 6x^2 + 56 = -1(2x-7)(3x+8)$$

57. $6x^2 + 5xy - 6y^2 = (\underline{} \ \underline{})(\underline{} \ \underline{})$

Possible factors	Middle term
$(x-y)(6x+6y)$	$6xy - 6xy = 0$
$(x+y)(6x-6y)$	$-6xy + 6xy = 0$
$(x-6y)(6x+y)$	$xy - 36xy = -35xy$
$(x+6y)(6x-y)$	$-xy + 36xy = 35xy$
$(x-2y)(6x+3y)$	$3xy - 12xy = -9xy$
$(x+2y)(6x-3y)$	$-3xy + 12xy = 9xy$
$(x-3y)(6x+2y)$	$2xy - 18xy = -16xy$
$(x+3y)(6x-2y)$	$-2xy + 18xy = 16xy$
$(2x-y)(3x+6y)$	$12xy - 3xy = 9xy$
$(2x+y)(3x-6y)$	$-12xy + 3xy = -9xy$
$(2x-6y)(3x+y)$	$2xy - 18xy = -16xy$
$(2x+6y)(3x-y)$	$-2xy + 18xy = 16xy$
$(2x-2y)(3x+3y)$	$6xy - 6xy = 0$
$(2x+2y)(3x-3y)$	$-6xy + 6xy = 0$
$(2x-3y)(3x+2y)$	$4xy - 9xy = -5xy$
$(2x+3y)(3x-2y)$	$-4xy + 9xy = 5xy$ ← bx

$$6x^2 + 5xy - 6y^2 = (2x+3y)(3x-2y)$$

59. $4u^2 - 39uv + 56v^2 = (\underline{} \ \underline{})(\underline{} \ \underline{})$

Possible factors	Middle term
$(u-v)(4u-56v)$	$-56uv - 4uv = -60uv$
$(u-56v)(4u-v)$	$-uv - 224uv = -225uv$
$(u-2v)(4u-28v)$	$-28uv - 8uv = -36uv$
$(u-28v)(4u-2v)$	$-2uv - 112uv = -14uv$
$(u-4v)(4u-14v)$	$-14uv - 16uv = -30uv$
$(u-14v)(4u-4v)$	$-4uv - 56uv = -60uv$
$(u-7v)(4u-8v)$	$-8uv - 28uv = -36uv$
$(u-8v)(4u-7v)$	$-7uv - 32uv = -39uv$ ← bx
$(2u-v)(2u-56v)$	$-112uv - 2uv = -114uv$
$(2u-2v)(2u-28v)$	$-56uv - 4uv = -60uv$
$(2u-4v)(2u-14v)$	$-28uv - 8uv = -36uv$
$(2u-7v)(2u-8v)$	$-16uv - 14uv = -30uv$

$$4u^2 - 39uv + 56v^2 = (u-8v)(4u-7v)$$

61. $9x^4 + 13x^2y^2 + 4y^4 = (\underline{} \ \underline{})(\underline{} \ \underline{})$

Possible factors	Middle term
$(x^2+y^2)(9x^2+4y^2)$	$4x^2y^2 + 9x^2y^2 = 13x^2y^2$ ← bx
$(x^2+4y^2)(9x^2+y^2)$	$x^2y^2 + 36x^2y^2 = 37x^2y^2$
$(x^2+2y^2)(9x^2+2y^2)$	$x^2y^2 + 36x^2y^2 = 37x^2y^2$
$(3x^2+y^2)(3x^2+4y^2)$	$12x^2y^2 + 3x^2y^2 = 15x^2y^2$
$(3x^2+2y^2)(3x^2+2y^2)$	$6x^2y^2 + 6x^2y^2 = 12x^2y^2$

$$9x^4 + 13x^2y^2 + 4y^4 = (x^2+y^2)(9x^2+4y^2)$$

63. $10x^2y^2 + xy - 21 = (\underline{} \ \underline{})(\underline{} \ \underline{})$

Possible factors	Middle term
$(xy-1)(10xy+21)$	$21xy - 10xy = 11xy$
$(xy+1)(10xy-21)$	$-21xy + 10xy = -11xy$
$(xy-21)(10xy+1)$	$xy - 210xy = -209xy$
$(xy+21)(10xy-1)$	$-xy + 210xy = 209xy$
$(xy-3)(10xy+7)$	$7xy - 30xy = -23xy$
$(xy+3)(10xy-7)$	$-7xy + 30xy = 23xy$
$(xy-7)(10xy+3)$	$3xy - 70xy = -67xy$
$(xy+7)(10xy-3)$	$-3xy + 70xy = 67xy$
$(2xy-1)(5xy+21)$	$21xy - 10xy = 11xy$
$(2xy+1)(5xy-21)$	$-21xy + 10xy = -11xy$
$(2xy-21)(5xy+1)$	$2xy - 105xy = -103xy$

(2xy + 21)(5xy – 1)	–2xy + 105xy = 103xy
(2xy – 3)(5xy + 7)	14xy – 15xy = – xy
(2xy + 3)(5xy – 7)	– 14xy + 15xy = xy	← bx
(2xy – 7)(5xy + 3)	6xy – 35xy = – 29xy
(2xy + 7)(5xy – 3)	– 6xy + 35xy = 29xy

$10x^2y^2 + xy - 21 = (2xy+3)(5xy-7)$

65. Let w = the original width of the rectangle.
$w + 8$ = the original length of the rectangle.

a. $w^2 + 14w + 24$
$a = 1, b = 14, c = 24$

Factor	Factor	Sum of factors
1	24	25
2	12	14	← b
3	8	11
4	6	10

$w^2 + 14w + 24 = (w+2)(w+12)$
The increased width is $(w+2)$ inches and the increased length is $(w+12)$ inches.

b. $(w+2) - w = 2$
The width was increased by 2 inches.

c. $(w+12) - (w+8) = 4$
The length was increased by 4 inches.

67. a. $x^2 + \frac{21}{2}x + 20 = \frac{1}{2}(2x^2 + 21x + 40)$
$= \frac{1}{2}(\_\_\_\_)(\_\_\_\_)$

Possible factors	Middle term
(x + 1)(2x + 40)	40x + 2x = 42x
(x + 40)(2x + 1)	x + 80x = 81x
(x + 2)(2x + 20)	20x + 4x = 24x
(x + 20)(2x + 2)	2x + 40x = 42x
(x + 4)(2x + 10)	10x + 8x = 18x

(x + 10)(2x + 4)	4x + 20x = 24x
(x + 5)(2x + 8)	8x + 10 = 18x
(x + 8)(2x + 5)	5x + 16x = 21x	← bx

$x^2 + \frac{21}{2}x + 20 = \frac{1}{2}(x+8)(2x+5)$

The lengths of the legs of the right triangle are $(x+8)$ inches and $(2x+5)$ inches.

b. $(x+8) - x = 8$
Each leg was increased by 8 inches.

c. $(2x+5) - 8 = 2x - 3$
The expression for the length of the long leg of the original triangle is $(2x-3)$ inches.

69. $4x^2 + 228x + 2808 = 4(x^2 + 57x + 702)$
$a = 1, b = 57, c = 702$

Factor	Factor	Sum of factors
1	702	702
2	351	353
3	234	237
6	117	123
9	78	87
13	54	67
18	39	57	← b
26	27	53

$4x^2 + 228x + 2808 = 4(x+18)(x+39)$
$= 2(x+18) \cdot 2(x+39)$
$= (2x+36)(2x+78)$
$= (36+2x)(78+2x)$

Width: $(36+2x) - 2x = 36$ feet.
Length: $(78+2x) - 2x = 78$ feet.
The dimensions of the playing area of the tennis court are 36 feet by 78 feet.

Chapter 10: Factoring

10.3 Calculator Exercises

1. Let $Y1 = 40x^2 + 21x - 180$.

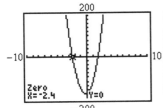 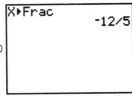

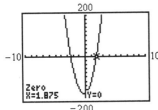 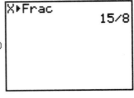

$x = -\dfrac{12}{5}$ $x = \dfrac{15}{8}$

$5x = -12$ $8x = 15$

$5x + 12 = 0$ and $8x - 15 = 0$

$40x^2 + 21x - 180 = (5x + 12)(8x - 15)$

2. Let $Y1 = 80x^2 - 232x + 117$.

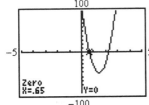 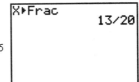

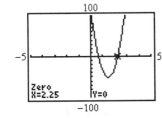 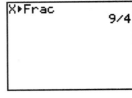

$x = \dfrac{13}{20}$ $x = \dfrac{9}{4}$

$20x = 13$ $4x = 9$

$20x - 13 = 0$ and $4x - 9 = 0$

$80x^2 - 232x + 117 = (20x - 13)(4x - 9)$

3. Let $Y1 = 108x^2 - 177x + 55$.

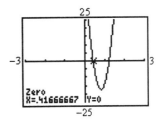 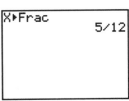

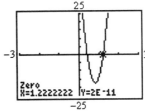 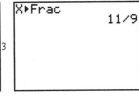

$x = \dfrac{5}{12}$ $x = \dfrac{11}{9}$

$12x = 5$ $9x = 11$

$12x - 5 = 0$ and $9x - 11 = 0$

$108x^2 - 177x + 55 = (12x - 5)(9x - 11)$

4. Let $Y1 = 180x^2 + 327x + 143$.

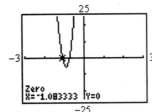 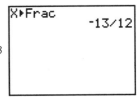

SSM: Experiencing Introductory and Intermediate Algebra

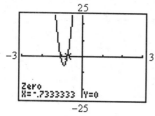
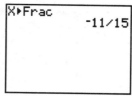

$x = -\dfrac{13}{12}$ \qquad $x = -\dfrac{11}{15}$

$12x = -13$ \qquad $15x = -11$

$12x + 13 = 0$ \; and \; $15x + 11 = 0$

$180x^2 + 327x + 143 = (12x+13)(15x+11)$

10.4 Exercises

1. $a = 3$, $b = 19$, $c = 20$, $ac = 60$
$$3x^2 + 19x + 20 = (3x^2 + 15x) + (4x + 20)$$
$$= 3x(x+5) + 4(x+5)$$
$$= (x+5)(3x+4)$$

3. $a = 6$, $b = 31$, $c = 35$, $ac = 210$
$$6x^2 + 31x + 35 = (6x^2 + 21x) + (10x + 35)$$
$$= 3x(2x+7) + 5(2x+7)$$
$$= (2x+7)(3x+5)$$

5. $a = 10$, $b = 27$, $c = 11$, $ac = 110$
$$10q^2 + 27q + 11 = (10q^2 + 22q) + (5q + 11)$$
$$= 2q(5q+11) + 1(5q+11)$$
$$= (5q+11)(2q+1)$$

7. $a = 4$, $b = 14$, $c = 45$, $ac = 180$
Because no two factors of 180 have a sum of 14, $4x^2 + 14x + 45$ does not factor.

9. $a = 20$, $b = -51$, $c = 12$, $ac = 240$
Because no two factors of 240 have a sum of -51, $20z^2 - 51z + 12$ does not factor.

11. $a = 6$, $b = -13$, $c = 6$, $ac = 36$
$$6m^2 - 13m + 6 = (6m^2 - 9m) + (-4m + 6)$$
$$= 3m(2m-3) - 2(2m-3)$$
$$= (2m-3)(3m-2)$$

13. $a = 56$, $b = 13$, $c = -3$, $ac = -168$
$$56p^2 + 13p - 3 = (56p^2 + 21p) + (-8p - 3)$$
$$= 7p(8p+3) - 1(8p+3)$$
$$= (8p+3)(7p-1)$$

15. $a = 15$, $b = -26$, $c = 8$, $ac = 120$
$$15p^2 - 26p + 8 = (15p^2 - 20p) + (-6p + 8)$$
$$= 5p(3p-4) - 2(3p-4)$$
$$= (3p-4)(5p-2)$$

17. $a = 32$, $b = -12$, $c = -5$, $ac = -160$
$$32x^2 - 12x - 5 = (32x^2 - 20x) + (8x - 5)$$
$$= 4x(8x-5) + 1(8x-5)$$
$$= (8x-5)(4x+1)$$

19. $a = 28$, $b = -19$, $c = -99$, $ac = -2772$
$$28k^2 - 19k - 99 = (28k^2 - 63k) + (44k - 99)$$
$$= 7k(4k-9) + 11(4k-9)$$
$$= (4k-9)(7k+11)$$

21. $a = 24$, $b = 77$, $c = -117$, $ac = -2808$
$$24m^2 + 77m - 117 = (24m^2 + 104m) + (-27m - 117)$$
$$= 8m(3m+13) - 9(3m+13)$$
$$= (3m+13)(8m-9)$$

23. $40x^2 - 148x + 28 = 4(10x^2 - 37x + 7)$
$a = 10$, $b = -37$, $c = 7$, $ac = 70$
$$40x^2 - 148x + 28 = 4(10x^2 - 37x + 7)$$
$$= 4\left[(10x^2 - 35x) + (-2x + 7)\right]$$
$$= 4\left[5x(2x-7) - 1(2x-7)\right]$$
$$= 4(2x-7)(5x-1)$$

426

25. $30x^2 + 87x + 30 = 3(10x^2 + 29x + 10)$
$a = 10$, $b = 29$, $c = 10$, $ac = 100$
$30x^2 + 87x + 30 = 3(10x^2 + 29x + 10)$
$ = 3\left[(10x^2 + 25x) + (4x + 10)\right]$
$ = 3\left[5x(2x + 5) + 2(2x + 5)\right]$
$ = 3(2x + 5)(5x + 2)$

27. $16x^2 + 42x - 18 = 2(8x^2 + 21x - 9)$
$a = 8$, $b = 21$, $c = -9$, $ac = -72$
$16x^2 + 42x - 18 = 2(8x^2 + 21x - 9)$
$ = 2\left[(8x^2 + 24x) + (-3x - 9)\right]$
$ = 2\left[8x(x + 3) - 3(x + 3)\right]$
$ = 2(x + 3)(8x - 3)$

29. $-18x^2 + 75x - 72 = -3(6x^2 - 25x + 24)$
$a = 6$, $b = -25$, $c = 24$, $ac = 144$
$-18x^2 + 75x - 72 = -3(6x^2 - 25x + 24)$
$ = -3\left[(6x^2 - 16x) + (-9x + 24)\right]$
$ = -3\left[2x(3x - 8) - 3(3x - 8)\right]$
$ = -3(3x - 8)(2x - 3)$

31. $16x^2 - 52x - 80 = 4(4x^2 - 13x - 20)$
$a = 4$, $b = -13$, $c = -20$, $ac = -80$
Because no two factors of -80 have a sum of -13, $4x^2 - 13x - 20$ does not factor. Thus,
$16x^2 - 52x - 80 = 4(4x^2 - 13x - 20)$

33. $6x^3 + 25x^2 + 25x = x(6x^2 + 25x + 25)$
$a = 6$, $b = 25$, $c = 25$, $ac = 150$

$6x^3 + 25x^2 + 25x = x(6x^2 + 25x + 25)$
$ = x\left[(6x^2 + 15x) + (10x + 25)\right]$
$ = x\left[3x(2x + 5) + 5(2x + 5)\right]$
$ = x(2x + 5)(3x + 5)$

35. $a = 6$, $b = 7$, $c = 2$, $ac = 12$
$6x^4 + 7x^2 + 2 = (6x^4 + 4x^2) + (3x^2 + 2)$
$ = 2x^2(3x^2 + 2) + 1(3x^2 + 2)$
$ = (3x^2 + 2)(2x^2 + 1)$

37. $a = 8$, $b = 46$, $c = 63$, $ac = 504$
$8x^4 + 46x^2 + 63 = (8x^4 + 28x^2) + (18x^2 + 63)$
$ = 4x^2(2x^2 + 7) + 9(2x^2 + 7)$
$ = (2x^2 + 7)(4x^2 + 9)$

39. $a = 6$, $b = -19$, $c = 15$, $ac = 90$
$6x^4 - 19x^2 + 15 = (6x^4 - 10x^2) + (-9x^2 + 15)$
$ = 2x^2(3x^2 - 5) - 3(3x^2 - 5)$
$ = (3x^2 - 5)(2x^2 - 3)$

41. $a = 12$, $b = -17$, $c = 6$, $ac = 72$
$12x^4 - 17x^2 + 6 = (12x^4 - 9x^2) + (-8x^2 + 6)$
$ = 3x^2(4x^2 - 3) - 2(4x^2 - 3)$
$ = (4x^2 - 3)(3x^2 - 2)$

43. $a = 8$, $b = -2$, $c = -15$, $ac = -120$
$8m^4 - 2m^2 - 15 = (8m^4 - 12m^2) + (10m^2 - 15)$
$ = 4m^2(2m^2 - 3) + 5(2m^2 - 3)$
$ = (2m^2 - 3)(4m^2 + 5)$

45. $a = 15$, $b = 14$, $c = -16$, $ac = -240$
$$15x^4 + 14x^2 - 16 = \left(15x^4 + 24x^2\right) + \left(-10x^2 - 16\right)$$
$$= 3x^2\left(5x^2 + 8\right) - 2\left(5x^2 + 8\right)$$
$$= \left(5x^2 + 8\right)\left(3x^2 - 2\right)$$

47. $60y^4 + 57y^2 - 84 = 3\left(20y^4 + 19y^2 - 28\right)$
$a = 20$, $b = 19$, $c = -28$, $ac = -560$
$$60y^4 + 57y^2 - 84$$
$$= 3\left(20y^4 + 19y^2 - 28\right)$$
$$= 3\left[\left(20y^4 + 35y^2\right) + \left(-16y^2 - 28\right)\right]$$
$$= 3\left[5y^2\left(4y^2 + 7\right) - 4\left(4y^2 + 7\right)\right]$$
$$= 3\left(4y^2 + 7\right)\left(5y^2 - 4\right)$$

49. $-24x^4 + 32x^2 - 10 = -2\left(12x^4 - 16x^2 + 5\right)$
$a = 12$, $b = -16$, $c = 5$, $ac = 60$
$$-24x^4 + 32x^2 - 10$$
$$= -2\left(12x^4 - 16x^2 + 5\right)$$
$$= -2\left[\left(12x^4 - 10x^2\right) + \left(-6x^2 + 5\right)\right]$$
$$= -2\left[2x^2\left(6x^2 - 5\right) - 1\left(6x^2 - 5\right)\right]$$
$$= -2\left(6x^2 - 5\right)\left(2x^2 - 1\right)$$

51. $a = 6$, $b = -7$, $c = -20$, $ac = -120$
$$6x^2 - 7xy - 20y^2 = \left(6x^2 - 15xy\right) + \left(8xy - 20y^2\right)$$
$$= 3x(2x - 5y) + 4y(2x - 5y)$$
$$= (2x - 5y)(3x + 4y)$$

53. $a = 40$, $b = -67$, $c = 28$, $ac = 1120$
$$40p^2 - 67pq + 28q^2$$
$$= \left(40p^2 - 35pq\right) + \left(-32pq + 28q^2\right)$$
$$= 5p(8p - 7q) - 4q(8p - 7q)$$
$$= (8p - 7q)(5p - 4q)$$

55. $6x^2 + \dfrac{49}{2}x + \dfrac{49}{2} = \dfrac{1}{2}\left(12x^2 + 49x + 49\right)$
$$= \dfrac{1}{2}\left[\left(12x^2 + 21x\right) + (28x + 49)\right]$$
$$= \dfrac{1}{2}\left[3x(4x + 7) + 7(4x + 7)\right]$$
$$= \dfrac{1}{2}(4x + 7)(3x + 7)$$

New base: $(4x + 7)$ feet
New height: $(3x + 7)$ feet
Shannon increased both the base and the height by 7 feet.

57. $2x^2 + 19x + 42 = \left(2x^2 + 12x\right) + (7x + 42)$
$$= 2x(x + 6) + 7(x + 6)$$
$$= (x + 6)(2x + 7)$$
$$= (x + 6)\left[(2x + 1) + 6\right]$$

The width of the free space is $6 \div 2 = 3$ meters. If the width of the court is 6 meters, then the width of the total area is $6 + 6 = 12$ meters and the length of the total area is $2(6) + 7 = 19$ meters. The total area of the badminton court and the free space is $12(19) = 228$ square meters.

10.4 Calculator Exercises

1. $96x^2 - 16x - 2 = 2\left(48x^2 - 8x - 1\right)$
$a = 48$, $b = -8$, $c = -1$

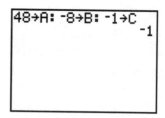

Let $Y1 = AC/x$ and $Y2 = x + Y1 = B$.

$$96x^2 - 16x - 2 = 2(48x^2 - 8x - 1)$$
$$= 2\left[(48x^2 + 4x) + (-12x - 1)\right]$$
$$= 2\left[4x(12x+1) - 1(12x+1)\right]$$
$$= 2(12x+1)(4x-1)$$

2. $a = 24$, $b = 7$, $c = -55$

Let $Y1 = AC/x$ and $Y2 = x + Y1 = B$.

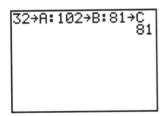

$$24x^2 + 7x - 55 = (24x^2 + 40x) + (-33x - 55)$$
$$= 8x(3x+5) - 11(3x+5)$$
$$= (3x+5)(8x-11)$$

3. $a = 32$, $b = 102$, $c = 81$

Let $Y1 = AC/x$ and $Y2 = x + Y1 = B$.

$$32x^2 + 102x + 81 = (32x^2 + 48x) + (54x + 81)$$
$$= 16x(2x+3) + 27(2x+3)$$
$$= (2x+3)(16x+27)$$

4. $a = 72$, $b = -99$, $c = 34$

Let $Y1 = AC/x$ and $Y2 = x + Y1 = B$.

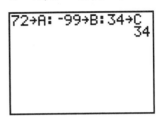

$$72x^2 - 99x + 34 = (72x^2 - 48x) + (-51x + 34)$$
$$= 24x(3x-2) - 17(3x-2)$$
$$= (3x-2)(24x-17)$$

5. $a = 4$, $b = 109$, $c = 225$

Let $Y1 = AC/x$ and $Y2 = x + Y1 = B$.

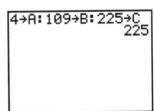

$$4p^4 + 109p^2 + 225$$
$$= \left(4p^4 + 9p^2\right) + \left(100p^2 + 225\right)$$
$$= p^2\left(4p^2 + 9\right) + 25\left(4p^2 + 9\right)$$
$$= \left(4p^2 + 9\right)\left(p^2 + 25\right)$$

10.5 Exercises

1. $3y^3 + 3y^2 + 3y = 3y\left(y^2 + y + 1\right)$

3. $10abc^2 + 15abc - 20ab = 5ab\left(2c^2 + 3c - 4\right)$

5. $-40a^2 - 24ab + 48ac = -8a\left(5a + 3b - 6c\right)$

7. $108x^5 - 75x^3 = 3x^3\left(36x^2 - 25\right)$
$$= 3x^3\left[(6x)^2 - 5^2\right]$$
$$= 3x^3(6x+5)(6x-5)$$

9. $-200x^3y + 32xy^3 = -8xy\left(25x^2 - 4y^2\right)$
$$= -8xy\left[(5x)^2 - (2y)^2\right]$$
$$= -8xy(5x+2y)(5x-2y)$$

11. $x^3 - 16x^2 + 64x = x\left(x^2 - 16x + 64\right)$
$$= x\left(x^2 - 2\cdot x\cdot 8 + 8^2\right)$$
$$= x(x-8)^2$$

13. $12u^2v + 36uv^2 + 27v^3$
$$= 3v\left(4u^2 + 12uv + 9v^2\right)$$
$$= 3v\left[(2u)^2 + 2(2u)(3v) + (3v)^2\right]$$
$$= 3v(2u+3v)^2$$

15. $3x^4 - 48x^2 + 7x^2 - 112 = 3x^2\left(x^2 - 16\right) + 7\left(x^2 - 16\right)$
$$= \left(x^2 - 16\right)\left(3x^2 + 7\right)$$
$$= \left(x^2 - 4^2\right)\left(3x^2 + 7\right)$$
$$= (x+4)(x-4)\left(3x^2 + 7\right)$$

17. $a = 4$, $b = -37$, $c = 9$, $ac = 36$
$$4p^4 - 37p^2 + 9 = \left(4p^4 - 36p^2\right) + \left(-p^2 + 9\right)$$
$$= 4p^2\left(p^2 - 9\right) - 1\left(p^2 - 9\right)$$
$$= \left(p^2 - 9\right)\left(4p^2 - 1\right)$$
$$= \left(p^2 - 3^2\right)\left[(2p)^2 - 1^2\right]$$
$$= (p+3)(p-3)(2p+1)(2p-1)$$

19. $2 + 8y - 42y^2 = -42y^2 + 8y + 2$
$$= -2\left(21y^2 - 4y - 1\right)$$

$a = 21$, $b = -4$, $c = -1$, $ac = -21$
$$= -2\left[\left(21y^2 - 7y\right) + (3y - 1)\right]$$
$$= -2\left[7y(3y - 1) + 1(3y - 1)\right]$$
$$= -2(3y - 1)(7y + 1)$$

21. $4u^2v^2 + 36uv + 56 = 4\left(u^2v^2 + 9uv + 14\right)$

$a = 1$, $b = 9$, $c = 14$

Factor	Factor	Sum of factors
1	14	15
2	7	9	← b

$4u^2v^2 + 36uv + 56 = 4(uv+2)(uv+7)$

23. $32x^3 - 64x^2 - 28x^2 + 56x$
$$= 4x\left(8x^2 - 16x - 7x + 14\right)$$
$$= 4x\left[8x(x-2) - 7(x-2)\right]$$
$$= 4x(x-2)(8x-7)$$

25. $12x^4 + 26x^3 - 30x^2$
$$= 2x^2\left(6x^2 + 13x - 15\right)$$

$a = 6$, $b = 13$, $c = -15$, $ac = -90$

$= 2x^2\left[(6x^2+18x)+(-5x-15)\right]$

$= 2x^2\left[6x(x+3)-5(x+3)\right]$

$= 2x^2(x+3)(6x-5)$

27. $x^6 - 5x^3y^3 + 3x^3y^3 - 15y^6$

$= x^3(x^3-5y^3)+3y^3(x^3-5y^3)$

$= (x^3-5y^3)(x^3+3y^3)$

29. $x^4 - 5x^3 + 8x^2 - 40x$

$= x(x^3-5x^2+8x-40)$

$= x\left[x^2(x-5)+8(x-5)\right]$

$= x(x-5)(x^2+8)$

31. $1-k^8 = 1^2 - (k^4)^2$

$= (1+k^4)(1-k^4)$

$= (1+k^4)\left[1^2-(k^2)^2\right]$

$= (1+k^4)(1+k^2)(1-k^2)$

$= (1+k^4)(1+k^2)(1^2-k^2)$

$= (1+k^4)(1+k^2)(1+k)(1-k)$

33. $14.75x - 0.95x^2 = x(14.75 - 0.95x)$

To figure the discount, Big Ed reduced the price of $14.75 for a pizza by $0.95 per pizza ordered.

Yes, Big Ed should limit the number of pizzas an individual may order to at the very most 8. If an individual were allowed to order 9 or more pizzas, the total charge would be less than the charge 8 pizzas. This would not be a good business move. (Plus, if an individual were allowed to order 16 or more pizzas, the charged would be negative.)

35. $19.95x + 2.99x^2 = x(19.95 + 2.99x)$

Tallmart adds a surcharge of $2.99 per doll purchased to the regular price of $19.95.

37. $135x - 48x^2 + 4x^3$

$= x(135 - 48x + 4x^2)$

$= x\left[(135-30x)+(-18x+4x^2)\right]$

$= x\left[15(9-2x)-2x(9-2x)\right]$

$= x(9-2x)(15-2x)$

The dimensions of the box are x inches, by $(9-2x)$ inches, by $(15-2x)$ inches.

If $x = 2$, then $9-2x = 5$ and $15-2x = 11$. The dimensions of the box will be 2 inches, by 5 inches, by 11 inches. The volume will be $2(5)(11) = 110$ in^3.

39. $\pi(4x^3 - 20x^2 + 25x)$

$= \pi x(4x^2 - 20x + 25)$

$= \pi x\left[(2x)^2 - 2(2x)(5) + 5^2\right]$

$= \pi x(2x-5)^2$

The radius is $(2x-5)$ inches.

If the height is $x = 8$ inches, then the radius is $2(8)-5 = 11$ inches and the volume is

$\pi(8)(2\cdot 8 - 5)^2 = 968\pi$ in^3 or about 3041 in^3.

41. $x^3 - 216 = x^3 - 6^3$

$= (x-6)(x^2 + x\cdot 6 + 6^2)$

$= (x-6)(x^2 + 6x + 36)$

No, the factors have no meaningful interpretation.

The carved-out cube is 6 inches on each edge.

10.5 Calculator Exercises

1. Let $Y1 = 16x^3 + 4x^2 - 264x + 189$.

SSM: Experiencing Introductory and Intermediate Algebra

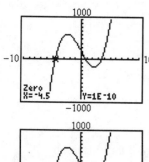
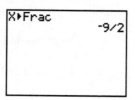
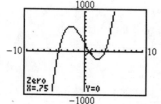
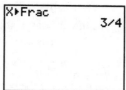
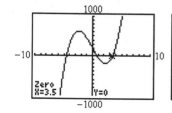
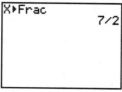

$x = -\dfrac{9}{2}$ $\quad x = \dfrac{3}{4}$ $\quad x = \dfrac{7}{2}$

$2x = -9$ $\quad\quad 4x = 3$ $\quad\quad 2x = 7$

$2x + 9 = 0$ $\quad 4x - 3 = 0$ $\quad 2x - 7 = 0$

$16x^3 + 4x^2 - 264x + 189$
$= (2x+9)(4x-3)(2x-7)$

2. Let $Y1 = 50x^3 + 25x^2 - 18x - 9$.

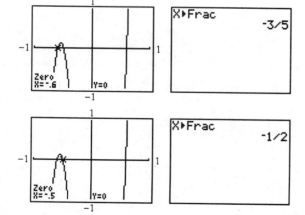

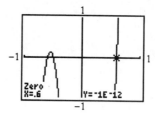
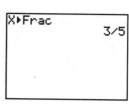

$x = -\dfrac{3}{5}$ $\quad x = -\dfrac{1}{2}$ $\quad x = \dfrac{3}{5}$

$5x = -3$ $\quad\quad 2x = -1$ $\quad\quad 5x = 3$

$5x + 3 = 0$ $\quad 2x + 1 = 0$ $\quad 5x - 3 = 0$

$50x^3 + 25x^2 - 18x - 9 = (5x+3)(2x+1)(5x-3)$

3. Let $Y1 = 90x^3 + 213x^2 - 53x - 140$.

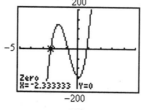
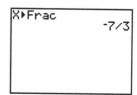

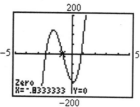
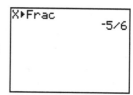

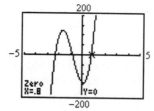
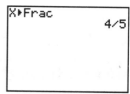

$x = -\dfrac{7}{3}$ $\quad x = -\dfrac{5}{6}$ $\quad x = \dfrac{4}{5}$

$3x = -7$ $\quad\quad 6x = -5$ $\quad\quad 5x = 4$

$3x + 7 = 0$ $\quad 6x + 5 = 0$ $\quad 5x - 4 = 0$

$90x^3 + 213x^2 - 53x - 140$
$= (3x+7)(6x+5)(5x-4)$

4. Let $Y1 = 18x^3 - 45x^2 - 128x + 320$.

432

Chapter 10: Factoring

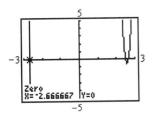 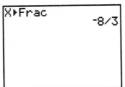

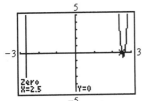 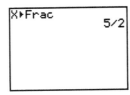

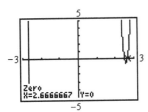 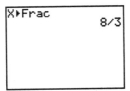

$x = -\dfrac{8}{3}$ $x = \dfrac{5}{2}$ $x = \dfrac{8}{3}$

$3x = -8$ $2x = 5$ $8x = 3$

$3x + 8 = 0$ $2x - 5 = 0$ $8x - 3 = 0$

$18x^3 - 45x^2 - 128x + 320$
$= (3x + 8)(2x - 5)(3x - 8)$

Chapter 10 Section-By-Section Review

1. $20a^6 - 28a^4 + 44a^2 = 4a^2(5a^4 - 7a^2 + 11)$

2. $22u^3v^2 + 22u^2v^3 = 22u^2v^2(u + v)$

3. $3x^3 + 3x + x^2 + 1 = (3x^3 + 3x) + (x^2 + 1)$
$= 3x(x^2 + 1) + 1(x^2 + 1)$
$= (x^2 + 1)(3x + 1)$

4. $7a^4 + 7a^2b^2 + 7a^2b^2 + 7b^4$
$= 7(a^4 + a^2b^2 + a^2b^2 + b^4)$
$= 7[a^2(a^2 + b^2) + b^2(a^2 + b^2)]$
$= 7(a^2 + b^2)(a^2 + b^2)$
$= 7(a^2 + b^2)^2$

5. $15ac + 18ad + 20bc + 24bd$
$= (15ac + 18ad) + (20bc + 24bd)$
$= 3a(5c + 6d) + 4b(5c + 6d)$
$= (5c + 6d)(3a + 4b)$

6. a. $\dfrac{1}{2}n^2 + \dfrac{1}{2}n = \dfrac{1}{2}n(n+1)$

   b. $\dfrac{1}{2}(12)^2 + \dfrac{1}{2}(12) = 78$
   $\dfrac{1}{2}(12)(12+1) = \dfrac{1}{2}(12)(13) = 78$
   The sum is 78.

   c. The factored expression was easier to evaluate. Reasons provided will vary.

7. $2x^2 - 6x + 5x - 15 = (2x^2 - 6x) + (5x - 15)$
$= 2x(x - 3) + 5(x - 3)$
$= (x - 3)(2x + 5)$

   The width is $(x - 3)$ units and the length is $(2x + 5)$ units.

8. $x + (x + 10) + (x + 20) + (x + 30) + (x + 40)$
$+ (x + 50) + (x + 60)$
$= 7x + 210$
   The employee's total pay for the 7 months was $7x + 210$.
   $7x + 210 = 7(x + 30)$

   The employee earned an average of $(x + 30)$ dollars per month for the 7-month period.

   $7(960 + 30) = 6930$
   The total pay for the 7 months was $6930.

433

SSM: Experiencing Introductory and Intermediate Algebra

9. $(24+2x)(15+2x)-(24)(15)$
$= 360+48x+30x+4x^2-360$
$= 4x^2+78x$

The area of the brick border is $(4x^2+78x)$ ft^2.

Factoring: $4x^2+78x = 2x(2x+39)$

Meaning: The area of the green strip is equivalent to a rectangle that measures $(2x)$ feet by $(2x+39)$ feet.

If $x = 3$: $2x(2x+39) = 2 \cdot 3(2 \cdot 3+39)$
$= 2 \cdot 3 \cdot 45$
$= 270$

If the brick boarder is 3 feet wide, then there are 270 ft^2 of boarder.

10. $p^2+12p+36 = p^2+2 \cdot p \cdot 6+6^2$
$= (p+6)^2$

11. $q^2-16q+64 = q^2-2 \cdot q \cdot 8+8^2$
$= (q-8)^2$

12. $9x^2+30x+25 = (3x)^2+2(3x)(5)+5^2$
$= (3x+5)^2$

13. $49y^2-112y+64 = (7y)^2-2(7y)(8)+8^2$
$= (7y-8)^2$

14. $a^2-9a+81$ does not factor. The first and last terms are both perfect squares, x^2 and 9^2, but the middle term is not $2 \cdot x \cdot 9$.

15. $x^7+6x^5y+9x^3y^2$
$= x^3(x^4+6x^2y+9y^2)$
$= x^3\left[(x^2)^2+2(x^2)(3y)+(3y)^2\right]$
$= x^3(x^2+3y)^2$

16. $x^2-169 = x^2-13^2$
$= (x+13)(x-13)$

17. $625-a^2 = 25^2-a^2$
$= (25+a)(25-a)$

18. $12x^2-75 = 3(4x^2-25)$
$= 3\left[(2x)^2-5^2\right]$
$= 3(2x+5)(2x-5)$

19. $p^2-q^2 = (p+q)(p-q)$

20. p^2+q^2 will not factor. This is a sum of two squares, not a difference of two squares.

21. $9x^2-25y^2 = (3x)^2-(5y)^2$
$= (3x+5y)(3x-5y)$

22. $16x^4-81 = (4x^2)^2-9^2$
$= (4x^2+9)(4x^2-9)$
$= (4x^2+9)\left[(2x)^2-3^2\right]$
$= (4x^2+9)(2x+3)(2x-3)$

23. $x^8-1 = (x^4)^2-1^2$
$= (x^4+1)(x^4-1)$
$= (x^4+1)\left[(x^2)^2-1^2\right]$
$= (x^4+1)(x^2+1)(x^2-1)$
$= (x^4+1)(x^2+1)(x^2-1^2)$
$= (x^4+1)(x^2+1)(x+1)(x-1)$

24. $c^3+27 = c^3+3^3$
$= (c+3)(c^2-c \cdot 3+3^2)$
$= (c+3)(c^2-3c+9)$

25. $c^3-27 = c^3-3^3$
$= (c-3)(c^2+c \cdot 3+3^2)$
$= (c-3)(c^2+3c+9)$

26. $8z^3 - 125$
$= (2z)^3 - 5^3$
$= (2z-5)\left[(2z)^2 + (2z)(5) + 5^2\right]$
$= (2z-5)(4z^2 + 10z + 25)$

27. $5h^3 + 40k^3 = 5(h^3 + 8k^3)$
$= 5\left[h^3 + (2k)^3\right]$
$= 5(h+2k)\left[h^2 - h(2k) + (2k)^2\right]$
$= 5(h+2k)(h^2 - 2hk + 4k^2)$

28. a. Let x = the length of each side of the flag.
Navy-blue area: $x^2 - 2^2 = x^2 - 4$

b. $x^2 - 4 = x^2 - 2^2 = (x+2)(x-2)$
A rectangle with an equivalent area will have dimensions of $(x + 2)$ ft by $(x - 2)$ ft.

c. $x^2 - 4 = (5)^2 - 4 = 21$
$3(8) = 24$
No, the area of the 3 ft by 8 ft rectangular piece of material square has a larger area than is needed for the flag.

29. $\pi(x^2 + 200x + 10000) = \pi(x^2 + 2 \cdot x \cdot 100 + 100^2)$
$= \pi(x+100)^2$
The radius of the pond is $(x + 100) - x = 100$ ft.
Total area: $\pi(x+100)^2 = \pi(20+100)^2$
$= \pi(120)^2$
$= 14400\pi$ ft^2
$\approx 45{,}239$ ft^2
Area of pond: $\pi \cdot 100^2 = 10000\pi$ ft^2.
$\approx 31{,}416$ ft^2
Area of exercise space:
$14400\pi - 10000\pi = 4400\pi$ ft^2
$\approx 13{,}823$ ft^2

30. $z^2 + 2z - 99$
$a = 1, b = 2, c = -99$

Factor	Factor	Sum of factors	
−1	99	98	
1	−99	−98	
−3	33	30	
3	−33	−30	
−9	11	2	←b
9	−11	−2	

$z^2 + 2z - 99 = (z-9)(u+11)$

31. $p^2 + 5pq - 66q^2$
$a = 1, b = 5, c = -66$

Factor	Factor	Sum of factors	
−1	66	65	
1	−66	−65	
−2	33	31	
2	−33	−31	
−3	22	19	
3	−22	−19	
−6	11	5	←b
6	−11	−5	

$p^2 + 5pq - 66q^2 = (p-6q)(p+11q)$

32. $6a^2 + 96a + 234 = 6(a^2 + 16a + 39)$
$a = 1, b = 16, c = 39$

Factor	Factor	Sum of factors	
1	13	14	
3	13	16	←b

$6a^2 + 96a + 234 = 6(a+3)(a+13)$

33. $x^4 + 8x^2 + 15$
$a = 1, b = 8, c = 15$

SSM: Experiencing Introductory and Intermediate Algebra

Factor	Factor	Sum of factors
1	15	16
3	5	8	←b

$x^4 + 8x^2 + 15 = (x^2 + 3)(x^2 + 5)$

34. $4q^3 - 28q^2 - 240q = 4q(q^2 - 7q - 60)$
    $a = 1, b = -7, c = -60$

Factor	Factor	Sum of factors
−1	60	59
1	−60	−59
−2	30	28
2	−30	−28
−3	20	17
3	−20	−17
−4	15	11
4	−15	−11
−5	12	7
5	−12	−7	←b
−6	10	4
6	−10	−4

$4q^3 - 28q^2 - 240q = 4q(q+5)(q-12)$

35. $x^2 y^2 - 4xy - 117$
    $a = 1, b = -4, c = -117$

Factor	Factor	Sum of factors
−1	117	116
1	−117	−116
−3	39	36
3	−39	−36
−9	13	4
9	−13	−4	←b

$x^2 y^2 - 4xy - 117 = (xy + 9)(xy - 13)$

36. $-7x^2 + 98x - 168 = -7(x^2 - 14x + 24)$
    $a = 1, b = -14, c = 24$

Factor	Factor	Sum of factors
−1	−24	−25
−2	−12	−14	←b
−3	−8	−11
−4	−6	−10

$-7x^2 + 98x - 168 = -7(x - 2)(x - 12)$

37. $2x^2 - 11x + 5 = (\_\_\ \_\_)(\_\_\ \_\_)$

Possible factors	Middle term
$(x - 1)(2x - 5)$	$-5x - 2x = -7x$
$(x - 5)(2x - 1)$	$-x - 10x = -11x$	←bx

$2x^2 - 11x + 5 = (x - 5)(2x - 1)$

38. $6x^2 + 17x + 5 = (\_\_\ \_\_)(\_\_\ \_\_)$

Possible factors	Middle term
$(x + 1)(6x + 5)$	$5x + 6x = 11x$
$(x + 5)(6x + 1)$	$x + 30x = 31x$
$(2x + 1)(3x + 5)$	$10x + 3x = 13x$
$(2x + 5)(3x + 1)$	$2x + 15x = 17x$	←bx

$6x^2 + 17x + 5 = (2x + 5)(3x + 1)$

39. $28a^2 b^2 + 91ab + 21 = 7(4a^2 b^2 + 13ab + 3)$
    $= 7(\_\_\ \_\_)(\_\_\ \_\_)$

Possible factors	Middle term
$(ab + 1)(4ab + 3)$	$3ab + 4ab = 7ab$
$(ab + 3)(4ab + 1)$	$ab + 12ab = 13ab$	←bx
$(2ab + 1)(2ab + 3)$	$6ab + 2ab = 8ab$

$28a^2 b^2 + 91ab + 21 = 7(ab + 3)(4ab + 1)$

Chapter 10: Factoring

40. $-45x^3 - 102x^2 + 48x = -3x(15x^2 + 34x - 16)$
$= -3x(\_\_ \_\_)(\_\_ \_\_)$

Possible factors	Middle term
$(x-1)(15x+16)$	$16x - 15x = x$
$(x+1)(15x-16)$	$-16x + 15x = -x$
$(x-16)(15x+1)$	$x - 240x = -239x$
$(x+16)(15x-1)$	$-x + 240x = 239x$
$(x-2)(15x+8)$	$8x - 30x = -22x$
$(x+2)(15x-8)$	$-8x + 30x = 22x$
$(x-8)(15x+2)$	$2x - 120x = -118x$
$(x+8)(15x-2)$	$-2x + 120x = 118x$
$(x-4)(15x+4)$	$4x - 60x = -56x$
$(x+4)(15x-4)$	$-4x + 60x = 56x$
$(3x-1)(5x+16)$	$48x - 5x = 43x$
$(3x+1)(5x-16)$	$-48x + 5x = -43x$
$(3x-16)(5x+1)$	$3x - 80x = -77x$
$(3x+16)(5x-1)$	$-3x + 80x = 77x$
$(3x-2)(5x+8)$	$24x - 10x = 14x$
$(3x+2)(5x-8)$	$-24x + 10x = -14x$
$(3x-8)(5x+2)$	$6x - 40x = -34x$
$(3x+8)(5x-2)$	$-6x + 40x = 34x$ ← bx
$(3x-4)(5x+4)$	$12x - 20x = -8x$
$(3x+4)(5x-4)$	$-12x + 20x = 8x$

$-45x^3 - 102x^2 + 48x = -3x(3x+8)(5x-2)$

41. a. $2x^2 + 10x + \dfrac{21}{2} = \dfrac{1}{2}(4x^2 + 20x + 21)$
$= \dfrac{1}{2}(\_\_ \_\_)(\_\_ \_\_)$

Possible factors	Middle term
$(x+1)(4x+21)$	$21x + 4x = 25x$
$(x+21)(4x+1)$	$x + 84x = 85x$
$(x+3)(4x+7)$	$7x + 12x = 19x$
$(x+7)(4x+3)$	$3x + 28x = 31x$
$(2x+1)(2x+21)$	$42x + 2x = 44x$
$(2x+3)(2x+7)$	$14x + 6x = 20x$ ← bx

$2x^2 + 10x + \dfrac{21}{2} = \dfrac{1}{2}(2x+3)(2x+7)$

The base is $(2x+7)$ units and the height is $(2x+3)$ units.

b. $(2x+7) - 2x = 7$
The base was increased by 7 units.

c. $(2x+3) - x = x+3$
The height was increased by $(x+3)$ units.

42. a. $x^2 + 17x + 30$
$a = 1,\ b = 17,\ c = 30$

Factor	Factor	Sum of factors
1	30	31
2	15	17 ← b
3	10	13
5	6	11

$x^2 + 17x + 30 = (x+2)(x+15)$

The new length is $(x+15)$ inches and the new width is $(x+2)$ inches.

b. $(x+15) - x = 15$
The length was increased by 15 inches.

c. $(x+2) - (x-6) = 8$
The width was increased by 8 inches.

43. $420 + 86x + 4x^2 = 4x^2 + 86x + 420$
$= 2(2x^2 + 43x + 210)$

Possible factors	Middle term
$(x+1)(2x+210)$	$210x + 2x = 212x$
$(x+210)(2x+1)$	$x + 420x = 421x$
$(x+2)(2x+105)$	$210x + 4x = 214x$

SSM: Experiencing Introductory and Intermediate Algebra

$(x+105)(2x+2)$	$2x+210x=212x$
$(x+3)(2x+70)$	$70x+6x=76x$
$(x+70)(2x+3)$	$3x+140x=143x$
$(x+5)(2x+42)$	$42x+10x=52x$
$(x+42)(2x+5)$	$5x+84x=89x$
$(x+6)(2x+35)$	$35x+12x=47x$
$(x+35)(2x+6)$	$6x+70x=76x$
$(x+7)(2x+30)$	$30x+14x=44x$
$(x+30)(2x+7)$	$7x+60x=67x$
$(x+10)(2x+21)$	$21x+20x=41x$
$(x+21)(2x+10)$	$10x+42x=52x$
$(x+14)(2x+15)$	$15x+28x=43x$ ← bx
$(x+15)(2x+14)$	$14x+30x=44x$

$420+86x+4x^2 = 4x^2+86x+420$
$\qquad = 2(2x^2+43x+210)$
$\qquad = 2(x+14)(2x+15)$
$\qquad = (2x+28)(2x+15)$
$\qquad = (28+2x)(15+2x)$

Length: $(28+2x)-2x = 28$ meters.
Width: $(15+2x)-2x = 15$ meters.
The dimensions of the playing area of the basketball court are 28 meters by 15 meters.

44. $a=3$, $b=23$, $c=4$, $ac=12$
Because no two factors of 12 have a sum of 23, $3x^2+23x+4$ does not factor.

45. $a=8$, $b=22$, $c=9$, $ac=72$
$8x^2+22x+9 = (8x^2+18x)+(4x+9)$
$\qquad = 2x(4x+9)+1(4x+9)$
$\qquad = (4x+9)(2x+1)$

46. $a=15$, $b=-29$, $c=12$, $ac=180$
$15y^2-29y+12 = (15y^2-20y)+(-9y+12)$
$\qquad = 5y(3y-4)-3(3y-4)$
$\qquad = (3y-4)(5y-3)$

47. $a=10$, $b=29$, $c=24$, $ac=240$
Because no two factors of 240 have a sum of 29, $10x^2+29x+24$ does not factor.

48. $a=12$, $b=-13$, $c=-14$, $ac=-168$
$12y^2-13y-14 = (12y^2-21y)+(8y-14)$
$\qquad = 3y(4y-7)+2(4y-7)$
$\qquad = (4y-7)(3y+2)$

49. $a=8$, $b=22$, $c=15$, $ac=120$
$8x^4+22x^2+15 = (8x^4+12x^2)+(10x^2+15)$
$\qquad = 4x^2(2x^2+3)+5(2x^2+3)$
$\qquad = (2x^2+3)(4x^2+5)$

50. $a=10$, $b=9$, $c=-9$, $ac=-90$
$10x^4+9x^2-9 = (10x^4+15x^2)+(-6x^2-9)$
$\qquad = 5x^2(2x^2+3)-3(2x^2+3)$
$\qquad = (2x^2+3)(5x^2-3)$

51. $a=12$, $b=-35$, $c=9$, $ac=108$
Because no two factors of 108 have a sum of -35, $12x^2-35x+9$ does not factor.

52. $a=6$, $b=-29$, $c=-28$, $ac=-168$
Because no two factors of -168 have a sum of -29, $6p^4-29p^2-28$ does not factor.

53. $24x^2+76x+32 = 4(6x^2+19x+8)$
$a=6$, $b=19$, $c=8$, $ac=48$
$24x^2+76x+32 = 4(6x^2+19x+8)$
$\qquad = 4\left[(6x^2+16x)+(3x+8)\right]$
$\qquad = 4\left[2x(3x+8)+1(3x+8)\right]$
$\qquad = 4(3x+8)(2x+1)$

54. $-40x^4 - 30x^2 + 175 = -5(8x^4 + 6x^2 - 35)$

$a = 8,\ b = 6,\ c = -35,\ ac = -280$

$-40x^4 - 30x^2 + 175$
$= -5(8x^4 + 6x^2 - 35)$
$= -5\left[(8x^4 + 20x^2) + (-14x^2 - 35)\right]$
$= -5\left[4x^2(2x^2 + 5) - 7(2x^2 + 5)\right]$
$= -5(2x^2 + 5)(4x^2 - 7)$

55. $a = 6,\ b = -7,\ c = -20,\ ac = -120$

$6x^2 - 7xy - 20y^2 = (6x^2 - 15xy) + (8xy - 20y^2)$
$= 3x(2x - 5y) + 4y(2x - 5y)$
$= (2x - 5y)(3x + 4y)$

56. $a = 12,\ b = 40,\ c = 25,\ ac = 300$

$12x^2 + 40xy + 25y^2$
$= (12x^2 + 30xy) + (10xy + 25y^2)$
$= 6x(2x + 5y) + 5y(2x + 5y)$
$= (2x + 5y)(6x + 5y)$

57. $a = 20,\ b = -27,\ c = 9,\ ac = 180$

$20x^2 - 27xy + 9y^2$
$= (20x^2 - 15xy) + (-12xy + 9y^2)$
$= 5x(4x - 3y) - 3y(4x - 3y)$
$= (4x - 3y)(5x - 3y)$

58. $19x^2 + 66x + 27 = (19x^2 + 57x) + (9x + 27)$
$= 19x(x + 3) + 9(x + 3)$
$= (x + 3)(19x + 9)$

The total area is $(x + 3)$ meters wide and $(19x + 9)$ meters long.

If the width of the lane is 1 meter, then the width of the total area is $1 + 3 = 4$ meters and the length of the total area is $19(1) + 9 = 28$ meters. The total area is $4 \cdot 28 = 112$ square meters.

59. $-12x^3 + 60x^2 y - 75xy^2$
$= -3x(4x^2 - 20xy + 25y^2)$
$= -3x\left[(2x)^2 - 2(2x)(5y) + (5y)^2\right]$
$= -3x(2x - 5y)^2$

60. $7x^4 + 7x^3 + 7x^2 = 7x^2(x^2 + x + 1)$

61. $12x^3 - 243x = 3x(4x^2 - 81)$
$= 3x\left[(2x)^2 - 9^2\right]$
$= 3x(2x + 9)(2x - 9)$

62. $32x^3 + 32x^2 + 8x = 8x(4x^2 + 4x + 1)$
$= 8x\left[(2x)^2 + 2(4x)(1) + 1^2\right]$
$= 8x(2x + 1)^2$

63. $24x^3 - 14x^2 - 90x = 2x(12x^2 - 7x - 45)$

$a = 12,\ b = -7,\ c = -45,\ ac = -540$

$24x^3 - 14x^2 - 90x = 2x(12x^2 - 7x - 45)$
$= 2x(12x^2 - 27x + 20x - 45)$
$= 2x\left[3x(4x - 9) + 5(4x - 9)\right]$
$= 2x(4x - 9)(3x + 5)$

64. $256x^4 - 288x^2 + 81$
$= (16x^2)^2 - 2(16x^2)(9) + 9^2$
$= (16x^2 - 9)^2$
$= (16x^2 - 9)(16x^2 - 9)$
$= \left[(4x)^2 - 3^2\right]\left[(4x)^2 - 3^2\right]$
$= (4x + 3)(4x - 3)(4x + 3)(4x - 3)$
$= (4x + 3)^2 (4x - 3)^2$

SSM: Experiencing Introductory and Intermediate Algebra

65. $a = 36, b = -25, c = 4, ac = 144$

$36x^4 - 25x^2 + 4$
$= \left(36x^4 - 16x^2\right) + \left(-9x^2 + 4\right)$
$= 4x^2\left(9x^2 - 4\right) - 1\left(9x^2 - 4\right)$
$= \left(9x^2 - 4\right)\left(4x^2 - 1\right)$
$= \left[(3x)^2 - 2^2\right]\left[(2x)^2 - 1^2\right]$
$= (3x+2)(3x-2)(2x+1)(2x-1)$

66. $2x^4 + 14x^3 - 8x^2 - 56x$
$= 2x\left(x^3 + 7x^2 - 4x - 28\right)$
$= 2x\left[x^2(x+7) - 4(x+7)\right]$
$= 2x(x+7)\left(x^2 - 4\right)$
$= 2x(x+7)\left(x^2 - 2^2\right)$
$= 2x(x+7)(x+2)(x-2)$

67. a. $x(4x) - 5^2 = 4x^2 - 25$

   The land area not covered by the statue is $\left(4x^2 - 25\right)$ ft².

   b. $4x^2 - 25 = (2x)^2 - 5^2 = (2x+5)(2x-5)$
   The dimensions of a rectangle with area equal to the uncovered area are $(2x+5)$ feet by $(2x-5)$ feet.

   c. If $x = 80$, then $2x + 5 = 2 \cdot 80 + 5 = 165$ and $2x - 5 = 2 \cdot 80 - 5 = 155$.
   The dimensions of a rectangle would be 165 feet by 155 feet.

68. $19.95x - 0.10x^2 = x(19.95 - 0.10x)$

   The regular price of $19.95 per person is reduced by $0.10 per person attending the dinner.

Chapter 10 Chapter Review

1. $z^2 + 9z - 90$
   $a = 1, b = 9, c = -90$

Factor	Factor	Sum of factors	
−1	90	89	
1	−90	−89	
−2	45	43	
2	−45	−43	
−3	30	27	
3	−30	−27	
−5	18	13	
5	−18	−13	
−6	15	9	←b
6	−15	−9	
−9	10	1	
9	−10	−1	

$z^2 + 9z - 90 = (z-6)(z+15)$

2. $a^2 - 18a + 72$
   $a = 1, b = -18, c = 72$

Factor	Factor	Sum of factors	
−1	−72	−73	
−2	−36	−38	
−3	−24	−27	
−4	−18	−22	
−6	−12	−18	←b
−8	−9	−17	

$a^2 - 18a + 72 = (a-6)(a-12)$

3. $x^2 + 14xy + 45y^2$
   $a = 1, b = 14, c = 45$

Factor	Factor	Sum of factors	
1	45	46	
3	15	18	
5	9	14	←b

$x^2 + 14xy + 45y^2 = (x+5y)(x+9y)$

440

4. $5a^2 + 70a + 245 = 5(a^2 + 14a + 49)$
$= 5(a^2 + 2 \cdot a \cdot 7 + 7^2)$
$= 5(a+7)^2$

5. $2a^2 + 8ab + 12b^2 = 2(a^2 + 4ab + 6b^2)$

6. $x^4 + 10x^2 + 21$
$a = 1, b = 10, c = 21$

Factor	Factor	Sum of factors
1	21	22
3	7	10	←b

$x^4 + 10x^2 + 21 = (x^2 + 3)(x^2 + 7)$

7. $3q^3 - 33q^2 - 126q = 3q(q^2 - 11q - 42)$
$a = 1, b = -11, c = -42$

Factor	Factor	Sum of factors
−1	42	41
1	−42	−41
−2	21	19
2	−21	−19
−3	14	11
3	−14	−11	←b
−6	7	1
6	−7	−1

$3q^3 - 33q^2 - 126q = 3q(q+3)(q-14)$

8. $-6x^2 + 42x + 360 = -6(x^2 - 7x - 60)$
$a = 1, b = -7, c = -60$

Factor	Factor	Sum of factors
−1	60	59
1	−60	−59
−2	30	28
2	−30	−28
−3	20	17
3	−20	−17
−4	15	11
4	−15	−11
−5	12	7
5	−12	−7	←b
−6	10	4
6	−10	−4

$-6x^2 + 42x + 360 = -6(x+5)(x-12)$

9. $10 + 7x + x^2 = x^2 + 7x + 10$
$a = 1, b = 7, c = 10$

Factor	Factor	Sum of factors
1	10	11
2	5	7	←b

$10 + 7x + x^2 = (x+2)(x+5)$

10. $x^2 - 289 = x^2 - 17^2$
$= (x+17)(x-17)$

11. $x^3 - 1 = x^3 - 1^3$
$= (x-1)(x^2 + x \cdot 1 + 1^2)$
$= (x-1)(x^2 + x + 1)$

12. $4x^2 - 64 = 4(x^2 - 16)$
$= 4(x+4)(x-4)$

13. $u^2 + v^2$ will not factor. This is a sum of two squares, not a difference of two squares.

14. $36x^2 - 49y^2 = (6x)^2 - (7y)^2$
$= (6x + 7y)(6x - 7y)$

SSM: Experiencing Introductory and Intermediate Algebra

15. $81x^4 - 1 = (9x^2)^2 - 1^2$
$= (9x^2 + 1)(9x^2 - 1)$
$= (9x^2 + 1)\left[(3x)^2 - 1^2\right]$
$= (9x^2 + 1)(3x + 1)(3x - 1)$

16. $p^2 + 22p + 121 = p^2 + 2 \cdot p \cdot 11 + 11^2$
$= (p + 11)^2$

17. $q^2 - 30q + 225 = q^2 - 2 \cdot q \cdot 15 + 15^2$
$= (q - 15)^2$

18. $27a^3 + 64b^3$
$= (3a)^3 + (4b)^3$
$= (3a + 4b)\left[(3a)^2 - (3a)(4b) + (4b)^2\right]$
$= (3a + 4b)(9a^2 - 12ab + 16b^2)$

19. $27a^2b^2 - 72ab + 48$
$= 3(9a^2b^2 - 24ab + 16)$
$= 3\left[(3ab)^2 - 2(3ab)(4) + 4^2\right]$
$= 3(3ab - 4)^2$

20. $-50x^3 + 120x^2y - 72xy^2$
$= -2x(25x^2 - 60xy + 36y^2)$
$= -2x\left[(5x)^2 - 2(5x)(6y) + (6y)^2\right]$
$= -2x(5x - 6y)^2$

21. $8x^4 - 2x^3 + 6x^2 - 12x = 2x(4x^3 - x^2 + 3x - 6)$

22. $35u^3v^2 + 25u^2v^3 = 5u^2v^2(7u + 5v)$

23. $2x^3 + 10x + x^2 + 5$
$= 2x(x^2 + 5) + 1(x^2 + 5)$
$= (x^2 + 5)(2x + 1)$

24. $m^2 - 2mn - 8mn + 16n^2$
$= m(m - 2n) - 8n(m - 2n)$
$= (m - 2n)(m - 8n)$

25. $4a^4 + 8a^2b^2 + 8a^2b^2 + 16b^4$
$= 4(a^4 + 2a^2b^2 + 2a^2b^2 + 4b^4)$
$= 4\left[a^2(a^2 + 2b^2) + 2b^2(a^2 + 2b^2)\right]$
$= 4(a^2 + 2b^2)(a^2 + 2b^2)$
$= 4(a^2 + 2b^2)^2$

26. $2x^2 - 13x + 11$
$a = 2, b = -13, c = 11, ac = 22$
$2x^2 - 13x + 11 = (2x^2 - 2x) + (-11x + 11)$
$= 2x(x - 1) - 11(x - 1)$
$= (x - 1)(2x - 11)$

27. $24ac + 20ad + 18bc + 15bd$
$= 4a(6c + 5d) + 3b(6c + 5d)$
$= (6c + 5d)(4a + 3b)$

28. $7x^2 - 19x - 6$
$a = 7, b = -19, c = -6, ac = -42$
$7x^2 - 19x - 6 = (7x^2 - 21x) + (2x - 6)$
$= 7x(x - 3) + 2(x - 3)$
$= (x - 3)(7x + 2)$

29. $10x^2 - 11x - 6$
$a = 10, b = -11, c = -6, ac = -60$
$10x^2 - 11x - 6 = (10x^2 - 15x) + (4x - 6)$
$= 5x(2x - 3) + 2(2x - 3)$
$= (2x - 3)(5x + 2)$

30. $36a^2 + 66a + 24 = 6(6a^2 + 11a + 4)$
$a = 6, b = 11, c = 4, ac = 24$
$36a^2 + 66a + 24 = 6(6a^2 + 11a + 4)$
$= 6\left[(6a^2 + 8a) + (3a + 4)\right]$
$= 6\left[2a(3a + 4) + 1(3a + 4)\right]$
$= 6(3a + 4)(2a + 1)$

31. $-30x^2 - 28x + 32 = -2(15x^2 + 14x - 16)$
$a = 15, b = 14, c = -16, ac = -240$
$-30x^2 - 28x + 32$
$= -2(15x^2 + 14x - 16)$
$= -2\left[(15x^2 + 10x) + (-24x - 16)\right]$
$= -2\left[5x(3x + 2) - 8(3x + 2)\right]$
$= -2(3x + 2)(5x - 8)$

32. $12x^4 + 13x^2 + 3$
$a = 12, b = 13, c = 3, ac = 36$
$12x^4 + 13x^2 + 3 = (12x^4 + 9x^2) + (4x^2 + 3)$
$= 3x^2(4x^2 + 3) + 1(4x^2 + 3)$
$= (4x^2 + 3)(3x^2 + 1)$

33. $54x^3 + 36x^2 + 6x = 6x(9x^2 + 6x + 1)$
$= 6x\left[(3x)^2 + 2(3x)(1) + 1^2\right]$
$= 6x(3x + 1)^2$

34. $81x^4 - 72x^2 + 16 = (9x^2)^2 - 2(9x^2)(4) + 4^2$
$= (9x^2 - 4)^2 = \left[(3x)^2 - 2^2\right]^2$
$= \left[(3x + 2)(3x - 2)\right]^2$
$= (3x + 2)^2 (3x - 2)^2$

35. $4x^4 - 61x^2 + 225$
$= (4x^4 - 25x^2) + (-36x^2 + 225)$
$= x^2(4x^2 - 25) - 9(4x^2 - 25)$
$= (4x^2 - 25)(x^2 - 9)$
$= \left[(2x)^2 - 5^2\right](x^2 - 3^2)$
$= (2x + 5)(2x - 5)(x + 3)(x - 3)$

36. $12x^3 + 18x^2 - 30x^2 - 45x$
$= 3x(4x^2 + 6x - 10x - 15)$
$= 3x\left[2x(2x + 3) - 5(2x + 3)\right]$
$= 3x(2x + 3)(2x - 5)$

37. $3x^4 + 15x^3 - 27x^2 - 135x$
$= 3x(x^3 + 5x^2 - 9x - 45)$
$= 3x\left[x^2(x + 5) - 9(x + 5)\right]$
$= 3x(x + 5)(x^2 - 9)$
$= 3x(x + 5)(x^2 - 3^2)$
$= 3x(x + 5)(x + 3)(x - 3)$

38. $8x^2 - 2x - 3$
$a = 8, b = -2, c = -3, ac = -24$
$8x^2 - 2x - 3 = (8x^2 - 6x) + (4x - 3)$
$= 2x(4x - 3) + 1(4x - 3)$
$= (4x - 3)(2x + 1)$
The width is $(2x + 1)$ feet and the length is $(4x - 3)$ feet.

SSM: Experiencing Introductory and Intermediate Algebra

39. a. $6x^2 - 10x - 4 = 2(3x^2 - 5x - 2)$
$a = 3, b = -5, c = -2, ac = -6$
$6x^2 - 10x - 4 = 2(3x^2 - 5x - 2)$
$= 2\left[(6x^2 + 2x) + (-12x - 4)\right]$
$= 2\left[x(3x+1) - 2(3x+1)\right]$
$= 2(3x+1)(x-2)$
$= (3x+1)(2x-4)$
The new length is $(3x+1)$ inches and the new width is $(2x-4)$ inches.

b. $(3x+1) - x = 2x+1$
The length was increased by $(2x+1)$ inches.

c. $(2x-4) - (x-2) = x - 2$
The width was increased by $(x-2)$ inches.

40. a. $a = 1, b = 12, c = 32$
$x^2 + 12x + 32 = (x+8)(x+4)$
$= \frac{1}{2} \cdot 2 \cdot (x+8)(x+4)$
$= \frac{1}{2}(2x+16)(x+4)$
The base is $(2x+16)$ feet and the height is $(x+4)$ feet.

b. $(2x+16) - 2x = 16$
The base was increased by 16 feet.

c. $(x+4) - x = 4$
The height was increased by 4 feet.

41. $50x + 25x^2 = 25x(2+x) = x(50+25x)$
The fine for the first offense is \$75. After that, an additional \$25 per offense is added to each subsequent fine amount.

42. a. Let x = length of each side of the calendar.
$x^2 - 3^2 = x^2 - 9$

b. $x^2 - 9 = x^2 - 3^2 = (x+3)(x-3)$
The dimensions of a rectangle with area equal to the area not covered by the cartoon are $(x+3)$ inches by $(x-3)$ inches.

c. Yes. If $x = 5$, then the area not covered by the cartoon is $x^2 - 9 = 5^2 - 9 = 16$ square inches. The area covered by the cartoon is $3^2 = 9$ square inches.

43. $\pi(372100 + 1220x + x^2)$
$= \pi(610^2 + 2 \cdot 610 \cdot x + x^2)$
$= \pi(610 + x)^2$
The radius of the face and border of the target is $(610+x)$ mm.

The radius of the face of the target (not including the border) is $(610+x) - x = 610$ mm.

The area of the face of the target (not including the border) is $\pi \cdot 610^2 = 372,100\pi$ mm^2, or approximately 1,168,987 mm^2.

44. $(68+2x)(88+2x) - (68)(88) = 312x + 4x^2$
$312x + 4x^2 = 4x(78+x)$.
The area of the border is equivalent to the area of a rectangle measuring $4x$ inches by $(78+x)$ inches.

If $x = 16$, then the area of the border is
$312x + 4x^2 = 312(16) + 4(16)^2 = 6016$ square inches.

Chapter 10 Test

1. $81a^3 + 54a^2 + 9a = 9a(9a^2 + 6a + 1)$
$= 9a\left[(3a)^2 + 2(3a)(1) + 1^2\right]$
$= 9a(3a+1)^2$

2. $p^3 + 125 = p^3 + 5^3$
$= (p+5)(p^2 - p \cdot 5 + 5^2)$
$= (p+5)(p^2 - 5p + 25)$

3. $-8a^4b^2 - 36a^3b^3 - 16a^2b^4$
$= -4a^2b^2(2a^2 + 9ab + 4b^2)$
$a = 2, b = 9, c = 4, ac = 8$
$-8a^4b^2 - 36a^3b^3 - 16a^2b^4$
$= -4a^2b^2(2a^2 + 9ab + 4b^2)$
$= -4a^2b^2\left[(2a^2 + 8ab) + (ab + 4b^2)\right]$
$= -4a^2b^2\left[2a(a+4b) + b(a+4b)\right]$
$= -4a^2b^2(a+4b)(2a+b)$

4. $a(a^2+b^2) - 5b(a^2+b^2) = (a^2+b^2)(a-5b)$

5. $15x^2 - 21xy + 10xy - 14y^2$
$= 3x(5x-7y) + 2y(5x-7y)$
$= (5x-7y)(3x+2y)$

6. $64a^2 - 49b^2 = (8a)^2 - (7b)^2$
$= (8a+7b)(8a-7b)$

7. $25x^2 - 70x + 49 = (5x)^2 - 2(5x)(7) + 7^2$
$= (5x-7)^2$

8. $3x^3 - 27x^2 + 24x = 3x(x^2 - 9x + 8)$
$a = 1, b = -9, c = 8$

Factor	Factor	Sum of factors	
−1	−8	−9	←b
−2	−4	−6	

$3x^3 - 27x^2 + 24x = 3x(x-1)(x-8)$

9. $a = 1, b = -4, c = -21$

Factor	Factor	Sum of factors	
−1	21	20	
1	−21	−20	
−3	7	4	
3	−7	−4	←b

$x^2 - 4xy - 21y^2 = (x-7y)(x+3y)$

10. $a = 14, b = 25, c = 9, ac = 126$
$14x^2 + 25x + 9 = (14x^2 + 18x) + (7x + 9)$
$= 2x(7x+9) + 1(7x+9)$
$= (7x+9)(2x+1)$

11. $a = 4, b = 27, c = -7, ac = -28$
$4x^4 + 27x^2 - 7 = (4x^4 + 28x^2) + (-x^2 - 7)$
$= 4x^2(x^2+7) - 1(x^2+7)$
$= (x^2+7)(4x^2-1)$
$= (x^2+7)\left[(2x)^2 - 1^2\right]$
$= (x^2+7)(2x+1)(2x-1)$

12. $a = 1, b = 8, c = 14$
Because no two factors of 14 have a sum of 8, $x^2 + 8x + 14$ does not factor.

13. $2x^3 + 5x^2 - 3x = x(2x^2 + 5x - 3)$
$a = 2, b = 5, c = -3, ac = -6$
$2x^3 + 5x^2 - 3x = x(2x^2 + 5x - 3)$
$= x\left[(2x^2 + 6x) + (-x - 3)\right]$
$= x\left[2x(x+3) - 1(x+3)\right]$
$= x(x+3)(2x-1)$

The length of the box is $(x+3)$ inches and the width of the box is $(2x-1)$ inches.

14. a. $\frac{1}{2}x^2 + \frac{17}{2}x + 30 = \frac{1}{2}(x^2 + 17x + 60)$
$a = 1, b = 17, c = 60$

SSM: Experiencing Introductory and Intermediate Algebra

Factor	Factor	Sum of factors
1	60	61
2	30	32
3	20	23
2	15	19
5	12	17 ← b
6	10	16

$$\frac{1}{2}x^2 + \frac{17}{2}x + 30 = \frac{1}{2}(x+5)(x+12)$$

The base of the new triangle is $(x+12)$ inches and the height of the new triangle is $(x+5)$ inches.

b. $(x+5) - x = 5$
$(x+12) - (x+7) = 5$
The base and height were both increased by 5 inches.

c. The base of the original triangle was $(x+7)$ inches.

15. $20x + 5x^2 = 5x(4+x) = x(20+5x)$
Edgar's parents gave him $25 for passing the first test. After that, they added $5 per passed test to the each subsequent reward amount.

16. a. Let s = the length of each edge of the crate.
$s^3 - 2^3 = s^3 - 8$

b. $s^3 - 8 = s^3 - 2^3$
$= (s-2)(s^2 + s \cdot 2 + 2^2)$
$= (s-2)(s^2 + 2s + 4)$
No. The factors have no meaningful physical interpretation. Because the volume of a cube is given by the formula $V = s^3$, the expression would have to factor into a perfect cube in order for the factors to be the length of each edge of the cube. The expression above is not a perfect cube.

d. If $x = 12$ inches, then the amount of gelatin required to fill the crate along with the boxed egg is $x^3 - 8 = 12^3 - 8 = 1720$ cubic inches.

17. $\pi(x^2 + 18x + 81) = \pi(x^2 + 2 \cdot x \cdot 9 + 9^2)$
$= \pi(x+9)^2$
The radius of the center of the tray (not including the border) is $(x+9) - x = 9$ inches.
If the width of the border is $x = 2$ inches, then the area of the tray and border is
$\pi(2+9)^2 = 121\pi$ in^2 or approximately 380 in^2.
The area of the center of the tray (not including the border) is $\pi \cdot 9^2 = 81\pi$ in^2, or approximately 254 in^2.
The area or the border is $121\pi - 81\pi = 40\pi$ in^2, or approximately $380 - 254 = 126$ in^2.

18. a. David factored by grouping.

b. He has not factored completely. The expression $(4x^2 - 9)$ can be factored further.

c. $12x^3 + 28x^2 - 27x - 63$
$= (3x+7)(4x^2 - 9)$
$= (3x+7)[(2x)^2 - 3^2]$
$= (3x+7)(2x+3)(2x-3)$

Chapter 11

11.1 Exercises

1. $3x^3 - 2x^2 + x = 5$ or $3x^3 - 2x^2 + x - 5 = 0$ is a cubic polynomial equation.

3. $3\sqrt{y} + y - 4 = 0$ is not a polynomial equation because the square root has y as its radicand.

5. $\frac{1}{4}x^4 + 3x^2 - \frac{3}{4} = 0$ is a polynomial equation.

7. $4(x-2)(x+7) = 16$ or $4x^2 + 20x - 72 = 0$ is a quadratic polynomial equation.

9. $3x^{-2} - 5x = 4x^2$ is not a polynomial equation because x has an exponent of -2.

11. $1.7x^2 + 3.2x = 5.7$ or $1.7x^2 + 3.2x - 5.7 = 0$ is a quadratic polynomial equation.

13. Let $Y1 = x^2 + 8$ and $Y2 = 6x$.

X	Y1	Y2
0	8	0
1	9	6
2	12	12
3	17	18
4	24	24
5	33	30
6	44	36

    X=2

    The solutions are 2 and 4.

15. Let $Y1 = 4x^3$ and $Y2 = x + 1$.

X	Y1	Y2
-2	-32	-1
-1	-4	0
0	0	1
1	4	2
2	32	3
3	108	4
4	256	5

    X=0

    Because $Y1 < Y2$ when $x = 0$ and $Y1 > Y2$ when $x = 1$, the solution is a noninteger number between 1 and 2.

17. Let $Y1 = x^2 - 7$ and $Y2 = x^2 + 3$.

X	Y1	Y2
-3	2	12
-2	-3	7
-1	-6	4
0	-7	3
1	-6	4
2	-3	7
3	2	12

    X=-3

    Because $Y1 - Y2 = -10$ for all values of x, the expressions will never be equal. The equation has no solution.

19. Let $Y1 = x^2 + 5x + 1$ and $Y2 = 1 + x(5 + x)$.

X	Y1	Y2
-3	-5	-5
-2	-5	-5
-1	-3	-3
0	1	1
1	7	7
2	15	15
3	25	25

    X=-3

    Because $Y1 = Y2$ for all values of x, the solution is the set of all real numbers.

21. Let $Y1 = x^2 - 2x + 6$ and $Y2 = 12 - 4x + 2x^2$.

X	Y1	Y2
-3	21	42
-2	14	28
-1	9	18
0	6	12
1	5	10
2	6	12
3	9	18

    X=-3

    Because $Y1 < Y2$ for all values of x, the expressions will never be equal. The equation has no real-number solution.

23. Let $Y1 = \frac{1}{2}x^2 - x$ and $Y2 = 6 - 3x$.

X	Y1	Y2
-6	24	24
-5	17.5	21
-4	12	18
-3	7.5	15
-2	4	12
-1	1.5	9
0	0	6

    X=-6

X	Y1	Y2
-4	12	18
-3	7.5	15
-2	4	12
-1	1.5	9
0	0	6
1	-.5	3
2	0	0

    X=2

    The solutions are -6 and 2.

25. Let $Y1 = 2 - 0.2x^2$ and $Y2 = 0.6x$.

X	Y₁	Y₂
-5	-3	-3
-4	-1.2	-2.4
-3	.2	-1.8
-2	1.2	-1.2
-1	1.8	-.6
0	2	0
1	1.8	.6

X= -5

X	Y₁	Y₂
-4	-1.2	-2.4
-3	.2	-1.8
-2	1.2	-1.2
-1	1.8	-.6
0	2	0
1	1.8	.6
2	1.2	1.2

X=2

The solutions are −5 and 2.

27. Let $Y1 = x^2 + 8$ and $Y2 = 6x$.

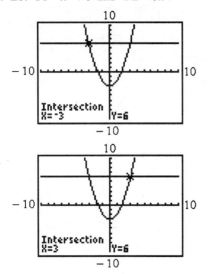

The solutions are −3 and 3.

29. Let $Y1 = x^2 - 3$ and $Y2 = 2x$.

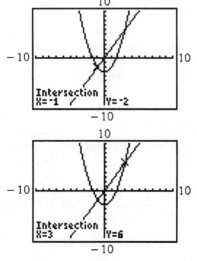

The solutions are −1 and 3.

31. Let $Y1 = x^3$ and $Y2 = 4x$.

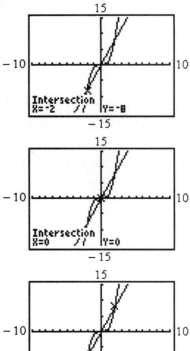

The solutions are −2, 0, and 2.

33. Let $Y1 = x^2 - 3x - 10$ and $Y2 = 0$.

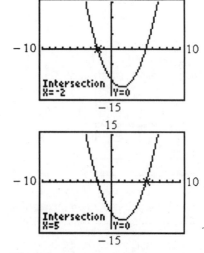

The solutions are −2 and 5.

35. Let $Y1 = x^2 - 2x + 1$ and $Y2 = x^2 - 2x - 3$.

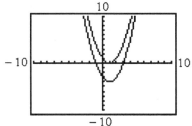

Because the graphs do not intersect, the equation has no solution.

37. Let $Y1 = x^2 + 1$ and $Y2 = 3x^2 + 3$.

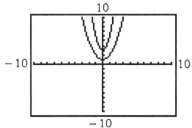

Because the graphs do not intersect, the equation has no real-number solution.

39. Let $Y1 = x(x+3)$ and $Y2 = x^2 + 3x$.

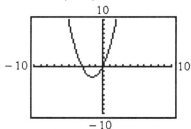

Because the graphs are the same, the solution is the set of all real numbers.

41. Let $Y1 = 4x^2 - x^3$ and $Y2 = x^2 - 4x$.

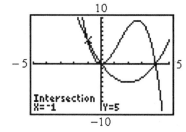

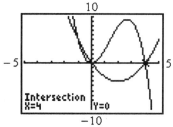

The solutions are -1, 0, and 4.

43. Let $Y1 = x^2 - 10x + 30$ and $Y2 = \frac{1}{2}x^2 - 5x + 15$.

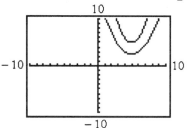

Because the graphs do not intersect, the equation has no real-number solution.

45. Let $Y1 = x^3 - 2x^2 + 1$ and $Y2 = x^3 - 2x^2 + 9$.

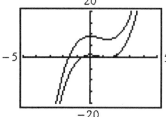

Because the graphs do not intersect, the equation has no solution.

47. Let $Y1 = x(x^2 - 3) - 5(x+1)$ and $Y2 = x^3 - 8x - 5$.

SSM: Experiencing Introductory and Intermediate Algebra

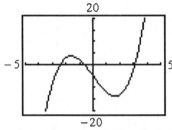

Because the graphs are the same, the solution is the set of all real numbers.

49. Let $Y1 = 4x^2$ and $Y2 = 9$.

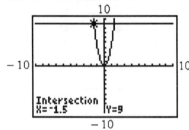

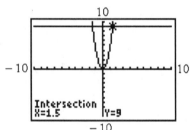

The solutions are -1.5 and 1.5, or as fractions, the solutions are $-\dfrac{3}{2}$ and $\dfrac{3}{2}$.

51. Let $Y1 = 10x^3 - 7x^2 - 4x$ and $Y2 = 3x - 4$.

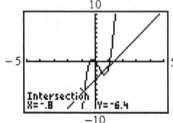

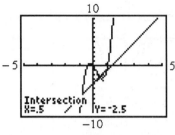

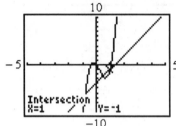

The solutions are -0.8, 0.5, and 1.

53. Let $Y1 = x^2 - 0.9x - 10.36$ and $Y2 = 0$.

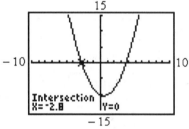

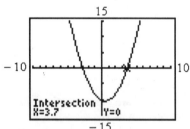

The solutions are -2.8 and 3.7.

55. Let $Y1 = 4x^3 + 3.7x^2$ and $Y2 = 1.74x + 7.56$.

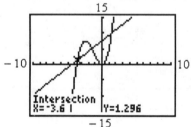

Chapter 11: Quadratic and Other Polynomial Equations and Inequalities

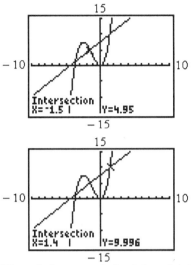

The solutions are −3.6, −1.5, and 1.4.

57. $s = -16t^2 + v_0 t + s_0$; $s = 0$; $v_0 = 0$; $s_0 = 40$

$0 = -16t^2 + 0t + 40$ or $0 = -16t^2 + 40$

Let $Y1 = 0$ and $Y2 = -16x^2 + 40$.

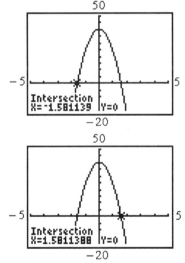

The solutions to the equation are approximately −1.58 and 1.58. Time cannot be negative, so it will take approximately 1.58 seconds for the hat to hit the ground.

59. $s = -16t^2 + v_0 t + s_0$; $s = 0$; $v_0 = -5$; $s_0 = 40$

$0 = -16t^2 - 5t + 40$

Let $Y1 = 0$ and $Y2 = -16x^2 - 5x + 40$.

The solutions to the equation are approximately −1.75 and 1.43. Time cannot be negative, so it will take approximately 1.43 seconds for the dagger to hit the ground.

61. $s = -16t^2 + v_0 t + s_0$; $s = 0$; $v_0 = 5$; $s_0 = 40$

$0 = -16t^2 + 5t + 40$

Let $Y1 = 0$ and $Y2 = -16x^2 + 5x + 40$.

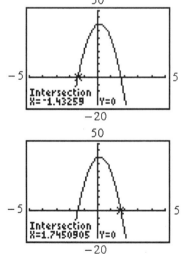

The solutions to the equation are approximately −1.43 and 1.75. Time cannot be negative, so it will take approximately 1.75 seconds for the baton to hit the ground.

SSM: Experiencing Introductory and Intermediate Algebra

63. $175 - 25 = 150$

$150 = -2.143x^2 + 19.371x + 173.314$

Let $Y1 = 150$ and

$Y2 = -2.143x^2 + 19.371x + 173.314$.

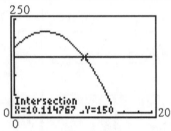

The solution to the equation is approximately 10.11. Now, $1995 + 10.11 = 2005.11$ or about 2006. According to the function, the collections from corporate income taxes will be $25 billion less than those of 1995 in about the year 2006.

No, this relation will not continue to apply. If it did, collections would eventually be negative.

65. $2(25,700) = 51,400 = 51.4$ thousand

$51.4 = 0.175x^2 + 0.165x + 25.715$

Let $Y1 = 51.4$ and

$Y2 = 0.175x^2 + 0.165x + 25.715$.

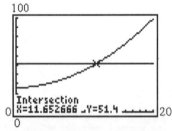

The solution to the equation is approximately 11.65. Now, $1997 + 11.65 = 2008.65$ or about 2009. According to the function, the Hawaiian per capita personal income will be twice that of 1997 in about 2009.

Yes, this result seems reasonable. Incomes doubling over a 12-year period seems feasible.

67. $100 = 11.55 + 20.67x - 1.35x^2$

Let $Y1 = 100$ and $Y2 = 11.55 + 20.67x - 1.35x^2$.

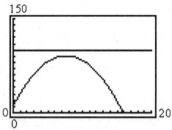

The graphs do not intersect, so the equation has no real solution. No possible number of sales representatives can meet the goal of $100,000.

The solutions to the equation are approximately 4.25 and 11.06. In order to realize the $75,000 goal, 5 to 11 sales representatives should be assigned to the territory.

11.1 Calculator Exercises

Students should verify the given solutions with the calculator.

11.2 Exercises

1. $(x+6)(x+11) = 0$

$x + 6 = 0$ or $x + 11 = 0$

$x = -6$ $\qquad x = -11$

The solutions are -11 and -6.

3. $\left(\dfrac{3}{5}x - \dfrac{9}{20}\right)\left(x + \dfrac{2}{3}\right) = 0$

$\dfrac{3}{5}x - \dfrac{9}{20} = 0$ or $x + \dfrac{2}{3} = 0$

$\dfrac{3}{5}x = \dfrac{9}{20}$ $\qquad x = -\dfrac{2}{3}$

$x = \dfrac{3}{4}$

The solutions are $-\dfrac{2}{3}$ and $\dfrac{3}{4}$.

5. $3x(x+9)(2x-5)=0$

$3x=0$ or $x+9=0$ or $2x-5=0$

$x=0$ $\quad\quad x=-9 \quad\quad 2x=5$

$\quad\quad\quad\quad\quad\quad\quad\quad\quad\quad x=\dfrac{5}{2}$

The solutions are -9, 0, and $\dfrac{5}{2}$.

7. $(7x-49)(49x-7)=0$

$7x-49=0$ or $49x-7=0$

$7x=49 \quad\quad\quad 49x=7$

$x=7 \quad\quad\quad\quad x=\dfrac{1}{7}$

The solutions are $\dfrac{1}{7}$ and 7.

9. $(4x+3)(2x-9)(x+6)=0$

$4x+3=0$ or $2x-9=0$ or $x+6=0$

$4x=-3 \quad\quad 2x=9 \quad\quad\quad x=-6$

$x=-\dfrac{3}{4} \quad\quad x=\dfrac{9}{2}$

The solutions are -6, $-\dfrac{3}{4}$, and $\dfrac{9}{2}$.

11. $(0.2x+6.8)(1.3x-1.69)=0$

$0.2x+6.8=0$ or $1.3x-1.69=0$

$0.2x=-6.8 \quad\quad 1.3x=1.69$

$x=-34 \quad\quad\quad x=1.3$

The solutions are -34 and 1.3.

13. $0=x^2+10x+24$

$0=(x+6)(x+4)$

$x+6=0$ or $x+4=0$

$x=-6 \quad\quad x=-4$

The solutions are -6 and -4.

15. $x^2+33=14x$

$x^2-14x+33=0$

$(x-11)(x-3)=0$

$x-11=0$ or $x-3=0$

$x=11 \quad\quad x=3$

The solutions are 3 and 11.

17. $4x^2+5x+24x+30=0$

$x(4x+5)+6(4x+5)=0$

$(4x+5)(x+6)=0$

$4x+5=0$ or $x+6=0$

$4x=-5 \quad\quad x=-6$

$x=-\dfrac{5}{4}$

The solutions are -6 and $-\dfrac{5}{4}$.

19. $5x^2+3x=8$

$5x^2+3x-8=0$

$(5x+8)(x-1)=0$

$5x+8=0$ or $x-1=0$

$5x=-8 \quad\quad x=1$

$x=-\dfrac{8}{5}$

The solutions are $-\dfrac{8}{5}$ and 1.

21. $15x^2=35x$

$15x^2-35x=0$

$5x(3x-7)=0$

$5x=0$ or $3x-7=0$

$x=0 \quad\quad 3x=7$

$\quad\quad\quad\quad x=\dfrac{7}{3}$

The solutions are 0 and $\dfrac{7}{3}$.

23. $18x^2-3x=5-30x$

$18x^2+27x-5=0$

$(6x-1)(3x+5)=0$

$6x-1=0$ or $3x+5=0$

$6x=1 \quad\quad 3x=-5$

$x=\dfrac{1}{6} \quad\quad x=-\dfrac{5}{3}$

The solutions are $-\dfrac{5}{3}$ and $\dfrac{1}{6}$.

25. $16x^2 + 72x + 81 = 0$

$(4x+9)^2 = 0$

$4x+9 = 0$

$4x = -9$

$x = -\dfrac{9}{4}$

The solution is $-\dfrac{9}{4}$ (which is a double root).

27. $4x^2 + 25x + 18 = 5x - 7$

$4x^2 + 20x + 25 = 0$

$(2x+5)^2 = 0$

$2x + 5 = 0$

$2x = -5$

$x = -\dfrac{5}{2}$

The solution is $-\dfrac{5}{2}$ (which is a double root).

29. $b^2 + 7 = 71$

$b^2 - 64 = 0$

$(b+8)(b-8) = 0$

$b+8 = 0$ or $b-8 = 0$

$b = -8 \qquad b = 8$

The solutions are -8 and 8.

31. $9z^2 = 25$

$9z^2 - 25 = 0$

$(3z+5)(3z-5) = 0$

$3z+5 = 0$ or $3z-5 = 0$

$3z = -5 \qquad 3z = 5$

$z = -\dfrac{5}{3} \qquad z = \dfrac{5}{3}$

The solutions are $-\dfrac{5}{3}$ and $\dfrac{5}{3}$.

33. $(x+1)^2 = 49$

$x^2 + 2x + 1 = 49$

$x^2 + 2x - 48 = 0$

$(x+8)(x-6) = 0$

$x+8 = 0$ or $x-6 = 0$

$x = -8 \qquad x = 6$

The solutions are -8 and 6.

35. $(x-3)(x-2) = 42$

$x^2 - 5x + 6 = 42$

$x^2 - 5x - 36 = 0$

$(x-9)(x+4) = 0$

$x-9 = 0$ or $x+4 = 0$

$x = 9 \qquad x = -4$

The solutions are -4 and 9.

37. $x^2 + (x+3)^2 = 225$

$x^2 + x^2 + 6x + 9 = 225$

$2x^2 + 6x - 216 = 0$

$2(x^2 + 3x - 108)$

$2(x+12)(x-9) = 0$

$x+12 = 0$ or $x-9 = 0$

$x = -12 \qquad x = 9$

The solutions are -12 and 9.

39. $x^3 + 7x^2 - 9x - 63 = 0$

$x^2(x+7) - 9(x+7) = 0$

$(x+7)(x^2 - 9) = 0$

$(x+7)(x+3)(x-3) = 0$

$x+7 = 0$ or $x+3 = 0$ or $x-3 = 0$

$x = -7 \qquad x = -3 \qquad x = 3$

The solutions are -7, -3, and 3.

Chapter 11: Quadratic and Other Polynomial Equations and Inequalities

41. $18x^3 + 45x^2 - 50x - 125 = 0$
$9x^2(2x+5) - 25(2x+5) = 0$
$(2x+5)(9x^2 - 25) = 0$
$(2x+5)(3x+5)(3x-5) = 0$
$2x+5=0$ or $3x+5=0$ or $3x-5=0$
$2x=-5$ $\quad$ $3x=-5$ $\quad$ $3x=5$
$x=-\dfrac{5}{2}$ $\quad$ $x=-\dfrac{5}{3}$ $\quad$ $x=\dfrac{5}{3}$

The solutions are $-\dfrac{5}{2}$, $-\dfrac{5}{3}$, and $\dfrac{5}{3}$.

43. $3x^3 - 3x^2 + 12x - 12 = 0$
$3(x^3 - x^2 + 4x - 4) = 0$
$3\left[x^2(x-1) + 4(x-1)\right] = 0$
$3(x-1)(x^2+4) = 0$
$x-1=0$ or $x^2+4=0$
$x=1$ $\qquad$ $x^2=-4$ (not real)

The solution is 1.

45. a. Let x = the length of the trough
$\dfrac{1}{2}x + 1$ = the width of the trough
3 = the height of the trough
$V = x\left(\dfrac{1}{2}x+1\right)(3) = \dfrac{3}{2}x^2 + 3x$

The volume is $\left(\dfrac{3}{2}x^2 + 3x\right)$ ft^3.

b. $72 = \dfrac{3}{2}x^2 + 3x$
$144 = 3x^2 + 6x$
$0 = 3x^2 + 6x - 144$
$0 = 3(x^2 + 2x - 48)$
$0 = 3(x+8)(x-6)$
$x+8=0$ or $x-6=0$
$x=-8$ $\qquad$ $x=6$

Because length cannot be negative, the length of the trough must be $x = 6$ feet. The width is $\dfrac{1}{2}(6)+1 = 4$ feet.

Thus, the dimensions of the trough are 6 feet by 4 feet by 3 feet.

47. a. Let x = the height of each drawer
$3x$ = the width of each drawer
$4x+2$ = the length of each drawer
$S = 2(x)(4x+2) + 2(x)(3x) + (3x)(4x+2)$
$= 8x^2 + 4x + 6x^2 + 12x^2 + 6x$
$= 26x^2 + 10x$

The surface area is $\left(26x^2 + 10x\right)$ in^2.

b. $800 = 26x^2 + 10x$
$0 = 26x^2 + 10x - 800$
$0 = 2(13x^2 + 5x - 400)$
$0 = 2(13x+70)(x-5)$
$13x+70=0$ or $x-5=0$
$13x=-70$ $\qquad$ $x=5$
$x=-\dfrac{70}{13}$

Because length cannot be negative, the height of each drawer must be $x = 5$ inches. The width is $3(5) = 15$ inches and the length is $4(5) + 2 = 22$ inches.

Thus, the dimensions of each drawer will be 22 inches by 15 inches by 5 inches.

49. a. Let x = the base of the sail
$x+4$ = the height of the sail
$A = \dfrac{1}{2}(x)(x+4) = \dfrac{1}{2}x^2 + 2x$

The area is $\left(\dfrac{1}{2}x^2 + 2x\right)$ ft^2.

b. $30 = \frac{1}{2}x^2 + 2x$

$60 = x^2 + 4x$

$0 = x^2 + 4x - 60$

$0 = (x+10)(x-6)$

$x + 10 = 0$ or $x - 6 = 0$

$x = -10 \qquad x = 6$

Because length cannot be negative, the base of the sail must be $x = 6$ feet. The height will be $6 + 4 = 10$ feet.

51. $V^2 = 30FS$

$V^2 = 30(.40)(243)$

$V^2 = 2916$

$V^2 - 2196 = 0$

$(V + 54)(v - 54) = 0$

$V + 54 = 0$ or $V - 54 = 0$

$V = -54 \qquad V = 54$

The vehicle was traveling at 54 mph.

53. a. $y = -0.12x^2 - 0.06x + 2.9$

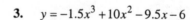

b. $2 = -0.12x^2 - 0.06x + 2.9$

$0.12x^2 + 0.06x - 0.9 = 0$

$12x^2 + 6x - 90 = 0$

$6(2x^2 + x - 15) = 0$

$6(2x - 5)(x + 3) = 0$

$2x - 5 = 0$ or $x + 3 = 0$

$2x = 5 \qquad x = -3$

$x = \frac{5}{2} = 2.5$

Disregarding –3, the solution is 2.5. Now, 1998 + 2.5 = 2000.5 or about 2001.

According to the function, profits will drop to $2 million in about 2001.

c. Answers will vary.

11.2 Calculator Exercises

1. $y = x^3 - 2x^2 + 3x - 4$

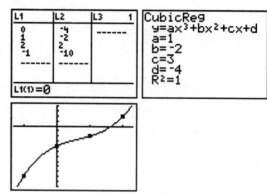

2. $y = x^4 - x^3 + x^2 - x + 1$

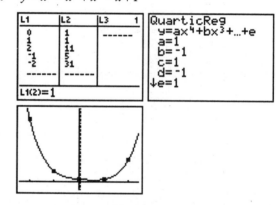

3. $y = -1.5x^3 + 10x^2 - 9.5x - 6$

Chapter 11: Quadratic and Other Polynomial Equations and Inequalities

4. $y = 0.5x^4 - 0.2x^3 - 7x + 6$

11.3 Exercises

1. $\sqrt{63} = \sqrt{9 \cdot 7} = 3\sqrt{7}$

3. $\sqrt{243} = \sqrt{81 \cdot 3} = 9\sqrt{3}$

5. $\sqrt{147} = \sqrt{49 \cdot 3} = 7\sqrt{3}$

7. $-\sqrt{125} = -\sqrt{25 \cdot 5} = -5\sqrt{5}$

9. $-3 + \sqrt{20} = -3 + \sqrt{4 \cdot 5} = -3 + 2\sqrt{5}$

11. $2 - \sqrt{50} = 2 - \sqrt{25 \cdot 2} = 2 - 5\sqrt{2}$

13. $\dfrac{4 - \sqrt{48}}{4} = \dfrac{4 - \sqrt{16 \cdot 3}}{4} = \dfrac{4 - 4\sqrt{3}}{4} = 1 - \sqrt{3}$

15. $\dfrac{-6 + \sqrt{45}}{3} = \dfrac{-6 + \sqrt{9 \cdot 5}}{3} = \dfrac{-6 + 3\sqrt{5}}{3} = -2 + \sqrt{5}$

17. $\dfrac{-5 - \sqrt{18}}{-3} = \dfrac{-5 - \sqrt{9 \cdot 2}}{-3} = \dfrac{-5 - 3\sqrt{2}}{-3} = \dfrac{5}{3} + \sqrt{2}$

19. $\dfrac{-6 - \sqrt{54}}{-3} = \dfrac{-6 - \sqrt{9 \cdot 6}}{-3} = \dfrac{-6 - 3\sqrt{6}}{-3} = 2 + \sqrt{6}$

21. $\sqrt{\dfrac{16}{5}} = \dfrac{\sqrt{16}}{\sqrt{5}} = \dfrac{4}{\sqrt{5}} = \dfrac{4}{\sqrt{5}} \cdot \dfrac{\sqrt{5}}{\sqrt{5}} = \dfrac{4\sqrt{5}}{5}$

23. $\sqrt{\dfrac{50}{48}} = \dfrac{\sqrt{25 \cdot 2}}{\sqrt{16 \cdot 3}} = \dfrac{5\sqrt{2}}{4\sqrt{3}} = \dfrac{5\sqrt{2}}{4\sqrt{3}} \cdot \dfrac{\sqrt{3}}{\sqrt{3}} = \dfrac{5\sqrt{6}}{4 \cdot 3} = \dfrac{5\sqrt{6}}{12}$

25. $x^2 = 144$
$x = \pm\sqrt{144}$
$x = \pm 12$
The solutions are ± 12.

27. $a^2 = 13$
$a = \pm\sqrt{13}$
The solutions are $\pm\sqrt{13}$.

29. $q^2 = 98$
$q = \pm\sqrt{98}$
$q = \pm\sqrt{49 \cdot 2}$
$q = \pm 7\sqrt{2}$
The solutions are $\pm 7\sqrt{2}$.

31. $2x^2 - 32 = 0$
$2x^2 = 32$
$x^2 = 16$
$x = \pm\sqrt{16}$
$x = \pm 4$
The solutions are ± 4.

33. $4x^2 - 25 = 0$
$4x^2 = 25$
$x^2 = \dfrac{25}{4}$
$x = \pm\sqrt{\dfrac{25}{4}}$
$x = \pm\dfrac{5}{2}$
The solutions are $\pm\dfrac{5}{2}$.

35. $9x^2 = 2$

$x^2 = \dfrac{2}{9}$

$x = \pm\sqrt{\dfrac{2}{9}}$

$x = \pm\dfrac{\sqrt{2}}{3}$

The solutions are $\pm\dfrac{\sqrt{2}}{3}$.

37. $3x^2 + 4 = 6$

$3x^2 = 2$

$x^2 = \dfrac{2}{3}$

$x = \pm\sqrt{\dfrac{2}{3}}$

$x = \pm\dfrac{\sqrt{2}}{\sqrt{3}} \cdot \dfrac{\sqrt{3}}{\sqrt{3}}$

$x = \pm\dfrac{\sqrt{6}}{3}$

The solutions are $\pm\dfrac{\sqrt{6}}{3}$.

39. $m^2 + 7 = 5$

$m^2 = -2$

The equation has no real-number solution.

41. $(x-5)^2 = 0$

$x - 5 = \pm\sqrt{0}$

$x - 5 = 0$

$x = 5$

The solution is 5 (which is a double root).

43. $(z-7)^2 = 4$

$z - 7 = \pm\sqrt{4}$

$z - 7 = \pm 2$

$z = 7 \pm 2$

$z = 7 + 2$ or $z = 7 - 2$

$z = 9$ $\qquad z = 5$

The solutions are 5 and 9.

45. $(4a-3)^2 = 4$

$4a - 3 = \pm\sqrt{4}$

$4a - 3 = \pm 2$

$4a = 3 \pm 2$

$4a = 3 + 2$ or $4a = 3 - 2$

$4a = 5$ $\qquad 4a = 1$

$a = \dfrac{5}{4}$ $\qquad a = \dfrac{1}{4}$

The solutions are $\dfrac{1}{4}$ and $\dfrac{5}{4}$.

47. $x^2 = 1.69$

$x = \pm\sqrt{1.69}$

$x = \pm 1.3$

The solutions are ± 1.3.

49. $(x+3)^2 - 1 = 3$

$(x+3)^2 = 4$

$x + 3 = \pm\sqrt{4}$

$x + 3 = \pm 2$

$x = -3 \pm 2$

$x = -3 + 2$ or $x = -3 - 2$

$x = -1$ $\qquad x = -5$

The solutions are -5 and -1.

51. $2(m-4)^2 - 6 = 12$

$2(m-4)^2 = 18$

$(m-4)^2 = 9$

$m - 4 = \pm\sqrt{9}$

$m - 4 = \pm 3$

$m = 4 \pm 3$

$m = 4 + 3$ or $m = 4 - 3$

$m = 7$ $\qquad m = 1$

The solutions are 1 and 7.

Chapter 11: Quadratic and Other Polynomial Equations and Inequalities

53. $x^2 + 10x + 25 = 9$
$(x+5)^2 = 9$
$x+5 = \pm\sqrt{9}$
$x+5 = \pm 3$
$x = -5 \pm 3$
$n = -5+3$ or $n = -5-3$
$n = -2$ $n = -8$
The solutions are -8 and -2.

55. $9x^2 - 6x + 1 = 144$
$(3x-1)^2 = 144$
$3x-1 = \pm\sqrt{144}$
$3x-1 = \pm 12$
$3x = 1 \pm 12$
$3x = 13$ or $3x = -11$
$x = \dfrac{13}{3}$ $x = -\dfrac{11}{3}$
The solutions are $-\dfrac{11}{3}$ and $\dfrac{13}{3}$.

57. $(x-7)^2 - 5 = 1$
$(x-7)^2 = 6$
$x-7 = \pm\sqrt{6}$
$x = 7 \pm \sqrt{6}$
The solutions are $7 \pm \sqrt{6}$.

59. $(2x+1)^2 - 3 = 7$
$(2x+1)^2 = 10$
$2x+1 = \pm\sqrt{10}$
$2x = -1 \pm \sqrt{10}$
$x = \dfrac{-1 \pm \sqrt{10}}{2}$
The solutions are $\dfrac{-1 \pm \sqrt{10}}{2}$.

61. $(x+3)^2 - 6 = 6$
$(x+3)^2 = 12$
$x+3 = \pm\sqrt{12}$
$x+3 = \pm 2\sqrt{3}$
$x = -3 \pm 2\sqrt{3}$
The solutions are $-3 \pm 2\sqrt{3}$.

63. $5x^2 - 4 = 0$
$5x^2 = 4$
$x^2 = \dfrac{4}{5}$
$x = \pm\sqrt{\dfrac{4}{5}}$
$x = \pm\dfrac{2}{\sqrt{5}} \cdot \dfrac{\sqrt{5}}{\sqrt{5}}$
$x = \pm\dfrac{2\sqrt{5}}{5}$
The solutions are $\pm\dfrac{2\sqrt{5}}{5}$.

65. $2a^2 - 13 = 12$
$2a^2 = 25$
$a^2 = \dfrac{25}{2}$
$a = \pm\sqrt{\dfrac{25}{2}}$
$x = \pm\dfrac{5}{\sqrt{2}} \cdot \dfrac{\sqrt{2}}{\sqrt{2}}$
$x = \pm\dfrac{5\sqrt{2}}{2}$
The solutions are $\pm\dfrac{5\sqrt{2}}{2}$.

67. Let a = the vertical distance of the ramp.
$a^2 + 60^2 = 61^2$
$a^2 + 3600 = 3721$
$a^2 = 121$
$a = \pm\sqrt{121}$
$a = \pm 11$
Distance cannot be negative, so the ramp is 11 inches high.

69. Let a = the distance along the ground.
$a^2 + 3000^2 = 5000^2$
$a^2 + 9,000,000 = 25,000,000$
$a^2 = 16,000,000$
$a = \pm\sqrt{16,000,000}$
$a = \pm 4000$
Distance cannot be negative, so the distance along the ground is 4000 feet.

71. $x^2 + x^2 = 50^2$
$2x^2 = 2500$
$x^2 = 1250$
$x = \pm\sqrt{1250}$
$x = \pm\sqrt{625 \cdot 2}$
$x = \pm 25\sqrt{2}$
Distance cannot be negative, so the distance is $25\sqrt{2}$ feet, or approximately 35.36 feet.

73. Let x = the amount the center would rise.
$2640^2 + x^2 = 2640.5^2$
$6,969,600 + x^2 = 6,972,240.25$
$x^2 = 2640.25$
$x = \pm\sqrt{2640.25}$
$x \approx 51.38$
Distance cannot be negative, so the center of the bridge would rise approximately 51.38 feet.

75. Let x = the height of the ball.
$x^2 + 25^2 = 26^2$
$x^2 + 625 = 676$
$x^2 = 51$
$x = \pm\sqrt{51}$
$x \approx \pm 7.14$
Distance cannot be negative, so the height of the ball as it crossed the goal line was 7.14 feet.

$8 - 7.14 = 0.86$; the ball was 0.86 feet below the goal's horizontal bar.

77. $4708.90 = 4000(1+r)^2$
$1.177225 = (1+r)^2$
$\pm\sqrt{1.177225} = 1+r$
$1.085 = 1+r$
$-1 \pm 1.085 = r$
$r = 0.085$ or $r = -2.085$
Disregarding -2.085, the annual interest rate of the investment is $0.085 = 8.5\%$.

79. $6000 = 5000(1+r)^2$
$1.2 = (1+r)^2$
$\pm\sqrt{1.2} = 1+r$
$\pm 1.0954 \approx 1+r$
$-1 \pm 1.0954 \approx r$
$r \approx 0.0954$ or $r \approx -2.0954$
Disregarding -2.0954, the annual interest rate must be approximately $0.0954 = 9.54\%$.

81. $9500 = 7500(1+r)^2$
$\dfrac{19}{15} = (1+r)^2$
$\pm\sqrt{\dfrac{19}{15}} = 1+r$
$\pm 1.1255 \approx 1+r$
$-1 \pm 1.1255 \approx r$
$r \approx 0.1255$ or $r \approx -2.1255$
Disregarding -2.1255, Jean would need an annual compound interest rate of approximately $0.1255 = 12.55\%$.

83. $s = -16t^2 + v_0 t + s_0$; $s = 0$; $v_0 = 0$; $s_0 = 400$

$0 = -16t^2 + 0t + 400$

$16t^2 = 400$

$t^2 = 25$

$t = \pm\sqrt{25}$

$t = \pm 5$

Time cannot be negative, so it will take 5 seconds for the balloon to reach the ground.

85. $s = -16t^2 + v_0 t + s_0$; $s = 5000$; $v_0 = 0$; $s_0 = 12,000$

$5000 = -16t^2 + 0t + 12000$

$16t^2 = 7000$

$t^2 = 437.5$

$t = \pm\sqrt{437.5}$

$t \approx \pm 20.92$

Time cannot be negative, so it will take about 20.92 seconds for the skydiver to descend from 12,000 feet to 5000 feet.

87. $9000 = 0.044x^2 + 4700$

$4300 = 0.044x^2$

$\dfrac{4300}{0.044} = x^2$

$\pm\sqrt{\dfrac{4300}{0.044}} = x$

$\pm 312.61 \approx x$

Disregarding −312.61, the solution to the equation is 312.61. The salesperson manages about 313 accounts.

11.3 Calculator Exercises

Original Pizza Diameter	Twice-as-Large Pizza Diameter
5 inches	7 inches
6 inches	8 inches
7 inches	10 inches
8 inches	11 inches
9 inches	13 inches
10 inches	14 inches
11 inches	16 inches
12 inches	17 inches

11.4 Exercises

1. $\left(\dfrac{1}{2} \cdot 18\right)^2 = 9^2 = 81$

Add 81 to the expression.

3. $\left[\dfrac{1}{2}(-9)\right]^2 = \left(-\dfrac{9}{2}\right)^2 = \dfrac{81}{4}$

Add $\dfrac{81}{4}$ to the expression.

5. $\left(\dfrac{1}{2} \cdot \dfrac{3}{4}\right)^2 = \left(\dfrac{3}{8}\right)^2 = \dfrac{9}{64}$

Add $\dfrac{9}{64}$ to the expression.

7. $\left(\dfrac{1}{2} \cdot 1\right)^2 = \left(\dfrac{1}{2}\right)^2 = \dfrac{1}{4}$

Add $\dfrac{1}{4}$ to the expression.

9. $\left[\dfrac{1}{2}(-6)\right]^2 = (-3)^2 = 9$

Add 9 to the expression.

11. $\left(\dfrac{1}{2} \cdot 9\right)^2 = \left(\dfrac{9}{2}\right)^2 = \dfrac{81}{4}$

Add $\dfrac{81}{4}$ to the expression.

13. $\left(\dfrac{1}{2} \cdot \dfrac{8}{9}\right)^2 = \left(\dfrac{4}{9}\right)^2 = \dfrac{16}{81}$

Add $\dfrac{16}{81}$ to the expression.

SSM: Experiencing Introductory and Intermediate Algebra

15. $\left[\dfrac{1}{2}(-14)\right]^2 = (-7)^2 = 49$

    Add 49 to the expression.

17. $x^2 + 6x - 20 = 35$
    $x^2 + 6x = 55$
    $x^2 + 6x + 9 = 55 + 9$
    $(x+3)^2 = 64$
    $x + 3 = \pm\sqrt{64}$
    $x + 3 = \pm 8$
    $x = -3 \pm 8$
    $x = -3 + 8$ or $x = -3 - 8$
    $x = 5 \qquad\quad x = -11$
    The solutions are -11 and 5.

19. $x^2 - 3x = 28$
    $x^2 - 3x + \dfrac{9}{4} = 28 + \dfrac{9}{4}$
    $\left(x - \dfrac{3}{2}\right)^2 = \dfrac{121}{4}$
    $x - \dfrac{3}{2} = \pm\sqrt{\dfrac{121}{4}}$
    $x - \dfrac{3}{2} = \pm\dfrac{11}{2}$
    $x = \dfrac{3}{2} \pm \dfrac{11}{2}$
    $x = \dfrac{3}{2} + \dfrac{11}{2}$ or $x = \dfrac{3}{2} - \dfrac{11}{2}$
    $x = \dfrac{14}{2} \qquad\quad x = -\dfrac{8}{2}$
    $x = 7 \qquad\qquad x = -4$
    The solutions are -4 and 7.

21. $x^2 + \dfrac{4}{7}x + \dfrac{3}{49} = 0$
    $x^2 + \dfrac{4}{7}x = -\dfrac{3}{49}$

$x^2 + \dfrac{4}{7}x + \dfrac{4}{49} = -\dfrac{3}{49} + \dfrac{4}{49}$
$\left(x + \dfrac{2}{7}\right)^2 = \dfrac{1}{49}$
$x + \dfrac{2}{7} = \pm\sqrt{\dfrac{1}{49}}$
$x + \dfrac{2}{7} = \pm\dfrac{1}{7}$
$x = -\dfrac{2}{7} \pm \dfrac{1}{7}$
$x = -\dfrac{2}{7} + \dfrac{1}{7}$ or $x = -\dfrac{2}{7} - \dfrac{1}{7}$
$x = -\dfrac{1}{7} \qquad\quad x = -\dfrac{3}{7}$

The solutions are $-\dfrac{3}{7}$ and $-\dfrac{1}{7}$.

23. $x^2 + x - 30 = 60$
    $x^2 + x = 90$
    $x^2 + x + \dfrac{1}{4} = 90 + \dfrac{1}{4}$
    $\left(x + \dfrac{1}{2}\right)^2 = \dfrac{361}{4}$
    $x + \dfrac{1}{2} = \pm\sqrt{\dfrac{361}{4}}$
    $x + \dfrac{1}{2} = \pm\dfrac{19}{2}$
    $x = -\dfrac{1}{2} \pm \dfrac{19}{2}$
    $x = -\dfrac{1}{2} + \dfrac{19}{2}$ or $x = -\dfrac{1}{2} - \dfrac{19}{2}$
    $x = \dfrac{18}{2} \qquad\quad x = -\dfrac{20}{2}$
    $x = 9 \qquad\qquad x = -10$
    The solutions are -10 and 9.

Chapter 11: Quadratic and Other Polynomial Equations and Inequalities

25. $x^2 - 6x = 2$
$x^2 - 6x + 9 = 2 + 9$
$(x-3)^2 = 11$
$x - 3 = \pm\sqrt{11}$
$x = 3 \pm \sqrt{11}$
The solutions are $3 \pm \sqrt{11}$.

27. $x^2 + 9x = 1$
$x^2 + 9x + \frac{81}{4} = 1 + \frac{81}{4}$
$\left(x + \frac{9}{2}\right)^2 = \frac{85}{4}$
$x + \frac{9}{2} = \pm\sqrt{\frac{85}{4}}$
$x + \frac{9}{2} = \pm\frac{\sqrt{85}}{2}$
$x = \frac{-9 \pm \sqrt{85}}{2}$
The solutions are $\frac{-9 \pm \sqrt{85}}{2}$.

29. $x^2 + \frac{8}{9}x = 2$
$x^2 + \frac{8}{9}x + \frac{16}{81} = 2 + \frac{16}{81}$
$\left(x + \frac{4}{9}\right)^2 = \frac{178}{81}$
$x + \frac{4}{9} = \pm\sqrt{\frac{178}{81}}$
$x + \frac{4}{9} = \pm\frac{\sqrt{178}}{9}$
$x = \frac{-4 \pm \sqrt{178}}{9}$
The solutions are $\frac{-4 \pm \sqrt{178}}{9}$.

31. $x^2 - x - 5 = 0$
$x^2 - x = 5$
$x^2 - x + \frac{1}{4} = 5 + \frac{1}{4}$
$\left(x - \frac{1}{2}\right)^2 = \frac{21}{4}$
$x - \frac{1}{2} = \pm\sqrt{\frac{21}{4}}$
$x - \frac{1}{2} = \pm\frac{\sqrt{21}}{2}$
$x = \frac{1 \pm \sqrt{21}}{2}$
The solutions are $\frac{1 \pm \sqrt{21}}{2}$.

33. $x^2 + x + 10 = 0$
$x^2 + x = -10$
$x^2 + x + \frac{1}{4} = -10 + \frac{1}{4}$
$\left(x + \frac{1}{2}\right)^2 = -\frac{39}{4}$
The equation has no real-number solution.

35. $2x^2 + 6x - 1 = 0$
$2x^2 + 6x = 1$
$x^2 + 3x = \frac{1}{2}$
$x^2 + 3x + \frac{9}{4} = \frac{1}{2} + \frac{9}{4}$
$\left(x + \frac{3}{2}\right)^2 = \frac{11}{4}$
$x + \frac{3}{2} = \pm\sqrt{\frac{11}{4}}$
$x + \frac{3}{2} = \pm\frac{\sqrt{11}}{2}$
$x = \frac{-3 \pm \sqrt{11}}{2}$
The solutions are $\frac{-3 \pm \sqrt{11}}{2}$.

SSM: Experiencing Introductory and Intermediate Algebra

37. $3x^2 + x - 7 = 0$
$3x^2 + x = 7$
$x^2 + \frac{1}{3}x = \frac{7}{3}$
$x^2 + \frac{1}{3}x + \frac{1}{36} = \frac{7}{3} + \frac{1}{36}$
$\left(x + \frac{1}{6}\right)^2 = \frac{85}{36}$
$x + \frac{1}{6} = \pm\sqrt{\frac{85}{36}}$
$x + \frac{1}{6} = \pm\frac{\sqrt{85}}{6}$
$x = \frac{-1 \pm \sqrt{85}}{6}$

The solutions are $\frac{-1 \pm \sqrt{85}}{6}$.

39. $x^2 - 14x + 55 = 6$
$x^2 - 14x = -49$
$x^2 - 14x + 49 = -49 + 49$
$(x-7)^2 = 0$
$(x-7)^2 = \pm\sqrt{0}$
$x - 7 = 0$
$x = 7$
The solution is 7 (which is a double root).

41. $4x^2 - 20x + 30 = 5$
$4x^2 - 20x = -25$
$x^2 - 5x = -\frac{25}{4}$
$x^2 - 5x + \frac{25}{4} = -\frac{25}{4} + \frac{25}{4}$
$\left(x - \frac{5}{2}\right)^2 = 0$

$x - \frac{5}{2} = \pm\sqrt{0}$
$x - \frac{5}{2} = 0$
$x = \frac{5}{2}$

The solution is $\frac{5}{2}$ (which is a double root).

43. $\frac{1}{2}x^2 + 5x - 2 = 0$
$\frac{1}{2}x^2 + 5x = 2$
$x^2 + 10x = 4$
$x^2 + 10x + 25 = 4 + 25$
$(x+5)^2 = 29$
$x + 5 = \pm\sqrt{29}$
$x = -5 \pm \sqrt{29}$
The solution is $-5 \pm \sqrt{29}$.

45. $\frac{1}{3}x^2 + \frac{1}{9}x - \frac{1}{6} = 0$
$\frac{1}{3}x^2 + \frac{1}{9}x = \frac{1}{6}$
$x^2 + \frac{1}{3}x = \frac{1}{2}$
$x^2 + \frac{1}{3}x + \frac{1}{36} = \frac{1}{2} + \frac{1}{36}$
$\left(x + \frac{1}{6}\right)^2 = \frac{19}{36}$
$x + \frac{1}{6} = \pm\sqrt{\frac{19}{36}}$
$x + \frac{1}{6} = \pm\frac{\sqrt{19}}{6}$
$x = \frac{-1 \pm \sqrt{19}}{6}$

The solution is $\frac{-1 \pm \sqrt{19}}{6}$.

47. $\dfrac{2}{3}x^2 - 2x - \dfrac{5}{6} = 0$

$\dfrac{2}{3}x^2 - 2x = \dfrac{5}{6}$

$x^2 - 3x = \dfrac{5}{4}$

$x^2 - 3x + \dfrac{9}{4} = \dfrac{5}{4} + \dfrac{9}{4}$

$\left(x - \dfrac{3}{2}\right)^2 = \dfrac{14}{4}$

$x - \dfrac{3}{2} = \pm\sqrt{\dfrac{14}{4}}$

$x - \dfrac{3}{2} = \pm\dfrac{\sqrt{14}}{2}$

$x = \dfrac{3 \pm \sqrt{14}}{2}$

The solution is $\dfrac{3 \pm \sqrt{14}}{2}$.

49. Let x = the number of tickets to be sold.
$36 - x$ = the price per ticket

$x(36 - x) = 250$

$36x - x^2 = 250$

$x^2 - 36x = -250$

$x^2 - 36x + 324 = -250 + 324$

$(x - 18)^2 = 74$

$x - 18 = \pm\sqrt{74}$

$x = 18 \pm \sqrt{74}$

$x \approx 26.60 \quad \text{or} \quad x \approx 9.40$

The theater will approximately break even if the group is sold 10 tickets or 27 tickets.

51. Let x = the number of people to be in the group.
$19.50 - 0.25x$ = the price per person.

$x(19.50 - 0.25x) = 200$

$19.5x - 0.25x^2 = 200$

$0.25x^2 - 19.5x = -200$

$x^2 - 78x = -800$

$x^2 - 78x + 1521 = -800 + 1521$

$(x - 39)^2 = 721$

$x - 39 = \pm\sqrt{721}$

$x = 39 \pm \sqrt{721}$

$x \approx 65.85 \quad \text{or} \quad x \approx 12.15$

The restaurant will approximately break even if 13 people or 66 people are in the group.

53. Let x = the number of cars to be purchased.
$20 - x$ = the price per car.

$x(20 - x) = 75$

$20x - x^2 = 75$

$x^2 - 20x = -75$

$x^2 - 20x + 100 = -75 + 100$

$(x - 10)^2 = 25$

$x - 10 = \pm\sqrt{25}$

$x - 10 = \pm 5$

$x = 10 \pm 5$

$x = 15 \quad \text{or} \quad x = 5$

The number of cars that can be purchased is 5. (Note: if 15 were purchased the price would be below $12 each.)

55. $\dfrac{1}{3}x^2 - 64x + 3100 = 2000$

$\dfrac{1}{3}x^2 - 64x = -1100$

$x^2 - 192x = -3300$

$x^2 - 192x + 9216 = -3300 + 9216$

$(x - 96)^2 = 5916$

$x - 96 = \pm\sqrt{5916}$

$x = 96 \pm \sqrt{5916}$

$x \approx 173 \quad \text{or} \quad x \approx 19$

Disregarding 173 because it is above 50, the price should be set at $19 per unit.

SSM: Experiencing Introductory and Intermediate Algebra

57. $55,000 = $55 thousand
$$0.11x^2 + 3.08x + 11 = 55$$
$$0.11x^2 + 3.08x = 44$$
$$x^2 + 28x = 400$$
$$x^2 + 28x + 196 = 400 + 196$$
$$(x+14)^2 = 596$$
$$x + 14 = \pm\sqrt{596}$$
$$x = -14 \pm \sqrt{596}$$
$$x \approx 10.41 \quad \text{or} \quad x \approx -38.41$$
Disregarding -38.41, the size of the contract should be about $10.41 million.

59. Let w = the width of the rectangle.
$w + 4$ = the length of the rectangle.
$$w(w+4) = 117$$
$$w^2 + 4w = 117$$
$$w^2 + 4w + 4 = 117 + 4$$
$$(w+2)^2 = 121$$
$$w + 2 = \pm\sqrt{121}$$
$$w + 2 = \pm 11$$
$$w = -2 \pm 11$$
$$w = 9 \quad \text{or} \quad w = -13$$
Because distance cannot be negative, the width of the rectangle is 9 inches. The length is $9 + 4 = 13$ inches.

61. Let x = the length of the foundation.
$x - 5$ = the width of the foundation.
$$x(x-5) = 85$$
$$x^2 - 5x = 85$$
$$x^2 - 5x + \frac{25}{4} = 85 + \frac{25}{4}$$
$$\left(x - \frac{5}{2}\right)^2 = \frac{365}{4}$$
$$x - \frac{5}{2} = \pm\sqrt{\frac{365}{4}}$$

$$x - \frac{5}{2} = \pm\frac{\sqrt{365}}{2}$$
$$x = \frac{5 \pm \sqrt{365}}{2}$$
$$x \approx 12.1 \quad \text{or} \quad x \approx -7.1$$
Because length cannot be negative, the length of the foundation is 12.1 feet. The width of the foundation is $12.1 - 5 = 7.1$ feet.

63. a. $s = 0;\ v_0 = 32;\ s_0 = 16$
$$0 = -16t^2 + 32t + 16$$

b. $16t^2 - 32t = 16$
$$t^2 - 2t = 1$$
$$t^2 - 2t + 1 = 1 + 1$$
$$(t-1)^2 = 2$$
$$t - 1 = \pm\sqrt{2}$$
$$t = 1 \pm \sqrt{2}$$
Time cannot be negative, so the ball will hit the ground in $1 + \sqrt{2}$ seconds.

c. $1 + \sqrt{2} \approx 2.4$
It will hit the ground in about 2.4 seconds.

c. $\dfrac{11 + \sqrt{129}}{4} \approx 5.6$
It will hit the ground in about 5.6 seconds.

65. a. $y = -0.2x^2 + 1.5x + 5.4$

L1	L2	L3	1
0	5.4		
1	6.7		
2	7.6		

L1(2)=1

QuadReg
y=ax²+bx+c
a=-.2
b=1.5
c=5.4
R²=1

Chapter 11: Quadratic and Other Polynomial Equations and Inequalities

b. $5.4 = -0.2x^2 + 1.5x + 5.4$
$0.2x^2 - 1.5x = 0$
$x^2 - 7.5x = 0$
$x^2 - 7.5x + 14.0625 = 0 + 14.0625$
$(x - 3.75)^2 = 14.0625$
$x - 3.75 = \pm\sqrt{14.0625}$
$x - 3.75 = \pm 3.75$
$x = 3.75 \pm 3.75$
$x = 7.5 \quad \text{or} \quad x = 0$

Disregarding 0, the number of employees will be the same as in 1997 after about 7.5 years. Now, 1997 + 7.5 = 2004.5, or 2005. That is, according to the function, the number of employees will return to the 1997 level in the year 2005.

11.4 Calculator Exercises

1. $x^2 + 8x + 15 = 0$
$x^2 + 8x = -15$
$x^2 + 8x + 16 = -15 + 16$
$(x + 4)^2 = 1$
$x + 4 = \pm\sqrt{1}$
$x + 4 = \pm 1$
$x = -4 \pm 1$
$x = -4 + 1 \quad \text{or} \quad x = -4 - 1$
$x = -3 \qquad\qquad x = -5$

Let $Y1 = x^2 + 8x + 15$ and $Y1 = 0$.

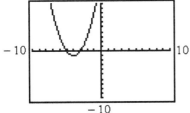

The solutions are -5 and -3.

2. $x^2 + 4x + 7 = 0$
$x^2 + 4x = -7$
$x^2 + 4x + 4 = -7 + 4$
$(x + 2)^2 = -3$

No real-number solution.

Let $Y1 = x^2 + 4x + 7$ and $Y1 = 0$.

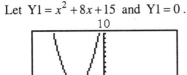

The equation has no real solution.

3. $x^2 + x - 1 = 0$
$x^2 + x = 1$
$x^2 + x + \dfrac{1}{4} = 1 + \dfrac{1}{4}$
$\left(x + \dfrac{1}{2}\right)^2 = \dfrac{5}{4}$
$x + \dfrac{1}{2} = \pm\sqrt{\dfrac{5}{4}}$
$x + \dfrac{1}{2} = \pm\dfrac{\sqrt{5}}{2}$
$x = \dfrac{-1 \pm \sqrt{5}}{2}$
$x \approx 0.618 \quad \text{or} \quad x \approx -1.618$

Let $Y1 = x^2 + x - 1$ and $Y1 = 0$.

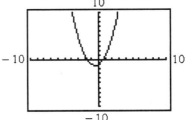

The solutions are approximately -1.618 and 0.618.

4. $2x^2 + 11x + 12 = 0$
$2x^2 + 11x = -12$
$x^2 + \dfrac{11}{2}x = -6$
$x^2 + \dfrac{11}{2}x + \dfrac{121}{16} = -6 + \dfrac{121}{16}$
$\left(x + \dfrac{11}{4}\right)^2 = \dfrac{25}{16}$
$x + \dfrac{11}{4} = \pm\sqrt{\dfrac{25}{16}}$
$x + \dfrac{11}{4} = \pm\dfrac{5}{4}$
$x = -\dfrac{11}{4} \pm \dfrac{5}{4}$
$x = -\dfrac{6}{4}$ or $x = -\dfrac{16}{4}$
$x = -\dfrac{3}{2}$ $x = -4$

Let $Y1 = 2x^2 + 11x + 12$ and $Y1 = 0$.

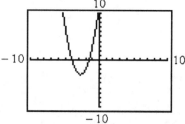

The solutions are -4 and -1.5.

5. $2x^2 + 6x + 7 = 0$
$2x^2 + 6x = -7$
$x^2 + 3x = -\dfrac{7}{2}$
$x^2 + 3x + \dfrac{9}{4} = -\dfrac{7}{2} + \dfrac{9}{4}$
$\left(x + \dfrac{3}{2}\right)^2 = -\dfrac{5}{4}$
No real solution.

Let $Y1 = 2x^2 + 6x + 7$ and $Y1 = 0$.

The equation has no real-number solution.

6. $8x^2 + 8x - 5 = 0$
$8x^2 + 8x = 5$
$x^2 + x = \dfrac{5}{8}$
$x^2 + x + \dfrac{1}{4} = \dfrac{5}{8} + \dfrac{1}{4}$
$\left(x + \dfrac{1}{2}\right)^2 = \dfrac{7}{8}$
$x + \dfrac{1}{2} = \pm\sqrt{\dfrac{7}{8}}$
$x + \dfrac{1}{2} = \pm\dfrac{\sqrt{7}}{\sqrt{8}} \cdot \dfrac{\sqrt{2}}{\sqrt{2}}$
$x + \dfrac{1}{2} = \pm\dfrac{\sqrt{14}}{4}$
$x = -\dfrac{1}{2} \pm \dfrac{\sqrt{14}}{4}$
$x = \dfrac{-2 \pm \sqrt{14}}{4}$
$x \approx 0.435$ or $x \approx -1.435$

Let $Y1 = 8x^2 + 8x - 5$ and $Y1 = 0$.

The solutions are approximately -1.435 and 0.435.

11.5 Exercises

1. $x^2 - 12x + 27 = 0$

$x = \dfrac{-(-12) \pm \sqrt{(-12)^2 - 4(1)(27)}}{2(1)}$

$x = \dfrac{12 \pm \sqrt{36}}{2}$

$x = \dfrac{12 \pm 6}{2}$

$x = \dfrac{18}{2}$ or $x = \dfrac{6}{2}$

$x = 9 \qquad x = 3$

The solutions are 3 and 9.

3. $2x^2 + 3x - 15 = 2x + 6$

$2x^2 + x - 21 = 0$

$x = \dfrac{-1 \pm \sqrt{(1)^2 - 4(2)(-21)}}{2(2)}$

$x = \dfrac{-1 \pm \sqrt{169}}{4}$

$x = \dfrac{-1 \pm 13}{4}$

$x = \dfrac{12}{4}$ or $x = \dfrac{-14}{4}$

$x = 3 \qquad x = -\dfrac{7}{2}$

The solutions are $-\dfrac{7}{2}$ and 3.

5. $2z^2 + 11z + 5 = 0$

$z = \dfrac{-11 \pm \sqrt{(11)^2 - 4(2)(5)}}{2(2)}$

$z = \dfrac{-11 \pm \sqrt{81}}{4}$

$z = \dfrac{-11 \pm 9}{4}$

$z = \dfrac{-2}{4}$ or $z = \dfrac{-20}{4}$

$z = -\dfrac{1}{2} \qquad z = -5$

The solutions are -5 and $-\dfrac{1}{2}$.

7. $8(2p^2 - p + 1) = 7$

$16p^2 - 8p + 8 = 7$

$16p^2 - 8p + 1 = 0$

$p = \dfrac{-(-8) \pm \sqrt{(-8)^2 - 4(16)(1)}}{2(16)}$

$p = \dfrac{8 \pm \sqrt{0}}{32}$

$p = \dfrac{-8 \pm 0}{32}$

$p = \dfrac{-8}{32}$

$p = -\dfrac{1}{4}$

The solution is $-\dfrac{1}{4}$ (which is a double root).

9. $x^2 + 3x + 4 = 0$

$x = \dfrac{-3 \pm \sqrt{(3)^2 - 4(1)(4)}}{2(1)}$

$x = \dfrac{-3 \pm \sqrt{-7}}{6}$

The equation has no real-number solution.

11. $-5a^2 + 4a - 7 = 0$

$a = \dfrac{-4 \pm \sqrt{(4)^2 - 4(-5)(-7)}}{2(-5)}$

$a = \dfrac{-4 \pm \sqrt{-124}}{-10}$

The equation has no real-number solution.

SSM: Experiencing Introductory and Intermediate Algebra

13. $v^2 - 5v + 2 = 0$

$v = \dfrac{-(-5) \pm \sqrt{(-5)^2 - 4(1)(2)}}{2(1)}$

$v = \dfrac{5 \pm \sqrt{17}}{2}$

The solutions are $\dfrac{5 \pm \sqrt{17}}{2}$.

15. $x(x-4) + 1 = 0$

$x^2 - 4x + 1 = 0$

$x = \dfrac{-(-4) \pm \sqrt{(-4)^2 - 4(1)(1)}}{2(1)}$

$x = \dfrac{4 \pm \sqrt{12}}{2}$

$x = \dfrac{4 \pm 2\sqrt{3}}{2}$

$x = 2 \pm \sqrt{3}$

The solutions are $2 \pm \sqrt{3}$.

17. $4x(x+1) + 6 = 6 - x$

$4x^2 + 5x = 0$

$x = \dfrac{-5 \pm \sqrt{(5)^2 - 4(4)(0)}}{2(4)}$

$x = \dfrac{-5 \pm \sqrt{25}}{8}$

$x = \dfrac{-5 \pm 5}{8}$

$x = \dfrac{-10}{8}$ or $x = \dfrac{0}{8}$

$x = -\dfrac{5}{4}$ $\quad x = 0$

The solutions are $-\dfrac{5}{4}$ and 0.

19. $3d^2 + 10 = 17$

$3d^2 - 7 = 0$

$d = \dfrac{-0 \pm \sqrt{(0)^2 - 4(3)(-7)}}{2(3)}$

$d = \dfrac{0 \pm \sqrt{84}}{6}$

$d = \dfrac{0 \pm 2\sqrt{21}}{6}$

$d = \pm \dfrac{\sqrt{21}}{3}$

The solutions are $\pm \dfrac{\sqrt{21}}{3}$.

21. $16m = m^2 + 55$

$0 = m^2 - 16m + 55$

$m = \dfrac{-(-16) \pm \sqrt{(-16)^2 - 4(1)(55)}}{2(1)}$

$m = \dfrac{16 \pm \sqrt{36}}{4}$

$m = \dfrac{16 \pm 6}{2}$

$m = \dfrac{22}{2}$ or $m = \dfrac{10}{2}$

$m = 11 \quad\quad m = 5$

The solutions are 5 and 11.

23. $(x-4)(x+4) = 2(x-4)$

$x^2 - 2x - 8 = 0$

$x = \dfrac{-(-2) \pm \sqrt{(-2)^2 - 4(1)(-8)}}{2(1)}$

$x = \dfrac{2 \pm \sqrt{36}}{2}$

$x = \dfrac{2 \pm 6}{2}$

$x = \dfrac{8}{2}$ or $x = \dfrac{-4}{2}$

$x = 4 \quad\quad x = -2$

The solutions are -2 and 4.

25. $x^2 - 6.3x + 7.2 = 0$

$x = \dfrac{-(-6.3) \pm \sqrt{(-6.3)^2 - 4(1)(7.2)}}{2(1)}$

$x = \dfrac{6.3 \pm \sqrt{10.89}}{2}$

$x = \dfrac{6.3 \pm 3.3}{2}$

$x = \dfrac{9.6}{2}$ or $x = \dfrac{3}{2}$
$x = 4.8 \qquad x = 1.5$
The solutions are 1.5 and 4.8.

27. $1.8 - 5.6x - x^2 = 0$

$x^2 + 5.6x - 1.8 = 0$

$x = \dfrac{-5.6 \pm \sqrt{(5.6)^2 - 4(1)(-1.8)}}{2(1)}$

$x = \dfrac{-5.6 \pm \sqrt{38.56}}{2}$

$x = \dfrac{-5.6 \pm 2\sqrt{9.64}}{2}$

$x = -2.8 \pm \sqrt{9.64}$

The solutions are $-2.8 \pm \sqrt{9.64}$.

29. $a^2 + 24.01 = 9.8a$

$a^2 - 9.8a + 24.01 = 0$

$a = \dfrac{-(-9.8) \pm \sqrt{(-9.8)^2 - 4(1)(24.01)}}{2(1)}$

$a = \dfrac{9.8 \pm \sqrt{0}}{2}$

$a = \dfrac{9.8 \pm 0}{2}$

$a = \dfrac{9.8}{2}$

$a = 4.9$

The solution is 4.9 (which is a double root).

31. $1.7z^2 + 1.3z + 5.6 = 0$

$z = \dfrac{-1.3 \pm \sqrt{(1.3)^2 - 4(1.7)(5.6)}}{2(1.7)}$

$z = \dfrac{-1.3 \pm \sqrt{-36.39}}{3.4}$

The equation has no real-number solution.

33. $x^2 - 11x + 24 = 0$

$b^2 - 4ac = (-11)^2 - 4(1)(24) = 25$

Because 25 is a perfect square, the equation has two rational solutions.

35. $a^2 + 12a + 36 = 0$

$b^2 - 4ac = 12^2 - 4(1)(36) = 0$

Because the discriminant is 0, the equation has one rational solution.

37. $z^2 = 4z - 5$

$z^2 - 4z + 5 = 0$

$b^2 - 4ac = (-4)^2 - 4(1)(5) = -4$

Because the discriminant is negative, the equation has no real-number solution.

39. $6x^2 - 11x - 7 = 0$

$b^2 - 4ac = (-11)^2 - 4(6)(-7) = 289$

Because 289 is a perfect square, the equation has two rational solutions.

41. $7p^2 - 15 = 0$

$b^2 - 4ac = 0^2 - 4(7)(-15) = 420$

Because 420 is positive but not a perfect square, the equation has two irrational solutions.

43. $1 - 5x - 4x^2 = 0$

$4x^2 + 5x - 1 = 0$

$b^2 - 4ac = 5^2 - 4(4)(-1) = 41$

Because 41 is positive but not a perfect square, the equation has two irrational solutions.

45. $z - 1.25 = 0.2z^2$

$0.2z^2 - z + 1.25 = 0$

$b^2 - 4ac = (-1)^2 - 4(0.2)(1.25) = 0$

Because the discriminant is 0, the equation has one rational solution.

47. $0.3x - 2.8 = 1.7x^2$

$1.7x^2 - 0.3x + 2.8 = 0$

$b^2 - 4ac = (-0.3)^2 - 4(1.7)(2.8) = -18.95$

Because the discriminant is negative, the equation has no real-number solution.

49. Let x = the amount of steak to be purchased.
$9 - 0.10x$ = price per pound.

a. $R(x) = x(9 - 0.10x)$

b. $C(x) = 6x + 7.50$

c. $P(x) = R(x) - C(x)$
$= x(9 - 0.10x) - (6x + 7.50)$
$= 9x - 0.1x^2 - 6x - 7.5$
$= -0.1x^2 + 3x - 7.5$

d. $-0.1x^2 + 3x - 7.5 = 0$

$x = \dfrac{-3 \pm \sqrt{3^2 - 4(-0.1)(-7.5)}}{2(-0.1)}$

$x = \dfrac{-3 \pm \sqrt{6}}{-0.2}$

$x \approx 2.75$ or $x \approx 27.25$
$9 - 0.1x \approx 8.73 \quad 9 - 0.1x \approx 6.28$

The shop will approximately break even if it sells 2.75 pounds at $8.73 per pound or 27.25 pounds sold at $6.28 per pound.

51. Let x = the number of people attending the party.
$20 - 0.25x$ = price per person charged by caterer.

a. $R(x) = 50x$

b. $C(x) = x(20 - 0.25x) + 600$

c. $P(x) = R(x) - C(x)$
$= 50x - [x(20 - 0.25x) + 600]$
$= 50x - (20x - 0.25x^2 + 600)$
$= 50x - 20x + 0.25x^2 - 600$
$= 0.25x^2 + 30x - 600$

d. $0.25x^2 + 30x - 600 = 0$

$x = \dfrac{-30 \pm \sqrt{30^2 - 4(0.25)(-600)}}{2(0.25)}$

$x = \dfrac{-30 \pm \sqrt{1500}}{0.5}$

$x = \dfrac{-30 \pm 10\sqrt{15}}{0.5}$

$x = -60 \pm 20\sqrt{15}$
$x \approx 18$ or $x \approx -137$
$20 - 0.25x \approx 15.50$

Disregarding -137, the promoter will approximately break even if 18 people attend, paying $15.50 each.

53. $0.00026x^2 - 0.028x + 1.13 = 7$
$0.00026x^2 - 0.028x - 5.87 = 0$

$x = \dfrac{-(-0.028) \pm \sqrt{(-0.028)^2 - 4(0.00026)(-5.87)}}{2(0.00026)}$

$x = \dfrac{0.028 \pm \sqrt{0.0068888}}{0.00052}$

$x \approx 214$ or $x \approx -106$

It will take about 214 years after 1800 for the world population to reach 7 billion.

Now, $1800 + 214 = 2014$. Thus, the model predicts that the world population will reach 7 billion is about the year 2014.

55. $0.066x^2 - 0.145x + 22.81 = 100$
$0.066x^2 - 0.145x - 77.19 = 0$

$x = \dfrac{-(-0.145) \pm \sqrt{(-0.145)^2 - 4(0.066)(-77.19)}}{2(0.066)}$

$x = \dfrac{0.145 \pm \sqrt{20.399185}}{0.132}$

$x \approx 36$ or $x \approx -33$

Disregarding -33, the CPI was 100% about 36 years after 1950.

Now, $1950 + 36 = 1986$. Thus, according to the model, the CPI was 100% in about 1986.

57. Let x = the height of the ramp.
$x + 6$ = the base of the ramp.
$x + 12$ = the hypotenuse of the ramp.
$x^2 + (x + 6)^2 = (x + 12)^2$
$x^2 + x^2 + 12x + 36 = x^2 + 24x + 144$
$x^2 - 12x - 108 = 0$

$x = \dfrac{-(-12) \pm \sqrt{(-12)^2 - 4(1)(-108)}}{2(1)}$

$x = \dfrac{12 \pm \sqrt{576}}{2}$

$x = \dfrac{12 \pm 24}{2}$

$x = \dfrac{36}{2}$ or $x = \dfrac{-12}{2}$

$x = 18 \qquad x = -6$

Because length cannot be negative, the height of the ramp is 18 feet. The base is 18 + 6 = 24 feet long and the hypotenuse is 18 + 12 = 30 feet long.

11.5 Calculator Exercises

	Equation	Value of Discriminant	Types of Roots	Number of Unlike Roots	Roots
1.	$x^2 + 6 = 5x$	1	rational	2	2, 3
2.	$9x^2 + 6x = -1$	0	rational	1	$-\dfrac{1}{3}$
3.	$2x^2 + 1 = 7x$	41	irrational	2	0.149, 3.351
4.	$x^2 + 6x = -10$	-4	not real	2	not real
5.	$x^2 = 6 - x$	25	rational	2	$-3, 2$
6.	$5x^2 - 6x = 0$	36	rational	2	0, 1.2
7.	$x^2 + 0.36 = 1.2x$	0	rational	1	0.6
8.	$1.7x^2 + x + 1.9 = 0$	-11.92	not real	2	not real
9.	$1.5x^2 + 1.2x = 3.6$	23.04	rational	2	$-2, 1.2$
10.	$\dfrac{1}{4}x^2 + x = \dfrac{1}{8}$	1.125	irrational	2	$-4.121, 0.121$
11.	$x^2 - \dfrac{1}{6}x = \dfrac{1}{6}$	$0.69\overline{4}$	rational	2	$-0.\overline{3}, 0.5$
12.	$\dfrac{1}{5}x^2 + \dfrac{2}{3}x = -\dfrac{7}{8}$	$-0.2\overline{5}$	not real	2	not real

11.6 Exercises

1. $E = mc^2$

    $\dfrac{E}{m} = c^2$

    $c = \pm\sqrt{\dfrac{E}{m}}$

    $c = \pm\dfrac{\sqrt{E}}{\sqrt{m}} \cdot \dfrac{\sqrt{m}}{\sqrt{m}}$

    $c = \pm\dfrac{\sqrt{Em}}{m}$

3. $V = \pi r^2 h$

    $\dfrac{V}{\pi h} = r^2$

    $r = \pm\sqrt{\dfrac{V}{\pi h}}$

    $r = \pm\dfrac{\sqrt{V}}{\sqrt{\pi h}} \cdot \dfrac{\sqrt{\pi h}}{\sqrt{\pi h}}$

    $r = \pm\dfrac{\sqrt{V\pi h}}{\pi h}$

5. $E = \dfrac{1}{2}mv^2$

    $2E = mv^2$

    $\dfrac{2E}{m} = v^2$

    $v = \pm\sqrt{\dfrac{2E}{m}}$

    $v = \pm\dfrac{\sqrt{2E}}{\sqrt{m}} \cdot \dfrac{\sqrt{m}}{\sqrt{m}}$

    $v = \pm\dfrac{\sqrt{2Em}}{m}$

7. $C = \dfrac{n(n-1)}{2}$

    $C = \dfrac{n^2 - n}{2}$

    $2C = n^2 - n$

    $0 = n^2 - n - 2C$

    $n = \dfrac{-(-1) \pm \sqrt{(-1)^2 - 4(1)(-2C)}}{2(1)}$

    $n = \dfrac{1 \pm \sqrt{1 + 8C}}{2}$

9. $x^2 - y^2 = c^2$

    $x^2 = y^2 + c^2$

    $x = \pm\sqrt{y^2 + c^2}$

11. $A = P(1+r)^2$

    $\dfrac{A}{P} = (1+r)^2$

    $1 + r = \pm\sqrt{\dfrac{A}{P}}$

    $1 + r = \pm\dfrac{\sqrt{A}}{\sqrt{P}} \cdot \dfrac{\sqrt{P}}{\sqrt{P}}$

    $1 + r = \pm\dfrac{\sqrt{AP}}{P}$

    $r = -1 \pm \dfrac{\sqrt{AP}}{P}$

13. $D^2 = L^2 + W^2 + H^2$

    $D^2 - W^2 - H^2 = L^2$

    $L = \pm\sqrt{D^2 - W^2 - H^2}$

    $L = \pm\sqrt{24^2 - 16^2 - 8^2}$

    $= \pm\sqrt{256}$

    $= \pm 16$

    Because length cannot be negative, the length of the box is 16 inches.

15. $c = a\sqrt{2} = 3.75\sqrt{2} \approx 5.30$

    The hypotenuse measures approximately 5.30 m.

17. $c = a\sqrt{2}$

$7.6 = a\sqrt{2}$

$a = \dfrac{7.6}{\sqrt{2}} = \dfrac{7.6}{\sqrt{2}} \cdot \dfrac{\sqrt{2}}{\sqrt{2}} = \dfrac{7.6\sqrt{2}}{2} = 3.8\sqrt{2} \approx 5.37$

Each leg measures approximately 5.37 ft.

19. $c = a\sqrt{2} = 20\sqrt{2} \approx 28.28$

The wire is approximately 28.28 feet long.

21. $c = a\sqrt{2}$

$42.5 = a\sqrt{2}$

$a = \dfrac{42.5}{\sqrt{2}} = \dfrac{42.5}{\sqrt{2}} \cdot \dfrac{\sqrt{2}}{\sqrt{2}} = \dfrac{42.5\sqrt{2}}{2} = 21.25\sqrt{2} \approx 30.05$

The building is approximately 30.05 feet high.

23. $c = a\sqrt{2}$

$40 = a\sqrt{2}$

$a = \dfrac{40}{\sqrt{2}} = \dfrac{40}{\sqrt{2}} \cdot \dfrac{\sqrt{2}}{\sqrt{2}} = \dfrac{40\sqrt{2}}{2} = 20\sqrt{2} \approx 28.28$

Yes, the diver is about 28.28 feet below the surface which is less than 30 feet.

25. $b = a\sqrt{3}$

$18.6 = a\sqrt{3}$

$a = \dfrac{18.6}{\sqrt{3}} = \dfrac{18.6}{\sqrt{3}} \cdot \dfrac{\sqrt{3}}{\sqrt{3}} = \dfrac{18.6\sqrt{3}}{3} = 6.2\sqrt{3} \approx 10.7$

$c = 2a = 2(6.2\sqrt{3}) = 12.4\sqrt{3} \approx 21.5$

The lengths of the sides are approximately 10.7 inches and 21.5 inches.

27. $c = 2a$

$22 = 2a$

$a = \dfrac{22}{2} = 11$

$b = a\sqrt{3} = 11\sqrt{3} \approx 19.1$

The lengths of the sides are 11 inches and approximately 19.1 inches.

29. $c = 2a = 2(21) = 42$

$b = a\sqrt{3} = 21\sqrt{3} \approx 36.4$

The lengths of the sides are 42 cm and approximately 36.4 cm.

31. $b = a\sqrt{3} = 4000\sqrt{3} \approx 6928.2$

$c = 2a = 2(4000) = 8000$

The horizontal distance to the set-down point is approximately 6928.2 feet. The slanted distance to the set-down point is 8000 feet.

33. $b = a\sqrt{3} = 8\sqrt{3} \approx 13.9$

$c = 2a = 2(8) = 16$

The pole is approximately 13.9 feet high and the wire is 16 feet long.

11.6 Calculator Exercises

1. a.
```
8→A: (8√(2))→C: A²
+A²=C²
                    1
```

Yes, $a = 8$ and $c = 8\sqrt{2}$ can be the legs and hypotenuse of an isosceles right triangle.

b.
```
12→A: (12√(3))→C:
A²+A²=C²
                    0
```

No, $a = 12$ and $c = 12\sqrt{3}$ cannot be the legs and hypotenuse of an isosceles right triangle.

c.
```
(6√(2))→A: 12→C: A
²+A²=C²
                    1
```

Yes, $a = 6\sqrt{2}$ and $c = 12$ can be the legs and hypotenuse of an isosceles right triangle.

d.
```
(8√(3)/3)→A: 8→C:
A²+A²=C²
                    0
```

No, $a = \frac{8\sqrt{3}}{3}$ and $c = 8$ cannot be the legs and hypotenuse of an isosceles right triangle.

2. a.

Yes, $a = 5$, $b = 5\sqrt{3}$, and $c = 10$ can be the sides of a 30°-60°-90° triangle.

b.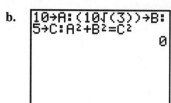

No, $a = 10$, $b = 10\sqrt{3}$, and $c = 5$ cannot be the sides of a 30°-60°-90° triangle.

c. [5→A:(5√(2))→B:10
→C:A²+B²=C²
               0]

No, $a = 5$, $b = 5\sqrt{2}$, and $c = 10$ cannot be the sides of a 30°-60°-90° triangle.

d. [5→A:10→B:(5√(3))
→C:A²+B²=C²
               0]

No, $a = 5$, $b = 10$, and $c = 5\sqrt{3}$ cannot be the sides of a 30°-60°-90° triangle.

e. [10→A:(10√(3))→B:
20→C:A²+B²=C²
               1]

Yes, $a = 10$, $b = 10\sqrt{3}$, and $c = 20$ can be the sides of a 30°-60°-90° triangle.

11.7 Exercises

1. $2x^2 - 5 < 4x + 1$ is a quadratic inequality, because it simplifies to $2x^2 - 4x - 6 < 0$.

3. $1.3z - 2.8 > 4.3z^2$ is a quadratic inequality, because it simplifies to $4.3z^2 - 1.3z + 2.8 < 0$.

5. $4x^{-2} + 5x^{-1} \geq 2x + 1$ is not a quadratic inequality because it has variables with negative exponents.

7. $\frac{1}{3}x^2 - \frac{5}{6}x > \frac{3}{4}$ is a quadratic inequality, because it simplifies to $4x^2 - 10x - 9 > 0$.

9. $2x^3 + 4 \leq x^2 + x$ is not a quadratic inequality, because it has a third-degree term.

11. $3\sqrt{x} + x < x^2 - 5$ is not a quadratic inequality, because it has a variable under a radical.

13. Let $Y1 = x^2 + 3x - 7$ and $Y2 = x + 8$.

The solutions are all integers less than or equal to – 5 or greater than or equal to 3.

15. Let $Y1 = x^2 + 4x - 3$ and $Y2 = 2x + 5$.

The solutions are all integers greater than – 4 and less than 1.

Chapter 11: Quadratic and Other Polynomial Equations and Inequalities

17. Let $Y1 = 4x^2 - 13x - 7$ and $Y2 = 18x + 1$.

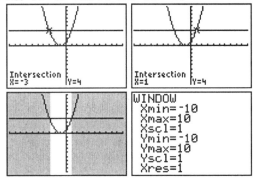

The solutions are all integers less than 0 or greater than 8.

19. Let $Y1 = x^2 + 2x + 1$ and $Y2 = 4$.

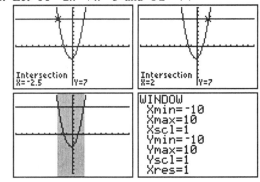

The solution is $(-\infty, -3) \cup (1, \infty)$.

21. Let $Y1 = 2x^2 + x - 3$ and $Y2 = 7$.

The solution is $(-2.5, 2)$.

23. Let $Y1 = 24 + 5x - x^2$ and $Y2 = -x + 8$.

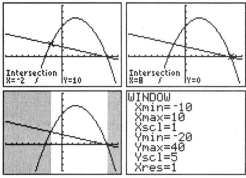

The solution is $(-\infty, -2) \cup (8, \infty)$.

25. Let $Y1 = 2x^2 - 5$ and $Y2 = x^2 + 4$.

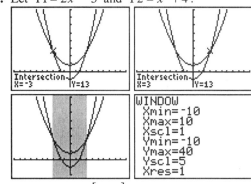

The solution is $[-3, 3]$.

27. Let $Y1 = \dfrac{1}{4}x^2 - 3$ and $Y2 = x^2 + 2$.

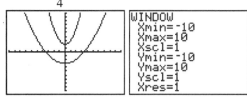

Because the graph of Y1 is always below the graph of Y2, the inequality has no solution.

29. Let $Y1 = x^2 + 6x + 10$ and $Y2 = -x^2 - 6x - 8$.

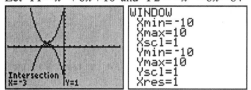

Because $Y1 \geq Y2$ for all values of x, the solution is the set of all real numbers, or $(-\infty, \infty)$.

31. Let $Y1 = x^2 - 4x + 8$ and $Y2 = \dfrac{1}{4}x^2 - x + 5$.

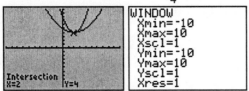

Because $Y1 \geq Y2$ for all values of x, the solution is the set of all real numbers, or $(-\infty, \infty)$.

33. $x^2 - 2x - 15 \geq 9$
Determine the critical points.
$x^2 - 2x - 15 = 9$
$x^2 - 2x - 24 = 0$
$(x+4)(x-6) = 0$
$x = -4$ or $x = 6$

Possible intervals:
$(-\infty, -4]$, $[-4, 6]$, and $[6, \infty)$
Use a test point in each interval to determine that the solution is $(-\infty, -4] \cup [6, \infty)$.

35. $8 + 2x - x^2 > -7$
Determine the critical points.
$8 + 2x - x^2 = -7$
$0 = x^2 - 2x - 15$
$0 = (x+3)(x-5)$
$x + 3 = 0$ or $x - 5 = 0$
$x = -3$ $\quad$ $x = 5$
Possible intervals:
$(-\infty, -3)$, $(-3, 5)$, and $(5, \infty)$
Use a test point in each interval to determine that the solution is $(-3, 5)$.

37. $4x^2 - 12x + 9 \leq 4$
Determine the critical points.
$4x^2 - 12x + 9 = 4$
$4x^2 - 12x + 5 = 0$
$(2x - 1)(2x - 5) = 0$

$2x - 1 = 0$ or $2x - 5 = 0$
$x = \dfrac{1}{2}$ $\quad$ $x = \dfrac{5}{2}$
Possible intervals:
$\left(-\infty, \dfrac{1}{2}\right]$, $\left[\dfrac{1}{2}, \dfrac{5}{2}\right]$, and $\left[\dfrac{5}{2}, \infty\right)$
Use a test point in each interval to determine that the solution is $\left[\dfrac{1}{2}, \dfrac{5}{2}\right]$.

39. $6 + 5x - x^2 < -8$
Determine the critical points.
$6 + 5x - x^2 = -8$
$0 = x^2 - 5x - 14$
$(x+2)(x-7) = 0$
$x + 2 = 0$ or $x - 7 = 0$
$x = -2$ $\quad$ $x = 7$
Possible intervals:
$(-\infty, -2)$, $(-2, 7)$, and $(7, \infty)$
Use a test point in each interval to determine that the solution is $(-\infty, -2) \cup (7, \infty)$.

41. $x^2 + 3 \geq -x + 15$
Determine the critical points.
$x^2 + 3 = -x + 15$
$x^2 + x - 12 = 0$
$(x-3)(x+4) = 0$
$x - 3 = 0$ or $x + 4 = 0$
$x = 3$ $\quad$ $x = -4$
Possible intervals:
$(-\infty, -4]$, $[-4, 3]$, and $[3, \infty)$
Use a test point in each interval to determine that the solution is $(-\infty, -4] \cup [3, \infty)$.

43. $10 - 2x^2 < 4x + 4$
Determine the critical points.
$10 - 2x^2 = 4x + 4$
$0 = 2x^2 + 4x - 6$
$0 = 2(x^2 + 2x - 3)$
$0 = 2(x - 1)(x + 3)$

Chapter 11: Quadratic and Other Polynomial Equations and Inequalities

$x - 1 = 0$ or $x + 3 = 0$
$x = 1 \qquad x = -3$
Possible intervals:
$(-\infty, -3)$, $(-3, 1)$, and $(1, \infty)$
Use a test point in each interval to determine that the solution is $(-\infty, -3) \cup (1, \infty)$.

45. $x^2 - x - 6 \leq 12 + 2x - 2x^2$
Determine the critical points.
$x^2 - x - 6 = 12 + 2x - 2x^2$
$3x^2 - 3x - 18 = 0$
$3(x^2 - x - 6) = 0$
$3(x + 2)(x - 3) = 0$
$x + 2 = 0$ or $x - 3 = 0$
$x = -2 \qquad x = 3$
Possible intervals:
$(-\infty, -2]$, $[-2, 3]$, and $[3, \infty)$
Use a test point in each interval to determine that the solution is $[-2, 3]$.

47. $2x^2 - 3x + \dfrac{21}{4} \geq 2x^2 - 6x + \dfrac{15}{2}$
Determine the critical points.
$x^2 - 3x + \dfrac{21}{4} = 2x^2 - 6x + \dfrac{15}{2}$
$4x^2 - 12x + 21 = 8x^2 - 24x + 30$
$0 = 4x^2 - 12x + 9$
$(2x - 3)^2 = 0$
$2x - 3 = 0$
$x = \dfrac{3}{2}$
Possible intervals:
$\left(-\infty, \dfrac{3}{2}\right]$ and $\left[\dfrac{3}{2}, \infty\right)$
Use a test point in each interval to determine that neither interval is a solution. Thus, the solution is only $\dfrac{3}{2}$.

49. $x^2 - x - \dfrac{7}{4} > -2x^2 + 2x - \dfrac{5}{2}$
Determine the critical points.
$x^2 - x - \dfrac{7}{4} = -2x^2 + 2x - \dfrac{5}{2}$
$4x^2 - 4x - 7 = -8x^2 + 8x - 10$
$12x^2 - 12x + 3 = 0$
$3(4x^2 - 4x + 1) = 0$
$3(2x - 1)^2 = 0$
$2x - 1 = 0$
$x = \dfrac{1}{2}$
Possible intervals:
$\left(-\infty, \dfrac{1}{2}\right)$ and $\left(\dfrac{1}{2}, \infty\right)$
Use a test point in each interval to determine that both intervals are solutions. Thus, the solution is $\left(-\infty, \dfrac{1}{2}\right) \cup \left(\dfrac{1}{2}, \infty\right)$.

51. $x^2 + 4x + 8 \geq 4$
Determine the critical points.
$x^2 + 4x + 8 = 4$
$x^2 + 4x + 4 = 0$
$(x + 2)^2 = 0$
$x + 2 = 0$
$x = -2$
Possible intervals:
$(-\infty, -2]$ and $[-2, \infty)$
Use a test point in each interval to determine that both intervals are solutions. Thus, the solution is the set of all real numbers, or $(-\infty, \infty)$.

53. $2 + 4x - x^2 > 10 - 4x + x^2$
Determine the critical points.
$x^2 + 6x + 5 = -x^2 - 6x - 13$
$2x^2 + 12x + 18 = 0$

$2(x^2+6x+9)=0$

$2(x+3)^2=0$

$x+3=0$

$x=-3$

Possible intervals:

$(-\infty,-3)$ and $(-3,\infty)$

Use a test point in each interval to determine that neither interval is a solution. Thus, the inequality has no solution.

55. $s=-16t^2+v_0t+s_0$; $s=25$; $v_0=30$; $s_0=25$

$25 < -16t^2+30t+25$

Determine the critical points.

$25=-16t^2+30t+25$

$16t^2-30t=0$

$2t(8t-15)=0$

$2t=0$ or $8t-15=0$

$t=0$ $\qquad t=\dfrac{15}{8}=1.875$

Possible intervals:

$(-\infty,0)$, $(0,1.875)$, and $(1.875,\infty)$

Use a test point in each interval to determine that the solution is $(0,1.875)$.

The egg is higher than the stand between 0 and 1.875 seconds.

57. $s=-16t^2+v_0t+s_0$; $s=800$; $v_0=75$; $s_0=1200$

$800 \geq -16t^2+75t+1200$

Determine the critical points.

$800=-16t^2+75t+1200$

$16t^2-75t-400=0$

$t=\dfrac{-(-75)\pm\sqrt{(-75)^2-4(16)(-400)}}{2(16)}$

$t=\dfrac{75\pm\sqrt{31,225}}{32}$

$t\approx 7.866$ or $t\approx -3.178$

Since time cannot be negative, the possible intervals are approximately $[0,7.866]$ and

$[7.866,\infty)$.

Use a test point in each interval to determine that the solution is approximately $[7.866,\infty)$.

Recognize that the rock will hit the ground when $0=-16t^2+75t+1200$. Solving this equation yields $t\approx 11.316$ seconds.
Thus, the rock will be no more than 800 feet high in the air from about 7.866 seconds to 11.316 seconds. After 11.316 seconds, the rock will be on the ground.

59. $3x^2+x+35 \leq 100$

Determine the critical points.

$3x^2+x+35=100$

$3x^2+x-65=0$

$x=\dfrac{-1\pm\sqrt{1^2-4(3)(-65)}}{2(3)}$

$x=\dfrac{-1\pm\sqrt{781}}{6}$

$x\approx 4.491$ or $x\approx -4.824$

Since the number of days must be positive, the possible intervals are approximately $[0,4.491]$ and $[4.491,\infty)$.

Use a test point in each interval to determine that the solution is approximately $[0,4.491]$.

The booth can be operated from 1 to 4 days.

61. $0.01x^2+0.57x+2.27 > 10$

Determine the critical points.

$0.01x^2+0.57x+2.27=10$

$0.01x^2+0.57x-7.73=0$

$x=\dfrac{-0.57\pm\sqrt{(0.57)^2-4(0.01)(-7.73)}}{2(0.01)}$

$x=\dfrac{-0.57\pm\sqrt{0.6341}}{0.02}$

$x\approx 11.315$ or $x\approx -68.315$

Since the number of years must be positive, the possible intervals are approximately $(0,11.315)$ and $(11.315,\infty)$.

Use a test point in each interval to determine that the solution is approximately $(11.315,\infty)$.

Now $1975 + 11.315 = 1986.315$. Thus, the hourly costs exceed \$10 from 1987 and beyond.

63. $-0.007x^2 + 0.68x + 6.51 > 15$
Determine the critical points.
$-0.007x^2 + 0.68x + 6.51 = 15$
$0 = 0.007x^2 - 0.68x + 8.49$
$x = \dfrac{-(-0.68) \pm \sqrt{(-0.68)^2 - 4(0.007)(8.49)}}{2(0.007)}$
$x = \dfrac{0.68 \pm \sqrt{0.22468}}{0.014}$
$x \approx 82.429$ or $x \approx 14.714$

Disregarding 82.429, the possible intervals are $(0, 14.713)$ and $(14.713, \infty)$.

Use a test point in each interval to determine that the solution is $(14.713, \infty)$.

Now $1975 + 14.713 = 1989.713$. Thus, the hourly costs exceed \$15 from 1990 and beyond.

11.7 Calculator Exercises

Students should use the described procedures in order to check the solutions of the appropriate exercises in the 11.7 Exercises.

Chapter 11 Section-By-Section Review

1. Let $Y1 = 2x^2 + 5x - 16$ and $Y2 = 7x + 8$.

X	Y1	Y2		X	Y1	Y2
-3	-13	-13		-2	-18	-6
-2	-18	-6		-1	-19	1
-1	-19	1		0	-16	8
0	-16	8		1	-9	15
1	-9	15		2	2	22
2	2	22		3	17	29
3	17	29		4	36	36
X=-3				X=4		

The solutions are −3 and 4.

2. Let $Y1 = \dfrac{1}{3}x^2 + 5x + 6$ and $Y2 = x - 3$.

X	Y1	Y2
-9	-12	-12
-8	-12.67	-11
-7	-12.67	-10
-6	-12	-9
-5	-10.67	-8
-4	-8.667	-7
-3	-6	-6
X=-9		

The solutions are −9 and −3.

3. Let $Y1 = 0.7x^2 - 2.5x + 4$ and $Y2 = 1.7x + 1.1$.

X	Y1	Y2
0	4.6	1.1
1	2.8	2.8
2	2.4	4.5
3	3.4	6.2
4	5.8	7.9
5	9.6	9.6
6	14.8	11.3
X=1		

The solutions are 1 and 5.

4. Let $Y1 = x^2 - 4x + 4$ and $Y2 = 9$.

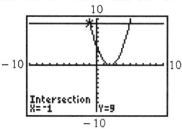

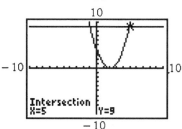

The solutions are −1 and 5.

5. Let $Y1 = x^2 - 6$ and $Y2 = \dfrac{1}{4}x^2 + 6$.

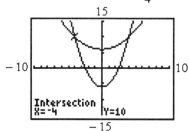

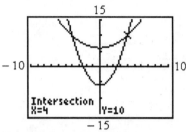
The solutions are –4 and 4.

6. Let $Y1 = x^2 + 3x - 5$ and $Y2 = 7 - 2x - x^2$.

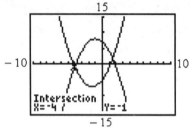

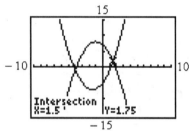

The solutions are –4 and 1.5.

7. Let $Y1 = -0.3x^2 + 4x + 5$ and $Y2 = 2.2x - 11.5$.

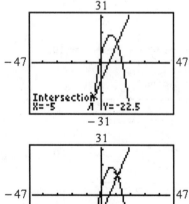

The solutions are –5 and 11.

8. Let $Y1 = x^2 + 8x + 8$ and $Y2 = 0$.

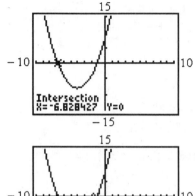

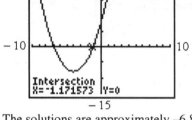

The solutions are approximately –6.83 and –1.17.

9. Let $Y1 = 2x^3 - 18x$ and $Y2 = 3x^2 - 27$.

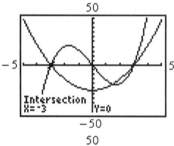

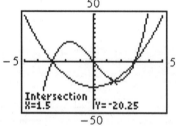

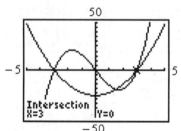

The solutions are –3, 1.5, and 3.

Chapter 11: Quadratic and Other Polynomial Equations and Inequalities

10. $s = -16t^2 + v_0 t + s_0$; $s = 0$; $v_0 = 0$; $s_0 = 96$

    $0 = -16t^2 + 0t + 96$ or $0 = -16t^2 + 96$

    Let $Y1 = 0$ and $Y2 = -16x^2 + 96$.

    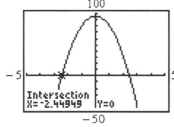

    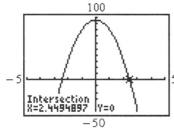

    The solutions to the equation are approximately -2.45 and 2.45. Time cannot be negative, so it will take approximately 2.45 seconds for the lead weight to hit the ground.

11. $2(5.5) = 11$; $11 = 0.014x^2 + 0.42x + 5.49$

    Let $Y1 = 11$ and $Y2 = 0.014x^2 + 0.42x + 5.49$.

    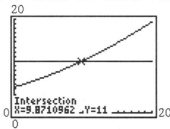

    The solution to the equation is approximately 9.87. Now, $1995 + 9.87 = 2004.87$ or about 2005. According to the function, the box office receipts will be twice that of 1995 in about 2005.

12. $x^2 - 5 = x + 1$

    $x^2 - x - 6 = 0$

    $(x-3)(x+2) = 0$

    $x - 3 = 0$ or $x + 2 = 0$

    $x = 3$ $\qquad$ $x = -2$

    The solutions are -2 and 3.

13. $6x^2 - x - 77 = 0$

    $(2x+7)(3x-11) = 0$

    $2x + 7 = 0$ or $3x - 11 = 0$

    $2x = -7$ $\qquad$ $3x = 11$

    $x = -\dfrac{7}{2}$ $\qquad$ $x = \dfrac{11}{3}$

    The solutions are $-\dfrac{7}{2}$ and $\dfrac{11}{3}$.

14. $2x + 10 = x^2 - 5x + 2$

    $0 = x^2 - 7x - 8 = 0$

    $(x-8)(x+1) = 0$

    $x - 8 = 0$ or $x + 1 = 0$

    $x = 8$ $\qquad$ $x = -1$

    The solutions are -1 and 8.

15. $x^2 - 7x - 60 = 0$

    $(x-12)(x+5) = 0$

    $x - 12 = 0$ or $x + 5 = 0$

    $x = 12$ $\qquad$ $x = -5$

    The solutions are -5 and 12.

16. $3x^2 + 5x - 3 = 1 + 5x + 3x^2$

    $-3 = 1$

    This is a false equation (a contradiction). It has no solution.

17. $6x^2 - 8x = 9x - 12$

    $6x^2 - 17x + 12 = 0$

    $(2x-3)(3x-4) = 0$

    $2x - 3 = 0$ or $3x - 4 = 0$

    $2x = 3$ $\qquad$ $3x = 4$

    $x = \dfrac{3}{2}$ $\qquad$ $x = \dfrac{4}{3}$

    The solutions are $\dfrac{4}{3}$ and $\dfrac{3}{2}$.

18. $\dfrac{1}{4}x^2 + \dfrac{3}{2}x - \dfrac{19}{8} = \dfrac{3}{8}x + 2$

    $2x^2 + 12x - 19 = 3x + 16$

SSM: Experiencing Introductory and Intermediate Algebra

$2x^2 + 9x - 35 = 0$
$(2x-5)(x+7) = 0$
$2x-5 = 0$ or $x+7 = 0$
$2x = 5 \qquad x = -7$
$x = \dfrac{5}{2}$

The solutions are -7 and $\dfrac{5}{2}$.

19. $9x^2 - 49 = 0$
$(3x+7)(3x-7) = 0$
$9x^2 - 49 = 0$
$(3x+7)(3x-7) = 0$
$3x+7 = 0$ or $3x-7 = 0$
$3x = -7 \qquad 3x = 7$
$x = -\dfrac{7}{3} \qquad x = \dfrac{7}{3}$

The solutions are $-\dfrac{7}{3}$ and $\dfrac{7}{3}$.

20. Let w = the width of the carpeted area
$w+5$ = the length of the carpeted area
$w(w+5) = 300$
$w^2 + 5w = 300$
$w^2 + 5w - 300 = 0$
$(w+20)(w-15) = 0$
$w+20 = 0$ or $w-15 = 0$
$w = -20 \qquad w = 15$

Because length cannot be negative, the width of the carpeted area must be $w = 15$ feet. The length is $15 + 5 = 20$ feet. Thus, the dimensions of the carpeted area are 15 feet by 20 feet.

21. a. $y = -0.06x^2 + 0.12x + 6.44$

L1	L2	L3	1	QuadReg
0	6.44	------		y=ax²+bx+c
1	6.5			a=-.06
2	6.44			b=.12
------	------			c=6.44
				R²=1

L1(2)=1

b. $5 = -0.06x^2 + 0.12x + 6.44$
$0.06x^2 - 0.12x - 1.44 = 0$
$0.06(x^2 - 2x - 24) = 0$
$0.06(x-6)(x+4) = 0$
$x-6 = 0$ or $x+4 = 0$
$x = 6 \qquad x = -4$

Disregarding -4, the solution is 6. Now, $1998 + 6 = 2004$. Thus, according to the function, donations will drop to \$5 million in about 2004.

22. $-\sqrt{49} = -7$

23. $\sqrt{200} = \sqrt{100 \cdot 2} = 10\sqrt{2}$

24. $\sqrt{32} = \sqrt{16 \cdot 2} = 4\sqrt{2}$

25. $\sqrt{8} = \sqrt{4 \cdot 2} = 2\sqrt{2}$

26. $-\sqrt{72} = -\sqrt{36 \cdot 2} = -6\sqrt{2}$

27. $\dfrac{-6 - \sqrt{50}}{5} = \dfrac{-6 - \sqrt{25 \cdot 2}}{5}$
$= \dfrac{-6 - 5\sqrt{2}}{5}$ or $-\dfrac{6}{5} - \sqrt{2}$

28. $\dfrac{14 + \sqrt{98}}{14} = \dfrac{14 + \sqrt{49 \cdot 2}}{14}$
$= \dfrac{14 + 7\sqrt{2}}{14}$
$= \dfrac{2 + \sqrt{2}}{2}$ or $1 + \dfrac{\sqrt{2}}{2}$

29. $\dfrac{3 - \sqrt{27}}{6} = \dfrac{3 - \sqrt{9 \cdot 3}}{6} = \dfrac{3 - 3\sqrt{3}}{6} = \dfrac{1 - \sqrt{3}}{2}$

30. $\dfrac{-4 + \sqrt{20}}{-2} = \dfrac{-4 + \sqrt{4 \cdot 5}}{-2} = \dfrac{-4 + 2\sqrt{5}}{-2} = 2 - \sqrt{5}$

31. $\sqrt{\dfrac{36}{5}} = \dfrac{\sqrt{36}}{\sqrt{5}} = \dfrac{6}{\sqrt{5}} \cdot \dfrac{\sqrt{5}}{\sqrt{5}} = \dfrac{6\sqrt{5}}{5}$

32. $\sqrt{\dfrac{40}{32}} = \sqrt{\dfrac{5}{4}} = \dfrac{\sqrt{5}}{\sqrt{4}} = \dfrac{\sqrt{5}}{2}$

Chapter 11: Quadratic and Other Polynomial Equations and Inequalities

33. $4x^2 - 100 = 0$
    $4x^2 = 100$
    $x^2 = 25$
    $x = \pm\sqrt{25}$
    $x = \pm 5$
    The solutions are ± 5.

34. $x^2 + 5 = 9$
    $x^2 = 4$
    $x = \pm\sqrt{4}$
    $x = \pm 2$
    The solutions are ± 2.

35. $x^2 + 3 = 15$
    $x^2 = 12$
    $x = \pm\sqrt{12}$
    $x = \pm\sqrt{4 \cdot 3}$
    $x = \pm 2\sqrt{3}$
    The solutions are $\pm 2\sqrt{3}$.

36. $15 + b^2 = 7$
    $b^2 = -8$
    The equation has no real-number solution.

37. $(x-4)^2 = 9$
    $x - 4 = \pm\sqrt{9}$
    $x - 4 = \pm 3$
    $x = 4 \pm 3$
    $x = 4 + 3$ or $x = 4 - 3$
    $x = 7$ $x = 1$
    The solutions are 1 and 7.

38. $x^2 + 18x + 81 = 16$
    $(x+9)^2 = 16$
    $x + 9 = \pm\sqrt{16}$
    $x + 9 = \pm 4$
    $x = -9 \pm 4$
    $x = -9 + 4$ or $x = -9 - 4$
    $x = -5$ $x = -13$
    The solutions are -13 and -5.

39. $(z-2)^2 + 5 = 7$
    $(z-2)^2 = 2$
    $z - 2 = \pm\sqrt{2}$
    $z = 2 \pm \sqrt{2}$
    The solutions are $2 \pm \sqrt{2}$.

40. Let x = the horizontal distance of the ramp.
    $x^2 + 15^2 = 35^2$
    $x^2 + 225 = 1225$
    $x^2 = 1000$
    $x = \pm\sqrt{1000}$
    $x = \pm 10\sqrt{10}$
    Length cannot be negative, so the horizontal distance of the ramp is $10\sqrt{10}$ feet, or approximately 31.62 feet.

41. $A = P(1+r)^t$; $P = 2000$, $A = 2332.80$, $t = 2$
    $2332.80 = 2000(1+r)^2$
    $1.1664 = (1+r)^2$
    $\pm\sqrt{1.1664} = 1 + r$
    $\pm 1.08 = 1 + r$
    $-1 \pm 1.08 = r$
    $r = 0.08$ or $r = -2.08$
    Disregarding -2.08, Stephanie's annual compound interest rate was $0.08 = 8\%$.

42. $s = -16t^2 + v_0 t + s_0$; $s = 2500$; $v_0 = 0$; $s_0 = 12,000$
    $2500 = -16t^2 + 0t + 12000$
    $16t^2 = 9500$
    $t^2 = 593.75$
    $t = \pm\sqrt{593.75}$
    $t \approx \pm 24.37$
    Time cannot be negative, so it will take about 24.37 seconds for skydivers to descend from 12,000 feet to 2500 feet.

43. $p^2 - 4p - 96 = 0$
    $p^2 - 4p = 96$
    $p^2 - 4p + 4 = 96 + 4$

485

$(p-2)^2 = 100$
$p-2 = \pm\sqrt{100}$
$p-2 = \pm 10$
$p = 2 \pm 10$
$p = 12$ or $p = -8$
The solutions are -8 and 12.

44. $2x^2 - 5x - 12 = 0$
$2x^2 - 5x = 12$
$x^2 - \frac{5}{2}x = 6$
$x^2 - \frac{5}{2}x + \frac{25}{16} = 6 + \frac{25}{16}$
$\left(x - \frac{5}{4}\right)^2 = \frac{121}{16}$
$x - \frac{5}{4} = \pm\sqrt{\frac{121}{16}}$
$x - \frac{5}{4} = \pm\frac{11}{4}$
$x = \frac{5}{4} \pm \frac{11}{4}$
$x = \frac{16}{4}$ or $x = -\frac{6}{4}$
$x = 4$ $\qquad x = -\frac{3}{2}$
The solutions are $-\frac{3}{2}$ and 4.

45. $x^2 + \frac{1}{2}x = 1$
$x^2 + \frac{1}{2}x + \frac{1}{16} = 1 + \frac{1}{16}$
$\left(x + \frac{1}{4}\right)^2 = \frac{17}{16}$
$x + \frac{1}{4} = \sqrt{\frac{17}{16}}$

$x + \frac{1}{4} = \pm\frac{\sqrt{17}}{\sqrt{16}}$
$x = -\frac{1}{4} \pm \frac{\sqrt{17}}{4}$
$x = \frac{-1 \pm \sqrt{17}}{4}$

The solutions are $\frac{-1 \pm \sqrt{17}}{4}$.

46. a. $y = -0.15x^2 + 0.25x + 2.6$

L1	L2	L3	1	QuadReg
0	2.6			y=ax²+bx+c
1	2.7			a=-.15
2	2.5			b=.25
				c=2.6
				R²=1

L1(2)=1

b. $-0.15x^2 + 0.25x + 2.6 = 2$
$-0.15x^2 + 0.25x = -0.6$
$x^2 - \frac{5}{3}x = 4$
$x^2 - \frac{5}{3}x + \frac{25}{36} = 4 + \frac{25}{36}$
$\left(x - \frac{5}{6}\right)^2 = \frac{169}{36}$
$x - \frac{5}{6} = \pm\sqrt{\frac{169}{36}}$
$x - \frac{5}{6} = \pm\frac{13}{6}$
$x = \frac{5}{6} \pm \frac{13}{6}$
$x = \frac{18}{6}$ or $x = -\frac{8}{6}$
$x = 3$ $\qquad x = -\frac{4}{3}$

Disregarding $-\frac{4}{3}$, the cost of goods will be $2 billion 3 years after 1999. That is, according to the function, the cost of goods will be $2 billion in the year 2002.

47. Let $x =$ the number of figurines to be purchased.
$20 - x =$ the price per figurine.

$x(20-x) = 100$

$20x - x^2 = 100$

$x^2 - 20x = -100$

$x^2 - 20x + 100 = -100 + 100$

$(x-10)^2 = 0$

$x - 10 = \pm\sqrt{0}$

$x - 10 = 0$

$x = 10$

The retailer will break even if each purchaser buys 10 figurines.

48. Let x = the width of the rectangle.
$2x + 8$ = the length of the rectangle.

$x(2x+8) = 90$

$2x^2 + 8x = 90$

$x^2 + 4x = 45$

$x^2 + 4x + 4 = 45 + 4$

$(x+2)^2 = 49$

$x + 2 = \pm\sqrt{49}$

$x + 2 = \pm 7$

$x = -2 \pm 7$

$x = 5$ or $x = -9$

Because distance cannot be negative, the width of the rectangle is 5 inches. The length is $2(5) + 8 = 18$ inches. Thus, the rectangle is 18 inches by 5 inches.

49. $x^2 - 2x - 63 = 0$

$x = \dfrac{-(-2) \pm \sqrt{(-2)^2 - 4(1)(-63)}}{2(1)}$

$x = \dfrac{2 \pm \sqrt{256}}{2}$

$x = \dfrac{2 \pm 16}{2}$

$x = \dfrac{18}{2}$ or $x = \dfrac{-14}{2}$

$x = 9 \qquad x = -7$

The solutions are -7 and 9.

50. $x^2 = 3x + 3$

$x^2 - 3x - 3 = 0$

$x = \dfrac{-(-3) \pm \sqrt{(-3)^2 - 4(1)(-3)}}{2(1)}$

$x = \dfrac{3 \pm \sqrt{21}}{2}$

The solutions are $\dfrac{3 \pm \sqrt{21}}{2}$.

51. $25x^2 + 1 = 10x$

$25x^2 - 10x + 1 = 0$

$x = \dfrac{-(-10) \pm \sqrt{(-10)^2 - 4(25)(1)}}{2(25)}$

$x = \dfrac{10 \pm \sqrt{0}}{50}$

$x = \dfrac{10 \pm 0}{50}$

$x = \dfrac{10}{50}$

$x = \dfrac{1}{5}$

The solution is $\dfrac{1}{5}$ (which is a double root).

52. $z^2 + \dfrac{7}{20}z - \dfrac{3}{10} = 0$

$20\left(z^2 + \dfrac{7}{20}z - \dfrac{3}{10}\right) = 20(0)$

$20z^2 + 7z - 6 = 0$

$z = \dfrac{-7 \pm \sqrt{7^2 - 4(20)(-6)}}{2(20)}$

$z = \dfrac{-7 \pm \sqrt{529}}{40}$

$z = \dfrac{-7 \pm 23}{40}$

$z = \dfrac{16}{40}$ or $z = \dfrac{-30}{40}$

$z = \dfrac{2}{5}$ $\qquad z = -\dfrac{3}{4}$

The solutions are $-\dfrac{3}{4}$ and $\dfrac{2}{5}$.

53. $x^2 + 2.1x - 10.8 = 0$

$x = \dfrac{-2.1 \pm \sqrt{(2.1)^2 - 4(1)(-10.8)}}{2(1)}$

$x = \dfrac{-2.1 \pm \sqrt{47.61}}{2}$

$x = \dfrac{-2.1 \pm 6.9}{2}$

$x = \dfrac{4.8}{2}$ or $x = \dfrac{-9}{2}$

$x = 2.4 \qquad x = -4.5$

The solutions are -4.5 and 2.4.

54. $y^2 - 5y + 12 = 0$

$y = \dfrac{-(-5) \pm \sqrt{(-5)^2 - 4(1)(12)}}{2(1)}$

$y = \dfrac{5 \pm \sqrt{-23}}{2}$

The equation has no real-number solution.

55. $x^2 - 10x + 6 = 0$

$x = \dfrac{-(-10) \pm \sqrt{(-10)^2 - 4(1)(6)}}{2(1)}$

$x = \dfrac{10 \pm \sqrt{76}}{2}$

$x = \dfrac{10 \pm 2\sqrt{19}}{2}$

$x = 5 \pm \sqrt{19}$

The solutions are $5 \pm \sqrt{19}$.

56. $3a^2 - 4a - 12 = 0$

$a = \dfrac{-(-4) \pm \sqrt{(-4)^2 - 4(3)(-12)}}{2(3)}$

$a = \dfrac{4 \pm \sqrt{160}}{6}$

$a = \dfrac{4 \pm 4\sqrt{10}}{6}$

$a = \dfrac{2 \pm 2\sqrt{10}}{3}$

The solutions are $\dfrac{2 \pm 2\sqrt{10}}{3}$.

57. $x^2 + 2x + 10 = 0$

$b^2 - 4ac = 2^2 - 4(1)(10) = -36$

Because the discriminant is negative, the equation has no real-number solution.

58. $x^2 + 20x + 55 = 2x - 26$

$x^2 + 18x + 81 = 0$

$b^2 - 4ac = 18^2 - 4(1)(81) = 0$

Because the discriminant is 0, the equation has one rational solution.

59. $x^2 = 10x + 75$

$x^2 - 10x - 75 = 0$

$b^2 - 4ac = (-10)^2 - 4(1)(-75) = 400$

Because 400 is a perfect square, the equation has two rational solutions.

60. $x = x^2 - 11$

$0 = x^2 - x - 11$

$b^2 - 4ac = (-1)^2 - 4(1)(-11) = 45$

Because 45 is positive but not a perfect square, the equation has two irrational solutions.

61. Let $x =$ the number of cakes ordered.
$6.50 - 0.50x =$ price per cake.

a. $R(x) = x(6.50 - 0.50x)$

b. $C(x) = 2.50x + 3.95$

c. $P(x) = R(x) - C(x)$
$= x(6.50 - 0.50x) - (2.50x + 3.95)$
$= 6.5x - 0.5x^2 - 2.5x - 3.95$
$= -0.5x^2 + 4x - 3.95$

d. $-0.5x^2 + 4x - 3.95 = 0$
$0.5x^2 - 4x + 3.95 = 0$
$x = \dfrac{-(-4) \pm \sqrt{(-4)^2 - 4(0.5)(3.95)}}{2(0.5)}$
$x = \dfrac{5 \pm \sqrt{8.1}}{1}$
$x = 5 \pm \sqrt{8.1}$
$x \approx 8 \quad \text{or} \quad x \approx 2$
$6.50 - 0.50x \approx 2.50 \quad 6.50 - 0.50x \approx 5.50$
Mary will approximately break even if she sells 2 cakes at $5.50 each and if she sells 8 cakes at $2.50 each.

62. $2000 = -4.1x^2 + 163x + 693$
$4.1x^2 - 163x + 1307 = 0$
$x = \dfrac{-(-163) \pm \sqrt{(-163)^2 - 4(4.1)(1307)}}{2(4.1)}$
$x = \dfrac{163 \pm \sqrt{5134.2}}{8.2}$
$x \approx 28.6 \quad \text{or} \quad x \approx 11.1$

The world's total generation first reached 2000 billion kilowatt-hours approximately 11.1 years after 1980. Now $1980 + 11.1 = 1991.1$, or about 1992. Thus, according to the model, the world's total nuclear power generation first reached 2000 billion kilowatt-hours in about the year 1992.

63. $s = -16t^2 + 50$
$16t^2 = 50 - s$
$t^2 = \dfrac{50 - s}{16}$
$t = \pm\sqrt{\dfrac{50 - s}{16}}$
$t = \pm\dfrac{\sqrt{50 - s}}{\sqrt{16}}$
$t = \pm\dfrac{\sqrt{50 - s}}{4}$

64. $A = \dfrac{1}{4}\pi d^2$
$4A = \pi d^2$
$\dfrac{4A}{\pi} = d^2$
$d = \pm\sqrt{\dfrac{4A}{\pi}}$
$d = \pm\dfrac{\sqrt{4A}}{\sqrt{\pi}} \cdot \dfrac{\sqrt{\pi}}{\sqrt{\pi}}$
$d = \pm\dfrac{2\sqrt{A\pi}}{\pi}$

65. $a = bx^2 + c$
$a - c = bx^2$
$\dfrac{a - c}{b} = x^2$
$x = \pm\sqrt{\dfrac{a - c}{b}}$
$x = \pm\dfrac{\sqrt{a - c}}{\sqrt{b}} \cdot \dfrac{\sqrt{b}}{\sqrt{b}}$
$x = \pm\dfrac{\sqrt{b(a - c)}}{b}$

66. $s = -16t^2 + v_0 t + s_0$
$0 = -16t^2 + v_0 t + (s_0 - s)$
$t = \dfrac{-v_0 \pm \sqrt{(v_0)^2 - 4(-16)(s_0 - s)}}{2(-16)}$
$t = \dfrac{-v_0 \pm \sqrt{(v_0)^2 + 64(s_0 - s)}}{-32}$

67. $D^2 = L^2 + W^2 + H^2$

$D^2 - L^2 - W^2 = H^2$

$H = \pm\sqrt{D^2 - L^2 - W^2}$

$H = \pm\sqrt{12^2 - 8^2 - 4^2}$

$= \pm\sqrt{64}$

$= \pm 8$

Because length cannot be negative, the height of the box is 8 inches.

68. $c = a\sqrt{2} = 7\sqrt{2} \approx 9.90$

The rope is $7\sqrt{2}$ feet long, or approximately 9.90 feet long.

69. $c = 2a$

$20 = 2a$

$a = \dfrac{20}{2} = 10$

$b = a\sqrt{3} = 10\sqrt{3} \approx 17.32$

The lengths of the sides are 10 inches and $10\sqrt{3}$ inches, or approximately 17.32 inches.

70. Let $Y1 = x^2 + 6x - 9$ and $Y2 = 4x + 15$.

The solutions are all integers greater than or equal to -6 and less than or equal to 4.

71. Let $Y1 = 2x^2 + 3x - 8$ and $Y2 = 6 + 4x - x^2$.

The solutions are all integers less than -2 or greater than 2.

72. Let $Y1 = x^2 + 2x - 3$ and $Y2 = 5$.

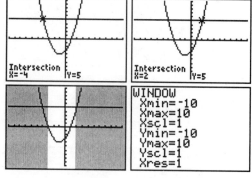

The solution is $(-\infty, -4] \cup [2, \infty)$.

73. Let $Y1 = 5 - x^2$ and $Y2 = -4$.

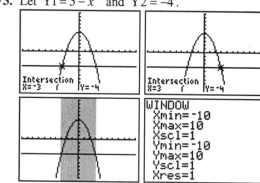

The solution is $[-3, 3]$.

74. Let $Y1 = x^2 - 2x - 1$ and $Y2 = x + 3$.

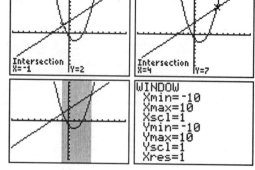

The solution is $(-1, 4)$.

75. Let $Y1 = x^2 + 3$ and $Y2 = 1 - x^2$.

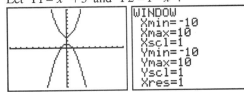

Because the graph of Y1 is always above the graph of Y2, the inequality has no solution.

76. Let $Y1 = x^2 - 6x + 11$ and $Y2 = -\frac{1}{2}x^2 + 3x - \frac{5}{2}$.

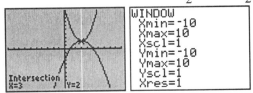

Because the graph of Y1 is above the graph of Y2 everywhere except at the intersection point, the solution is $(-\infty, 3) \cup (3, \infty)$.

77. Let $Y1 = x^2 - 8x + 18$ and $Y2 = \frac{1}{4}x^2 - 2x + 6$.

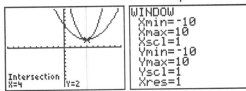

Because the graph of Y1 is above the graph of Y2 everywhere except at the intersection point, the only solution to the inequality is 4.

78. Let $Y1 = 10 - 6x + x^2$ and $Y2 = -4 + 6x - x^2$.

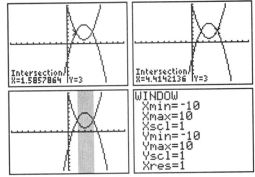

The solution is $(1.586, 4.414)$.

79. $x^2 - 4x + 4 < \frac{1}{4}x^2 - x + 3$

Determine the critical points.

$x^2 - 4x + 4 = \frac{1}{4}x^2 - x + 3$

$4x^2 - 16x + 16 = x^2 - 4x + 12$

$3x^2 - 12x + 4 = 0$

$x = \dfrac{-(-12) \pm \sqrt{(-12)^2 - 4(3)(4)}}{2(3)}$

$x = \dfrac{12 \pm \sqrt{96}}{6}$

$x = \dfrac{12 \pm 4\sqrt{6}}{6}$

$x = 2 \pm \dfrac{2}{3}\sqrt{6}$

Possible intervals:

$\left(-\infty, 2 - \dfrac{2}{3}\sqrt{6}\right)$, $\left(2 - \dfrac{2}{3}\sqrt{6}, 2 + \dfrac{2}{3}\sqrt{6}\right)$, and

$\left(2 \pm \dfrac{2}{3}\sqrt{6}, \infty\right)$

Use a test point in each interval to determine that the solution is $\left(2 - \dfrac{2}{3}\sqrt{6}, 2 + \dfrac{2}{3}\sqrt{6}\right)$.

80. $x^2 + x - 6 \leq 12 - 2x - 2x^2$

Determine the critical points.

$x^2 + x - 6 = 12 - 2x - 2x^2$

$3x^2 + 3x - 18 = 0$

$3(x^2 + x - 6) = 0$

$3(x - 2)(x + 3) = 0$

$x - 2 = 0$ or $x + 3 = 0$

$x = 2 \qquad x = -3$

Possible intervals:

$(-\infty, -3]$, $[-3, 2]$, and $[2, \infty)$

Use a test point in each interval to determine that the solution is $[-3, 2]$.

SSM: Experiencing Introductory and Intermediate Algebra

81. $3 - z^2 > -2x$
Determine the critical points.
$3 - z^2 = -2x$
$0 = z^2 - 2x - 3$
$0 = (z+1)(z-3)$
$z + 1 = 0$ or $z - 3 = 0$
$z = -1 \quad\quad z = 3$
Possible intervals:
$(-\infty, -1)$, $(-1, 3)$, and $(3, \infty)$
Use a test point in each interval to determine that the solution is $(-1, 3)$.

82. $x^2 + 6x + 11 > -x^2 - 6x - 7$
Determine the critical points.
$x^2 + 6x + 11 = -x^2 - 6x - 7$
$2x^2 + 12x + 18 = 0$
$2(x^2 + 6x + 9) = 0$
$2(x+3)^2 = 0$
$x + 3 = 0$
$x = -3$
Possible intervals:
$(-\infty, -3)$ and $(-3, \infty)$
Use a test point in each interval to determine that both intervals are solutions. Thus, the solution is $(-\infty, -3) \cup (-3, \infty)$.

83. $1 - 2x^2 \leq 1 - x^2$
Determine the critical points.
$1 - 2x^2 = 1 - x^2$
$0 = x^2$
$x = 0$
Possible intervals:
$(-\infty, 3]$ and $[3, \infty)$
Use a test point in each interval to determine that both intervals are solutions. Thus, the solution is the set of all real numbers, or $(-\infty, \infty)$.

84. $x^2 + 4x - 2 < -\dfrac{1}{2}x^2 - 2x - 8$
Determine the critical points.

$x^2 + 4x - 2 = -\dfrac{1}{2}x^2 - 2x - 8$
$2x^2 + 8x - 4 = -x^2 - 4x - 16$
$3x^2 + 12x + 12 = 0$
$3(x^2 + 4x + 4) = 0$
$3(x+2)^2 = 0$
$x + 2 = 0$
$x = -2$
Possible intervals:
$(-\infty, -2)$ and $(-2, \infty)$
Use a test point in each interval to determine that neither interval is a solution. Thus, the inequality has no solution.

85. $p^2 - 6p - 1 < 0$
Determine the critical points.
$p^2 - 6p - 1 = 0$
$p = \dfrac{-(-6) \pm \sqrt{(-6)^2 - 4(1)(-1)}}{2(1)}$
$x = \dfrac{6 \pm \sqrt{40}}{2}$
$x = \dfrac{6 \pm 2\sqrt{10}}{2}$
$x = 3 \pm \sqrt{10}$
Possible intervals:
$(-\infty, 3-\sqrt{10})$, $(3-\sqrt{10}, 3+\sqrt{10})$, $(3+\sqrt{10}, \infty)$
Use a test point in each interval to determine that the solution is $(3-\sqrt{10}, 3+\sqrt{10})$.

86. $3x^2 + 25x + 50 \leq 200$
Determine the critical points.
$3x^2 + 25x + 50 = 200$
$3x^2 + 25x - 150 = 0$
$x = \dfrac{-25 \pm \sqrt{25^2 - 4(3)(-150)}}{2(3)}$
$x = x = \dfrac{-25 \pm \sqrt{2425}}{6}$
$x \approx 4.041$ or $x \approx -12.374$

Since the number of days must be positive, the possible intervals are $[0, 4.041]$ and $[4.041, \infty)$. Use a test point in each interval to determine that the solution is $[0, 4.041]$.

The booth can be rented from 1 to 4 days.

87. $s = -16t^2 + v_0 t + s_0$; $s = 0$; $v_0 = 25$; $s_0 = 30$

$0 < -16t^2 + 25t + 30$

Determine the critical points.

$0 = -16t^2 + 25t + 30$

$16t^2 - 25t - 30 = 0$

$t = \dfrac{-(-25) \pm \sqrt{(-25)^2 - 4(16)(-30)}}{2(16)}$

$t = \dfrac{25 \pm \sqrt{2545}}{32}$

$t \approx 2.357$ or $t \approx -0.795$

Since time must be positive, the possible intervals are $(0, 2.358)$ and $(2.358, \infty)$.

Use a test point in each interval to determine that the solution is $(0, 2.358)$.

The dart is above the ground between 0 and approximately 2.358 seconds.

Chapter 11 Chapter Review

1. $\sqrt{50} = \sqrt{25 \cdot 2} = 5\sqrt{2}$

2. $-\sqrt{48} = -\sqrt{16 \cdot 3} = -4\sqrt{3}$

3. $\dfrac{-3 + \sqrt{18}}{3} = \dfrac{-3 + \sqrt{9 \cdot 2}}{3} = \dfrac{-3 + 3\sqrt{2}}{3} = -1 + \sqrt{2}$

4. $\dfrac{4 - \sqrt{80}}{-2} = \dfrac{4 - \sqrt{16 \cdot 5}}{-2} = \dfrac{4 - 4\sqrt{5}}{-2} = -2 + 2\sqrt{5}$

5. $\sqrt{\dfrac{14}{25}} = \dfrac{\sqrt{14}}{\sqrt{25}} = \dfrac{\sqrt{14}}{5}$

6. $\sqrt{\dfrac{75}{18}} = \sqrt{\dfrac{25}{6}} = \dfrac{\sqrt{25}}{\sqrt{6}} = \dfrac{5}{\sqrt{6}} = \dfrac{5}{\sqrt{6}} \cdot \dfrac{\sqrt{6}}{\sqrt{6}} = \dfrac{5\sqrt{6}}{6}$

7. Let $Y1 = x^2 - 4$ and $Y2 = 2x - 1$.

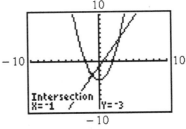
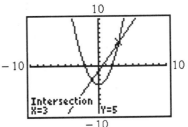

The solutions are -1 and 3.

8. Let $Y1 = 3 - x^2$ and $Y2 = -3 - \dfrac{1}{3}x^2$.

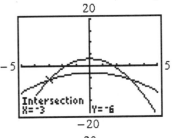
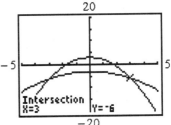

The solutions are -3 and 3.

9. Let $Y1 = x^2 + 6x + 9$ and $Y2 = 4$.

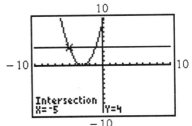

SSM: Experiencing Introductory and Intermediate Algebra

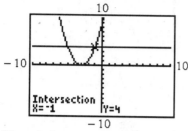

The solutions are −5 and −1.

10. Let $Y1 = 15x^2 + 13x - 72$ and $Y2 = 0$.

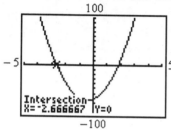

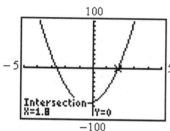

The solutions are $-2.\overline{6}$ and 1.8. As mixed numbers, the solutions are $-2\frac{2}{3}$ and $1\frac{4}{5}$.

11. Let $Y1 = 2x + 10$ and $Y2 = x^2 + 4x + 7$.

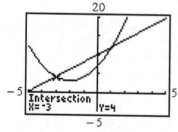

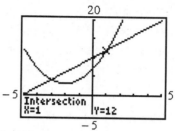

The solutions are −3 and 1.

12. Let $Y1 = x^2 + 9x + 7$ and $Y2 = 0$.

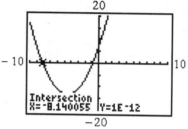

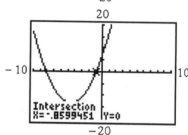

The solutions are approximately −8.14 and −0.86.

13. Let $Y1 = \frac{1}{6}x^2 - 2x + 7$ and $Y2 = 9 - \frac{1}{4}x^2$.

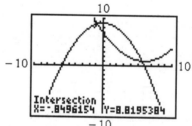

Chapter 11: Quadratic and Other Polynomial Equations and Inequalities

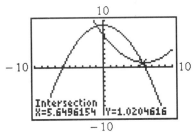

The solutions are approximately −0.85 and 5.65.

14. Let $Y1 = x^2 - 6x - 40$ and $Y2 = 0$.

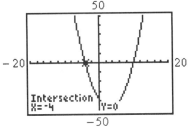

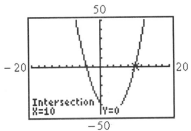

The solutions are approximately −4 and 10.

15. $Y1 = 2x^2 - x - 28$ and $Y2 = x^2 + 3x + 17$.

X	Y1	Y2
-5	27	27
-4	8	21
-3	-7	17
-2	-18	15
-1	-25	15
0	-28	17
1	-27	21

X=-5

X	Y1	Y2
3	-13	35
4	0	45
5	17	57
6	38	71
7	63	87
8	92	105
9	125	125

X=9

The solutions are −5 and 9.

16. $Y1 = 3x^2 - 5x - 10$ and $Y2 = 7x + 5$.

X	Y1	Y2
-1	-2	-2
0	-10	5
1	-12	12
2	-8	19
3	2	26
4	18	33
5	40	40

X=-1

The solutions are −1 and 5.

17. $12x^2 + 9x = 8x + 6$

$12x^2 + x - 6 = 0$

$(3x - 2)(4x + 3) = 0$

$3x - 2 = 0$ or $4x + 3 = 0$

$3x = 2 \qquad 4x = -3$

$x = \dfrac{2}{3} \qquad x = -\dfrac{3}{4}$

The solutions are $-\dfrac{3}{4}$ and $\dfrac{2}{3}$.

18. $x^2 + 3x - 88 = 0$

$(x + 11)(x - 8) = 0$

$x + 11 = 0$ or $x - 8 = 0$

$x = -11 \qquad x = 8$

The solutions are −11 and 8.

19. $5x^2 + 14x - 6 = 8 - 19x$

$5x^2 + 33x - 14 = 0$

$(5x - 2)(x + 7) = 0$

$5x - 2 = 0$ or $x + 7 = 0$

$5x = 2 \qquad x = -7$

$x = \dfrac{2}{5}$

The solutions are −7 and $\dfrac{2}{5}$.

20. $49x^2 - 16 = 0$

$(7x + 4)(7x - 4) = 0$

$7x + 4 = 0$ or $7x - 4 = 0$

$7x = -4 \qquad 7x = 4$

$x = -\dfrac{4}{7} \qquad x = \dfrac{4}{7}$

The solutions are $\pm \dfrac{4}{7}$.

21. $5x^2 - 180 = 0$

$5(x^2 - 36) = 0$

$5(x + 6)(x - 6) = 0$

$x + 6 = 0$ or $x - 6 = 0$

$x = -6 \qquad x = 6$

The solutions are ± 6.

495

22. $11 + b^2 = 2$
    $b^2 = -9$
    The equation has no real-number solution.

23. $(x-9)^2 = 16$
    $x - 9 = \pm\sqrt{16}$
    $x - 9 = \pm 4$
    $x = 9 \pm 4$
    $x = 9 + 4$ or $x = 9 - 4$
    $x = 13$ $\qquad x = 5$
    The solutions are 5 and 13.

24. $(p-7)^2 + 6 = 9$
    $(p-7)^2 = 3$
    $p - 7 = \pm\sqrt{3}$
    $p = 7 \pm \sqrt{3}$
    The solutions are $7 \pm \sqrt{3}$.

25. $z^2 + 12z = -33$
    $z^2 + 12z + 36 = -33 + 36$
    $(z+6)^2 = 3$
    $z + 6 = \pm\sqrt{3}$
    $z = -6 \pm \sqrt{3}$
    The solutions are $-6 \pm \sqrt{3}$.

26. $m^2 + 7m + 12 = 9$
    $m^2 + 7m = -3$
    $m^2 + 7m + \dfrac{49}{4} = -3 + \dfrac{49}{4}$
    $\left(m + \dfrac{7}{2}\right)^2 = \dfrac{37}{4}$
    $m + \dfrac{7}{2} = \pm\sqrt{\dfrac{37}{4}}$
    $m + \dfrac{7}{2} = \pm\dfrac{\sqrt{37}}{2}$

$m = -\dfrac{7}{2} \pm \dfrac{\sqrt{37}}{2} = \dfrac{-7 \pm \sqrt{37}}{2}$

The solutions are $\dfrac{-7 \pm \sqrt{37}}{2}$.

27. $b^2 + 2.4b - 4.32 = 0$
    $b^2 + 2.4b = 4.32$
    $b^2 + 2.4b + 1.44 = 4.32 + 1.44$
    $(b+1.2)^2 = 5.76$
    $b + 1.2 = \pm\sqrt{5.76}$
    $b + 1.2 = \pm 2.4$
    $b = -1.2 \pm 2.4$
    $b = -1.2 + 2.4$ or $b = -1.2 - 2.4$
    $b = 1.2$ $\qquad b = -3.6$
    The solutions are -3.6 and 1.2.

28. $7q^2 - 7q - 78 = (8q^2 - 3q - 50) - (q^2 + 4q + 28)$
    $7q^2 - 7q - 78 = 8q^2 - 3q - 50 - q^2 - 4q - 28$
    $7q^2 - 7q - 78 = 7q^2 - 7q - 78$
    $-78 = -78$
    This equation must always be true (an identity).
    The solution is the set of all real numbers.

29. $x^2 + 20x + 84 = 0$
    $(x+14)(x+6) = 0$
    $x + 14 = 0$ or $x + 6 = 0$
    $x = -14$ $\qquad x = -6$
    The solutions are -14 and -6.

30. $2y^2 + 3y - 5 = 0$
    $(2y+5)(y-1) = 0$
    $2y + 5 = 0$ or $y - 1 = 0$
    $2y = -5$ $\qquad y = 1$
    $y = -\dfrac{5}{2}$
    The solutions are $-\dfrac{5}{2}$ and 1.

Chapter 11: Quadratic and Other Polynomial Equations and Inequalities

31. $4x^2 + 16x - 2 = 0$

$4x^2 + 16x = 2$

$x^2 + 4x = \dfrac{1}{2}$

$x^2 + 4x + 4 = \dfrac{1}{2} + 4$

$(x+2)^2 = \dfrac{9}{2}$

$x + 2 = \pm\sqrt{\dfrac{9}{2}}$

$x + 2 = \pm\dfrac{3}{\sqrt{2}}$

$x + 2 = \pm\dfrac{3}{\sqrt{2}} \cdot \dfrac{\sqrt{2}}{\sqrt{2}}$

$x + 2 = \pm\dfrac{3\sqrt{2}}{2}$

$x = -2 \pm \dfrac{3\sqrt{2}}{2} = \dfrac{-4 \pm 3\sqrt{2}}{2}$

The solutions are $\dfrac{-4 \pm 3\sqrt{2}}{2}$.

32. $4 + 8x + x^2 = 0$

$x^2 + 8x + 4 = 0$

$x = \dfrac{-8 \pm \sqrt{8^2 - 4(1)(4)}}{2(1)}$

$x = \dfrac{-8 \pm \sqrt{48}}{2}$

$x = \dfrac{-8 \pm 4\sqrt{3}}{2}$

$x = -4 \pm 2\sqrt{3}$

The solutions are $-4 \pm 2\sqrt{3}$.

33. $x^2 = 7x + 2$

$x^2 - 7x - 2 = 0$

$x = \dfrac{-(-7) \pm \sqrt{(-7)^2 - 4(1)(-2)}}{2(1)}$

$x = \dfrac{7 \pm \sqrt{57}}{2}$

The solutions are $\dfrac{7 \pm \sqrt{57}}{2}$.

34. $36q^2 + 25 = 60q$

$36q^2 - 60q + 25 = 0$

$q = \dfrac{-(-60) \pm \sqrt{(-60)^2 - 4(36)(25)}}{2(36)}$

$q = \dfrac{60 \pm \sqrt{0}}{72}$

$q = \dfrac{60 \pm 0}{72}$

$q = \dfrac{60}{72}$

$q = \dfrac{5}{6}$

The solution is $\dfrac{5}{6}$ (which is a double root).

35. $b^2 - 7b + 16 = 0$

$b = \dfrac{-(-7) \pm \sqrt{(-7)^2 - 4(1)(16)}}{2(1)}$

$b = \dfrac{7 \pm \sqrt{-15}}{2}$

The equation has no real-number solution.

36. $z^2 + 6z - 55 = 0$

$z = \dfrac{-6 \pm \sqrt{6^2 - 4(1)(-55)}}{2(1)}$

$z = \dfrac{-6 \pm \sqrt{256}}{2}$

$z = \dfrac{-6 \pm 16}{2}$

$z = \dfrac{10}{2}$ or $z = \dfrac{-22}{2}$

$z = 5$ $z = -11$

The solutions are -11 and 5.

SSM: Experiencing Introductory and Intermediate Algebra

37. $4m^2 + 7m = 36$

$4m^2 + 7m - 36 = 0$

$m = \dfrac{-7 \pm \sqrt{7^2 - 4(4)(-36)}}{2(4)}$

$m = \dfrac{-7 \pm \sqrt{625}}{8}$

$m = \dfrac{-7 \pm 25}{8}$

$m = \dfrac{18}{8}$ or $m = \dfrac{-32}{8}$

$m = \dfrac{9}{4}$ $\quad\quad m = -4$

The solutions are -4 and $\dfrac{9}{4}$.

38. $a = \dfrac{x(x+1)}{2}$

$a = \dfrac{x^2 + x}{2}$

$2a = x^2 + x$

$0 = x^2 + x - 2a$

$x = \dfrac{-1 \pm \sqrt{1^2 - 4(1)(-2a)}}{2(1)}$

$x = \dfrac{-1 \pm \sqrt{1 + 8a}}{2}$

39. $x^2 + y^2 + z^2 = r^2$

$y^2 = r^2 - x^2 - z^2$

$y = \pm\sqrt{r^2 - x^2 - z^2}$

40. $A = 4\pi r^2$

$\dfrac{A}{4\pi} = r^2$

$r = \pm\sqrt{\dfrac{A}{4\pi}}$

$r = \pm\dfrac{\sqrt{A}}{\sqrt{4\pi}} \cdot \dfrac{\sqrt{\pi}}{\sqrt{\pi}}$

$r = \pm\dfrac{\sqrt{A\pi}}{2\pi}$

41. Let $Y1 = x^2 + 2x - 7$ and $Y2 = 4x + 8$.

The solutions are all integers greater than or equal to -3 and less than or equal to 5.

42. Let $Y1 = x^2 + 9x - 11$ and $Y2 = 4x + 3$.

The solutions are all integers greater than or equal to -7 and less than or equal to 2.

43. $z^2 + 5z - 6 > 12 - 10z - 2z^2$

Determine the critical points.

$z^2 + 5z - 6 = 12 - 10z - 2z^2$

$3z^2 + 15z - 18 = 0$

$3(z^2 + 5z - 6) = 0$

$3(z - 1)(z + 6) = 0$

$z - 1 = 0$ or $z + 6 = 0$

$z = 1$ $\quad\quad z = -6$

Possible intervals:

$(-\infty, -6)$, $(-6, 1)$, and $(1, \infty)$

Use a test point in each interval to determine that the solution is $(-6, 1)$.

44. $m^2 - 15 > 2m$

Determine the critical points.

$m^2 - 15 = 2m$

$m^2 - 2m - 15 = 0$

$(m + 3)(m - 5) = 0$

$m = -3$ or $m = 5$

Possible intervals:

$(-\infty, -3)$, $(-3, 5)$ and $(5, \infty)$

Use a test point in each interval to determine that both intervals are solutions. Thus, the solution is $(-\infty, -3) \cup (5, \infty)$.

45. $x^2 + 2x - 1 > -x^2 - 2x + 1$
Determine the critical points.
$x^2 + 2x - 1 = -x^2 - 2x + 1$
$2x^2 + 4x - 2 = 0$
$2(x^2 + 2x - 1) = 0$
$x = \dfrac{-2 \pm \sqrt{2^2 - 4(1)(-1)}}{2(1)}$
$x = \dfrac{-2 \pm \sqrt{8}}{2}$
$x = \dfrac{-2 \pm 2\sqrt{2}}{2}$
$x = -1 \pm \sqrt{2}$

Possible intervals:
$(-\infty, -1 - \sqrt{2})$, $(-1 - \sqrt{2}, -1 + \sqrt{2})$, $(-1 + \sqrt{2}, \infty)$
Use a test point in each interval to determine that the solution is $(-\infty, -1 - \sqrt{2}) \cup (-1 + \sqrt{2}, \infty)$.

46. $p^2 - 8p - 2 > 0$
Determine the critical points.
$p^2 - 8p - 2 = 0$
$p = \dfrac{-(-8) \pm \sqrt{(-8)^2 - 4(1)(-2)}}{2(1)}$
$p = \dfrac{8 \pm \sqrt{72}}{2}$
$p = \dfrac{8 \pm 6\sqrt{2}}{2}$
$p = 4 \pm 3\sqrt{2}$

Possible intervals:
$(-\infty, -4 - 3\sqrt{2})$, $(-4 - 3\sqrt{2}, -4 + 3\sqrt{2})$, and $(-4 + 3\sqrt{2}, \infty)$.
Use a test point in each interval to determine that the solution is $(-\infty, -4 - 3\sqrt{2}) \cup (-4 + 3\sqrt{2}, \infty)$.

47. $2x^2 - 5x - 11 \le 19 + 8x - x^2$
Determine the critical points.
$2x^2 - 5x - 11 = 19 + 8x - x^2$
$3x^2 - 13x - 30 = 0$
$(3x + 5)(x - 6) = 0$
$3x + 5 = 0$ or $x - 6 = 0$
$x = -\dfrac{5}{3}$ \quad $x = 6$

Possible intervals:
$\left(-\infty, -\dfrac{5}{3}\right]$, $\left[-\dfrac{5}{3}, 6\right]$, and $[6, \infty)$
Use a test point in each interval to determine that the solution is $\left[-\dfrac{5}{3}, 6\right]$.

48. $3q^2 + 2q - 17 \ge 5q + 19$
Determine the critical points.
$3q^2 + 2q - 17 = 5q + 19$
$3q^2 - 3q - 36 = 0$
$3(q^2 - q - 12) = 0$
$3(q + 3)(q - 4) = 0$
$q + 3 = 0$ or $q - 4 = 0$
$q = -3$ \quad $q = 4$

Possible intervals:
$(-\infty, -3]$, $[-3, 4]$, and $[4, \infty)$.
Use a test point in each interval to determine that the solution is $(-\infty, -3] \cup [4, \infty)$.

49. Let $Y1 = x^2 - 4x - 1$ and $Y2 = x + 5$.

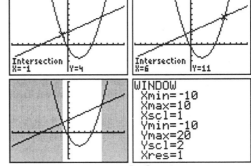

The solution is $(-\infty, -1] \cup [6, \infty)$.

SSM: Experiencing Introductory and Intermediate Algebra

50. Let $Y1 = 2 - x^2$ and $Y2 = 5 + x^2$.

Because the graph of Y1 is always below the graph of Y2, the inequality has no solution.

51. Let $Y1 = x^2 - 2x - 3$ and $Y2 = 12$.

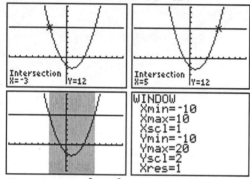

The solution is $[-3, 5]$.

52. Let $Y1 = 7 - x^2$ and $Y2 = -9$.

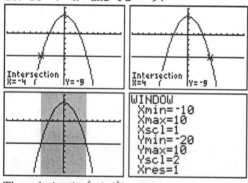

The solution is $(-4, 4)$.

53. Let $Y1 = x^2 + 4x - 2$ and $Y2 = -5 - 4x - x^2$.

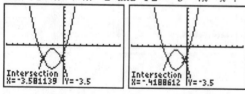

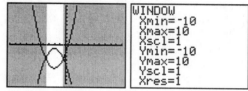

The solution is $(-\infty, -3.581) \cup (-0.419, \infty)$.

54. Let $Y1 = -x^2 - 10x - 27$ and $Y2 = x^2 + 10x + 23$.

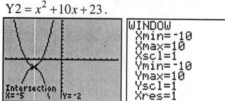

Because the graph of Y1 is below the graph of Y2 everywhere except at the intersection point, the solution is $(-\infty, -5) \cup (-5, \infty)$.

55. Let $Y1 = -x^2 - 10x - 27$ and $Y2 = x^2 + 10x + 23$.

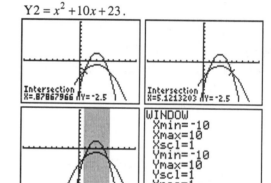

The solution is $(-0.879, 5.121)$.

56. Let $Y1 = -x^2 + 6x - 6$ and $Y2 = -\frac{1}{3}x^2 + 2x$.

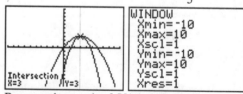

Because the graph of Y1 is below the graph of Y2 everywhere except at the intersection point, the only solution to the inequality is 3.

500

57. $s = -16t^2 + v_0t + s_0$; $s = 0$; $v_0 = 0$; $s_0 = 160$

$0 = -16t^2 + 0t + 160$

$16t^2 = 160$

$t^2 = 10$

$t = \pm\sqrt{10}$

$t \approx \pm 3.16$

Time cannot be negative, so the notebook will hit the ground in $\sqrt{10}$ seconds, or approximately 3.16 seconds.

58. $5x^2 + 75x + 875 = 3500$

$5x^2 + 75x - 2625 = 0$

$5(x^2 + 15x - 525) = 0$

$x = \dfrac{-15 \pm \sqrt{15^2 - 4(1)(-525)}}{2(1)}$

$x = \dfrac{-15 \pm \sqrt{2325}}{2}$

$x \approx 16.6$ or $x \approx -31.6$

Disregarding −31.6, about 16 items can be produced for $3500.

59. $s = -16t^2 + v_0t + s_0$; $s = 0$; $v_0 = 64$; $s_0 = 32$

$0 = -16t^2 + 64t + 32$

$0 = -16(t^2 - 4t - 2)$

$t = \dfrac{-(-4) \pm \sqrt{(-4)^2 - 4(1)(-2)}}{2(1)}$

$t = \dfrac{4 \pm \sqrt{24}}{2}$

$t = \dfrac{4 \pm 2\sqrt{6}}{2}$

$t = 2 \pm \sqrt{6}$

$t \approx 4.45$ or $t \approx -0.45$

Time cannot be negative, so the notebook will hit the ground in $2 + \sqrt{6}$ seconds, or approximately 4.45 seconds.

60. $s = -16t^2 + v_0t + s_0$; $s = 4000$; $v_0 = 0$; $s_0 = 11,000$

$4000 = -16t^2 + 0t + 11,000$

$16t^2 = 7000$

$t^2 = 437.5$

$t = \pm\sqrt{437.5}$

$t \approx \pm 20.92$

Time cannot be negative, so the sky diver will freefall for $\sqrt{437.5}$ seconds, or approximately 20.92 seconds.

61. Let x = the length of the other leg of the triangle.

$x^2 + 34^2 = 42.5^2$

$x^2 + 1156 = 1806.25$

$x^2 = 650.25$

$x = \pm\sqrt{650.25}$

$x = \pm 25.5$

Length cannot be negative, so the length of the other leg of the triangle is 25.5 meters long.

62. Let w = the width of the rectangle.
$3w - 6$ = the length of the rectangle.

$w(3w - 6) = 144$

$3w^2 - 6w - 144 = 0$

$3(w^2 - 2w - 48) = 0$

$3(w - 8)(w + 6) = 0$

$w - 8 = 0$ or $w + 6 = 0$

$w = 8$ $w = -6$

$3w - 6 = 18$

Distance cannot be negative, so the width of the rectangle is 8 cm and the length is 18 cm.

63. $1323 = 1200(1 + r)^2$

$1.1025 = (1 + r)^2$

$1 + r = \pm\sqrt{1.1025}$

$1 + r = \pm 1.05$

$r = -1 \pm 1.05$

$r = 0.05$ or $r = -2.05$

Disregarding −2.05, the annual interest rate must be $0.05 = 5\%$.

SSM: Experiencing Introductory and Intermediate Algebra

64. $1500 = 1200(1+r)^2$

$(1+r)^2 = 1.25$

$1+r = \pm\sqrt{1.25}$

$1+r \approx \pm 1.118$

$r \approx -1 \pm 1.118$

$r \approx 0.118$ or $r \approx -2.118$

Disregarding -2.118, the annual interest rate must be approximately $0.118 = 11.8\%$.

65. Let $x =$ the number of baskets to be purchased.
$10.50 - 0.50x =$ price per basket.

a. $R(x) = x(10.50 - 0.50x)$

b. $C(x) = 7.50x + 3.00$

c. $P(x) = R(x) - C(x)$
$= x(10.50 - 0.50x) - (7.50x + 3.00)$
$= 10.5x - 0.5x^2 - 7.5x - 3$
$= -0.5x^2 + 3x - 3$

d. $0 = -0.5x^2 + 3x - 3$

$0.5x^2 - 3x + 3 = 0$

$x = \dfrac{-(-3) \pm \sqrt{(-3)^2 - 4(0.5)(3)}}{2(0.5)}$

$x = \dfrac{3 \pm \sqrt{3}}{1}$

$x = 3 \pm \sqrt{3}$

$x \approx 4.7 \approx 5$ or $x \approx 1.3 \approx 2$

$10.50 - 0.50x \approx 8$ $10.50 - 0.50x \approx 9.50$

Angela will approximately break even if she sells 2 baskets at $9.50 each and if she sells 5 baskets at $8.00 each.

66. Let $x =$ the length of the hypotenuse.

$x^2 = 20^2 + 48^2$

$x^2 = 2704$

$x = \pm\sqrt{2704}$

$x = \pm 52$

Length cannot be negative, so the length of the hypotenuse is 52 inches.

67. $1 = -0.01x^2 + 0.01x + 2.8$

$0.01x^2 - 0.01x - 1.8 = 0$

$x = \dfrac{-(-0.01) \pm \sqrt{(-0.01)^2 - 4(0.01)(-1.8)}}{2(0.01)}$

$x = \dfrac{0.01 \pm \sqrt{0.0721}}{0.02}$

$x \approx 13.9$ or $x \approx -12.9$

Disregarding -12.9, the per capita consumption of cigarettes will drop to 1000 approximately 13.9 years after 1990. Now, $1990 + 13.9 = 2003.9$, or about 2004. Thus, according to the model, the per capita consumption of cigarettes will be 1000 in about the year 2004.

68. a. $y = -0.18x^2 + 0.99x + 3.55$

L1	L2	L3	1
0	3.55		
1	4.36		
2	4.81		

L1(2)=1

QuadReg
y=ax²+bx+c
a=-.18
b=.99
c=3.55
R²=1

b. $4 = -0.18x^2 + 0.99x + 3.55$

$0.18x^2 - 0.99x + 0.45 = 0$

$0.18(x^2 - 5.5x + 2.5) = 0$

$0.18(x - 5)(x - 0.5) = 0$

$x = 5$ or $x = 0.5$

Disregarding 0.5, the sales will drop to 4 million units 5 years after 1998, which would be the year 2003.

69. $c = a\sqrt{2}$

$14.21 = a\sqrt{2}$

$a = \dfrac{14.21}{\sqrt{2}}$

$= \dfrac{14.21}{\sqrt{2}} \cdot \dfrac{\sqrt{2}}{\sqrt{2}}$

$= \dfrac{14.21\sqrt{2}}{2}$

$= 7.105\sqrt{2}$

≈ 10.05

Each leg measures $7.105\sqrt{2}$ cm, or approximately 10.05 cm.

70. $c = a\sqrt{2} = 11\sqrt{2} \approx 15.56$

The rope is $11\sqrt{2}$ feet long, or approximately 15.56 feet long.

71. $c = 2a$

$32 = 2a$

$a = \dfrac{32}{2} = 16$

$b = a\sqrt{3} = 16\sqrt{3} \approx 27.7$

The lengths of the sides are 16 inches and $16\sqrt{3}$ inches, or approximately 27.7 inches.

72. $5x^2 + 20x + 100 \le 500$

Determine the critical points.

$5x^2 + 20x + 100 = 500$

$5x^2 + 20x - 400 = 0$

$5(x^2 + 4x - 80) = 0$

$x = \dfrac{-4 \pm \sqrt{4^2 - 4(1)(-80)}}{2(1)}$

$x = \dfrac{-4 \pm \sqrt{336}}{2}$

$x \approx 7.165$ or $x \approx -11.165$

Since the number of days must be positive, the possible intervals are $[0, 7.165]$ and $[7.165, \infty)$. Use a test point in each interval to determine that the solution is $[0, 7.165]$.

The booth can be rented from 1 to 7 days.

Chapter 11 Test

1. $x^2 - 4x - 9 = 2x + 7$

$x^2 - 6x - 16 = 0$

$(x - 8)(x + 2) = 0$

$x - 8 = 0$ or $x + 2 = 0$

$x = 8 \qquad\quad x = -2$

The solutions are -2 and 8.

2. Let $Y1 = 2x^3 + 9x^2 - 23x - 66$ and $Y2 = 0$.

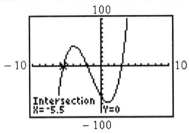

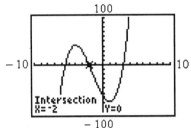

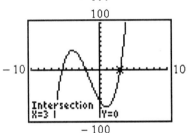

The solutions are -5.5, -2, and 3.

3. $a^2 + 2a - 5 = 2(a^2 + a - 2)$

$a^2 + 2a - 5 = 2a^2 + 2a - 4$

$0 = a^2 + 1$

$-1 = a^2$

The equation has no real-number solution.

4. $x^2 - 6x + 13 = 0$

$x^2 - 6x = -13$

$x^2 - 6x + 9 = -13 + 9$

$(x - 3)^2 = -4$

The equation has no real-number solution.

5. $2x^2 - 7x + 3 = 2x - 1$
   $2x^2 - 9x + 4 = 0$
   $(2x-1)(x-4) = 0$
   $2x - 1 = 0$ or $x - 4 = 0$
   $2x = 1$ $\quad\quad x = 4$
   $x = \dfrac{1}{2}$
   The solutions are $\dfrac{1}{2}$ and 4.

6. $x^2 - 5x + 4 = 7 - 3x - x^2$
   $2x^2 - 2x - 3 = 0$
   $x = \dfrac{-(-2) \pm \sqrt{(-2)^2 - 4(2)(-3)}}{2(2)}$
   $= \dfrac{2 \pm \sqrt{28}}{4}$
   $= \dfrac{2 \pm \sqrt{4 \cdot 7}}{4}$
   $= \dfrac{2 \pm 2\sqrt{7}}{4}$
   $= \dfrac{1 \pm \sqrt{7}}{2}$
   The solutions are $\dfrac{1 \pm \sqrt{7}}{2}$, or approximately -0.82 and 1.82.

7. $x^2 - x - 6 \le 6$
   Determine the critical points.
   $x^2 - x - 6 = 6$
   $x^2 - x - 12 = 0$
   $(x+3)(x-4) = 0$
   $x + 3 = 0$ or $x - 4 = 0$
   $x = -3$ $\quad\quad x = 4$
   Possible intervals:
   $(-\infty, -3]$, $[-3, 4]$, and $[4, \infty)$
   Use a test point in each interval to determine that the solution is $[-3, 4]$.

8. $(x+3)(x-2) < 3(x+3)$
   Determine the critical points.
   $(x+3)(x-2) = 3(x+3)$
   $x^2 - 2x + 3x - 6 = 3x + 9$
   $x^2 - 2x - 15 = 0$
   $(x+3)(x-5) = 0$
   $x + 3 = 0$ or $x - 5 = 0$
   $x = -3$ $\quad\quad x = 5$
   Possible intervals:
   $(-\infty, -3)$, $(-3, 5)$, and $(5, \infty)$
   Use a test point in each interval to determine that the solution is $(-3, 5)$.

9. $2x^2 + 7x - 9 \ge 4x + 11$
   Determine the critical points.
   $2x^2 + 7x - 9 = 4x + 11$
   $2x^2 + 3x - 20 = 0$
   $(2x-5)(x+4) = 0$
   $2x - 5 = 0$ or $x + 4 = 0$
   $x = \dfrac{5}{2}$ $\quad\quad x = -4$
   Possible intervals:
   $(-\infty, -4]$, $\left[-4, \dfrac{5}{2}\right]$, and $\left[\dfrac{5}{2}, \infty\right)$
   Use a test point in each interval to determine that the solution is $(-\infty, -4] \cup \left[\dfrac{5}{2}, \infty\right)$.

10. $x^2 - 9x + 9 = 7x - 5$
    $x^2 - 16x + 14 = 0$
    $b^2 - 4ac = (-16)^2 - 4(1)(14) = 200$
    Because 200 is positive but not a perfect square, the equation has two irrational solutions.
    $x = \dfrac{-(-16) \pm \sqrt{200}}{2(1)} = \dfrac{16 \pm \sqrt{100 \cdot 2}}{2} = \dfrac{16 \pm 10\sqrt{2}}{2}$
    $= 8 \pm \sqrt{2}$
    The exact roots are $8 \pm \sqrt{2}$.

11. $x^2 - 3x + 12 = 0$

$b^2 - 4ac = (-3)^2 - 4(1)(12) = -39$

Because the discriminant is negative, the equation has no real-number solution.

12. $s = at^2 + c$

$s - c = at^2$

$\dfrac{s-c}{a} = t^2$

$t = \pm\sqrt{\dfrac{s-c}{a}}$

$= \pm\sqrt{\dfrac{s-c}{a} \cdot \dfrac{a}{a}}$

$= \pm\dfrac{\sqrt{a(s-c)}}{a}$

13. **a.** Let $Y1 = x^2 - 2$ and $Y2 = x + 4$.

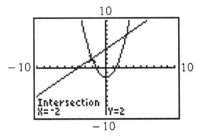

The solutions are -2 and 3.

b. Let $Y1 = x^2 - 2$ and $Y2 = x + 4$.

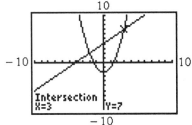

The solutions are -2 and 3.

c. $x^2 - 2 = x + 4$

$x^2 - x - 6 = 0$

$(x-3)(x+2) = 0$

$x - 3 = 0$ or $x + 2 = 0$

$x = 3 \qquad x = -2$

The solutions are -2 and 3.

14. $c = 2a$

$28 = 2a$

$a = \dfrac{28}{2} = 14$

$b = a\sqrt{3} = 14\sqrt{3} \approx 24.2$

The lengths of the sides are 14 inches and $14\sqrt{3}$ inches, or approximately 24.2 inches.

15. $2600 = -16t^2 + 0t + 11{,}500$

$16t^2 = 8900$

$t^2 = 556.25$

$t = \pm\sqrt{556.25}$

$t \approx \pm 23.58$

Time cannot be negative, so Keanu will freefall for $\sqrt{556.25}$ seconds, or approximately 23.58 seconds.

16. $1573 = 1400(1+r)^2$

$\dfrac{1573}{1400} = (1+r)^2$

$1 + r = \pm\sqrt{\dfrac{1573}{1400}}$

$r = -1 \pm \sqrt{\dfrac{1573}{1400}}$

$r \approx 0.06$ or $r \approx -2.06$

Disregarding -2.06, the annual interest rate must be approximately $0.06 = 6\%$.

17. Let $x =$ the number of flower arrangements.
$12.50 - 0.50x =$ price per basket.

a. $R(x) = x(12.50 - 0.50x)$

b. $C(x) = 9.50x + 3.95$

SSM: Experiencing Introductory and Intermediate Algebra

c. $P(x) = R(x) - C(x)$
$= x(12.50 - 0.50x) - (9.50x + 3.95)$
$= 12.5x - 0.5x^2 - 9.5x - 3.95$
$= -0.5x^2 + 3x - 3.95$

d. $0 = -0.5x^2 + 3x - 3.95$
$0.5x^2 - 3x + 3.95 = 0$
$x = \dfrac{-(-3) \pm \sqrt{(-3)^2 - 4(0.5)(3.95)}}{2(0.5)}$
$x = \dfrac{3 \pm \sqrt{1.1}}{1}$
$x = 3 \pm \sqrt{1.1}$
$x \approx 4.05 \approx 4$ or $x \approx 1.95 \approx 2$
$12.50 - 0.50x \approx 10.50$ $12.50 - 0.50x \approx 11.50$
Nicole will approximately break even if she sells 2 flower arrangements at $11.50 each and if she sells 4 flower arrangements at $10.50 each.

18. a. $y = 0.35x^2 + 0.65x + 0.4$

L1	L2	L3		QuadReg
0	.4			y=ax²+bx+c
1	1.4			a=.35
2	3.1			b=.65
				c=.4
				R²=1

    L1(2)=1

    b. $5 = 0.35x^2 + 0.65x + 0.4$
    $0 = 0.35x^2 + 0.65x - 4.6$
    $x = \dfrac{-0.65 \pm \sqrt{(0.65)^2 - 4(0.35)(-4.6)}}{2(0.35)}$
    $x = \dfrac{-0.65 \pm \sqrt{6.8265}}{0.7}$
    $x \approx 2.8$ or $x \approx -4.7$
    Disregarding –4.7, the net income will be $5 billion approximately 2.8 years after 1998. Now, 1998 + 2.8 = 2000.8, or about 2001. Thus, the net income will be $5 billion in about the year 2001.

19. Let x = the horizontal distance.
    $x^2 + 10^2 = 40^2$
    $x^2 + 100 = 1600$
    $x^2 = 1500$
    $x = \pm\sqrt{1500}$
    $x = \pm 10\sqrt{15}$
    $x \approx 38.73$
    The horizontal distance between the ship and the end of the gangplank is $10\sqrt{15}$ feet, or approximately 38.73 feet.

20. First calculate the value of the discriminant, $b^2 - 4ac$, for the equation at hand. If the value of the discriminant is negative, then the equation will have no real-number solution. If the value of the discriminant is zero, then the equation will have one real rational solution (a double root). If the value of the discriminant is positive and also a perfect square, then the equation will have two different rational solutions. Finally, if the value of the discriminant is positive but not a perfect square, then the equation will have two different irrational solutions.

Cumulative Review Chapters 1 – 11

1. $(2a^2 + 3ab - 4b^2) + (a^2 + ab - b^2)$
   $= 3a^2 + 4ab - 5b^2$

2. $(1.5x^2 + 2.3xy - y^2) - (0.5x^2 + 1.3xy - y^2)$
   $= 1.5x^2 + 2.3xy - y^2 - 0.5x^2 - 1.3xy + y^2$
   $= x^2 + xy$

3. $(3x+1)(2x-6) = 6x^2 - 18x + 2x - 6$
   $= 6x^2 - 16x - 6$

4. $(5x+1)^2 = (5x)^2 + 2(5x)(1) + 1^2$
   $= 25x^2 + 10x + 1$

5. $(2x+3)(2x-3) = 4x^2 - 6x + 6x - 9$
   $= 4x^2 - 9$

Chapter 11: Quadratic and Other Polynomial Equations and Inequalities

6. $(2.1xy^2z)(3xy^3z^2) = 2.1(3)x^{1+1}y^{2+3}z^{1+2}$
$= 6.3x^2y^5z^3$

7. $6x^2y^{-1}z^0 = \dfrac{6x^2z^0}{y^1} = \dfrac{6x^2}{y}$

8. $\dfrac{2x^{-1}y}{4xy^2} = \dfrac{1}{2x^{1-(-1)}y^{2-1}} = \dfrac{1}{2x^2y}$

9. $\left[\dfrac{(-2a)^3}{5b}\right]^{-2} = \left(\dfrac{-8a^3}{5b}\right)^{-2}$
$= \left(\dfrac{5b}{-8a^3}\right)^2$
$= \dfrac{(5b)^2}{(-8a^3)^2}$
$= \dfrac{25b^2}{64a^6}$

10. $(-2p^2q^{-3}r)(4p^3q^5r) = -2 \cdot 4 p^{2+3} q^{-3+5} r^{1+1}$
$= -8p^5q^2r^2$

11. $\dfrac{2x^2+4x-6}{2x} = \dfrac{2x^2}{2x} + \dfrac{4x}{2x} - \dfrac{6}{2x} = x+2-\dfrac{3}{x}$

12. $\dfrac{x^2+4x-5}{x+5} = \dfrac{(x-1)(x+5)}{x+5} = x-1$

13. $f(-3) = -(-3)^2 + 2(-3) - 1 = -9-6-1 = -16$

14. $5.6 \times 10^{-3} = 0.0056$

15. $-3.4\text{E}6 = -3,400,000$

16. $A = \dfrac{1}{2}bh$
$2A = bh$
$\dfrac{2A}{h} = b$

17. $s = -16t^2 + 5$
$16t^2 + s = 5$
$16t^2 = 5 - s$
$t^2 = \dfrac{5-s}{16}$
$t = \pm\sqrt{\dfrac{5-s}{16}}$
$t = \pm\dfrac{\sqrt{5-s}}{4}$

18. $x^2 + 2x - 15 = (x-3)(x+5)$

19. $a = 10,\ b = 7,\ c = -12,\ ac = -120$
$10s^2 + 7s - 12 = (10s^2 + 15s) + (-8s - 12)$
$= 5s(2s+3) - 4(2s+3)$
$= (2s+3)(5s-4)$

20. $9x^2 - 16 = (3x)^2 - 4^2 = (3x+4)(3x-4)$

21. $4x^2 + 12x + 9 = (2x)^2 + 2(2x)(3) + 3^2$
$= (2x+3)^2$

22. $a^3 + 27 = a^3 + 3^3$
$= (a+3)(a^2 - a \cdot 3 + 3^2)$
$= (a+3)(a^2 - 3a + 9)$

23. $2a^4 - 32 = 2(a^4 - 16)$
$= 2(a^2+4)(a^2-4)$
$= 2(a^2+4)(a+2)(a-2)$

24. $2x - 4y = 1 \Rightarrow 2x - 4y = 1$
$x + y = 6 \Rightarrow \underline{4x - 4y = 24}$
$6x = 25$
$x = \dfrac{25}{6}$

$\dfrac{25}{6} + y = 6$

$y = \dfrac{11}{6}$

The solution is $\left(\dfrac{25}{6}, \dfrac{11}{6}\right)$.

25. $y < 2x + 5$
$y \geq -2x - 1$

Find the intersection point (using substitution).
$2x + 5 = -2x - 1$
$4x = -6$
$x = \dfrac{-6}{4} = -\dfrac{3}{2}$
$y = 2\left(-\dfrac{3}{2}\right) + 5 = -3 + 5 = 2$

The intersection point is $\left(-\dfrac{3}{2}, 2\right)$.

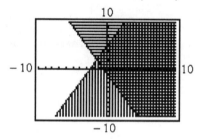

One possible solution is (1, 2).

26. $x^2 - 4x + 7 = 2x - 1$
$x^2 - 6x + 8 = 0$
$(x-2)(x-4) = 0$
$x - 2 = 0$ or $x - 4 = 0$
$x = 2 \qquad x = 4$
The solutions are 2 and 4.

27. $x^2 - 5x - 12 = 0$
$x = \dfrac{-(-5) \pm \sqrt{(-5)^2 - 4(1)(-12)}}{2(1)}$
$x = \dfrac{5 \pm \sqrt{73}}{2}$
The solutions are $\dfrac{5 \pm \sqrt{73}}{2}$.

28. $3x^2 + 6x = 9$
$3x^2 + 6x - 9 = 0$
$3(x^2 + 2x - 3) = 0$
$3(x-1)(x+3) = 0$
$x - 1 = 0$ or $x + 3 = 0$
$x = 1 \qquad x = -3$
The solutions are −3 and 1.

29. $2x^2 + 4x - 5 = 2x^2 + 4x + 4$
$-5 = 4$
The equation is always false (a contradiction). It has no solution.

30. $2x^2 - 5x + 10 = x^2 - 2$
$x^2 - 5x + 12 = 0$
$x = \dfrac{-(-5) \pm \sqrt{(-5)^2 - 4(1)(12)}}{2(1)}$
$x = \dfrac{5 \pm \sqrt{-73}}{2}$
The equation has no real-number solution.

31. $(x-2)(2x+1) = 4$
$2x^2 + x - 4x - 2 = 4$
$2x^2 - 3x - 6 = 0$
$x = \dfrac{-(-3) \pm \sqrt{(-3)^2 - 4(2)(-6)}}{2(2)}$
$x = \dfrac{3 \pm \sqrt{57}}{4}$
The solutions are $\dfrac{3 \pm \sqrt{57}}{4}$.

32. $2x + 4 = 5x - 1$
$5 = 3x$
$\dfrac{5}{3} = x$
The solution is $\dfrac{5}{3}$.

33. $2(x+3)+4=(x+4)+(x+6)$
$2x+10=2x+10$
$10=10$
The equation is always true (an identity). The solution is the set of all real numbers.

34. $5x-2(2.5x-1)=4$
$5x-5x+2=4$
$2=4$
The equation is always false (a contradiction). It has no solution.

35. $|2x-1|=4$
$2x-1=4$ or $2x-1=-4$
$2x=5$ $\qquad$ $2x=-3$
$x=\dfrac{5}{2}$ $\qquad$ $x=-\dfrac{3}{2}$
The solutions are $-\dfrac{3}{2}$ and $\dfrac{5}{2}$.

36. $3|x|-2=7$
$3|x|=9$
$|x|=3$
$x=3$ or $x=-3$
The solutions are -3 and 3.

37. $2x-5>3x+4$
$-x-5>4$
$-x>9$
$x<-9$
The solution is $(-\infty,-9)$.

38. $x^2-4x-5\geq x+1$
Determine the critical points.
$x^2-4x-5=x+1$
$x^2-5x-6=0$
$(x+1)(x-6)=0$
$x=-1$ or $x=6$
Possible intervals:
$(-\infty,-1]$, $[-1,6]$, and $[6,\infty)$
Use a test point in each interval to determine that the solution is $(-\infty,-1]\cup[6,\infty)$.

39. $3t^2-4t<2t^2+1$
Determine the critical points.
$3t^2-4t=2t^2+1$
$t^2-4t-1=0$
$t=\dfrac{-(-4)\pm\sqrt{(-4)^2-4(1)(-1)}}{2(1)}$
$x=\dfrac{4\pm\sqrt{20}}{2}$
$x=\dfrac{4\pm 2\sqrt{5}}{2}$
$x=2\pm\sqrt{5}$
Possible intervals:
$(-\infty,2-\sqrt{5})$, $(2-\sqrt{5},2+\sqrt{5})$, $(2+\sqrt{5},\infty)$
Use a test point in each interval to determine that the solution is $(2-\sqrt{5},\ 2+\sqrt{5})$.

40. a.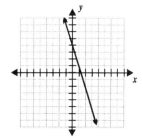

b. Yes, the relation $y=-3x+4$ is a function. It passes the vertical line test.

c. It is decreasing all x.

d. The domain is all real numbers, or $(-\infty,\infty)$. The range is all real numbers, or $(-\infty,\infty)$.

e. x-intercept: $0=-3x+4$
$3x=4$
$x=\dfrac{4}{3}$
y-intercept: $y=-3(0)+4=4$
The x-intercept is $\left(\dfrac{4}{3},0\right)$ and the y-intercept is $(0,4)$.

41. a. *x*-intercept: $0 = x^2 + 4x - 1$

$$x = \frac{-4 \pm \sqrt{4^2 - 4(1)(-1)}}{2(1)}$$

$$x = \frac{-4 \pm \sqrt{20}}{2}$$

$$x = \frac{-4 \pm 2\sqrt{5}}{2}$$

$$x = -2 \pm \sqrt{5}$$

y-intercept: $y = 0^2 + 4(0) - 1 = -1$

The *x*-intercepts are $(-2 \pm \sqrt{5}, 0)$ and the *y*-intercept is $(0, -1)$.

b. $x = \dfrac{-b}{2a} = \dfrac{-4}{2(1)} = -2$

$y = (-2)^2 + 4(-2) - 1 = -5$

The vertex is $(-2, -5)$.

c.

x	$y = x^2 + 4x - 1$	y
-5	$y = (-5)^2 + 4(-5) - 1 = 4$	4
-4	$y = (-4)^2 + 4(-4) - 1 = -1$	-1
-2	$y = (-2)^2 + 4(-2) - 1 = -5$	-5
0	$y = 0^2 + 4(0) - 1 = -1$	-1
1	$y = 1^2 + 4(1) - 1 = 4$	4

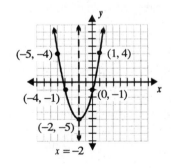

d. The domain is all real numbers, or $(-\infty, \infty)$.
The range is $y \geq -5$, or $[-5, \infty)$.

42. $m = \dfrac{1-(-2)}{4-3} = \dfrac{1+2}{4-3} = \dfrac{3}{1} = 3$

$y - (-2) = 3(x - 3)$

$y + 2 = 3x - 9$

$y = 3x - 11$

43. $2x + 3y = 1$

$3y = -2x + 1$

$y = -\dfrac{2}{3}x + \dfrac{1}{3}$

The slope of the given line is $-\dfrac{2}{3}$. The slope of the line perpendicular to the given line is $\dfrac{3}{2}$.

$y - 2 = \dfrac{3}{2}[x - (-1)]$

$y - 2 = \dfrac{3}{2}(x + 1)$

$y - 2 = \dfrac{3}{2}x + \dfrac{3}{2}$

$y = \dfrac{3}{2}x + \dfrac{7}{2}$

44. Let x = the number of pounds of peanuts.

$7 - x$ = the number of pounds of cashews.

$5x + 12(7 - x) = 10(7)$

$5x + 84 - 12x = 70$

$-7x = -14$

$x = 2$

$7 - x = 5$

Raynoc should mix 2 pounds of peanuts and 5 pounds of cashews.

45. $s = -16t^2 + v_0 t + s_0$; $s = 0$; $v_0 = 0$; $s_0 = 77$

$0 = -16t^2 + 0t + 77$

$16t^2 = 77$

$t^2 = \dfrac{77}{16}$

$t = \pm\sqrt{\dfrac{77}{16}}$

$t = \pm\dfrac{\sqrt{77}}{4}$

$t \approx \pm 2.194$

Time cannot be negative, so the ball would hit the ground in $\dfrac{\sqrt{77}}{4}$ seconds, or approximately 2.194 seconds.

46. Let $x =$ the amount Georgette should invest.

$x + 0.12x = 16,800$

$1.12x = 16,800$

$x = \dfrac{16,800}{1.12}$

$x = 15,000$

Georgette should invest $15,000.

47. a. $y = -0.6x^2 + 0.9x + 9.5$

L1	L2	L3	1	QuadReg
0	9.5	---		y=ax²+bx+c
1	9.8			a=-.6
2	8.9			b=.9
---	---			c=9.5
				R²=1
L1(4)=				

b. $3.5 = -0.6x^2 + 0.9x + 9.5$

$0.6x^2 - 0.9x - 6 = 0$

$0.6(x^2 - 1.5x - 10) = 0$

$0.6(x + 2.5)(x - 4) = 0$

$x = -2.5$ or $x = 4$

Disregarding −2.5, the profit will drop to $3.5 million 4 years after 1999, which would be the year 2003.

48. Let $x =$ the number of packages purchased. $3.50 - 0.25x =$ price per package.

a. $R(x) = x(3.50 - 0.25x)$

b. $C(x) = 2.00x + 1.50$

c. $P(x) = R(x) - C(x)$

$= x(3.50 - 0.25x) - (2.00x + 1.50)$

$= 3.5x - 0.25x^2 - 2x - 1.5$

$= -0.25x^2 + 1.5x - 1.5$

d. $0 = -0.25x^2 + 1.5x - 1.5$

$0 = -0.25(x^2 - 6x + 6)$

$x = \dfrac{-(-6) \pm \sqrt{(-6)^2 - 4(1)(6)}}{2(1)}$

$x = \dfrac{6 \pm \sqrt{12}}{2}$

$x = \dfrac{6 \pm 2\sqrt{3}}{2}$

$x = 3 \pm \sqrt{3}$

$x \approx 4.7 \approx 5$ or $x \approx 1.3 \approx 2$

$3.5 - 0.25x \approx 1.5$ $3.5 - 0.25x \approx 3$

Sam will approximately break even if he sells 5 packages at $1.50 each and if he sells 3 packages at $3.00 each.

49. $c = a\sqrt{2}$

$8 = a\sqrt{2}$

$a = \dfrac{8}{\sqrt{2}} = \dfrac{8}{\sqrt{2}} \cdot \dfrac{\sqrt{2}}{\sqrt{2}} = \dfrac{8\sqrt{2}}{2} = 4\sqrt{2} \approx 5.657$

Each leg measures $4\sqrt{2}$ meters, or about 5.657 meters.

50. Let $x =$ Janet's score on the last test.

$85 \leq \dfrac{96 + 82 + 94 + x}{4} \leq 90$

$85 \leq \dfrac{272 + x}{4} \leq 90$

$340 \leq 272 + x \leq 360$

$68 \leq x \leq 88$

Janet must score between 68 and 88 points, inclusive, in order to earn a B in history.

Chapter 12

12.1 Exercises

1. $\dfrac{x^2+10x+25}{x+5}$ is a rational expression because both $x^2+10x+25$ and $x+5$ are polynomial expressions.

3. $z-4-\dfrac{25}{z-4}$ is a rational expression because is can be written as $\dfrac{z}{1}-\dfrac{4}{1}-\dfrac{25}{z-4}$.

5. $\dfrac{x+3}{\sqrt{x}-2}$ is not a rational expression because $\sqrt{x}-2$ is not a polynomial expression.

7. $6x^2+17x-3$ is a rational expression because is can be written as $\dfrac{6x^2+17x-3}{1}$.

9. $\dfrac{11}{3xy}$ is a rational expression because both 11 and $3xy$ are polynomial expressions.

11. $3x^{-2}+4x^{-1}+3-x$ is a rational expression because is can be written as $\dfrac{3}{x^2}+\dfrac{4}{x}+\dfrac{3}{1}-\dfrac{x}{1}$.

13. $\dfrac{0.5x+7}{1.2x-4}$ is a rational expression because both $0.5x+7$ and $1.2x-4$ are polynomial expressions.

15. $\dfrac{2-3\sqrt{x}}{2x}$ is not a rational expression because $2-3\sqrt{x}$ is not a polynomial expression.

17. $y=\dfrac{3-x}{x}$
    $x=0$
    Restricted value: 0
    Domain: $x\ne 0$
    $(-\infty,0)\cup(0,\infty)$

19. $y=\dfrac{x-5}{x^2-11x+30}$
    $x^2-11x+30=0$
    $(x-5)(x-6)=0$
    $x=5$ or $x=6$
    Restricted values: 5 and 6
    Domain: $x\ne 5, x\ne 6$
    $(-\infty,5)\cup(5,6)\cup(6,\infty)$

21. $y=\dfrac{x+7}{x^2-4}$
    $x^2-4=0$
    $(x-2)(x+2)=0$
    $x=2$ or $x=-2$
    Restricted values: -2 and 2
    Domain: $x\ne -2, x\ne 2$
    $(-\infty,-2)\cup(-2,2)\cup(2,\infty)$

23. $y=\dfrac{2x^2+7}{x^2-5x}$
    $x^2-5x=0$
    $x(x-5)=0$
    $x=0$ or $x=5$
    Restricted values: 0 and 5
    Domain: $x\ne 0, x\ne 5$
    $(-\infty,0)\cup(0,5)\cup(5,\infty)$

25. $y=\dfrac{x+2}{x^3-2x^2+4x-8}$
    $x^3-2x^2+4x-8=0$
    $x^2(x-2)+4(x-2)=0$
    $(x-2)(x^2+4)=0$
    $x=2$
    Restricted value: 2
    Domain: $x\ne 2$
    $(-\infty,2)\cup(2,\infty)$

Chapter 12: Rational Expressions, Functions, and Equations

27. $y = \dfrac{7}{2x^2 + 17x + 35}$

    $2x^2 + 17x + 35 = 0$

    $(2x + 7)(x + 5) = 0$

    $x = -\dfrac{7}{2}$ or $x = -5$

    Restricted values: -5 and $-\dfrac{7}{2}$

    Domain: $x \ne -5, x \ne -\dfrac{7}{2}$

    $(-\infty, -5) \cup \left(-5, -\dfrac{7}{2}\right) \cup \left(-\dfrac{7}{2}, \infty\right)$

29. $y = \dfrac{x^2 - 3x}{8x^2 + 6x - 9}$

    $8x^2 + 6x - 9 = 0$

    $(2x + 3)(4x - 3) = 0$

    $x = -\dfrac{3}{2}$ or $x = \dfrac{3}{4}$

    Restricted values: $-\dfrac{3}{2}$ and $\dfrac{3}{4}$

    Domain: $x \ne -\dfrac{3}{2}, x \ne \dfrac{3}{4}$

    $\left(-\infty, -\dfrac{3}{2}\right) \cup \left(-\dfrac{3}{2}, \dfrac{3}{4}\right) \cup \left(\dfrac{3}{4}, \infty\right)$

31. $y = \dfrac{4x}{4x^2 + 36x + 81}$

    $4x^2 + 36x + 81 = 0$

    $(2x + 9)^2 = 0$

    $x = -\dfrac{9}{2}$

    Restricted value: $-\dfrac{9}{2}$

    Domain: $x \ne -\dfrac{9}{2}$

    $\left(-\infty, -\dfrac{9}{2}\right) \cup \left(-\dfrac{9}{2}, \infty\right)$

33. $y = \dfrac{6 - x}{x^2 - 22x + 121}$

    $x^2 - 22x + 121 = 0$

    $(x - 11)^2 = 0$

    $x = 11$

    Restricted value: 11

    Domain: $x \ne 11$

    $(-\infty, 11) \cup (11, \infty)$

35. $y = \dfrac{3x^2}{9x^2 - 25}$

    $9x^2 - 25 = 0$

    $(3x + 5)(3x - 5) = 0$

    $x = -\dfrac{5}{3}$ or $x = \dfrac{5}{3}$

    Restricted values: $-\dfrac{5}{3}$ and $\dfrac{5}{3}$

    Domain: $x \ne -\dfrac{5}{3}, x \ne \dfrac{5}{3}$

    $\left(-\infty, -\dfrac{5}{3}\right) \cup \left(-\dfrac{5}{3}, \dfrac{5}{3}\right) \cup \left(\dfrac{5}{3}, \infty\right)$

37. $y = \dfrac{5x^3 + 1}{2x^3 + 4x^2 - 18x - 36}$

    $2x^3 + 4x^2 - 18x - 36 = 0$

    $2(x^3 + 2x^2 - 9x - 18) = 0$

    $2[x^2(x + 2) - 9(x + 2)] = 0$

    $2(x + 2)(x^2 - 9) = 0$

    $2(x + 2)(x + 3)(x - 3) = 0$

    $x = -2, \ x = -3, \text{ or } x = 3$

    Restricted values: $-3, -2,$ and 3

    Domain: $x \ne -3, x \ne -2, x \ne 3$

    $(-\infty, -3) \cup (-3, -2) \cup (-2, 3) \cup (3, \infty)$

39. $y = \dfrac{3x + 8}{2x^2 + 5x + 7}$

    $2x^2 + 5x + 7 = 0$

513

SSM: Experiencing Introductory and Intermediate Algebra

$$x = \frac{-5 \pm \sqrt{5^2 - 4(2)(7)}}{2(2)}$$

$$x = \frac{-5 \pm \sqrt{-31}}{4} \text{ (Not Real)}$$

Restricted values: None
Domain: All real numbers
$$(-\infty, \infty)$$

41. $y = \dfrac{x+4}{2x+3}$

$2x + 3 = 0$

$x = -\dfrac{3}{2}$

Restricted value: $-\dfrac{3}{2}$

Domain: $x \neq -\dfrac{3}{2}$

$$\left(-\infty, -\frac{3}{2}\right) \cup \left(-\frac{3}{2}, \infty\right)$$

43. $y = \dfrac{6}{x} + \dfrac{9}{x^2}$

The restricted value is 0. Set up a table of *x*-values being sure to use both *x*-values that are less than 0 and *x*-values that are greater than 0. Connect the points with a smooth curve. Do not connect across the vertical line $x = 0$.

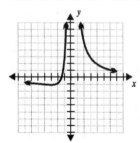

45. $h(x) = \dfrac{x^2 - 49}{x - 7}$

The restricted value is 7. Set up a table of *x*-values being sure to use both *x*-values that are less than 7 and *x*-values that are greater than 7. Connect the points with a smooth curve. Do not

connect across the vertical line $x = 7$.

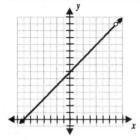

47. $y = \dfrac{2x}{x^2 + 3}$

The function has no restricted values. Set up a table of *x*-values. Connect the points with a smooth curve.

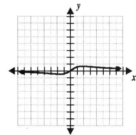

49. $y = \dfrac{6x}{x^3 + 8}$

The restricted value is –2. Set up a table of *x*-values being sure to use both *x*-values that are less than –2 and *x*-values that are greater than –2. Connect the points with a smooth curve. Do not connect across the vertical line $x = -2$.

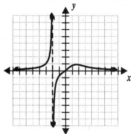

51. $f(x) = \dfrac{x^2 + 1}{x - 3}$

The restricted value is 3. Set up a table of *x*-values being sure to use both *x*-values that are less than 3 and *x*-values that are greater than 3. Connect the points with a smooth curve. Do not connect across the vertical line $x = 3$.

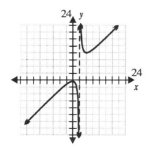

53. $R(x) = \dfrac{2x^2 + 7x + 6}{x + 2}$

The restricted value is –2. Set up a table of x-values being sure to use both x-values that are less than –2 and x-values that are greater than –2. Connect the points with a smooth curve. Do not connect across the vertical line $x = -2$.

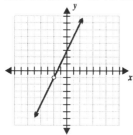

55. $y = \dfrac{8}{x}$

The restricted value is 0. Set up a table of x-values being sure to use both x-values that are less than 0 and x-values that are greater than 0. Connect the points with a smooth curve. Do not connect across the vertical line $x = 0$.

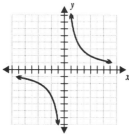

57. $y = \dfrac{x - 5}{x + 3}$

The restricted value is –3. Set up a table of x-values being sure to use both x-values that are less than –3 and x-values that are greater than –3. Connect the points with a smooth curve. Do not connect across the vertical line $x = -3$.

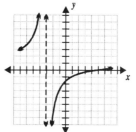

59. $f(x) = \dfrac{2x + 3}{x^2 - 2x - 3} = \dfrac{2x + 3}{(x - 3)(x + 1)}$

The restricted values are –1 and 3. Set up a table of x-values being sure to use x-values that are less than –1, x-values that are between –1 and 3, and x-values that are greater than 3. Connect the points with a smooth curve. Do not connect across the vertical lines $x = -1$ and $x = 3$.

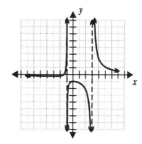

61. $g(x) = \dfrac{2x^2 - 5x + 2}{x - 2}$

The restricted value is 2. Set up a table of x-values being sure to use both x-values that are less than 2 and x-values that are greater than 2. Connect the points with a smooth curve. Do not connect across the vertical line $x = 2$.

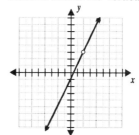

SSM: Experiencing Introductory and Intermediate Algebra

63. a. $C_{ave}(x) = \dfrac{6500 + 100x + 25x^2}{x}$

b. The minimum average cost per year would be approximately $906.25, when the number of years is $x = 16$.

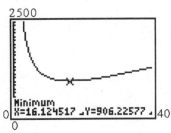

c. The customer should replace the heating- and-cooling system after 16 years of use.

65. a.

Year	Proportion from domestic
1995	$\dfrac{2394}{5037} \approx 0.475$
1998	$\dfrac{2282}{5037} \approx 0.418$
1999	$\dfrac{2147}{5334} \approx 0.403$

b. $P(x) = \dfrac{D(x)}{T(x)} = \dfrac{-56.12x + 2405}{89.65x + 5068}$

c.

Year	$P(x)$
1995	$P(0) = \dfrac{-56.12(0) + 2405}{89.65(0) + 5068} \approx 0.475$
1998	$P(3) = \dfrac{-56.12(3) + 2405}{89.65(3) + 5068} \approx 0.419$
1999	$P(4) = \dfrac{-56.12(4) + 2405}{89.65(4) + 5068} \approx 0.402$

d. The approximations from part c are very close to the proportions in part a. The function is a good predictor of the proportion of domestic crude-oil production.

12.1 Calculator Exercises

1. Answers will vary. One possibility is shown below. Let $Y1 = \dfrac{x^3 + 5}{x^2 - 3}$

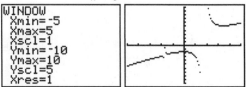

2. Answers will vary. One possibility is shown below. Let $Y1 = \dfrac{1}{x^3 - 1} + \dfrac{1}{x^2}$

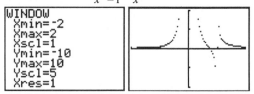

3. Answers will vary. One possibility is shown next. Let $Y1 = \dfrac{3}{x - 2} + \dfrac{5}{x^2 + 1}$

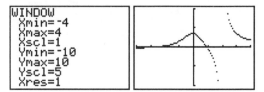

12.2 Exercises

1. $\dfrac{36x^3y^2z}{54x^2y^5} = \dfrac{(18x^2y^2)(2xz)}{(18x^2y^2)(3y^3)} = \dfrac{2xz}{3y^3}$

3. $\dfrac{-56a^3b^2}{84a^2b^2c} = \dfrac{(28a^2b^2)(-2a)}{(28a^2b^2)(3c)} = -\dfrac{2a}{3c}$

5. $\dfrac{9x - 27}{18x - 6} = \dfrac{9(x - 3)}{6(3x - 1)} = \dfrac{3 \cdot 3(x - 3)}{3 \cdot 2(3x - 1)} = \dfrac{3(x - 3)}{2(3x - 1)}$

7. $\dfrac{8 - z}{z - 8} = \dfrac{-z + 8}{z - 8} = \dfrac{-1(z - 8)}{z - 8} = -1$

9. $\dfrac{5x^2y^3 + 10xy^4}{15xy^2 - 20x^2y^2} = \dfrac{5xy^3(x + 2y)}{5xy^2(3 - 4x)} = \dfrac{y(x + 2y)}{(3 - 4x)}$

11. $\dfrac{-6x+9y-3z}{3x-21y+3z} = \dfrac{3(-2x+3y-z)}{3(x-7y+z)} = \dfrac{-2x+3y-z}{x-7y+z}$

13. $\dfrac{x^2-8x+12}{x^2-10x+24} = \dfrac{(x-2)(x-6)}{(x-4)(x-6)} = \dfrac{x-2}{x-4}$

15. $\dfrac{x^2-x-6}{x^2+9x+14} = \dfrac{(x-3)(x+2)}{(x+7)(x+2)} = \dfrac{x-3}{x+7}$

17. $\dfrac{3x^2+7x+2}{3x^2-11x-4} = \dfrac{(3x+1)(x+2)}{(3x+1)(x-4)} = \dfrac{x+2}{x-4}$

19. $\dfrac{6x^2-x-2}{4x^2-4x-3} = \dfrac{(2x+1)(3x-2)}{(2x+1)(2x-3)} = \dfrac{3x-2}{2x-3}$

21. $\dfrac{8x^2+12x+4}{24x^2-40x-16} = \dfrac{4(2x^2+3x+1)}{8(3x^2-5x-2)}$
$= \dfrac{4(x+1)(2x+1)}{8(x-2)(3x+1)}$
$= \dfrac{(x+1)(2x+1)}{2(x-2)(3x+1)}$

23. $\dfrac{2x^2+xy-y^2}{3x^2+4xy+y^2} = \dfrac{(x+y)(2x-y)}{(x+y)(3x+y)} = \dfrac{2x-y}{3x+y}$

25. $\dfrac{64-x^2}{x^3+2x^2-64x-128} = \dfrac{-x^2+64}{x^2(x+2)-64(x+2)}$
$= \dfrac{-1(x^2-64)}{(x+2)(x^2-64)}$
$= -\dfrac{1}{x+2}$

27. $\dfrac{2x^2-7x-15}{x^3-5x^2+11x-55} = \dfrac{(x-5)(2x+3)}{x^2(x-5)+11(x-5)}$
$= \dfrac{(x-5)(2x+3)}{(x-5)(x^2+11)}$
$= \dfrac{2x+3}{x^2+11}$

29. $\dfrac{12a^2b^2}{35cd^2} \cdot \dfrac{49c^2d^2}{27ab^3} = \dfrac{588a^2b^2c^2d^2}{945ab^3cd^2}$
$= \dfrac{(21ab^2cd^2)(35ac)}{(21ab^2cd^2)(45b)}$
$= \dfrac{28ac}{45b}$

31. $\dfrac{-2x^3}{5y^2} \cdot \dfrac{15xy^2}{8x^2} = \dfrac{-30x^4y^2}{40x^2y^2}$
$= \dfrac{(10x^2y^2)(-3x^2)}{(10x^2y^2)(4)}$
$= -\dfrac{3x^2}{4}$

33. $(2.6x^2) \cdot \dfrac{3.1x}{y^2} = \dfrac{2.6x^2}{1} \cdot \dfrac{3.1x}{y^2} = \dfrac{8.06x^3}{y^2}$

35. $\dfrac{1}{3} \cdot \dfrac{2a^2}{3b} = \dfrac{2a^2}{9b}$

37. $\dfrac{x+5}{x} \cdot \dfrac{x^3}{x+4} = \dfrac{x^3(x+5)}{x(x+4)} = \dfrac{x^2(x+5)}{(x+4)}$

39. $\dfrac{3x+2}{x+4} \cdot \dfrac{x-4}{3x+2} = \dfrac{(3x+2)(x-4)}{(3x+2)(x+4)} = \dfrac{x-4}{x+4}$

41. $\dfrac{x+7}{2x+1} \cdot \dfrac{5x+2}{x} = \dfrac{(x+7)(5x+2)}{x(2x+1)}$

43. $\dfrac{x^2-x-6}{2x^2+3x+1} \cdot \dfrac{2x^2-x-1}{x^2+5x+6}$
$= \dfrac{(x^2-x-6)(2x^2-x-1)}{(2x^2+3x+1)(x^2+5x+6)}$
$= \dfrac{(x+2)(x-3)(x-1)(2x+1)}{(x+1)(2x+1)(x+2)(x+3)}$
$= \dfrac{(x-3)(x-1)}{(x+1)(x+3)}$

45. $\dfrac{x^2-xy-2y^2}{x^2-9y^2} \cdot \dfrac{x^2-3xy}{x^2-y^2}$

$= \dfrac{(x^2-xy-2y^2)(x^2-3xy)}{(x^2-9y^2)(x^2-y^2)}$

$= \dfrac{(x+y)(x-2y)(x)(x-3y)}{(x+3y)(x-3y)(x+y)(x-y)}$

$= \dfrac{x(x-2y)}{(x+3y)(x-y)}$

47. $\dfrac{54a^3b^2}{c} \div \dfrac{36ab^2}{c} = \dfrac{54a^3b^2}{c} \cdot \dfrac{c}{36ab^2}$

$= \dfrac{54a^3b^2c}{36ab^2c}$

$= \dfrac{(18ab^2c)(3a^2)}{(18ab^2c)(2)}$

$= \dfrac{3a^2}{2}$

49. $\dfrac{-21x^2y^3}{z} \div \dfrac{7xy^2}{z^2} = \dfrac{-21x^2y^3}{z} \cdot \dfrac{z^2}{7xy^2}$

$= \dfrac{-21x^2y^3z^2}{7xy^2z}$

$= \dfrac{(7xy^2z)(-3xyz)}{7xy^2z}$

$= -3xyz$

51. $(32ab^2) \div \dfrac{8ab^3}{c} = \dfrac{32ab^2}{1} \cdot \dfrac{c}{8ab^3}$

$= \dfrac{32ab^2c}{8ab^3}$

$= \dfrac{(8ab^2)(4c)}{(8ab^2)(b)}$

$= \dfrac{4c}{b}$

53. $\dfrac{21x^2y^2}{z^3} \div (-3xy^3) = \dfrac{21x^2y^2}{z^3} \cdot \dfrac{1}{-3xy^3}$

$= \dfrac{21x^2y^2}{-3xy^3z^3}$

$= \dfrac{(3xy^2)(7x)}{(3xy^2)(-yz^3)}$

$= -\dfrac{7x}{yz^3}$

55. $\dfrac{a+4}{9a^2-1} \div \dfrac{5a+20}{3a+1} = \dfrac{a+4}{9a^2-1} \cdot \dfrac{3a+1}{5a+20}$

$= \dfrac{(a+4)(3a+1)}{(9a^2-1)(5a+20)}$

$= \dfrac{(a+4)(3a+1)}{(3a+1)(3a-1)(5)(a+4)}$

$= \dfrac{1}{5(3a-1)}$

57. $\dfrac{x^2-16}{5y-2} \div \dfrac{x+4}{25y^2-4} = \dfrac{x^2-16}{5y-2} \cdot \dfrac{25y^2-4}{x+4}$

$= \dfrac{(x^2-16)(25y^2-4)}{(5y-2)(x+4)}$

$= \dfrac{(x+4)(x-4)(5y+2)(5y-2)}{(5y-2)(x+4)}$

$= (x-4)(5y+2)$

59. $\dfrac{2x^2-5x-3}{x^2-5x-14} \div \dfrac{3x^2+10x+3}{x^2-4x-21}$

$= \dfrac{2x^2-5x-3}{x^2-5x-14} \cdot \dfrac{x^2-4x-21}{3x^2+10x+3}$

$= \dfrac{(2x^2-5x-3)(x^2-4x-21)}{(x^2-5x-14)(3x^2+10x+3)}$

$= \dfrac{(x-3)(2x+1)(x+3)(x-7)}{(x+2)(x-7)(x+3)(3x+1)}$

$= \dfrac{(x-3)(2x+1)}{(x+2)(3x+1)}$

61. $\dfrac{8x^2+2x-3}{4x^2-1} \div (4x+3) = \dfrac{8x^2+2x-3}{4x^2-1} \cdot \dfrac{1}{4x+3}$

$= \dfrac{8x^2+2x-3}{(4x^2-1)(4x+3)}$

$= \dfrac{(2x-1)(4x+3)}{(2x+1)(2x-1)(4x+3)}$

$= \dfrac{1}{2x+1}$

63. $(x+9) \div \dfrac{2x^2+19x+9}{3x+4} = \dfrac{x+9}{1} \cdot \dfrac{3x+4}{2x^2+19x+9}$

$= \dfrac{(x+9)(3x+4)}{2x^2+19x+9}$

$= \dfrac{(x+9)(3x+4)}{(x+9)(2x+1)}$

$= \dfrac{3x+4}{2x+1}$

65. $\dfrac{1}{3x-5} \div \dfrac{1}{x+7} = \dfrac{1}{3x-5} \cdot \dfrac{x+7}{1} = \dfrac{x+7}{3x-5}$

67. $\dfrac{x^2+3xy}{7xy+2y^2} \div \dfrac{x^2+4xy+3y^2}{7x^2-5xy-2y^2}$

$= \dfrac{x^2+3xy}{7xy+2y^2} \cdot \dfrac{7x^2-5xy-2y^2}{x^2+4xy+3y^2}$

$= \dfrac{(x^2+3xy)(7x^2-5xy-2y^2)}{(7xy+2y^2)(x^2+4xy+3y^2)}$

$= \dfrac{x(x+3y)(7x+2y)(x-y)}{y(7x+2y)(x+y)(x+3y)}$

$= \dfrac{x(x-y)}{y(x+y)}$

69. $\dfrac{4x+8y}{3x-12y} \div \dfrac{4x+4y}{6x+6y} = \dfrac{4x+8y}{3x-12y} \cdot \dfrac{6x+6y}{4x+4y}$

$= \dfrac{(4x+8y)(6x+6y)}{(3x-12y)(4x+4y)}$

$= \dfrac{4(x+2y)(6)(x+y)}{3(x-4y)(4)(x+y)}$

$= \dfrac{2(x+2y)}{x-4y}$

71. $t = \dfrac{d}{r} = \dfrac{170}{x}$

The expression for the Katarina's time of travel is $\dfrac{170}{x}$ hours.

$d = rt = (x-10)\left(\dfrac{170}{x}\right)$

The expression for Katarina's new distance is $(x-10)\left(\dfrac{170}{x}\right)$ miles.

If $x = 50$, then $d = (50-10)\left(\dfrac{170}{50}\right) = 136$.

Katarina would have traveled 136 miles at the reduced speed if her original speed was 50 mph.

73. $m(x) = \dfrac{500 - 500x^3}{1-x}$

$= \dfrac{500(1-x^3)}{1-x}$

$= \dfrac{500(1-x)(1+x+x^2)}{1-x}$

$= 500(1+x+x^2)$

$m(1.20) = \dfrac{500 - 500(1.20)^3}{1-1.20} = 1820$

$m(1.20) = 500(1+1.20+1.20^2) = 1820$

Sharon's grandson received a total of $1820.

75. $V = LWH$; $L = \dfrac{V}{WH}$

If $V = 15x^3 + 7x^2 - 2x$, $H = x$, and $W = 3x+2$, then

$L = \dfrac{15x^3+7x^2-2x}{x(3x+2)}$

$= \dfrac{x(15x^2+7x-2)}{x(3x+2)} = \dfrac{x(3x+2)(5x-1)}{x(3x+2)}$

$= 5x - 1$

The expression for the length is $(5x-1)$ inches. If the height is $x = 4$ inches, then the width will be $3(4)+2 = 14$ inches and the length will be $5(4)-1 = 19$ inches.

SSM: Experiencing Introductory and Intermediate Algebra

77. a. $C(x) = 300 + 5x$
$R(x) = 25x$

b. $\dfrac{25x}{300+5x} = \dfrac{25x}{5(60+x)} = \dfrac{5x}{60+x}$

c. $\dfrac{5(40)}{60+40} = \dfrac{200}{100} = 2$
The revenue-to-cost ratio is 2.

12.2 Calculator Exercises

Value for n	Original function	Simplified function
2	$y = \dfrac{(1-x^2)}{(1-x)}$	$y = 1 + x$
3	$y = \dfrac{(1-x^3)}{(1-x)}$	$y = 1 + x + x^2$
4	$y = \dfrac{(1-x^4)}{(1-x)}$	$y = 1 + x + x^2 + x^3$
5	$y = \dfrac{(1-x^5)}{(1-x)}$	$y = 1 + x + x^2 + x^3 + x^4$
6	$y = \dfrac{(1-x^6)}{(1-x)}$	$y = 1 + x + x^2 + x^3 + x^4 + x^5$

The pattern appears to be as follows:
$$\dfrac{(1-x^n)}{(1-x)} = 1 + x + x^2 + \cdots + x^{n-1}$$

12.3 Exercises

1. $\dfrac{5}{x} + \dfrac{8}{x} = \dfrac{5+8}{x} = \dfrac{13}{x}$

3. $\dfrac{2x}{5xy} + \dfrac{3x}{5xy} = \dfrac{2x+3x}{5xy} = \dfrac{5x}{5xy} = \dfrac{1}{y}$

5. $\dfrac{x-1}{x+6} + \dfrac{x+1}{x+6} = \dfrac{(x-1)+(x+1)}{x+6} = \dfrac{2x}{x+6}$

7. $\dfrac{2b}{b^2-25} + \dfrac{10}{b^2-25} = \dfrac{2b+10}{b^2-25} = \dfrac{2(b+5)}{(b+5)(b-5)} = \dfrac{2}{b-5}$

9. $\dfrac{x-5}{x+3} + \dfrac{x+5}{x+3} + \dfrac{x+9}{x+3} = \dfrac{(x-5)+(x+5)+(x+9)}{x+3}$
$= \dfrac{3x+9}{x+3}$
$= \dfrac{3(x+3)}{x+3}$
$= 3$

11. $\dfrac{4}{d} - \dfrac{9}{d} = \dfrac{4-9}{d} = -\dfrac{5}{d}$

13. $\dfrac{9xy}{7x^2y} - \dfrac{2xy}{7x^2y} = \dfrac{9xy - 2xy}{7x^2y} = \dfrac{7xy}{7x^2y} = \dfrac{1}{x}$

15. $\dfrac{2x-3}{x+9} - \dfrac{x-3}{x+9} = \dfrac{(2x-3)-(x-3)}{x+9}$
$= \dfrac{2x-3-x+3}{x+9}$
$= \dfrac{x}{x+9}$

17. $\dfrac{4x-13}{x-5} - \dfrac{2x-3}{x-5} = \dfrac{(4x-13)-(2x-3)}{x-5}$
$= \dfrac{4x-13-2x+3}{x-5}$
$= \dfrac{2x-10}{x-5}$
$= \dfrac{2(x-5)}{x-5}$
$= 2$

19. $\dfrac{x+3}{x+11} - \dfrac{2x-8}{x+11} = \dfrac{(x+3)-(2x-8)}{x+11}$
$= \dfrac{x+3-2x+8}{x+11}$
$= \dfrac{-x+11}{x+11}$

21. $\dfrac{z^2}{z+4} - \dfrac{16}{z+4} = \dfrac{z^2-16}{z+4} = \dfrac{(z+4)(z-4)}{z+4} = z-4$

23. $\dfrac{3c}{c^2-9} - \dfrac{9}{c^2-9} = \dfrac{3c-9}{c^2-9} = \dfrac{3(c-3)}{(c+3)(c-3)} = \dfrac{3}{c+3}$

25. $\dfrac{5x+1}{2x+3} - \dfrac{x+2}{2x+3} - \dfrac{2x-5}{2x+3} = \dfrac{(5x+1)-(x+2)-(2x-5)}{2x+3}$

$= \dfrac{5x+1-x-2-2x+5}{2x+3}$

$= \dfrac{2x+4}{2x+3}$

$= \dfrac{2(x+2)}{2x+3}$

27. $\dfrac{5}{x} + \dfrac{8}{-x} = \dfrac{5}{x} + \dfrac{-8}{x} = \dfrac{5+(-8)}{x} = -\dfrac{3}{x}$

29. $\dfrac{3x-2}{x-5} + \dfrac{x+8}{5-x} = \dfrac{3x-2}{x-5} + \dfrac{-(x+8)}{x-5}$

$= \dfrac{3x-2-x-8}{x-5}$

$= \dfrac{2x-10}{x-5}$

$= \dfrac{2(x-5)}{x-5}$

$= 2$

31. $\dfrac{a^2}{a-3} + \dfrac{9}{3-a} = \dfrac{a^2}{a-3} + \dfrac{-9}{a-3}$

$= \dfrac{a^2-9}{a-3}$

$= \dfrac{(a+3)(a-3)}{a-3}$

$= a+3$

33. $\dfrac{5x}{x^2-9} + \dfrac{15}{9-x^2} = \dfrac{5x}{x^2-9} + \dfrac{-15}{x^2-9}$

$= \dfrac{5x-15}{x^2-9}$

$= \dfrac{5(x-3)}{(x+3)(x-3)}$

$= \dfrac{5}{x+3}$

35. $\dfrac{3}{2x} + \dfrac{7}{6x^2} = \dfrac{3(3x)}{2x(3x)} + \dfrac{7}{6x^2} = \dfrac{9x}{6x^2} + \dfrac{7}{6x^2} = \dfrac{9x+7}{6x^2}$

37. $\dfrac{3x+1}{x-6} + \dfrac{x-4}{2x-12} = \dfrac{2(3x+1)}{2(x-6)} + \dfrac{x-4}{2(x-6)}$

$= \dfrac{6x+2+x-4}{2(x-6)}$

$= \dfrac{7x-2}{2(x-6)}$

39. $\dfrac{2x}{2x-3} + \dfrac{6x+9}{4x^2-9}$

$= \dfrac{2x(2x+3)}{(2x+3)(2x-3)} + \dfrac{6x+9}{(2x+3)(2x-3)}$

$= \dfrac{4x^2+6x+6x+9}{(2x+3)(2x-3)}$

$= \dfrac{4x^2+12x+9}{(2x+3)(2x-3)}$

$= \dfrac{(2x+3)(2x+3)}{(2x+3)(2x-3)}$

$= \dfrac{2x+3}{2x-3}$

41. $\dfrac{7}{3x-5} + \dfrac{4}{3x+5} = \dfrac{7(3x+5)}{(3x+5)(3x-5)} + \dfrac{4(3x-5)}{(3x+5)(3x-5)}$

$= \dfrac{21x+35+12x-20}{(3x+5)(3x-5)}$

$= \dfrac{33x+15}{(3x+5)(3x-5)}$

$= \dfrac{3(11x+5)}{(3x+5)(3x-5)}$

43. $\dfrac{7}{x-5} + \dfrac{x+1}{x^2-10x+25} = \dfrac{7(x-5)}{(x-5)^2} + \dfrac{x+1}{(x-5)^2}$

$= \dfrac{7x-35+x+1}{(x-5)^2}$

$= \dfrac{8x-34}{(x-5)^2}$

$= \dfrac{2(4x-17)}{(x-5)^2}$

45. $\dfrac{b+8}{b} + \dfrac{b}{b+8} = \dfrac{(b+8)^2}{b(b+8)} + \dfrac{b^2}{b(b+8)}$

$= \dfrac{b^2+16b+64+b^2}{b(b+8)}$

$= \dfrac{2b^2+16b+64}{b(b+8)}$

$= \dfrac{2(b^2+8b+32)}{b(b+8)}$

47. $\dfrac{x-2}{4x+12} + \dfrac{2x+1}{x^2-9} = \dfrac{x-2}{4(x+3)} + \dfrac{2x+1}{(x+3)(x-3)}$

$= \dfrac{(x-2)(x-3)}{4(x+3)(x-3)} + \dfrac{4(2x+1)}{4(x+3)(x-3)}$

$= \dfrac{x^2-3x-2x+6+8x+4}{4(x+3)(x-3)}$

$= \dfrac{x^2+3x+10}{4(x+3)(x-3)}$

49. $\dfrac{x+7}{x^2+2x-15} + \dfrac{8-x}{x^2+9x+20}$

$= \dfrac{x+7}{(x-3)(x+5)} + \dfrac{8-x}{(x+4)(x+5)}$

$= \dfrac{(x+7)(x+4)}{(x-3)(x+4)(x+5)} + \dfrac{(8-x)(x-3)}{(x-3)(x+4)(x+5)}$

$= \dfrac{x^2+4x+7x+28+8x-24-x^2+3x}{(x-3)(x+4)(x+5)}$

$= \dfrac{22x+4}{(x-3)(x+4)(x+5)}$

$= \dfrac{2(11x+2)}{(x-3)(x+4)(x+5)}$

51. $\dfrac{3a-b}{a^2b} + \dfrac{a+2b}{ab^2} = \dfrac{b(3a-b)}{a^2b^2} + \dfrac{a(a+2b)}{a^2b^2}$

$= \dfrac{3ab-b^2+a^2+2ab}{a^2b^2}$

$= \dfrac{a^2+5ab-b^2}{a^2b^2}$

53. $\dfrac{3x}{x-2y} + \dfrac{4y}{2y-x} = \dfrac{3x}{x-2y} + \dfrac{-4y}{x-2y}$

$= \dfrac{3x-4y}{x-2y}$

55. $\dfrac{5x+2y}{x^2-9y^2} + \dfrac{6}{x+3y}$

$= \dfrac{5x+2y}{(x+3y)(x-3y)} + \dfrac{6(x-3y)}{(x+3y)(x-3y)}$

$= \dfrac{5x+2y+6x-18y}{(x+3y)(x-3y)}$

$= \dfrac{11x-16y}{(x+3y)(x-3y)}$

57. $\dfrac{8}{x^2+6xz+9z^2} + \dfrac{2}{x^2-9z^2}$

$= \dfrac{8}{(x+3z)^2} + \dfrac{2}{(x+3z)(x-3z)}$

$= \dfrac{8(x-3z)}{(x+3z)^2(x-3z)} + \dfrac{2(x+3z)}{(x+3z)^2(x-3z)}$

$= \dfrac{8x-24z+2x+6z}{(x+3z)^2(x-3z)}$

$= \dfrac{10x-18z}{(x+3z)^2(x-3z)}$

$= \dfrac{2(5x-9z)}{(x+3z)^2(x-3z)}$

59. $\dfrac{7x+5}{x-2} - \dfrac{x-1}{2-x} = \dfrac{7x+5}{x-2} - \dfrac{-(x-1)}{x-2}$

$= \dfrac{(7x+5)-[-(x-1)]}{x-2}$

$= \dfrac{7x+5+x-1}{x-2}$

$= \dfrac{8x+4}{x-2}$

$= \dfrac{4(2x+1)}{x-2}$

61. $\dfrac{3x}{x^2-4} - \dfrac{6}{4-x^2} = \dfrac{3x}{x^2-4} - \dfrac{-6}{x^2-4}$

$= \dfrac{3x-(-6)}{x^2-4}$

$= \dfrac{3x+6}{x^2-4}$

$= \dfrac{3(x+2)}{(x+2)(x-2)}$

$= \dfrac{3}{x-2}$

63. $\dfrac{6x-3}{3x+15} - \dfrac{x-1}{x+5}$

$= \dfrac{6x-3}{3(x+5)} - \dfrac{3(x-1)}{3(x+5)}$

$$= \frac{6x-3-3(x-1)}{3(x+5)}$$

$$= \frac{6x-3-3x+3}{3(x+5)}$$

$$= \frac{3x}{3(x+5)}$$

$$= \frac{x}{x+5}$$

65. $\dfrac{7}{2x-5} - \dfrac{9}{2x+5} = \dfrac{7(2x+5)}{(2x+5)(2x-5)} - \dfrac{9(2x-5)}{(2x+5)(2x-5)}$

$$= \frac{7(2x+5)-9(2x-5)}{(2x+5)(2x-5)}$$

$$= \frac{14x+35-18x+45}{(2x+5)(2x-5)}$$

$$= \frac{-4x+80}{(2x+5)(2x-5)}$$

$$= \frac{-4(x-20)}{(2x+5)(2x-5)}$$

67. $\dfrac{6}{x^2+12x+36} - \dfrac{2x+1}{x+6} = \dfrac{6}{(x+6)^2} - \dfrac{(2x+1)(x+6)}{(x+6)^2}$

$$= \frac{6-(2x^2+12x+x+6)}{(x+6)^2}$$

$$= \frac{6-2x^2-12x-x-6}{(x+6)^2}$$

$$= \frac{-2x^2-13x}{(x+6)^2}$$

$$= \frac{-x(2x+13)}{(x+6)^2}$$

69. $\dfrac{x}{8-x} - \dfrac{8-x}{x} = \dfrac{x^2}{x(8-x)} - \dfrac{(8-x)^2}{x(8-x)}$

$$= \frac{x^2-(64-16x+x^2)}{x(8-x)}$$

$$= \frac{x^2-64+16x-x^2}{x(8-x)}$$

$$= \frac{16x-64}{x(8-x)}$$

$$= \frac{16(x-4)}{x(8-x)}$$

71. $\dfrac{x+2}{3x+12} - \dfrac{x-4}{x^2-16} = \dfrac{x+2}{3(x+4)} - \dfrac{x-4}{(x+4)(x-4)}$

$$= \frac{(x+2)(x-4)-3(x-4)}{3(x+4)(x-4)}$$

$$= \frac{x^2-4x+2x-8-3x+12}{3(x+4)(x-4)}$$

$$= \frac{x^2-5x+4}{3(x+4)(x-4)}$$

$$= \frac{(x-1)(x-4)}{3(x+4)(x-4)}$$

$$= \frac{x-1}{3(x+4)}$$

73. $\dfrac{11}{b^2-9} - \dfrac{7}{2b^2-b-15}$

$$= \frac{11}{(b+3)(b-3)} - \frac{7}{(b-3)(2b+5)}$$

$$= \frac{11(2b+5)-7(b+3)}{(b+3)(b-3)(2b+5)}$$

$$= \frac{22b+55-7b-21}{(b+3)(b-3)(2b+5)}$$

$$= \frac{15b+34}{(b+3)(b-3)(2b+5)}$$

75. $\dfrac{7}{5ab^2} - \dfrac{15}{10a^2b} - \dfrac{3}{2b}$

$$= \frac{7(2a)}{5ab^2(2a)} - \frac{15(b)}{10a^2b(b)} - \frac{3(5a^2b)}{2b(5a^2b)}$$

$$= \frac{14a}{10a^2b^2} - \frac{15b}{10a^2b^2} - \frac{15a^2b}{10a^2b^2}$$

$$= \frac{14a-15b-15a^2b}{10a^2b^2}$$

77. $\dfrac{4p-q}{p+3q} - \dfrac{p+2q}{p+3q} = \dfrac{(4p-q)-(p+2q)}{p+3q}$

$$= \frac{4p-q-p-2q}{p+3q}$$

$$= \frac{3p-3q}{p+3q}$$

$$= \frac{3(p-q)}{p+3q}$$

SSM: Experiencing Introductory and Intermediate Algebra

79. $\dfrac{7}{x^2+6xy+9y^2} - \dfrac{2}{x^2-9y^2}$
$= \dfrac{7}{(x+3y)^2} - \dfrac{2}{(x+3y)(x-3y)}$
$= \dfrac{7(x-3y)-2(x+3y)}{(x+3y)^2(x-3y)}$
$= \dfrac{7x-21y-2x-6y}{(x+3y)^2(x-3y)}$
$= \dfrac{5x-27y}{(x+3y)^2(x-3y)}$

81. $\dfrac{x}{x+5} - \dfrac{4}{x-5} + \dfrac{10x}{x^2-25}$
$= \dfrac{x(x-5)}{(x+5)(x-5)} - \dfrac{4(x+5)}{(x+5)(x-5)} + \dfrac{10x}{(x+5)(x-5)}$
$= \dfrac{x(x-5)-4(x+5)+10x}{(x+5)(x-5)}$
$= \dfrac{x^2-5x-4x-20+10x}{(x+5)(x-5)}$
$= \dfrac{x^2+x-20}{(x+5)(x-5)}$
$= \dfrac{(x-4)(x+5)}{(x+5)(x-5)}$
$= \dfrac{x-4}{x-5}$

83. $1 + \dfrac{x-9}{x+10} = \dfrac{x+10}{x+10} + \dfrac{x-9}{x+10}$
$= \dfrac{x+10+x-9}{x+10}$
$= \dfrac{2x+1}{x+10}$

85. $5 - \dfrac{4}{2x+7} = \dfrac{5(2x+7)}{2x+7} - \dfrac{4}{2x+7}$
$= \dfrac{10x+35-4}{2x+7}$
$= \dfrac{10x+31}{2x+7}$

87. $\dfrac{2x-3}{x+7} - 1 = \dfrac{2x-3}{x+7} - \dfrac{x+7}{x+7}$
$= \dfrac{2x-3-(x+7)}{x+7}$
$= \dfrac{2x-3-x-7}{x+7}$
$= \dfrac{x-10}{x+7}$

89. $\dfrac{3}{x+3} + x + 5 = \dfrac{3}{x+3} + \dfrac{(x+5)(x+3)}{x+3}$
$= \dfrac{3+x^2+3x+5x+15}{x+3}$
$= \dfrac{x^2+8x+18}{x+3}$

91. $2x+1 - \dfrac{3}{x-4} = \dfrac{(2x+1)(x-4)}{x-4} - \dfrac{3}{x-4}$
$= \dfrac{2x^2-8x+x-4-3}{x-4}$
$= \dfrac{2x^2-7x-7}{x-4}$

93. **a.** Width of smaller rectangle $= \dfrac{20}{L}$

   Width of larger rectangle $= \dfrac{45}{L}$

   $\dfrac{20}{L} + \dfrac{45}{L} = \dfrac{20+45}{L} = \dfrac{65}{L}$

   The total width is $\dfrac{65}{L}$ feet

   b. Width of smaller rectangle $= \dfrac{20}{L}$

   Width of larger rectangle $= \dfrac{45}{2L-5}$

   $\dfrac{20}{L} + \dfrac{45}{2L-5} = \dfrac{20(2L-5)}{L(2L-5)} + \dfrac{45L}{L(2L-5)}$
   $= \dfrac{40L-100+45L}{L(2L-5)}$
   $= \dfrac{85L-100}{L(2L-5)}$
   $= \dfrac{5(17L-20)}{L(2L-5)}$

   The total width is $\dfrac{5(17L-20)}{L(2L-5)}$ feet

Chapter 12: Rational Expressions, Functions, and Equations

95. $2\left(\dfrac{24}{x-1}\right)+2\left(\dfrac{72}{x+1}\right)=\dfrac{48}{x-1}+\dfrac{144}{x+1}$

$=\dfrac{48(x+1)}{(x+1)(x-1)}+\dfrac{144(x-1)}{(x+1)(x-1)}$

$=\dfrac{48x+48+144x-144}{(x+1)(x-1)}$

$=\dfrac{192x-96}{(x+1)(x-1)}$

$=\dfrac{96(2x-1)}{(x+1)(x-1)}$

The perimeter is $\dfrac{96(2x-1)}{(x+1)(x-1)}$ feet.

97. **a.** $t=\dfrac{d}{r}$

running time: $\dfrac{26}{x}$

bicycle time: $\dfrac{110}{x+20}$

total time: $T(x)=\dfrac{26}{x}+\dfrac{110}{x+20}$

b. Let $Y1=\dfrac{26}{x}+\dfrac{110}{x+20}$

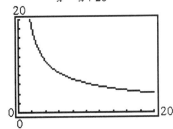

c. When $x=9$, $y\approx 6.7$.

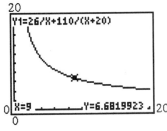

The time will be approximately 6.7 hours.

d. $x\approx 6.7$ when $y=8$

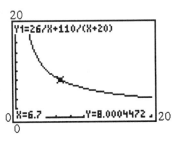

The average speed for running is approximately 6.7 mph and the average speed for bicycling is approximately $6.7+20=26.7$ mph.

99. $t=\dfrac{d}{r}$

Let $x=$ the distance traveled upriver.
$2x=$ total distance for the round-trip.

Time to travel upriver $=\dfrac{x}{25}$.

Time to travel downriver $=\dfrac{x}{35}$.

Total time of travel $=\dfrac{x}{25}+\dfrac{x}{35}=\dfrac{7x}{175}+\dfrac{5x}{175}=\dfrac{12x}{175}$.

Average speed $=$ total distance $\div$ total time

$=2x\div\dfrac{12x}{175}$

$=2x\cdot\dfrac{175}{12x}$

$=\dfrac{175}{6}$

≈ 29.2

The average speed is approximately 29.2 mph.

12.3 Calculator Exercise

Students should use the method described to check the equivalent expressions.

SSM: Experiencing Introductory and Intermediate Algebra

12.4 Exercises

1. $x^{-3} + 2x^{-2} - x^{-1} = 5$ is a rational equation because it can be written in the form $\frac{1}{x^3} + \frac{2}{x^2} - \frac{1}{x} - 5 = 0$.

3. $\frac{1}{x+7} + \frac{1}{x} = 5$ is a rational equation because it can be written in the form $\frac{1}{x+7} + \frac{1}{x} - 5 = 0$.

5. $v^{1/2} + 3v = 10$ is not a rational equation because a variable base raised to a fractional exponent cannot be written as a rational expression.

7. $\frac{\sqrt{2x+7}}{x} = \frac{1}{3-x}$ is not a rational equation because it has a variable in the radicand of a square root.

9. $\frac{1}{3x^2} + \frac{5}{6x} = \frac{8}{9}$ is a rational equation because it can be written in the form $\frac{1}{3x^2} + \frac{5}{6x} - \frac{8}{9} = 0$.

11. $\frac{\sqrt{3}}{x^2} + \frac{\sqrt{2}}{x} = 7$ is a rational equation because it can be written in the form $\frac{\sqrt{3}}{x^2} + \frac{\sqrt{2}}{x} - 7 = 0$. No variables are in the radicands of the square roots.

13. $\frac{1}{a} = \frac{5}{9} + \frac{2}{3}$

    $9a\left(\frac{1}{a}\right) = 9a\left(\frac{5}{9}\right) + 9a\left(\frac{2}{3}\right)$

    $9 = 5a + 6a$

    $9 = 11a$

    $\frac{9}{11} = a$

    The solution is $\frac{9}{11}$.

15. $\frac{11}{x} - \frac{3}{4} = \frac{5}{8}$

    $8x\left(\frac{11}{x}\right) - 8x\left(\frac{3}{4}\right) = 8x\left(\frac{5}{8}\right)$

    $88 - 6x = 5x$

    $88 = 11x$

    $8 = x$

    The solution is 8.

17. $1 - \frac{8}{x} = \frac{10}{x}$

    $x(1) - x\left(\frac{8}{x}\right) = x\left(\frac{10}{x}\right)$

    $x - 8 = 10$

    $x = 18$

    The solution is 18.

19. $\frac{1}{2} = \frac{5}{2x} - \frac{6}{x}$

    $2x\left(\frac{1}{2}\right) = 2x\left(\frac{5}{2x}\right) - 2x\left(\frac{6}{x}\right)$

    $x = 5 - 12$

    $x = -7$

    The solution is -7.

21. $2 = \frac{6}{x-2}$

    $(x-2)(2) = (x-2)\left(\frac{6}{x-2}\right)$

    $2x - 4 = 6$

    $2x = 10$

    $x = 5$

    The solution is 5.

23. $\frac{x-4}{2x+3} = \frac{5}{21}$

    $5(2x+3) = 21(x-4)$

    $10x + 15 = 21x - 84$

    $99 = 11x$

    $9 = x$

    The solution is 9.

25. $\frac{1}{x-6} = \frac{17}{2x+3}$

    $17(x-6) = 1(2x+3)$

    $17x - 102 = 2x + 3$

    $15x = 105$

    $x = 7$

    The solution is 7.

27. $\frac{8}{2x-3} + \frac{4}{x} = 0$

    $\frac{8}{2x-3} = \frac{-4}{x}$

$-4(2x-3) = 8x$
$-8x+12 = 8x$
$12 = 16x$
$x = \dfrac{12}{16} = \dfrac{3}{4}$

The solution is $\dfrac{3}{4}$.

29. $\dfrac{2x+3}{x+4} = \dfrac{5}{x+4}$

$(x+4)\left(\dfrac{2x+3}{x+4}\right) = (x+4)\left(\dfrac{5}{x+4}\right)$

$2x+3 = 5$
$2x = 2$
$x = 1$

The solution is 1.

31. $\dfrac{16}{x} = x$

$x^2 = 16$
$x = \pm\sqrt{16}$
$x = \pm 4$

The solutions are ± 4.

33. $\dfrac{z}{7} - \dfrac{7}{z} = 0$

$\dfrac{z}{7} = \dfrac{7}{z}$

$49 = z^2$
$z = \pm\sqrt{49}$
$z = \pm 7$

The solutions are ± 7.

35. $\dfrac{10}{k} + 3 = k$

$k\left(\dfrac{10}{k}\right) + k(3) = k(k)$

$10 + 3k = k^2$
$0 = k^2 - 3k - 10$
$0 = (k+2)(k-5)$
$k+2 = 0$ or $k-5 = 0$
$k = -2$ $k = 5$

The solutions are -2 and 5.

37. $\dfrac{x}{4} + \dfrac{4}{x} = 3 - \dfrac{4}{x}$

$4x\left(\dfrac{x}{4}\right) + 4x\left(\dfrac{4}{x}\right) = 4x(3) - 4x\left(\dfrac{4}{x}\right)$

$x^2 + 16 = 12x - 16$
$x^2 - 12x + 32 = 0$
$(x-4)(x-8) = 0$
$x-4 = 0$ or $x-8 = 0$
$x = 4$ $x = 8$

The solutions are 4 and 8.

39. $\dfrac{y}{5} - \dfrac{21}{5y} = 1 + \dfrac{3}{y}$

$5y\left(\dfrac{y}{5}\right) - 5y\left(\dfrac{21}{5y}\right) = 5y(1) + 5y\left(\dfrac{3}{y}\right)$

$y^2 - 21 = 5y + 15$
$y^2 - 5y - 36 = 0$
$(y+4)(y-9) = 0$
$y+4 = 0$ or $y-9 = 0$
$y = -4$ $y = 9$

The solutions are -4 and 9.

41. $\dfrac{x+7}{x+9} = \dfrac{x}{x+1}$

$x(x+9) = (x+7)(x+1)$
$x^2 + 9x = x^2 + x + 7x + 7$
$9x = 8x + 7$
$x = 7$

The solution is 7.

43. $\dfrac{x-6}{x+5} = \dfrac{x-4}{3x+2}$

$(x-4)(x+5) = (x-6)(3x+2)$
$x^2 + 5x - 4x - 20 = 3x^2 + 2x - 18x - 12$
$0 = 2x^2 - 17x + 8$
$0 = (x-8)(2x-1)$
$x-8 = 0$ or $2x-1 = 0$
$x = 8$ $x = \dfrac{1}{2}$

The solutions are $\dfrac{1}{2}$ and 8.

45. $x^{-1} + \dfrac{2}{3} = \dfrac{3}{5}$

$\dfrac{1}{x} + \dfrac{2}{3} = \dfrac{3}{5}$

$15x\left(\dfrac{1}{x}\right) + 15x\left(\dfrac{2}{3}\right) = 15x\left(\dfrac{3}{5}\right)$

$15 + 10x = 9x$

$x = -15$

The solution is -15.

47. $x^{-2} - 6 = 3$

$\dfrac{1}{x^2} - 6 = 3$

$x^2\left(\dfrac{1}{x^2}\right) - x^2(6) = x^2(3)$

$1 - 6x^2 = 3x^2$

$1 = 9x^2$

$\dfrac{1}{9} = x^2$

$x = \pm\sqrt{\dfrac{1}{9}}$

$x = \pm\dfrac{1}{3}$

The solutions are $\pm\dfrac{1}{3}$.

49. $1 - 4x^{-1} = 60x^{-2}$

$1 - \dfrac{4}{x} = \dfrac{60}{x^2}$

$x^2(1) - x^2\left(\dfrac{4}{x}\right) = x^2\left(\dfrac{60}{x^2}\right)$

$x^2 - 4x = 60$

$x^2 - 4x - 60 = 0$

$(x+6)(x-10) = 0$

$x + 6 = 0$ or $x - 10 = 0$

$x = -6 \qquad x = 10$

The solutions are -6 and 10.

51. $\dfrac{9}{x-7} + 3 = \dfrac{2x-5}{x-7}$

$(x-7)\left(\dfrac{9}{x-7}\right) + (x-7)(3) = (x-7)\left(\dfrac{2x-5}{x-7}\right)$

$9 + 3x - 21 = 2x - 5$

$3x - 12 = 2x - 5$

$x = 7$

Now 7 is a restricted value (extraneous), so the equation has no solution.

53. $\dfrac{x+2}{3} = \dfrac{x^2 - x + 7}{3x - 6}$

$3(x-2)\left(\dfrac{x+2}{3}\right) = 3(x-2)\left[\dfrac{x^2 - x + 7}{3(x-2)}\right]$

$(x-2)(x+2) = x^2 - x + 7$

$x^2 - 4 = x^2 - x + 7$

$-4 = -x + 7$

$x = 11$

The solution is 11.

55. $\dfrac{x-3}{x+4} = \dfrac{14}{x^2 + 6x + 8}$

$(x+2)(x+4)\left(\dfrac{x-3}{x+4}\right) = (x+2)(x+4)\left[\dfrac{14}{(x+2)(x+4)}\right]$

$(x+2)(x-3) = 14$

$x^2 - 3x + 2x - 6 = 14$

$x^2 - x - 20 = 0$

$(x+4)(x-5) = 0$

$x + 4 = 0$ or $x - 5 = 0$

$x = -4 \qquad x = 5$

Now -4 is a restricted value (extraneous), so the solution is 5.

Chapter 12: Rational Expressions, Functions, and Equations

57. $\dfrac{25}{p^2+p-12} = \dfrac{p+4}{p-3}$

$(p-3)(p+4)\left[\dfrac{25}{(p-3)(p+4)}\right]$

$= (p-3)(p+4)\left(\dfrac{p+4}{p-3}\right)$

$25 = (p+4)(p+4)$
$25 = p^2 + 4p + 4p + 16$
$0 = p^2 + 8p - 9$
$0 = (p-1)(p+9)$
$p-1=0$ or $p+9=0$
$p=1 \qquad p=-9$

The solutions are –9 and 1.

59. $\dfrac{m-2}{m+1} = \dfrac{7m+22}{m^2+6m+5}$

$(m+1)(m+5)\left(\dfrac{m-2}{m+1}\right)$

$= (m+1)(m+5)\left[\dfrac{7m+22}{(m+1)(m+5)}\right]$

$(m+5)(m-2) = 7m+22$
$m^2 - 2m + 5m - 10 = 7m + 22$
$m^2 - 4m - 32 = 0$
$(m+4)(m-8) = 0$
$m+4=0$ or $m-8=0$
$m=-4 \qquad m=8$

The solutions are –4 and 8.

61. $\dfrac{1-4x}{1-x} + 2 = \dfrac{6x-3}{x-1}$

$\dfrac{4x-1}{x-1} + 2 = \dfrac{6x-3}{x-1}$

$(x-1)\left(\dfrac{4x-1}{x-1}\right) + (x-1)(2) = (x-1)\left(\dfrac{6x-3}{x-1}\right)$

$4x - 1 + 2x - 2 = 6x - 3$
$6x - 3 = 6x - 3$
$-3 = -3$

The equation is true for all values of x except the restricted value(s). The solution is the set of all real numbers not equal to 1.

63. $2 + \dfrac{z+2}{2z+1} = \dfrac{3z+6}{2z+1} + 1$

$(2z+1)(2) + (2z+1)\left(\dfrac{z+2}{2z+1}\right)$

$= (2z+1)\left(\dfrac{3z+6}{2z+1}\right) + (2z+1)(1)$

$4z + 2 + z + 2 = 3z + 6 + 2z + 1$
$5z + 4 = 5z + 7$
$4 = 7$

The equation is always false (a contradiction). It has no solution.

65. $\dfrac{1}{x+5} + \dfrac{21}{x^2-25} = 2$

$(x+5)(x-5)\left[\dfrac{1}{x+5} + \dfrac{21}{(x+5)(x-5)}\right]$

$= (x+5)(x-5)(2)$

$x - 5 + 21 = (x^2 - 25)(2)$
$x + 16 = 2x^2 - 50$
$0 = 2x^2 - x = 66$
$0 = (x-6)(2x+11)$
$(x-6)(2x+11) = 0$
$x-6=0$ or $2x+11=0$
$x=6 \qquad x=-\dfrac{11}{2}$

The solutions are $-\dfrac{11}{2}$ and 6.

67. $\dfrac{x+5}{x-3} = \dfrac{1+x}{2-x}$

$(1+x)(x-3) = (x+5)(2-x)$
$x - 3 + x^2 - 3x = 2x - x^2 + 10 - 5x$
$2x^2 + x - 13 = 0$

$x = \dfrac{-1 \pm \sqrt{1^2 - 4(2)(-13)}}{2(2)}$

$x = \dfrac{-1 \pm \sqrt{105}}{4}$

$x \approx 2.3117$ or $x \approx -2.8117$

The solutions are $\dfrac{-1 \pm \sqrt{105}}{4}$, or approximately –2.8117 and 2.3117.

SSM: Experiencing Introductory and Intermediate Algebra

69. $\dfrac{1}{x+3} + \dfrac{1}{x} = 2$

$x(x+3)\left(\dfrac{1}{x+3}\right) + x(x+3)\left(\dfrac{1}{x}\right) = x(x+3)(2)$

$x + x + 3 = 2x^2 + 6x$

$0 = 2x^2 + 4x - 3$

$x = \dfrac{-4 \pm \sqrt{4^2 - 4(2)(-3)}}{2(2)}$

$x = \dfrac{-4 \pm \sqrt{40}}{4}$

$x = \dfrac{-4 \pm 2\sqrt{10}}{4}$

$x = \dfrac{-2 \pm \sqrt{10}}{2}$

$x \approx 0.5811$ or $x \approx -2.5811$

The solutions are $\dfrac{-2 \pm \sqrt{10}}{2}$, or approximately -2.5811 and 0.5811.

71. $2w^{-3} + 2w^{-2} - w^{-1} = 1$

$\dfrac{2}{w^3} + \dfrac{2}{w^2} - \dfrac{1}{w} = 1$

$w^3\left(\dfrac{2}{w^3}\right) + w^3\left(\dfrac{2}{w^2}\right) - w^3\left(\dfrac{1}{w}\right) = w^3(1)$

$2 + 2w - w^2 = w^3$

$0 = w^3 + w^2 - 2w - 2$

$0 = w^2(w+1) - 2(w+1)$

$0 = (w+1)(w^2 - 2)$

$w = -1$ or $w^2 = 2$

$\quad\quad\quad\quad\quad w = \pm\sqrt{2}$

The solutions are -1 and $\pm\sqrt{2}$.

73. $\dfrac{1}{3u^2} + \dfrac{5}{6u} = \dfrac{8}{9}$

$18u^2\left(\dfrac{1}{3u^2}\right) + 18u^2\left(\dfrac{5}{6u}\right) = 18u^2\left(\dfrac{8}{9}\right)$

$6 + 15u = 16u^2$

$0 = 16u^2 - 15u - 6$

$u = \dfrac{-(-15) \pm \sqrt{(-15)^2 - 4(16)(-6)}}{2(16)}$

$u = \dfrac{15 \pm \sqrt{609}}{32}$

$u \approx 1.2399$ or $u \approx -0.3024$

The solutions are $\dfrac{15 \pm \sqrt{609}}{32}$, or approximately -0.3024 and 1.2399.

75. $\dfrac{3}{x-5} + \dfrac{3}{x^2-25} = \dfrac{3x+1}{x^2-25}$

$(x+5)(x-5)\left[\dfrac{3}{x-5} + \dfrac{3}{(x+5)(x-5)}\right]$

$\quad = (x+5)(x-5)\left[\dfrac{3x+1}{(x+5)(x-5)}\right]$

$3(x+5) + 3 = 3x + 1$

$3x + 15 + 3 = 3x + 1$

$18 = 1$

The equation is always false (a contradiction). It has no solution.

77. $\dfrac{3x+7}{x-1} = \dfrac{3x-3}{x-1}$

$(x-1)\left(\dfrac{3x+7}{x-1}\right) = (x-1)\left(\dfrac{3x-3}{x-1}\right)$

$3x + 7 = 3x - 3$

$7 = -3$

The equation is always false (a contradiction). It has no solution.

79. $\dfrac{b+4}{b-2} = \dfrac{b^2+3b-4}{b^2-3b+2}$

$(b-1)(b-2)\left(\dfrac{b+4}{b-2}\right) = (b-1)(b-2)\left[\dfrac{b^2+3b-4}{(b-1)(b-2)}\right]$

$(b-1)(b+4) = b^2 + 3b - 4$

$b^2 + 4b - b - 4 = b^2 + 3b - 4$

$-4 = -4$

The equation is true for all values of x except the restricted value(s). The solution is the set of all real numbers except 1 and 2.

Chapter 12: Rational Expressions, Functions, and Equations

81. $\dfrac{a-3}{a-1} + \dfrac{3}{a+1} = \dfrac{(a+3)(a-2)}{a^2-1}$

$(a+1)(a-1)\left(\dfrac{a-3}{a-1} + \dfrac{3}{a+1}\right)$

$\qquad = (a+1)(a-1)\left[\dfrac{(a+3)(a-2)}{(a+1)(a-1)}\right]$

$(a+1)(a-3) + 3(a-1) = (a+3)(a-2)$

$a^2 - 3a + a - 3 + 3a - 3 = a^2 - 2a + 3a - 6$

$a^2 + a - 6 = a^2 + a - 6$

$-6 = -6$

The equation is true for all values of x except the restricted value(s). The solution is the set of all real numbers except -1 and 1.

83. $\dfrac{x}{2} - 1 + \dfrac{4}{x} = \dfrac{3}{2x} - \dfrac{1}{2}$

$2x\left(\dfrac{x}{2} - 1 + \dfrac{4}{x}\right) = 2x\left(\dfrac{3}{2x} - \dfrac{1}{2}\right)$

$x^2 - 2x + 8 = 3 - x$

$x^2 - x + 5 = 0$

$x = \dfrac{-(-1) \pm \sqrt{(-1)^2 - 4(1)(5)}}{2(1)}$

$x = \dfrac{1 \pm \sqrt{-19}}{2}$

The equation has no real-number solution.

85. Let $x =$ the interest rate on the second account.
$x + 0.02 =$ the interest rate on the first account.

$\dfrac{286}{(x+0.02)2} + \dfrac{117}{x(2)} = 3500$

$2x(x+0.02)\left[\dfrac{286}{(x+0.02)2} + \dfrac{117}{x(2)}\right]$

$\qquad = 2x(x+0.02)(3500)$

$286x + 117(x+0.02) = 7000x(x+0.02)$

$286x + 117x + 2.34 = 7000x^2 + 140x$

$0 = 7000x^2 - 263x - 2.34$

$x = \dfrac{-(-263) \pm \sqrt{(-263)^2 - 4(7000)(-2.34)}}{2(7000)}$

$x = \dfrac{263 \pm \sqrt{134{,}689}}{14{,}000}$

$x = 0.045 \qquad$ or $\qquad x \approx -0.0074$

$x + 0.02 = 0.065$

Disregarding -0.0074, Marianne received 6.5% interest on the first account and 4.5% interest on the second account.

87. $\dfrac{1}{R} = \dfrac{1}{r_1} + \dfrac{1}{r_2}$

$\dfrac{1}{11.5} = \dfrac{1}{20.5} + \dfrac{1}{r_2}$

$235.75 r_2\left(\dfrac{1}{11.5}\right) = 235.75 r_2\left(\dfrac{1}{20.5} + \dfrac{1}{r_2}\right)$

$20.5 r_2 = 11.5 r_2 + 235.75$

$9 r_2 = 235.75$

$r_2 \approx 26.2$

The resistance of the second blood vessel is about 26.2 dynes.

89. $\dfrac{360}{x} = 6$

$6x = 360$

$x = 60$

The focal length needed is 60 millimeters.

12.4 Calculator Exercises

1. The solution is 7.

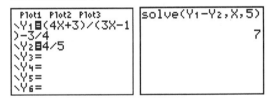

2. The solutions are 10 and $0.5 = \dfrac{1}{2}$.

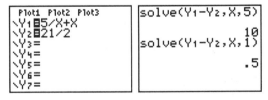

3. The solution is 2.

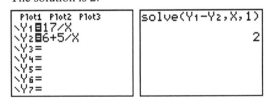

531

4. The solution is $0.8 = \frac{4}{5}$.

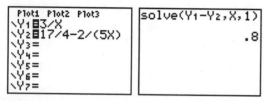

5. The solution is $1.\overline{6} = \frac{5}{3}$.

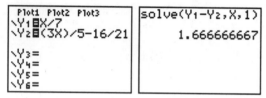

6. The solutions are -4 and -1.

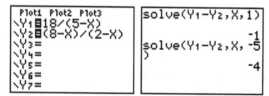

12.5 Exercises

1. Let x = the time to write the report together.

   $\frac{1}{3}$ is Miriam's rate.

   $\frac{1}{4.2}$ is Saul's rate.

   $\frac{1}{x}$ is their combined rate.

   $\frac{1}{3} + \frac{1}{4.2} = \frac{1}{x}$

   $12.6x\left(\frac{1}{3}\right) + 12.6x\left(\frac{1}{4.2}\right) = 12.6x\left(\frac{1}{x}\right)$

   $4.2x + 3x = 12.6$

   $7.2x = 12.6$

   $x = 1.75$

   It will take them 1.75 hours to write the report if they work together.

3. Let x = Simone's time to clean the house alone.
   $2x$ = Jacque's time to clean the house alone.

   $\frac{1}{x}$ is Simone's rate.

   $\frac{1}{2x}$ is Jacque's rate.

   $\frac{1}{3\frac{1}{3}} = \frac{1}{\frac{10}{3}} = \frac{3}{10}$ is their combined rate.

   $\frac{1}{x} + \frac{1}{2x} = \frac{3}{10}$

   $10x\left(\frac{1}{x}\right) + 10x\left(\frac{1}{2x}\right) = 10x\left(\frac{3}{10}\right)$

   $10 + 5 = 3x$

   $15 = 3x$

   $5 = x$

   $2x = 10$

   It will take Jacque 10 hours to clean the house if he works alone; it will take Simone 5 hours if she works alone.

5. Let x = the time to do the job together.

   $\frac{1}{9}$ is Felix's rate.

   $\frac{1}{6}$ is Oscar's rate.

   $\frac{1}{x}$ is their combined rate.

   $\frac{1}{9} + \frac{1}{6} = \frac{1}{x}$

   $18x\left(\frac{1}{9}\right) + 18x\left(\frac{1}{6}\right) = 18x\left(\frac{1}{x}\right)$

   $2x + 3x = 18$

   $5x = 18$

   $x = 3.6$

   It will take them 3.6 hours to do the job if they work together.

7. Let x = the time to fill the tank if both pipelines are used together.

   $\frac{1}{4}$ is the red pipe's rate.

   $\frac{1}{7}$ is the blue pipe's rate.

   $\frac{1}{x}$ is the combined rate.

   $\frac{1}{4} + \frac{1}{7} = \frac{1}{x}$

   $28x\left(\frac{1}{4}\right) + 28x\left(\frac{1}{7}\right) = 28x\left(\frac{1}{x}\right)$

$7x + 4x = 28$

$11x = 28$

$x = \dfrac{28}{11} \approx 2.55$

It will take approximately 2.55 hours to fill the tank if both pipes are used together.

9. Let x = the time for line B to fill the tank alone.

   $\dfrac{2}{3}x$ = the time for line A to fill the tank alone.

   $\dfrac{1}{x}$ is line B's rate.

   $\dfrac{1}{\frac{2}{3}x} = \dfrac{3}{2x}$ is line A's rate.

   $\dfrac{1}{4\frac{4}{5}} = \dfrac{1}{\frac{24}{5}} = \dfrac{5}{24}$ is the combined rate.

   $\dfrac{1}{x} + \dfrac{3}{2x} = \dfrac{5}{24}$

   $24x\left(\dfrac{1}{x}\right) + 24x\left(\dfrac{3}{2x}\right) = 24x\left(\dfrac{5}{24}\right)$

   $24 + 3x = 5x$

   $24 = 2x$

   $12 = x$

   $\dfrac{2}{3}x = 8$

   It will take line B 12 hours to fill the tank working alone; it will take line A 8 hours working alone.

11. Let x = the number of additional points.

    $\dfrac{45 + x}{60 + x} = \dfrac{80}{100}$

    $80(60 + x) = 100(45 + x)$

    $4800 + 80x = 4500 + 100x$

    $300 = 20x$

    $15 = x$

    Ricki would need an additional 15 points.

13. Let x = additional money to be raised and spent.

    $\dfrac{x + 120}{x + 300} = \dfrac{45}{100}$

    $45(x + 300) = 100(x + 120)$

    $45x + 13500 = 100x + 12000$

    $1500 = 55x$

    $x = \dfrac{1500}{55} = \dfrac{300}{11} \approx 27.27$

    The club should raise and spend an additional $27.27.

15. $\dfrac{PQ}{XY} = \dfrac{QR}{YZ}$

    $\dfrac{24}{x} = \dfrac{12}{2}$

    $12x = 24(2)$

    $12x = 48$

    $x = 4$

    $\overline{XY}$ measures 4 inches.

    $\dfrac{PR}{XZ} = \dfrac{QR}{YZ}$

    $\dfrac{x}{3} = \dfrac{12}{2}$

    $3(12) = 2x$

    $36 = 2x$

    $18 = x$

    $\overline{PR}$ measures 18 inches.

17. $\dfrac{PQ}{XY} = \dfrac{PR}{XZ}$

    $\dfrac{x}{4.03} = \dfrac{5.875}{2.35}$

    $4.03(5.875) = 2.35x$

    $23.67625 = 2.35x$

    $10.075 = x$

    $\overline{PQ}$ measures 10.075 cm.

    $\dfrac{QR}{YZ} = \dfrac{PR}{XZ}$

    $\dfrac{4.4}{x} = \dfrac{5.875}{2.35}$

    $5.875x = 4.4(2.35)$

    $5.875x = 10.34$

    $x = 1.76$

    $\overline{YZ}$ measures 1.76 cm.

19. $\dfrac{QR}{YZ} = \dfrac{PQ}{XY}$

    $\dfrac{x}{1\frac{3}{4}} = \dfrac{10\frac{5}{16}}{4\frac{1}{8}}$

    $\left(10\frac{5}{16}\right)\left(1\frac{3}{4}\right) = \left(4\frac{1}{8}\right)x$

    $18\frac{23}{24} = \left(4\frac{1}{8}\right)x$

    $4\frac{3}{8} = x$

$\overline{QR}$ measures $4\frac{3}{8}$ feet.

$\dfrac{PR}{XZ} = \dfrac{PQ}{XY}$

$\dfrac{5\frac{5}{8}}{x} = \dfrac{10\frac{5}{16}}{4\frac{1}{8}}$

$\left(10\frac{5}{16}\right)x = \left(5\frac{5}{8}\right)\left(4\frac{1}{8}\right)$

$\left(10\frac{5}{16}\right)x = 23\frac{13}{64}$

$x = 2\frac{1}{4}$

$\overline{XZ}$ measures $2\frac{1}{4}$ feet.

21. Let x = the height of the cliff.

$\dfrac{x}{35} = \dfrac{5.5}{4}$

$35(5.5) = 4x$

$192.5 = 4x$

$48.125 = x$

The estimated height of the cliff is 48.125 feet.

23. $F = \dfrac{kq_1q_2}{D^2}$

$FD^2 = kq_1q_2$

$D^2 = \dfrac{kq_1q_2}{F}$

$D = \pm\sqrt{\dfrac{kq_1q_2}{F}}$

Distance cannot be negative, so $D = \sqrt{\dfrac{kq_1q_2}{F}}$.

25. Let r = the resistance of one resistor.
$2r - 5$ = the resistance of the other resistor.

$\dfrac{1}{R} = \dfrac{1}{r_1} + \dfrac{1}{r_2}$

$\dfrac{1}{6} = \dfrac{1}{r} + \dfrac{1}{2r-5}$

$6r(2r-5)\left(\dfrac{1}{6}\right) = 6r(2r-5)\left(\dfrac{1}{r} + \dfrac{1}{2r-5}\right)$

$r(2r-5) = 6(2r-5) + 6r$

$2r^2 - 5r = 12r - 30 + 6r$

$2r^2 - 23r + 30 = 0$

$(2r-3)(r-10) = 0$

$2r - 3 = 0$ or $r - 10 = 0$

$r = \dfrac{3}{2}$ \qquad $r = 10$

$2r - 5 = -2$ \qquad $2r - 5 = 15$

Resistance cannot be negative, so the resistances of the two resistors are 10 ohms and 15 ohms.

12.5 Calculator Exercises

1. Let $Y1 = \dfrac{2x+1}{x}$, $Y2 = 1 + \dfrac{12}{x^2}$, and $Y3 = Y1 - Y2$.
The graph of Y3 is shown below.

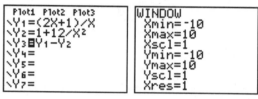

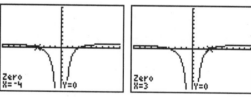

The solutions are –4 and 3.

2. Let $Y1 = \dfrac{x+1}{x+3}$, $Y2 = \dfrac{8}{2x}$, and $Y3 = Y1 - Y2$. The graph of Y3 is shown below.

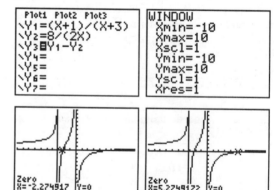

The solutions are approximately –2.27 and 5.27.

Chapter 12: Rational Expressions, Functions, and Equations

3. Let $Y1 = \dfrac{1}{x} + \dfrac{1}{x+2}$, $Y2 = \dfrac{5}{12}$, and $Y3 = Y1 - Y2$.
   The graph of Y3 is shown below.

   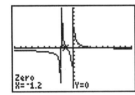 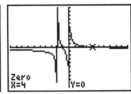

   The solutions are $-1.2 = -\dfrac{6}{5}$ and 4.

Chapter 12 Section-By-Section Review

1. $3 + \dfrac{x-5}{x^2+6x+9}$ is a rational expression because it can be written as $\dfrac{3}{1} + \dfrac{x-5}{x^2+6x+9}$.

2. $x^2 - \dfrac{\sqrt{x}+3x}{x+1}$ is a not rational expression because $\sqrt{x}+3x$ is not a polynomial expression.

3. $5x^2 - 3x + 2 - 4x^{-1}$ is a rational expression because is can be written as $\dfrac{5x^2}{1} - \dfrac{3x}{1} + \dfrac{2}{1} - \dfrac{4}{x}$.

4. $2x+5$ is a rational expression because is can be written as $\dfrac{2x}{1} + \dfrac{5}{1}$.

5. $y = \dfrac{-3}{x^2}$
   $x^2 = 0$
   $x = 0$
   Restricted value: 0
   Domain: $x \neq 0$
   $(-\infty, 0) \cup (0, \infty)$

6. $y = \dfrac{2x+7}{x^2+5x-24}$
   $x^2 + 5x - 24 = 0$
   $(x-3)(x+8) = 0$
   $x = 3$ or $x = -8$
   Restricted values: -8 and 3
   Domain: $x \neq -8, x \neq 3$
   $(-\infty, -8) \cup (-8, 3) \cup (3, \infty)$

7. $y = \dfrac{5}{4x^2+9}$
   $4x^2 + 9 = 0$
   $4x^2 = -9$
   $x^2 = -\dfrac{9}{4}$
   $x = \pm\sqrt{-\dfrac{9}{4}}$ (Not Real)
   Restricted values: None
   Domain: All real numbers
   $(-\infty, \infty)$

8. $y = \dfrac{5}{4x^2-9}$
   $4x^2 - 9 = 0$
   $4x^2 = 9$
   $x^2 = \dfrac{9}{4}$
   $x = \pm\sqrt{\dfrac{9}{4}}$
   $x = \pm\dfrac{3}{2}$
   Restricted values: $\pm\dfrac{3}{2}$
   Domain: $x \neq -\dfrac{3}{2}, x \neq \dfrac{3}{2}$
   $\left(-\infty, -\dfrac{3}{2}\right) \cup \left(-\dfrac{3}{2}, \pm\dfrac{3}{2}\right) \cup \left(\dfrac{3}{2}, \infty\right)$

9. $y = \dfrac{180}{x^2} + \dfrac{30}{x}$
   The restricted value is 0. Set up a table of x-values being sure to use both x-values that are

less than 0 and x-values that are greater than 0. Connect the points with a smooth curve. Do not connect across the vertical line $x = 0$.

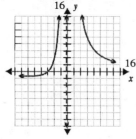

10. $g(x) = \dfrac{x^2 - x - 2}{x - 2}$

The restricted value is 2. Set up a table of x-values being sure to use both x-values that are less than 2 and x-values that are greater than 2. Connect the points with a smooth curve. Do not connect across the vertical line $x = 2$.

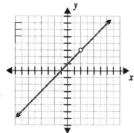

11. $p(x) = 1 - \dfrac{4}{x} - \dfrac{21}{x^2}$

The restricted value is 0. Set up a table of x-values being sure to use both x-values that are less than 0 and x-values that are greater than 0. Connect the points with a smooth curve. Do not connect across the vertical line $x = 0$.

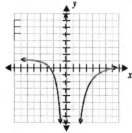

12. $y = \dfrac{x^2 - 3x - 10}{x^2 - x - 12} = \dfrac{x^2 - 3x - 10}{(x + 3)(x - 4)}$

The restricted values are –3 and 4. Set up a table

of x-values being sure to use x-values that are less than –3, x-values that are between –3 and 4, and x-values that are greater than 4. Connect the points with a smooth curve. Do not connect across the vertical lines $x = -3$ and $x = 4$.

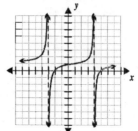

13. a. $C_{ave}(x) = \dfrac{1800 + 100x + 25x^2}{x}$

b. The minimum average cost per year would be approximately $524.27, when the number of years is $x \approx 8.5$.

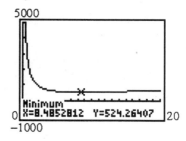

c. The customer should replace the cash register after about 8.5 years of use.

14. $\dfrac{-56x^3 yz}{126 x^2 y^3 z} = \dfrac{(14x^2 yz)(-4x)}{(14x^2 yz)(9y^2)} = -\dfrac{4x}{9y^2}$

15. $\dfrac{5p - 15}{15p + 25} = \dfrac{5(p - 3)}{5(3p + 5)} = \dfrac{p - 3}{3p + 5}$

16. $\dfrac{15x - 3x^2}{6x^3 - 30x^2} = \dfrac{-3x^2 + 15x}{6x^3 - 30x^2} = \dfrac{-3x(x - 5)}{6x^2(x - 5)} = -\dfrac{1}{2x}$

17. $\dfrac{2x^2 + x - 15}{4x^2 + 13x + 3} = \dfrac{(2x - 5)(x + 3)}{(4x + 1)(x + 3)} = \dfrac{2x - 5}{4x + 1}$

18. $\dfrac{b^2-49}{b^3+7b^2+2b+14} = \dfrac{(b+7)(b-7)}{b^2(b+7)+2(b+7)}$
$= \dfrac{(b+7)(b-7)}{(b+7)(b^2+2)}$
$= \dfrac{b-7}{b^2+2}$

19. $\dfrac{9x^3}{35y} \cdot \dfrac{-5y^2}{6x^5} = \dfrac{-45x^3y^2}{210x^5y}$
$= \dfrac{(15x^3y)(-3y)}{(15x^3y)(14x^2)}$
$= -\dfrac{3y}{14x^2}$

20. $\dfrac{a+5}{a-4} \cdot (a+1) = \dfrac{(a+5)(a+1)}{a-4} = \dfrac{a^2+6a+5}{a-4}$

21. $\dfrac{4m-4}{m^2-25} \cdot \dfrac{m-5}{2m^2+5m-7}$
$= \dfrac{(4m-4)(m-5)}{(m^2-25)(2m^2+5m-7)}$
$= \dfrac{4(m-1)(m-5)}{(m+5)(m-5)(2m+7)(m-1)}$
$= \dfrac{4}{(m+5)(2m+7)}$

22. $\dfrac{a-5}{a+4} \cdot \dfrac{a+4}{5-a} = \dfrac{(a-5)(a+4)}{(a+4)(5-a)}$
$= \dfrac{(a-5)(a+4)}{(a+4)(-1)(a-5)}$
$= \dfrac{1}{-1}$
$= -1$

23. $\dfrac{7x^2y}{5y-3x} \cdot \dfrac{3x-5y}{42xy^3} = \dfrac{7x^2y(3x-5y)}{42xy^3(5y-3x)}$
$= \dfrac{7x^2y(3x-5y)}{-42xy^3(3x-5y)}$
$= -\dfrac{x}{6y^2}$

24. $\dfrac{2m+6}{9m+45} \cdot \dfrac{-3m-15}{10m+30} = \dfrac{(2m+6)(-3m-15)}{(9m+45)(10m+30)}$
$= \dfrac{2(m+3)(-3)(m+5)}{9(m+5)(10)(m+3)}$
$= -\dfrac{1}{15}$

25. $\dfrac{14z^5}{25x^2y^3} \div \dfrac{-7z^3}{15xy^4} = \dfrac{14z^5}{25x^2y^3} \cdot \dfrac{15xy^4}{-7z^3}$
$= \dfrac{210xy^4z^5}{-175x^2y^3z^3}$
$= \dfrac{(35xy^3z^3)(6yz^2)}{(35xy^3z^3)(-5x)}$
$= -\dfrac{6yz^2}{5x}$

26. $\dfrac{27a^2b}{8cd} \div \dfrac{9c^2d}{16ab^2} = \dfrac{27a^2b}{8cd} \cdot \dfrac{16ab^2}{9c^2d}$
$= \dfrac{432a^3b^3}{72c^3d^2}$
$= \dfrac{72(6a^3b^3)}{72c^3d^2}$
$= \dfrac{6a^3b^3}{c^3d^2}$

27. $\dfrac{x^2+4x+3}{5x^2+5} \div \dfrac{x^2-3x+2}{x^4-1}$
$= \dfrac{x^2+4x+3}{5x^2+5} \cdot \dfrac{x^4-1}{x^2-3x+2}$
$= \dfrac{(x^2+4x+3)(x^4-1)}{(5x^2+5)(x^2-3x+2)}$
$= \dfrac{(x+1)(x+3)(x^2+1)(x+1)(x-1)}{5(x^2+1)(x-1)(x-2)}$
$= \dfrac{(x+1)(x+3)(x+1)}{5(x-2)}$
$= \dfrac{(x+1)^2(x+3)}{5(x-2)}$

28. $\dfrac{3x^2y}{z^3} \div (2xyz) = \dfrac{3x^2y}{z^3} \cdot \dfrac{1}{2xyz}$

$= \dfrac{3x^2y}{2xyz^4}$

$= \dfrac{(xy)(3x)}{(xy)(2z^4)}$

$= \dfrac{3x}{2z^4}$

29. $(21a^2bc) \div \dfrac{3ab}{2c} = \dfrac{21a^2bc}{1} \cdot \dfrac{2c}{3ab}$

$= \dfrac{42a^2bc^2}{3ab}$

$= \dfrac{(3ab)(14ac^2)}{3ab}$

$= 14ac^2$

30. $\dfrac{x^2-4}{x^2+4} \div (x-2) = \dfrac{x^2-4}{x^2+4} \cdot \dfrac{1}{x-2}$

$= \dfrac{(x+2)(x-2)}{(x^2+4)(x-2)}$

$= \dfrac{x+2}{x^2+4}$

31. $(2x^2-x-3) \div \dfrac{2x^2-7x+6}{2x^2-x-6}$

$= \dfrac{2x^2-x-3}{1} \cdot \dfrac{2x^2-x-6}{2x^2-7x+6}$

$= \dfrac{(2x-3)(x+1)(2x+3)(x-2)}{(2x-3)(x-2)}$

$= (x+1)(2x+3)$

32. $\dfrac{a+3}{a+6} \div \dfrac{a+2}{a+4} = \dfrac{a+3}{a+6} \cdot \dfrac{a+4}{a+2} = \dfrac{(a+3)(a+4)}{(a+6)(a+2)}$

33. $\dfrac{x-11}{x-3} \div \dfrac{11-x}{3-x} = \dfrac{x-11}{x-3} \cdot \dfrac{3-x}{11-x}$

$= \dfrac{(x-11)(-1)(x-3)}{(x-3)(-1)(x-11)}$

$= 1$

34. $r = \dfrac{d}{t} = \dfrac{220}{t}$

The expression for the Mario's average speed is

$\dfrac{220}{t}$ mph.

$d = rt = \left(\dfrac{220}{t} \cdot \dfrac{1}{2}\right)(t+2) = \dfrac{110}{t}(t+2) = \dfrac{110(t+2)}{t}$

The expression for Mario's new distance is $\dfrac{110(t+2)}{t}$ miles.

35. $V = LWH$; $H = \dfrac{V}{LH}$

If $V = 4x^3 + 12x^2$ and $H = L = 2x$, then

$H = \dfrac{4x^3 + 12x^2}{2x(2x)}$

$= \dfrac{4x^2(x+3)}{4x^2}$

$= x+3$

The expression for the height is $(x+3)$ inches.

36. a. $C(x) = 500 + 40x$

$R(x) = 80x$

b. $\dfrac{500+40x}{80x} = \dfrac{20(25+2x)}{20(4x)} = \dfrac{25+2x}{4x}$

The cost-to-revenue ratio is $\dfrac{25+2x}{4x}$.

37. $\dfrac{3y}{5x} + \dfrac{7y^2}{5x} = \dfrac{3y+7y^2}{5x} = \dfrac{y(3+7y)}{5x}$

38. $\dfrac{5b^3}{7a} - \dfrac{2b}{7a} = \dfrac{5b^3-2b}{7a} = \dfrac{b(5b^2-2)}{7a}$

39. $\dfrac{x-8}{2x-5} + \dfrac{3x-2}{2x-5} = \dfrac{x-8+3x-2}{2x-5}$

$= \dfrac{4x-10}{2x-5}$

$= \dfrac{2(2x-5)}{2x-5}$

$= 2$

40. $\dfrac{10x^2+3x+5}{4x^2-9} - \dfrac{2-7x}{4x^2-9} = \dfrac{10x^2+3x+5-2+7x}{4x^2-9}$

$= \dfrac{10x^2+10x+3}{(2x-3)(2x+3)}$

41. $\dfrac{4}{5x^3y} + \dfrac{7}{15x^2y^2} = \dfrac{4(3y)}{5x^3y(3y)} + \dfrac{7(x)}{15x^2y^2(x)}$

$= \dfrac{12y + 7x}{15x^3y^2}$

42. $\dfrac{5x}{x-8} + \dfrac{2x}{x+4} = \dfrac{5x(x+4)}{(x-8)(x+4)} + \dfrac{2x(x-8)}{(x-8)(x+4)}$

$= \dfrac{5x^2 + 20x + 2x^2 - 16x}{(x-8)(x+4)}$

$= \dfrac{7x^2 + 4x}{(x-8)(x+4)}$

$= \dfrac{x(7x+4)}{(x-8)(x+4)}$

43. $\dfrac{8}{2x-1} - \dfrac{4}{x+5} = \dfrac{8(x+5)}{(2x-1)(x+5)} - \dfrac{4(2x-1)}{(2x-1)(x+5)}$

$= \dfrac{8x + 40 - 8x + 4}{(2x-1)(x+5)}$

$= \dfrac{44}{(2x-1)(x+5)}$

44. $\dfrac{7}{x-y} + \dfrac{4}{y-x} = \dfrac{7}{x-y} + \dfrac{-4}{x-y} = \dfrac{7-4}{x-y} = \dfrac{3}{x-y}$

45. $\dfrac{4x}{9x^2 - 16} + \dfrac{2}{15x - 20}$

$= \dfrac{4x}{(3x+4)(3x-4)} + \dfrac{2}{5(3x-4)}$

$= \dfrac{5(4x)}{5(3x+4)(3x-4)} + \dfrac{2(3x+4)}{5(3x+4)(3x-4)}$

$= \dfrac{20x + 6x + 8}{5(3x+4)(3x-4)}$

$= \dfrac{26x + 8}{5(3x+4)(3x-4)}$

$= \dfrac{2(13x+4)}{5(3x+4)(3x-4)}$

46. $\dfrac{5}{2x^2 - 5x - 3} + \dfrac{7}{3x^2 - 11x + 6}$

$= \dfrac{5}{(2x+1)(x-3)} + \dfrac{7}{(3x-2)(x-3)}$

$= \dfrac{5(3x-2)}{(2x+1)(x-3)(3x-2)} + \dfrac{7(2x+1)}{(2x+1)(x-3)(3x-2)}$

$= \dfrac{15x - 10 + 14x + 7}{(2x+1)(x-3)(3x-2)}$

$= \dfrac{29x - 3}{(2x+1)(x-3)(3x-2)}$

47. $\dfrac{7}{2x - 18} - \dfrac{3x+4}{x^2 - 81}$

$= \dfrac{7}{2(x-9)} - \dfrac{3x+4}{(x+9)(x-9)}$

$= \dfrac{7(x+9)}{2(x+9)(x-9)} - \dfrac{2(3x+4)}{2(x+9)(x-9)}$

$= \dfrac{7x + 63 - 6x - 8}{2(x+9)(x-9)}$

$= \dfrac{x + 55}{2(x+9)(x-9)}$

48. $\dfrac{8}{2x^2 + x - 6} - \dfrac{2}{3x^2 + 4x - 4}$

$= \dfrac{8}{(2x-3)(x+2)} - \dfrac{2}{(3x-2)(x+2)}$

$= \dfrac{8(3x-2)}{(2x-3)(x+2)(3x-2)} - \dfrac{2(2x-3)}{(2x-3)(x+2)(3x-2)}$

$= \dfrac{24x - 16 - 4x + 6}{(2x-3)(x+2)(3x-2)}$

$= \dfrac{20x - 10}{(2x-3)(x+2)(3x-2)}$

$= \dfrac{10(2x-1)}{(2x-3)(x+2)(3x-2)}$

49. $\dfrac{7}{x+5}+\dfrac{6}{x-5}+\dfrac{8}{5-x}=\dfrac{7}{x+5}+\dfrac{6}{x-5}+\dfrac{-8}{x-5}$

$=\dfrac{7}{x+5}+\dfrac{6-8}{x-5}$

$=\dfrac{7}{x+5}+\dfrac{-2}{x-5}$

$=\dfrac{7(x+5)}{(x+5)(x-5)}+\dfrac{-2(x-5)}{(x+5)(x-5)}$

$=\dfrac{7x+35-2x+10}{(x+5)(x-5)}$

$=\dfrac{5x+45}{(x+5)(x-5)}$

$=\dfrac{5(x+9)}{(x+5)(x-5)}$

50. $\dfrac{7x}{x+2}-\dfrac{11}{3x}-\dfrac{5}{6}$

$=\dfrac{7x(6x)}{6x(x+2)}-\dfrac{11(2)(x+2)}{6x(x+2)}-\dfrac{5(x)(x+2)}{6x(x+2)}$

$=\dfrac{42x^2-22x-44-5x^2-10x}{6x(x+2)}$

$=\dfrac{37x^2-32x-44}{6x(x+2)}$

51. $\dfrac{x-1}{x+1}+\dfrac{x-3}{x}+\dfrac{x+1}{2x}$

$=\dfrac{2x(x-1)}{2x(x+1)}+\dfrac{2(x+1)(x-3)}{2x(x+1)}+\dfrac{(x+1)(x+1)}{2x(x+1)}$

$=\dfrac{2x^2-2x+2x^2-4x-6+x^2+2x+1}{2x(x+1)}$

$=\dfrac{5x^2-4x-5}{2x(x+1)}$

The perimeter is $\dfrac{5x^2-4x-5}{2x(x+1)}$ units.

52. $\dfrac{1}{2}\cdot\dfrac{14x^2+14x+6}{x(x+1)}-\dfrac{3x}{x+1}$

$=\dfrac{14x^2+14x+6}{2x(x+1)}-\dfrac{2x(3x)}{2x(x+1)}$

$=\dfrac{14x^2+14x+6-6x^2}{2x(x+1)}$

$=\dfrac{8x^2+14x+6}{2x(x+1)}$

$=\dfrac{2(4x+3)(x+1)}{2x(x+1)}$

$=\dfrac{4x+3}{x}$

The width $\dfrac{4x+3}{x}$ units.

53. a. $t=\dfrac{d}{r}$

Bob's time: $\dfrac{125}{x+10}$

Gretchen's: $\dfrac{80}{x}$

total time: $t(x)=\dfrac{125}{x+10}+\dfrac{80}{x}$

b. Let $Y1=\dfrac{125}{x+10}+\dfrac{80}{x}$

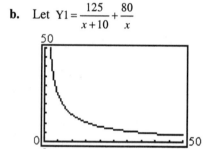

c. $t(30)=\dfrac{125}{30+10}+\dfrac{80}{30}\approx 5.8$

The total time of the trip will be approximately 5.8 hours.

d. $4.5=\dfrac{125}{x+10}+\dfrac{80}{x}$

$x(x+10)(4.5)=x(x+10)\left(\dfrac{125}{x+10}+\dfrac{80}{x}\right)$

$4.5x(x+10)=125x+80(x+10)$

$4.5x^2+45x=125x+80x+800$

$4.5x^2 - 160x - 800 = 0$

$x = \dfrac{-(-160) \pm \sqrt{(-160)^2 - 4(4.5)(-800)}}{2(4.5)}$

$x = \dfrac{160 \pm \sqrt{40{,}000}}{9} = \dfrac{160 \pm 200}{9}$

$x = 40$ or $x = -\dfrac{40}{9}$

Disregarding $-\dfrac{40}{9}$, Gretchen's average speed was 40 mph and Bob's was 50 mph.

54. $\dfrac{2x+3}{6} = \dfrac{5}{\sqrt{x}-1}$ is not a rational equation because it has a variable in the radicand of a square root..

55. $3x - 7 = 4x^{1/3}$ is not a rational equation because a variable base raised to a fractional exponent cannot be written as a rational expression.

56. $5x^{-2} + 3x^{-1} + 4 = 18$ is a rational equation because it can be written in the form $\dfrac{5}{x^2} + \dfrac{3}{x} - 14 = 0$.

57. $\dfrac{2}{x} - \dfrac{5}{x-3} = 17$ is a rational equation because it can be written in the form $\dfrac{2}{x} - \dfrac{5}{x-3} - 17 = 0$.

58. $x + 3 = \dfrac{3x-26}{2-x}$

$3x - 26 = (x+3)(2-x)$
$3x - 26 = 2x - x^2 + 6 - 3x$
$x^2 + 4x - 32 = 0$
$(x-4)(x+8) = 0$
$x - 4 = 0$ or $x + 8 = 0$
$x = 4$ $\quad\quad x = -8$
The solutions are -8 and 4.

59. $\dfrac{3}{x+3} = \dfrac{6}{2x+6}$

$2(x+3)\left[\dfrac{3}{x+3}\right] = 2(x+3)\left[\dfrac{6}{2(x+3)}\right]$

$2(3) = 6$
$6 = 6$
The equation is true for all values of x except the restricted value(s). The solution is the set of all real numbers not equal to -3.

60. $\dfrac{7-x}{2-x} = \dfrac{x+9}{x-2}$

$(x+9)(2-x) = (7-x)(x-2)$
$2x - x^2 + 18 - 9x = 7x - 14 - x^2 + 2x$
$-x^2 - 7x + 18 = -x^2 + 9x - 14$
$32 = 16x$
$2 = x$
Now 2 is a restricted value (extraneous), so the equation has no solution.

61. $\dfrac{4}{3x} = 1 - \dfrac{17}{3x}$

$(3x)\left(\dfrac{4}{3x}\right) = (3x)(1) - (3x)\left(\dfrac{17}{3x}\right)$

$4 = 3x - 17$
$21 = 3x$
$7 = x$
The solution is 7.

62. $\dfrac{5}{2x+4} = \dfrac{2x+1}{2}$

$(2x+1)(2x+4) = 5(2)$
$4x^2 + 8x + 2x + 4 = 10$
$4x^2 + 10x - 6 = 0$
$2(2x^2 + 5x - 3) = 0$
$2(x+3)(2x-1) = 0$
$x + 3 = 0$ or $2x - 1 = 0$
$x = -3$ $\quad\quad x = \dfrac{1}{2}$

The solutions are -3 and $\dfrac{1}{2}$.

63. $\dfrac{3x+7}{2x-4} = \dfrac{11}{3}$

$11(2x-4) = 3(3x+7)$
$22x - 44 = 9x + 21$
$13x = 65$
$x = 5$
The solution is 5.

64. $\dfrac{z}{50} - \dfrac{2}{z} = 0$

$\dfrac{z}{50} = \dfrac{2}{z}$

$100 = z^2$

$z = \pm\sqrt{100}$

$z = \pm 10$

The solutions are ± 10.

65. $b = 1 + \dfrac{30}{b}$

$b(b) = b\left(1 + \dfrac{30}{b}\right)$

$b^2 = b + 30$

$b^2 - b - 30 = 0$

$(b+5)(b-6) = 0$

$b + 5 = 0$ or $b - 6 = 0$

$b = -5 \qquad b = 6$

The solutions are -5 and 6.

66. $a + \dfrac{13}{a} = -\dfrac{3}{a} - 8$

$a\left(a + \dfrac{13}{a}\right) = a\left(-\dfrac{3}{a} - 8\right)$

$a^2 + 13 = -3 - 8a$

$a^2 + 8a + 16 = 0$

$(a+4)^2 = 0$

$a + 4 = 0$

$a = -4$

The solution is -4.

67. $1 = 12p^{-2} + p^{-1}$

$1 = \dfrac{12}{p^2} + \dfrac{1}{p}$

$p^2(1) = p^2\left(\dfrac{12}{p^2} + \dfrac{1}{p}\right)$

$p^2 = 12 + p$

$p^2 - p - 12 = 0$

$(p+3)(p-4) = 0$

$p + 3 = 0$ or $p - 4 = 0$

$p = -3 \qquad p = 4$

The solutions are -3 and 4.

68. $\dfrac{x-2}{x-6} + 3 = \dfrac{3x-14}{x-6}$

$(x-6)\left(\dfrac{x-2}{x-6} + 3\right) = (x-6)\left(\dfrac{3x-14}{x-6}\right)$

$x - 2 + 3x - 18 = 3x - 14$

$4x - 20 = 3x - 14$

$x = 6$

Now 6 is a restricted value (extraneous), so the equation has no solution.

69. $\dfrac{3}{m+4} + \dfrac{7}{m^2-16} = 2$

$(m+4)(m-4)\left(\dfrac{3}{m+4} + \dfrac{7}{(m+4)(m-4)}\right)$
$\qquad\qquad = (m+4)(m-4)(2)$

$3(m-4) + 7 = 2(m^2 - 16)$

$3m - 12 + 7 = 2m^2 - 32$

$0 = 2m^2 - 3m - 27$

$0 = (m+3)(2m-9) = 0$

$m + 3 = 0$ or $2m - 9 = 0$

$m = -3 \qquad m = \dfrac{9}{2}$

The solutions are -3 and $\dfrac{9}{2}$.

70. $\dfrac{2}{(3x+5)-(2x+1)} = \dfrac{8}{3(x+5)+(x+1)}$

$\dfrac{2}{3x+5-2x-1} = \dfrac{8}{3x+15+x+1}$

$\dfrac{2}{x+4} = \dfrac{8}{4x+16}$

$4(x+4)\left(\dfrac{2}{x+4}\right) = 4(x+4)\left[\dfrac{8}{4(x+4)}\right]$

$8 = 8$

The equation is true for all values of x except the restricted value(s). The solution is the set of all real numbers not equal to -4.

71. $\dfrac{7}{x+3} + \dfrac{2x-1}{x^2+x-6} = \dfrac{5x-2}{x^2-4}$

$(x-2)(x+3)(x+2)\left[\dfrac{7}{x+3} + \dfrac{2x-1}{(x-2)(x+3)}\right]$

$\qquad = (x-2)(x+3)(x+2)\left[\dfrac{5x-2}{(x+2)(x-2)}\right]$

$7(x-2)(x+2)+(x+2)(2x-1)=(x+3)(5x-2)$
$7x^2-28+2x^2+3z-2=5x^2+13x-6$
$4x^2-10x-24=0$
$2(x-4)(2x+3)=0$
$x-4=0$ or $2x+3=0$
$x=4 \qquad x=-\dfrac{3}{2}$

The solutions are $-\dfrac{3}{2}$ and 4.

72. $1+\dfrac{1}{x^2}=\dfrac{5}{x}$

$x^2\left(1+\dfrac{1}{x^2}\right)=x^2\left(\dfrac{5}{x}\right)$

$x^2+1=5x$

$x^2-5x+1=0$

$x=\dfrac{-(-5)\pm\sqrt{(-5)^2-4(1)(1)}}{2(1)}$

$x=\dfrac{5\pm\sqrt{21}}{2}$

$x\approx 4.7913$ or $x\approx 0.2087$

The solutions are $\dfrac{5\pm\sqrt{21}}{2}$, or approximately 0.2087 and 4.7913.

73. $\dfrac{5x}{3}+\dfrac{2}{3}=\dfrac{x-1}{x}$

$3x\left(\dfrac{5x}{3}+\dfrac{2}{3}\right)=3x\left(\dfrac{x-1}{x}\right)$

$5x^2+2x=3(x-1)$

$5x^2+2x=3x-3$

$5x^2-x+3=0$

$x=\dfrac{-(-1)\pm\sqrt{(-1)^2-4(5)(3)}}{2(1)}$

$x=\dfrac{1\pm\sqrt{-59}}{2}$

The equation has no real-number solution.

74. Let x = interest rate on the first account.
$x+0.01$ = interest rate on the second account.

$\dfrac{300}{x}+\dfrac{560}{x+0.01}=13000$

$x(x+0.01)\left[\dfrac{300}{x}+\dfrac{560}{x+0.01}\right]=x(x+0.01)(13000)$

$300(x+0.01)+560x=13000x(x+0.01)$

$300x+3+560x=13000x^2+130x$

$0=13000x^2-730x-3$

$x=\dfrac{-(-730)\pm\sqrt{(-730)^2-4(13000)(-3)}}{2(13000)}$

$x=\dfrac{730\pm\sqrt{688900}}{26000}=\dfrac{730\pm 830}{26000}$

$x=0.06$ or $x\approx -0.0038$

$x+0.01=0.07$

Disregarding -0.0038, Hope received 6% interest on the first account and 7% interest on the second account.

75. Let x = the time to do the room together.

$\dfrac{1}{6}$ is Norman's rate.

$\dfrac{1}{8}$ is his assistant's rate.

$\dfrac{1}{x}$ is their combined rate.

$\dfrac{1}{6}+\dfrac{1}{8}=\dfrac{1}{x}$

$24x\left(\dfrac{1}{6}\right)+24x\left(\dfrac{1}{8}\right)=24x\left(\dfrac{1}{x}\right)$

$4x+3x=24$

$7x=24$

$x=\dfrac{24}{7}\approx 3.4$

It will take them $\dfrac{24}{7}$ hours, or approximately 3.4 hours, to do the job together.

76. Let x = time it takes Ethel to pack a crate alone.
$x+10$ = time it takes Lucy alone.

$\dfrac{1}{x}$ is Ethel's rate.

$\dfrac{1}{x+10}$ is Lucy's.

$\dfrac{1}{30}$ is their combined rate.

$$\frac{1}{x} + \frac{1}{x+10} = \frac{1}{30}$$

$$30x(x+10)\left(\frac{1}{x} + \frac{1}{x+10}\right) = 30x(x+10)\left(\frac{1}{30}\right)$$

$$30(x+10) + 30x = x(x+10)$$

$$30x + 300 + 30x = x^2 + 10x$$

$$0 = x^2 - 50x - 300$$

$$x = \frac{-(-50) \pm \sqrt{(-50)^2 - 4(1)(-300)}}{2(1)}$$

$$x = \frac{50 \pm \sqrt{3700}}{2} = \frac{50 \pm 10\sqrt{37}}{2} = 25 \pm 5\sqrt{37}$$

$$x \approx 55.4 \quad \text{or} \quad x \approx -5.4$$

$$x + 10 = 65.4$$

It takes about 55.4 minutes for Ethel to pack a crate alone. It takes Lucy 65.4 minutes.

77. Let x = the time it takes to fill the barrel together.

$\frac{1}{10}$ is one line's rate.

$\frac{1}{15}$ is the other line's rate.

$\frac{1}{x}$ is the combined rate.

$$\frac{1}{10} + \frac{1}{15} = \frac{1}{x}$$

$$30x\left(\frac{1}{10} + \frac{1}{15}\right) = 30x\left(\frac{1}{x}\right)$$

$$3x + 2x = 30$$

$$5x = 30$$

$$x = 6$$

It takes 6 minutes to fill the barrel if both lines work together.

78. Let x = time for high-speed line working alone.
$2x$ = time for the other line working alone.

$\frac{1}{x}$ is the high-speed line's rate.

$\frac{1}{2x}$ is the other line's rate.

$\frac{1}{3\frac{1}{2}} = \frac{1}{\frac{7}{2}} = \frac{2}{7}$ is the combined rate.

$$\frac{1}{x} + \frac{1}{2x} = \frac{2}{7}$$

$$14x\left(\frac{1}{x} + \frac{1}{2x}\right) = 14x\left(\frac{2}{7}\right)$$

$$14 + 7 = 4x$$
$$21 = 4x$$
$$5.25 = x$$
$$2x = 10.5$$

It will take the high-speed 5.25 hours working alone. It will take the other line 10.5 hours.

79. Let x = the additional amount Maya should add.

$$\frac{x+85}{x+210} = \frac{50}{100}$$

$$50(x+210) = 100(x+85)$$

$$50x + 10500 = 100x + 8500$$

$$2000 = 50x$$

$$40 = x$$

She should add another $40 to the account.

80. Let x = the additional amount of pure alcohol.

$$\frac{1.5 + x}{12 + x} = \frac{25}{100}$$

$$25(12 + x) = 100(1.5 + x)$$

$$300 + 25x = 150 + 100x$$

$$150 = 75x$$

$$2 = x$$

Another 2 cups of pure alcohol should be added.

81. a. $\frac{GI}{JL} = \frac{GH}{JK}$

$$\frac{x}{21} = \frac{4}{7}$$

$$4(21) = 7x$$

$$84 = 7x$$

$$12 = x$$

$\overline{GI}$ measures 12 inches.

$$\frac{HI}{KL} = \frac{GH}{JK}$$

$$\frac{8}{x} = \frac{4}{7}$$

$$4x = 8(7)$$

$$4x = 56$$

$$x = 14$$

$\overline{KL}$ measures 14 inches

Chapter 12: Rational Expressions, Functions, and Equations

b. $\dfrac{HI}{KL} = \dfrac{GH}{JK}$

$\dfrac{x}{5\frac{13}{16}} = \dfrac{3\frac{1}{2}}{2\frac{5}{8}}$

$\left(3\frac{1}{2}\right)\left(5\frac{13}{16}\right) = \left(2\frac{5}{8}\right)x$

$20\frac{11}{32} = \left(2\frac{5}{8}\right)x$

$7\frac{3}{4} = x$

$\overline{HI}$ measures $7\frac{3}{4}$ feet.

$\dfrac{GI}{JL} = \dfrac{GH}{JK}$

$\dfrac{13}{x} = \dfrac{3\frac{1}{2}}{2\frac{5}{8}}$

$\left(3\frac{1}{2}\right)x = 13\left(2\frac{5}{8}\right)$

$\left(3\frac{1}{2}\right)x = 34\frac{1}{8}$

$x = 9\frac{3}{4}$

$\overline{JL}$ measures $9\frac{3}{4}$ feet.

c. $\dfrac{GH}{JK} = \dfrac{HI}{KL}$

$\dfrac{x}{1.88} = \dfrac{9.22}{2.305}$

$9.22(1.88) = 2.305x$

$17.3336 = 2.305x$

$7.52 = x$

$\overline{GH}$ measures 7.52 cm.

$\dfrac{GI}{UL} = \dfrac{HI}{KL}$

$\dfrac{14.08}{x} = \dfrac{9.22}{2.305}$

$9.22x = 14.08(2.305)$

$9.22x = 32.4544$

$x = 3.52$

$\overline{JL}$ measures 3.52 cm.

82. Let x = the height of the bluff.

$\dfrac{x}{120} = \dfrac{6}{15}$

$6(120) = 15x$

$720 = 15x$

$48 = x$

The bluff is 48 feet high.

83. Let x = the length of the tree's shadow.

$\dfrac{x}{9} = \dfrac{64}{6}$

$64(9) = 6x$

$576 = 6x$

$96 = x$

The tree's shadow was 96 feet long.

84. Let r = the resistance of one resistor.
$2r + 5$ = the resistance of the other resistor.

$\dfrac{1}{R} = \dfrac{1}{r_1} + \dfrac{1}{r_2}$

$\dfrac{1}{30} = \dfrac{1}{r} + \dfrac{1}{2r+5}$

$30r(2r+5)\left(\dfrac{1}{30}\right) = 30r(2r+5)\left(\dfrac{1}{r} + \dfrac{1}{2r+5}\right)$

$r(2r+5) = 30(2r+5) + 30r$

$2r^2 + 5r = 60r + 150 + 30r$

$2r^2 - 85r - 150 = 0$

$r = \dfrac{-(-85) \pm \sqrt{(-85)^2 - 4(2)(-150)}}{2(2)}$

$r = \dfrac{85 \pm \sqrt{8425}}{4}$

$r \approx 44$ or $r \approx -2$

$2r + 5 \approx 93$

Resistance cannot be negative, so the resistances of the two resistors are approximately 44 ohms and 93 ohms.

85. $\dfrac{1}{f} = \dfrac{1}{f_1} + \dfrac{1}{f_2}$

$ff_1f_2\left(\dfrac{1}{f}\right) = ff_1f_2\left(\dfrac{1}{f_1} + \dfrac{1}{f_2}\right)$

$f_1f_2 = ff_2 + ff_1$

$f_1f_2 = f(f_2 + f_1)$

$f = \dfrac{f_1f_2}{f_2 + f_1}$

SSM: Experiencing Introductory and Intermediate Algebra

Chapter 12 Chapter Review

1. $y = \dfrac{45}{x^2} + \dfrac{15}{x}$

   The restricted value is 0. Set up a table of x-values being sure to use both x-values that are less than 0 and x-values that are greater than 0. Connect the points with a smooth curve. Do not connect across the vertical line $x = 0$.

   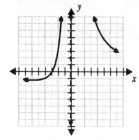

2. $g(x) = \dfrac{x+5}{x^2+5x+4} = \dfrac{x+5}{(x+1)(x+4)}$

   The restricted values are -4 and -1. Set up a table of x-values being sure to use x-values that are less than -4, x-values that are between -4 and -1, and x-values that are greater than -1. Connect the points with a smooth curve. Do not connect across the vertical lines $x = -4$ and $x = -1$.

   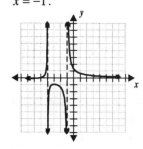

3. $p(x) = \dfrac{2x^2 - x - 3}{x+1}$

   The restricted value is -1. Set up a table of x-values being sure to use both x-values that are less than -1 and x-values that are greater than -1. Connect the points with a smooth curve. Do not connect across the vertical line $x = -1$.

   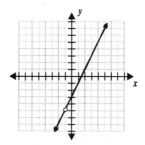

4. $y = \dfrac{x^2 - 5x - 6}{x^2 + 4x - 5} = \dfrac{x^2 - 5x - 6}{(x-1)(x+5)}$

   The restricted values are -5 and 1. Set up a table of x-values being sure to use x-values that are less than -5, x-values that are between -5 and 1, and x-values that are greater than 1. Connect the points with a smooth curve. Do not connect across the vertical lines $x = -5$ and $x = 1$.

   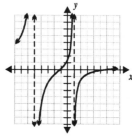

5. $y = \dfrac{5}{9x^2 + 25}$

   $9x^2 + 25 = 0$

   $9x^2 = -25$

   $x^2 = -\dfrac{25}{9}$

   $x = \pm\sqrt{-\dfrac{25}{9}}$ (Not Real)

   Restricted values: None
   Domain: All real numbers
   $(-\infty, \infty)$

6. $y = \dfrac{5}{9x^2 - 25}$

   $9x^2 - 25 = 0$

   $9x^2 = 25$

Chapter 12: Rational Expressions, Functions, and Equations

$$x^2 = \frac{25}{9}$$

$$x = \pm\sqrt{\frac{25}{9}} = \pm\frac{5}{3}$$

Restricted values: $\pm\frac{5}{3}$

Domain: $x \neq -\frac{5}{3}, x \neq \frac{5}{3}$

$$\left(-\infty, -\frac{5}{3}\right) \cup \left(-\frac{5}{3}, \frac{5}{3}\right) \cup \left(\frac{5}{3}, \infty\right)$$

7. $g(x) = \dfrac{x+5}{3x}$

$3x = 0$

$x = 0$

Restricted values: 0
Domain: $x \neq 0$

$$(-\infty, 0) \cup (0, \infty)$$

8. $y = \dfrac{5}{9x^2 - 25}$

$3x^2 + 2x - 8 = 0$

$(3x - 4)(x + 2) = 0$

$3x - 4 = 0$ or $x + 2 = 0$

$x = \dfrac{4}{3}$ $\qquad x = -2$

Restricted values: -2 and $\dfrac{4}{3}$

Domain: $x \neq -2, x \neq \dfrac{4}{3}$

$$\left(-\infty, -2\right) \cup \left(-2, \frac{4}{3}\right) \cup \left(\frac{4}{3}, \infty\right)$$

9. $\dfrac{3x^2 + 19x + 20}{2x^2 + 13x + 15} = \dfrac{(x+5)(3x+4)}{(x+5)(2x+3)} = \dfrac{3x+4}{2x+3}$

10. $\dfrac{k^2 - 25}{k^3 - 5k^2 + 6k - 30} = \dfrac{(k+5)(k-5)}{k^2(k-5) + 6(k-5)}$

$$= \dfrac{(k+5)(k-5)}{(k-5)(k^2+6)}$$

$$= \dfrac{k+5}{k^2+6}$$

11. $\dfrac{-72x^4y^2z}{96xy^6z} = \dfrac{(24xy^2z)(-3x^3)}{(24xy^2z)(4y^4)} = -\dfrac{3x^3}{4y^4}$

12. $\dfrac{7q - 21}{14q + 28} = \dfrac{7(q-3)}{14(q+2)} = \dfrac{q-3}{2(q+2)}$

13. $\dfrac{12x^2 - 6x^3}{9x^3 - 36x} = \dfrac{6x^2(2-x)}{9x(x^2-4)}$

$$= \dfrac{-6x^2(x-2)}{9x(x+2)(x-2)}$$

$$= \dfrac{-2x}{3(x+2)}$$

14. $\dfrac{7}{3x+2} - \dfrac{2}{x-3} = \dfrac{7(x-3)}{(3x+2)(x-3)} - \dfrac{2(3x+2)}{(3x+2)(x-3)}$

$$= \dfrac{7x - 21 - 6x - 4}{(3x+2)(x-3)}$$

$$= \dfrac{x - 25}{(3x+2)(x-3)}$$

15. $\dfrac{14}{a-b} + \dfrac{9}{b-a} = \dfrac{14}{a-b} + \dfrac{-9}{a-b} = \dfrac{14-9}{a-b} = \dfrac{5}{a-b}$

16. $\dfrac{7b}{3a} + \dfrac{5b^2}{3a} = \dfrac{7b + 5b^2}{3a} = \dfrac{b(7 + 5b)}{3a}$

17. $\dfrac{3p^2}{4m} - \dfrac{2p}{4m} = \dfrac{3p^2 - 2p}{4m} = \dfrac{p(3p - 2)}{4m}$

18. $\dfrac{5}{4x^2y^3} + \dfrac{3}{6x^2y} = \dfrac{5(3)}{4x^2y^3(3)} + \dfrac{3(2y^2)}{6x^2y(2y^2)}$

$$= \dfrac{5(3) + 3(2y^2)}{12x^2y^3}$$

$$= \dfrac{3(5 + 2y^2)}{12x^2y^3}$$

$$= \dfrac{5 + 2y^2}{4x^2y^3}$$

19. $\dfrac{x}{x+3} + \dfrac{4x}{x-5} = \dfrac{x(x-5)}{(x+3)(x-5)} + \dfrac{4x(x+3)}{(x+3)(x-5)}$

$$= \dfrac{x^2 - 5x + 4x^2 + 12x}{(x+3)(x-5)}$$

$$= \frac{5x^2 + 7x}{(x+3)(x-5)}$$
$$= \frac{x(5x+7)}{(x+3)(x-5)}$$

20. $\dfrac{2x}{4x^2 - 25} + \dfrac{1}{6x - 15}$

$$= \frac{2x}{(2x+5)(2x-5)} + \frac{1}{3(2x-5)}$$
$$= \frac{2x(3)}{3(2x+5)(2x-5)} + \frac{2x+5}{3(2x+5)(2x-5)}$$
$$= \frac{6x + 2x + 5}{3(2x+5)(2x-5)}$$
$$= \frac{8x + 5}{3(2x+5)(2x-5)}$$

21. $\dfrac{7}{2x^2 + 5x - 3} + \dfrac{5}{3x^2 + 11x + 6}$

$$= \frac{7(3x+2)}{(2x-1)(x+3)(3x+2)} + \frac{5(2x-1)}{(3x+2)(x+3)(3x+2)}$$
$$= \frac{21x + 14 + 10x - 5}{(2x-1)(x+3)(3x+2)}$$
$$= \frac{31x + 9}{(2x-1)(x+3)(3x+2)}$$

22. $\dfrac{3z-5}{2z-7} + \dfrac{z-9}{2z-7} = \dfrac{3z-5+z-9}{2z-7}$

$$= \frac{4z - 14}{2z - 7} = \frac{2(2z-7)}{2z-7}$$
$$= 2$$

23. $\dfrac{10x^2 - 8x}{25x^2 - 16} - \dfrac{5x - 4}{25x^2 - 16} = \dfrac{10x^2 - 8x - 5x + 4}{25x^2 - 16}$

$$= \frac{10x^2 - 13x + 4}{(5x+4)(5x-4)} = \frac{(2x-1)(5x-4)}{(5x+4)(5x-4)}$$
$$= \frac{2x-1}{5x+4}$$

24. $\dfrac{3}{x+6} + \dfrac{2}{x-6} + \dfrac{4}{6-x} = \dfrac{3}{x+6} + \dfrac{2}{x-6} + \dfrac{-4}{x-6}$

$$= \frac{3}{x+6} + \frac{2-4}{x-6} = \frac{3}{x+6} + \frac{-2}{x-6}$$

$$= \frac{3(x-6)}{(x+6)(x-6)} + \frac{-2(x+6)}{(x+6)(x-6)}$$
$$= \frac{3x - 18 - 2x - 12}{(x+6)(x-6)}$$
$$= \frac{x - 30}{(x+6)(x-6)}$$

25. $\dfrac{x}{x+1} - \dfrac{3}{2x} - \dfrac{3}{4}$

$$= \frac{4x(x)}{4x(x+1)} - \frac{3(2)(x+1)}{4x(x+1)} - \frac{3(x)(x+1)}{4x(x+1)}$$
$$= \frac{4x^2 - 6x - 6 - 3x^2 - 3x}{4x(x+1)}$$
$$= \frac{x^2 - 9x - 6}{4x(x+1)}$$

26. $\dfrac{5}{2x+10} - \dfrac{2x+1}{x^2 - 25}$

$$= \frac{5}{2(x+5)} - \frac{2x+1}{(x+5)(x-5)}$$
$$= \frac{5(x-5)}{2(x+5)(x-5)} - \frac{2(2x+1)}{2(x+5)(x-5)}$$
$$= \frac{5x - 25 - 4x - 2}{2(x+5)(x-5)}$$
$$= \frac{x - 27}{2(x+5)(x-5)}$$

27. $\dfrac{3}{4x^2 - 17x + 15} - \dfrac{2}{5x^2 - 19x + 12}$

$$= \frac{3}{(4x-5)(x-3)} - \frac{2}{(5x-4)(x-3)}$$
$$= \frac{3(5x-4)}{(4x-5)(x-3)(5x-4)} - \frac{2(4x-5)}{(4x-5)(x-3)(5x-4)}$$
$$= \frac{15x - 12 - 8x + 10}{(4x-5)(x-3)(5x-4)}$$
$$= \frac{7x - 2}{(4x-5)(x-3)(5x-4)}$$

28. $\dfrac{m^2 - 36}{2m^2 + 9m + 10} \cdot \dfrac{3m + 6}{m - 6} = \dfrac{(m+6)(m-6)(3)(m+2)}{(2m+5)(m+2)(m-6)}$

$$= \frac{3(m+6)}{2m+5}$$

29. $\dfrac{z-3}{z-7} \cdot \dfrac{7-z}{z-3} = \dfrac{(z-3)(-1)(z-7)}{(z-7)(z-3)} = -1$

30. $\dfrac{8a^4}{21b} \cdot \dfrac{-35b^3}{12a^7} = \dfrac{-280a^4b^3}{252a^7b} = \dfrac{(28a^4b)(-10b^2)}{(28a^4b)(9a^3)} = -\dfrac{10b^2}{9a^3}$

31. $(2c+3) \cdot \dfrac{c+7}{c+3} = \dfrac{(2c+3)(c+7)}{c+3}$

32. $\dfrac{3b-4a}{24ab^3} \cdot \dfrac{8a^2b}{4a-3b} = \dfrac{-1(4a-3b)(8a^2b)}{24ab^3(4a-3b)} = -\dfrac{a}{3b^2}$

33. $\dfrac{5p+20}{12p+96} \cdot \dfrac{-9p-72}{15p+60} = \dfrac{5(p+4)(-9)(p+8)}{12(p+8)(15)(p+4)} = -\dfrac{1}{4}$

34. $\dfrac{22g^8}{15h^3k^2} \div \dfrac{-11g^4}{10hk^3} = \dfrac{22g^8}{15h^3k^2} \cdot \dfrac{10hk^3}{-11g^4}$

    $= \dfrac{220g^8hk^3}{-165g^4h^3k^2}$

    $= -\dfrac{4g^4k}{3h^2}$

35. $\dfrac{18p^2q^2}{25m^2n} \div \dfrac{3mn^2}{5pq^5} = \dfrac{18p^2q^2}{25m^2n} \cdot \dfrac{5pq^5}{3mn^2}$

    $= \dfrac{90p^3q^7}{75m^3n^4}$

    $= \dfrac{6p^3q^7}{5m^3n^4}$

36. $(15m^3np^2) \div \dfrac{2mn}{5p} = \dfrac{15m^3np^2}{1} \cdot \dfrac{5p}{2mn}$

    $= \dfrac{75m^3np^3}{2mn}$

    $= \dfrac{75m^2p^3}{2}$

37. $\dfrac{4x^2-9}{4x^2+9} \div (2x+3) = \dfrac{4x^2-9}{4x^2+9} \cdot \dfrac{1}{2x+3}$

    $= \dfrac{(2x+3)(2x-3)}{(4x^2+9)(2x+3)}$

    $= \dfrac{2x-3}{4x^2+9}$

38. $\dfrac{x^2-x-6}{2x^2+8} \div \dfrac{x^2-5x+6}{x^4-16}$

    $= \dfrac{x^2-x-6}{2x^2+8} \cdot \dfrac{x^4-16}{x^2-5x+6}$

    $= \dfrac{(x+2)(x-3)(x^2+4)(x+2)(x-2)}{2(x^2+4)(x-2)(x-3)}$

    $= \dfrac{(x+2)^2}{2}$

39. $\dfrac{5a^2b^3}{c^5} \div (3a^2bc^2) = \dfrac{5a^2b^3}{c^5} \cdot \dfrac{1}{3a^2bc^2} = \dfrac{5a^2b^3}{3a^2bc^7} = \dfrac{5b^2}{3c^7}$

40. $\dfrac{z-3}{z-6} \div \dfrac{z+4}{z+8} = \dfrac{z-3}{z-6} \cdot \dfrac{z+8}{z+4} = \dfrac{(z-3)(z+8)}{(z-6)(z+4)}$

41. $\dfrac{k-7}{k-12} \div \dfrac{7-k}{12-k} = \dfrac{k-7}{k-12} \cdot \dfrac{12-k}{7-k}$

    $= \dfrac{(k-7)(-1)(k-12)}{(k-12)(-1)(k-7)}$

    $= 1$

42. $(2x^2+3x-2) \div \dfrac{2x^2+3x-2}{2x^2+5x+2}$

    $= \dfrac{2x^2+3x-2}{1} \cdot \dfrac{2x^2+5x+2}{2x^2+3x-2}$

    $= \dfrac{(2x-1)(x+5)(2x+1)(x+2)}{(2x-1)(x+2)}$

    $= (x+5)(2x+1)$

43. $\dfrac{4}{7} - \dfrac{1}{y} = \dfrac{9}{28}$

    $28y\left(\dfrac{4}{7} - \dfrac{1}{y}\right) = 28y\left(\dfrac{9}{28}\right)$

    $16y - 28 = 9y$

    $7y = 28$

    $y = 4$

    The solution is 4.

44. $\dfrac{4}{x-1} = \dfrac{5x-16}{7}$

    $(5x-16)(x-1) = 4(7)$

SSM: Experiencing Introductory and Intermediate Algebra

$5x^2 - 5x - 16x + 16 = 28$
$5x^2 - 21x - 12 = 0$
$x = \dfrac{-(-21) \pm \sqrt{(-21)^2 - 4(5)(-12)}}{2(5)}$
$x = \dfrac{21 \pm \sqrt{681}}{10}$
$x \approx 4.71$ or $x \approx -0.51$
The solutions are $\dfrac{21 \pm \sqrt{681}}{10}$, or approximately -0.51 and 4.71.

45. $2 - \dfrac{4}{x+3} = \dfrac{x+2}{x+3}$
$(x+3)\left(2 - \dfrac{4}{x+3}\right) = (x+3)\left(\dfrac{x+2}{x+3}\right)$
$2(x+3) - 4 = x+2$
$2x + 6 - 4 = x + 2$
$2x + 2 = x + 2$
$x = 0$
The solution is 0.

46. $\dfrac{5}{m+3} + \dfrac{8}{m^2 - 9} = \dfrac{3}{5}$
$5(m+3)(m-3)\left[\dfrac{5}{m+3} + \dfrac{8}{(m+3)(m-3)}\right]$
$\qquad = 5(m+3)(m-3)\left(\dfrac{3}{5}\right)$
$5(5)(m-3) + 5(8) = 3(m+3)(m-3)$
$25m - 75 + 40 = 3m^2 - 27$
$0 = 3m^2 - 25m + 8$
$0 = (3m - 1)(m - 8)$
$m = \dfrac{1}{3}$ or $m = 8$
The solutions are $\dfrac{1}{3}$ and 8.

47. $\dfrac{16}{2x+6} = \dfrac{5}{x}$
$5(2x+6) = 16x$
$10x + 30 = 16x$
$30 = 6x$
$5 = x$
The solution is 5.

48. $35 = p^{-2} + 2p^{-1}$
$35 = \dfrac{1}{p^2} + \dfrac{2}{p}$
$p^2(35) = p^2\left(\dfrac{1}{p^2} + \dfrac{2}{p}\right)$
$35p^2 = 1 + 2p$
$35p^2 - 2p - 1 = 0$
$(7p + 1)(5p - 1) = 0$
$p = -\dfrac{1}{7}$ or $p = \dfrac{1}{5}$
The solutions are $-\dfrac{1}{7}$ and $\dfrac{1}{5}$.

49. $\dfrac{x-6}{x-4} = \dfrac{x+7}{2(x+3) - (x+10)}$
$\dfrac{x-6}{x-4} = \dfrac{x+7}{2x+6-x-10}$
$\dfrac{x-6}{x-4} = \dfrac{x+7}{x-4}$
$(x-4)\left(\dfrac{x-6}{x-4}\right) = (x-4)\left(\dfrac{x+7}{x-4}\right)$
$x - 6 = x + 7$
$-6 = 7$
The equation is always false (a contradiction). It has no solution.

50. $\dfrac{a}{a-7} = \dfrac{a+1}{a-2}$
$(a+1)(a-7) = a(a-2)$
$a^2 - 6a - 7 = a^2 - 2a$
$-4a = 7$
$a = -\dfrac{7}{4}$
The solution is $-\dfrac{7}{4}$.

51. $1 - \dfrac{6}{x} = \dfrac{6}{x} - \dfrac{1}{2} + \dfrac{21}{2x}$
$2x\left(1 - \dfrac{6}{x}\right) = 2x\left(\dfrac{6}{x} - \dfrac{1}{2} + \dfrac{21}{2x}\right)$
$2x - 12 = 12 - x + 21$
$3x = 45$
$x = 15$
The solution is 15.

52. $1 + 7x^{-1} = 3$

$1 + \dfrac{7}{x} = 3$

$x\left(1 + \dfrac{7}{x}\right) = x(3)$

$x + 7 = 3x$

$7 = 2x$

$\dfrac{7}{2} = x$

The solution is $\dfrac{7}{2}$.

53. $\dfrac{10}{x+5} - \dfrac{24}{x^2+x-20} = \dfrac{1}{x^2-16}$

$\dfrac{10}{x+5} - \dfrac{24}{(x-4)(x+5)} = \dfrac{1}{(x+4)(x-4)}$

$(x-4)(x+5)(x+4)\left[\dfrac{10}{x+5} - \dfrac{24}{(x-4)(x+5)}\right]$

$= (x-4)(x+5)(x+4)\left[\dfrac{1}{(x+4)(x-4)}\right]$

$10(x-4)(x+4) - 24(x+4) = x+5$

$10x^2 - 160 - 24x - 96 = x + 5$

$10x^2 - 25x - 261 = 0$

$x = \dfrac{-(-25) \pm \sqrt{(-25)^2 - 4(10)(-261)}}{2(10)}$

$x = \dfrac{25 \pm \sqrt{11{,}065}}{20}$

$x \approx 6.51$ or $x \approx -4.01$

The solutions are $\dfrac{25 \pm \sqrt{11{,}065}}{20}$, or approximately -4.01 and 6.51.

54. $\dfrac{2}{(x+3)+(x-9)} = \dfrac{3}{2(x+4)+(x-17)}$

$\dfrac{2}{2x-6} = \dfrac{3}{3x-9}$

$6(x-3)\left[\dfrac{2}{2(x-3)}\right] = 6(x-3)\left[\dfrac{3}{3(x-3)}\right]$

$6 = 6$

The equation is true for all values of x except the restricted value(s). The solution is the set of all real numbers except 3.

55. $x + 5 = \dfrac{x-11}{2-x}$

$x - 11 = (x+5)(2-x)$

$x - 11 = 2x - x^2 + 10 - 5x$

$x^2 + 4x - 21 = 0$

$(x-3)(x+7) = 0$

$x = 3$ or $x = -7$

The solutions are -7 and 3.

56. $5 = \dfrac{3(x+5)}{x+3} + \dfrac{4x}{2x+6}$

$2(x+3)(5) = 2(x+3)\left[\dfrac{3(x+5)}{x+3} + \dfrac{4x}{2(x+3)}\right]$

$10x + 30 = 6x + 60 + 4x$

$30 = 30$

The equation is true for all values of x except the restricted value(s). The solution is the set of all real numbers except -3.

57. $\dfrac{1}{2} - \dfrac{3}{2x} = \dfrac{1}{2} + \dfrac{1}{2x} - \dfrac{2}{x}$

$2x\left(\dfrac{1}{2} - \dfrac{3}{2x}\right) = 2x\left(\dfrac{1}{2} + \dfrac{1}{2x} - \dfrac{2}{x}\right)$

$x - 3 = x + 1 - 4$

$-3 = -3$

The equation is true for all values of x except the restricted value(s). The solution is the set of all real numbers except 0.

58. $5x = \dfrac{9x+4}{x+2}$

$9x + 4 = 5x(x+2)$

$9x + 4 = 5x^2 + 10x$

$0 = 5x^2 + x - 4$

$0 = (5x-4)(x+1) = 0$

$x = \dfrac{4}{5}$ or $x = -1$

The solutions are -1 and $\dfrac{4}{5}$.

SSM: Experiencing Introductory and Intermediate Algebra

59. $\dfrac{x}{3} = \dfrac{27}{x}$

$81 = x^2$

$x = \pm\sqrt{81}$

$x = \pm 9$

The solutions are –9 and 9.

60. $\dfrac{4-x}{3-x} = \dfrac{x+11}{x-3}$

$(x+11)(3-x) = (4-x)(x-3)$

$3x - x^2 + 33 - 11x = 4x - 12 - x^2 + 3x$

$-x^2 - 8x + 33 = -x^2 + 7x - 12$

$45 = 15x$

$3 = x$

Now 3 is a restricted value (extraneous), so the equation has no solution.

61. $\dfrac{x+8}{2} = \dfrac{2x+11}{x+1}$

$2(2x+11) = (x+8)(x+1)$

$4x + 22 = x^2 + 9x + 8$

$0 = x^2 + 5x - 14 = 0$

$0 = (x-2)(x+7) = 0$

$x = 2 \quad \text{or} \quad x = -7$

The solutions are 2 and –7.

62. $x + 1 = \dfrac{x^2 + 5x - 12}{3x - 1}$

$x^2 + 5x - 12 = (x+1)(3x-1)$

$x^2 + 5x - 12 = 3x^2 + 2x - 1$

$0 = 2x^2 - 3x + 11$

$x = \dfrac{-(-3) \pm \sqrt{(-3)^2 - 4(2)(11)}}{2(2)}$

$x = \dfrac{3 \pm \sqrt{-79}}{4}$

The equation has no real-number solution.

63. $\dfrac{2x}{x+4} = 1 - \dfrac{8}{x+4}$

$(x+4)\left(\dfrac{2x}{x+4}\right) = (x+4)\left(1 - \dfrac{8}{x+4}\right)$

$2x = x + 4 - 8$

$x = -4$

The solution is –4.

64. $\dfrac{x-4}{x-8} = \dfrac{6x}{x^2 - 4x - 32}$

$(x+4)(x-8)\left(\dfrac{x-4}{x-8}\right) = (x+4)(x-8)\left[\dfrac{6x}{(x+4)(x-8)}\right]$

$(x+4)(x-4) = 6x$

$x^2 - 16 = 6x$

$x^2 - 6x - 16 = 0$

$(x+2)(x-8) = 0$

$x = -2 \quad \text{or} \quad x = 8$

The solutions are –2 and 8.

65. Let x = the time to do the room together.

$\dfrac{1}{10}$ is Dave's rate.

$\dfrac{1}{7}$ is Joan's rate.

$\dfrac{1}{x}$ is their combined rate.

$\dfrac{1}{10} + \dfrac{1}{7} = \dfrac{1}{x}$

$70x\left(\dfrac{1}{10}\right) + 70x\left(\dfrac{1}{7}\right) = 70x\left(\dfrac{1}{x}\right)$

$7x + 10x = 70$

$17x = 70$

$x = \dfrac{70}{17} \approx 4.1$

It will take them $\dfrac{70}{17}$ hours, or approximately 4.1 hours, to do the job together.

66. Let x = the additional amount to be added.

$\dfrac{2+x}{7+x} = \dfrac{30}{100}$

$30(7+x) = 100(2+x)$

$210 + 30x = 200 + 100x$

$10 = 70x$

$x = \dfrac{10}{70}$

$x = \dfrac{1}{7}$

Another $\dfrac{1}{7}$ pint of glue should be added.

Chapter 12: Rational Expressions, Functions, and Equations

67. $r = \dfrac{d}{t} = \dfrac{120}{t}$

    The expression for the Antonio's average rate is $\dfrac{120}{t}$ mph.

    $d = rt = \dfrac{2}{3}\left(\dfrac{120}{t}\right)(t+1) = \dfrac{80(t+1)}{t}$

    The expression for Antonio's new distance is $\dfrac{80(t+1)}{t}$ miles.

68. Let x = the height of the tree.
    $\dfrac{x}{4} = \dfrac{18}{3}$
    $72 = 3x$
    $24 = x$
    The tree is 24 feet high.

69. $V = LWH$; $H = \dfrac{V}{LW}$

    If $V = 10x^3 + 5x^2$, $W = x$, and $L = 5x$, then
    $H = \dfrac{10x^3 + 5x^2}{5x(x)} = \dfrac{5x^2(2x+1)}{5x^2} = 2x+1$

    The expression for the height is $(2x+1)$ inches.

70. Let x = time for first line working alone.
    $x + 1.5$ = time for the other line working alone.
    $\dfrac{1}{x}$ is the first line's rate.
    $\dfrac{1}{x+1.5}$ is the second line's rate.
    $\dfrac{1}{3\frac{1}{3}} = \dfrac{1}{\frac{10}{3}} = \dfrac{3}{10}$ is the combined rate.
    $\dfrac{1}{x} + \dfrac{1}{x+1.5} = \dfrac{3}{10}$
    $10x(x+1.5)\left(\dfrac{1}{x} + \dfrac{1}{x+1.5}\right) = 10x(x+1.5)\left(\dfrac{3}{10}\right)$
    $10(x+1.5) + 10x = 3x(x+1.5)$
    $10x + 15 + 10x = 3x^2 + 4.5x$
    $0 = 3x^2 - 15.5x - 15$

    $x = \dfrac{-(-15.5) \pm \sqrt{(-15.5)^2 - 4(3)(-15)}}{2(3)}$
    $x = \dfrac{15.5 \pm \sqrt{420.25}}{6} = \dfrac{15.5 \pm 20.25}{6}$
    $x = 6$ or $x = -\dfrac{5}{6}$
    $x + 1.5 = 7.5$
    It takes the first line 6 hours to fill an order when working alone. It takes the second line 7.5 hours.

71. $\dfrac{6}{x-3} + 3 + \dfrac{x-1}{x}$
    $= \dfrac{6x}{x(x-3)} + \dfrac{3x(x-3)}{x(x-3)} + \dfrac{(x-1)(x-3)}{x(x-3)}$
    $= \dfrac{6x + 3x^2 - 9x + x^2 - 3x - x + 3}{x(x-3)}$
    $= \dfrac{4x^2 - 7x + 3}{x(x-3)}$
    $= \dfrac{(4x-3)(x-1)}{x(x-3)}$

    The perimeter is $\dfrac{(4x-3)(x-1)}{x(x-3)}$ units.

72. Let x = the additional amount.
    $\dfrac{70+x}{250+x} = \dfrac{1}{3}$
    $1(250+x) = 3(70+x)$
    $250 + x = 210 + 3x$
    $40 = 2x$
    $20 = x$
    Latoya must add another \$20.

73. Let r = the resistance of one resistor.
    $r + 12$ = the resistance of the other resistor.
    $\dfrac{1}{R} = \dfrac{1}{r_1} + \dfrac{1}{r_2}$
    $\dfrac{1}{25} = \dfrac{1}{r} + \dfrac{1}{r+12}$
    $25r(r+12)\left(\dfrac{1}{25}\right) = 25r(r+12)\left(\dfrac{1}{r} + \dfrac{1}{r+12}\right)$
    $r(r+12) = 25(r+12) + 25r$
    $r^2 + 12r = 25r + 300 + 25r$
    $r^2 - 38r - 300 = 0$

$$r = \frac{-(-38) \pm \sqrt{(-38)^2 - 4(1)(-300)}}{2(1)}$$

$$r = \frac{38 \pm \sqrt{2644}}{2}$$

$r \approx 44.7$ or $r = -6.7$

$r + 12 = 56.7$

Resistance cannot be negative, so the resistances of the two resistors are approximately 45 ohms and 57 ohms.

74. $\dfrac{1}{C} = \dfrac{1}{C_1} + \dfrac{1}{C_2}$

$CC_1C_2 \left(\dfrac{1}{C}\right) = CC_1C_2 \left(\dfrac{1}{C_1} + \dfrac{1}{C_2}\right)$

$C_1C_2 = CC_2 + CC_1$

$C_1C_2 = C(C_2 + C_1)$

$C = \dfrac{C_1C_2}{C_2 + C_1}$

Chapter 12 Test

1. $y = \dfrac{x-7}{2x^2 + 6x}$

    $2x^2 + 6x = 0$

    $2x(x+3) = 0$

    $x = 0$ or $x = -3$

    Restricted values: -3 and 0

    Domain: $x \neq -3, x \neq 0$

    $(-\infty, -3) \cup (-3, 0) \cup (0, \infty)$

2. $g(x) = \dfrac{x^2 - 4x - 21}{x^2 - 5x - 36}$

    $x^2 - 5x - 36 = 0$

    $(x+4)(x-9) = 0$

    $x = -4$ or $x = 9$

    Restricted values: -4 and 9

    Domain: $x \neq -4, x \neq 9$

    $(-\infty, -4) \cup (-4, 9) \cup (9, \infty)$

3. $f(x) = \dfrac{x+4}{x-2}$

    The restricted value is 2. Set up a table of x-values being sure to use both x-values that are less than 2 and x-values that are greater than 2. Connect the points with a smooth curve. Do not connect across the vertical line $x = 2$.

    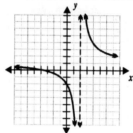

4. $\dfrac{24x + 56}{9x^2 - 49} = \dfrac{8(3x+7)}{(3x+7)(3x-7)} = \dfrac{8}{3x-7}$

5. $\dfrac{2z^3 - 18z}{z^3 + 3z^2 + 9z + 27} = \dfrac{2z(z^2 - 9)}{z^2(z+3) + 9(z+3)}$

    $= \dfrac{2z(z+3)(z-3)}{(z+3)(z^2 + 9)}$

    $= \dfrac{2z(z-3)}{(z^2 + 9)}$

6. $\dfrac{5}{x^2 + 4x} + \dfrac{2}{x^2 + x - 12}$

    $= \dfrac{5}{x(x+4)} + \dfrac{2}{(x-3)(x+4)}$

    $= \dfrac{5(x-3)}{x(x-3)(x+4)} + \dfrac{2x}{x(x-3)(x+4)}$

    $= \dfrac{5x - 15 + 2x}{x(x-3)(x+4)}$

    $= \dfrac{7x - 15}{x(x-3)(x+4)}$

Chapter 12: Rational Expressions, Functions, and Equations

7. $\dfrac{5x^2-40x}{20x^2+30x} \div \dfrac{x^2-64}{2x^2+19x+24}$

$= \dfrac{5x^2-40x}{20x^2+30x} \cdot \dfrac{2x^2+19x+24}{x^2-64}$

$= \dfrac{(5x^2-40x)(2x^2+19x+24)}{(20x^2+30x)(x^2-64)}$

$= \dfrac{5x(x-8)(2x+3)(x+8)}{10x(2x+3)(x+8)(x-8)}$

$= \dfrac{1}{2}$

8. $\dfrac{2x+1}{x^2+5x} - \dfrac{x-4}{x^2+10x+25}$

$= \dfrac{2x+1}{x(x+5)} - \dfrac{x-4}{(x+5)(x+5)}$

$= \dfrac{(2x+1)(x+5)}{x(x+5)^2} - \dfrac{x(x-4)}{x(x+5)^2}$

$= \dfrac{2x^2+10x+x+5-x^2+4x}{x(x+5)^2}$

$= \dfrac{x^2+15x+5}{x(x+5)^2}$

9. $\dfrac{24x^2y}{x^2-xy-2y^2} \cdot \dfrac{3x^2-12xy+12y^2}{16x^3y}$

$= \dfrac{24x^2y(3x^2-12xy+12y^2)}{16x^3y(x^2-xy-2y^2)}$

$= \dfrac{24x^2y(3)(x-2y)(x-2y)}{(x+y)(x-2y)16x^3y}$

$= \dfrac{9(x-2y)}{2x(x+y)}$

10. $x-5 = \dfrac{5-x}{x}$

$5-x = x(x-5)$

$5-x = x^2-5x$

$0 = x^2-4x-5$

$0 = (x+1)(x-5)$

$x = -1$ or $x = 5$

The solutions are -1 and 5.

11. $\dfrac{16}{x-3} = \dfrac{x-3}{x}$

$(x-3)(x-3) = 16x$

$x^2-9 = 16x$

$x^2-16x-9 = 0$

$x = \dfrac{-(-22)\pm\sqrt{(-22)^2-4(1)(9)}}{2(1)}$

$x = \dfrac{22\pm\sqrt{448}}{2}$

$x = \dfrac{22\pm 8\sqrt{7}}{4}$

$x = 11\pm 4\sqrt{7}$

$x \approx 21.5830$ or $x \approx 0.4170$

The solutions are $11\pm 4\sqrt{7}$, or approximately 0.4170 and 21.5830.

12. $\dfrac{b}{5}+\dfrac{1}{2} = \dfrac{1}{5}-\dfrac{11}{10b}$

$10b\left(\dfrac{b}{5}+\dfrac{1}{2}\right) = 10b\left(\dfrac{1}{5}-\dfrac{11}{10b}\right)$

$2b^2+5b = 2b-11$

$2b^2+3b+11 = 0$

$b = \dfrac{-3\pm\sqrt{3^2-4(2)(11)}}{2(2)}$

$b = \dfrac{-3\pm\sqrt{-79}}{4}$

The equation has no real-number solution.

13. $\dfrac{3}{m}-\dfrac{m}{75} = 0$

$\dfrac{3}{m} = \dfrac{m}{75}$

$m^2 = 225$

$m = \pm\sqrt{225}$

$m = \pm 15$

The solutions are -15 and 15.

14. $6 + 14x^{-1} + 4x^{-2} = 0$

$6 + \dfrac{14}{x} + \dfrac{4}{x^2} = 0$

$x^2\left(6 + \dfrac{14}{x} + \dfrac{4}{x^2} = 0\right) = x^2(0)$

$6x^2 + 14x + 4 = 0$

$2(3x^2 + 7x + 2) = 0$

$2(3x+1)(x+2) = 0$

$(3x+1)(x+2) = 0$

$p = -\dfrac{1}{3}$ or $p = -2$

The solutions are -2 and $-\dfrac{1}{3}$.

15. $\dfrac{x^2 + 6x + 9}{4x + 20} = \dfrac{x+6}{x+5}$

$4(x+5)\left[\dfrac{x^2+6x+9}{4(x+5)}\right] = 4(x+5)\left(\dfrac{x+6}{x+5}\right)$

$x^2 + 6x + 9 = 4(x+6)$

$x^2 + 6x + 9 = 4x + 24$

$x^2 + 2x - 15 = 0$

$(x-3)(x+5) = 0$

$x = 3$ or $x = -5$

Now -5 is a restricted value (extraneous), so the solution is 3.

16. $\dfrac{2}{x+2} = \dfrac{3}{x+3}$

$3(x+2) = 2(x+3)$

$3x + 6 = 2x + 6$

$x = 0$

The solution is 0.

17. $\dfrac{x-5}{x+1} = 1 - \dfrac{6}{x+1}$

$(x+1)\left(\dfrac{x-5}{x+1}\right) = (x+1)\left(1 - \dfrac{6}{x+1}\right)$

$x - 5 = x + 1 - 6$

$-5 = -5$

The equation is true for all values of x except the restricted value(s). The solution is the set of all real numbers except -1.

18. $\dfrac{5-x}{3-x} = \dfrac{x+9}{x-3}$

$(x+9)(3-x) = (5-x)(x-3)$

$3x - x^2 + 27 - 9x = 5x - 15 - x^2 + 3x$

$-x^2 - 6x + 27 = -x^2 + 8x - 15$

$-14x = -42$

$x = 3$

Now 3 is a restricted value (extraneous), so the equation has no solution.

19. $z = \dfrac{x-m}{s}$

$x - m = zs$

$s = \dfrac{x-m}{z}$

20. Let x = the height of the tree.

$\dfrac{x}{7} = \dfrac{32.5}{10.5}$

$227.5 = 10.5x$

$\dfrac{227.5}{10.5} = x$

$\dfrac{65}{3} = x$

$x = 21\dfrac{2}{3}$

The height of the tree is $21\dfrac{2}{3}$ feet.

21. Let x = time it takes Jill to rake the leaves alone.

$\dfrac{1}{x}$ is Jill's rate.

$\dfrac{1}{6}$ is Jack's rate.

$\dfrac{1}{2\frac{2}{5}} = \dfrac{1}{\frac{12}{5}} = \dfrac{5}{12}$ is the combined rate.

$\dfrac{1}{x} + \dfrac{1}{6} = \dfrac{5}{12}$

$12x\left(\dfrac{1}{x} + \dfrac{1}{6}\right) = 12x\left(\dfrac{5}{12}\right)$

$12 + 2x = 5x$

$12 = 3x$

$4 = x$

It will take Jill 4 hours to rake the leaves when working alone.

22. Answers will vary.

Chapter 13

13.1 Exercises

1. $-\sqrt[4]{1296} = -(6) = -6$

3. $\sqrt[4]{-1296}$ is not a real number because the radicand is negative and the root is even.

5. $\sqrt[3]{216} = 6$ because $(6)^3 = 216$.

7. $\sqrt[3]{-216} = -6$ because $(-6)^3 = -216$.

9. $-\sqrt[3]{216} = -(6) = -6$

11. $\sqrt{110} \approx 10.488$

13. $-\sqrt{110} \approx -10.488$

15. $\sqrt{-110}$ is not a real number because the radicand is negative and the root is even.

17. $\sqrt[3]{110} \approx 4.791$

19. $-\sqrt[3]{110} \approx -4.791$

21. $\sqrt[5]{20} \approx 1.821$

23. $\sqrt[3]{12.5} \approx 2.321$

25. $y = \sqrt{5}x + 7$ is not radical. The variable is not in the radicand.

27. $f(x) = 2\sqrt{3x+1} - 5$ is radical. The variable is in the radicand.

29. $y = \sqrt{x-3}$
    $x - 3 < 0$
    $x < 3$
    The restricted values are all real numbers less than 3.
    Domain: $[3, \infty)$

31. $y = \sqrt[3]{x-3}$
    There are no restricted values because the index of the radical is odd.
    Domain: $(-\infty, \infty)$

33. $y = 2\sqrt{x} + 3$
    $x < 0$
    The restricted values are all real numbers less than 0.
    Domain: $[0, \infty)$

35. $y = \sqrt{2x+1} - \sqrt{x-3}$
    $2x - 1 < 0$ or $x - 3 < 0$
    $2x < 1$ or $x < 3$
    $x < \dfrac{1}{2}$ or $x < 3$
    The restricted values are all real numbers less than 3.
    Domain: $[3, \infty)$

37. $y = \left(\sqrt[3]{5x+9}\right)^4$
    There are no restricted values because the index of the radical is odd.
    Domain: $(-\infty, \infty)$

39. $y = \left(\sqrt[4]{3x+2}\right)^5$
    $3x + 2 < 0$
    $3x < -2$
    $x < -\dfrac{2}{3}$
    The restricted values are all real numbers less than $-\dfrac{2}{3}$.
    Domain: $\left[-\dfrac{2}{3}, \infty\right)$

41. Table values will vary; possible values follow.

x	$y = \sqrt{x}$	y
0	$y = \sqrt{0} = 0$	0
1	$y = \sqrt{1} = 1$	1
4	$y = \sqrt{4} = 2$	2
9	$y = \sqrt{9} = 3$	3

SSM: Experiencing Introductory and Intermediate Algebra

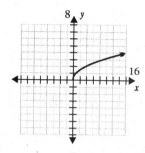

43. Table values will vary; possible values follow.

x	$y = \sqrt[3]{x} - 5$	y
-1	$y = \sqrt[3]{-1} - 5 = -6$	-6
0	$y = \sqrt[3]{0} - 5 = -5$	-5
1	$y = \sqrt[3]{1} - 5 = -4$	-4
8	$y = \sqrt[3]{8} - 5 = -3$	-3

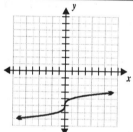

45. Table values will vary; possible values follow.

x	$y = 5\sqrt{x} - 10$	y
0	$y = 5\sqrt{0} - 10 = -10$	-10
1	$y = 5\sqrt{1} - 10 = -5$	-5
4	$y = 5\sqrt{4} - 10 = 0$	0
9	$y = 5\sqrt{9} - 10 = 5$	5

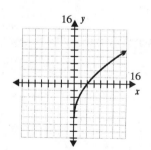

47. Table values will vary; possible values follow.

x	$f(x) = \sqrt{x - 5}$	$f(x)$
5	$f(5) = \sqrt{5 - 5} = 0$	0
6	$f(6) = \sqrt{6 - 5} = 1$	1
9	$f(9) = \sqrt{9 - 5} = 2$	2
14	$f(14) = \sqrt{14 - 5} = 3$	3

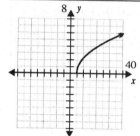

49. Table values will vary; possible values follow.

x	$g(x) = \sqrt[3]{8x}$	$g(x)$
$-\dfrac{1}{8}$	$g\left(-\dfrac{1}{8}\right) = \sqrt[3]{8\left(-\dfrac{1}{8}\right)} = -1$	-1
0	$g(0) = \sqrt[3]{8(0)} = 0$	0
$\dfrac{1}{8}$	$g\left(\dfrac{1}{8}\right) = \sqrt[3]{8\left(\dfrac{1}{8}\right)} = 1$	1
1	$g(1) = \sqrt[3]{8(1)} = 2$	2

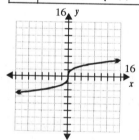

51. Let s = length of a side, and A = area.
$$A = s^2$$
$$s = \sqrt{A}$$

a. $s = \sqrt{289}$
$s = 17$
The length of the side is 17 feet.

b. $s = \sqrt{90.25} = 9.5$
The length of the side is 9.5 inches.

c. $s = \sqrt{115} \approx 10.724$
The length of the side is about 10.724 mm.

53. Let r = radius of circle, and A = area.
$$A = \pi r^2$$
$$r = \sqrt{\frac{A}{\pi}}$$

a. $r = \sqrt{\frac{196\pi}{\pi}} = \sqrt{196} = 14$
The radius is 14 inches.

b. $r = \sqrt{\frac{49}{\pi}} \approx 3.949$
The radius is about 3.949 yards.

c. $r = \sqrt{\frac{108}{\pi}} \approx 5.863$
The radius is about 5.863 meters.

55. Let x = length of a side, and V = volume.
$$V = s^3$$
$$s = \sqrt[3]{V}$$

a. $s = \sqrt[3]{1728} = 12$
The length of the side is 12 inches.

b. $s = \sqrt[3]{120} \approx 4.932$
The length of the side is about 4.932 meters.

57. Let r = radius of sphere, and V = volume.
$$V = \frac{4}{3}\pi r^3$$
$$r = \sqrt[3]{\frac{3V}{4\pi}} \;;\; d = 2\sqrt[3]{\frac{3V}{4\pi}}$$

a. $r = \sqrt[3]{\frac{3(288\pi)}{4\pi}} = \sqrt[3]{216} = 6$

$d = 2 \cdot \sqrt[3]{\frac{3(288\pi)}{4\pi}} = 2 \cdot \sqrt[3]{216} = 2(6) = 12$

The balloon has a radius of 6 decimeters and a diameter of 12 decimeters.

b. $r = \sqrt[3]{\frac{3(250)}{4\pi}} \approx 3.908$

$d = 2 \cdot \sqrt[3]{\frac{3(250)}{4\pi}} \approx 7.816$

The balloon has a radius of about 3.908 yards and a diameter of about 7.816 yards.

59. Circular area covered by dome = πr^2
$$A = \pi r^2$$
$$r = \sqrt{\frac{A}{\pi}}$$
$$r = \sqrt{\frac{8200}{\pi}} \approx 51.09$$
$$d = 2r = 2\sqrt{\frac{8200}{\pi}} \approx 102.179$$

The radius of the dome is about 51.09 feet and the diameter is about 102.179 feet.

61. $(8,-3),(8,2)$
$$d = \sqrt{(x_2 - x_1)^2 + (y_2 - y_1)^2}$$
$$= \sqrt{(8-8)^2 + (2-(-3))^2}$$
$$= \sqrt{0^2 + 5^2} = \sqrt{25}$$
$$= 5$$

The distance between the points is 5 units.

63. $(-3,6),(5,6)$
$$d = \sqrt{(x_2 - x_1)^2 + (y_2 - y_1)^2}$$
$$= \sqrt{(5-(-3))^2 + (6-6)^2}$$
$$= \sqrt{8^2 + 0^2} = \sqrt{64}$$
$$= 8$$

The distance between the points is 8 units.

65. $(-3, -4), (2, 5)$

$$d = \sqrt{(x_2 - x_1)^2 + (y_2 - y_1)^2}$$
$$= \sqrt{(2-(-3))^2 + (5-(-4))^2}$$
$$= \sqrt{5^2 + 9^2} = \sqrt{25 + 81}$$
$$= \sqrt{106}$$

The distance between the points is $\sqrt{106}$ units.

67. $d(x) = 9.4\sqrt[4]{x}$, x = number of years after 1993.

a.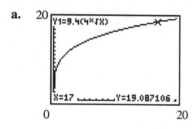

b. The function is increasing.

c. See graph. Assuming that the trend continues, the 2010 market value of data communications equipment will be about $19.1 million.

69. $r = \sqrt{n}$

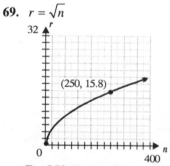

For 250 pieces of data, there should be about 16 rows in the table.

13.1 Calculator Exercises

The function $f(x) = \sqrt{x^2 + 6x + 9} = \sqrt{(x+3)^2}$ does not simplify to $x = 3$ due to the definition of $\sqrt{x}$. Recall that for even roots (such as the square root), the principal root is always non-negative. That is,

$\sqrt{x} \geq 0$. Thus, $f(x) = \sqrt{x^2 + 6x + 9}$ will always be non-negative. However, $x + 3$ can be negative so it is not equivalent to $f(x)$. The simplification is correct as long as $x + 3 \geq 0$. The correct simplification would be:

$$f(x) = \sqrt{x^2 + 6x + 9} = \sqrt{(x+3)^2} = |x+3|$$

1. $f(x) = \sqrt{x^2 - 4x + 4} = \sqrt{(x-2)^2} = |x-2|$

2. $f(x) = \sqrt{4x^2 - 12x + 9} = \sqrt{(2x-3)^2} = |2x-3|$

13.2 Exercises

1. $36^{1/2} = \sqrt{36} = 6$

3. $216^{1/3} = \sqrt[3]{216} = 6$

5. $-81^{1/4} = -\sqrt[4]{81} = -3$

7. $(-81)^{1/4} = \sqrt[4]{-81}$ not a real number.

9. $-8^{1/3} = -\sqrt[3]{8} = -2$

11. $(-8)^{1/3} = \sqrt[3]{-8} = -2$

13. $36^{-1/2} = \dfrac{1}{36^{1/2}} = \dfrac{1}{\sqrt{36}} = \dfrac{1}{6}$

15. $216^{-1/3} = \dfrac{1}{216^{1/3}} = \dfrac{1}{\sqrt[3]{216}} = \dfrac{1}{6}$

17. $81^{-1/4} = \dfrac{1}{81^{1/4}} = \dfrac{1}{\sqrt[4]{81}} = \dfrac{1}{3}$

19. $-81^{-1/4} = -\dfrac{1}{81^{1/4}} = -\dfrac{1}{\sqrt[4]{81}} = -\dfrac{1}{3}$

21. $8^{-1/3} = \dfrac{1}{8^{1/3}} = \dfrac{1}{\sqrt[3]{8}} = \dfrac{1}{2}$

23. $(-8)^{-1/3} = \dfrac{1}{(-8)^{1/3}} = \dfrac{1}{\sqrt[3]{-8}} = \dfrac{1}{-2} = -\dfrac{1}{2}$

25. $522^{1/2} = \sqrt{522} \approx 22.847$

27. $522^{1/4} = \sqrt[4]{522} \approx 4.780$

29. $522^{-1/3} = \dfrac{1}{522^{1/3}} = \dfrac{1}{\sqrt[3]{522}} \approx 0.124$

31. $522^{-1/5} = \dfrac{1}{522^{1/5}} = \dfrac{1}{\sqrt[5]{522}} \approx 0.286$

33. $27^{4/3} = \left(\sqrt[3]{27}\right)^4 = 3^4 = 81$

35. $(-27)^{4/3} = \left(\sqrt[3]{-27}\right)^4 = (-3)^4 = 81$

37. $27^{-4/3} = \dfrac{1}{27^{4/3}} = \dfrac{1}{\left(\sqrt[3]{27}\right)^4} = \dfrac{1}{(3)^4} = \dfrac{1}{81}$

39. $-27^{-4/3} = -\dfrac{1}{27^{4/3}} = -\dfrac{1}{\left(\sqrt[3]{27}\right)^4} = -\dfrac{1}{3^4} = -\dfrac{1}{81}$

41. $4^{5/2} = \left(\sqrt{4}\right)^5 = 2^5 = 32$

43. $8^{7/3} = \left(\sqrt[3]{8}\right)^7 = 2^7 = 128$

45. $4^{-7/2} = \dfrac{1}{4^{7/2}} = \dfrac{1}{\left(\sqrt{4}\right)^7} = \dfrac{1}{2^7} = \dfrac{1}{128}$

47. $-16^{3/2} = -\left(\sqrt{16}\right)^3 = -4^3 = -64$

49. $(-9)^{3/2} = \left(\sqrt{-9}\right)^3$ not a real number.

51. $-81^{-3/2} = -\dfrac{1}{81^{3/2}} = -\dfrac{1}{\left(\sqrt{81}\right)^3} = -\dfrac{1}{9^3} = -\dfrac{1}{729}$

53. $28^{5/4} = \left(\sqrt[4]{28}\right)^5 \approx 64.409$

55. $(-21)^{4/3} = \left(\sqrt[3]{-21}\right)^4 \approx 57.937$

57. $5^{-2/3} = \dfrac{1}{5^{2/3}} = \dfrac{1}{\left(\sqrt[3]{5}\right)^2} \approx 0.342$

59. $-42^{-2/3} = -\dfrac{1}{42^{2/3}} = -\dfrac{1}{\left(\sqrt[3]{42}\right)^2} \approx -0.083$

61. $(-88)^{3/8} = \left(\sqrt[8]{-88}\right)^3$ not a real number.

63. $f(x) = (5-4x)^{3/4} = \left(\sqrt[4]{5-4x}\right)^3$
$5-4x < 0$
$5 < 4x$
$\dfrac{5}{4} < x$
The restricted values are all real numbers greater than $\dfrac{5}{4}$.
Domain: $\left(-\infty, \dfrac{5}{4}\right]$.

65. $g(x) = (5-4x)^{2/3} = \left(\sqrt[3]{5-4x}\right)^2$
There are no restricted values because the index of the radical is odd.
Domain: $(-\infty, \infty)$

67. $y = x^{3/4} + 2$
Domain: $[0, \infty)$
Values will vary; possible solutions follow.

x	$y = x^{3/4} + 2$	y
0	$y = (0)^{3/4} + 2 = 2$	2
1	$y = (1)^{3/4} + 2 = 3$	3
16	$y = (2)^{3/4} + 2 = 10$	10

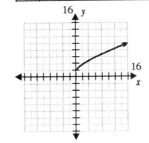

SSM: Experiencing Introductory and Intermediate Algebra

69. $y = 3 - x^{2/3}$

Domain: $(-\infty, \infty)$

Values will vary; possible solutions follow.

x	$y = 3 - x^{2/3}$	y
-8	$y = 3 - (-8)^{2/3} = -1$	-1
0	$y = 3 - (0)^{2/3} = 3$	3
8	$y = 3 - (8)^{2/3} = -1$	-1

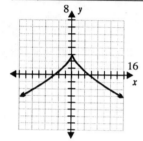

71. $y = x^{4/5} + 1$

Domain: $(-\infty, \infty)$

Values will vary; possible solutions follow.

x	$y = x^{4/5} + 1$	y
-32	$y = (-32)^{4/5} + 1 = 17$	17
0	$y = (0)^{4/5} + 1 = 1$	1
32	$y = (32)^{4/5} + 1 = 17$	17

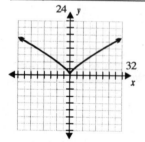

73. $y = 5(x-4)^{2/3}$

Domain: $(-\infty, \infty)$

Values will vary; possible solutions follow.

x	$y = 5(x-4)^{2/3}$	y
3	$y = 5((3)-4)^{2/3} = 5$	5
4	$y = 5((4)-4)^{2/3} = 0$	0
5	$y = 5((5)-4)^{2/3} = 5$	5

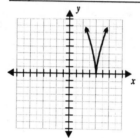

75. $V = (12S)^{1/2}$

A car that leaves a skid mark that is 250 feet long was traveling at about 55 miles per hour.

13.2 Calculator Exercises

1. $(-27)^{2/3} = \left(\sqrt[3]{-27}\right)^2 = (-3)^2 = 9$

2. $\left[(-27)^{1/3}\right]^2$

$\left[(-27)^2\right]^{1/3}$

$\left(\sqrt[3]{-27}\right)^2$

$\sqrt[3]{(-27)^2}$

Chapter 13: Radical Expressions, Functions, and Equations

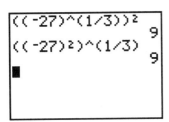

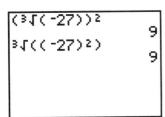

All four ways yield the same correct result.

13.3 Exercises

1. $9^{-1/2} = \dfrac{1}{9^{1/2}} = \dfrac{1}{3}$

3. $\dfrac{1}{8^{-1/3}} = 8^{1/3} = 2$

5. $27^{1/3} \cdot 27^{4/3} = 27^{1/3+4/3} = 27^{5/3} = 243$

7. $\dfrac{32^{4/5}}{32^{1/5}} = 32^{(4/5)-(1/5)} = 32^{3/5} = 8$

9. $\left(125^{1/2}\right)^{2/3} = 125^{\frac{1}{2}\cdot\frac{2}{3}} = 125^{1/3} = 5$

11. $3^{1/2} \cdot 12^{1/2} = (3 \cdot 12)^{1/2} = 36^{1/2} = 6$

13. $\left(\dfrac{8}{27}\right)^{1/3} = \dfrac{8^{1/3}}{27^{1/3}} = \dfrac{2}{3}$

15. $\left(\dfrac{9}{16}\right)^{-1/2} = \left(\dfrac{16}{9}\right)^{1/2} = \dfrac{16^{1/2}}{9^{1/2}} = \dfrac{4}{3}$

17. $\left[\left(2^{1/2}\right)\left(8^{1/2}\right)\right]^{-3/2} = \left[(2\cdot 8)^{1/2}\right]^{-3/2}$
$= \left[16^{1/2}\right]^{-3/2} = 4^{-3/2} = \dfrac{1}{4^{3/2}} = \dfrac{1}{8}$

19. $x^{-2/3} = \dfrac{1}{x^{2/3}}$

21. $\dfrac{12}{y^{-3/4}} = 12y^{3/4}$

23. $z^{2/3} \cdot z^{3/4} = z^{\frac{2}{3}+\frac{3}{4}} = z^{\frac{8}{12}+\frac{9}{12}} = z^{17/12}$

25. $\dfrac{p^{5/6}}{p^{2/3}} = p^{\frac{5}{6}-\frac{2}{3}} = p^{\frac{5}{6}-\frac{4}{6}} = p^{1/6}$

27. $\left(b^{3/4}\right)^{8/9} = b^{\frac{3}{4}\cdot\frac{8}{9}} = b^{2/3}$

29. $\left(x^2 y\right)^{3/4} = x^{2\cdot 3/4} y^{3/4} = x^{3/2} y^{3/4}$

31. $\left(\dfrac{x^2}{y^3}\right)^{5/6} = \dfrac{x^{2\cdot 5/6}}{y^{3\cdot 5/6}} = \dfrac{x^{5/3}}{y^{5/2}}$

33. $\left(\dfrac{a}{b}\right)^{-1/3} = \left(\dfrac{b}{a}\right)^{1/3} = \dfrac{b^{1/3}}{a^{1/3}}$

35. $c^{3/5} \cdot c^{-4/5} = c^{\frac{3}{5}+\left(-\frac{4}{5}\right)} = c^{-1/5} = \dfrac{1}{c^{1/5}}$

37. $\left(8a^5 b^6\right)^{2/3} = 8^{2/3} a^{5\cdot 2/3} b^{6\cdot 2/3} = 4a^{10/3} b^4$

39. $\left(5a^{3/4} b^{2/5}\right)\left(2a^{1/3} b^{2/5}\right) = 5\cdot 2 a^{\frac{3}{4}+\frac{1}{3}} b^{\frac{2}{5}+\frac{2}{5}}$
$= 10 a^{\frac{9}{12}+\frac{4}{12}} b^{4/5} = 10 a^{13/12} b^{4/5}$

41. $\left(\dfrac{m^{3/7}}{2m^{-2/7}}\right)^2 = \left(\dfrac{m^{\frac{3}{7}-\left(-\frac{2}{7}\right)}}{2}\right)^2 = \left(\dfrac{m^{5/7}}{2}\right)^2$
$= \dfrac{m^{2\cdot 5/7}}{2^2} = \dfrac{m^{10/7}}{4}$

43. $\left(3a^{1/3}b^2c^{1/6}\right)^{-2}\left(2a^{4/3}b^{1/4}c^{5/6}\right)^3$

$= \left(3^{-2}a^{-2\cdot 1/3}b^{-2\cdot 2}c^{-2\cdot 1/6}\right)\left(2^3 a^{3\cdot 4/3}b^{3\cdot 1/4}c^{3\cdot 5/6}\right)$

$= \left(\dfrac{1}{9}a^{-2/3}b^{-4}c^{-1/3}\right)\left(8a^4 b^{3/4} c^{5/2}\right)$

$= \dfrac{8}{9}a^{-\frac{2}{3}+4}b^{-4+\frac{3}{4}}c^{-\frac{1}{3}+\frac{5}{2}} = \dfrac{8}{9}a^{10/3}b^{-13/4}c^{13/6}$

$= \dfrac{8a^{10/3}c^{13/6}}{9b^{13/4}}$

45. $x^{1/4}\left(x^{2/3} - x^{4/5}\right) = x^{\frac{1}{4}+\frac{2}{3}} - x^{\frac{1}{4}+\frac{4}{5}}$

$= x^{11/12} - x^{21/25}$

47. $2x^{2/5}\left(x^{1/5} + 3y^{2/5}\right) = 2x^{\frac{2}{5}+\frac{1}{5}} + 2\cdot 3x^{2/5}y^{2/5}$

$= 2x^{3/5} + 6x^{2/5}y^{2/5}$

49. $a^{1/4}b^{1/3}\left(3 - a^{1/2}b^{5/6}\right)$

$= 3a^{1/4}b^{1/3} - a^{\frac{1}{4}+\frac{1}{2}}b^{\frac{1}{3}+\frac{5}{6}}$

$= 3a^{1/4}b^{1/3} - a^{3/4}b^{7/6}$

51. $\left(x^{1/2} - y^{1/2}\right)\left(x^{1/3} + y^{1/3}\right)$

$= x^{\frac{1}{2}+\frac{1}{3}} - x^{1/3}y^{1/2} + x^{1/2}y^{1/3} - y^{\frac{1}{2}+\frac{1}{3}}$

$= x^{5/6} - x^{1/3}y^{1/2} + x^{1/2}y^{1/3} - y^{5/6}$

53. $\left(x^{1/4} + y^{1/4}\right)\left(x^{1/4} - y^{1/4}\right) = \left(x^{1/4}\right)^2 - \left(y^{1/4}\right)^2$

$= x^{2\cdot 1/4} - y^{2\cdot 1/4} = x^{1/2} - y^{1/2}$

55. $\left(x^2 + y^2\right)\left(x^{1/2} - y^{1/2}\right)$

$= x^{2+\frac{1}{2}} + x^{1/2}y^2 - x^2 y^{1/2} - y^{2+\frac{1}{2}}$

$= x^{5/2} - x^2 y^{1/2} + x^{1/2}y^2 - y^{5/2}$

57. $\left(x^{1/3} + 2\right)\left(x^{1/3} - 2\right) = \left(x^{1/3}\right)^2 - (2)^2$

$= x^{2\cdot 1/3} - 2^2 = x^{2/3} - 4$

59. $\left(x^{1/4} + 2\right)^2 = \left(x^{1/4}\right)^2 + 2\cdot 2x^{1/4} + (2)^2$

$= x^{1/2} + 4x^{1/4} + 4$

61. $\left(x - y^{1/2}\right)^2 = (x)^2 - 2\cdot xy^{1/2} + \left(y^{1/2}\right)^2$

$= x^2 - 2xy^{1/2} + y$

63. $\left(a^{1/2} + b^{1/2}\right)^2 = \left(a^{1/2}\right)^2 + 2\cdot a^{1/2}b^{1/2} + \left(b^{1/2}\right)^2$

$= a + 2a^{1/2}b^{1/2} + b$

65. Let s_1 = original length of a side

s_2 = length of the larger side

$s_1 = V^{1/3}$

$s_2 = (8V)^{1/3} = 8^{1/3}V^{1/3} = 2V^{1/3} = 2s_1$

The length of each side should be doubled in order to increase the volume to 8 times its original size.

67. Let

t_1 = time to fall from original height

t_2 = time to fall from 3 times the original height

$t_1 = \left(\dfrac{d}{16.1}\right)^{1/2}$

$t_2 = \left(\dfrac{3d}{16.1}\right)^{1/2} = 3^{1/2}\left(\dfrac{d}{16.1}\right)^{1/2} = 3^{1/2}t_1 = \sqrt{3}t_1$

Raising the height to three times its original height will increase the time to fall by a factor of $\sqrt{3}$.

69. Let

t_1 = time to fall from original height

t_2 = time to fall from 3 times the original height

$t_1 = \left(\dfrac{d}{4.9}\right)^{1/2}$

$t_2 = \left(\dfrac{3d}{4.9}\right)^{1/2} = 3^{1/2}\left(\dfrac{d}{4.9}\right)^{1/2} = 3^{1/2}t_1 = \sqrt{3}t_1$

Raising the height to three times its original height will increase the time to fall by a factor of $\sqrt{3}$. The answers are the same; no, changing the units from feet to meters did not affect the time.

13.3 Calculator Exercises

1. $\left(x^3\right)^{2/3} = x^{3 \cdot 2/3} = x^2$

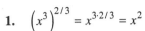

2. $\left(x^3\right)^{-2/3} = x^{3 \cdot (-2/3)} = x^{-2} = \dfrac{1}{x^2}$

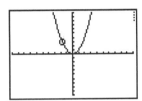

3. $\dfrac{x^4}{x^{9/4}} = x^{4-\frac{9}{4}} = x^{7/4}$

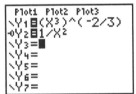
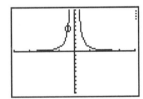

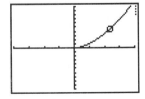

4. $\dfrac{x^2}{x^{7/2}} = x^{2-\frac{7}{2}} = x^{-3/2} = \dfrac{1}{x^{3/2}}$

5. $x^{3/4}\left(x^{3/2} - x^{2/3}\right) = x^{\frac{3}{4}+\frac{3}{2}} - x^{\frac{3}{4}+\frac{2}{3}}$
$= x^{9/4} - x^{17/12}$

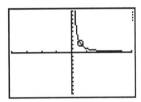

6. $\left(x^{1/3} + 1\right)\left(x^{1/3} - 1\right) = \left(x^{1/3}\right)^2 - (1)^2$
$= x^{2 \cdot 1/3} - 1^2 = x^{2/3} - 1$

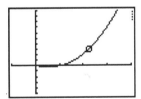

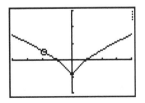

SSM: Experiencing Introductory and Intermediate Algebra

13.4 Exercises

1. $\sqrt{28} = \sqrt{4 \cdot 7} = \sqrt{4} \cdot \sqrt{7} = 2\sqrt{7}$

3. $\sqrt[3]{-686} = \sqrt[3]{-343 \cdot 2} = \sqrt[3]{-343} \cdot \sqrt[3]{2} = -7\sqrt[3]{2}$

5. $\sqrt[4]{112} = \sqrt[4]{16 \cdot 7} = \sqrt[4]{16} \cdot \sqrt[4]{7} = 2\sqrt[4]{7}$

7. $\sqrt[5]{-6250} = \sqrt[5]{-3125 \cdot 2} = \sqrt[5]{-3125} \cdot \sqrt[5]{2} = -5\sqrt[5]{2}$

9. $\sqrt{20x^4 y^3 z^2}$
$= \sqrt{4 \cdot 5 \cdot x^4 \cdot y^2 \cdot y \cdot z^2}$
$= \sqrt{4x^4 y^2 z^2} \cdot \sqrt{5y}$
$= 2x^2 yz\sqrt{5y}$

11. $\sqrt[3]{72m^5 n^9} = \sqrt[3]{8 \cdot 9 \cdot m^3 \cdot m^2 \cdot n^9}$
$= \sqrt[3]{8m^3 n^9} \cdot \sqrt[3]{9m^2}$
$= 2mn^2 \sqrt[3]{9m^2}$

13. $\sqrt[4]{162x^4 y^5} = \sqrt[4]{81 \cdot 2 \cdot x^4 \cdot y^4 \cdot y}$
$= \sqrt[4]{81x^4 y^4} \cdot \sqrt[4]{2y}$
$= 3xy\sqrt[4]{2y}$

15. $\sqrt[5]{486a^6 b^{10} c^2} = \sqrt[5]{243 \cdot 2 \cdot a^5 \cdot a \cdot b^{10} \cdot c^2}$
$= \sqrt[5]{243a^5 b^{10}} \cdot \sqrt[5]{2ac^2}$
$= 3ab^2 \sqrt[5]{2ac^2}$

17. $\sqrt{5x^2 + 30x + 45} = \sqrt{5(x^2 + 6x + 9)}$
$= \sqrt{5(x+3)^2} = \sqrt{5} \cdot \sqrt{(x+3)^2}$
$= (x+3)\sqrt{5}$

19. $\sqrt{7x} \cdot \sqrt{14y} = \sqrt{7x \cdot 14y}$
$= \sqrt{98xy} = \sqrt{49 \cdot 2xy}$
$= 7\sqrt{2xy}$

21. $2\sqrt{7xy} \cdot x\sqrt{14y} = 2x\sqrt{7xy \cdot 14y}$
$= 2x\sqrt{98xy^2} = 2x\sqrt{49 \cdot 2 \cdot x \cdot y^2}$
$= 2x(7y)\sqrt{2x}$
$= 14xy\sqrt{2x}$

23. $\sqrt{7x + 14y} \cdot \sqrt{3x + 6y}$
$= \sqrt{(7x + 14y)(3x + 6y)}$
$= \sqrt{7(x+2y)(3)(x+2y)}$
$= \sqrt{21(x+2y)^2}$
$= (x+2y)\sqrt{21}$

25. $\sqrt{x+2} \cdot \sqrt{x^2 + 3x + 2}$
$= \sqrt{(x+2)(x^2 + 3x + 2)}$
$= \sqrt{(x+2)(x+2)(x+1)}$
$= \sqrt{(x+2)^2 (x+1)}$
$= (x+2)\sqrt{x+1}$

27. $\sqrt{x^2 + x - 2} \cdot \sqrt{x^2 + 3x - 4}$
$= \sqrt{(x^2 + x - 2)(x^2 + 3x - 4)}$
$= \sqrt{(x+2)(x-1)(x+4)(x-1)}$
$= \sqrt{(x-1)^2 (x+2)(x+4)}$
$= (x-1)\sqrt{(x+2)(x+4)}$

29. $\sqrt[3]{-4x^4 y^2} \cdot \sqrt[3]{6xy} = \sqrt[3]{-4x^4 y^2 \cdot 6xy}$
$= \sqrt[3]{-24x^5 y^3} = \sqrt[3]{-8 \cdot 3 \cdot x^3 \cdot x^2 \cdot y^3}$
$= -2xy\sqrt[3]{3x^2}$

31. $-\sqrt{\dfrac{36}{49}} = -\dfrac{\sqrt{36}}{\sqrt{49}} = -\dfrac{6}{7}$

33. $\sqrt{\dfrac{1210}{1440}} = \sqrt{\dfrac{121 \cdot 10}{144 \cdot 10}} = \sqrt{\dfrac{121}{144}} = \dfrac{\sqrt{121}}{\sqrt{144}} = \dfrac{11}{12}$

Chapter 13: Radical Expressions, Functions, and Equations

35. $-\sqrt{\dfrac{8}{27}} = -\dfrac{\sqrt{8}}{\sqrt{27}} = -\dfrac{\sqrt{4\cdot 2}}{\sqrt{9\cdot 3}}$
$= -\dfrac{2\sqrt{2}}{3\sqrt{3}} = -\dfrac{2\sqrt{2}}{3\sqrt{3}} \cdot \dfrac{\sqrt{3}}{\sqrt{3}} = -\dfrac{2\sqrt{2\cdot 3}}{3\cdot 3}$
$= -\dfrac{2\sqrt{6}}{9}$

37. $\sqrt[3]{\dfrac{4}{125}} = \dfrac{\sqrt[3]{4}}{\sqrt[3]{125}} = \dfrac{\sqrt[3]{4}}{5}$

39. $\sqrt[3]{-\dfrac{9}{25}} = \dfrac{\sqrt[3]{-9}}{\sqrt[3]{25}} = \dfrac{\sqrt[3]{-9}}{\sqrt[3]{25}} \cdot \dfrac{\sqrt[3]{5}}{\sqrt[3]{5}} = \dfrac{\sqrt[3]{-9\cdot 5}}{5} = \dfrac{-\sqrt[3]{45}}{5}$

41. $\sqrt[5]{\dfrac{1}{16}} = \dfrac{\sqrt[5]{1}}{\sqrt[5]{16}} = \dfrac{\sqrt[5]{1}}{\sqrt[5]{16}} \cdot \dfrac{\sqrt[5]{2}}{\sqrt[5]{2}} = \dfrac{\sqrt[5]{2}}{2}$

43. $\sqrt{\dfrac{4x^2}{9y^2}} = \dfrac{\sqrt{4x^2}}{\sqrt{9y^2}} = \dfrac{2x}{3y}$

45. $\sqrt{\dfrac{3x}{25y^2}} = \dfrac{\sqrt{3x}}{\sqrt{25y^2}} = \dfrac{\sqrt{3x}}{5y}$

47. $\sqrt{\dfrac{4xy}{5z}} = \dfrac{\sqrt{4xy}}{\sqrt{5z}} = \dfrac{2\sqrt{xy}}{\sqrt{5z}} \cdot \dfrac{\sqrt{5z}}{\sqrt{5z}} = \dfrac{2\sqrt{5xyz}}{5z}$

49. $\sqrt{\dfrac{2x^2}{6}} = \sqrt{\dfrac{x^2}{3}} = \dfrac{\sqrt{x^2}}{\sqrt{3}} = \dfrac{x}{\sqrt{3}} \cdot \dfrac{\sqrt{3}}{\sqrt{3}} = \dfrac{x\sqrt{3}}{3}$

51. $\sqrt[3]{\dfrac{27x^3}{y^6}} = \dfrac{\sqrt[3]{27x^3}}{\sqrt[3]{y^6}} = \dfrac{3x}{y^2}$

53. $\sqrt[3]{\dfrac{3}{25x^2}} = \dfrac{\sqrt[3]{3}}{\sqrt[3]{25x^2}} = \dfrac{\sqrt[3]{3}}{\sqrt[3]{25x^2}} \cdot \dfrac{\sqrt[3]{5x}}{\sqrt[3]{5x}} = \dfrac{\sqrt[3]{15x}}{5x}$

55. $\sqrt[3]{\dfrac{3x^2 y}{5xy^2 z}} = \sqrt[3]{\dfrac{3x}{5yz}} = \dfrac{\sqrt[3]{3x}}{\sqrt[3]{5yz}} = \dfrac{\sqrt[3]{3x}}{\sqrt[3]{5yz}} \cdot \dfrac{\sqrt[3]{25y^2 z^2}}{\sqrt[3]{25y^2 z^2}}$
$= \dfrac{\sqrt[3]{75xy^2 z^2}}{5yz}$

57. $\dfrac{\sqrt{5}}{\sqrt{x}} = \dfrac{\sqrt{5}}{\sqrt{x}} \cdot \dfrac{\sqrt{x}}{\sqrt{x}} = \dfrac{\sqrt{5x}}{x}$

59. $\dfrac{\sqrt{4x}}{\sqrt{8xy}} = \sqrt{\dfrac{4x}{8xy}} = \sqrt{\dfrac{1}{2y}} = \dfrac{\sqrt{1}}{\sqrt{2y}} = \dfrac{\sqrt{1}}{\sqrt{2y}} \cdot \dfrac{\sqrt{2y}}{\sqrt{2y}}$
$= \dfrac{\sqrt{2y}}{2y}$

61. $\dfrac{\sqrt{z^3}}{\sqrt{8}} = \dfrac{\sqrt{z^2 \cdot z}}{\sqrt{4\cdot 2}} = \dfrac{z\sqrt{z}}{2\sqrt{2}} = \dfrac{z\sqrt{z}}{2\sqrt{2}} \cdot \dfrac{\sqrt{2}}{\sqrt{2}} = \dfrac{z\sqrt{2z}}{4}$

63. $\dfrac{\sqrt{ab^4}}{\sqrt{a^2 b}} = \sqrt{\dfrac{ab^4}{a^2 b}} = \sqrt{\dfrac{b^3}{a}} = \dfrac{\sqrt{b^3}}{\sqrt{a}} = \dfrac{b\sqrt{b}}{\sqrt{a}} \cdot \dfrac{\sqrt{a}}{\sqrt{a}}$
$= \dfrac{b\sqrt{ab}}{a}$

65. $\dfrac{3x\sqrt{5x^2 y}}{6\sqrt{15x^3 y}} = \dfrac{x}{2}\sqrt{\dfrac{5x^2 y}{15x^3 y}} = \dfrac{x}{2}\sqrt{\dfrac{1}{3x}} = \dfrac{x}{2\sqrt{3x}}$
$= \dfrac{x}{2\sqrt{3x}} \cdot \dfrac{\sqrt{3x}}{\sqrt{3x}} = \dfrac{x\sqrt{3x}}{6x} = \dfrac{\sqrt{3x}}{6}$

67. $\dfrac{\sqrt[3]{3}}{\sqrt[3]{x^2}} = \dfrac{\sqrt[3]{3}}{\sqrt[3]{x^2}} \cdot \dfrac{\sqrt[3]{x}}{\sqrt[3]{x}} = \dfrac{\sqrt[3]{3x}}{x}$

69. $\dfrac{\sqrt[3]{3x}}{\sqrt[3]{12xy}} = \sqrt[3]{\dfrac{3x}{12xy}} = \sqrt[3]{\dfrac{1}{4y}} = \dfrac{\sqrt[3]{1}}{\sqrt[3]{4y}} = \dfrac{\sqrt[3]{1}}{\sqrt[3]{4y}} \cdot \dfrac{\sqrt[3]{2y^2}}{\sqrt[3]{2y^2}}$
$= \dfrac{\sqrt[3]{2y^2}}{2y}$

71. $\dfrac{\sqrt[3]{-8xy^2 z^2}}{\sqrt[3]{3x^2 y^2 z}} = \sqrt[3]{\dfrac{-8xy^2 z^2}{3x^2 y^2 z}} = \sqrt[3]{\dfrac{-8z}{3x}} = \dfrac{\sqrt[3]{-8z}}{\sqrt[3]{3x}}$
$= \dfrac{-2\sqrt[3]{z}}{\sqrt[3]{3x}} \cdot \dfrac{\sqrt[3]{9x^2}}{\sqrt[3]{9x^2}} = \dfrac{-2\sqrt[3]{9x^2 z}}{3x}$

73. $T = 2\pi\sqrt{\dfrac{L}{32}}$
$= 2\pi\sqrt{\dfrac{7}{32}} \approx 2.94$

The period of the pendulum is about 2.94 seconds.

SSM: Experiencing Introductory and Intermediate Algebra

75. $T = 2\pi\sqrt{\dfrac{L}{32}}$

$= 2\pi\sqrt{\dfrac{3}{32}} \approx 1.92$

It will take the man about 1.92 seconds to complete one forward-and-back motion.

77. $T = 2\pi\sqrt{\dfrac{L}{32}}$

$= 2\pi\sqrt{\dfrac{2.5}{32}} \approx 1.76$

It takes Eydie about 1.76 seconds to complete one forward-and-back motion.

$d = r \cdot t$

$r = \dfrac{d}{t} = \dfrac{2(4)}{1.76} \approx 4.56 \;\dfrac{\text{ft}}{\text{sec}}$

$\dfrac{4.56 \text{ ft}}{1 \text{ sec}} \cdot \dfrac{3600 \text{ sec}}{1 \text{ hr}} \cdot \dfrac{1 \text{ mile}}{5280 \text{ ft}} \approx 3.11 \text{ miles/hr}$

Eydie's walking speed is about 4.55 ft/sec, or 3.11 miles per hour.

13.4 Calculator Exercises

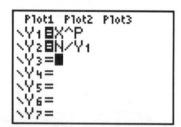

1.

$13718 = 6859 \cdot 2 = 19^3 \cdot 2$

$\sqrt[3]{13718} = \sqrt[3]{19^3 \cdot 2} = 19\sqrt[3]{2}$

2.

$199927 = 28561 \cdot 7 = 13^4 \cdot 7$

$\sqrt[4]{199927} = \sqrt[4]{13^4 \cdot 7} = 13\sqrt[4]{7}$

3.

$-1664 = -128 \cdot 13 = (-2)^7 \cdot 13$

$\sqrt[7]{-1664} = \sqrt[7]{(-2)^7 \cdot 13} = -2\sqrt[7]{13}$

4.

$9251 = 841 \cdot 11 = 29^2 \cdot 11$

$\sqrt{9251} = \sqrt{29^2 \cdot 11} = 29\sqrt{11}$

13.5 Exercises

1. $2\sqrt{28} - \sqrt{63}$
$= 2\sqrt{4 \cdot 7} - \sqrt{9 \cdot 7}$
$= 4\sqrt{7} - 3\sqrt{7} = (4-3)\sqrt{7}$
$= \sqrt{7}$

3. $\sqrt{10} - \sqrt{\dfrac{1}{10}}$
$= \sqrt{10} - \dfrac{\sqrt{10}}{10} = \left(1 - \dfrac{1}{10}\right)\sqrt{10}$
$= \dfrac{9}{10}\sqrt{10}$

5. $\sqrt[3]{24} - 4\sqrt[3]{3}$
$= 2\sqrt[3]{3} - 4\sqrt[3]{3} = (2-4)\sqrt[3]{3}$
$= -2\sqrt[3]{3}$

7. $5\sqrt{75} - 2\sqrt{27} + \sqrt{48}$
$= 5\sqrt{25 \cdot 3} - 2\sqrt{9 \cdot 3} + \sqrt{16 \cdot 3}$
$= 25\sqrt{3} - 6\sqrt{3} + 4\sqrt{3} = (25-6+4)\sqrt{3}$
$= 23\sqrt{3}$

9. $5\sqrt{x} + 9\sqrt{x}$
$= (5+9)\sqrt{x}$
$= 14\sqrt{x}$

11. $\sqrt{25x} + \sqrt{36x}$
$= 5\sqrt{x} + 6\sqrt{x} = (5+6)\sqrt{x}$
$= 11\sqrt{x}$

13. $\sqrt{8x^3} - \sqrt{50x^3}$
$= \sqrt{4 \cdot 2 \cdot x^2 \cdot x} - \sqrt{25 \cdot 2 \cdot x^2 \cdot x}$
$= 2x\sqrt{2x} - 5x\sqrt{2x} = (2x-5x)\sqrt{2x}$
$= -3x\sqrt{2x}$

15. $\sqrt{9a} + \sqrt{16a^3}$
$= 3\sqrt{a} + 4a\sqrt{a} = (3+4a)\sqrt{a}$

17. $7a\sqrt{b} + 9a\sqrt{b} - 2a\sqrt{b}$
$= (7a + 9a - 2a)\sqrt{b}$
$= 14a\sqrt{b}$

19. $5\sqrt{pq} - 4\sqrt[3]{pq} + 2\sqrt{pq} + 11\sqrt[3]{pq}$
$= (5+2)\sqrt{pq} + (-4+11)\sqrt[3]{pq}$
$= 7\sqrt{pq} + 7\sqrt[3]{pq}$

21. $7y\sqrt{x^3y} + 3x\sqrt{xy^3} - 4xy\sqrt{xy}$
$= 7xy\sqrt{xy} + 3xy\sqrt{xy} - 4xy\sqrt{xy}$
$= (7+3-4)xy\sqrt{xy}$
$= 6xy\sqrt{xy}$

23. $2x\sqrt[3]{x^2y^4z} - 3y\sqrt[3]{x^5yz}$
$= 2xy\sqrt[3]{x^2yz} - 3xy\sqrt[3]{x^2yz}$
$= (2-3)xy\sqrt[3]{x^2yz}$
$= -xy\sqrt[3]{x^2yz}$

25. $\sqrt{7}\left(\sqrt{5} - \sqrt{7}\right)$
$= \sqrt{35} - \sqrt{49} = \sqrt{35} - 7$

27. $\sqrt{3}\left(\sqrt{x} - \sqrt{5}\right) = \sqrt{3x} - \sqrt{15}$

29. $3\sqrt{a}\left(2\sqrt{a} - 5\right) = 6\sqrt{a^2} - 15\sqrt{a}$
$= 6a - 15\sqrt{a}$

31. $2\sqrt[3]{x}\left(4\sqrt[3]{x^2} - 6\sqrt[3]{x}\right) = 8\sqrt[3]{x^3} - 12\sqrt[3]{x^2}$
$= 8x - 12\sqrt[3]{x^2}$

33. $\left(\sqrt{3} - 5\sqrt{6}\right)\left(2\sqrt{3} + \sqrt{8}\right)$
$= 2\sqrt{9} - 10\sqrt{18} + \sqrt{24} - 5\sqrt{48}$
$= 6 - 30\sqrt{2} + 2\sqrt{6} - 20\sqrt{3}$

35. $\left(\sqrt{3} - \sqrt{x}\right)\left(\sqrt{2} + \sqrt{x}\right)$
$= \sqrt{6} - \sqrt{2x} + \sqrt{3x} - \sqrt{x^2}$
$= \sqrt{6} - \sqrt{2x} + \sqrt{3x} - x$

SSM: Experiencing Introductory and Intermediate Algebra

37. $(5-\sqrt{6})(5+\sqrt{6})$
$= (5)^2 - (\sqrt{6})^2 = 25 - 6$
$= 19$

39. $(12+\sqrt{p})(12-\sqrt{p})$
$= (12)^2 - (\sqrt{p})^2$
$= 144 - p$

41. $(\sqrt{2x}+\sqrt{3y})(\sqrt{2x}-\sqrt{3y})$
$= (\sqrt{2x})^2 - (\sqrt{3y})^2$
$= 2x - 3y$

43. $(\sqrt{a}+4)^2 = (\sqrt{a})^2 + 2\cdot 4\sqrt{a} + (4)^2$
$= a + 8\sqrt{a} + 16$

45. $(3\sqrt{b}-2)^2 = (3\sqrt{b})^2 - 2(3\sqrt{b})(2) + (2)^2$
$= 9b - 12\sqrt{b} + 4$

47. $(\sqrt{x}-\sqrt{y})^2 = (\sqrt{x})^2 - 2(\sqrt{x})(\sqrt{y}) + (\sqrt{y})^2$
$= x - 2\sqrt{xy} + y$

49. $\dfrac{\sqrt{21x}-\sqrt{14}}{\sqrt{7}} = \dfrac{\sqrt{21x}}{\sqrt{7}} - \dfrac{\sqrt{14}}{\sqrt{7}} = \sqrt{3x} - \sqrt{2}$

51. $\dfrac{\sqrt{a}-12}{\sqrt{a}} = \dfrac{\sqrt{a}}{\sqrt{a}} - \dfrac{12}{\sqrt{a}} = 1 - \dfrac{12}{\sqrt{a}} \cdot \dfrac{\sqrt{a}}{\sqrt{a}}$
$= 1 - \dfrac{12\sqrt{a}}{a}$

53. $\dfrac{\sqrt{x}+\sqrt{y}+\sqrt{z}}{\sqrt{x}}$
$= \dfrac{\sqrt{x}}{\sqrt{x}} + \dfrac{\sqrt{y}}{\sqrt{x}} + \dfrac{\sqrt{z}}{\sqrt{x}}$
$= 1 + \dfrac{\sqrt{y}}{\sqrt{x}} \cdot \dfrac{\sqrt{x}}{\sqrt{x}} + \dfrac{\sqrt{z}}{\sqrt{x}} \cdot \dfrac{\sqrt{x}}{\sqrt{x}}$
$= 1 + \dfrac{\sqrt{xy}}{x} + \dfrac{\sqrt{xz}}{x}$

55. $\dfrac{18}{\sqrt{6}+\sqrt{3}}$
$= \dfrac{18}{\sqrt{6}+\sqrt{3}} \cdot \dfrac{\sqrt{6}-\sqrt{3}}{\sqrt{6}-\sqrt{3}} = \dfrac{18(\sqrt{6}-\sqrt{3})}{(\sqrt{6})^2 - (\sqrt{3})^2}$
$= \dfrac{18(\sqrt{6}-\sqrt{3})}{6-3} = \dfrac{18(\sqrt{6}-\sqrt{3})}{3}$
$= 6(\sqrt{6}-\sqrt{3})$

57. $\dfrac{\sqrt{3}+\sqrt{2}}{\sqrt{3}-\sqrt{2}}$
$= \dfrac{\sqrt{3}+\sqrt{2}}{\sqrt{3}-\sqrt{2}} \cdot \dfrac{\sqrt{3}+\sqrt{2}}{\sqrt{3}+\sqrt{2}}$
$= \dfrac{(\sqrt{3})^2 + 2\cdot\sqrt{3}\cdot\sqrt{2} + (\sqrt{2})^2}{(\sqrt{3})^2 - (\sqrt{2})^2}$
$= \dfrac{3 + 2\sqrt{6} + 2}{3 - 2}$
$= 5 + 2\sqrt{6}$

59. $\dfrac{3x}{\sqrt{x}-2}$
$= \dfrac{3x}{\sqrt{x}-2} \cdot \dfrac{\sqrt{x}+2}{\sqrt{x}+2} = \dfrac{3x(\sqrt{x}+2)}{(\sqrt{x})^2 - 2^2}$
$= \dfrac{3x(\sqrt{x}+2)}{x-4}$
$= \dfrac{3x\sqrt{x} + 6\sqrt{x}}{x-4}$

61. $\dfrac{3b-4}{\sqrt{3b}-2}$
$= \dfrac{3b-4}{\sqrt{3b}-2} \cdot \dfrac{\sqrt{3b}+2}{\sqrt{3b}+2} = \dfrac{(3b-4)(\sqrt{3b}+2)}{(\sqrt{3b})^2 - (2)^2}$
$= \dfrac{(3b-4)(\sqrt{3b}+2)}{3b-4}$
$= \sqrt{3b} + 2$

63. $\dfrac{\sqrt{x}+3}{\sqrt{x}-3}$

$= \dfrac{\sqrt{x}+3}{\sqrt{x}-3} \cdot \dfrac{\sqrt{x}+3}{\sqrt{x}+3} = \dfrac{\left(\sqrt{x}\right)^2 + 2 \cdot 3\sqrt{x} + (3)^2}{\left(\sqrt{x}\right)^2 - (3)^2}$

$= \dfrac{x + 6\sqrt{x} + 9}{x - 9}$

65. $\dfrac{3\sqrt{x}-4}{4-3\sqrt{x}} = \dfrac{-\left(4-3\sqrt{x}\right)}{4-3\sqrt{x}} = -1$

67. $A = s^2$

$P = 4s$

Let s_1 = length of side for 1st painting.

$s_1 = \sqrt{A_1}$

$s_1 = \sqrt{490} = 7\sqrt{10}$

Perimeter of first painting:

$4s = 4\left(7\sqrt{10}\right) = 28\sqrt{10}$

Let s_2 = length of side of 2nd painting.

$s_2 = \sqrt{A_2}$

$s_2 = \sqrt{810} = 9\sqrt{10}$

Perimeter of second painting:

$4s = 4\left(9\sqrt{10}\right) = 36\sqrt{10}$

Difference:

$36\sqrt{10} - 28\sqrt{10} = 8\sqrt{10} \approx 25.3$

The larger frame will require about 25.3 inches more than for the smaller frame.

69. $V = s^3$

$P = 4s$

Let s_1 = length of a side of the larger box.

$s_1 = \sqrt[3]{V_1}$

$s_1 = \sqrt[3]{1750} = 5\sqrt[3]{14}$

Perimeter of large base:

$4s = 4\left(5\sqrt[3]{14}\right) = 20\sqrt[3]{14}$

Let s_2 = length of a side of the smaller box.

$s_2 = \sqrt[3]{V_2}$

$s_2 = \sqrt[3]{896} = 4\sqrt[3]{14}$

Perimeter of smaller base:

$4s = 4\left(4\sqrt[3]{14}\right) = 16\sqrt[3]{14}$

Difference:

$20\sqrt[3]{14} - 16\sqrt[3]{14} = 4\sqrt[3]{14} \approx 9.64$

The larger box has a perimeter that is about 9.64 inches more than for the smaller box.

13.5 Calculator Exercises

1. $\left(3\sqrt[3]{x} + 5\sqrt{x}\right) + \left(\sqrt[3]{x} - 3\sqrt{x}\right) + \left(4\sqrt{x} + 7\right)$

$= 3\sqrt[3]{x} + 5\sqrt{x} + \sqrt[3]{x} - 3\sqrt{x} + 4\sqrt{x} + 7$

$= 4\sqrt[3]{x} + 6\sqrt{x} + 7$

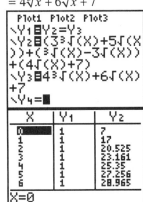

2. $\left(2\sqrt{x} + 7\right) - \left(\sqrt{x} + 9\right)$

$= 2\sqrt{x} + 7 - \sqrt{x} - 9$

$= \sqrt{x} - 2$

3. $2\sqrt[3]{x}\left(3\sqrt[3]{2x}+5\right)$

$= 6\sqrt[3]{2x^2} + 10\sqrt[3]{x}$

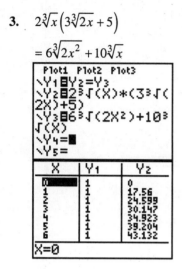

4. $\left(2\sqrt{x}+3\sqrt{2x}\right)\left(\sqrt{x}+\sqrt{2x}\right)$

$= 2\sqrt{x^2} + 3\sqrt{2x^2} + 2\sqrt{2x^2} + 3\sqrt{4x^2}$

$= 2x + 3x\sqrt{2} + 2x\sqrt{2} + 6x$

$= 8x + 5x\sqrt{2}$

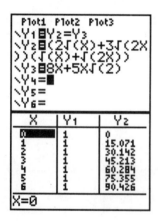

5. $\dfrac{x-16}{\sqrt{x}+4}$

$= \dfrac{x-16}{\sqrt{x}+4} \cdot \dfrac{\sqrt{x}-4}{\sqrt{x}-4}$

$= \dfrac{(x-16)(\sqrt{x}-4)}{\left(\sqrt{x}\right)^2 - (4)^2} = \dfrac{(x-16)(\sqrt{x}-4)}{x-16}$

$= \sqrt{x}-4$

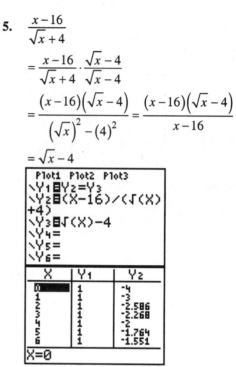

6. $\dfrac{21}{\sqrt{x}-7}$

$= \dfrac{21}{\sqrt{x}-7} \cdot \dfrac{\sqrt{x}+7}{\sqrt{x}+7} = \dfrac{21\left(\sqrt{x}+7\right)}{\left(\sqrt{x}\right)^2 - (7)^2}$

$= \dfrac{21\left(\sqrt{x}+7\right)}{x-49}$

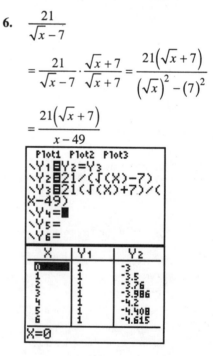

Chapter 13: Radical Expressions, Functions, and Equations

7. $(3\sqrt{x}+5)(\sqrt{x}-7)+3\sqrt{x}-6$
$= 3(\sqrt{x})^2 + 5\sqrt{x} - 21\sqrt{x} - 35 + 3\sqrt{x} - 6$
$= 3x - 13\sqrt{x} - 41$

```
Plot1 Plot2 Plot3
\Y1=Y2=Y3
\Y2=(3√(X)+5)(√(
X)-7)+3√(X)-6
\Y3=3X-13√(X)-41
\Y4=
\Y5=

 X  | Y1 | Y2
 0  | 1  | -41
 1  | 1  | -51
 2  | 1  | -53.38
 3  | 1  | -54.52
 4  | 1  | -55
 5  | 1  | -55.07
 6  | 1  | -54.84
X=0
```

13.6 Exercises

1. $\sqrt{x}-3=5$
$\sqrt{x}=8$
$(\sqrt{x})^2 = 8^2$
$x = 64$
The solution is $x = 64$.

3. $\sqrt{x}+1.7=4.5$
$\sqrt{x}=2.8$
$(\sqrt{x})^2 = 2.8^2$
$x = 7.84$
The solution is $x = 7.84$.

5. $\sqrt{x}+8=5$
$\sqrt{x}=-3$
This has no real solution since the principal root is always non-negative.

7. $\sqrt{3x}-4=2$
$\sqrt{3x}=6$
$(\sqrt{3x})^2 = 6^2$
$3x = 36$
$x = 12$
The solution is $x = 12$.

9. $9-3\sqrt{2z}=0$
$9 = 3\sqrt{2z}$
$3 = \sqrt{2z}$
$3^2 = (\sqrt{2z})^2$
$9 = 2z$
$z = \dfrac{9}{2}$
The solution is $z = \dfrac{9}{2}$.

11. $3\sqrt{6x}+1=10$
$3\sqrt{6x}=9$
$\sqrt{6x}=3$
$(\sqrt{6x})^2 = 3^2$
$6x = 9$
$x = \dfrac{3}{2}$
The solution is $x = \dfrac{3}{2}$.

13. $\sqrt{x+5}=3$
$(\sqrt{x+5})^2 = 3^2$
$x+5 = 9$
$x = 4$
The solution is $x = 4$.

15. $\sqrt{2x+5} + 4 = 9$
$\sqrt{2x+5} = 5$
$\left(\sqrt{2x+5}\right)^2 = 5^2$
$2x+5 = 25$
$2x = 20$
$x = 10$
The solution is $x = 10$.

17. $\sqrt{3x+4} - 2 = 6$
$\sqrt{3x+4} = 8$
$\left(\sqrt{3x+4}\right)^2 = 8^2$
$3x+4 = 64$
$3x = 60$
$x = 20$
The solution is $x = 20$.

19. $\sqrt{5-4x} = x$
$\left(\sqrt{5-4x}\right)^2 = x^2$
$5-4x = x^2$
$x^2 + 4x - 5 = 0$
$(x+5)(x-1) = 0$
$x = -5$ (extraneous) or $x = 1$
$x = -5$ is extraneous because the principal root cannot be negative.
The solution is $x = 1$.

21. $\sqrt{6x-5} = \sqrt{4x+5}$
$\left(\sqrt{6x-5}\right)^2 = \left(\sqrt{4x+5}\right)^2$
$6x-5 = 4x+5$
$2x = 10$
$x = 5$
The solution is $x = 5$.

23. $3\sqrt{x+2} = \sqrt{x+10}$
$\left(3\sqrt{x+2}\right)^2 = \left(\sqrt{x+10}\right)^2$
$9(x+2) = x+10$
$9x+18 = x+10$
$8x = -8$
$x = -1$
The solution is $x = -1$.

25. $x = 3 + 2\sqrt{x-4}$
$x - 3 = 2\sqrt{x-4}$
$(x-3)^2 = \left(2\sqrt{x-4}\right)^2$
$x^2 - 6x + 9 = 4(x-4)$
$x^2 - 6x + 9 = 4x - 16$
$x^2 - 10x + 25 = 0$
$(x-5)^2 = 0$
$x = 5$
The solution is $x = 5$.

27. $x = \sqrt{21-x-x^2} + 3$
$x - 3 = \sqrt{21-x-x^2}$
$(x-3)^2 = \left(\sqrt{21-x-x^2}\right)^2$
$x^2 - 6x + 9 = 21 - x - x^2$
$2x^2 - 5x - 12 = 0$
$(2x+3)(x-4) = 0$
$x = -\dfrac{3}{2}$ (extraneous) or $x = 4$
The value $x = -\dfrac{3}{2}$ is not a solution because the principal root cannot be negative.
The solution is $x = 4$.

Chapter 13: Radical Expressions, Functions, and Equations

29. $\sqrt{x^2 - 14x + 49} = 7 - x$
$\left(\sqrt{x^2 - 14x + 49}\right)^2 = (7-x)^2$
$x^2 - 14x + 49 = 49 - 14x + x^2$
$0 = 0$
The solution is all real numbers less than or equal to 7.

31. $x = \sqrt{2x + 7}$
$x^2 = \left(\sqrt{2x+7}\right)^2$
$x^2 = 2x + 7$
$x^2 - 2x - 7 = 0$
$a = 1, b = -2, c = -7$
$x = \dfrac{-(-2) \pm \sqrt{(-2)^2 - 4(1)(-7)}}{2(1)}$
$= \dfrac{2 \pm \sqrt{4 + 28}}{2} = \dfrac{2 \pm \sqrt{32}}{2} = \dfrac{2 \pm 4\sqrt{2}}{2}$
$= 1 \pm 2\sqrt{2}$
The value $x = 1 - 2\sqrt{2}$ is not a solution because the principal root cannot be negative.
The solution is $x = 1 + 2\sqrt{2} \approx 3.828$.

33. $5 + \sqrt{25 - 2x} = x$
$\sqrt{25 - 2x} = x - 5$
$\left(\sqrt{25 - 2x}\right)^2 = (x-5)^2$
$25 - 2x = x^2 - 10x + 25$
$x^2 - 8x = 0$
$x(x - 8) = 0$
$x = 0$ (extraneous) or $x = 8$
The value $x = 0$ is not a solution because the principal root cannot be negative.
The solution is $x = 8$.

35. $\sqrt{x} - 3 = \sqrt{x - 27}$
$\left(\sqrt{x} - 3\right)^2 = \left(\sqrt{x - 27}\right)^2$
$x - 6\sqrt{x} + 9 = x - 27$
$-6\sqrt{x} + 9 = -27$
$-6\sqrt{x} = -36$
$\sqrt{x} = 6$
$\left(\sqrt{x}\right)^2 = 6^2$
$x = 36$
The solution is $x = 36$.

37. $\sqrt{x - 7} = 7 - \sqrt{x}$
$\left(\sqrt{x - 7}\right)^2 = \left(7 - \sqrt{x}\right)^2$
$x - 7 = 49 - 14\sqrt{x} + x$
$-7 = 49 - 14\sqrt{x}$
$-56 = -14\sqrt{x}$
$4 = \sqrt{x}$
$4^2 = \left(\sqrt{x}\right)^2$
$16 = x$
The solution is $x = 16$.

39. $\sqrt{2x + 5} - 7 = \sqrt{5 + 2x} + 3$
$\sqrt{2x + 5} - 7 = \sqrt{2x + 5} + 3$
$-7 = 3$
This is a contradiction. There is no solution.

41. $\sqrt{2x + 3} = 1 - \sqrt{x + 5}$
$\left(\sqrt{2x + 3}\right)^2 = \left(1 - \sqrt{x+5}\right)^2$
$2x + 3 = 1 - 2\sqrt{x + 5} + x + 5$
$x - 3 = -2\sqrt{x + 5}$
$(x - 3)^2 = \left(-2\sqrt{x+5}\right)^2$
$x^2 - 6x + 9 = 4(x + 5)$

SSM: Experiencing Introductory and Intermediate Algebra

$x^2 - 6x + 9 = 4x + 20$
$x^2 - 10x - 11 = 0$
$(x-11)(x+1) = 0$
$x = 11$ (extraneous) or $x = -1$ (extraneous)
Both values are extraneous. There is no solution.

43. $\sqrt[3]{3x} = -6$
$\left(\sqrt[3]{3x}\right)^3 = (-6)^3$
$3x = -216$
$x = -72$
The solution is $x = -72$.

45. $\sqrt[4]{2x} = 10$
$\left(\sqrt[4]{2x}\right)^4 = (10)^4$
$2x = 10000$
$x = 5000$
The solution is $x = 5000$.

47. $\sqrt[5]{2x} = -2$
$\left(\sqrt[5]{2x}\right)^5 = (-2)^5$
$2x = -32$
$x = -16$
The solution is $x = -16$.

49. $\sqrt[3]{x} + 3 = 1$
$\sqrt[3]{x} = -2$
$\left(\sqrt[3]{x}\right)^3 = (-2)^3$
$x = -8$
The solution is $x = -8$.

51. $\sqrt[4]{x} + 1 = 5$
$\sqrt[4]{x} = 4$
$\left(\sqrt[4]{x}\right)^4 = (4)^4$
$x = 256$
The solution is $x = 256$.

53. $\sqrt[3]{2x+1} = 3$
$\left(\sqrt[3]{2x+1}\right)^3 = 3^3$
$2x + 1 = 27$
$2x = 26$
$x = 13$
The solution is $x = 13$.

55. $\sqrt[4]{x-5} = 2$
$\left(\sqrt[4]{x-5}\right)^4 = 2^4$
$x - 5 = 16$
$x = 21$
The solution is $x = 21$.

57. $\sqrt[5]{2x-5} = 3$
$\left(\sqrt[5]{2x-5}\right)^5 = 3^5$
$2x - 5 = 243$
$2x = 248$
$x = 124$
The solution is $x = 124$.

59. $\sqrt[3]{x^2 + 7x} + 7 = 9$
$\sqrt[3]{x^2 + 7x} = 2$
$\left(\sqrt[3]{x^2 + 7x}\right)^3 = 2^3$
$x^2 + 7x = 8$
$x^2 + 7x - 8 = 0$
$(x+8)(x-1) = 0$
$x = -8$ or $x = 1$
The solutions are $x = -8$ and $x = 1$.

61. $\sqrt[4]{3x-5} = \sqrt[4]{2x+4}$
$\left(\sqrt[4]{3x-5}\right)^4 = \left(\sqrt[4]{2x+4}\right)^4$
$3x - 5 = 2x + 4$
$x = 9$
The solution is $x = 9$.

63. $\sqrt[4]{5x-2} = 2\sqrt[4]{3}$
$\left(\sqrt[4]{5x-2}\right)^4 = \left(2\sqrt[4]{3}\right)^4$
$5x - 2 = 16(3)$
$5x = 50$
$x = 10$
The solution is $x = 10$.

65. $x = \sqrt[4]{18x^2 - 81}$
$x^4 = \left(\sqrt[4]{18x^2 - 81}\right)^4$
$x^4 = 18x^2 - 81$
$x^4 - 18x^2 - 81 = 0$
$\left(x^2 - 9\right)^2 = 0$
$x^2 - 9 = 0$
$(x+3)(x-3) = 0$
$x = -3$ (extraneous) or $x = 3$
The value $x = -3$ is extraneous.
The solution is $x = 3$.

67. $\sqrt[4]{2x} + \sqrt[4]{3x} = 0$
$\sqrt[4]{2x} = -\sqrt[4]{3x}$
$\left(\sqrt[4]{2x}\right)^4 = \left(-\sqrt[4]{3x}\right)^4$
$2x = 3x$
$x = 0$
The solution is $x = 0$.

69. $\sqrt[4]{3x-5} + \sqrt[4]{x-3} = 0$
$\sqrt[4]{3x-5} = -\sqrt[4]{x-3}$
$\left(\sqrt[4]{3x-5}\right)^4 = \left(-\sqrt[4]{x-3}\right)^4$
$3x - 5 = x - 3$
$2x = 2$
$x = 1$ (restricted value)
The value $x = 1$ is part of the set of restricted values.
There is no solution.

71. $\sqrt[3]{3x-5} + \sqrt[3]{x-3} = 0$
$\sqrt[3]{3x-5} = -\sqrt[3]{x-3}$
$\left(\sqrt[3]{3x-5}\right)^3 = \left(-\sqrt[3]{x-3}\right)^3$
$3x - 5 = -(x - 3)$
$3x - 5 = -x + 3$
$4x = 8$
$x = 2$
The solution is $x = 2$.

73. $\sqrt[4]{7x-1} - \sqrt[4]{x+11} = 0$
$\sqrt[4]{7x-1} = \sqrt[4]{x+11}$
$\left(\sqrt[4]{7x+1}\right)^4 = \left(\sqrt[4]{x+11}\right)^4$
$7x + 1 = x + 11$
$6x = 10$
$x = \dfrac{5}{3}$
The solution is $x = \dfrac{5}{3}$.

75. $x^{4/3} = 16$
$\left(x^{4/3}\right)^3 = 16^3$
$x^4 = 4096$
$x = \pm\sqrt[4]{4096}$
$x = \pm 8$
The solutions are $x = \pm 8$.

77. $(x+6)^{2/5} = 4$
$\left((x+6)^{2/5}\right)^5 = 4^5$
$(x+6)^2 = 1024$
$x + 6 = \pm\sqrt{1024}$
$x + 6 = \pm 32$
$x = -6 \pm 32$
The solutions are $x = 26$ and $x = -38$.

79. $(x-4)^{3/4} = 27$
$\left((x-4)^{3/4}\right)^4 = 27^4$
$(x-4)^3 = 531441$
$x-4 = \sqrt[3]{531441}$
$x-4 = 81$
$x = 85$
The solution is $x = 85$.

81. $x^{-2/3} = 4$
$\left(x^{-2/3}\right)^3 = 4^3$
$\dfrac{1}{x^2} = 64$
$x^2 = \dfrac{1}{64}$
$x = \pm\sqrt{\dfrac{1}{64}}$
$x = \pm\dfrac{1}{8}$
The solutions are $x = \pm\dfrac{1}{8}$.

83. $(x-5)^{-3/4} = 27$
$\left((x-5)^{-3/4}\right)^4 = 27^4$
$(x-5)^{-3} = 531441$
$\dfrac{1}{(x-5)^3} = 531441$
$(x-5)^3 = \dfrac{1}{531441}$
$x-5 = \sqrt[3]{\dfrac{1}{531441}}$
$x = 5 + \dfrac{1}{81}$
$x = \dfrac{406}{81}$
The solution is $x = \dfrac{406}{81}$.

85. $(5x-3)^{-2/3} = \dfrac{1}{25}$
$\left((5x-3)^{-2/3}\right)^3 = \left(\dfrac{1}{25}\right)^3$
$(5x-3)^{-2} = \dfrac{1}{15625}$
$\dfrac{1}{(5x-3)^2} = \dfrac{1}{15625}$
$(5x-3)^2 = 15625$
$5x-3 = \pm\sqrt{15625}$
$5x = 3 \pm 125$
$x = \dfrac{3 \pm 125}{5}$
$x = -\dfrac{122}{5}$ or $x = \dfrac{128}{5}$
The solutions are $x = -\dfrac{122}{5}$ and $x = \dfrac{128}{5}$.

87. $x^{2/3} + 5 = 3$
$x^{2/3} = -2$
$\left(x^{2/3}\right)^3 = (-2)^3$
$x^2 = -8$
There is no real number solution.

89. $x^{3/4} - 7 = 1$
$x^{3/4} = 8$
$\left(x^{3/4}\right)^4 = 8^4$
$x^3 = 4096$
$x = \sqrt[3]{4096}$
$x = 16$
The solution is $x = 16$.

Chapter 13: Radical Expressions, Functions, and Equations

91. $t = 2\sqrt{\dfrac{2d}{g}}$

$1.16 = 2\sqrt{\dfrac{2d}{32}}$

$0.58 = \sqrt{\dfrac{2d}{32}}$

$(0.58)^2 = \left(\sqrt{\dfrac{2d}{32}}\right)^2$

$0.3364 = \dfrac{2d}{32}$

$10.7648 = 2d$

$5.3824 = d$

His vertical distance is 5.3824 feet. Added to his height of about 6.5833 feet, he can reach a height of 11.9657 feet. This is about as high as a 12-foot basket.

93. $d = \sqrt{(x_2 - x_1)^2 + (y_2 - y_1)^2}$

$13 = \sqrt{(3-8)^2 + (4-y)^2}$

$(13)^2 = \left(\sqrt{(-5)^2 + 16 - 8y + y^2}\right)^2$

$169 = y^2 - 8y + 41$

$0 = y^2 - 8y - 128$

$0 = (y+8)(y-16)$

$y = -8$ or $y = 16$

The y-coordinates are -8 and 16.
The two possible points are $(8,-8)$ and $(8,16)$.

95. $d = \sqrt{(x_2 - x_1)^2 + (y_2 - y_1)^2}$

$10 = \sqrt{(-2-x)^2 + (5+1)^2}$

$10^2 = \left(\sqrt{4 + 4x + x^2 + 36}\right)^2$

$100 = 40 + 4x + x^2$

$0 = x^2 + 4x - 60$

$0 = (x+10)(x-6)$

$x = -10$ or $x = 6$

The x-coordinates are -10 and 6.
The two possible points are $(-10,-1)$ and $(6,-1)$.

97. $d = \sqrt{(x_2 - x_1)^2 + (y_2 - y_1)^2}$

$5 = \sqrt{(-2-6)^2 + (1-y)^2}$

$5 = \sqrt{64 + (1-y)^2}$

$25 = 64 + (1-y)^2$

$-39 = (1-y)^2$

There is no real solution.

99. $d = kE^{1/3}$

$13,200 = 0.02E^{1/3}$

$660,00 = E^{1/3}$

$(660,000)^3 = \left(E^{1/3}\right)^3$

$2.875 \times 10^{17} \approx E$

The energy was approximately 2.875×10^{17} ft-lb.

101. $T = \left(\dfrac{LH^2}{25}\right)^{1/4}$

$4 = \left(\dfrac{L \cdot 8}{25}\right)^{1/4}$

$4^4 = \dfrac{L \cdot 8}{25}$

$256 = \dfrac{64L}{25}$

$6400 = 64L$

$100 = L$

The crushing load was 100 tons.

SSM: Experiencing Introductory and Intermediate Algebra

13.6 Calculator Exercises

1. $\left(\sqrt[4]{2x+1}\right)\left(\sqrt[3]{x-13}\right) = 9$

   $Y_1 = \left(\sqrt[4]{2x+1}\right)\left(\sqrt[3]{x-13}\right)$

   $Y_2 = 9$

   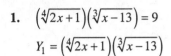

   The solution is $x = 40$.

2. $\left(\sqrt[5]{3x+2}\right) - \left(\sqrt[4]{8x+1}\right) = -1$

   $Y_1 = \left(\sqrt[5]{3x+2}\right) - \left(\sqrt[4]{8x+1}\right)$

   $Y_2 = -1$

   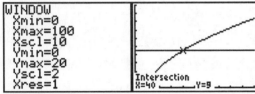

   The solution is $x = 10$.

3. $\dfrac{\sqrt{3x+4}}{\sqrt[4]{x-4}} = \sqrt[3]{4x-16}$

   $Y_1 = \dfrac{\sqrt{3x+4}}{\sqrt[4]{x-4}}$

   $Y_2 = \sqrt[3]{4x-16}$

   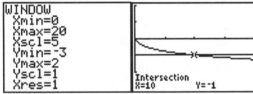

   The solution is $x = 20$.

4. $\sqrt[3]{2x+1} = \sqrt[4]{3x^2}$

   $Y_1 = \sqrt[3]{2x+1}$

   $Y_2 = \sqrt[4]{3x^2}$

   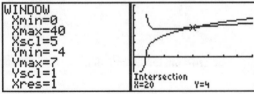

   The solutions are $x \approx -0.307$ and $x \approx 1.412$.

5. $x^{3/2} + 2x^{1/2} - 7 = 0$

   $Y_1 = x^{3/2} + 2x^{1/2} - 7$

   $Y_2 = 0$

   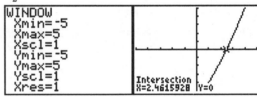

   The solution is $x \approx 2.462$.

6. $3x^{2/3} - 5x^{1/3} - 9 = 0$

   $Y_1 = 3x^{2/3} - 5x^{1/3} - 9$

   $Y_2 = 0$

   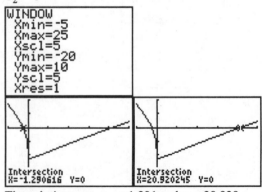

   The solutions are $x \approx -1.291$ and $x \approx 20.920$.

13.7 Exercises

1. $\sqrt{-100} = \sqrt{100} \cdot i = 10i$

3. $\sqrt{-\dfrac{16}{49}} = \sqrt{\dfrac{16}{49}} \cdot i = \dfrac{4}{7}i$

5. $\sqrt{-32} = \sqrt{32} \cdot i = 4\sqrt{2} \cdot i = 4i\sqrt{2}$

7. $2\sqrt{-50} = 2\sqrt{50} \cdot i = 2 \cdot 5\sqrt{2} \cdot i = 10i\sqrt{2}$

9. $7\sqrt{-36} + 9\sqrt{-4}$
   $= 7\sqrt{36} \cdot i + 9\sqrt{4} \cdot i$
   $= 7 \cdot 6 \cdot i + 9 \cdot 2 \cdot i$
   $= 42i + 18i$
   $= 60i$

11. $2\sqrt{-121} - 3\sqrt{-9}$
    $= 2\sqrt{121} \cdot i - 3\sqrt{9} \cdot i$
    $= 2 \cdot 11 \cdot i - 3 \cdot 3 \cdot i$
    $= 22i - 9i$
    $= 13i$

13. $\dfrac{\sqrt{-144}}{\sqrt{-225}}$
    $= \dfrac{\sqrt{144} \cdot i}{\sqrt{225} \cdot i} = \dfrac{12i}{15i}$
    $= \dfrac{4}{5}$

15. $\sqrt{-5}\sqrt{-8}\sqrt{-10}$
    $= \sqrt{5} \cdot i \cdot \sqrt{8} \cdot i \cdot \sqrt{10} \cdot i$
    $= \sqrt{400} \cdot i^3 = 20i^3$
    $= -20i$

17. $\left(\sqrt{-6} + \sqrt{-4}\right)\left(\sqrt{-3} - \sqrt{-8}\right)$
    $= \left(\sqrt{6} \cdot i + 2i\right)\left(\sqrt{3} \cdot i - 2\sqrt{2} \cdot i\right)$
    $= \sqrt{18} \cdot i^2 + 2\sqrt{3} \cdot i^2 - 2\sqrt{12} \cdot i^2 - 4\sqrt{2} \cdot i^2$
    $= 3\sqrt{2}(-1) + 2\sqrt{3}(-1) - 4\sqrt{3}(-1) - 4\sqrt{2}(-1)$
    $= -3\sqrt{2} - 2\sqrt{3} + 4\sqrt{3} + 4\sqrt{2}$
    $= \sqrt{2} + 2\sqrt{3}$

19. $\left(\sqrt{-5} - \sqrt{-7}\right)\left(\sqrt{-5} + \sqrt{-7}\right)$
    $= \left(\sqrt{5} \cdot i - \sqrt{7} \cdot i\right)\left(\sqrt{5} \cdot i + \sqrt{7} \cdot i\right)$
    $= \left(\sqrt{5} \cdot i\right)^2 - \left(\sqrt{7} \cdot i\right)^2$
    $= 5i^2 - 7i^2 = -2i^2$
    $= 2$

21. $(3 + 5i) + (-8 - i)$
    $= 3 + 5i - 8 - i$
    $= -5 + 4i$

23. $(6 + 2i) - (3 + 3i)$
    $= 6 + 2i - 3 - 3i$
    $= 3 - i$

25. $(4 - 5i) - (3 - 5i)$
    $= 4 - 5i - 3 + 5i$
    $= 1$

27. $\left(\dfrac{1}{2} + 5i\right) + \left(3 - \dfrac{2}{3}i\right)$
    $\dfrac{1}{2} + 5i + 3 - \dfrac{2}{3}i$
    $\dfrac{7}{2} + \dfrac{13}{3}i$

29. $(4.5 + 6.7i) - (2.88 - 4.68i)$
    $= 4.5 + 6.7i - 2.88 + 4.68i$
    $= 1.62 + 11.38i$

31. $\left(2\sqrt{3} - i\sqrt{2}\right) + \left(5\sqrt{3} + 2i\sqrt{2}\right)$
    $= 2\sqrt{3} - i\sqrt{2} + 5\sqrt{3} + 2i\sqrt{2}$
    $= 7\sqrt{3} + i\sqrt{2}$

33. $(4 - 3i)(6 + 5i)$
    $= 24 - 18i + 20i - 15i^2$
    $= 24 + 2i + 15$
    $= 39 + 2i$

SSM: Experiencing Introductory and Intermediate Algebra

35. $(5+7i)(5-7i)$
$= (5)^2 - (7i)^2 = 25 - 49i^2$
$= 25 + 49$
$= 74$

37. $(-6-i)(-6+i)$
$= (-6)^2 - (i)^2$
$= 36 + 1$
$= 37$

39. $(0.4 - 3.1i)(5.7 + 0.8i)$
$= 2.28 - 17.67i + 0.32i - 2.48i^2$
$= 2.28 - 17.35i + 2.48$
$= 4.76 - 17.35i$

41. $\left(\dfrac{3}{5} + \dfrac{1}{2}i\right)\left(\dfrac{2}{5} - \dfrac{2}{3}i\right)$
$= \dfrac{6}{25} + \dfrac{1}{5}i - \dfrac{2}{5}i - \dfrac{1}{3}i^2$
$= \dfrac{6}{25} - \dfrac{1}{5}i + \dfrac{1}{3}$
$= \dfrac{43}{75} - \dfrac{1}{5}i$

43. $i\sqrt{2}(\sqrt{2} + i\sqrt{3})$
$= 2i + i^2\sqrt{6} = 2i - \sqrt{6}$
$= -\sqrt{6} + 2i$

45. $(\sqrt{3} - \sqrt{5}i)(\sqrt{3} + i\sqrt{5})$
$= (\sqrt{3})^2 - (i\sqrt{5})^2$
$= 3 - 5i^2 = 3 + 5$
$= 8$

47. $\dfrac{7 + 21i}{7}$
$= \dfrac{7}{7} + \dfrac{21i}{7}$
$= 1 + 3i$

49. $\dfrac{-4 + 7i}{3 - i}$
$= \dfrac{(-4+7i)(3+i)}{(3-i)(3+i)} = \dfrac{-12 + 21i - 4i + 7i^2}{9 - i^2}$
$= \dfrac{-12 + 17i - 7}{9 + 1} = \dfrac{-19 + 17i}{10}$
$= -\dfrac{19}{10} + \dfrac{17}{10}i$

$= \dfrac{235 + 65i - 94i - 26i^2}{25 - 4i^2}$
$= \dfrac{235 - 29i + 26}{25 + 4}$
$= \dfrac{261 - 29i}{29}$
$= 9 - i$

51. $\dfrac{6 - 5i}{2i}$
$= \dfrac{(6-5i)i}{(2i)i} = \dfrac{6i - 5i^2}{2i^2}$
$= \dfrac{6i + 5}{-2}$
$= -\dfrac{5}{2} - 3i$

53. $\dfrac{16.2 - 13.5i}{2.7i}$
$= \dfrac{(16.2 - 13.5i)i}{(2.7i)i} = \dfrac{16.2i - 13.5i^2}{2.7i^2}$
$= \dfrac{16.2i + 13.5}{-2.7} = \dfrac{13.5 + 16.2i}{-2.7}$
$= -5 - 6i$

55. $\dfrac{21.2 - 10.4i}{4.5 + i}$
$= \dfrac{(21.2 - 10.4i)(4.5 - i)}{(4.5 + i)(4.5 - i)}$
$= \dfrac{95.4 - 46.8i - 21.2i + 10.4i^2}{(4.5)^2 - i^2}$
$= \dfrac{95.4 - 68i - 10.4}{20.25 + 1} = \dfrac{85 - 68i}{21.25}$
$= 4 - 3.2i$

Chapter 13: Radical Expressions, Functions, and Equations

57. $\dfrac{\sqrt{6}+i\sqrt{14}}{i\sqrt{2}}$

$= \dfrac{\left(\sqrt{6}+i\sqrt{14}\right)i}{\left(i\sqrt{2}\right)i} = \dfrac{i\sqrt{6}+i^2\sqrt{14}}{i^2\sqrt{2}}$

$= \dfrac{i\sqrt{6}-\sqrt{14}}{-\sqrt{2}} = \dfrac{-\sqrt{14}+i\sqrt{6}}{-\sqrt{2}}$

$= \sqrt{7}-i\sqrt{3}$

59. $\dfrac{2+5i\sqrt{6}}{2\sqrt{3}+i\sqrt{2}}$

$= \dfrac{\left(2+5i\sqrt{6}\right)\left(2\sqrt{3}-i\sqrt{2}\right)}{\left(2\sqrt{3}+i\sqrt{2}\right)\left(2\sqrt{3}-i\sqrt{2}\right)}$

$= \dfrac{4\sqrt{3}+10i\sqrt{18}-2i\sqrt{2}-5i^2\sqrt{12}}{\left(2\sqrt{3}\right)^2-\left(i\sqrt{2}\right)^2}$

$= \dfrac{4\sqrt{3}+30i\sqrt{2}-2i\sqrt{2}+10\sqrt{3}}{12+2}$

$= \dfrac{14\sqrt{3}+28i\sqrt{2}}{14}$

$= \sqrt{3}+2i\sqrt{2}$

61. $a^2+7=0$
$a^2=-7$
$a=\pm\sqrt{-7}$
$a=\pm i\sqrt{7}$

63. $z^2+5=1$
$z^2=-4$
$z=\pm\sqrt{-4}$
$z=\pm 2i$

65. $3p^2+75=0$
$3p^2=-75$
$p^2=-25$
$p=\pm\sqrt{-25}$
$p=\pm 5i$

67. $4d^2+12=0$
$4d^2=-12$
$d^2=-3$
$d=\pm\sqrt{-3}$
$d=\pm i\sqrt{3}$

69. $(t+1)^2+9=0$
$(t+1)^2=-9$
$(t+1)=\pm\sqrt{-9}$
$t+1=\pm 3i$
$t=-1\pm 3i$

71. $4(x-5)^2+22=2$
$4(x-5)^2=-20$
$(x-5)^2=-5$
$x-5=\pm\sqrt{-5}$
$x=5\pm i\sqrt{5}$

73. $4(2x+5)^2=-16$
$(2x+5)^2=-4$
$2x+5=\pm\sqrt{-4}$
$2x=-5\pm 2i$
$x=-\dfrac{5}{2}\pm i$

75. $2(z+2.5)^2+10.58=0$
$2(z+2.5)^2=-10.58$
$(z+2.5)^2=-5.29$
$z+2.5=\pm\sqrt{-5.29}$
$z=-2.5\pm 2.3i$

77. $\left(b-\frac{1}{2}\right)^2 + \frac{1}{4} = 0$

$\left(b-\frac{1}{2}\right)^2 = -\frac{1}{4}$

$b-\frac{1}{2} = \pm\sqrt{-\frac{1}{4}}$

$b = \frac{1}{2} \pm \frac{1}{2}i$

79. $x^2 + 2x + 4 = 0$

$x = \frac{-2 \pm \sqrt{(2)^2 - 4(1)(4)}}{2(1)}$

$= \frac{-2 \pm \sqrt{4-16}}{2} = \frac{-2 \pm \sqrt{-12}}{2}$

$= \frac{-2 \pm 2i\sqrt{3}}{2}$

$= -1 \pm i\sqrt{3}$

81. $b^2 - 10b + 27 = 0$

$b = \frac{-(-10) \pm \sqrt{(-10)^2 - 4(1)(27)}}{2(1)}$

$= \frac{10 \pm \sqrt{100-108}}{2} = \frac{10 \pm \sqrt{-8}}{2}$

$= \frac{10 \pm 2i\sqrt{2}}{2}$

$= 5 \pm i\sqrt{2}$

83. $4y^2 + 4y + 5 = 0$

$y = \frac{-4 \pm \sqrt{4^2 - 4(4)(5)}}{2(4)}$

$= \frac{-4 \pm \sqrt{16-80}}{8} = \frac{-4 \pm \sqrt{-64}}{8}$

$= \frac{-4 \pm 8i}{8}$

$= -\frac{1}{2} \pm i$

85. $9p^2 - 12p + 8 = 0$

$p = \frac{-(-12) \pm \sqrt{(-12)^2 - 4(9)(8)}}{2(9)}$

$= \frac{12 \pm \sqrt{144-288}}{18} = \frac{12 \pm \sqrt{-144}}{18}$

$= \frac{12 \pm 12i}{18}$

$= \frac{2}{3} \pm \frac{2}{3}i$

87. $x^2 - 2.4x + 3.44 = 0$

$x = \frac{-(-2.4) \pm \sqrt{(-2.4)^2 - 4(1)(3.44)}}{2(1)}$

$= \frac{2.4 \pm \sqrt{5.76-13.76}}{2} = \frac{2.4 \pm \sqrt{-8}}{2}$

$= \frac{2.4 \pm 2i\sqrt{2}}{2}$

$= 1.2 \pm i\sqrt{2}$

89. $2y^2 - 2y + 1.22 = 0$

$y = \frac{-(-2) \pm \sqrt{(-2)^2 - 4(2)(1.22)}}{2(2)}$

$= \frac{2 \pm \sqrt{4-9.76}}{4} = \frac{2 \pm \sqrt{-5.76}}{4}$

$= \frac{2 \pm 2.4i}{4}$

$= \frac{1}{2} \pm \frac{3}{5}i$

91. $x^2 - \frac{2}{3}x + \frac{13}{36} = 0$

$x = \frac{-\left(-\frac{2}{3}\right) \pm \sqrt{\left(-\frac{2}{3}\right)^2 - 4(1)\left(\frac{13}{36}\right)}}{2(1)}$

$= \frac{\frac{2}{3} \pm \sqrt{\frac{4}{9} - \frac{13}{9}}}{2} = \frac{\frac{2}{3} \pm \sqrt{-1}}{2}$

$= \frac{1}{3} \pm \frac{1}{2}i$

93. $4z^2 + \frac{4}{3}z + \frac{2}{9} = 0$

$36z^2 + 12z + 2 = 0$

$z = \dfrac{-12 \pm \sqrt{12^2 - 4(36)(2)}}{2(36)}$

$= \dfrac{-12 \pm \sqrt{144 - 288}}{72} = \dfrac{-12 \pm \sqrt{-144}}{72}$

$= \dfrac{-12 \pm 12i}{72}$

$= -\dfrac{1}{6} \pm \dfrac{1}{6}i$

95. $\dfrac{x}{x-2} = \dfrac{5}{x+3}$

$x(x+3) = 5(x-2)$

$x^2 + 3x = 5x - 10$

$x^2 - 2x + 10 = 0$

$x = \dfrac{-(-2) \pm \sqrt{(-2)^2 - 4(1)(10)}}{2(1)}$

$= \dfrac{2 \pm \sqrt{-36}}{2} = \dfrac{2 \pm 6i}{2}$

$= 1 \pm 3i$

97. $\dfrac{y+7}{y-3} = \dfrac{6}{y+4}$

$(y+7)(y+4) = 6(y-3)$

$y^2 + 11y + 28 = 6y - 18$

$y^2 + 5y + 46 = 0$

$y = \dfrac{-5 \pm \sqrt{5^2 - 4(1)(46)}}{2(1)}$

$= \dfrac{-5 \pm \sqrt{-159}}{2} = \dfrac{-5 \pm i\sqrt{159}}{2}$

$= -\dfrac{5}{2} \pm \dfrac{\sqrt{159}}{2}i$

99. $\dfrac{z-5}{z(z+3)} = \dfrac{4}{5}$

$5(z-5) = 4z(z+3)$

$5z - 25 = 4z^2 + 12z$

$4z^2 + 7z + 25 = 0$

$z = \dfrac{-7 \pm \sqrt{7^2 - 4(4)(25)}}{2(4)}$

$= \dfrac{-7 \pm \sqrt{-351}}{8} = \dfrac{-7 \pm 3i\sqrt{39}}{8}$

$= -\dfrac{7}{8} \pm \dfrac{3\sqrt{39}}{8}i$

101. $V = IZ$

$V = (3+5i)(5+4i)$

$= 15 + 12i + 25i + 20i^2$

$= 15 + 37i - 20$

$= -5 + 37i$

The total voltage is $-5 + 37i$ volts.

$|V| = \sqrt{(-5)^2 + (37)^2} \approx 37.336$

The magnitude of the total voltage is approximately 37.336 volts.

103. $I = \dfrac{V}{Z}$

$I = \dfrac{18+i}{2+3i}$

$= \dfrac{(18+i)(2-3i)}{(2+3i)(2-3i)} = \dfrac{36 + 2i - 54i - 3i^2}{4+9}$

$= \dfrac{36 - 52i + 3}{13} = \dfrac{39 - 52i}{13}$

$= 3 - 4i$

The current is $3 - 4i$ amperes.
The magnitude of the current is

$|I| = \sqrt{(3)^2 + (-4)^2} = 5$

The magnitude of the current is 5 amperes.

105. $K = 1.00 - 2.67r + r^2$

$-1.75 = 1 - 2.67r + r^2$

$0 = 2.75 - 2.67r + r^2$

$r = \dfrac{2.67 \pm \sqrt{(-2.67)^2 - 4(1)(2.75)}}{2(1)}$

$= \dfrac{2.67 \pm \sqrt{-3.8711}}{2}$

$= 1.335 \pm \dfrac{\sqrt{3.8711}}{2} i$

13.7 Calculator Exercises

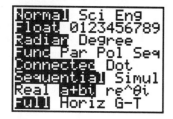

Problems **1.** to **3.**

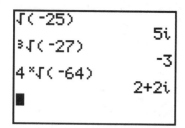

Problems **4.** to **6.**

```
5*√(-32)
               -2
4*√(-324)
             3+3i
4*√(-40000)
           10+10i
■
```

Problems **7.** to **9.**

```
1.5-3.5i▶Frac
          3/2-7/2i
0.4+3.2i▶Frac
          2/5+16/5i
-1.8i▶Frac
             -9/5i
```

Problems **10.** to **12.**

```
abs(6+8i)
               10
abs(6-8i)
               10
abs(2-5i)
       5.385164807
■
```

Problems **13.** to **15.**

```
abs(22)
               22
abs(-7i)
                7
abs(i√(3))
       1.732050808
```

Chapter 13 Section-By-Section Review

1. $\sqrt{225} = \sqrt{(15)^2} = 15$

2. $\sqrt{2.89} = \sqrt{(1.7)^2} = 1.7$

3. $\sqrt{\dfrac{49}{64}} = \sqrt{\left(\dfrac{7}{8}\right)^2} = \dfrac{7}{8}$

4. $\sqrt[3]{-\dfrac{27}{125}} = \sqrt[3]{\left(-\dfrac{3}{5}\right)^3} = -\dfrac{3}{5}$

Chapter 13: Radical Expressions, Functions, and Equations

5. $-\sqrt[4]{1296} = -\sqrt[4]{(6)^4} = -6$

6. $\sqrt[4]{-1296}$ is not a real number.

7. $\sqrt{150} \approx 12.247$

8. $\sqrt[3]{-35} \approx -3.271$

9. $\sqrt[5]{125} \approx 2.627$

10. $y = \sqrt{2x-7}$
    $2x - 7 < 0$
    $2x < 7$
    $x < \dfrac{7}{2}$
    The restricted values are all real numbers less than $\dfrac{7}{2}$.
    Domain: $\left[\dfrac{7}{2}, \infty\right)$

11. $y = \sqrt[3]{2x-7}$
    There are no restricted values.
    Domain: $(-\infty, \infty)$

12. $f(x) = 2\sqrt{x}$

x	$f(x) = 2\sqrt{x}$	$f(x)$
0	$f(0) = 2\sqrt{0} = 0$	0
1	$f(1) = 2\sqrt{1} = 2$	2
4	$f(4) = 2\sqrt{4} = 4$	4
9	$f(9) = 2\sqrt{9} = 6$	6

    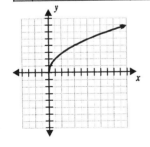

13. $y = \sqrt[3]{3x-4}$

x	$y = \sqrt[3]{3x-4}$	y
-4	$y = \sqrt[3]{3(-4)-4} \approx -2.52$	-2.52
1	$y = \sqrt[3]{3(1)-4} = -1$	-1
4	$y = \sqrt[3]{3(4)-4} = 2$	2
10	$y = \sqrt[3]{3(10)-4} \approx 2.96$	2.96

    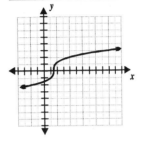

14. $s = \sqrt{\dfrac{n(n+1)}{12}}$
    $= \sqrt{\dfrac{23(23+1)}{12}} = \sqrt{\dfrac{23(24)}{12}}$
    $= \sqrt{46} \approx 6.78$
    The variability is about 6.78.

15. $d = \sqrt{(x_2 - x_1)^2 + (y_2 - y_1)^2}$
    $= \sqrt{(4-4)^2 + (7-(-3))^2}$
    $= \sqrt{0^2 + 10^2} = \sqrt{10^2}$
    $= 10$
    The distance is 10 units.

16. $d = \sqrt{(x_2 - x_1)^2 + (y_2 - y_1)^2}$
    $= \sqrt{(7-(-2))^2 + (-4-(-4))^2}$
    $= \sqrt{9^2 + 0^2} = \sqrt{9^2}$
    $= 9$
    The distance is 9 units.

SSM: Experiencing Introductory and Intermediate Algebra

17. $d = \sqrt{(x_2 - x_1)^2 + (y_2 - y_1)^2}$
$= \sqrt{(9-6)^2 + (6-2)^2}$
$= \sqrt{3^2 + 4^2} = \sqrt{25}$
$= 5$
The distance is 5 units.

18. $d = \sqrt{(x_2 - x_1)^2 + (y_2 - y_1)^2}$
$= \sqrt{(4-(-3))^2 + (8-(-1))^2}$
$= \sqrt{7^2 + 9^2} = \sqrt{49 + 81} = \sqrt{130}$
≈ 11.402
The distance is about 11.402 units.

19. $A = s^2$
$s = \sqrt{A}$
$s = \sqrt{121} = \sqrt{11^2} = 11$
The length of the side is 11 inches.

20. $V = s^3$
$s = \sqrt[3]{V}$
$s = \sqrt[3]{\frac{125}{216}} = \sqrt[3]{\left(\frac{5}{6}\right)^3} = \frac{5}{6}$
The length of the side is $\frac{5}{6}$ yards.

21. $V = \frac{4}{3}\pi r^3$
$r = \sqrt[3]{\frac{3V}{4\pi}}$
$r = \sqrt[3]{\frac{3(65.45)}{4\pi}} \approx 2.5$
$d = 2r = 2(2.5) = 5$
The radius of the ball is about 2.5 centimeters and its diameter is about 5 centimeters.

22. a. $v(x) = 13\sqrt[10]{x}$

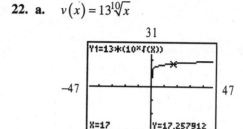

b. The function is increasing.

c. See graph.
In 2010, $x = 17$. If the trend continues, the market value of athletic footwear will be about $17.3 million.

23. $(-64)^{1/3} = \sqrt[3]{-64} = \sqrt[3]{(-4)^3} = -4$

24. $-16^{1/4} = -\sqrt[4]{16} = -\sqrt[4]{2^4} = -2$

25. $(-16)^{1/4} = \sqrt[4]{-16}$ is not a real number.

26. $-64^{-1/3} = -\frac{1}{\sqrt[3]{64}} = -\frac{1}{\sqrt[3]{4^3}} = -\frac{1}{4}$

27. $-64^{4/3} = -\left(\sqrt[3]{64}\right)^4 = -(4)^4 = -256$

28. $(-64)^{3/2} = \left(\sqrt{-64}\right)^3$ is not a real number.

29. $16^{-3/4} = \frac{1}{16^{3/4}} = \frac{1}{\left(\sqrt[4]{16}\right)^3} = \frac{1}{2^3} = \frac{1}{8}$

30. $(-16)^{-3/4} = \frac{1}{(-16)^{3/4}} = \frac{1}{\left(\sqrt[4]{-16}\right)^3}$ is not a real number.

31. $y = (4x+9)^{3/4}$
$4x + 9 < 0$
$4x < -9$
$x < -\frac{9}{4}$
The restricted values are all real numbers less

than $-\frac{9}{4}$.

Domain: $\left[-\frac{9}{4}, \infty\right)$

32. $f(x) = 2x^{1/3} + 2$

There are no restricted values.

Domain: $(-\infty, \infty)$

33. $g(x) = x^{3/2} + 1$

x	$f(x) = x^{3/2} + 1$	$f(x)$
0	$f(0) = 0^{3/2} + 1 = 1$	1
1	$f(1) = 1^{3/2} + 1 = 2$	2
2	$f(2) = 2^{3/2} + 1 \approx 3.828$	3.828
4	$f(4) = 4^{3/2} + 1 = 9$	9

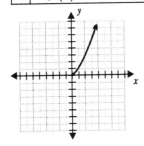

34. $y = (2x+1)^{1/3}$

x	$y = (2x+1)^{1/3}$	y
-14	$y = (2(-14)+1)^{1/3} = -3$	-3
-1	$y = (2(-1)+1)^{1/3} = -1$	-1
0	$y = (2(0)+1)^{1/3} = 1$	1
2	$y = (2(2)+1)^{1/3} \approx 1.71$	1.71

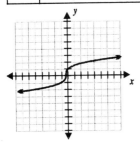

35. $V = (9s)^{1/2}$

[graph showing $Y_1 = f(9X)$ with $X = 150$, $Y = 36.742346$, range 0 to 2000, height to 150]

The car was traveling at a speed of about 36.7 miles per hour.

36. $\dfrac{1}{4^{-1/2}} = 4^{1/2} = \sqrt{4} = 2$

37. $8^{2/3} \cdot 8^{5/3} = 8^{\frac{2}{3} + \frac{5}{3}} = 8^{7/3}$
$= \left(\sqrt[3]{8}\right)^7 = 2^7 = 128$

38. $\dfrac{9^{7/3}}{9^{2/3}} = 9^{\frac{7}{3} - \frac{2}{3}} = 9^{5/3} = \left(\sqrt[3]{9}\right)^5 \approx 38.941$

39. $\left(\dfrac{4}{9}\right)^{-3/2} = \left(\dfrac{9}{4}\right)^{3/2} = \left(\sqrt{\dfrac{9}{4}}\right)^3$
$= \left(\dfrac{3}{2}\right)^3 = \dfrac{3^3}{2^3} = \dfrac{27}{8}$

40. $\left[\left(3^{1/2}\right)\left(27^{1/2}\right)\right]^{-3/2} = \left[(3 \cdot 27)^{1/2}\right]^{-3/2}$
$= \left[(81)^{1/2}\right]^{-3/2} = 9^{-3/2} = \dfrac{1}{9^{3/2}}$
$= \dfrac{1}{\left(\sqrt{9}\right)^3} = \dfrac{1}{3^3}$
$= \dfrac{1}{27}$

41. $\left(64^{2/3}\right)^{3/4} = 64^{\frac{2 \cdot 3}{3 \cdot 4}} = 64^{1/2} = \sqrt{64} = 8$

42. $\dfrac{x^{2/3}}{x^{5/6}} = x^{\frac{2}{3} - \frac{5}{6}} = x^{-1/6} = \dfrac{1}{x^{1/6}}$

43. $y^{2/5} \cdot y^{-3/10} = y^{\frac{2}{5} + \left(-\frac{3}{10}\right)} = y^{1/10}$

44. $\left(z^{3/5}\right)^{5/9} = z^{\frac{3\cdot 5}{5\cdot 9}} = z^{1/3}$

45. $\left(\dfrac{a^3}{b^6}\right)^{-5/12} = \left(\dfrac{b^6}{a^3}\right)^{5/12} = \dfrac{b^{6\cdot 5/12}}{a^{3\cdot 5/12}} = \dfrac{b^{5/2}}{a^{5/4}}$

46. $\left(8x^6 y^9\right)^{4/3} = 8^{4/3} x^{6\cdot 4/3} y^{9\cdot 4/3}$
$= \left(\sqrt[3]{8}\right)^4 x^8 y^{12} = 2^4 x^8 y^{12}$
$= 16 x^8 y^{12}$

47. $\left(2a^{3/4} b^{1/3}\right)\left(3a^{1/3} b^{2/3}\right)$
$= 2\cdot 3 a^{\frac{3}{4}+\frac{1}{3}} b^{\frac{1}{3}+\frac{2}{3}}$
$= 6a^{13/12} b$

48. $x^{1/3}\left(x^{2/3} - x^{1/3}\right)$
$= x^{\frac{1}{3}+\frac{2}{3}} - x^{\frac{1}{3}+\frac{1}{3}}$
$= x - x^{2/3}$

49. $A = s^2$
$s = \sqrt{A}$
Let s_1 = original length of side
s_2 = new length
$s_2 = \sqrt{A_2} = \sqrt{5A_1} = \sqrt{5s_1^2} = s_1\sqrt{5}$
The length of the sides should be increased by a factor of $\sqrt{5}$.

50. $\sqrt{6300} = \sqrt{900\cdot 7} = 30\sqrt{7}$

51. $\sqrt[3]{-320} = \sqrt[3]{-64\cdot 5} = -4\sqrt[3]{5}$

52. $\sqrt[4]{162} = \sqrt[4]{81\cdot 2} = 3\sqrt[4]{2}$

53. $\sqrt{45x^4 y^7 z^2} = \sqrt{9\cdot 5\cdot x^4 \cdot y^6 \cdot y \cdot z^2}$
$= 3x^2 y^3 z\sqrt{5y}$

54. $\sqrt[3]{-64 x^2 y^7} = \sqrt[3]{-64\cdot x^2 \cdot y^6 \cdot y}$
$= -4y^2 \sqrt[3]{x^2 y}$

55. $\sqrt{12x^2 - 36x + 27}$
$= \sqrt{3(4x^2 - 12x + 9)}$
$= \sqrt{3(2x-3)^2}$
$= (2x-3)\sqrt{3}$

56. $\sqrt{2x}\cdot\sqrt{8x^3}$
$= \sqrt{(2x)(8x^3)}$
$= \sqrt{16x^4}$
$= 4x^2$

57. $\sqrt[3]{-2a^2 b}\cdot\sqrt[3]{20a^2 b^5}$
$= \sqrt[3]{(-2a^2 b)(20a^2 b^5)}$
$= \sqrt[3]{-2\cdot 20 a^{2+2} b^{1+5}}$
$= \sqrt[3]{-40 a^4 b^6}$
$= -2ab^2 \sqrt[3]{5a}$

58. $\sqrt{x^2 - 2xy}\cdot\sqrt{3x - 6y}$
$= \sqrt{(x^2 - 2xy)(3x - 6y)}$
$= \sqrt{x(x-2y)\cdot 3(x-2y)}$
$= \sqrt{3x(x-2y)^2}$
$= (x-2y)\sqrt{3x}$

59. $\sqrt{\dfrac{25}{64}} = \dfrac{\sqrt{25}}{\sqrt{64}} = \dfrac{5}{8}$

60. $\sqrt{\dfrac{13}{289}} = \dfrac{\sqrt{13}}{\sqrt{289}} = \dfrac{\sqrt{13}}{17}$

61. $\sqrt{\dfrac{27}{343}} = \sqrt{\dfrac{9\cdot 3}{49\cdot 7}} = \dfrac{3\sqrt{3}}{7\sqrt{7}} = \dfrac{3\sqrt{3}}{7\sqrt{7}}\cdot\dfrac{\sqrt{7}}{\sqrt{7}}$
$= \dfrac{3\sqrt{21}}{49}$

Chapter 13: Radical Expressions, Functions, and Equations

62. $\sqrt[3]{\dfrac{6}{25}} = \dfrac{\sqrt[3]{6}}{\sqrt[3]{25}} \cdot \dfrac{\sqrt[3]{5}}{\sqrt[3]{5}}$

    $= \dfrac{\sqrt[3]{30}}{5}$

63. $\dfrac{\sqrt{50}}{\sqrt{60}} = \sqrt{\dfrac{50}{60}} = \sqrt{\dfrac{5}{6}} = \dfrac{\sqrt{5}}{\sqrt{6}} \cdot \dfrac{\sqrt{6}}{\sqrt{6}}$

    $= \dfrac{\sqrt{30}}{6}$

64. $\sqrt{\dfrac{25a^2}{64b^4}} = \dfrac{\sqrt{25a^2}}{\sqrt{64b^4}}$

    $= \dfrac{5a}{8b^2}$

65. $\sqrt{\dfrac{16m}{5}} = \dfrac{\sqrt{16m}}{\sqrt{5}} = \dfrac{4\sqrt{m}}{\sqrt{5}} \cdot \dfrac{\sqrt{5}}{\sqrt{5}}$

    $= \dfrac{4\sqrt{5m}}{5}$

66. $\sqrt[3]{\dfrac{3z^3}{4x^2y}} = \dfrac{\sqrt[3]{3z^3}}{\sqrt[3]{4x^2y}} = \dfrac{z\sqrt[3]{3}}{\sqrt[3]{4x^2y}} \cdot \dfrac{\sqrt[3]{2xy^2}}{\sqrt[3]{2xy^2}}$

    $= \dfrac{z\sqrt[3]{6xy^2}}{2xy}$

67. $\dfrac{4a\sqrt{5ab^2}}{12\sqrt{10ab^3}} = \dfrac{a}{3}\sqrt{\dfrac{5ab^2}{10ab^3}} = \dfrac{a}{3}\sqrt{\dfrac{1}{2b}}$

    $= \dfrac{a}{3\sqrt{2b}} = \dfrac{a}{3\sqrt{2b}} \cdot \dfrac{\sqrt{2b}}{\sqrt{2b}}$

    $= \dfrac{a\sqrt{2b}}{6b}$

68. $T = 2\pi\sqrt{\dfrac{L}{32}}$

    $T = 2\pi\sqrt{\dfrac{3}{32}} \approx 1.924$

    The period of the pendulum is about 1.924 seconds.

69. $3\sqrt{5} - 2\sqrt{5} + 7\sqrt{5} - \sqrt{5}$

    $= (3 - 2 + 7 - 1)\sqrt{5}$

    $= 7\sqrt{5}$

70. $7\sqrt{11} + 2\sqrt{44}$

    $= 7\sqrt{11} + 4\sqrt{11}$

    $= (7 + 4)\sqrt{11}$

    $= 11\sqrt{11}$

71. $\sqrt{15} + \sqrt{\dfrac{1}{15}}$

    $= \sqrt{15} + \dfrac{1}{\sqrt{15}} \cdot \dfrac{\sqrt{15}}{\sqrt{15}}$

    $= \sqrt{15} + \dfrac{1}{15}\sqrt{15}$

    $= \dfrac{16}{15}\sqrt{15}$

72. $5\sqrt[3]{16} - \sqrt[3]{54}$

    $= 10\sqrt[3]{2} - 3\sqrt[3]{2}$

    $= (10 - 3)\sqrt[3]{2}$

    $= 7\sqrt[3]{2}$

73. $\sqrt{49x} - \sqrt{25x}$

    $= 7\sqrt{x} - 5\sqrt{x}$

    $= (7 - 5)\sqrt{x}$

    $= 2\sqrt{x}$

74. $4b\sqrt{a^3b} - 7a\sqrt{ab^3} + 8ab\sqrt{ab}$

    $= 4ab\sqrt{ab} - 7ab\sqrt{ab} + 8ab\sqrt{ab}$

    $= (4 - 7 + 8)ab\sqrt{ab}$

    $= 5ab\sqrt{ab}$

75. $\sqrt{7}\left(\sqrt{14} - \sqrt{7}\right)$

    $= \sqrt{7} \cdot \sqrt{14} - \left(\sqrt{7}\right)^2$

    $= \sqrt{98} - 7$

    $= 7\sqrt{2} - 7$

76. $\sqrt{2a}(6+\sqrt{2a})$
$= \sqrt{2a}\cdot 6 + (\sqrt{2a})^2$
$= 6\sqrt{2a} + 2a$

77. $(2-\sqrt{5})(4+\sqrt{5})$
$= 8 - 4\sqrt{5} + 2\sqrt{5} - (\sqrt{5})^2$
$= 8 - 2\sqrt{5} - 5$
$= 3 - 2\sqrt{5}$

78. $(\sqrt{5x}+\sqrt{7y})(\sqrt{5x}-\sqrt{7y})$
$= (\sqrt{5x})^2 - (\sqrt{7y})^2$
$= 5x - 7y$

79. $(\sqrt{x}+8)^2$
$= (\sqrt{x})^2 + 2\cdot 8\sqrt{x} + 8^2$
$= x + 16\sqrt{x} + 64$

80. $3\sqrt[3]{x}(2\sqrt[3]{x^2}+\sqrt[3]{x})$
$= 6\sqrt[3]{x^3} + 3\sqrt[3]{x^2}$
$= 6x + 3\sqrt[3]{x^2}$

81. $\dfrac{\sqrt{15x}-\sqrt{30}}{\sqrt{5}}$
$= \dfrac{\sqrt{15x}}{\sqrt{5}} - \dfrac{\sqrt{30}}{\sqrt{5}}$
$= \sqrt{\dfrac{15x}{5}} - \sqrt{\dfrac{30}{5}}$
$= \sqrt{3x} - \sqrt{6}$

82. $\dfrac{\sqrt{z}-5}{\sqrt{z}}$
$= \dfrac{\sqrt{z}}{\sqrt{z}} - \dfrac{5}{\sqrt{z}}$
$= 1 - \dfrac{5}{\sqrt{z}}\cdot\dfrac{\sqrt{z}}{\sqrt{z}}$
$= 1 - \dfrac{5\sqrt{z}}{z}$

83. $\dfrac{24}{\sqrt{5}+2}$
$= \dfrac{24(\sqrt{5}-2)}{(\sqrt{5}+2)(\sqrt{5}-2)}$
$= \dfrac{24(\sqrt{5}-2)}{(\sqrt{5})^2 - 2^2}$
$= \dfrac{24\sqrt{5}-48}{(5-4)}$
$= 24\sqrt{5} - 48$

84. $\dfrac{\sqrt{x}+2}{\sqrt{x}-2}$
$= \dfrac{(\sqrt{x}+2)(\sqrt{x}+2)}{(\sqrt{x}-2)(\sqrt{x}+2)}$
$= \dfrac{(\sqrt{x})^2 + 2\cdot 2\sqrt{x} + 2^2}{(\sqrt{x})^2 - 2^2}$
$= \dfrac{x + 4\sqrt{x} + 4}{x - 4}$

85. $\dfrac{2\sqrt{x}-5}{5-2\sqrt{x}}$
$= \dfrac{-(5-2\sqrt{x})}{5-2\sqrt{x}}$
$= -1$

86. $\dfrac{2x-9}{\sqrt{2x}-3}$

$= \dfrac{(2x-9)(\sqrt{2x}+3)}{(\sqrt{2x}-3)(\sqrt{2x}+3)}$

$= \dfrac{(2x-9)(\sqrt{2x}+3)}{(\sqrt{2x})^2 - 3^2}$

$= \dfrac{(2x-9)(\sqrt{2x}+3)}{2x-9}$

$= \sqrt{2x}+3$

87. Let s_1 = length of the side of the first mirror.

$s_1 = \sqrt{A_1}$

$s_1 = \sqrt{720}$

$s_1 = 12\sqrt{5}$

Let s_2 = length of the side of the second mirror.

$s_2 = \sqrt{A_2}$

$s_2 = \sqrt{1620}$

$s_2 = 18\sqrt{5}$

Difference:

$4s_2 - 4s_1 = 4(18\sqrt{5}) - 4(12\sqrt{5})$

$= 72\sqrt{5} - 48\sqrt{5} = 24\sqrt{5}$

The second mirror will require $24\sqrt{5}$ inches more material.

88. $2\sqrt{3x} + 3 = 15$

$2\sqrt{3x} = 12$

$\sqrt{3x} = 6$

$(\sqrt{3x})^2 = 6^2$

$3x = 36$

$x = 12$

The solution is $x = 12$.

89. $\sqrt{4x-3} = x - 2$

$(\sqrt{4x-3})^2 = (x-2)^2$

$4x - 3 = x^2 - 4x + 4$

$0 = x^2 - 8x + 7$

$0 = (x-7)(x-1)$

$x = 7$ or $x = 1$ (extraneous)

The solution is $x = 7$.

90. $2\sqrt{x+5} = \sqrt{5x+9}$

$(2\sqrt{x+5})^2 = (\sqrt{5x+9})^2$

$4(x+5) = 5x + 9$

$4x + 20 = 5x + 9$

$11 = x$

The solution is $x = 11$.

91. $\sqrt{x+6} + 3 = \sqrt{5x-1}$

$(\sqrt{x+6}+3)^2 = (\sqrt{5x-1})^2$

$x + 6 + 6\sqrt{x+6} + 9 = 5x - 1$

$6\sqrt{x+6} = 4x - 16$

$3\sqrt{x+6} = 2x - 8$

$(3\sqrt{x+6})^2 = (2x-8)^2$

$9(x+6) = 4x^2 - 32x + 64$

$9x + 54 = 4x^2 - 32x + 64$

$0 = 4x^2 - 41x + 10$

$0 = (4x-1)(x-10)$

$x = \dfrac{1}{4}$ (extraneous) or $x = 10$

The solution is $x = 10$.

92. $\sqrt{x+8} = \sqrt{x-2}$

$(\sqrt{x+8})^2 = (\sqrt{x-2})^2$

$x + 8 = x - 2$

$8 = -2$

This is a contradiction. There is no solution.

SSM: Experiencing Introductory and Intermediate Algebra

93. $\sqrt[3]{2x} = -4$
$\left(\sqrt[3]{2x}\right)^3 = (-4)^3$
$2x = -64$
$x = -32$
The solution is $x = -32$.

94. $\sqrt[4]{x+3} = 3$
$\left(\sqrt[4]{x+3}\right)^4 = 3^4$
$x + 3 = 81$
$x = 78$
The solution is $x = 78$.

95. $(x+7)^{2/5} = 4$
$\left((x+7)^{2/5}\right)^5 = 4^5$
$(x+7)^2 = 1024$
$x + 7 = \pm\sqrt{1024}$
$x = -7 \pm 32$
$x = -39$ or $x = 25$
The solutions are $x = -39$ and $x = 25$.

96. $x^{-3/2} = 8$
$\left(x^{-3/2}\right)^2 = 8^2$
$x^{-3} = 64$
$\frac{1}{x^3} = 64$
$x^3 = \frac{1}{64}$
$\sqrt[3]{x^3} = \sqrt[3]{\frac{1}{64}}$
$x = \frac{1}{4}$
The solution is $x = \frac{1}{4}$.

97. $2x^{2/3} - 5 = 1$
$2x^{2/3} = 6$
$x^{2/3} = 3$
$\left(x^{2/3}\right)^3 = 3^3$
$x^2 = 27$
$x = \pm\sqrt{27}$
$x = \pm 3\sqrt{3}$
The solutions are $x = \pm 3\sqrt{3}$.

98. $d = \sqrt{(x_2 - x_1)^2 + (y_2 - y_1)^2}$
$5 = \sqrt{(6-3)^2 + (y+2)^2}$
$5 = \sqrt{3^2 + y^2 + 4y + 4}$
$5^2 = \left(\sqrt{y^2 + 4y + 13}\right)^2$
$25 = y^2 + 4y + 13$
$0 = y^2 + 4y - 12$
$0 = (y+6)(y-2)$
$y = -6$ or $y = 2$
The y-coordinates are $y = -6$ or $y = 2$.
The two possible points are $(6, -6)$ and $(6, 2)$.

99. $V = \sqrt{10.5s}$
$40 = \sqrt{10.5s}$
$40^2 = \left(\sqrt{10.5s}\right)^2$
$1600 = 10.5s$
$152.381 \approx s$
The skid mark would be approximately 152.4 feet long.

100. $t = \sqrt{\dfrac{2d}{32.2}}$

$6 = \sqrt{\dfrac{2d}{32.2}}$

$6 = \sqrt{\dfrac{d}{16.1}}$

$6^2 = \left(\sqrt{\dfrac{d}{16.1}}\right)^2$

$36 = \dfrac{d}{16.1}$

$579.6 = d$

The peach was dropped from a height of 579.6 feet.

101. $\sqrt{-64} = \sqrt{64} \cdot i = 8i$

102. $\sqrt{-\dfrac{25}{49}} = \sqrt{\dfrac{25}{49}} \cdot i = \dfrac{5}{7}i$

103. $\sqrt{-6.25} = \sqrt{6.25} \cdot i = 2.5i$

104. $\sqrt{-50} = \sqrt{50} \cdot i = 5i\sqrt{2}$

105. $\sqrt{-\dfrac{81}{2}} = \sqrt{\dfrac{81}{2}} \cdot i = \dfrac{9}{\sqrt{2}} \cdot \dfrac{\sqrt{2}}{\sqrt{2}} \cdot i$

$= \dfrac{9\sqrt{2}}{2}i$

106. $-5\sqrt{-32} = -5\sqrt{32} \cdot i = -5 \cdot 4\sqrt{2} \cdot i$

$= -20i\sqrt{2}$

107. $\sqrt{-25} + \sqrt{-49}$

$= \sqrt{25} \cdot i + \sqrt{49} \cdot i$

$= 5i + 7i$

$= 12i$

108. $\sqrt{-144} - \sqrt{-64}$

$= \sqrt{144} \cdot i - \sqrt{64} \cdot i$

$= 12i - 8i$

$= 4i$

109. $2\sqrt{-36} + 6\sqrt{-81}$

$= 2\sqrt{36} \cdot i + 6\sqrt{81} \cdot i$

$= 12i + 54i$

$= 66i$

110. $11\sqrt{-4} - 3\sqrt{-36}$

$= 11\sqrt{4} \cdot i - 3\sqrt{36} \cdot i$

$= 22i - 18i$

$= 4i$

111. $\dfrac{\sqrt{-169}}{\sqrt{-225}}$

$= \dfrac{\sqrt{169} \cdot i}{\sqrt{225} \cdot i}$

$= \dfrac{13}{15}$

112. $\sqrt{-2}\sqrt{-6}\sqrt{-75}$

$= \sqrt{2} \cdot i \cdot \sqrt{6} \cdot i \cdot \sqrt{75} \cdot i$

$= \sqrt{2}\sqrt{6}\sqrt{75} \cdot i^3$

$= \sqrt{900} \cdot i^3$

$= 30(-i)$

$= -30i$

113. $\sqrt{-2}\left(\sqrt{-6} + \sqrt{75}\right)$

$= \sqrt{2} \cdot i\left(\sqrt{6} \cdot i + 5\sqrt{3}\right)$

$= \sqrt{12} \cdot i^2 + 5i\sqrt{6}$

$= -2\sqrt{3} + 5i\sqrt{6}$

114. $\left(\sqrt{-2} + \sqrt{-18}\right)\left(\sqrt{-4} - \sqrt{-9}\right)$

$= \left(\sqrt{2} \cdot i + \sqrt{18} \cdot i\right)\left(\sqrt{4} \cdot i - \sqrt{9} \cdot i\right)$

$= \sqrt{8} \cdot i^2 + \sqrt{72} \cdot i^2 - \sqrt{18} \cdot i^2 - \sqrt{162} \cdot i^2$

$= -2\sqrt{2} - 6\sqrt{2} + 3\sqrt{2} + 9\sqrt{2}$

$= (-2 - 6 + 3 + 9)\sqrt{2}$

$= 4\sqrt{2}$

115. $\left(\sqrt{-7}+\sqrt{-11}\right)\left(\sqrt{-7}-\sqrt{-11}\right)$
$= \left(i\sqrt{7}+i\sqrt{11}\right)\left(i\sqrt{7}-i\sqrt{11}\right)$
$= \left(i\sqrt{7}\right)^2 - \left(i\sqrt{11}\right)^2$
$= 7i^2 - 11i^2 = -4i^2$
$= 4$

116. $(17+4i)+(12-6i)$
$= 17+4i+12-6i$
$= 29-2i$

117. $(12-3i)-(1+i)$
$= 12-3i-1-i$
$= 11-4i$

118. $(2+5i)(-3+4i)$
$= -6-15i+8i+20i^2$
$= -6-7i-20$
$= -26-7i$

119. $(3+11i)(3-11i)$
$= (3)^2 - (11i)^2$
$= 9+121$
$= 130$

120. $\dfrac{12+15i}{3}$
$= \dfrac{12}{3}+\dfrac{15i}{3}$
$= 4+5i$

121. $\dfrac{14-21i}{-7i}$
$= \dfrac{14}{-7i}+\dfrac{-21i}{-7i} = -\dfrac{2}{i}+3$
$= -\dfrac{2}{i}\cdot\dfrac{i}{i}+3 = -\dfrac{2i}{i^2}+3$
$= 3+2i$

122. $\dfrac{41+i}{5-2i}$
$= \dfrac{(41+i)(5+2i)}{(5-2i)(5+2i)} = \dfrac{205+5i+82i+2i^2}{25-4i^2}$
$= \dfrac{205+87i-2}{25+4} = \dfrac{203+87i}{29}$
$= 7+3i$

123. $\left(5\sqrt{7}+8i\sqrt{13}\right)+\left(3\sqrt{7}-2i\sqrt{13}\right)$
$= 5\sqrt{7}+8i\sqrt{13}+3\sqrt{7}-2i\sqrt{13}$
$= 8\sqrt{7}+6i\sqrt{13}$

124. $i\sqrt{6}\left(\sqrt{3}-i\sqrt{6}\right)$
$= i\sqrt{18}-i^2\sqrt{36}$
$= 3i\sqrt{2}+6$
$= 6+3i\sqrt{2}$

125. $\left(\dfrac{2}{3}-\dfrac{3}{5}i\right)\left(\dfrac{1}{2}+\dfrac{1}{3}i\right)$
$= \dfrac{1}{3}-\dfrac{3}{10}i+\dfrac{2}{9}i-\dfrac{1}{5}i^2$
$= \dfrac{1}{3}-\dfrac{7}{90}i+\dfrac{1}{5}$
$= \dfrac{8}{15}-\dfrac{7}{90}i$

126. $\dfrac{18-11i\sqrt{6}}{3\sqrt{2}+i\sqrt{3}}$
$= \dfrac{\left(18-11i\sqrt{6}\right)\left(3\sqrt{2}-i\sqrt{3}\right)}{\left(3\sqrt{2}+i\sqrt{3}\right)\left(3\sqrt{2}-i\sqrt{3}\right)}$
$= \dfrac{54\sqrt{2}-33i\sqrt{12}-18i\sqrt{3}+11i^2\sqrt{18}}{\left(3\sqrt{2}\right)^2-\left(i\sqrt{3}\right)^2}$
$= \dfrac{54\sqrt{2}-66i\sqrt{3}-18i\sqrt{3}-33\sqrt{2}}{18+3}$
$= \dfrac{21\sqrt{2}-84i\sqrt{3}}{21}$
$= \sqrt{2}-4i\sqrt{3}$

Chapter 13: Radical Expressions, Functions, and Equations

127. $z^2 + 10 = 1$
$z^2 = -9$
$z = \pm\sqrt{-9}$
$z = \pm 3i$

128. $5t^2 + 125 = 0$
$5t^2 = -125$
$t^2 = -25$
$t = \pm\sqrt{-25}$
$t = \pm 5i$

129. $3a^2 + 33 = 0$
$3a^2 = -33$
$a^2 = -11$
$a = \pm\sqrt{-11}$
$a = \pm i\sqrt{11}$

130. $(r-5)^2 + 36 = 0$
$(r-5)^2 = -36$
$r - 5 = \pm\sqrt{-36}$
$r - 5 = \pm 6i$
$r = 5 \pm 6i$

131. $7(x+2)^2 + 23 = 2$
$7(x+2)^2 = -21$
$(x+2)^2 = -3$
$x + 2 = \pm\sqrt{-3}$
$x = -2 \pm i\sqrt{3}$

132. $3(4x+1)^2 = -27$
$(4x+1)^2 = -9$
$4x + 1 = \pm\sqrt{-9}$
$4x = -1 \pm 3i$
$x = -\dfrac{1}{4} \pm \dfrac{3}{4}i$

133. $\left(m + \dfrac{2}{5}\right)^2 + \dfrac{16}{25} = 0$
$\left(m + \dfrac{2}{5}\right)^2 = -\dfrac{16}{25}$
$m + \dfrac{2}{5} = \pm\sqrt{-\dfrac{16}{25}}$
$m = -\dfrac{2}{5} \pm \dfrac{4}{5}i$

134. $x^2 - 10x + 29 = 0$
$x = \dfrac{-(-10) \pm \sqrt{(-10)^2 - 4(1)(29)}}{2(1)}$
$= \dfrac{10 \pm \sqrt{-16}}{2} = \dfrac{10 \pm 4i}{2}$
$= 5 \pm 2i$

135. $4y^2 + 4y + 10 = 0$
$y = \dfrac{-4 \pm \sqrt{(4)^2 - 4(4)(10)}}{2(4)}$
$= \dfrac{-4 \pm \sqrt{-144}}{8} = \dfrac{-4 \pm 12i}{8}$
$= -\dfrac{1}{2} \pm \dfrac{3}{2}i$

136. $z^2 + 16z + 17 = 0$
$z = \dfrac{-16 \pm \sqrt{16^2 - 4(1)(17)}}{2(1)}$
$= \dfrac{-16 \pm \sqrt{188}}{2} = \dfrac{-16 \pm 2\sqrt{47}}{2}$
$= -8 \pm \sqrt{47}$

137. $9x^2 - 12x + 40 = 0$
$x = \dfrac{-(-12) \pm \sqrt{(-12)^2 - 4(9)(40)}}{2(9)}$
$= \dfrac{12 \pm \sqrt{-1296}}{18} = \dfrac{12 \pm 36i}{18}$
$= \dfrac{2}{3} \pm 2i$

SSM: Experiencing Introductory and Intermediate Algebra

138. $4x^2 + 4x + 5.84 = 0$

$x = \dfrac{-4 \pm \sqrt{4^2 - 4(4)(5.84)}}{2(4)}$

$= \dfrac{-4 \pm \sqrt{-77.44}}{8} = \dfrac{-4 \pm 8.8i}{8}$

$= -0.5 \pm 1.1i$

139. $x^2 - \dfrac{3}{2}x + \dfrac{5}{8} = 0$

$8x^2 - 12x + 5 = 0$

$x = \dfrac{-(-12) \pm \sqrt{(-12)^2 - 4(8)(5)}}{2(8)}$

$= \dfrac{12 \pm \sqrt{-16}}{16} = \dfrac{12 \pm 4i}{16}$

$= \dfrac{3}{4} \pm \dfrac{1}{4}i$

140. $\dfrac{x}{x-5} = \dfrac{10}{x+1}$

$x(x+1) = 10(x-5)$

$x^2 + x = 10x - 50$

$x^2 - 9x + 50 = 0$

$x = \dfrac{-(-9) \pm \sqrt{(-9)^2 - 4(1)(50)}}{2(1)}$

$= \dfrac{9 \pm \sqrt{-119}}{2}$

$= \dfrac{9}{2} \pm \dfrac{\sqrt{119}}{2}i$

141. $\dfrac{y+4}{y-8} = \dfrac{9}{y+6}$

$(y+4)(y+6) = 9(y-8)$

$y^2 + 10y + 24 = 9y - 72$

$y^2 + y + 96 = 0$

$y = \dfrac{-1 \pm \sqrt{1^2 - 4(1)(96)}}{2(1)}$

$= \dfrac{-1 \pm \sqrt{-383}}{2}$

$= -\dfrac{1}{2} \pm \dfrac{\sqrt{383}}{2}i$

142. $\dfrac{b-8}{b(b+7)} = \dfrac{2}{3}$

$3(b-8) = 2b(b+7)$

$3b - 24 = 2b^2 + 14b$

$0 = 2b^2 + 11b + 24$

$b = \dfrac{-11 \pm \sqrt{(11)^2 - 4(2)(24)}}{2(2)}$

$= \dfrac{-11 \pm \sqrt{-71}}{4}$

$= -\dfrac{11}{4} \pm \dfrac{\sqrt{71}}{4}i$

143. $V = IZ$

$V = (5 + 4i)(9 - 2i)$

$= 45 + 36i - 10i - 8i^2$

$= 45 + 26i + 8$

$= 53 + 26i$

The total voltage is $53 + 26i$ volts.

$|V| = \sqrt{53^2 + 26^2} \approx 59.0$

The magnitude of the total voltage is about 59.0 volts.

Chapter 13: Radical Expressions, Functions, and Equations

144. $Z = \dfrac{V}{I}$

$Z = \dfrac{33 + 19i}{5 + 2i}$

$= \dfrac{(33 + 19i)(5 - 2i)}{(5 + 2i)(5 - 2i)}$

$= \dfrac{165 - 66i + 95i + 38}{29}$

$= \dfrac{203 + 29i}{29}$

$= 7 + i$

The impedance is $7 + i$ ohms.

$|Z| = \sqrt{7^2 + 1^2} = \sqrt{50} = 5\sqrt{2}$

The magnitude of the impedance is $5\sqrt{2}$ ohms.

Chapter 13 Chapter Review

1. $\sqrt{196} = \sqrt{14^2} = 14$

2. $\sqrt{4.41} = \sqrt{2.1^2} = 2.1$

3. $\sqrt[3]{-\dfrac{27}{64}} = \sqrt[3]{\left(-\dfrac{3}{4}\right)^3} = -\dfrac{3}{4}$

4. $\sqrt[4]{256} = \sqrt[4]{4^4} = 4$

5. $\sqrt[4]{-256}$ is not a real number.

6. $\sqrt[3]{-29} \approx -3.072$

7. $121^{1/2} = \sqrt{121} = \sqrt{11^2} = 11$

8. $(-125)^{1/3} = \sqrt[3]{-125} = \sqrt[3]{(-5)^3} = -5$

9. $-81^{1/4} = -\sqrt[4]{81} = -\sqrt[4]{3^4} = -3$

10. $(-81)^{1/4} = \sqrt[4]{-81}$ is not a real number.

11. $729^{5/6} = \left(\sqrt[6]{729}\right)^5 = 3^5 = 243$

12. $(-729)^{5/6} = \left(\sqrt[6]{-729}\right)^5$ is not a real number.

13. $(-8)^{-2/3} = \dfrac{1}{(-8)^{2/3}} = \dfrac{1}{\left(\sqrt[3]{-8}\right)^2} = \dfrac{1}{2^2} = \dfrac{1}{4}$

14. $(-81)^{-3/4} = \dfrac{1}{(-81)^{3/4}} = \dfrac{1}{\left(\sqrt[4]{-81}\right)^3}$ is not a real number.

15. $\dfrac{1}{9^{-1/2}} = 9^{1/2} = \sqrt{9} = 3$

16. $27^{2/3} \cdot 27^{5/3}$
$= 27^{\frac{2}{3} + \frac{5}{3}} = 27^{7/3}$
$= \left(\sqrt[3]{27}\right)^7 = 3^7$
$= 2187$

17. $\dfrac{6^{7/3}}{6^{2/3}}$
$= 6^{\frac{7}{3} - \frac{2}{3}} = 6^{5/3}$
$= \left(\sqrt[3]{6}\right)^5 \approx 19.812$

18. $\left(36^{2/3}\right)^{3/4}$
$= 36^{\frac{2}{3} \cdot \frac{3}{4}} = 36^{1/2}$
$= \sqrt{36} = 6$

19. $\left(\dfrac{9}{16}\right)^{-3/2}$
$= \left(\dfrac{16}{9}\right)^{3/2} = \left(\sqrt{\dfrac{16}{9}}\right)^3 = \left(\dfrac{4}{3}\right)^3$
$= \dfrac{64}{27}$

20. $\left[\left(3^{1/2}\right)\left(12^{1/2}\right)\right]^{-3/2}$
$= \left[(3 \cdot 12)^{1/2}\right]^{-3/2}$
$= \left[36^{1/2}\right]^{-3/2} = 6^{-3/2}$
$= \dfrac{1}{6^{3/2}} = \dfrac{1}{\left(\sqrt{6}\right)^3}$
≈ 0.068

21. $-\sqrt{\dfrac{72}{98}} = -\dfrac{\sqrt{72}}{\sqrt{98}} = -\dfrac{6\sqrt{2}}{7\sqrt{2}} = -\dfrac{6}{7}$

22. $\sqrt{\dfrac{15}{144}} = \dfrac{\sqrt{15}}{\sqrt{144}} = \dfrac{\sqrt{15}}{12}$

23. $\sqrt{\dfrac{50}{338}} = \sqrt{\dfrac{25}{169}} = \dfrac{\sqrt{25}}{\sqrt{169}} = \dfrac{5}{13}$

24. $\sqrt[3]{-\dfrac{64}{125}} = \sqrt[3]{\left(-\dfrac{4}{5}\right)^3} = -\dfrac{4}{5}$

25. $\sqrt{150} = \sqrt{25 \cdot 6} = 5\sqrt{6}$

26. $\sqrt[3]{-448} = \sqrt[3]{64 \cdot 7} = 4\sqrt[3]{7}$

27. $\sqrt[4]{48} = \sqrt[4]{16 \cdot 3} = 2\sqrt[4]{3}$

28. $\sqrt[5]{-486} = \sqrt[5]{243 \cdot 2} = 3\sqrt[5]{2}$

29. $\sqrt{15} \cdot \sqrt{35} = \sqrt{525} = \sqrt{25 \cdot 21} = 5\sqrt{21}$

30. $\sqrt{\dfrac{3}{7}} - \sqrt{\dfrac{7}{3}}$
$= \dfrac{\sqrt{3}}{\sqrt{7}} - \dfrac{\sqrt{7}}{\sqrt{3}} = \dfrac{\sqrt{3} \cdot \sqrt{7}}{\sqrt{7} \cdot \sqrt{7}} - \dfrac{\sqrt{7} \cdot \sqrt{3}}{\sqrt{3} \cdot \sqrt{3}}$
$= \dfrac{\sqrt{21}}{7} - \dfrac{\sqrt{21}}{3} = \left(\dfrac{1}{7} - \dfrac{1}{3}\right)\sqrt{21}$
$= -\dfrac{4}{21}\sqrt{21}$

31. $6\sqrt{13} + 3\sqrt{52}$
$= 6\sqrt{13} + 6\sqrt{13}$
$= 12\sqrt{13}$

32. $4\sqrt{7} - 3\sqrt{7} + 11\sqrt{7} - \sqrt{7}$
$= (4 - 3 + 11 - 1)\sqrt{7}$
$= 11\sqrt{7}$

33. $\sqrt{3}\left(\sqrt{6} - \sqrt{3}\right)$
$= \sqrt{3} \cdot \sqrt{6} - \sqrt{3} \cdot \sqrt{3}$
$= \sqrt{18} - \sqrt{9}$
$= 3\sqrt{2} - 3$

34. $\left(7 - \sqrt{3}\right)\left(9 + \sqrt{3}\right)$
$= 63 - 9\sqrt{3} + 7\sqrt{3} - \left(\sqrt{3}\right)^2$
$= 63 - 2\sqrt{3} - 3$
$= 60 - 2\sqrt{3}$

35. $\left(\sqrt{10} - \sqrt{5}\right)\left(\sqrt{10} + \sqrt{5}\right)$
$= \left(\sqrt{10}\right)^2 - \left(\sqrt{5}\right)^2$
$= 10 - 5$
$= 5$

36. $6\sqrt[3]{375} + 2\sqrt[3]{24}$
$= 6 \cdot 5\sqrt[3]{3} + 2 \cdot 2\sqrt[3]{3}$
$= 30\sqrt[3]{3} + 4\sqrt[3]{3}$
$= 34\sqrt[3]{3}$

37. $\dfrac{x^{2/3}}{x^{3/4}} = x^{\frac{2}{3} - \frac{3}{4}} = x^{-1/12} = \dfrac{1}{x^{1/12}}$

38. $z^{3/5} \cdot z^{-3/2} = z^{\frac{3}{5} + \left(-\frac{3}{2}\right)} = z^{-9/10} = \dfrac{1}{z^{9/10}}$

39. $\left(\dfrac{p^4}{q^8}\right)^{-5/16} = \left(\dfrac{q^8}{p^4}\right)^{5/16} = \dfrac{q^{8 \cdot \frac{5}{16}}}{p^{4 \cdot \frac{5}{16}}} = \dfrac{q^{5/2}}{p^{5/4}}$

40. $\left(27x^9 y^{12}\right)^{4/3} = 27^{4/3} x^{9 \cdot \frac{4}{3}} y^{12 \cdot \frac{4}{3}} = 81x^{12}y^{16}$

Chapter 13: Radical Expressions, Functions, and Equations

41. $\sqrt{72x^6y^3z^4} = \sqrt{36 \cdot 2 \cdot x^6 \cdot y^2 \cdot y \cdot z^4}$
$= 6x^3yz^2\sqrt{2y}$

42. $\sqrt[3]{-64x^2y^7} = \sqrt[3]{-64 \cdot x^2 \cdot y^6 \cdot y}$
$= -4y^2\sqrt[3]{x^2y}$

43. $2\sqrt{3xy} \cdot y\sqrt{6x}$
$= 2y\sqrt{18x^2y}$
$= 6xy\sqrt{2y}$

44. $\sqrt[3]{-3a^5b^5} \cdot \sqrt[3]{18ab^2}$
$= \sqrt[3]{-54a^6b^7} = \sqrt[3]{-27 \cdot 2a^6 \cdot b^6 \cdot b}$
$= -3a^2b^2\sqrt[3]{2b}$

45. $\sqrt{10x + 35y} \cdot \sqrt{2x + 7y}$
$= \sqrt{(10x + 35y)(2x + 7y)}$
$= \sqrt{5(2x + 7y)(2x + 7y)}$
$= \sqrt{5(2x + 7y)^2}$
$= (2x + 7y)\sqrt{5}$

46. $\sqrt{\dfrac{36x^2}{49y^4}} = \dfrac{\sqrt{36x^2}}{\sqrt{49y^4}} = \dfrac{6x}{7y^2}$

47. $\sqrt{\dfrac{25z}{3}} = \dfrac{\sqrt{25z}}{\sqrt{3}} = \dfrac{5\sqrt{z}}{\sqrt{3}} \cdot \dfrac{\sqrt{3}}{\sqrt{3}} = \dfrac{5\sqrt{3z}}{3}$

48. $\sqrt[3]{\dfrac{-27a}{b^3}} = \dfrac{\sqrt[3]{-27a}}{\sqrt[3]{b^3}} = \dfrac{-3\sqrt[3]{a}}{b}$

49. $\dfrac{\sqrt{6a}}{\sqrt{12ab}} = \sqrt{\dfrac{6a}{12ab}} = \sqrt{\dfrac{1}{2b}}$
$= \dfrac{1}{\sqrt{2b}} \cdot \dfrac{\sqrt{2b}}{\sqrt{2b}} = \dfrac{\sqrt{2b}}{2b}$

50. $\dfrac{\sqrt{x^3y^2}}{\sqrt{x^4y^5}}$
$= \sqrt{\dfrac{x^3y^2}{x^4y^5}} = \sqrt{\dfrac{1}{xy^3}} = \dfrac{\sqrt{1}}{\sqrt{xy^3}}$
$= \dfrac{1}{y\sqrt{xy}} \cdot \dfrac{\sqrt{xy}}{\sqrt{xy}} = \dfrac{\sqrt{xy}}{xy^2}$

51. $\sqrt{121x} - \sqrt{81x}$
$= 11\sqrt{x} - 9\sqrt{x}$
$= (11 - 9)\sqrt{x}$
$= 2\sqrt{x}$

52. $16d\sqrt{c^3d} - 11c\sqrt{cd^3} + 9cd\sqrt{cd}$
$= 16cd\sqrt{cd} - 11cd\sqrt{cd} + 9cd\sqrt{cd}$
$= (16 - 11 + 9)cd\sqrt{cd}$
$= 14cd\sqrt{cd}$

53. $(\sqrt{a} + 9)^2$
$= (\sqrt{a})^2 + 2 \cdot 9\sqrt{a} + 9^2$
$= a + 18\sqrt{a} + 81$

54. $\sqrt{6a}(3 + \sqrt{6a})$
$= 3\sqrt{6a} + (\sqrt{6a})^2$
$= 3\sqrt{6a} + 6a$

55. $(\sqrt{3x} + \sqrt{2y})(\sqrt{3x} - \sqrt{2y})$
$= (\sqrt{3x})^2 - (\sqrt{2y})^2$
$= 3x - 2y$

56. $2\sqrt[3]{x^2}(3\sqrt[3]{x} + 5\sqrt[3]{x^2})$
$= 2\sqrt[3]{x^2} \cdot 3\sqrt[3]{x} + 2\sqrt[3]{x^2} \cdot 5\sqrt[3]{x^2}$
$= 6\sqrt[3]{x^3} + 10\sqrt[3]{x^4}$
$= 6x + 10x\sqrt[3]{x}$

SSM: Experiencing Introductory and Intermediate Algebra

57. $\dfrac{\sqrt{18x}-\sqrt{42}}{\sqrt{6}}$

$= \dfrac{\sqrt{18x}}{\sqrt{6}} - \dfrac{\sqrt{42}}{\sqrt{6}}$

$= \sqrt{\dfrac{18x}{6}} - \sqrt{\dfrac{42}{6}}$

$= \sqrt{3x} - \sqrt{7}$

58. $\dfrac{\sqrt{m}-9}{\sqrt{m}}$

$= \dfrac{\sqrt{m}}{\sqrt{m}} - \dfrac{9}{\sqrt{m}}$

$= \sqrt{\dfrac{m}{m}} - \dfrac{9}{\sqrt{m}} \cdot \dfrac{\sqrt{m}}{\sqrt{m}}$

$= \sqrt{1} - \dfrac{9\sqrt{m}}{m}$

$= 1 - \dfrac{9\sqrt{m}}{m}$

59. $\dfrac{18}{\sqrt{7}+2}$

$= \dfrac{18}{\sqrt{7}+2} \cdot \dfrac{\sqrt{7}-2}{\sqrt{7}-2}$

$= \dfrac{18(\sqrt{7}-2)}{(\sqrt{7})^2-2^2} = \dfrac{18\sqrt{7}-36}{7-4}$

$= \dfrac{18\sqrt{7}-36}{3} = \dfrac{18\sqrt{7}}{3} - \dfrac{36}{3}$

$= 6\sqrt{7}-12$

$= 6(\sqrt{7}-2)$

60. $\dfrac{3+\sqrt{x}}{3-\sqrt{x}}$

$= \dfrac{3+\sqrt{x}}{3-\sqrt{x}} \cdot \dfrac{3+\sqrt{x}}{3+\sqrt{x}}$

$= \dfrac{3^2+2\cdot 3\sqrt{x}+(\sqrt{x})^2}{(3)^2-(\sqrt{x})^2}$

$= \dfrac{9+6\sqrt{x}+x}{9-x}$

61. $\dfrac{7-3\sqrt{a}}{3\sqrt{a}-7} = \dfrac{-(3\sqrt{a}-7)}{3\sqrt{a}-7} = -1$

62. $\dfrac{25-7x}{5-\sqrt{7x}}$

$= \dfrac{25-7x}{5-\sqrt{7x}} \cdot \dfrac{5+\sqrt{7x}}{5+\sqrt{7x}}$

$= \dfrac{(25-7x)(5+\sqrt{7x})}{(5)^2-(\sqrt{7x})^2}$

$= \dfrac{(25-7x)(5+\sqrt{7x})}{25-7x}$

$= 5+\sqrt{7x}$

63. $f(x) = 5 - \sqrt{x}$

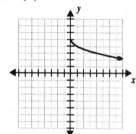

The restricted values are all real numbers less than 0.
Domain: $[0,\infty)$

64. $y = x^{3/2} - 3$

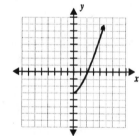

The restricted values are all real numbers less than 0.
Domain: $[0,\infty)$

Chapter 13: Radical Expressions, Functions, and Equations

65. $(x-1)^{3/2} = \sqrt[3]{2x+5}$

$Y1 = (x-1)^{3/2}$

$Y2 = \sqrt[3]{2x+5}$

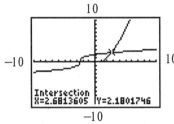

The approximate solution is the x-coordinate of the intersection point, 2.681.

66. $(x+4)^{3/5} = (5x-1)^{2/3}$

$Y1 = (x+4)^{3/5}$

$Y2 = (5x-1)^{2/3}$

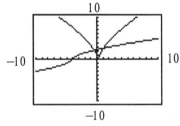

The approximate solutions are the x-coordinates of the intersection points, −0.429 and 1.061.

67. $\sqrt{x} + 2 = 11$

$\sqrt{x} = 9$

$(\sqrt{x})^2 = 9^2$

$x = 81$

68. $3\sqrt{5x} + 7 = 52$

$3\sqrt{5x} = 45$

$\sqrt{5x} = 15$

$(\sqrt{5x})^2 = 15^2$

$5x = 225$

$x = 45$

69. $\sqrt{3x+1} = x-3$

$(\sqrt{3x+1})^2 = (x-3)^2$

$3x+1 = x^2 - 6x + 9$

$0 = x^2 - 9x + 8$

$0 = (x-8)(x-1)$

$x = 8$ or $x = 1$ (extraneous)

The solution is $x = 8$.

70. $2\sqrt{x+9} = \sqrt{9x+1}$

$(2\sqrt{x+9})^2 = (\sqrt{9x+1})^2$

$4(x+9) = 9x+1$

$4x + 36 = 9x + 1$

$-5x = -35$

$x = 7$

The solution is $x = 7$.

71. $\sqrt{2x+1} + 2 = \sqrt{6x+1}$

$(\sqrt{2x+1} + 2)^2 = (\sqrt{6x+1})^2$

$2x + 1 + 4\sqrt{2x+1} + 4 = 6x + 1$

$4\sqrt{2x+1} = 4x - 4$

$\sqrt{2x+1} = x - 1$

$(\sqrt{2x+1})^2 = (x-1)^2$

$2x + 1 = x^2 - 2x + 1$

$0 = x^2 - 4x$

$0 = x(x-4)$

$x = 0$ (extraneous) or $x = 4$

The solution is $x = 4$.

72. $\sqrt{10-x} = \sqrt{3-x}$

$(\sqrt{10-x})^2 = (\sqrt{3-x})^2$

$10 - x = 3 - x$

$10 = 3$

This is a contradiction. There is no solution.

603

73. $\sqrt[3]{3x} = -6$
$\left(\sqrt[3]{3x}\right)^3 = (-6)^3$
$3x = -216$
$x = -72$

74. $\sqrt[4]{x+2} = 3$
$\left(\sqrt[4]{x+2}\right)^4 = 3^4$
$x + 2 = 81$
$x = 79$
The solution is $x = 79$.

75. $(x+5)^{2/5} = 9$
$\left((x+5)^{2/5}\right)^5 = 9^5$
$(x+5)^2 = 59049$
$x + 5 = \pm\sqrt{59049}$
$x = -5 \pm 243$
$x = -248$ or $x = 238$
The solutions are $x = -248$ and $x = 238$.

76. $x^{-3/2} = 27$
$\dfrac{1}{x^{3/2}} = 27$
$x^{3/2} = \dfrac{1}{27}$
$\left(x^{3/2}\right)^2 = \left(\dfrac{1}{27}\right)^2$
$x^3 = \dfrac{1}{729}$
$x = \sqrt[3]{\dfrac{1}{729}}$
$x = \dfrac{1}{9}$
The solution is $x = \dfrac{1}{9}$.

77. $3x^{2/3} - 7 = 2$
$3x^{2/3} = 9$
$x^{2/3} = 3$
$\left(x^{2/3}\right)^3 = 3^3$
$x^2 = 27$
$x = \pm\sqrt{27}$
$x = \pm 3\sqrt{3}$
The solutions are $x = \pm 3\sqrt{3}$.

78. $\sqrt{-169} = \sqrt{169} \cdot i = 13i$

79. $\sqrt{-\dfrac{64}{81}} = \sqrt{\dfrac{64}{81}} \cdot i = \dfrac{8}{9}i$

80. $\sqrt{-\dfrac{25}{6}} = \sqrt{\dfrac{25}{6}} \cdot i = \dfrac{\sqrt{25}}{\sqrt{6}}i$
$= \dfrac{5i}{\sqrt{6}} \cdot \dfrac{\sqrt{6}}{\sqrt{6}}$
$= \dfrac{5i\sqrt{6}}{6}$

81. $10\sqrt{-80} = 10\sqrt{80} \cdot i = 10 \cdot 4\sqrt{5} \cdot i = 40i\sqrt{5}$

82. $\sqrt{-6.25} = \sqrt{6.25} \cdot i = 2.5i$

83. $-\sqrt{-108} = -\sqrt{108} \cdot i = -6\sqrt{3} \cdot i = -6i\sqrt{3}$

84. $(22.3 - 1.33i) + (8.55 - 2.9i)$
$= 22.3 - 1.33i + 8.55 - 2.9i$
$= 30.85 - 4.23i$

85. $\sqrt{-64} + \sqrt{-4}$
$= \sqrt{64} \cdot i + \sqrt{4} \cdot i$
$= 8i + 2i$
$= 10i$

86. $\sqrt{-196} - \sqrt{-100}$
$= \sqrt{196} \cdot i - \sqrt{100} \cdot i$
$= 14i - 10i$
$= 4i$

Chapter 13: Radical Expressions, Functions, and Equations

87. $\left(\dfrac{3}{4}+\dfrac{1}{2}i\right)\left(\dfrac{4}{9}-2i\right)$

$= \dfrac{1}{3} + \dfrac{2}{9}i - \dfrac{3}{2}i - i^2$

$= \dfrac{1}{3} - \dfrac{23}{18}i + 1$

$= \dfrac{4}{3} - \dfrac{23}{18}i$

88. $-1.5i(2.9-4i)$

$= -4.35i + 6i^2$

$= -6 - 4.35i$

89. $7\sqrt{-25} + 3\sqrt{-64}$

$= 7\sqrt{25}\cdot i + 3\sqrt{64}\cdot i$

$= 7\cdot 5i + 3\cdot 8i$

$= 35i + 24i$

$= 59i$

90. $\sqrt{-2}\sqrt{-6}\sqrt{-48}$

$= \sqrt{2}\cdot i \cdot \sqrt{6}\cdot i \cdot \sqrt{48}\cdot i$

$= \sqrt{576}\cdot i^3$

$= 24\cdot(-i)$

$= -24i$

91. $\left(\sqrt{-3}-\sqrt{-10}\right)\left(\sqrt{-5}+\sqrt{-6}\right)$

$= \left(\sqrt{3}\cdot i - \sqrt{10}\cdot i\right)\left(\sqrt{5}\cdot i + \sqrt{6}\cdot i\right)$

$= \sqrt{15}\cdot i^2 - \sqrt{50}\cdot i^2 + \sqrt{18}\cdot i^2 - \sqrt{60}\cdot i^2$

$= -\sqrt{15} + 5\sqrt{2} - 3\sqrt{2} + 2\sqrt{15}$

$= \sqrt{15} + 2\sqrt{2}$

92. $\left(\sqrt{-13}+\sqrt{-17}\right)\left(\sqrt{-13}-\sqrt{-17}\right)$

$= \left(\sqrt{13}\cdot i + \sqrt{17}\cdot i\right)\left(\sqrt{13}\cdot i - \sqrt{17}\cdot i\right)$

$= \left(\sqrt{13}\cdot i\right)^2 - \left(\sqrt{17}\cdot i\right)^2$

$= 13i^2 - 17i^2 = -4i^2$

$= 4$

93. $(21-i)-(7+i)$

$= 21 - i - 7 - i$

$= 14 - 2i$

94. $(9-3i)(-9+3i)$

$= -81 + 27i + 27i - 9i^2$

$= -81 + 54i + 9$

$= -72 + 54i$

95. $\dfrac{18-27i}{9i}$

$= \dfrac{18}{9i} - \dfrac{27i}{9i} = \dfrac{2}{i} - 3$

$= \dfrac{2}{i}\cdot\dfrac{i}{i} - 3 = \dfrac{2i}{i^2} - 3$

$= -3 - 2i$

96. $\dfrac{32-9i}{2-3i}$

$= \dfrac{32-9i}{2-3i}\cdot\dfrac{2+3i}{2+3i}$

$= \dfrac{64 - 18i + 96i - 27i^2}{2^2 - (3i)^2}$

$= \dfrac{64 + 78i + 27}{4 - 9i^2} = \dfrac{91 + 78i}{4 + 9}$

$= \dfrac{91 + 78i}{13}$

$= 7 + 6i$

97. $\left(\dfrac{5}{6}-7i\right)+\left(\dfrac{5}{6}+i\right)$

$= \dfrac{5}{6} - 7i + \dfrac{5}{6} + i$

$= \dfrac{5}{3} - 6i$

SSM: Experiencing Introductory and Intermediate Algebra

98. $\dfrac{12i+\sqrt{10}}{2\sqrt{5}+i\sqrt{2}}$

$= \dfrac{12i+\sqrt{10}}{2\sqrt{5}+i\sqrt{2}} \cdot \dfrac{2\sqrt{5}-i\sqrt{2}}{2\sqrt{5}-i\sqrt{2}}$

$= \dfrac{24i\sqrt{5}+2\sqrt{50}-12i^2\sqrt{2}-i\sqrt{20}}{\left(2\sqrt{5}\right)^2-\left(i\sqrt{2}\right)^2}$

$= \dfrac{24i\sqrt{5}+10\sqrt{2}+12\sqrt{2}-2i\sqrt{5}}{20+2}$

$= \dfrac{22\sqrt{2}+22i\sqrt{5}}{22}$

$= \sqrt{2}+i\sqrt{5}$

99. $\dfrac{\sqrt{-289}}{\sqrt{-841}}$

$= \dfrac{\sqrt{289}\cdot i}{\sqrt{841}\cdot i} = \dfrac{\sqrt{289}}{\sqrt{841}}$

$= \dfrac{17}{29}$

100. $\dfrac{d-11}{d(d+9)} = \dfrac{5}{8}$

$8(d-11) = 5d(d+9)$

$8d-88 = 5d^2+45d$

$0 = 5d^2+37d+88$

$d = \dfrac{-37\pm\sqrt{(37)^2-4(5)(88)}}{2(5)}$

$= \dfrac{-37\pm\sqrt{-391}}{10}$

$= -\dfrac{37}{10}\pm\dfrac{\sqrt{391}}{10}i$

101. $(m+7)^2+25 = 0$

$(m+7)^2 = -25$

$m+7 = \pm\sqrt{-25}$

$m = -7\pm 5i$

102. $4(y-3)^2+29 = 5$

$4(y-3)^2 = -24$

$(y-3)^2 = -6$

$y-3 = \pm\sqrt{-6}$

$y = 3\pm i\sqrt{6}$

103. $x^2-16x+67 = 0$

$x = \dfrac{-(-16)\pm\sqrt{(-16)^2-4(1)(67)}}{2(1)}$

$= \dfrac{16\pm\sqrt{-12}}{2} = \dfrac{16\pm 2i\sqrt{3}}{2}$

$= 8\pm i\sqrt{3}$

104. $9y^2+30y+32 = 0$

$y = \dfrac{-30\pm\sqrt{(30)^2-4(9)(32)}}{2(9)}$

$= \dfrac{-30\pm\sqrt{-252}}{18} = \dfrac{-30\pm 6i\sqrt{7}}{18}$

$= -\dfrac{5}{3}\pm\dfrac{\sqrt{7}}{3}i$

105. $a^2+100 = 19$

$a^2 = -81$

$a = \pm\sqrt{-81}$

$a = \pm 9i$

106. $6b^2+78 = 0$

$6b^2 = -78$

$b^2 = -13$

$b = \pm\sqrt{-13}$

$b = \pm i\sqrt{13}$

107. $\dfrac{y+9}{y+1} = \dfrac{5}{y-7}$

$(y+9)(y-7) = 5(y+1)$

$y^2 + 9y - 7y - 63 = 5y + 5$

$y^2 - 3y - 68 = 0$

$y = \dfrac{-(-3) \pm \sqrt{(-3)^2 - 4(1)(-68)}}{2(1)}$

$= \dfrac{3 \pm \sqrt{281}}{2}$

$= \dfrac{3}{2} \pm \dfrac{\sqrt{281}}{2}$

108. $Z = \dfrac{V}{I}$ or $I = \dfrac{V}{Z}$

$I = \dfrac{30 + 52i}{9 + 5i}$

$= \dfrac{(30 + 52i)(9 - 5i)}{(9 + 5i)(9 - 5i)}$

$= \dfrac{270 - 150i + 468i - 260i^2}{81 + 25}$

$= \dfrac{530 + 318i}{106}$

$= 5 + 3i$

The current is $5 + 3i$ amperes.

$|I| = \sqrt{5^2 + 3^2} = \sqrt{34} \approx 5.8$

The magnitude of the current is approximately 5.8 amperes.

109. $s = \sqrt{A}$

$s = \sqrt{529}$

$s = 23$

The side is 23 inches.

110. $s = \sqrt{\dfrac{n(n+1)}{12}}$

$= \sqrt{\dfrac{27(27+1)}{12}}$

$= \sqrt{63} = 3\sqrt{7}$

≈ 7.937

The variability is approximately 7.937.

111. $t = \sqrt{\dfrac{2d}{32.2}}$

$4 = \sqrt{\dfrac{2d}{32.2}}$

$16 = \dfrac{2d}{32.2}$

$515.2 = 2d$

$257.6 = d$

The pear fell from a height of 257.6 feet.

112. $T = 2\pi\sqrt{\dfrac{L}{32}}$

$T = 2\pi\sqrt{\dfrac{3.5}{32}}$

≈ 2.08 seconds.

The period is approximately 2.08 seconds.

113. $d = \sqrt{(x_2 - x_1)^2 + (y_2 - y_1)^2}$

$= \sqrt{(6 - 6)^2 + (7 - -3)^2}$

$= \sqrt{10^2} = 10$

The distance is 10 units.

114. $d = \sqrt{(x_2 - x_1)^2 + (y_2 - y_1)^2}$

$= \sqrt{(2 + 4)^2 + (8 - 8)^2}$

$= \sqrt{6^2} = 6$

The distance is 6 units.

115. $d = \sqrt{(x_2 - x_1)^2 + (y_2 - y_1)^2}$

$= \sqrt{(9 - 5)^2 + (2 + 1)^2}$

$= \sqrt{4^2 + 3^2} = \sqrt{25}$

$= 5$

The distance is 5 units.

116. $d = \sqrt{(x_2 - x_1)^2 + (y_2 - y_1)^2}$

$= \sqrt{(-2 - 5)^2 + (10 - 7)^2}$

$= \sqrt{(-7)^2 + 3^2} = \sqrt{49 + 9}$

$= \sqrt{58}$

The distance is $\sqrt{58}$ units.

SSM: Experiencing Introductory and Intermediate Algebra

117. $d = \sqrt{(x_2 - x_1)^2 + (y_2 - y_1)^2}$

$4 = \sqrt{(2-x)^2 + (8-(-2))^2}$

$4 = \sqrt{(2-x)^2 + 100}$

$16 = (2-x)^2 + 100$

$-84 = (2-x)^2$

There is no solution.

118. $d = \sqrt{(x_2 - x_1)^2 + (y_2 - y_1)^2}$

$5 = \sqrt{(3-7)^2 + (2-y)^2}$

$25 = 16 + (2-y)^2$

$9 = (2-y)^2$

$\pm 3 = 2 - y$

$y = 2 \pm 3$

$y = 5$ or $y = -1$

The y-coordinates are 5 and –1. The possible points are $(7,5)$ and $(7,-1)$.

Chapter 13 Test

1. $\sqrt{196} = \sqrt{14^2} = 14$

2. $\sqrt[4]{21} \approx 2.141$

3. $\sqrt{-25}$ is not a real number.

4. $\sqrt[3]{\dfrac{-8}{125}} = \sqrt[3]{\left(\dfrac{-2}{5}\right)^3} = -\dfrac{2}{5}$

5. $16^{5/2} \cdot 16^{-1/4} = 16^{\frac{5}{2}+\left(-\frac{1}{4}\right)} = 16^{9/4}$

$= \left(\sqrt[4]{16}\right)^9 = 2^9 = 512$

6. $\left(81^{4/3}\right)^{3/8} = 81^{\frac{4}{3} \cdot \frac{3}{8}} = 81^{1/2} = \sqrt{81} = 9$

7. $\left(\dfrac{8}{27}\right)^{-2/3} = \left(\dfrac{27}{8}\right)^{2/3} = \left(\sqrt[3]{\dfrac{27}{8}}\right)^2 = \left(\dfrac{3}{2}\right)^2 = \dfrac{9}{4}$

8. $\dfrac{x^{4/5}}{x^{3/10}} = x^{\frac{4}{5}-\frac{3}{10}} = x^{1/2} = \sqrt{x}$

9. $\left(\sqrt{2x} - \sqrt{5y}\right)\left(\sqrt{2x} + \sqrt{5y}\right)$

$= \left(\sqrt{2x}\right)^2 - \left(\sqrt{5y}\right)^2$

$= 2x - 5y$

10. $\sqrt{6x^3 y} \cdot \sqrt{15xy^2}$

$= \sqrt{90x^4 y^3}$

$= \sqrt{9 \cdot 10 \cdot x^4 \cdot y^2 \cdot y}$

$= 3x^2 y\sqrt{10y}$

11. $\sqrt{\dfrac{3z^3}{2xy^2}}$

$= \dfrac{\sqrt{3 \cdot z^2 \cdot z}}{\sqrt{2 \cdot x \cdot y^2}}$

$= \dfrac{z\sqrt{3z}}{y\sqrt{2x}} \cdot \dfrac{\sqrt{2x}}{\sqrt{2x}}$

$= \dfrac{z\sqrt{6xz}}{2xy}$

12. $\dfrac{\sqrt{50xy^3}}{\sqrt{98x^3 y^2}}$

$= \sqrt{\dfrac{50xy^3}{98x^3 y^2}}$

$= \sqrt{\dfrac{25y}{49x^2}}$

$= \dfrac{\sqrt{25y}}{\sqrt{49x^2}}$

$= \dfrac{5\sqrt{y}}{7x}$

13. $\sqrt{81x} + 2\sqrt{25x} - 3\sqrt{16x}$

$= 9\sqrt{x} + 10\sqrt{x} - 12\sqrt{x}$

$= (9 + 10 - 12)\sqrt{x}$

$= 7\sqrt{x}$

Chapter 13: Radical Expressions, Functions, and Equations

14. $5\sqrt[3]{x}\left(2\sqrt[3]{x^2}+3\sqrt[3]{x}\right)$
$=5\sqrt[3]{x}\cdot 2\sqrt[3]{x^2}+5\sqrt[3]{x}\cdot 3\sqrt[3]{x}$
$=10\sqrt[3]{x^3}+15\sqrt[3]{x^2}$
$=10x+15\sqrt[3]{x^2}$

15. $\dfrac{\sqrt{x}+4}{\sqrt{x}+1}$
$=\dfrac{\sqrt{x}+4}{\sqrt{x}+1}\cdot\dfrac{\sqrt{x}-1}{\sqrt{x}-1}$
$=\dfrac{x+4\sqrt{x}-\sqrt{x}-4}{\left(\sqrt{x}\right)^2-1^2}$
$=\dfrac{x+3\sqrt{x}-4}{x-1}$

16. $(x+4)^{1/3}+3=2$
$(x+4)^{1/3}=-1$
$x+4=\sqrt[3]{-1}$
$x=-4-1$
$x=-5$
The solution is $x=-5$.

17. $3\sqrt{x}-5=13$
$3\sqrt{x}=18$
$\sqrt{x}=6$
$x=6^2$
$x=36$
The solution is $x=36$.

18. $\sqrt{3x+1}=x-1$
$\left(\sqrt{3x+1}\right)^2=(x-1)^2$
$3x+1=x^2-2x+1$
$0=x^2-5x$
$0=x(x-5)$
$x=0$ (extraneous) or $x=5$
The solution is $x=5$.

19. $\sqrt{5x+1}-2=\sqrt{x+1}$
$\left(\sqrt{5x+1}-2\right)^2=\left(\sqrt{x+1}\right)^2$
$5x+1-2\cdot 2\sqrt{5x+1}+4=x+1$
$-4\sqrt{5x+1}=-4x-4$
$\sqrt{5x+1}=x+1$
$\left(\sqrt{5x+1}\right)^2=(x+1)^2$
$5x+1=x^2+2x+1$
$0=x^2-3x$
$0=x(x-3)$
$x=0$ (extraneous) or $x=3$
The solution is $x=3$.

20. $f(x)=\sqrt{x+7}-3$
$x+7<0$
$x<-7$
The restricted values are all real numbers less than -7.
Domain: $[-7,\infty)$

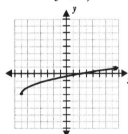

21. $3\sqrt{-100}-2\sqrt{-49}$
$=3\sqrt{100}\cdot i-2\sqrt{49}\cdot i$
$=30i-14i$
$=16i$

22. $\sqrt{-2}\sqrt{-7}\sqrt{-56}$
$=\sqrt{2}\cdot i\cdot\sqrt{7}\cdot i\cdot\sqrt{56}\cdot i$
$=\sqrt{784}\cdot i^3$
$=28(-i)$
$=-28i$

SSM: Experiencing Introductory and Intermediate Algebra

23. $\sqrt{-2}\left(\sqrt{-7}-\sqrt{56}\right)$
$= \sqrt{2}\cdot i\left(\sqrt{7}\cdot i - \sqrt{56}\right)$
$= \sqrt{14}\cdot i^2 - \sqrt{112}\cdot i$
$= -\sqrt{14} - 4i\sqrt{7}$

24. $\dfrac{\sqrt{-225}}{\sqrt{-289}}$
$= \dfrac{\sqrt{225}\cdot i}{\sqrt{289}\cdot i}$
$= \dfrac{15}{17}$

25. $\left(\sqrt{-6}-\sqrt{-13}\right)\left(\sqrt{-6}+\sqrt{-13}\right)$
$= \left(i\sqrt{6}-i\sqrt{13}\right)\left(i\sqrt{6}+i\sqrt{13}\right)$
$= \left(i\sqrt{6}\right)^2 - \left(i\sqrt{13}\right)^2$
$= 6i^2 - 13i^2 = -7i^2$
$= 7$

26. $(8+2i)(-5+i)$
$= -40 - 10i + 8i + 2i^2$
$= -40 - 2i - 2$
$= -42 - 2i$

27. $\dfrac{29+29i}{5-2i}$
$= \dfrac{29+29i}{5-2i}\cdot\dfrac{5+2i}{5+2i}$
$= \dfrac{145+145i+58i+58i^2}{5^2-(2i)^2}$
$= \dfrac{145+203i-58}{25+4}$
$= \dfrac{87+203i}{29}$
$= 3+7i$

28. $(22.47-13.6i)-(12.2-7.32i)$
$= -22.47 - 13.6i - 12.2 + 7.32i$
$= -34.67 - 6.28i$

29. $i\sqrt{15}\left(\sqrt{3}+i\sqrt{5}\right)$
$= i\sqrt{15}\cdot\sqrt{3} + i\sqrt{15}\cdot i\sqrt{5}$
$= i\sqrt{45} + i^2\sqrt{75}$
$= 3i\sqrt{5} - 5\sqrt{3}$

30. $\dfrac{44+16i}{2i}$
$= \dfrac{44}{2i}+\dfrac{16i}{2i} = \dfrac{22}{i}+8$
$= \dfrac{22}{i}\cdot\dfrac{i}{i}+8 = \dfrac{22i}{i^2}+8$
$= 8 - 22i$

31. $x^2 + 196 = 0$
$x^2 = -196$
$x = \pm\sqrt{-196}$
$x = \pm 14i$

32. $(t+2)^2 + 15 = 0$
$(t+2)^2 = -15$
$t+2 = \pm\sqrt{-15}$
$t = -2 \pm i\sqrt{15}$

33. $2z^2 + 5 = 2.12$
$2z^2 = -2.88$
$z^2 = -1.44$
$z = \pm\sqrt{-1.44}$
$z = \pm 1.2i$

34. $x^2 - 6x + 17 = 0$
$x = \dfrac{-(-6)+\sqrt{(-6)^2-4(1)(17)}}{2(1)}$
$= \dfrac{6\pm\sqrt{-32}}{2} = \dfrac{6\pm 4i\sqrt{2}}{2}$
$= 3 \pm 2i\sqrt{2}$

Chapter 13: Radical Expressions, Functions, and Equations

35. $9x^2 - 6x + 7 = 0$

$$x = \frac{-(-6) \pm \sqrt{(-6)^2 - 4(9)(7)}}{2(9)}$$

$$= \frac{6 \pm \sqrt{-216}}{18} = \frac{6 \pm 6i\sqrt{6}}{18}$$

$$= \frac{1}{3} \pm \frac{\sqrt{6}}{3}i$$

36. $\dfrac{y-2}{y(y+1)} = \dfrac{4}{7}$

$7(y-2) = 4y(y+1)$

$7y - 14 = 4y^2 + 4y$

$0 = 4y^2 - 3y + 14$

$$y = \frac{-(-3) \pm \sqrt{(-3)^2 - 4(4)(14)}}{2(4)}$$

$$= \frac{3 \pm \sqrt{-215}}{8}$$

$$= \frac{3}{8} \pm \frac{\sqrt{215}}{8}i$$

37. $d = \sqrt{(x_2 - x_1)^2 + (y_2 - y_1)^2}$

$= \sqrt{(3-3)^2 + (5+2)^2}$

$= \sqrt{0^2 + 7^2} = \sqrt{7^2}$

$= 7$

The distance is 7 units.

38. $d = \sqrt{(x_2 - x_1)^2 + (y_2 - y_1)^2}$

$= \sqrt{(5-2)^2 + (4+3)^2}$

$= \sqrt{3^2 + 7^2} = \sqrt{9 + 49}$

$= \sqrt{58}$

The distance is $\sqrt{58}$ units.

39. Let s_1 = length of the side of smaller tablecloth

$s_1 = \sqrt{A_1}$

$s_1 = \sqrt{2000}$

$= 20\sqrt{5}$

Let s_2 = length of the side of larger tablecloth

$s_2 = \sqrt{A_2}$

$s_2 = \sqrt{3125}$

$= 25\sqrt{5}$

Difference:

$4s_2 - 4s_1 = 4(25\sqrt{5}) - 4(20\sqrt{5})$

$= 100\sqrt{5} - 80\sqrt{5}$

$= 20\sqrt{5}$

The larger tablecloth will require $20\sqrt{5}$ inches more border.

40. The restricted values of a radical function, with an even index, are found by setting the radicand less than 0 and solving the resulting inequality. The domain is all real numbers that are not restricted values.
Answers will vary.

41. $V = IZ$

$V = (8 + 3i)(11 - 3i)$

$= 88 - 24i + 33i - 9i^2$

$= 88 + 9i + 9$

$= 97 + 9i$

The total voltage is $97 + 9i$ volts.

$|V| = \sqrt{97^2 + 9^2} = \sqrt{9490} \approx 97.4$

The magnitude of the total voltage is approximately 97.4 volts.

42. Answers will vary but should at least mention that the real-numbers are a subset of the complex numbers.

SSM: Experiencing Introductory and Intermediate Algebra

Chapters 1-13 Cumulative Review

1. $2^0 x^{-1} y^2 = \dfrac{y^2}{x}$

2. $\left(3x^{1/2} y^{3/4}\right)^2 = 3^2 x^{\frac{1}{2} \cdot 2} y^{\frac{3}{4} \cdot 2} = 9xy^{3/2}$

3. $\left[\dfrac{(2s)^2}{3t}\right]^{-3} = \left(\dfrac{3t}{(2s)^2}\right)^3 = \dfrac{3^3 t^3}{(2s)^6} = \dfrac{27t^3}{64s^6}$

4. $\left(2x^3 + 6x^2 y - 2xy^2 + y^3\right) + \left(3x^3 + 4xy^2 - y^3\right)$
   $= 2x^3 + 6x^2 y - 2xy^2 + y^3 + 3x^3 + 4xy^2 - y^3$
   $= 5x^3 + 6x^2 y + 2xy^2$

5. $\left(1.2a^2 - 3.6ab + b^2\right) - \left(4a^2 + 2.71ab - 3.4b^2\right)$
   $= 1.2a^2 - 3.6ab + b^2 - 4a^2 - 2.71ab + 3.4b^2$
   $= -2.8a^2 - 6.31ab + 4.4b^2$

6. $\left(6.8m^2 n\right)\left(-2mn^2 p\right)$
   $= -13.6 m^{2+1} n^{1+2} p$
   $= -13.6 m^3 n^3 p$

7. $(3a - b)(2a + 4b)$
   $= 6a^2 - 2ab + 12ab - 4b^2$
   $= 6a^2 + 10ab - 4b^2$

8. $(2x + 3)(2x - 3)$
   $= (2x)^2 - 3^2$
   $= 4x^2 - 9$

9. $(2x + 3)^2$
   $= (2x)^2 + 2 \cdot 3 \cdot 2x + 3^2$
   $= 4x^2 + 12x + 9$

10. $\dfrac{-15 x^2 y^3 z}{3xyz}$
    $= -5 x^{2-1} y^{3-1} z^{1-1}$
    $= -5xy^2 z^0$
    $= -5xy^2$

11. $\dfrac{2m^2 n + 4mn - 8n^2}{2m^2 n}$
    $= \dfrac{2m^2 n}{2m^2 n} + \dfrac{4mn}{2m^2 n} - \dfrac{8n^2}{2m^2 n}$
    $= 1 + \dfrac{2}{m} - \dfrac{4n}{m^2}$

12. $\dfrac{x^2 + 2x - 3}{x^2 - 9}$
    $= \dfrac{(x+3)(x-1)}{(x+3)(x-3)}$
    $= \dfrac{x-1}{x-3}$

13. $\dfrac{x}{x+2} + \dfrac{3}{x-2}$
    $= \dfrac{x(x-2)}{(x+2)(x-2)} + \dfrac{3(x+2)}{(x-2)(x+2)}$
    $= \dfrac{x^2 - 2x}{(x+2)(x-2)} + \dfrac{3x+6}{(x+2)(x-2)}$
    $= \dfrac{x^2 + x + 6}{(x+2)(x-2)}$

Chapter 13: Radical Expressions, Functions, and Equations

14. $\dfrac{1}{x+2} - \dfrac{3}{x-3} - \dfrac{x+3}{x^2+x-12}$

$= \dfrac{(x+4)(x-3)}{(x+2)(x+4)(x-3)} - \dfrac{3(x+2)(x+4)}{(x+2)(x+4)(x-3)} - \dfrac{(x+3)(x+2)}{(x+2)(x+4)(x-3)}$

$= \dfrac{(x^2+x-12) - 3(x^2+6x+8) - (x^2+5x+6)}{(x+2)(x+4)(x-3)}$

$= \dfrac{x^2+x-12-3x^2-18x-24-x^2-5x-6}{(x+2)(x+4)(x-3)}$

$= \dfrac{-3x^2-22x-42}{(x+2)(x+4)(x-3)} = -\dfrac{3x^2+22+42}{(x+2)(x+4)(x-3)}$

15. $\dfrac{2a-5}{a+2} \cdot \dfrac{2a^2+3a-2}{4a^2-25}$

$= \dfrac{(2a-5)(2a-1)(a+2)}{(a+2)(2a-5)(2a+5)}$

$= \dfrac{2a-1}{2a+5}$

16. $\dfrac{2x}{2x^2+9x+4} \div \dfrac{4x^2y}{x^2+9x+20}$

$= \dfrac{2x}{(2x+1)(x+4)} \cdot \dfrac{(x+5)(x+4)}{4x^2y}$

$= \dfrac{x+5}{2xy(2x+1)}$

17. $\sqrt{8y} + \sqrt{2y} - 2\sqrt{18y}$

$= 2\sqrt{2y} + \sqrt{2y} - 6\sqrt{2y}$

$= (2+1-6)\sqrt{2y}$

$= -3\sqrt{2y}$

18. $\sqrt[3]{25a^4b} \cdot \sqrt[3]{5ab}$

$= \sqrt[3]{25a^4b \cdot 5ab}$

$= \sqrt[3]{125a^5b^2}$

$= \sqrt[3]{5^3 a^3 a^2 b^2}$

$= 5a\sqrt[3]{a^2b^2}$

19. $(\sqrt{3}-\sqrt{5})(\sqrt{3}+\sqrt{5})$

$= (\sqrt{3})^2 - (\sqrt{5})^2$

$= 3 - 5$

$= -2$

20. $\dfrac{\sqrt{21abc^3}}{\sqrt{7a^3b}}$

$= \sqrt{\dfrac{21abc^3}{7a^3b}} = \sqrt{\dfrac{3c^3}{a^2}} = \dfrac{\sqrt{3c^3}}{\sqrt{a^2}} = \dfrac{\sqrt{3c^2 \cdot c}}{\sqrt{a^2}}$

$= \dfrac{c\sqrt{3c}}{a}$

21. $\dfrac{\sqrt{x}+2}{\sqrt{x}-1}$

$= \dfrac{\sqrt{x}+2}{\sqrt{x}-1} \cdot \dfrac{\sqrt{x}+1}{\sqrt{x}+1}$

$= \dfrac{x + 2\sqrt{x} + \sqrt{x} + 2}{(\sqrt{x})^2 - (1)^2}$

$= \dfrac{x + 3\sqrt{x} + 2}{x-1}$

613

SSM: Experiencing Introductory and Intermediate Algebra

22. $x^{1/2}\left(x^{2/3} - x^{3/4}\right)$
$= x^{1/2} \cdot x^{2/3} - x^{1/2} \cdot x^{3/4}$
$= x^{\frac{1}{2}+\frac{2}{3}} - x^{\frac{1}{2}+\frac{3}{4}}$
$= x^{7/6} - x^{5/4}$

23. $16a^2 - 25b^2$
$= (4a)^2 - (5b)^2$
$= (4a - 5b)(4a + 5b)$

24. $x^2 - 2x - 8 = (x-4)(x-2)$

25. $3x^2 - 9x - 30$
$= 3(x^2 - 3x - 10)$
$= 3(x-5)(x+2)$

26. $f(x) = -x^2 + 3x - 1$
$f(-2) = -(-2)^2 + 3(-2) - 1$
$ = -4 - 6 - 1$
$ = -11$

27. $f(x) = -3.2x + 1$

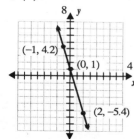

Domain: all real numbers
Range: all real numbers

28. $y = x^2 + 4x + 6$
y-intercept: $(0, 6)$
x-coordinate of vertex:
$x = -\dfrac{b}{2a} = -\dfrac{4}{2} = -2$
y-coordinate of vertex:
$y = (-2)^2 + 4(-2) + 6 = 2$

Vertex: $(-2, 2)$
Axis of symmetry: $x = -2$

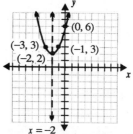

Domain: all real numbers
Range: all real numbers greater than or equal to -2

29. $g(x) = -2x^2 + x + 1$
y-intercept: $(0, 1)$
x-coordinate of vertex:
$x = -\dfrac{b}{2a} = -\dfrac{1}{2(-2)} = \dfrac{1}{4}$
y-coordinate of vertex:
$g\left(\dfrac{1}{4}\right) = -2\left(\dfrac{1}{4}\right)^2 + \left(\dfrac{1}{4}\right) + 1 = \dfrac{9}{8}$
Vertex: $\left(\dfrac{1}{4}, \dfrac{9}{8}\right)$
Axis of symmetry: $x = \dfrac{1}{4}$

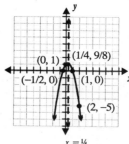

Domain: all real numbers
Range: all real numbers less than or equal to $\dfrac{9}{8}$.

30. $y = \dfrac{x^2 - 4}{x + 2}$

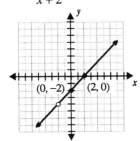

Domain: all real numbers $\neq -2$
Range: all real numbers $\neq -4$

31. $h(x) = \dfrac{2}{x}$

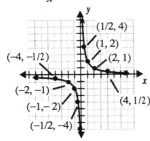

Domain: all real numbers $\neq 0$
Range: all real numbers $\neq 0$

32. $y = 2\sqrt{x}$

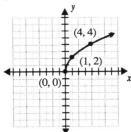

Domain: all real numbers ≥ 0
Range: all real numbers ≥ 0

33. $a(x) = x^{2/3} + 2$

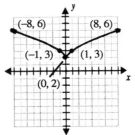

Domain: all real numbers
Range: all real numbers ≥ 2

34. $2(x + 3) - 4(x - 1) = x - 2$
$2x + 6 - 4x + 4 = x - 2$
$-2x + 10 = x - 2$
$-3x = -12$
$x = 4$

35. $x^2 - 3x + 9 = 0$
$a = 1, b = -3, c = 9$
$x = \dfrac{-(-3) \pm \sqrt{(-3)^2 - 4(1)(9)}}{2(1)}$
$= \dfrac{3 \pm \sqrt{-27}}{2}$
$= \dfrac{3}{2} \pm \dfrac{3i\sqrt{3}}{2}$

36. $x^2 + 2x - 15 = 0$
$(x + 5)(x - 3) = 0$
$x + 5 = 0$ or $x - 3 = 0$
$x = -5$ or $x = 3$
The solutions are -5 and 3.

37. $2t^2 + 3t = 4$
$2t^2 + 3t - 4 = 0$
$t = \dfrac{-3 \pm \sqrt{3^3 - 4(2)(-4)}}{2(2)} = \dfrac{-3 \pm \sqrt{41}}{4}$

38. $\dfrac{x+1}{x-1} = \dfrac{x+3}{x+2}$

$(x+1)(x+2) = (x-1)(x+3)$

$x^2 + x + 2x + 2 = x^2 - x + 3x - 3$

$x^2 + 3x + 2 = x^2 + 2x - 3$

$x = -5$

39. $\dfrac{2}{x^2} + \dfrac{3}{2x^2} = \dfrac{1}{8}$

$8x^2 \left(\dfrac{2}{x^2} + \dfrac{3}{2x^2} \right) = 8x^2 \left(\dfrac{1}{8} \right)$

$16 + 12 = x^2$

$28 = x^2$

$\pm\sqrt{28} = x$

$x = \pm 2\sqrt{7}$

40. $\sqrt{2x-3} + 5 = 9$

$\sqrt{2x-3} = 4$

$\left(\sqrt{2x-3} \right)^2 = 4^2$

$2x - 3 = 16$

$2x = 19$

$x = \dfrac{19}{2}$

41. $2x^2 - 9x - 18 < 0$

$(2x+3)(x-6) < 0$

The solution is all real numbers in the interval $\left(-\dfrac{3}{2}, 6 \right)$.

42. $3x - 4y = 3$

$x + y = 8$

Solve the second equation for x.

$x = 8 - y$

Substitute this result for x in the first equation.

$3(8-y) - 4y = 3$

$24 - 3y - 4y = 3$

$-7y = -21$

$y = 3$

Substitute this value for y in the second equation.

$x + 3 = 8$

$x = 5$

The solution is $(5, 3)$.

43. $m = \dfrac{y_2 - y_1}{x_2 - x_1} = \dfrac{2-(-5)}{-4-3} = \dfrac{7}{-7} = -1$

$y - y_1 = m(x - x_1)$

$y - (-5) = -1(x - 3)$

$y + 5 = -x + 3$

$y = -x - 2$

44. $2x + 3y = 5$

$3y = -2x + 5$

$y = -\dfrac{2}{3}x + \dfrac{5}{3}$

The line we want has slope $\dfrac{3}{2}$.

$y - y_1 = m(x - x_1)$

$y - 1 = \dfrac{3}{2}(x - (-2))$

$y - 1 = \dfrac{3}{2}x + 3$

$y = \dfrac{3}{2}x + 4$

45. A general quadratic equation has the form $y = ax^2 + bx + c$. From the given points we have:

$-4 = a(0)^2 + b(0) + c$

$0 = a(-4)^2 + b(-4) + c$

$0 = a(1)^2 + b(1) + c$

or

$-4 = c$

$0 = 16a - 4b + c$

$0 = a + b + c$

Substitute the value -4 for c in the second and third equations.

$0 = 16a - 4b - 4$
$0 = 4a + 4b - 16$
$0 = 20z - 20$
$20 = 20a$
$1 = a$
Since $a + b + c = 0$ and $a = 1, c = -4$, we have
$1 + b - 4 = 0$, or $b = 3$.
The quadratic equation is $y = x^2 + 3x - 4$.

46. Let l be the length of the rectangle and w be the width. Since the diagonal measures 20 feet, we have $\sqrt{l^2 + w^2} = 20$ or $l^2 + w^2 = 400$. Substitute $w + 4$ for l in this equation.
$(w+4)^2 + w^2 = 400$
$w^2 + 8w + 16 + w^2 = 400$
$2w^2 + 8w + 16 = 400$
$w^2 + 4w + 8 = 200$
$w^2 + 4w - 192 = 0$
$(w + 16)(w - 12) = 0$
$w = -16$ or $w = 12$
Since length cannot be negative, the width is 12 feet and the length is $12 + 4i + 16$ feet. The rectangle is 12 feet by 16 feet.

47. If the point $(0, y)$ is 5 units from the point $(-3, 4)$, then, by the distance formula,
$\sqrt{(-3-0)^2 + (4-y)^2} = 5$
$(-3)^2 + (4-y)^2 = 25$
$(4-y)^2 = 16$
$4 - y = \pm 4$
$y = 4 \pm 4$
$y = 0$ or $y = 8$
The possible y-coordinates are 0 and 8.

48. Since Kevin needs 5 hours to rake the yard, he rakes $\frac{1}{5}$ of the yard in 1 hour. Similarly, Elizabeth rakes $\frac{1}{4}$ of the yard in 1 hour. Let x be the number of hours that they work together raking leaves. The proportion of the yard that Kevin rakes is $\frac{1}{5}x$ and the proportion that Elizabeth rakes is $\frac{1}{4}x$.
$\frac{1}{5}x + \frac{1}{4}x = 1$
$20\left(\frac{1}{5}x\right) + 20\left(\frac{1}{4}x\right) = 20$
$4x + 5x = 20$
$9x = 20$
$x = \frac{20}{9} = 2\frac{2}{9}$
It will take Kevin and Elizabeth $2\frac{2}{9}$ hours (about 2 hours, 13 minutes) to rake the leaves. They will be done before 2:30 P.M.

49. Let x be the number of items produced and sold. The cost of producing x items is $1000 + 0.55x$, while the revenue from selling x items is $1.25x$.
$1.25x \geq 1000 + 0.55x$
$0.70x \geq 1000$
$x \geq \frac{1000}{0.70}$
$x \geq \frac{10,000}{7} \approx 1428.6$
The company must sell at least 1429 items in order to break even.

50. The position equation is $s = -16t^2 + v_0 t + s_0$. We have to find t when s is 2500 if $v_0 = 0$ and $s_0 = 10000$.
$2500 = -16t^2 + 0t + 10,000$
$16t^2 = 7500$
$t^2 = \frac{1875}{4}$
$t = \pm\sqrt{\frac{1875}{4}} = \pm\frac{25\sqrt{3}}{2}$
Since the amount of time will be positive, we need to use $t = \frac{25\sqrt{3}}{2} \approx 21.65$. The skydiver was free-falling for about 21.65 seconds.

Chapter 14

14.1 Exercises

1. $h = \{(-3,5),(-2,4),(-1,3),(0,2),(1,1),(2,0)\}$
   $h^{-1} = \{(5,-3),(4,-2),(3,-1),(2,0),(1,1),(0,2)\}$

3. $A = \{(3,6),(2,7),(1,8),(0,8),(-1,7),(-2,6)\}$
   $A^{-1} = \{(6,3),(7,2),(8,1),(8,0),(7,-1),(6,-2)\}$

5. $y = 2x - 8$
   $x = 2y - 8$
   $x + 8 = 2y$
   $\dfrac{x+8}{2} = y$
   $y = \dfrac{1}{2}x + 4$

7. $y = -3x + 2$
   $x = -3y + 2$
   $3y = -x + 2$
   $y = \dfrac{-x+2}{3}$
   $y = -\dfrac{1}{3}x + \dfrac{2}{3}$

9. $y = \dfrac{3}{4}x + 9$
   $x = \dfrac{3}{4}y + 9$
   $x - 9 = \dfrac{3}{4}y$
   $\dfrac{4}{3}(x-9) = \dfrac{4}{3}\left(\dfrac{3}{4}y\right)$
   $y = \dfrac{4}{3}x - 12$

11. $y = 0.125x - 2.5$
    $x = 0.125y - 2.5$
    $x + 2.5 = 0.125y$
    $\dfrac{x+2.5}{0.125} = y$
    $y = 8x + 20$

13. $y = x^2 - 2$
    $x = y^2 - 2$
    $x + 2 = y^2$
    $y = \pm\sqrt{x+2}$

15. $k = \{(5,2),(10,4),(15,8),(20,16),(25,32)\}$

    The relation k is a function, because each ordered pair has a different first coordinate.

    The function k is one-to-one, because each ordered pair has a different second coordinate.

17. $Q = \{(-2,0),(-4,2),(-6,4),(-8,2),(-10,0)\}$

    The relation Q is a function, because each ordered pair has a different first coordinate.

    The function Q is not one-to-one, because the second coordinates 0 and 2 are each repeated in two different ordered pairs.

19. $A = \{(-1,0),(0,1),(-1,2),(0,3)\}$

    The relation A is not a function, because the first coordinates -1 and 0 are each repeated in two different ordered pairs.

21. Yes, the graph represents a one-to-one function, because all possible vertical and horizontal lines cross the graph only once.

23. No, the graph does not represent a one-to-one function. The graph does represents a function, because all possible vertical lines cross the graph only once, but it is not one-to-one, because a horizontal line can be drawn that intersects the graph more than once.

25. Yes, the graph represents a one-to-one function, because all possible vertical and horizontal lines cross the graph only once.

27. $g(x) = 3x - 6$ $y = 3x - 6$
$x = 3y - 6$
$x + 6 = 3y$
$\dfrac{x+6}{3} = y$
$y = \dfrac{1}{3}x + 2$

Therefore, $g^{-1}(x) = \dfrac{1}{3}x + 2$.

Let $Y1 = 3x - 6$, $Y2 = \dfrac{1}{3}x + 2$, and $Y3 = x$.

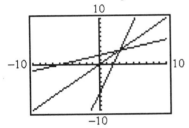

29. $y = \dfrac{2}{3}x - 4$ $x = \dfrac{2}{3}y - 4$
$x + 4 = \dfrac{2}{3}y$
$\dfrac{3}{2}(x+4) = \dfrac{3}{2}\left(\dfrac{2}{3}y\right)$
$\dfrac{3}{2}x + 6 = y$

Therefore, $y^{-1} = \dfrac{3}{2}x + 6$.

Let $Y1 = \dfrac{2}{3}x - 4$, $Y2 = \dfrac{3}{2}x + 6$, and $Y3 = x$.

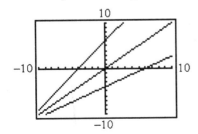

31. $h(x) = x^2 - 1$ $y = x^2 - 1$
$x = y^2 - 1$
$x + 1 = y^2$
$y = \pm\sqrt{x+1}$

Therefore, $h^{-1}(x) = \pm\sqrt{x+1}$.

Let $Y1 = x^2 - 1$, $Y2 = \sqrt{x+1}$, $Y3 = -\sqrt{x+1}$ and $Y4 = x$.

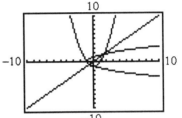

33. $y = \dfrac{1}{3}x^3 - 4$ $x = \dfrac{1}{3}y^3 - 4$
$3(x) = 3\left(\dfrac{1}{3}y^3 - 4\right)$
$3x = y^3 - 12$
$3x + 12 = y^3$
$y = \sqrt[3]{3x + 12}$

Therefore, $y^{-1} = \sqrt[3]{3x + 12}$.

Let $Y1 = \dfrac{1}{3}x^3 - 4$, $Y2 = \sqrt[3]{3x+12}$, and $Y3 = x$.

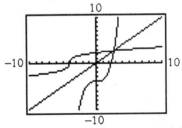

35. Let x = the value of stock transactions. Julia's monthly income can be represented by
$f(x) = 1500 + 0.02x$
$y = 1500 + 0.02x$
$x = 1500 + 0.02y$
$x - 1500 = 0.02y$

$$\frac{x-1500}{0.02} = y$$
$$50x - 75000 = y$$

Therefore, $f^{-1}(x) = 50x - 75000$.

The inverse function represents the amount of Julia's transactions in terms of her monthly income.

37. Let x = the amount time.
    The distance Antonio travels is represented by
    $f(x) = 55x$
    $y = 55x$
    $x = 55y$
    $$\frac{x}{55} = y$$

    Therefore, $f^{-1}(x) = \frac{x}{55}$.

    The inverse function represents the amount of time in terms of the distance traveled.

39. Let x = the number of months to repay the loan.
    $I = Prt = 2000(0.05)x = 10x$
    Total to be repaid = $P + I$
    The total to be repaid is represented by
    $f(x) = 2000 + 10x$
    $y = 2000 + 10x$
    $x = 2000 + 10y$
    $x - 2000 = 10y$
    $$\frac{x-2000}{10} = y$$
    $$y = \frac{1}{10}x - 200$$

    Therefore, $f^{-1}(x) = \frac{1}{10}x - 200$.

    The inverse function represents the number of months in terms of the total amount to repay.

14.1 Calculator Exercises

1. $y = 0.3x^2 - 4$
   $x = 0.3y^2 - 4$
   $x + 4 = 0.3y^2$
   $$\frac{x+4}{0.3} = y^2$$
   $$y = \pm\sqrt{\frac{x+4}{0.3}}$$

   Therefore, $y^{-1} = \pm\sqrt{\frac{x+4}{0.3}}$.

   Let $Y1 = 0.3x^2 - 4$, $Y2 = \sqrt{\frac{x+4}{0.3}}$, and $Y3 = -\sqrt{\frac{x+4}{0.3}}$.

   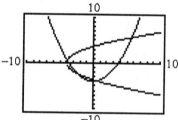

2. $f(x) = 0.1x^3 + 2$
   $y = 0.1x^3 + 2$
   $x = 0.1y^3 + 2$
   $x - 2 = 0.1y^3$
   $$\frac{x-2}{0.1} = y^3$$
   $$y = \sqrt[3]{\frac{x-2}{0.1}}$$

   Therefore, $f^{-1}(x) = \sqrt[3]{\frac{x-2}{0.1}}$.

Let Y1 $= 0.1x^3 + 2$ and Y2 $= \sqrt[3]{\dfrac{x-2}{0.1}}$.

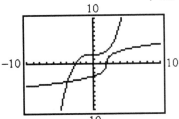

Let Y1 $= \dfrac{1}{x} + 2$ and Y2 $= \dfrac{1}{x-2}$.

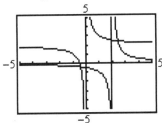

3. $g(x) = 5\sqrt{x} - 1.5$

$y = 5\sqrt{x} - 1.5$

$x = 5\sqrt{y} - 1.5$

$x + 1.5 = 5\sqrt{y}$

$\dfrac{x+1.5}{5} = \sqrt{y}$

$\left(\dfrac{x+1.5}{5}\right)^2 = y$

Therefore, $g^{-1}(x) = \left(\dfrac{x+1.5}{5}\right)^2$.

Let Y1 $= 5\sqrt{x} - 1.5$ and Y2 $= \left(\dfrac{x+1.5}{5}\right)^2$.

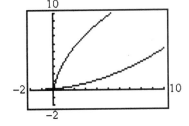

4. $y = \dfrac{1}{x} + 2$

$x = \dfrac{1}{y} + 2$

$x - 2 = \dfrac{1}{y}$

$y(x-2) = 1$

$y = \dfrac{1}{x-2}$

Therefore, $y^{-1} = \dfrac{1}{x-2}$.

14.2 Exercises

1. $y = 7^x$ is an exponential function because it has a rational-number base and a variable exponent.

3. $f(x) = 0.3^x$ is an exponential function because it has a rational-number base and a variable exponent.

5. $g(x) = x^{0.3}$ is not an exponential function because the base is a variable.

7. $y = 1.57^x$ is an exponential function because it has a rational-number base and a variable exponent.

9. $y = x^{-3}$ is not an exponential function because the base is a variable.

11. $R(x) = \left(\dfrac{2}{3}\right)^x$ is an exponential function because it has a rational-number base and a variable exponent.

13. $g(x) = 16^x$

$g(3) = 16^3 = 4096$

15. $g(x) = 16^x$

$g(-2) = 16^{-2} = \dfrac{1}{16^2} = \dfrac{1}{256}$

17. $g(x) = 16^x$

$g\left(\tfrac{1}{2}\right) = 16^{1/2} = \sqrt{16} = 4$

19. $g(x) = 16^x$
$g(\sqrt{2}) = 16^{\sqrt{2}} \approx 50.453$

21. $h(x) = 0.64^x$
$h(0.3) = 0.64^{0.3} \approx 0.875$

23. $h(x) = 0.64^x$
$h\left(-\frac{1}{3}\right) = 0.64^{-1/3} \approx 1.160$

25. $h(x) = 0.64^x$
$h(-\sqrt{2}) = 0.64^{-\sqrt{2}} \approx 1.880$

27. Let $Y1 = 4^x$

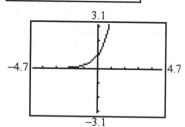

29. Let $Y1 = 4^x$

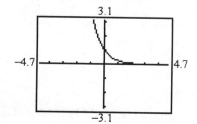

31. Let $Y1 = 4^{2x}$

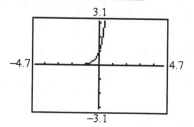

33. Let $Y1 = 4^{(1/2)x}$

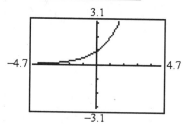

35. Let $Y1 = 4^{x-1}$

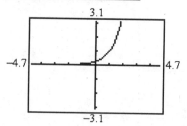

37. Let $Y1 = 4^x - 1$.

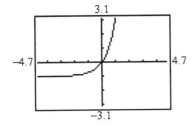

39. Let $Y1 = e^{(1/2)x}$

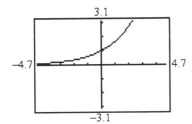

41. Let $Y1 = \frac{1}{2}e^x$

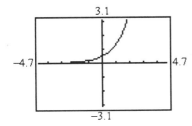

43. Let $Y1 = e^{(-1/2)x}$

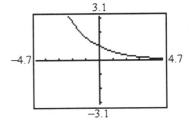

45. Let $Y1 = e^x + \frac{1}{2}$

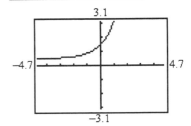

47. Let $Y1 = \frac{1}{2}e^{(1/2)x}$

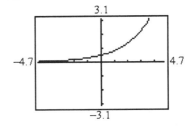

SSM: Experiencing Introductory and Intermediate Algebra

49. $A = P(1+r)^t$; $P = \$6000$; $r = 5.5\% = 0.055$

$A = 6000(1+0.055)^t$

$A = 6000(1.055)^t$

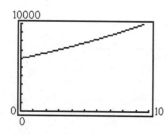

After seven years, Jerome's investment will be worth $A = 6000(1.055)^7 \approx \$8,728.08$

51. $A = Pe^t$; $P = \$8000$; $r = 4.8\% = 0.048$

$A = 8000e^{0.048t}$

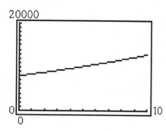

After five years, Jolene's investment will be worth $A = 8000e^{0.048(5)} \approx \$10,169.99$. At that point she will have earned approximately $\$10,169.99 - \$8000 = \$2,169.99$.

53. $y = 49.75(1.03)^x$

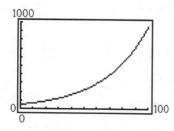

2010 is 20 years after 1990. According to the model, the number of subscribers will be in the year 2010 will be $y = 49.75(1.03)^{20} \approx 90$ million.

55. $y = 6.002e^{1.95x}$

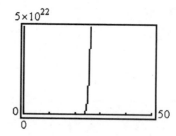

2015 is 25 years after 1990. According to the model, the operating revenue in the year 2015 will be $y = 6.002e^{1.95(25)} \approx \$8.92 \times 10^{21}$ million.

14.2 Calculator Exercises

1. a. Let $Y1 = e^x$

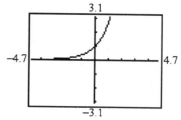

b. Let $Y1 = e^{-x}$

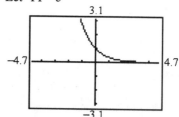

2. a. Let $Y1 = e^x + e^{-x}$

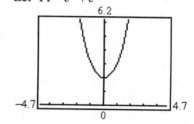

Chapter 14: Exponential and Logarithmic Functions and Equations

b. Let $Y1 = e^x - e^{-x}$

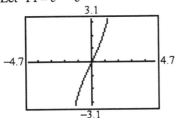

3. a. Let $Y1 = (e^x)(e^{-x})$

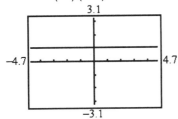

b. Let $Y1 = e^x \div e^{-x}$

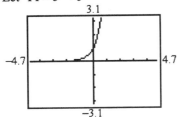

4. a. Let $Y1 = e^{(1/2)x}$

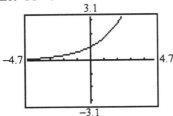

b. Let $Y1 = e^{(-1/2)x}$

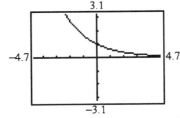

5. a. Let $Y1 = e^{(1/2)x} + e^{(-1/2)x}$

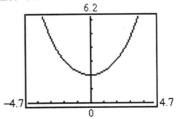

b. Let $Y1 = e^{(1/2)x} - e^{(-1/2)x}$

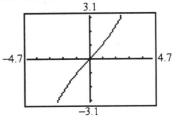

6. a. Let $Y1 = \left(e^{(1/2)x}\right)\left(e^{(-1/2)x}\right)$

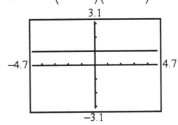

b. Let $Y1 = e^{(1/2)x} \div e^{(-1/2)x}$

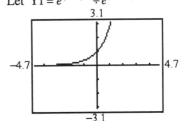

Answers to the questions will vary.

14.3 Exercises

1. $f(x) = 11^x$; $f^{-1}(x) = \log_{11} x$

3. $g(x) = 6^x$; $g^{-1}(x) = \log_6 x$

5. $H(x) = k^x$; $H^{-1}(x) = \log_k x$

SSM: Experiencing Introductory and Intermediate Algebra

7. $\log_2 64 = 6$ because $2^6 = 64$.

9. $\log_3 243 = 5$ because $3^5 = 243$.

11. $\log_4 256 = 4$ because $4^4 = 256$.

13. $\log_2\left(\frac{1}{8}\right) = -3$ because $2^{-3} = \frac{1}{2^3} = \frac{1}{8}$.

15. $\log_4 0.25 = -1$ because $4^{-1} = \frac{1}{4} = 0.25$.

17. $\log_3\left(\frac{1}{81}\right) = -4$ because $3^{-4} = \frac{1}{3^4} = \frac{1}{81}$.

19. $\log 10 = 1$

21. $\log 0.0001 = \log 10^{-4} = -4$

23. $\ln e^3 = 3$

25. $\ln e^{-5} = -5$

27. $\ln \frac{1}{e^5} = \ln e^{-5} = -5$

29. $\log 15 \approx 1.176$

31. $\log\left(\frac{1}{12}\right) \approx -1.079$

33. $\log 1.35 \approx 0.130$

35. $\ln 14 \approx 2.639$

37. $\ln 2.85 \approx 1.047$

39. $\ln \frac{1}{5} \approx -1.609$

41. $\log_4 12 = \frac{\log 12}{\log 4} \approx 1.792$

43. $\log_2 10 = \frac{\log 10}{\log 2} \approx 3.322$

45. $\log_5 2.88 = \frac{\log 2.88}{\log 5} \approx 0.657$

47. $\log_5 \frac{2}{3} = \frac{\log \frac{2}{3}}{\log 3} \approx -0.252$

49. $\log_8 15 = \frac{\ln 15}{\ln 8} \approx 1.302$

51. $\log_3 5.9 = \frac{\ln 5.9}{\ln 3} \approx 1.616$

53. $\log_5 \frac{1}{7} = \frac{\ln \frac{1}{7}}{\ln 5} \approx -1.209$

55. $f(x) = \log_5 x$

$Y1 = \frac{\log x}{\log 5}$

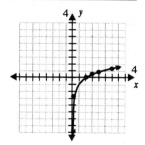

57. $g(x) = \log(x+2)$

$Y1 = \log(x+2)$

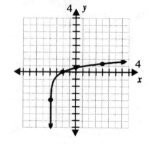

Chapter 14: Exponential and Logarithmic Functions and Equations

59. $h(x) = \ln(x+2)$

$Y1 = \ln(x+2)$

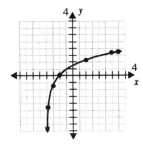

61. $pH = -\log[H^+]$

$6.2 = -\log[H^+]$

$\log[H^+] = -6.2$

$[H^+] = 10^{-6.2}$

$[H^+] \approx 6 \times 10^{-7} = 0.0000006$

The hydrogen ion concentration is approximately 0.0000006 moles per liter.

63. $pH = -\log[H^+]$

$pH = -\log[3.2 \times 10^{-9}]$

$pH \approx 8.49485$

The pH is approximately 8.5.

65. $R = \log \dfrac{I}{I_0}$

$3.5 = \log \dfrac{I}{I_0}$ $8.3 = \log \dfrac{I}{I_0}$

$10^{3.5} = \dfrac{I}{I_0}$ $10^{8.3} = \dfrac{I}{I_0}$

$I = 10^{3.5} I_0$ $I = 10^{8.3} I_0$

$\dfrac{10^{8.3} I_0}{10^{3.5} I_0} = \dfrac{10^{8.3}}{10^{3.5}} \approx 63,095.7$

The 1906 San Francisco earthquake' intensity was approximately 63,095.7 times as great as the smaller earthquake's intensity.

14.3 Calculator Exercises

1. Let $Y1 = \log x$.

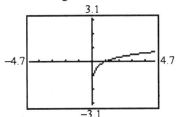

2. Let $Y1 = \log(x+1)$.

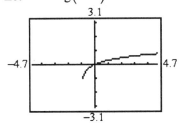

3. Let $Y1 = \log x + \log(x+1)$.

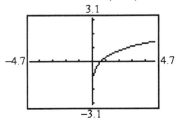

4. Let $Y1 = \log[x(x+1)]$.

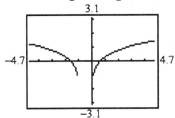

5. Let $Y1 = \log x - \log(x+1)$.

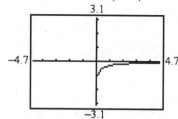

6. Let $Y1 = \log\left(\dfrac{x}{x+1}\right)$.

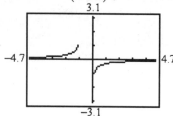

Answers to the questions will vary.

14.4 Exercises

1. $\log 12a = \log(2^2 \cdot 3 \cdot a)$
$= \log 2^2 + \log 3 + \log a$
$= 2\log 2 + \log 3 + \log a$

3. $\ln x^3 = 3\ln x$

5. $\log_5 \dfrac{x}{5} = \log_5 x - \log_5 5 = \log_5 x - 1$

7. $\log \dfrac{2x^2}{y} = \log 2x^2 - \log y$
$= \log 2 + \log x^2 - \log y$
$= \log 2 + 2\log x - \log y$

9. $\log_3 x^3 y^2 = \log_3 x^3 + \log_3 y^2$
$= 3\log_3 x + 2\log_3 y$

11. $\ln \sqrt[3]{xy^2} = \ln\left(xy^2\right)^{1/3}$
$= \dfrac{1}{3}\ln\left(xy^2\right)$
$= \dfrac{1}{3}\left(\ln x + \ln y^2\right)$
$= \dfrac{1}{3}\left(\ln x + 2\ln y\right)$
$= \dfrac{1}{3}\ln x + \dfrac{2}{3}\ln y$

13. $\log \dfrac{\sqrt{2x}}{\sqrt[3]{y}} = \log \sqrt{2x} - \log \sqrt[3]{y}$
$= \log(2x)^{1/2} - \log y^{1/3}$
$= \dfrac{1}{2}\log 2x - \dfrac{1}{3}\log y$
$= \dfrac{1}{2}(\log 2 + \log x) - \dfrac{1}{3}\log y$
$= \dfrac{1}{2}\log 2 + \dfrac{1}{2}\log x - \dfrac{1}{3}\log y$

15. $\log_3 3a = \log_3 3 + \log_3 a$
$= 1 + \log_3 a$

17. $\log_5 10xy = \log_5(5 \cdot 2 \cdot x \cdot y)$
$= \log_5 5 + \log_5 2 + \log_5 x + \log_5 y$
$= 1 + \log_5 2 + \log_5 x + \log_5 y$

19. $\log_a\left(ab^2\right) = \log_a a + \log_a b^2 = 1 + 2\log_a b$

21. $\log x + \log(x+5) = \log x(x+5) = \log\left(x^2 + 5x\right)$

23. $2\ln x + 3\ln y = \ln x^2 + \ln y^3 = \ln x^2 y^3$

25. $2\log_3(x+3) - \log_3(x-1)$
$= \log_3(x+3)^2 - \log_3(x-1)$
$= \log_3 \dfrac{(x+3)^2}{x-1}$

27. $\frac{1}{2}\ln x - \frac{1}{5}\ln(x+1) = \ln x^{1/2} - \ln(x+1)^{1/5}$
$= \ln \sqrt{x} - \ln \sqrt[5]{x+1}$
$= \ln \dfrac{\sqrt{x}}{\sqrt[5]{x+1}}$

29. $\log xy - \log xz = \log \dfrac{xy}{xz} = \log \dfrac{y}{z}$

31. $G = \log\left(\dfrac{P_o}{P_i}\right)^{10} = \log\left(\dfrac{20}{1.5}\right)^{10} \approx 11.249$

The power gain is about 11.249 decibels.

33. $y = c \ln \dfrac{I_0}{I} = 12 \ln(3.5)^{10} \approx 15.033$

The depth is approximately 15.033 units.

35. $t = \dfrac{\ln k}{r} = \dfrac{\ln 3}{0.05} \approx 21.972$

The amount will triple in about 22 years.

37. $t = \dfrac{\ln k}{r} = \dfrac{\ln 4}{0.09} \approx 15.403$

It will take approximately 15.4 years.

14.4 Calculator Exercises

1. Let $Y1 = \log 3x$.

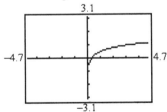

2. Let $Y1 = \log x + \log 2x$.

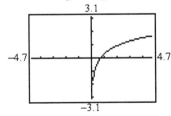

3. Let $Y1 = \log(x^2 - x)$.

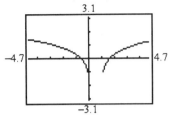

4. Let $Y1 = \log x^2 - \log x$.

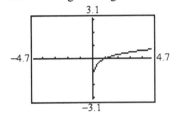

5. No, the graphs are not equivalent.

6. No, the graphs are not equivalent.

7. The logarithm of a sum is not equivalent to the sum of the logarithms.

8. The logarithm of a difference is not equivalent to the difference of the logarithms.

14.5 Exercises

1. $-3 + \log x^2 = 1$ is a logarithmic equation because it has a variable in the argument of the logarithm.

3. $e^3 + x = 2x - 1$ is neither an exponential nor a logarithmic equation. The exponential expression does not have a variable exponent.

5. $e^{2x} - 1 = 5$ is an exponential equation because it has a variable exponent.

7. $x = \log 4 + \log 7$ is neither a logarithmic nor an exponential equation. Neither of the logarithmic expressions has a variable in its argument.

9. $5^{2x} = 625$
$5^{2x} = 5^4$
$2x = 4$
$x = 2$
The solution is 2.

11. $e^t = 2$
$\ln e^t = \ln 2$
$t = \ln 2 \approx 0.693$
The solution is approximately 0.693.

13. $5^x = 1$
$5^x = 5^0$
$x = 0$
The solution is 0.

15. $3e^{-x} = 2$
$e^{-x} = \dfrac{2}{3}$
$\ln e^{-x} = \ln \dfrac{2}{3}$
$-x = \ln \dfrac{2}{3}$
$x = -\ln \dfrac{2}{3} \approx 0.405$
The solution is approximately 0.405.

17. $(2^{3x})(2^5) = 2^8$
$2^{3x+5} = 2^8$
$3x + 5 = 8$
$3x = 3$
$x = 1$
The solution is 1.

19. $2^x = 2^{3x-6}$
$x = 3x - 6$
$-2x = -6$
$x = 3$
The solution is 3.

21. $5^{2x(x-2)} = 5^{3(x-1)}$
$2x(x-2) = 3(x-1)$
$2x^2 - 4x = 3x - 3$
$2x^2 - 7x + 3 = 0$
$(2x-1)(x-3) = 0$
$x = \dfrac{1}{2}$ or $x = 3$
The solutions are $\dfrac{1}{2}$ and 3.

23. $e^{x(x-2)} = e^{x+10}$
$x(x-2) = x + 10$
$x^2 - 2x = x + 10$
$x^2 - 3x - 10 = 0$
$(x+2)(x-5) = 0$
$x = -2$ or $x = 5$
The solutions are -2 and 5.

25. $3^{x+x(x-3)} = 3^{(x-1)^2}$
$x + x(x-3) = (x-1)^2$
$x + x^2 - 3x = x^2 - 2x + 1$
$x^2 - 2x = x^2 - 2x + 1$
$0 = 1$
The equation is always false (a contradiction). It has no solution.

27. $8^{5(x-3)} = 8^{4(x-2)} 8^{(x-7)}$
$8^{5(x-3)} = 8^{4(x-2)+(x-7)}$
$5(x-3) = 4(x-2) + (x-7)$
$5x - 3 = 4x - 8 + x - 7$
$5x - 15 = 5x - 15$
$-15 = -15$
The equation is always true (an identity). The solution is the set of all real numbers.

Chapter 14: Exponential and Logarithmic Functions and Equations

29. $3^{3x} = 9^{x+1}$
$3^{3x} = (3^2)^{x+1}$
$3^{3x} = 3^{2x+2}$
$3x = 2x+2$
$x = 2$
The solution is 2.

31. $5^{2b+5} = 5^4$
$2b+5 = 4$
$2b = -1$
$b = -\dfrac{1}{2}$
The solution is $-\dfrac{1}{2}$.

32. $7^{3c-1} = 7^{c+9}$
$3c-1 = c+9$
$2c = 10$
$c = 5$
The solution is 5.

33. $3^{2x^2} \cdot 3^{5x} = 27$
$3^{2x^2+5x} = 3^3$
$2x^2 + 5x = 3$
$2x^2 + 5x - 3 = 0$
$(2x-1)(x+3) = 0$
$x = \dfrac{1}{2}$ or $x = -3$
The solutions are -3 and $\dfrac{1}{2}$.

35. $e^{0.5x} = 4$
$\ln e^{0.5x} = \ln 4$
$0.5x = \ln 4$
$x = \dfrac{\ln 4}{0.5} \approx 2.773$
The solution is approximately 2.773.

37. $(3.5)^x = 12$
$\log(3.5)^x = \log 12$
$x \log 3.5 = \log 12$
$x = \dfrac{\log 12}{\log 3.5} \approx 1.984$
The solution is approximately 1.984.

39. $\log_5 c = 5$
$5^5 = c$
$3125 = c$
The solution is approximately 3125.

41. $\log x = 0$
$10^0 = x$
$1 = x$
The solution is 1.

43. $\ln z = 0$
$e^0 = z$
$1 = z$
The solution is 1.

45. $\log_3 k = 2$
$3^2 = k$
$9 = k$
The solution is 9.

47. $\log_5 x = -3$
$5^{-3} = x$
$\dfrac{1}{5^3} = x$
$\dfrac{1}{125} = x$
The solution is $\dfrac{1}{125}$.

49. $\log_3(a+2) = 2$
$3^2 = a+2$
$9 = a+2$
$7 = a$
The solution is 7.

51. $2\ln x = 1$

$\ln x = \dfrac{1}{2}$

$e^{1/2} = x$

$\sqrt{e} = x$

$1.649 \approx x$

The solution is 1.649.

53. $\log x + 4 = 2$

$\log x = -2$

$10^{-2} = x$

$\dfrac{1}{10^2} = x$

$\dfrac{1}{100} = x$

The solution is $\dfrac{1}{100}$.

55. $5\ln z + 4 = 4$

$5\ln z = 0$

$\ln z = 0$

$e^0 = z$

$1 = z$

The solution is 1.

57. $\ln x + \ln(x+1) = \ln 12$

$\ln x(x+1) = \ln 12$

$x(x+1) = 12$

$x^2 + x - 12 = 0$

$(x-3)(x+4) = 0$

$x = 3$ or $x = -4$

Since –4 is a restricted value, the solution is 3.

59. $2\ln(p-5) = \ln 9$

$\ln(p-5)^2 = \ln 9$

$(p-5)^2 = 9$

$p - 5 = \pm\sqrt{9}$

$p - 5 = \pm 3$

$p = 5 \pm 3$

$p = 8$ or $x = 2$

Since 2 is a restricted value, the solution is 8.

61. $\log 2x + \log 3x = \log 6 + 2\log x$

$\log(2x)(3x) = \log 6 + \log x^2$

$\log 6x^2 = \log 6x^2$

$6x^2 = 6x^2$

The equation is true for all values of x except the restricted value(s). The solution is the set of all real numbers greater than 0.

63. $\ln x + \ln(x+2) = 2\ln(x+2)$

$\ln x(x+2) = \ln(x+2)^2$

$x(x+2) = (x+2)^2$

$x^2 + 2x = x^2 + 4x + 4$

$-2x = 4$

$x = -2$

Now –2 is a restricted value, so the equation has no solution.

65. $\log(x+4) - \log x = \log 2$

$\log\dfrac{x+4}{x} = \log 2$

$\dfrac{x+4}{x} = 2$

$2x = x + 4$

$x = 4$

The solution is 4.

67. $\log x - \log(x+6) = -1$

$\log\dfrac{x}{x+6} = -1$

$10^{-1} = \dfrac{x}{x+6}$

$\dfrac{1}{10} = \dfrac{x}{x+6}$

$10x = x + 6$

$9x = 6$

$x = \dfrac{6}{9} = \dfrac{2}{3}$

The solution is $\dfrac{2}{3}$.

Chapter 14: Exponential and Logarithmic Functions and Equations

69. $\log_3(x+16) - \log_3 x = 2$

$\log_3 \dfrac{x+16}{x} = 2$

$3^2 = \dfrac{x+16}{x}$

$9 = \dfrac{x+16}{x}$

$x + 16 = 9x$

$16 = 8x$

$2 = x$

The solution is 2.

71. Let t = the number of years since 1990.

$A(t) = A_0 e^{kt}$

$328 = 218 e^{k(8)}$

$\dfrac{328}{218} = e^{8k}$

$\ln\left(\dfrac{328}{218}\right) = 8k$

$\dfrac{1}{8} \ln\left(\dfrac{328}{218}\right) = k$

$0.051 \approx k$

$A(t) = 218 e^{0.051 t}$

$A(20) = 218 e^{0.051(20)} \approx 604.6$

In the year 2010, the estimated expenditures will be approximately $604.6 billion.

73. $A(t) = A_0 e^{0.0416 t}$

$2 A_0 = A_0 e^{0.0416 t}$

$2 = e^{0.0416 t}$

$\ln 2 = 0.0416 t$

$\dfrac{\ln 2}{0.0416} = t$

$16.7 \approx t$

$1990 + 16.7 = 2006.7$

Per capita income will double its 1990 value in approximately the year 2007.

75. $A(t) = A_0 e^{kt}$

$\dfrac{1}{2} A_0 = A_0 e^{k(4.5)}$

$\dfrac{1}{2} = e^{4.5k}$

$\ln\left(\dfrac{1}{2}\right) = 4.5 k$

$\dfrac{\ln(1/2)}{4.5} = k$

$-0.154 \approx k$

The decay factor is -0.154.

$0.98 A_0 = A_0 e^{-0.154 t}$

$0.98 = e^{-0.154 t}$

$\ln 0.98 = -0.154 t$

$\dfrac{\ln 0.98}{-0.154} = t$

$0.131 \approx t$

It will take approximately 0.131 billion years, or 131 million years.

14.5 Calculator Exercises

Students should read about graphing techniques.

Chapter 14 Section-By-Section Review

1. $h = \{(2,4.5),(4,3.5),(6,2.5),(8,1.5),(10,0.5)\}$

$h^{-1} = \{(4.5,2),(3.5,4),(2.5,6),(1.5,8),(0.5,10)\}$

SSM: Experiencing Introductory and Intermediate Algebra

2. $y = 3x - 9$ $x = 3y - 9$
$x + 9 = 3y$
$\dfrac{x+9}{3} = y$
$\dfrac{1}{3}x + 3 = y$
Therefore, $y^{-1} = \dfrac{1}{3}x + 3$

3. $y = x^2 + 1$ $x = y^2 + 1$
$x - 1 = y^2$
$y = \pm\sqrt{x-1}$
Therefore, $y^{-1} = \pm\sqrt{x-1}$

4. $y = x^3 - 1$ $x = y^3 - 1$
$x + 1 = y^3$
$y = \sqrt[3]{x+1}$
Therefore, $y^{-1} = \sqrt[3]{x+1}$

5. $P = \{(9,1),(8,2),(7,3),(6,1),(5,2),(4,3)\}$
The function P is not one-to-one, because the second coordinates 1, 2, and 3 are each repeated in two different ordered pairs. Therefore, the inverse of P is not a function.

6. $Q = \{(1,9),(2,8)(3,7),(4,6),(5,5),(6,4),(7,3)\}$
The function Q is one-to-one, because each ordered pair has a different second coordinate. Therefore the inverse of Q is a function.

7. No, the graph does not represent a function, because a vertical line can be drawn that intersects the graph more than once.

8. Yes, the graph represents a one-to-one function, because all possible vertical and horizontal lines cross the graph only once.

9. No, the graph does not represent a one-to-one function. The graph does represents a function, because all possible vertical lines cross the graph only once, but it is not one-to-one, because a horizontal line can be drawn that intersects the graph more than once.

10. $f(x) = \dfrac{3}{4}x - 3$ $y = \dfrac{3}{4}x - 3$
$x = \dfrac{3}{4}y - 3$
$x + 3 = \dfrac{3}{4}y$
$\dfrac{4}{3}(x+3) = \dfrac{4}{3}\left(\dfrac{3}{4}y\right)$
$\dfrac{4}{3}x + 4 = y$
Therefore, $f^{-1}(x) = \dfrac{4}{3}x + 4$.

Let $Y1 = \dfrac{3}{4}x - 3$, $Y2 = \dfrac{4}{3}x + 4$, and $Y3 = x$.

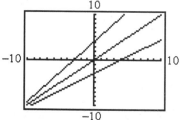

11. $y = x^3 + 8$ $x = y^3 + 8$
$x - 8 = y^3$
$\sqrt[3]{x-8} = y$
Therefore, $y^{-1} = \sqrt[3]{x-8}$

Let $Y1 = x^3 + 8$, $Y2 = \sqrt[3]{x-8}$, and $Y3 = x$.

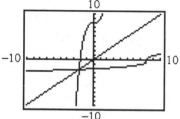

12. Let x = the value of all sales.
Motomo's weekly income is represented by $f(x) = 570 + 0.03x$.

Chapter 14: Exponential and Logarithmic Functions and Equations

$y = 570 + 0.03x$
$x = 570 + 0.03y$
$x - 570 = 0.03y$
$\dfrac{x - 570}{0.03} = y$
$33.\overline{3}x - 19000 = y$

Therefore, $f^{-1}(x) = 33.\overline{3}x - 19000$.

The inverse function represents the value of all sales in terms of Motomo's weekly salary.

13. Let $x =$ the number of hours.
The number of problem solved is represented by $f(x) = 15x$.
$y = 15x$
$x = 15y$
$\dfrac{x}{15} = y$

Therefore, $f^{-1}(x) = \dfrac{x}{15}$.

The inverse function represents the number of hours in terms of the number of problems solved.

14. $f(x) = 2^x$ is an exponential function because it has a rational-number base and a variable exponent.

15. $y = x^2$ is not an exponential function because the base is a variable.

16. $g(x) = 1.21^x$
$g(2) = 1.21^2 = 1.4641$

17. $g(-1) = 1.21^{-1} = \dfrac{1}{1.21} \approx 0.826$

18. $g(0) = 1.21^0 = 1$

19. $g\left(\tfrac{1}{2}\right) = 1.21^{1/2} = \sqrt{1.21} = 1.1$

20. $g(0.3) = 1.21^{0.3} \approx 1.059$

21. $g(\sqrt{2}) = 1.21^{\sqrt{2}} \approx 1.309$

22. $Y1 = 5^x - 1$

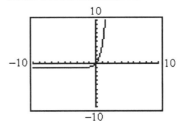

23. $Y1 = 9^{(1/2)x}$

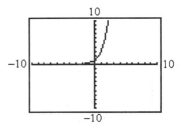

24. $Y1 = 2e^{0.5x}$

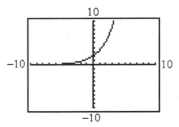

25. $A = P(1+r)^t$; $P = \$1000$; $r = 4\% = 0.04$

$A = 1000(1+0.04)^t$

$A = 1000(1.04)^t$

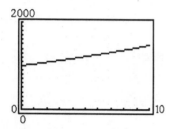

After four years, Paul's investment will be worth
$A = 1000(1.04)^4 \approx \$1,169.86$.

26. $y = 330.53(1.06)^x$

$x = 2010 - 1990 = 20$

$y = 330.53(1.06)^{20} \approx 1060.1$

The estimated expenditures predicted for the year 2010 will be about $1060.1 billion.

27. $h(x) = 8^x$; $h^{-1}(x) = \log_8 x$

28. $A(x) = Ae^{kx}$

$y = Ae^{kx}$

$x = Ae^{ky}$

$\dfrac{x}{A} = e^{ky}$

$\ln \dfrac{x}{A} = ky$

$\dfrac{1}{k} \ln \dfrac{x}{A} = y$

Therefore, $A^{-1}(x) = \dfrac{1}{k} \ln \dfrac{x}{A}$

29. $\log 100 = \log 10^2 = 2$

30. $\log_2 16 = \log_2 2^4 = 4$

31. $\log_3 \dfrac{1}{27} = \log_3 \dfrac{1}{3^3} = \log_3 3^{-3} = -3$

32. $\log 1.5 \approx 0.176$

33. $\ln e^4 = 4$

34. $\ln 10 \approx 2.303$

35. $\ln e^{-2} = -2$

36. $\log \dfrac{3}{5} \approx -0.222$

37. $\log_5 15 = \dfrac{\log 15}{\log 5} \approx 1.683$

38. $y = \log_3 x$

$Y1 = \dfrac{\log x}{\log 3}$

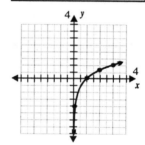

39. $f(x) = \ln(x-2)$

$Y1 = \ln(x-2)$

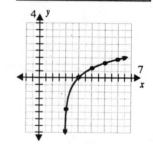

40. $\text{pH} = -\log\left[H^+\right]$

$\text{pH} = -\log\left[6.2\times 10^{-3}\right]$

$\text{pH} \approx 2.208$

The pH is approximately 2.208.

41. $\text{pH} = -\log\left[H^+\right]$

$7.4 = -\log\left[H^+\right]$

$\log\left[H^+\right] = -7.4$

$\left[H^+\right] = 10^{-7.4}$

$\left[H^+\right] \approx 3.98\times 10^{-8}$

The hydrogen ion concentration is approximately 3.98×10^{-8} moles per liter.

42. $\log 2x^4 = \log 2 + \log x^4 = \log 2 + 4\log x$

43. $\log_7 \dfrac{7x}{y} = \log_7 7x - \log_7 y$

$= \log_7 7 + \log_7 x - \log_7 y$

$= 1 + \log_7 x - \log_7 y$

44. $\ln 25x^2 y^3 z = \ln 5^2 x^2 y^3 z$

$= \ln 5^2 + \ln x^2 + \ln y^3 + \ln z$

$= 2\ln 5 + 2\ln x + 3\ln y + \ln z$

45. $\ln \dfrac{\sqrt[3]{2x^2 y}}{\sqrt{yz}} = \ln \sqrt[3]{2x^2 y} - \ln \sqrt{yz}$

$= \ln\left(2x^2 y\right)^{1/3} - \ln\left(yz\right)^{1/2}$

$= \dfrac{1}{3}\ln 2x^2 y - \dfrac{1}{2}\ln yz$

$= \dfrac{1}{3}\left(\ln 2 + \ln x^2 + \ln y\right) - \dfrac{1}{2}\left(\ln y + \ln z\right)$

$= \dfrac{1}{3}\left(\ln 2 + 2\ln x + \ln y\right) - \dfrac{1}{2}\left(\ln y + \ln z\right)$

$= \dfrac{1}{3}\ln 2 + \dfrac{2}{3}\ln x + \dfrac{1}{3}\ln y - \dfrac{1}{2}\ln y - \dfrac{1}{2}\ln z$

46. $\log(x-3) + \log(x+3) = \log(x-3)(x+3)$

$= \log(x^2 - 9)$

47. $5\ln x - 3\ln y = \ln x^5 - \ln y^3 = \ln \dfrac{x^5}{y^3}$

48. $\dfrac{1}{3}\log x - \dfrac{1}{2}\log y = \log x^{1/3} - \log y^{1/2}$

$= \log \sqrt[3]{x} - \log \sqrt{y}$

$= \log \dfrac{\sqrt[3]{x}}{\sqrt{y}}$

49. $\log ab - \log bc + \log cd = \log \dfrac{ab}{bc} + \log cd$

$= \log\left(\dfrac{ab}{bc}\cdot cd\right)$

$= \log ad$

50. $t = \dfrac{\ln k}{r} = \dfrac{\ln 3}{0.06} \approx 18.3$

It will take approximately 18.3 years.

51. $e^x + 5 = 7$ is an exponential equation because it has a variable exponent.

52. $\ln 7 - x = \ln 3$ is neither a logarithmic nor an exponential equation. Neither of the logarithmic expressions has a variable in its argument.

53. $\log x^2 + 2 = \log x$ is a logarithmic equation because it has a variable in the argument of the logarithms.

54. $x = e^3 - 1$ is neither an exponential nor a logarithmic equation. The exponential expression does not have a variable exponent.

55. $2^x + 7 = 2^{3x}$ is an exponential equation because it has variable exponents.

56. $y = \log_2 5 - e^2$ is neither a logarithmic nor an exponential equation. The logarithmic expression does not have a variable in its argument and the exponential expression does not have a variable exponent.

57. $3^{x+1} = 243$
$3^{x+1} = 3^5$
$x+1 = 5$
$x = 4$
The solution is 4.

58. $e^{-5t} - 2 = 3$
$e^{-5t} = 5$
$\ln e^{-5t} = \ln 5$
$-5t = \ln 5$
$t = \dfrac{\ln 5}{-5} \approx -0.322$
The solution is approximately -0.322.

59. $12^a = 1$
$12^a = 12^0$
$a = 0$
The solution is 0.

60. $(3^{2x})(3^5) = 3^8$
$3^{2x+5} = 3^8$
$2x + 5 = 8$
$2x = 3$
$x = \dfrac{3}{2}$
The solution is $\dfrac{3}{2}$.

61. $e^{2x(x-3)} = e^{2(x-2)(x-1)}$
$2x(x-3) = 2(x-2)(x-1)$
$2x^2 - 6x = 2x^2 - 6x + 4$
$0 = 4$
The equation is always false (a contradiction). It has no solution.

62. $4^{x^2+2x} = 64$
$4^{x^2+2x} = 4^3$
$x^2 + 2x = 3$
$x^2 + 2x - 3 = 0$
$(x-1)(x+3) = 0$
$x = 1$ or $x = -3$
The solutions are -3 and 1.

63. $5^{2x-3} = 5^{x+9} \cdot 5^{x-12}$
$5^{2x-3} = 5^{x+9+x-12}$
$2x - 3 = x + 9 + x - 12$
$2x - 3 = 2x - 3$
$-3 = -3$
The equation is always true (an identity). The solution is the set of all real numbers.

64. $10^{2.3x} = 4$
$\log 10^{2.3x} = \log 4$
$2.3x = \log 4$
$x = \dfrac{\log 4}{2.3} \approx 0.262$
The solution is approximately 0.262.

65. $\log_7 a = 49$
$7^{49} = a$
$2.569 \times 10^{41} \approx a$
The solution is approximately 2.569×10^{41}.

66. $5 \log_3 x = 45$
$\log_3 x = 9$
$3^9 = x$
$19{,}683 = x$
The solution is 19,683.

67. $\ln z = 1$
$e^1 = z$
$2.718 \approx z$
The solution is approximately 2.718.

Chapter 14: Exponential and Logarithmic Functions and Equations

68. $\log b = 0$
$10^0 = b$
$1 = b$
The solution is 1.

69. $\log k = -3$
$10^{-3} = k$
$k = \dfrac{1}{10^3} = \dfrac{1}{1000}$
The solution is $\dfrac{1}{1000}$.

70. $5 + \log_2 x = 3$
$\log_2 x = -2$
$2^{-2} = x$
$x = \dfrac{1}{2^2} = \dfrac{1}{4}$
The solution is $\dfrac{1}{4}$.

71. $3 \ln x = 1$
$\ln x = \dfrac{1}{3}$
$e^{1/3} = x$
$\sqrt[3]{e} = x$
$1.396 \approx x$
The solution is 1.396.

72. $\ln(2x^2 + 7x) = \ln 15$
$2x^2 + 7x = 15$
$2x^2 + 7x - 15 = 0$
$(2x - 3)(x + 5) = 0$
$x = \dfrac{3}{2}$ or $x = -5$
The solutions are -5 and $\dfrac{3}{2}$.

73. $2 \log(x - 2) = \log 9$
$\log(x - 2)^2 = \log 9$
$(x - 2)^2 = 9$
$x - 2 = \pm\sqrt{9}$
$x = 2 \pm 3$
$x = 5$ or $x = -1$
Since -1 is a restricted value, the solution is 5.

74. $\log x + \log(x + 4) = 2 \log(x + 2)$
$\log x(x + 4) = \log(x + 2)^2$
$x(x + 4) = (x + 2)^2$
$x^2 + 4x = x^2 + 4x + 4$
$0 = 4$
The equation is always false (a contradiction). It has no solution.

75. $7 \ln x - \ln x^3 = 4 \ln x$
$\ln x^7 - \ln x^3 = \ln x^4$
$\ln \dfrac{x^7}{x^3} = \ln x^4$
$\ln x^4 = \ln x^4$
$x^4 = x^4$
The equation is true for all values of x except the restricted value(s). The solution is the set of all real numbers greater than 0.

76. Let t = the number of years since 1990.
$A(t) = A_0 e^{kt}$ $4.177 = 3.022 e^{k(9)}$
$\dfrac{4.177}{3.022} = e^{9k}$
$\ln\left(\dfrac{4.177}{3.022}\right) = 9k$
$\dfrac{1}{9} \ln\left(\dfrac{4.177}{3.022}\right) = k$
$0.036 \approx k$
$A(t) = 3.022 e^{0.036t}$
$A(20) = 3.022 e^{0.036(20)} \approx 6.208$
The estimated resident population 85 years and older will be about 6.208 million in 2010. This is higher than the projected number.

639

77. $A(t) = A_0 e^{-0.000121t}$

$\frac{2}{3} A_0 = A_0 e^{-0.000121t}$

$\frac{2}{3} = e^{-0.000121t}$

$\ln\left(\frac{2}{3}\right) = -0.000121t$

$\frac{\ln(2/3)}{-0.000121} = t$

$3351 \approx t$

The time since the skeleton's demise is about 3351 years.

Chapter 14 Chapter Review

1. $\log 0.001 = -3$ because $10^{-3} = 0.001$.

2. $\log_7 343 = 3$ because $7^3 = 343$.

3. $\log_2 \frac{1}{64} = -6$ because $2^{-6} = \frac{1}{64}$.

4. $\log 5.1 \approx 0.708$

5. $\ln e^{-3} = -3$

6. $\ln 100 \approx 4.605$

7. $\ln e = 1$

8. $\ln \frac{4}{7} \approx -0.560$

9. $\log_3 36 = \frac{\log 36}{\log 3} \approx 3.262$

10. $\log 3x^2 y = \log 3 + \log x^2 + \log y$
    $= \log 3 + 2\log x + \log y$

11. $\log_3 \frac{9a}{b} = \log_3 9a - \log_3 b$
    $= \log_3 9 + \log_3 a - \log_3 b$
    $= 2 + \log_3 a - \log_3 b$

12. $\ln 100 p^3 q^2 r$
    $= \ln 2^2 5^2 p^3 q^2 r$
    $= \ln 2^2 + \ln 5^2 + \ln p^3 + \ln q^2 + \ln r$
    $= 2\ln 2 + 2\ln 5 + 3\ln p + 2\ln q + \ln r$

13. $\log \frac{\sqrt[5]{6x^3}}{\sqrt{xy}}$
    $= \log \sqrt[5]{2 \cdot 3x^3} - \log \sqrt{xy}$
    $= \log(2 \cdot 3x^3)^{1/5} - \log(xy)^{1/2}$
    $= \frac{1}{5}\log(2 \cdot 3x^3) - \frac{1}{2}\log xy$
    $= \frac{1}{5}(\log 2 + \log 3 + \log x^3) - \frac{1}{2}(\log x + \log y)$
    $= \frac{1}{5}(\log 2 + \log 3 + 3\log x) - \frac{1}{2}(\log x + \log y)$
    $= \frac{1}{5}\log 2 + \frac{1}{5}\log 3 + \frac{3}{5}\log x - \frac{1}{2}\log x - \frac{1}{2}\log y$

14. $\log(x^2 - 9) - \log(x + 3) = \log \frac{x^2 - 9}{x + 3}$
    $= \log \frac{(x+3)(x-3)}{x+3}$
    $= \log(x - 3)$

15. $2\log c - 5\log d = \log c^2 - \ln d^5 = \ln \frac{c^2}{d^5}$

16. $\frac{1}{2}\ln x - 3\ln y = \ln x^{1/2} - \ln y^3 = \ln \frac{\sqrt{x}}{y^3}$

17. $\log 2xy - \log xz + \log 3yz = \log \frac{2xy}{xz} + \log 3yz$
    $= \log\left(\frac{2xy}{xz} \cdot 3yz\right)$
    $= \log 6y^2$

18. $y = 5 - x$
    $x = 5 - y$
    $y = -x + 5$
    Therefore, $y^{-1} = -x + 5$.

19. $y = 1 - x^2$
$x = 1 - y^2$
$y^2 = 1 - x$
$y = \pm\sqrt{1-x}$
Therefore, $y^{-1} = \pm\sqrt{1-x}$.

20. $y = 4 - x^3$
$x = 4 - y^3$
$y^3 = 4 - x$
$y = \sqrt[3]{4-x}$
Therefore, $y^{-1} = \sqrt[3]{4-x}$.

21. $y = 2 - \dfrac{4}{5}x$
$x = 2 - \dfrac{4}{5}y$
$\dfrac{4}{5}y = -x + 2$
$\dfrac{5}{4}\left(\dfrac{4}{5}y\right) = \dfrac{5}{4}(-x+2)$
$y = -\dfrac{5}{4}x + \dfrac{5}{2}$
Therefore, $y^{-1} = -\dfrac{5}{4}x + \dfrac{5}{2}$.

22. $m(x) = 5^x$; $m^{-1}(x) = \log_5 x$

23. $G(x) = ab^x$ $y = ab^x$
$x = ab^y$
$\dfrac{x}{a} = b^y$
$\log_b\left(\dfrac{x}{a}\right) = y$
Therefore, $G^{-1}(x) = \log_b\left(\dfrac{x}{a}\right)$.

24. $g(t) = 1 + \left(\dfrac{4}{9}\right)^t$
$g(3) = 1 + \left(\dfrac{4}{9}\right)^3 \approx 1.088$

25. $g(-1) = 1 + \left(\dfrac{4}{9}\right)^{-1} = 1 + \dfrac{9}{4} = \dfrac{13}{4}$

26. $g(0) = 1 + \left(\dfrac{4}{9}\right)^0 = 1 + 1 = 2$

27. $g\left(\dfrac{1}{2}\right) = 1 + \left(\dfrac{4}{9}\right)^{1/2} = 1 + \dfrac{2}{3} = \dfrac{5}{3}$

28. $g(1.5) = 1 + \left(\dfrac{4}{9}\right)^{1.5} \approx 1.296$

29. $g(-\sqrt{3}) = 1 + \left(\dfrac{4}{9}\right)^{-\sqrt{3}} \approx 5.074$

30. Let $Y1 = 3^x - 2$

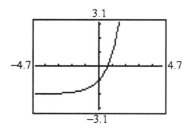

31. Let $Y1 = 8^{(1/3)x}$

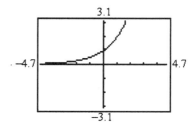

32. Let $Y1 = -2e^{0.4x}$

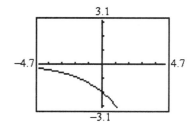

33. $y = \log_2 x - 2$; Let $Y1 = \dfrac{\log x}{\log 2} - 2$.

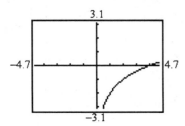

34. Let $Y1 = \log(x+1)$

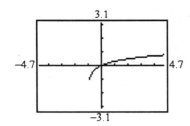

35. $\log_6 b = 216$
$6^{216} = b$
The solution is 6^{216}.

36. $2\log_3 z = 6$
$\log_3 z = 3$
$3^3 = z$
$27 = z$
The solution is 27.

37. $\log z = 1$
$10^1 = z$
$10 = z$
The solution is 10.

38. $\ln b = 0$
$e^0 = b$
$1 = b$
The solution is 1.

39. $\log 2m = -4$
$10^{-4} = 2m$
$m = \dfrac{1}{2 \cdot 10^4} = \dfrac{1}{20{,}000}$
The solution is $\dfrac{1}{20{,}000}$.

40. $-1 + \log_3 x = 4$
$\log_3 x = 5$
$3^5 = x$
$243 = z$
The solution is 243.

41. $2\log x = 1$
$\log x = \dfrac{1}{2}$
$10^{1/2} = x$
$x = \sqrt{10} \approx 3.162$
The solution is approximately 3.162.

42. $\log(2x^2 + 3x) = \log 9$
$2x^2 + 3x = 9$
$2x^2 + 3x - 9 = 0$
$(2x - 3)(x + 3) = 0$
$x = \dfrac{3}{2}$ or $x = -3$
The solutions are -3 and $\dfrac{3}{2}$.

43. $2\ln(x+3) = \ln 4$
$\ln(x+3)^2 = \ln 4$
$(x+3)^2 = 4$
$x + 3 = \pm\sqrt{4}$
$x + 3 = \pm 2$
$x = -3 \pm 2$
$x = -1$ or $x = -5$
Since -5 is a restricted value, the solution is -1.

44. $\log 4x + \log(x+1) = 2\log(2x+1)$
$\log 4x(x+1) = \log(2x+1)^2$
$4x(x+1) = (2x+1)^2$
$4x^2 + 4x = 4x^2 + 4x + 1$
$0 = 1$
The equation is always false (a contradiction). It has no solution.

Chapter 14: Exponential and Logarithmic Functions and Equations

45. $2\log a + \log a^4 = 6\log a$

$\log a^2 + \log a^4 = \log a^6$

$\log a^2 \cdot a^4 = \log a^6$

$\log a^6 = \log a^6$

$a^6 = a^6$

The equation is true for all values of x except the restricted value(s). The solution is the set of all real numbers greater than 0.

46. $5^{x-3} = 125$

$5^{x-3} = 5^3$

$x - 3 = 3$

$x = 6$

The solution is 6.

47. $e^{-t} + 4 = 5$

$e^{-t} = 1$

$\ln e^{-t} = \ln 1$

$-t = 0$

$t = 0$

The solution is 0.

48. $7^a = e$

$\ln 7^a = \ln e$

$a \ln 7 = 1$

$a = \dfrac{1}{\ln 7} \approx 0.514$

The solution is approximately 0.514.

49. $\left(6^7\right)\left(6^{3x}\right) = 6^2$

$6^{7+3x} = 6^2$

$7 + 3x = 2$

$3x = -5$

$x = -\dfrac{5}{3}$

The solution is $-\dfrac{5}{3}$.

50. $e^{x(x+6)} = e^{(x+5)(x+1)}$

$x(x+6) = (x+5)(x+1)$

$x^2 + 6x = x^2 + 6x + 5$

$0 = 5$

The equation is always false (a contradiction). It has no solution.

51. $2^{x^2} \cdot 2^{-3x} = 16$

$2^{x^2 - 3x} = 2^4$

$x^2 - 3x = 4$

$x^2 - 3x - 4 = 0$

$(x+1)(x-4) = 0$

$x = -1$ or $x = 4$

The solutions are -1 and 4.

52. $7^{2x-6} = \left(7^{3x-11}\right)\left(7^{5-x}\right)$

$7^{2x-6} = 7^{3x-11+5-x}$

$2x - 6 = 3x - 11 + 5 - x$

$2x - 6 = 2x - 6$

$-6 = -6$

The equation is true for all values of x. The solution is the set of all real numbers.

53. $10^{1.8x} = 9$

$\log 9 = 1.8x$

$x = \dfrac{\log 9}{1.8} \approx 0.530$

The solution is approximately 0.530.

54. Let x = the amount of sales.

Diana monthly income can be represented by

$f(x) = 1200 + 0.025x$

$y = 1200 + 0.025x$

$x = 1200 + 0.025y$

$x - 1200 = 0.025y$

$\dfrac{x - 1200}{0.025} = y$

$40x - 48000 = y$

Therefore, $f^{-1}(x) = 40x - 48000$.

The inverse function represents Diana's total sales in terms of her monthly income.

55. Let x = the number of hours.
The situation can be represented by
$f(x) = 95x$
$y = 95x$
$x = 95y$
$\dfrac{x}{95} = y$

Therefore, $f^{-1}(x) = \dfrac{x}{95}$.

The inverse function represents the number of hours in term of the distance traveled.

56. $A = P(1+r)^t$; $P = \$8500$; $r = 6.5\% = 0.065$
$A = 8500(1+0.065)^t$
$A = 8500(1.065)^t$

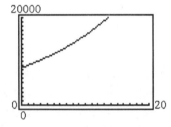

After 10 years, Kristin's investment will be worth $A = 8500(1.065)^{10} \approx \$15{,}955.67$

57. $2010 - 1990 = 20$
$y = 52.28(1.03)^x = 52.28(1.03)^{20} \approx 94.4$

According to the model, about 94.4 million U.S. households will have cable TV in the year 2010. Yes, it does seem reasonable that the same rate of growth would occur between 1990 and 2010.

58. $\text{pH} = -\log[\text{H}^+]$
$\text{pH} = -\log[1.6 \times 10^{-4}]$
$\text{pH} \approx 3.796$
The pH is approximately 3.796.

59. $\text{pH} = -\log[\text{H}^+]$
$1.6 = -\log[\text{H}^+]$
$\log[\text{H}^+] = -1.6$
$[\text{H}^+] = 10^{-1.6}$
$[\text{H}^+] \approx 0.025 = 2.5 \times 10^{-2}$

The hydrogen ion concentration is approximately 2.5×10^{-2} moles per liter.

60. $t = \dfrac{\ln k}{r} = \dfrac{\ln 2}{0.056} \approx 12.4$

It will take approximately 12.4 years.

61. Let t = the number of years since 1990.
$A(t) = A_0 e^{kt}$ $343 = 218e^{k(8)}$
$\dfrac{343}{218} = e^{8k}$
$\ln\left(\dfrac{343}{218}\right) = 8k$
$\dfrac{1}{8}\ln\left(\dfrac{343}{218}\right) = k$
$0.057 \approx k$

$A(t) = 218e^{0.057t}$
$A(15) = 218e^{0.057(15)} \approx 513$
$A(20) = 218e^{0.057(20)} \approx 682$

In 2005, the estimated revenue would be about $513 billion. In 2010, it would be about $682 billion.

62. $A(t) = A_0 e^{-0.000121t}$
$0.4A_0 = A_0 e^{-0.000121t}$
$0.4 = e^{-0.000121t}$
$\ln 0.4 = -0.000121t$
$\dfrac{\ln 0.4}{-0.000121} = t$
$7573 \approx t$

It has been about 7573 years since the organism died.

Chapter 14: Exponential and Logarithmic Functions and Equations

Chapter 14 Test

1. $f(x) = x^3 \quad y = x^3$
   $x = y^3$
   $\sqrt[3]{x} = y$
   Therefore, $f^{-1}(x) = \sqrt[3]{x}$
   Let $Y1 = x^3$, $Y2 = \sqrt[3]{x}$, and $Y3 = x$.

   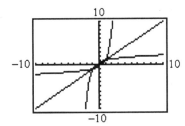

2. Let $Y1 = 2^x + 1$.

   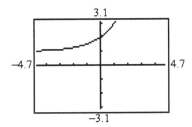

3. Let $Y1 = \ln(x+2)$.

   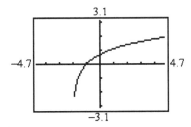

4. $F(x) = 2x - 8 \quad y = 2x - 8$
   $x = 2y - 8$
   $x + 8 = 2y$
   $\dfrac{x+8}{2} = y$
   $\dfrac{1}{2}x + 4 = y$
   Therefore, $F^{-1}(x) = \dfrac{1}{2}x + 4$

5. a. $g(x) = 16^{(1/2)x}$
   $g(-1) = 16^{(1/2)(-1)} = 16^{-1/2} = \dfrac{1}{16^{1/2}} = \dfrac{1}{\sqrt{16}} = \dfrac{1}{4}$

   b. $g(0) = 16^{(1/2)(0)} = 16^0 = 1$

   c. $g(4) = 16^{(1/2)(4)} = 16^2 = 256$

   d. $g(0.3) = 16^{(1/2)(0.3)} = 16^{0.15} \approx 1.516$

   e. $g(\sqrt{3}) = 16^{(1/2)(\sqrt{3})} \approx 11.036$

6. $\log 3x^2 y^3 z = \log 3 + \log x^2 + \log y^3 + \log z$
   $= \log 3 + 2\log x + 3\log y + \log z$

7. $\log_3 \dfrac{9a^2}{b} = \log_3 9a^2 - \log_3 b$
   $= \log_3 9 + \log_3 a^2 - \log_3 b$
   $= 2 + 2\log_3 a - \log_3 b$

8. $2\log x + \log(x-4) = \log x^2 + \log(x-4)$
   $\log x^2(x-4)$

9. $\dfrac{1}{2}\ln x - 2\ln y = \ln x^{1/2} - \ln y^2 = \ln \dfrac{\sqrt{x}}{y^2}$

10. a. Let $Y1 = 9^x$ and $Y2 = 3^{x+1}$.

X	Y1	Y2
-2	.01235	.33333
-1	.11111	1
0	1	3
1	9	9
2	81	27
3	729	81
4	6561	243

    X=1

    The solution is 1.

    b. Let $Y1 = 9^x$ and $Y2 = 3^{x+1}$.

    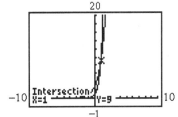

    The solution is 1.

c. $9^x = 3^{x+1}$
$(3^2)^x = 3^{x+1}$
$3^{2x} = 3^{x+1}$
$2x = x+1$
$x = 1$
The solution is 1.

11. $(2^{3x})(2^{-2}) = 8$
$2^{3x-2} = 2^3$
$3x - 2 = 3$
$3x = 5$
$x = \dfrac{5}{3}$
The solution is $\dfrac{5}{3}$.

12. $2\ln a = 3.1$
$\ln a = 1.55$
$a = e^{1.55}$
$a \approx 4.711$
The solution is approximately 4.711.

13. $2\log x + \log x^4 = 6\log x$
$\log x^2 + \log x^4 = \log x^6$
$\log x^2 \cdot x^4 = \log x^6$
$\log x^6 = \log x^6$
$x^6 = x^6$
The equation is true for all values of x except the restricted value(s). The solution is the set of all real numbers greater than 0.

14. $9^x = 5$
$\log 9^x = \log 5$
$x \log 9 = \log 5$
$x = \dfrac{\log 5}{\log 9} \approx 0.732$
The solution is approximately 0.732.

15. Let x = the amount of weekly sales.
$f(x)$ = the total weekly income.

a. $f(x) = 400 + 0.03x$

b. $y = 400 + 0.03x$
$x = 400 + 0.03y$
$x - 400 = 0.03y$
$\dfrac{x - 400}{0.03} = y$
$33.\overline{3}x - 13{,}333.\overline{3} = y$
Therefore, $f^{-1}(x) = 33.\overline{3}x - 13{,}333.\overline{3}$

c. The inverse function represents the value of weekly sales in terms of Pedro's weekly income.

16. $A = P(1+r)^x$; $P = \$2000$; $r = 5\% = 0.05$
$A = 2000(1+0.05)^x$
$A = 2000(1.05)^x$
After 20 years, the investment will be worth
$A = 2000(1.05)^{20} \approx \$5{,}306.60$

17. $t = \dfrac{\ln k}{r} = \dfrac{\ln 2}{0.075} \approx 9.24$
It will take approximately 9.24 years.

18. Let t = the number of years since 1990.
$A(t) = A_0 e^{kt}$ $28525 = 19614 e^{k(8)}$
$\dfrac{28525}{19614} = e^{8k}$
$\ln\left(\dfrac{28525}{19614}\right) = 8k$
$\dfrac{1}{8}\ln\left(\dfrac{28525}{19614}\right) = k$
$0.047 \approx k$
$A(t) = 19614 e^{0.047t}$
$A(20) = 19614 e^{0.047(20)} \approx 50211$
In 2010, the personal per capita income will be about \$50,211.

19. Answers will vary.

Chapter 14: Exponential and Logarithmic Functions and Equations

Cumulative Review 1 – 14

1. $5^{-2} x^0 y^2 = \dfrac{y^2}{5^2} = \dfrac{y^2}{25}$

2. $(-2x^{-3} y^2)^2 = (-2)^2 (x^{-3})^2 (y^2)^2$
   $= 4x^{-6} y^4$
   $= \dfrac{4y^4}{x^6}$

3. $\left(\dfrac{4x^{3/4}}{x^{-1/4}}\right)^2 = \left(4x^{3/4-(-1/4)}\right)^2 = (4x)^2 = 16x^2$

4. $(3.2a^2 - 2.6ab + 1.7b^2) - (4a^2 + 2.6ab - b^2)$
   $= 3.2a^2 - 2.6ab + 1.7b^2 - 4a^2 - 2.6ab + b^2$
   $= -0.8a^2 - 5.2ab + 2.7b^2$

5. $(4.5x^2 y)(-2y^2 z) = -9x^2 y^3 z$

6. $(3x + 2y)(3x - 2y) = (3x)^2 - (2y)^2 = 9x^2 - 4y^2$

7. $(2x + 6)(2x - 3) = 4x^2 - 6x + 12x - 18$
   $= 4x^2 + 6x - 18$

8. $(x - 4)^2 = x^2 - 2 \cdot x \cdot 4 + 4^2 = x^2 - 8x + 16$

9. $\dfrac{25x^2 y^4 z^2}{-5xyz} = -5x^{2-1} y^{4-1} z^{2-1} = -5xy^3 z$

10. $\dfrac{2r^2 s + 4rs - 8s^2}{2rs} = \dfrac{2r^2 s}{2rs} + \dfrac{4rs}{2rs} - \dfrac{8s^2}{2rs}$
    $= r + 2 - \dfrac{4s}{r}$

11. $\dfrac{x^2 - x - 6}{x^2 - 9} = \dfrac{(x+2)(x-3)}{(x+3)(x-3)} = \dfrac{x+2}{x+3}$

12. $\dfrac{x}{x+2} + \dfrac{3}{x-2} + \dfrac{x-1}{x^2 - 4}$
    $= \dfrac{x(x-2)}{(x+2)(x-2)} + \dfrac{3(x+2)}{(x+2)(x-2)} + \dfrac{x-1}{(x+2)(x-2)}$
    $= \dfrac{x^2 - 2x + 3x + 6 + x - 1}{(x+2)(x-2)}$
    $= \dfrac{x^2 + 2x + 5}{(x+2)(x-2)}$

13. $\dfrac{1}{x+4} - \dfrac{x}{x-2} = \dfrac{1(x-2)}{(x+4)(x-2)} - \dfrac{x(x+4)}{(x+4)(x-2)}$
    $= \dfrac{x - 2 - x^2 - 4x}{(x+4)(x-2)}$
    $= \dfrac{-x^2 - 3x - 2}{(x+4)(x-2)}$
    $= \dfrac{-1(x+1)(x+2)}{(x+4)(x-2)}$

14. $\dfrac{2m - 3}{m + 1} \cdot \dfrac{2m^2 + 5m + 3}{4m^2 - 9} = \dfrac{2m - 3}{m + 1} \cdot \dfrac{(m+1)(2m+3)}{(2m+3)(2m-3)}$
    $= \dfrac{(2m-3)(m+1)(2m+3)}{(m+1)(2m+3)(2m-3)}$
    $= 1$

15. $\dfrac{16x}{y^3 + y^2 - 12y} \div \dfrac{4x^2 y}{y^2 - y - 20}$
    $= \dfrac{16x}{y(y-3)(y+4)} \div \dfrac{4x^2 y}{(y+4)(y-5)}$
    $= \dfrac{16x}{y(y-3)(y+4)} \cdot \dfrac{(y+4)(y-5)}{4x^2 y}$
    $= \dfrac{16x(y+4)(y-5)}{4x^2 y^2 (y-3)(y+4)}$
    $= \dfrac{4(y-5)}{xy^2 (y-3)}$

16. $\sqrt{18x} - \sqrt{50x} + 4\sqrt{8x} = \sqrt{9 \cdot 2x} - \sqrt{25 \cdot 2x} + 4\sqrt{4 \cdot 2x}$
    $= 3\sqrt{2x} - 5\sqrt{2x} + 8\sqrt{2x}$
    $= 6\sqrt{2x}$

SSM: Experiencing Introductory and Intermediate Algebra

17. $\sqrt[3]{16a^2b^2} \cdot \sqrt[3]{8ab^2} = \sqrt[3]{16a^2b^2 \cdot 8ab^2}$
$= \sqrt[3]{128a^3b^4}$
$= \sqrt[3]{64a^3b^3 \cdot 2b}$
$= 4ab\sqrt[3]{2b}$

18. $(\sqrt{2}-\sqrt{3})(\sqrt{2}+\sqrt{3}) = (\sqrt{2})^2 - (\sqrt{3})^2 = 2 - 3 = -1$

19. $\dfrac{\sqrt{32xy^2z}}{\sqrt{8x^3y}} = \sqrt{\dfrac{32xy^2z}{8x^3y}} = \sqrt{\dfrac{4yz}{x^2}} = \dfrac{2}{x}\sqrt{yz}$

20. $\dfrac{\sqrt{x}+4}{\sqrt{x}-3} = \dfrac{\sqrt{x}+4}{\sqrt{x}-3} \cdot \dfrac{\sqrt{x}+3}{\sqrt{x}+3} = \dfrac{x+7\sqrt{x}+12}{x-9}$

21. $x^{2/3}\left(x^{3/4} - x^{1/3}\right) = x^{2/3+3/4} - x^{2/3+1/3}$
$= x^{17/12} - x$

22. $25m^2 - 36n^2 = (5m)^2 - (6n)^2$
$= (5m+6n)(5m-6n)$

23. $2x^2 - 4x - 16 = 2(x^2 - 2x - 8) = 2(x+2)(x-4)$

24. $x^2 + 6x - 6$ does not factor.

25. $f(x) = \dfrac{2}{3}x + 1$

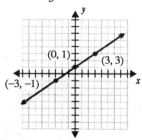

Domain: all real numbers
Range: all real numbers

26. $y = x^2 + 5x + 4$
$a = 1, b = 5, c = 4$
$-\dfrac{b}{2a} = \dfrac{-5}{2(1)} = -\dfrac{5}{2}$
$y = \left(-\dfrac{5}{2}\right)^2 + 5\left(-\dfrac{5}{2}\right) + 4 = -\dfrac{9}{4}$
Vertex: $\left(-\dfrac{5}{2}, -\dfrac{9}{4}\right)$

Axis of symmetry: $x = -\dfrac{5}{2}$
x-intercepts:
$x^2 + 5x + 4 = 0$
$(x+4)(x+1) = 0$
$x = -4$ or $x = -1$
The x-intercepts are $(-4,0)$ and $(-1,0)$.
y-intercept: $(0,4)$

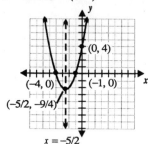

Domain: all real numbers
Range: all real numbers $\geq -\dfrac{9}{4}$

27. $y = \dfrac{x^2+4}{x+2}$

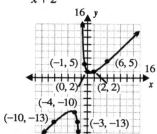

Domain: all real numbers $\neq -2$
Range: all real numbers ≤ -9.66 or ≥ 1.66

28. $h(x) = \sqrt{x-5}$

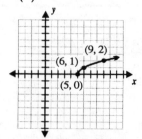

Domain: all real numbers $x \geq 5$
Range: all real numbers ≥ 0

29. $g(x) = x^{2/3} + 3$

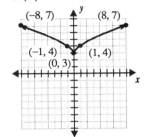

Domain: all real numbers
Range: all real numbers ≥ 3

30. $y = 3^x$

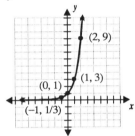

Domain: all real numbers
Range: all real numbers > 0

31. $f(x) = \ln(x-3)$

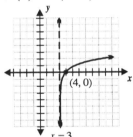

Domain: all real numbers > 3
Range: all real numbers

32. $2(x-5)-(x+4) = 2x-8$
$2x-10-x-4 = 2x-8$
$x-14 = 2x-8$
$-14 = x-8$
$-6 = x$
The solution is –6.

33. $2(x+3.1) = (x-4.2)+(x-6)$
$2x-6.2 = 2x-10.2$
$-6.2 = -10.2$
The equation is always false (a contraction). It has no solution.

34. $x^2 - 2x - 12 = 12$
$x^2 - 2x - 24 = 0$
$(x+4)(x-6) = 0$
$x = -4$ or $x = 6$
The solutions are –4 and 6.

35. $x^2 - 5x + 9 = 0$
$x = \dfrac{-(-5) \pm \sqrt{(-5)^2 - 4(1)(6)}}{2(1)}$
$x = \dfrac{5 \pm \sqrt{-11}}{2}$
$x = \dfrac{5 \pm i\sqrt{11}}{2} = \dfrac{5}{2} \pm \dfrac{\sqrt{11}}{2}i$
The solutions are $\dfrac{5}{2} \pm \dfrac{\sqrt{11}}{2}i$.

36. $\dfrac{x+1}{x+2} = \dfrac{x-1}{x+3}$
$(x-1)(x+2) = (x+1)(x+3)$
$x^2 + x - 2 = x^2 + 4x + 3$
$-3x = 5$
$x = -\dfrac{5}{3}$
The solutions is $-\dfrac{5}{3}$.

37. $4\sqrt{x+1} - 14 = -2$
$4\sqrt{x+1} = 12$
$\sqrt{x+1} = 3$
$x+1 = 9$
$x = 8$
The solutions is 8.

38. $e^{2x-4} = e^{x(x-2)}$
$2x - 4 = x(x-2)$
$2x - 4 = x^2 - 2x$
$0 = x^2 - 4x + 4$
$0 = (x-2)^2$
$0 = x - 2$
$2 = x$
The solution 2.

39. $2\log_3 x = \log_3 5$
$\log_3 x^2 = \log_3 5$
$x^2 = 5$
$x = \pm\sqrt{5}$
Since $-\sqrt{5}$ is a restricted value, the solution is $\sqrt{5}$

40. $2x - 3 < 3(x-4)$
$2x - 3 < 3x - 12$
$9 < x$
The solution is $(9, \infty)$.

41. $x^2 + 2x \geq 2$
$x^2 + 2x - 2 \geq 0$
Solve $x^2 + 2x - 2 = 0$
$x = \dfrac{-2 \pm \sqrt{2^2 - 4(1)(-2)}}{2(1)}$
$x = \dfrac{-2 \pm \sqrt{12}}{2} = \dfrac{-2 \pm 2\sqrt{3}}{2} = -1 \pm \sqrt{3}$

Possible intervals:
$(-\infty, -1-\sqrt{3}], [-1-\sqrt{3}, -1+\sqrt{3}], [-1+\sqrt{3}, \infty)$
Use a test point in each interval to determine the solution is $(-\infty, -1-\sqrt{3}] \cup [-1+\sqrt{3}, \infty)$

42. $3x - 2y = 4$
$x + y = 8$
From the second equation, $y = 8 - x$, substitute $8 - x$ for y in the first equation.

$3x - 2(8 - x) = 4$
$3x - 16 + 2x = 4$
$5x = 20$
$x = 4$

$y = 8 - x$
$y = 8 - 4 = y$
The solution is $(4, 4)$.

43. $m = \dfrac{y_2 - y_1}{x_2 - x_1} = \dfrac{2 - (-1)}{-2 - 3} = \dfrac{3}{-5} = -\dfrac{3}{5}$
$y - y_1 = m(x - x_1)$
$y - (-1) = -\dfrac{3}{5}(x - 3)$
$y + 1 = -\dfrac{3}{5}x + \dfrac{9}{5}$
$y = -\dfrac{3}{5}x + \dfrac{4}{5}$

44. $x + 4y = 2.3$
$4y = -x + 2.3$
$y = -\dfrac{1}{4}x + \dfrac{2.3}{4}$
The slope of the desired line is $\dfrac{-1}{-\frac{1}{4}} = 4$.
$y - y_1 = m(x - x_1)$
$y - 1 = 4[x - (-2)]$
$y - 1 = 4(x + 2)$
$y - 1 = 4x + 8$
$y = 4x + 9$

45. $y = 1.5x^2 + 4.5x - 6$

```
L1    L2    L3    2   QuadReg
 0    -6    ------    y=ax²+bx+c
-4     0              a=1.5
 1     0              b=4.5
      ------          c=-6
                      R²=1
L2(4) =
```

Chapter 14: Exponential and Logarithmic Functions and Equations

46. $f(x) = \frac{2}{3}x + 6 \qquad y = \frac{2}{3}x + 6$

$$x = \frac{2}{3}y + 6$$

$$x - 6 = \frac{2}{3}y$$

$$\frac{3}{2}(x-6) = \frac{3}{2}\left(\frac{2}{3}y\right)$$

$$\frac{3}{2}x - 9 = y$$

Therefore, $f^{-1}(x) = \frac{3}{2}x - 9$

47. Let x = the number of books printed.
$C(x) = 525 + 5x$
$R(x) = 15x$
$525 + 5x = 15x$
$525 = 10x$
$52.5 = x$
The company must sell 53 books in order to break even.

48. Let w = the width of Nathan's garden.
$w + 3$ = the length of the garden.
$$w^2 + (w+3)^2 = 15^2$$
$$w^2 + w^2 + 6w + 9 = 225$$
$$2w^2 + 6w - 216 = 0$$
$$2(w^2 + 3w - 108) = 0$$
$$2(w - 9)(w + 12) = 0$$
$w = 9 \qquad$ or $\qquad w = -12$
$w + 3 = 12$
The width cannot be negative, so the garden is 9 feet by 12 feet.

49. Let x = time to drain the tank using both drains.

$\frac{1}{4}$ is the left drain's rate.

$\frac{1}{6.5} = \frac{2}{13}$ is the right drain's rate.

$\frac{1}{x}$ is the combined rate.

$$\frac{1}{4} + \frac{2}{13} = \frac{1}{x}$$

$$52x\left(\frac{1}{4} + \frac{2}{13}\right) = 52x\left(\frac{1}{x}\right)$$

$$13x + 8x = 52$$

$$21x = 52$$

$$x = \frac{52}{21} \approx 2.5$$

It will take approximately 2.5 hours to drain the tank if both drains are used.

50. $t = \frac{\ln k}{r} = \frac{\ln 3}{0.045} \approx 24.4$

It will take about 24.4 years for the money to triple.

Calculator Appendix

Absolute Value

Absolute value expressions are entered in the calculator using the ABS function. To evaluate an absolute value expression, enter [MATH] [▶] [1] followed by the expression and [)].

Example: Evaluate $|-16|$.
[MATH] [▶] [1] [(-)] [1] [6] [)] [ENTER]
The result is 6.

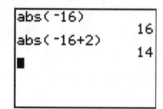

Example: Evaluate $|-16+2|$.
[MATH] [▶] [1] [(-)] [1] [6] [+] [2] [)] [ENTER]
The result is 14.

CALC function

Intersect under the CALC function determines the coordinates of the intersection point of two graphs on the graphics screen.

To find the intersection of two graphs, enter [2nd] [CALC] [5]; use the up and down arrow keys to move the cursor to the first graph and then press [ENTER]; use the up and down arrow keys again to move the cursor to the second graph and press [ENTER]; then use the left and right arrow keys to move the cursor as close to the intersection point as possible for a guess and press [ENTER]. The coordinates of the intersection point are displayed at the bottom of the screen.

Example: Determine the intersection point of the graphs of the equations $y = 3x + 5$ and $y = -x$.
[Y=] [3] [X,T,θ,n] [+] [5] [▼] [(-)] [X,T,θ,n] [GRAPH]
[2nd] [CALC] [5]
Use the arrow keys to select the first graph and press [ENTER]; move the cursor to the second graph and press [ENTER]; then make sure the cursor is near the intersection point and press [ENTER].
The result is $x = -1.25$ and $y = 1.25$.

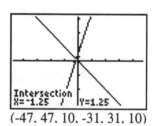

(-47, 47, 10, -31, 31, 10)

Maximum under the CALC function determines the coordinates of the maximum point of a graph between two given points on the graph that have been chosen by the user.

To find the maximum point of a graph of a graph, enter [2nd] [CALC] [4]; use the left and right arrow keys to move the cursor to the left of the maximum point and press [ENTER]; move the cursor to the right of the maximum point and pres [ENTER]; then move the cursor as close as possible to the maximum point (this should be between the previously chosen points) and press [ENTER]. The coordinates of the maximum point will be displayed at the bottom of the screen.

Example: Determine the coordinates of the maximum point of the graph of the equation $y = -x^2 - 5$.

Calculator Appendix

$\boxed{Y=}\ \boxed{(-)}\ \boxed{X,T,\theta,n}\ \boxed{x^2}\ \boxed{-}\ \boxed{5}\ \boxed{GRAPH}\ \boxed{2^{nd}}\ \boxed{CALC}\ \boxed{4}$
Use the arrow keys to move the cursor to the left of the maximum point and press $\boxed{ENTER}$ for the left bound; move the cursor to the right of the maximum point and press $\boxed{ENTER}$ for the right bound; then move the cursor as close to the maximum point as possible (this should be between the two previous points) and press $\boxed{ENTER}$ for the guess.
The result is $x = 2.4237E{-}6$ (this value may vary because it should be 0 but rounding error causes it to differ slightly) and $y = -5$.

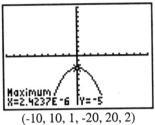
(-10, 10, 1, -20, 20, 2)

Minimum under the CALC function determines the coordinates of the minimum point of a graph between two given points on the graph that have been chosen by the user.
To find the minimum point of a graph, enter $\boxed{2^{nd}}\ \boxed{CALC}\ \boxed{3}$; use the left and right arrow keys to move the cursor to the left of the minimum point and press $\boxed{ENTER}$ for the left bound; move the cursor to the right of the minimum point and press $\boxed{ENTER}$ for the right bound; then move the cursor as close to the minimum point as possible (this should be between the two previous points) and press $\boxed{ENTER}$ for a guess.

Example: Determine the coordinates of the minimum point of the graph of the equation $y = x^2 - 5$.
$\boxed{Y=}\ \boxed{X,T,\theta,n}\ \boxed{x^2}\ \boxed{-}\ \boxed{5}\ \boxed{GRAPH}\ \boxed{2^{nd}}\ \boxed{CALC}\ \boxed{3}$.
Use the left and right arrow keys to move the cursor to the left of the minimum point and press $\boxed{ENTER}$ for the left bound; move the cursor to the right of the minimum point and press $\boxed{ENTER}$ for the right bound; then move the cursor as close to the minimum point as possible and press $\boxed{ENTER}$ for a guess. The result is $x = -1.075E{-}6$ (this value is close to 0 and may vary due to rounding error in the calculator) and $y = -5$.

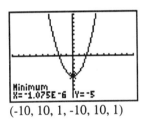
(-10, 10, 1, -10, 10, 1)

Value under the CALC function determines the y-coordinate of a point on the graphics screen given an x-coordinate chosen by the user.
To find the y-coordinate for a given x-coordinate on a graph, enter $\boxed{2^{nd}}\ \boxed{CALC}\ \boxed{1}$; then enter the given x-coordinate and press $\boxed{ENTER}$. The x-coordinate and the y-coordinate are displayed at the bottom of the screen.

Example: Determine the y-coordinate for the x-coordinate 3 on the graph of the equation $y = x^2 - 5$.
$\boxed{Y=}\ \boxed{X,T,\theta,n}\ \boxed{x^2}\ \boxed{-}\ \boxed{5}\ \boxed{GRAPH}\ \boxed{2^{nd}}\ \boxed{CALC}\ \boxed{1}$ $\boxed{3}\ \boxed{ENTER}$.
The x-coordinate, 3, and the y-coordinate, 4, are displayed at the bottom of the screen.

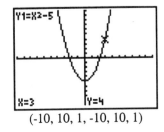
(-10, 10, 1, -10, 10, 1)

SSM: Experiencing Introductory and Intermediate Algebra

Zero under the CALC function determines the coordinates of the x–intercept(s) of a graph on the graphics screen.

To find an x-intercept of a graph, enter $\boxed{2^{nd}}$ $\boxed{CALC}$ $\boxed{1}$; use the left and right arrow keys to move the cursor to the left of the x-intercept and press $\boxed{ENTER}$ for a left bound; move the cursor to the right of the x-intercept and press $\boxed{ENTER}$ for a right bound; then move the cursor as close as possible to the intercept and press $\boxed{ENTER}$. The coordinates of the x-intecept are displayed at the bottom of the screen.

Example: Determine the coordinates of the x-intercept of the graph of the equation $y = 3x + 5$.

$\boxed{Y=}$ $\boxed{3}$ $\boxed{X,T,\theta,n}$ $\boxed{+}$ $\boxed{5}$ $\boxed{GRAPH}$ $\boxed{2^{nd}}$ $\boxed{CALC}$ $\boxed{2}$.
Use the left and right arrow keys to move to the left of the x-intercept and press $\boxed{ENTER}$; move the cursor to the right of the x-intercept and press $\boxed{ENTER}$; then use the left and right arrow keys to move the cursor as close as possible to the x-intercept (between the last two entries) and press $\boxed{ENTER}$.
The result is $x = -1.666667$ and $y = 0$.

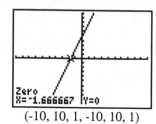
(-10, 10, 1, -10, 10, 1)

Colon

To enter more than one command within one entry, enter $\boxed{ALPHA}$ $\boxed{:}$.

Example: Evaluate $2x + 5$ for $x = 3$.
$\boxed{3}$ $\boxed{STO\blacktriangleright}$ $\boxed{X,T,\theta,n}$ $\boxed{ALPHA}$ $\boxed{:}$ $\boxed{2}$ $\boxed{X,T,\theta,n}$ $\boxed{+}$ $\boxed{5}$ $\boxed{ENTER}$.
The result is 11.

Evaluate

To evaluate an expression involving variables, first store the value of each variable. Enter the value, followed by $\boxed{STO\blacktriangleright}$, followed by the name of the variable ($\boxed{ALPHA}$ and the letter of the variable name), followed by a line entry ($\boxed{ALPHA}$ $\boxed{:}$). Repeat this process until all variables are entered with their values. After the last $\boxed{:}$, enter the expression to be evaluated.

Example: Evaluate $4a$ for $a = 3$.
$\boxed{3}$ $\boxed{STO\blacktriangleright}$ $\boxed{ALPHA}$ $\boxed{A}$ $\boxed{ALPHA}$ $\boxed{:}$ $\boxed{4}$ $\boxed{ALPHA}$ $\boxed{A}$ $\boxed{ENTER}$.
The result is 12.

To change an entry without reentering the entire entry, use $\boxed{2^{nd}}$ $\boxed{ENTRY}$ to recall the last entry. Change the entry by using the arrow keys to locate the position of the desired change and enter the change.

Calculator Appendix

[DEL] and [2nd] [INS] will allow you to delete and insert characters for entries that are shorter and longer than the original entry.

Example: Recall the previous entry and evaluate the expression $4a$ for $a = -2$.
Enter [2nd] [ENTRY]; then use the left arrow key to move over the 3. Enter [(-)] and then insert the 2 by pressing [2nd] [INS] [2] [ENTER].
The result is 8.

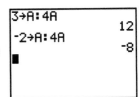

Exponent

To evaluate a squared expression, enter the expression followed by $[x^2]$. Be sure to use parentheses correctly.

Example: Evaluate 9^2.
[9] $[x^2]$ [ENTER]
The result is 81.

Evaluate $(9+5)^2$.
[(] [9] [+] [5] [)] $[x^2]$ [ENTER]
The result is 196.

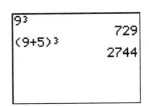

To evaluate a cubed expression, enter the expression followed by [MATH] [3]. Be sure to use parentheses correctly.

Example: Evaluate 9^3.
[9] [MATH] [3] [ENTER]
The result is 729.

Evaluate $(9+5)^3$.
[(] [9] [+] [5] [)] [MATH] [3]
The result is 2744.

To evaluate an exponential expression other than a squared or cubed expression, enter the expression followed by [^] followed by the value of the exponent. Be sure to use parentheses correctly.

Example: Evaluate 9^5.
[9] [^] [5] [ENTER]
The result is 59,049.

SSM: Experiencing Introductory and Intermediate Algebra

Evaluate $9^{1/2}$.
[9] [^] [(] [1] [÷] [2] [)] [ENTER]
The result is 3.

Evaluate $(9+5)^5$.
[(] [9] [+] [5] [)] [^] [5] [ENTER]
The result is 537,824.

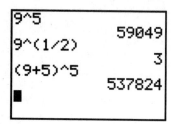

Fraction

Proper fractions are entered as division expressions. To obtain an answer in fractional notation, enter the expression followed by [MATH] [1] [ENTER]. To ensure the correct order of operations, place fractions inside parentheses.

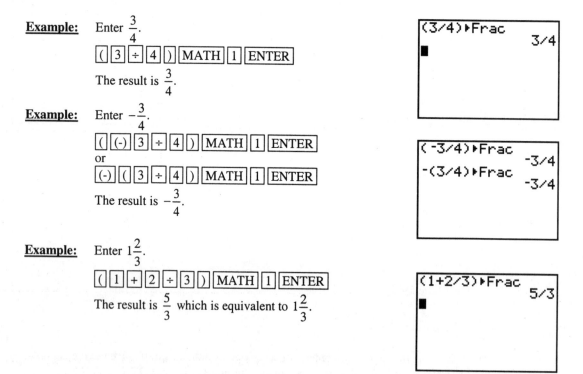

Example: Enter $\frac{3}{4}$.
[(] [3] [÷] [4] [)] [MATH] [1] [ENTER]
The result is $\frac{3}{4}$.

Example: Enter $-\frac{3}{4}$.
[(] [(-)] [3] [÷] [4] [)] [MATH] [1] [ENTER]
or
[(-)] [(] [3] [÷] [4] [)] [MATH] [1] [ENTER]
The result is $-\frac{3}{4}$.

Example: Enter $1\frac{2}{3}$.
[(] [1] [+] [2] [÷] [3] [)] [MATH] [1] [ENTER]
The result is $\frac{5}{3}$ which is equivalent to $1\frac{2}{3}$.

Negative mixed numbers are entered as the opposite of a sum of an integer and a proper fraction. The entire sum should be in parentheses to ensure proper order of operations. The opposite sign precedes the parentheses. There is no need to place the proper-fraction addend in a second set of parentheses.

Calculator Appendix

Example: Enter $-1\frac{2}{3}$.
[(-)][(][1][+][2][÷][3][)][MATH][1][ENTER]
The result is $-\frac{5}{3}$ which is equivalent to $-1\frac{2}{3}$.

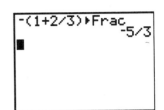

To evaluate an expression and obtain the results in fractional notation, enter the expression followed by [MATH][1][ENTER].

Example: Evaluate $\frac{3}{4} \div \frac{2}{3}$.
[(][3][÷][4][)][÷][(][2][÷][3][)][MATH][1][ENTER]
The result is $\frac{9}{8}$.

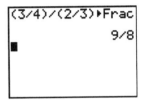

To change a real number to a fraction in lowest terms, enter the number followed by [MATH][1][ENTER].

Example: Write 0.25 as a fraction in lowest terms.
[.][2][5][MATH][1][ENTER]
The result is $\frac{1}{4}$.

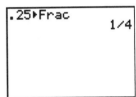

Irrational numbers will not convert to a fraction. Therefore, the result is a decimal approximation of the number.

Example: Write π as a fraction in lowest terms.
[2nd][π][MATH][1][ENTER]
The result is 3.141592654.

Graphing

To graph points on the calculator, first set the screen to the desired setting (see Screen). Enter [2nd][QUIT] to return to the home or text screen. Enter [2nd][DRAW][▶][1], followed by the x-coordinate of the point and [,], followed by the y-coordinate of the point and [)][ENTER]. To clear the screen of the graphed points, enter [2nd][DRAW][1].

Example: Graph the point $(3, 2)$ on a decimal screen.
[ZOOM][4][2nd][QUIT][2nd][DRAW][▶][1][3][,][2][)][ENTER]

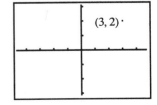

SSM: Experiencing Introductory and Intermediate Algebra

To graph an equation on the calculator, first set the screen to the desired setting (see Screens). Enter the equation that is solved for *y* by entering $\boxed{Y=}$, followed by the expression in terms of *x*. If more than one equation is to be entered, use $\boxed{\text{ENTER}}$ or $\boxed{\blacktriangledown}$ to move to the next *y*. Finally, enter $\boxed{\text{GRAPH}}$.

Example: Graph $y = 2x + 5$ on an integer screen.

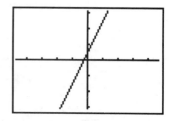

To determine the coordinates of points on the screen, use the arrow keys. The coordinates will be displayed on the bottom of the screen.

To determine the coordinates of points graphed on the screen, enter $\boxed{\text{TRACE}}$ and then use the left and right arrow keys to move along the graph. The coordinates will be displayed on the bottom of the screen, and the equation will be displayed in the upper left-hand corner of the screen.

Example: Determine whether $(4, 13)$ is on the graph of the equation $y = 2x + 5$.

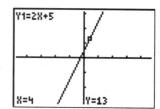

To graph an inequality on the calculator, first set the screen to the desired setting (see Screens). Solve the inequality for *y*, exchange the inequality symbol for an equal sign, and enter the equation. Press $\boxed{Y=}$ followed by the expression in terms of *x*. Move the cursor to the left of Y1 with the left arrow key. For a "greater than" inequality, press ENTER twice: $\boxed{\text{ENTER}}\boxed{\text{ENTER}}$. For a "less than" inequality, press ENTER three times: $\boxed{\text{ENTER}}\boxed{\text{ENTER}}\boxed{\text{ENTER}}$. Then press $\boxed{\text{GRAPH}}$.

Example: Graph $y > 2x + 5$ on an integer screen.

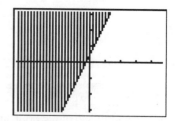

Display elements can be turned on and off using the FORMAT menu. Press $\boxed{2^{\text{nd}}}\boxed{\text{FORMAT}}$ and you can turn on (or off) the coordinates that display at the bottom, a background grid for graphs, the *x* and *y* axes, the axis label, and the expression being graphed. If your graph display is missing an element listed above, check the FORMAT menu to see if the feature is turned off.

Calculator Appendix

Grouping

To enter parentheses, brackets, and braces, use $\boxed{(}$ and $\boxed{)}$.

Example: Evaluate $2 + \{3[2 + (5 - 1)]\}$.
$\boxed{2}\boxed{+}\boxed{(}\boxed{(}\boxed{3}\boxed{(}\boxed{(}\boxed{2}\boxed{+}\boxed{(}\boxed{(}\boxed{5}\boxed{-}\boxed{1}\boxed{)}\boxed{)}\boxed{)}\boxed{)}$ ENTER
The result is 20.

```
2+(3(2+(5-1)))
              20
```

To evaluate division problems with an expression for the dividend and/or divisor, enter the expressions in parentheses.

Example: Evaluate $\dfrac{9+5}{3+4}$.
$\boxed{(}\boxed{9}\boxed{+}\boxed{5}\boxed{)}\boxed{\div}\boxed{(}\boxed{3}\boxed{+}\boxed{4}\boxed{)}$ ENTER
The result is 2.

```
(9+5)/(3+4)
            2
```

Negative sign

$\boxed{(\text{-})}$ means *negative* or *opposite of*
NOTE: Do **not** use the subtraction operation symbol, $\boxed{-}$, for a negative sign, $\boxed{(\text{-})}$.

Example: Evaluate $-5 - 3$.
$\boxed{(\text{-})}\boxed{5}\boxed{-}\boxed{3}$ ENTER
The result is -8.

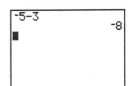

Opposite

Opposite expressions are entered in the calculator using the $\boxed{(\text{-})}$ key. To evaluate an opposite expression, enter $\boxed{(\text{-})}$ followed by the expression enclosed in parentheses.

Example: Evaluate $-(16)$.
$\boxed{(\text{-})}\boxed{(}\boxed{1}\boxed{6}\boxed{)}$ ENTER
The result is -16.

Evaluate $-(-16 + 2)$.
$\boxed{(\text{-})}\boxed{(}\boxed{(\text{-})}\boxed{1}\boxed{6}\boxed{+}\boxed{2}\boxed{)}$ ENTER
The result is 14.

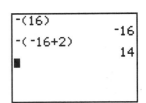

SSM: Experiencing Introductory and Intermediate Algebra

Radical

To evaluate a **square-root** expression, enter $\boxed{2^{nd}}$ $\boxed{\sqrt{\ }}$ followed by the expression and $\boxed{)}$.

Example: Evaluate $\sqrt{9}$.
$\boxed{2^{nd}}$ $\boxed{\sqrt{\ }}$ $\boxed{9}$ $\boxed{)}$ $\boxed{\text{ENTER}}$
The result is 3.

Evaluate $\sqrt{9+5}$.
$\boxed{2^{nd}}$ $\boxed{\sqrt{\ }}$ $\boxed{9}$ $\boxed{+}$ $\boxed{5}$ $\boxed{)}$ $\boxed{\text{ENTER}}$
The result is 3.741657387.

```
√(9)
            3
√(9+5)
   3.741657387
■
```

To evaluate a **cube-root** expression, enter $\boxed{\text{MATH}}$ $\boxed{4}$ followed by the term and $\boxed{)}$.

Example: Evaluate $\sqrt[3]{27}$.
$\boxed{\text{MATH}}$ $\boxed{4}$ $\boxed{2}$ $\boxed{7}$ $\boxed{)}$ $\boxed{\text{ENTER}}$
The result is 3.

Evaluate $\sqrt[3]{9+5}$.
$\boxed{\text{MATH}}$ $\boxed{4}$ $\boxed{9}$ $\boxed{+}$ $\boxed{5}$ $\boxed{)}$ $\boxed{\text{ENTER}}$
The result is 2.410142264.

```
³√(27)
            3
³√(9+5)
    2.410142264
■
```

To evaluate a **radical expression** other than a square or cube root, enter the value of the index, followed by $\boxed{\text{MATH}}$ $\boxed{5}$, followed by the term. Be sure to enclose the expression in parentheses, as needed.

Example: Evaluate $\sqrt[4]{81}$.
$\boxed{4}$ $\boxed{\text{MATH}}$ $\boxed{5}$ $\boxed{8}$ $\boxed{1}$ $\boxed{\text{ENTER}}$
The result is 3.

Evaluate $\sqrt[4]{9+5}$.
$\boxed{4}$ $\boxed{\text{MATH}}$ $\boxed{5}$ $\boxed{(}$ $\boxed{9}$ $\boxed{+}$ $\boxed{5}$ $\boxed{)}$ $\boxed{\text{ENTER}}$
The result is 1.93433642.

```
4*√81
            3
4*√(9+5)
    1.93433642
■
```

Screens

The **home** or **text screen** is a blank screen. If you want to return to this screen from another screen, enter $\boxed{2^{nd}}$ $\boxed{\text{QUIT}}$ as many times as needed.

The **integer graphics screen** used in this text is a centered integer screen set up by entering $\boxed{\text{ZOOM}}$ $\boxed{6}$ $\boxed{\text{ZOOM}}$ $\boxed{8}$ $\boxed{\text{ENTER}}$.
NOTE: Do **not** forget to press $\boxed{\text{ENTER}}$!
The window settings are $(-47, 47, 10, -31, 31, 10, 1)$.

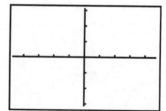

Calculator Appendix

The **decimal graphics screen** is the window setting $(-4.7, 4.7, 1, -3.1, 3.1, 1)$ and is set up by entering [ZOOM] [4].

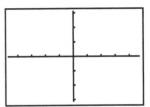

The **standard graphics screen** is the window setting $(-10, 10, 1, -10, 10, 1)$ and is set up by entering [ZOOM] [6].

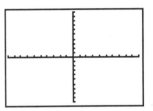

A **custom graphics screen** is set up by entering [WINDOW] [▼], followed by reentering values and using the down arrow until the desired setting is entered. To view the screen, enter [GRAPH].

Example: Enter the window $(0, 10, 1, 0, 25, 10, 1)$.

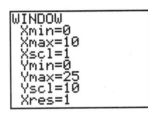
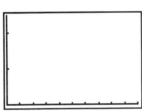

Subtraction

[−] means subtraction.

NOTE: Do **not** use the subtraction operation symbol, [−], for a negative sign, [(-)].

Example: Evaluate $-5 - 3$.
[(-)] [5] [−] [3] [ENTER]
The result is -8.

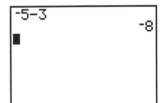

SSM: Experiencing Introductory and Intermediate Algebra

Table of Values

The calculator will display a table of values for expressions entered in the $\boxed{Y=}$ menu. First, enter $\boxed{Y=}$ followed by the expression. To set up the table, enter $\boxed{2^{nd}}$ $\boxed{\text{TBLSET}}$, followed by the least desired number wanted in the table and $\boxed{\blacktriangledown}$, followed by the increments desired and $\boxed{\blacktriangledown}$, followed by $\boxed{\text{ENTER}}$ $\boxed{\blacktriangledown}$ $\boxed{\text{ENTER}}$ to set the calculator to automatically calculate values. To view the table, enter $\boxed{2^{nd}}$ $\boxed{\text{TABLE}}$.

Example: Set up a table of values for the expression $2x$ for $x = 3, 4, 5, 6, 7, 8,$ and 9.

To set up the calculator to display only a limited number of values, first enter the expression in the $\boxed{Y=}$ menu. Then enter $\boxed{2^{nd}}$ $\boxed{\text{TBLSET}}$ and ignore the first two entries by pressing $\boxed{\blacktriangledown}$ $\boxed{\blacktriangledown}$; then set the calculator to ASK by entering $\boxed{\blacktriangleright}$ $\boxed{\text{ENTER}}$ $\boxed{\blacktriangledown}$ $\boxed{\text{ENTER}}$. To view the table, enter $\boxed{2^{nd}}$ $\boxed{\text{TABLE}}$, followed by the desired values for x.

Example: Set up a table of values for the expression $2x$ when $x = 3, 5,$ and 8.

$\boxed{Y=}$ $\boxed{2}$ $\boxed{X,T,\theta,n}$ $\boxed{2^{nd}}$ $\boxed{\text{TBLSET}}$ $\boxed{\blacktriangledown}$ $\boxed{\blacktriangledown}$ $\boxed{\blacktriangleright}$ $\boxed{\text{ENTER}}$ $\boxed{\blacktriangledown}$ $\boxed{\text{ENTER}}$ $\boxed{2^{nd}}$ $\boxed{\text{TABLE}}$ $\boxed{3}$ $\boxed{\text{ENTER}}$ $\boxed{5}$ $\boxed{\text{ENTER}}$ $\boxed{8}$ $\boxed{\text{ENTER}}$

TEST

The calculator tests whether an inequality or equation is true or false and returns 0 for a false statement and a 1 for a true statement. Enter the statement in the calculator and then press $\boxed{\text{ENTER}}$. The inequality symbols and the equal symbol are under the TEST menu. You will find these by entering $\boxed{2^{nd}}$ $\boxed{\text{TEST}}$ and then entering the number in front of the symbol.

Calculator Appendix

Example: Determine whether $3 < 4$.
$\boxed{3}\ \boxed{2^{nd}}\ \boxed{TEST}\ \boxed{5}\ \boxed{4}\ \boxed{ENTER}$
The result is 1 for true.

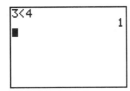

Since the calculator rounds decimal numbers, a 0 for false may result when the statement entered is true and a 1 for true when the statement is false. To recheck, enter each expression separately and check the results yourself.

Example: Determine whether $2.00000000001 = 2$ is true or false.
$\boxed{2}\ \boxed{.}\ \boxed{0}\ \boxed{0}\ \boxed{0}\ \boxed{0}\ \boxed{0}\ \boxed{0}\ \boxed{0}\ \boxed{0}\ \boxed{0}\ \boxed{1}\ \boxed{2^{nd}}\ \boxed{TEST}\ \boxed{1}\ \boxed{2}\ \boxed{ENTER}$
The result is 1 for true. However, we can see without a calculator that 2.00000000001 and 2 are not equal values.